D1237683

SOURCE BOOKS IN THE HISTORY OF THE SCIENCES

Edward H. Madden, *General Editor*

A SOURCE BOOK IN CHEMISTRY, 1400–1900
Henry M. Leicester and Herbert S. Klickstein

A SOURCE BOOK IN GREEK SCIENCE
Morris R. Cohen and I. E. Drabkin

SOURCE BOOK IN ASTRONOMY, 1900–1950
Harlow Shapley

A SOURCE BOOK IN PHYSICS
William Francis Magie

A SOURCE BOOK IN THE HISTORY OF PSYCHOLOGY
Richard J. Herrnstein and Edwin G. Boring

SOURCE BOOK IN GEOLOGY, 1900–1950
Kirtley F. Mather

GOTTLOB FREGE

1848–1925

FROM FREGE TO GÖDEL

A Source Book in Mathematical Logic, 1879–1931

Jean van Heijenoort

PROFESSOR OF PHILOSOPHY, BRANDEIS UNIVERSITY

HARVARD UNIVERSITY PRESS

CAMBRIDGE, MASSACHUSETTS · 1967

Distributed in Great Britain by Oxford University Press, London
Library of Congress Catalog Card Number 67-10905
Printed in the United States of America

General Editor's Preface

The *Source Books* in this series are collections of classical papers that have shaped the structure of the various sciences. Some of these classics are not readily available and many of them have never been translated into English, thus being lost to the general reader and in many cases to the scientist himself. The point of this series is to make these texts easily accessible and to provide good translations of the ones that have not been translated at all, or only poorly.

The series was planned to include volumes in all the major sciences from the Renaissance through the nineteenth century. It has been extended to include ancient and medieval Western science and the development of the sciences in the first half of the present century. Many of these books have been published already and others are in various stages of preparation.

The Carnegie Corporation originally financed the series by a grant to the American Philosophical Association. The History of Science Society and the American Association for the Advancement of Science have approved the project and are represented on the Editorial Advisory Board. This Board at present consists of the following members:

The series was begun and sustained by the devoted labors of Gregory D. Walcott and Everett W. Hall, the first two General Editors. I am indebted to them, to the members of the Advisory Board, and to Joseph D. Elder, Science Editor of Harvard University Press, for their indispensable aid in guiding the course of the *Source Books*.

EDWARD H. MADDEN

Department of Philosophy,
State University of New York at Buffalo

Preface

The second half of the nineteenth century saw a rebirth of logic. That science—which, many felt, had reached its completion and lacked any future—entered a renaissance that was to transform it radically. Though it had been heralded by Leibniz, the new development did not actually start till the middle of the nineteenth century. Boole, De Morgan, and Jevons are regarded as the initiators of modern logic, and rightly so. This first phase, however, suffered from a number of limitations. It tried to copy mathematics too closely, and often artificially. The multiplicity of interpretations of what became known as Boolean algebra created confusion and for a time was a hindrance rather than an advantage. Considered by itself, the period would, no doubt, leave its mark upon the history of logic, but it would not count as a great epoch.

A great epoch in the history of logic did open in 1879, when Gottlob Frege's *Begriffsschrift* was published. This book freed logic from an artificial connection with mathematics but at the same time prepared a deeper interrelation between these two sciences. It presented to the world, in full-fledged form, the propositional calculus and quantification theory. Although Frege's work was slow in winning recognition, the next decades saw striking advances in logic. Two new fields, set theory and foundations of mathematics, emerged on the borders of logic, mathematics, and philosophy.

The texts printed below have been chosen so as to depict this development. Their selection had to satisfy various conditions, sometimes hard to reconcile. The main constraint was that the texts to be selected had to fit between the covers of a single volume. This precluded encyclopedic completeness; the book had to hold to the main lines of the development. Some texts were included in the volume because they have become classics, others because they are perhaps not as well known as they deserve. Some works had to be omitted because of their length, some because of copyright restrictions. If the main strands were to be adequately covered, a number of accessory topics, such as modal logic, had to be left aside.

The volume opens with Frege's *Begriffsschrift*, which is followed by Peano's *Arithmetices principia*. These two works, each in its own manner, initiate the era of the logical reconstruction of mathematics that led to *Principia mathematica*. Dedekind's letter (*1890a*) fits into this trend. But soon difficulties arise; the modern paradoxes appear. Burali-Forti's paper, Cantor's letter (*1899*) to Dedekind, Russell's letter (*1902*) to Frege and Frege's answer, Richard's (*1905*) and König's (*1905a*) papers mark the troublesome spots. Hilbert (*1904*), Russell (*1908a*), Zermelo (*1908a*) present various responses to these difficulties and, with their different methods of overcoming the paradoxes, initiate three important developments in logic: proof theory and the search for a consistency proof, the theory of types, and axiomatic set theory.

Löwenheim's paper (*1915*), which links up with the work of Schröder, brings to the fore notions (validity, decision methods) that had remained in the background (Padoa's paper (*1900*), however, already deals with semantic questions) and presents the first proof—incomplete—of a very disquieting theorem. The work of Skolem belongs to several trends. In two papers (*1920, 1922*) he gives new proofs of Löwenheim's theorem and generalizes it; in another (*1923*) he studies primitive recursive arithmetic, which will become an important tool in foundations of mathematics, and in a fourth (*1928*) he deals with a proof procedure and decision problems. Fraenkel (*1922b*) and Skolem (*1922*) amend and broaden Zermelo's axiomization of set theory, while von Neumann (*1923, 1925*) offers a somewhat different system.

The twenties heard a fiery dialogue between two conceptions of mathematics, Hilbert's and Brouwer's. Hilbert's doctrine is expounded in two papers (*1925, 1927*), and with them are connected those of Weyl, Bernays, and Ackermann. One of Brouwer's papers (*1923b*) presents an aspect of his critique of classical mathematics, another (*1927a*) tries to draw the balance of the controversy with Hilbert, while a third (*1927*) sets forth some positive contributions of intuitionism (fan theorem, bar theorem). Kolmogorov's paper (*1925*) is the first systematic study of intuitionistic logic.

The last chapter of Herbrand's thesis (*1930*) presents his important theorem. Gödel's proof of the completeness of the first-order predicate calculus (*1930*) marks the end of a period in the history of quantification theory.

Gödel's incompleteness paper (*1931*) is here, of course. Herbrand's paper on the consistency of arithmetic (*1931a*), although published a few months after Gödel's, belongs to the preceding period. The importance of Gödel's results and methods, their direct impact and their indirect influence, the profound revision that they imposed upon Hilbert's program, the development of the theory of recursive functions, finally, the emergence in the thirties of an American school of logicians—all that makes the year 1931 a fitting terminal date for the selection presented here.

The volume is a source book in "mathematical logic". "Mathematical" here means that derivations in logic proceed according to definite and explicit rules, the way we do sums (Frege's system, as he says, is "modeled upon the language of arithmetic"). Thus "mathematical" perhaps expresses a deeper feature of present-day logic than does "symbolic". Logic today is symbolic indeed, but the use of symbols, though convenient and perhaps even indispensable, is possible because logic is mathematical in the sense just mentioned. "Formal logic", too, would still be a perfectly appropriate name. But there is no point in searching for an epithet. There are no two logics. Mathematical logic is what logic, through twenty-five centuries and a few transformations, has become today.

The papers selected are reproduced *in extenso*, except for five: those of Peano, Brouwer (*1927, 1927a*), Skolem (*1920*), and Herbrand (*1930*). For each of these papers the reasons for the excision and its extent are indicated in the corresponding introductory note.

Bibliographical references have been given a uniform style, and the list of them will be found on pages 629–655. Footnotes are numbered consecutively in each paper, even when they were not so in the original; when they were, the author's numeration has not been disturbed. Trifling mistakes have been corrected without further ado, but when a mistake, misprint, or inconsistency could provoke a misunderstanding, an editorial comment has been inserted. These editorial interpolations are given in "⟦ ⟧".

Professors Burton Dreben, W. V. Quine, and Hao Wang generously contributed their time in helping me to select the texts. They also lent a hand with clarifying passages in the originals or smoothing out phrases in the translations. Quine wrote the introductory notes to *Russell 1908*, *Whitehead and Russell 1910*, and *Schönfinkel 1924*. Wang wrote that to *Kolmogorov 1925*. Dreben wrote a number of footnotes and ten Notes for *Herbrand 1930*, and we jointly wrote the introductory note to *Skolem 1928*. Professor Charles Parsons wrote the introductory note to *Brouwer 1927*. The other introductory notes were written by me.

The volume contains forty-six papers, originally written in seven languages. Some authors were not writing in their mother tongues; some papers were hastily written, and many have a fair sprinkling of ambiguities, minor errors, and misprints; before 1930 the standards of editorship were not what they are today. All this means that the translators and the editor had their share of problems. They strove above all to produce a clear and unambiguous text, but they have endeavored to follow that narrow path where accuracy combines with good style. The translations were made by Mr. Stefan Bauer-Mengelberg (of the IBM Systems Research Institute), Dr. Dagfinn Føllesdal, Miss Beverly Woodward, and the editor; moreover, several other persons were enlisted to assist in translating one paper or another. For each paper the credit for the translation will be found in the introductory note. As a rule, the authors have not seen the translations of their texts. The exceptions are Lord Russell, who saw and approved the translation of his 1902 letter to Frege (written in German), Professor Fraenkel, who read and approved the translation of his paper, Professor Finsler, who approved the translation of his paper after suggesting a few changes, and Professor Gödel, who read the translations of his papers and approved them after introducing a number of changes in *1931*.

A teaser for translators of German texts on foundations of mathematics is the word "inhaltlich". Mr. Bauer-Mengelberg coined the neologism "contentual" and used it at a number of places. Elsewhere various periphrases were adopted; in particular, Professor Gödel suggested those that are used in the translation of his *1931*.

As the reader will see, a large share of the translations from the German devolved on Mr. Bauer-Mengelberg. Moreover, he went over the whole manuscript and suggested many changes for the better. Miss Paula Thibault read a good part of the volume and wielded her red pencil on many phrases that had been left awkward by the translators' desire for accuracy. Mr. Joseph D. Elder, Science Editor of Harvard University Press, has been at all times a dispenser of sound editorial advice.

JEAN VAN HEIJENOORT

The Springs, East Hampton, New York
23 July 1966

Contents

FROM FREGE TO GÖDEL

A Source Book in Mathematical Logic, 1879–1931

Begriffsschrift, *a formula language, modeled upon that of arithmetic, for pure thought*

GOTTLOB FREGE

(*1879*)

This is the first work that Frege wrote in the field of logic, and, although a mere booklet of eighty-eight pages, it is perhaps the most important single work ever written in logic. Its fundamental contributions, among lesser points, are the truth-functional propositional calculus, the analysis of the proposition into function and argument(s) instead of subject and predicate, the theory of quantification, a system of logic in which derivations are carried out exclusively according to the form of the expressions, and a logical definition of the notion of mathematical sequence. Any single one of these achievements would suffice to secure the book a permanent place in the logician's library.

Frege was a mathematician by training;[a] the point of departure of his investigations in logic was a mathematical question, and mathematics left its mark upon his logical accomplishments. In studying the concept of number, Frege was confronted with difficulties when he attempted to give a logical analysis of the notion of sequence. The imprecision and ambiguity of ordinary language led him to look for a more appropriate tool; he devised a new mode of expression, a language that deals with the "conceptual content" and that he came to call "Begriffsschrift".[b] This ideography is a "formula language", that is, a *lingua characterica*, a language written with special symbols, "for pure thought", that is, free from rhetorical embellishments, "modeled upon that of arithmetic", that is, constructed from specific symbols that are manipulated according to definite rules. The last phrase does not mean that logic mimics arithmetic, and the analogies, uncovered by Boole and others, between logic and arithmetic are useless for Frege, precisely because he wants to employ logic in

[a] See his *Inaugural-Dissertation* (*1873*) and his thesis for *venia docendi* (*1874*).

[b] In the translation below this term is rendered by "ideography", a word used by Jourdain in a paper (*1912*) read and annotated by Frege; that Frege acquiesced in its use was the reason why ultimately it was adopted here. Another acceptable rendition is "concept writing", used by Austin (*Frege 1950*, p. 92e).

Professor Günther Patzig was so kind as to report in a private communication that a student of his, Miss Carmen Diaz, found an occurrence of the word "Begriffsschrift" in Trendelenburg (*1867*, p. 4, line 1), a work that Frege quotes in his preface to *Begriffsschrift* (see below, p. 6). Frege used the word in other writings, and in particular in his major work (*1893, 1903*), but subsequently he seems to have become dissatisfied with it. In an unpublished fragment dated 26 July 1919 he writes: "I do not start from concepts in order to build up thoughts or propositions out of them; rather, I obtain the components of a thought by decomposition [[Zerfällung]] of the thought. In this respect my Begriffsschrift differs from the similar creations of Leibniz and his successors—in spite of its name, which perhaps I did not chose very aptly".

order to provide a foundation for arithmetic. He carefully keeps the logical symbols distinct from the arithmetic ones. Schröder (*1880*) criticized him for doing just that and thus wrecking a tradition established in the previous thirty years. Frege (*1882*, pp. 1–2) answered that his purpose had been quite different from that of Boole: "My intention was not to represent an abstract logic in formulas, but to express a content through written signs in a more precise and clear way than it is possible to do through words. In fact, what I wanted to create was not a mere *calculus ratiocinator* but a *lingua characterica* in Leibniz's sense".

Mathematics led Frege to an innovation that was to have a profound influence upon modern logic. He observes that we would do violence to mathematical statements if we were to impose upon them the distinction between subject and predicate. After a short but pertinent critique of that distinction, he replaces it by another, borrowed from mathematics but adapted to the needs of logic, that of function and argument. Frege begins his analysis by considering an ordinary sentence and remarks that the expression remains meaningful when certain words are replaced by others. A word for which we can make such successive substitutions occupies an argument place, and the stable component of the sentence is the function. This, of course, is not a definition, because in his system Frege deals not with ordinary sentences but with formulas; it is merely an explanation, after which he introduces functional letters and gives instructions for handling them and their arguments. Nowhere in the present text does Frege state what a function is or speak of the value of a function. He simply says that a judgment is obtained when the argument places between the parentheses attached to a functional letter have been properly filled (and, should the case so require, quantifiers have been properly used). It is only in his subsequent writings (*1891* and thereafter) that Frege will devote a great deal of attention to the nature of a function.

Frege's booklet presents the propositional calculus in a version that uses the conditional and negation as primitive connectives. Other connectives are examined for a moment, and their intertranslatability with the conditional and negation is shown. Mostly to preserve the simple formulation of the rule of detachment, Frege decides to use these last two. The notation that he introduces for the conditional has often been criticized, and it has not survived. It presents difficulties in printing and takes up a large amount of space. But, as Frege himself (*1896*, p. 364) says, "the comfort of the typesetter is certainly not the *summum bonum*", and the notation undoubtedly allows one to perceive the structure of a formula at a glance and to perform substitutions with ease. Frege's definition of the conditional is purely truth-functional, and it leads him to the rule of detachment, stated in § 6. He notes the discrepancy between this truth-functional definition and ordinary uses of the word "if". Frege dismisses modal considerations from his logic with the remark that they concern the grounds for accepting a judgment, not the content of the judgment itself. Frege's use of the words "affirmed" and "denied", with his listing of all possible cases in the assignment of these terms to propositions, in fact amounts to the use of the truth-table method. His axioms for the propositional calculus (they are not independent) are formulas (1), (2), (8), (28), (31), and (41). His rules of inference are the rule of detachment and an unstated rule of substitution. A number of theorems of the propositional calculus are proved, but no question of completeness, consistency, or independence is raised.

Quantification theory is introduced in § 11. Frege's instructions how to use italic and German letters contain, in effect, the rule of generalization and the rule that allows us to infer $A \supset (x)F(x)$ from $A \supset F(x)$ when x does not occur free in A. There are three new axioms: (58) for instantiation, (52) and (54) for identity. No rule of substitution is explicitly stated, and one has to examine Frege's practice in his derivations to see what he allows. The substitutions are indicated by tables on the left of the derivations. These substitutions are simultaneous substitutions. When a substitution is specified with the help of "Γ", which plays the role of what we would today call a syntactic variable, particular care should be exercised, and it proves convenient to perform the substitutions that do not involve "Γ" before that involving "Γ" is carried out. The point will become clear to the reader if he compares, for example, the derivation of (51) with that of (98). Frege's derivations are quite detailed and, even in the absence of an explicit rule of substitution, can be unambiguously reconstructed.

Frege allows a functional letter to occur in a quantifier (p. 24 below). This license is not a necessary feature of quantification theory, but Frege has to admit it in his system for the definitions and derivations of the third part of the book. The result is that the difference between function and argument is blurred. In fact, even before coming to quantification over functions, Frege states (p. 24 below) that we can consider $\Phi(A)$ to be a function of the argument Φ as well as of the argument A. (This is precisely the point that Russell will seize upon to make it bear the brunt of his paradox—see below, p. 125). It is true that Frege writes (p. 24 below) that, if a functional letter occurs in a quantifier, "this circumstance must be taken into account". But the phrase remains vague. The most generous interpretation would be that, in the scope of the quantifier in which it occurs, a functional letter has to be treated as such, that is, must be provided with a pair of parentheses and one or more arguments. Frege, however, does not say as much, and in the derivation of formula (77) he substitutes $\mathfrak{F}$ for $\mathfrak{a}$ in $f(\mathfrak{a})$, at least as an intermediate step. If we also observe that in the derivation of formula (91) he substitutes $\mathfrak{F}$ for f, we see that he is on the brink of a paradox. He will fall into the abyss when (*1891*) he introduces the course-of-values of a function as something "complete in itself", which "may be taken as an argument". For the continuation of the story see pages 124–128.

This flaw in Frege's system should not make us lose sight of the greatness of his achievement. The analysis of the proposition into function and argument, rather than subject and predicate, and quantification theory, which became possible only after such an analysis, are the very foundations of modern logic. The problems connected with quantification over functions could be approached only after a quantification theory had already been established. When the slowness and the wavering that marked the development of the propositional calculus are remembered, one cannot but marvel at seeing quantification theory suddenly coming full-grown into the world. Many years later (*1894*, p. 21) Peano still finds quantification theory "abstruse" and prefers to deal with it by means of just a few examples. Frege can proudly answer (*1896*, p. 376) that in 1879 he had already given all the laws of quantification theory; "these laws are few in number, and I do not know why they should be said to be abstruse".

In distinguishing his work from that of his predecessors and contemporaries, Frege repeatedly opposes a *lingua characterica* to a *calculus ratiocinator*. He uses these terms, suggested by Leibniz, to bring out an important feature of his system, in fact, one of the greatest achievements of his *Begriffsschrift*. In the pre-Fregean calculus of propositions and classes, logic, translated into formulas, is studied by means of arguments resting upon an intuitive logic. What Frege does is to construct logic as a language that need not be supplemented by any intuitive reasoning. Thus he is very careful to describe his system in purely *formal* terms (he even speaks of letters—Latin, German, and so on—rather than of variables, because of the imprecision of the latter term). He is fully aware that any system requires rules that cannot be expressed in the system; but these rules are void of any intuitive logic; they are "rules for the use of our signs" (p. 28 below): the rule of detachment, the rules for dealing with quantifiers. This is one of the great lessons of Frege's book. It was a new one in 1879, and it did not at once pervade the world of logic.

The third part of the book introduces a theory of mathematical sequences. Frege is moving toward his goal, the logical reconstruction of arithmetic. He defines the relation that Whitehead and Russell (*1910*, part II, sec. E) came to call the ancestral relation and that later (*1940*) Quine called the ancestral. The proper ancestral appears in § 26 and the ancestral proper in § 29. Subsequently Frege will use the notion for the justification of mathematical induction (*1884*, p. 93). Dedekind (below, p. 101, and *1893*, XVII) recognized that the ancestral agrees in essence with his own notion of chain, which was publicly introduced nine years after Frege's notion.

At times *Begriffsschrift* begs for a clarification of linguistic usage, for a distinction between expressions and what these expressions refer to. In his subsequent writings Frege will devote a great deal of attention to this problem. On one point, however, the book touches upon them, and not too happily. In § 8 identity of content is introduced as a relation between names, not their contents. "$\vdash\!\!\!-\!\!-\, A = B$" means that the signs "$A$" and "$B$" have the same conceptual content and, according to Frege, is a statement about signs.[c] There are strong arguments against such a conception, and Frege will soon recognize them. This will lead him to split the notion of conceptual content into sense ("Sinn") and reference ("Bedeutung") (*1892a*, but see also *1891*, p. 14; these two papers can be viewed as long emendations to *Begriffsschrift*).

In 1910 Jourdain sent to Frege the manuscript of a long paper that he had written on the history of logic and that contained a summary of *Begriffsschrift*. Frege answered with comments on a number of points, and Jourdain incorporated Frege's remarks in footnotes to his paper (*1912*). Some of these footnotes are reproduced below, at their appropriate places, with slight revisions in Jourdain's translation of Frege's comments (moreover, the German text used here is Frege's copy, and there are indications that the text that he sent to Jourdain and the copy that he preserved are not identical).

A few words should be said about Frege's use of the term "Verneinung". In a first use, "Verneinung" is opposed to "Bejahung", "verneinen" to "bejahen", and what these words express is, in fact, the ascription of truth values to contents of judgments; they are translated, respectively, by "denial" and "affirmation", "to deny" and "to affirm". The second use of "Verneinung" is for the connective, and when so used it is translated by "negation".

A number of misprints in the original were discovered during the translation. Most of them are included in the errata list that the reader will find in the reprint of Frege's booklet (*1964*, pp. 122–123). Those that are not in that list are the following:

(1) On page XV, lines 6u, 5u, and 3u of the German text, "A" and "B" (which are alpha and beta) are not of the same font as "Φ" and "Ψ", while they should be;

---

[c] On the nature of identity see comments in the present volume by Whitehead and Russell (below, pp. 218–219) and by Skolem (below, pp. 304–305).

(2) On page 29 of the German text, in § 15, the letters to the left of the long vertical line under (1) should be "$\mathfrak{a}$" and "$\mathfrak{b}$", not "a" and "b";

(3) The misprint indicated in footnote 18, p. 57 below;

(4) The misprint indicated in footnote 21, p. 65 below.

Moreover, Misprint 3 in the reprint's list does not occur in the German text used for the present translation; apparently, it is not a misprint at all but is simply due to the poor printing of some copies. The reprint also introduces misprints of its own: on page 1, line 4u, we find "——" where there should be "|——"; on page 62, near the top of the page, "$\genfrac{}{}{0pt}{}{\gamma}{\beta}$" should be "$\genfrac{}{}{0pt}{}{\gamma}{\tilde{\beta}}$".; on page 65 there should be a vertical negation stroke attached to the stroke preceding the first occurrence of "$h(y)$"; on page 39 an unreadable broken "c" has been left uncorrected.

The translation is by Stefan Bauer-Mengelberg, and it is published here by arrangement with Georg Olms Verlagsbuchhandlung.

PREFACE

In apprehending a scientific truth we pass, as a rule, through various degrees of certitude. Perhaps first conjectured on the basis of an insufficient number of particular cases, a general proposition comes to be more and more securely established by being connected with other truths through chains of inferences, whether consequences are derived from it that are confirmed in some other way or whether, conversely, it is seen to be a consequence of propositions already established. Hence we can inquire, on the one hand, how we have gradually arrived at a given proposition and, on the other, how we can finally provide it with the most secure foundation. The first question may have to be answered differently for different persons; the second is more definite, and the answer to it is connected with the inner nature of the proposition considered. The most reliable way of carrying out a proof, obviously, is to follow pure logic, a way that, disregarding the particular characteristics of objects, depends solely on those laws upon which all knowledge rests. Accordingly, we divide all truths that require justification into two kinds, those for which the proof can be carried out purely by means of logic and those for which it must be supported by facts of experience. But that a proposition is of the first kind is surely compatible with the fact that it could nevertheless not have come to consciousness in a human mind without any activity of the senses.[1] Hence it is not the psychological genesis but the best method of proof that is at the basis of the classification. Now, when I came to consider the question to which of these two kinds the judgments of arithmetic belong, I first had to ascertain how far one could proceed in arithmetic by means of inferences alone, with the sole support of those laws of thought that transcend all particulars. My initial step was to attempt to reduce the concept of ordering in a sequence to that of *logical* consequence, so as to proceed from there to the concept of number. To prevent anything intuitive ⟦Anschauliches⟧ from penetrating here unnoticed, I had to bend every effort to keep the chain of inferences free of gaps. In attempting to comply with this requirement in the strictest possible way I found the inadequacy of language to be an

[1] Since without sensory experience no mental development is possible in the beings known to us, that holds of all judgments.

obstacle; no matter how unwieldy the expressions I was ready to accept, I was less and less able, as the relations became more and more complex, to attain the precision that my purpose required. This deficiency led me to the idea of the present ideography. Its first purpose, therefore, is to provide us with the most reliable test of the validity of a chain of inferences and to point out every presupposition that tries to sneak in unnoticed, so that its origin can be investigated. That is why I decided to forgo expressing anything that is without significance for the *inferential sequence.* In § 3 I called what alone mattered to me the *conceptual content* ⟦*begrifflichen Inhalt*⟧. Hence this definition must always be kept in mind if one wishes to gain a proper understanding of what my formula language is. That, too, is what led me to the name "Begriffsschrift". Since I confined myself for the time being to expressing relations that are independent of the particular characteristics of objects, I was also able to use the expression "formula language for pure thought". That it is modeled upon the formula language of arithmetic, as I indicated in the title, has to do with fundamental ideas rather than with details of execution. Any effort to create an artificial similarity by regarding a concept as the sum of its marks ⟦Merkmale⟧ was entirely alien to my thought. The most immediate point of contact between my formula language and that of arithmetic is the way in which letters are employed.

I believe that I can best make the relation of my ideography to ordinary language ⟦Sprache des Lebens⟧ clear if I compare it to that which the microscope has to the eye. Because of the range of its possible uses and the versatility with which it can adapt to the most diverse circumstances, the eye is far superior to the microscope. Considered as an optical instrument, to be sure, it exhibits many imperfections, which ordinarily remain unnoticed only on account of its intimate connection with our mental life. But, as soon as scientific goals demand great sharpness of resolution, the eye proves to be insufficient. The microscope, on the other hand, is perfectly suited to precisely such goals, but that is just why it is useless for all others.

This ideography, likewise, is a device invented for certain scientific purposes, and one must not condemn it because it is not suited to others. If it answers to these purposes in some degree, one should not mind the fact that there are no new truths in my work. I would console myself on this point with the realization that a development of method, too, furthers science. Bacon, after all, thought it better to invent a means by which everything could easily be discovered than to discover particular truths, and all great steps of scientific progress in recent times have had their origin in an improvement of method.

Leibniz, too, recognized—and perhaps overrated—the advantages of an adequate system of notation. His idea of a universal characteristic, of a *calculus philosophicus* or *ratiocinator*,[2] was so gigantic that the attempt to realize it could not go beyond the bare preliminaries. The enthusiasm that seized its originator when he contemplated the immense increase in the intellectual power of mankind that a system of notation directly appropriate to objects themselves would bring about led him to underestimate the difficulties that stand in the way of such an enterprise. But, even if this worthy goal cannot be reached in one leap, we need not despair of a slow, step-by-step approximation. When a problem appears to be unsolvable in its full generality, one should

[2] On that point see *Trendelenburg 1867* ⟦pp. 1–47, *Ueber Leibnizens Entwurf einer allgemeinen Charakteristik*⟧.

temporarily restrict it; perhaps it can then be conquered by a gradual advance. It is possible to view the signs of arithmetic, geometry, and chemistry as realizations, for specific fields, of Leibniz's idea. The ideography proposed here adds a new one to these fields, indeed the central one, which borders on all the others. If we take our departure from there, we can with the greatest expectation of success proceed to fill the gaps in the existing formula languages, connect their hitherto separated fields into a single domain, and extend this domain to include fields that up to now have lacked such a language.[3]

I am confident that my ideography can be successfully used wherever special value must be placed on the validity of proofs, as for example when the foundations of the differential and integral calculus are established.

It seems to me to be easier still to extend the domain of this formula language to include geometry. We would only have to add a few signs for the intuitive relations that occur there. In this way we would obtain a kind of *analysis situs.*

The transition to the pure theory of motion and then to mechanics and physics could follow at this point. The latter two fields, in which besides rational necessity ⟦Denknothwendigkeit⟧ empirical necessity ⟦Naturnothwendigkeit⟧ asserts itself, are the first for which we can predict a further development of the notation as knowledge progresses. That is no reason, however, for waiting until such progress appears to have become impossible.

If it is one of the tasks of philosophy to break the domination of the word over the human spirit by laying bare the misconceptions that through the use of language often almost unavoidably arise concerning the relations between concepts and by freeing thought from that with which only the means of expression of ordinary language, constituted as they are, saddle it, then my ideography, further developed for these purposes, can become a useful tool for the philosopher. To be sure, it too will fail to reproduce ideas in a pure form, and this is probably inevitable when ideas are represented by concrete means; but, on the one hand, we can restrict the discrepancies to those that are unavoidable and harmless, and, on the other, the fact that they are of a completely different kind from those peculiar to ordinary language already affords protection against the specific influence that a particular means of expression might exercise.

The mere invention of this ideography has, it seems to me, advanced logic. I hope that logicians, if they do not allow themselves to be frightened off by an initial impression of strangeness, will not withhold their assent from the innovations that, by a necessity inherent in the subject matter itself, I was driven to make. These deviations from what is traditional find their justification in the fact that logic has hitherto always followed ordinary language and grammar too closely. In particular, I believe that the replacement of the concepts *subject* and *predicate* by *argument* and *function*, respectively, will stand the test of time. It is easy to see how regarding a content as a function of an argument leads to the formation of concepts. Furthermore, the demonstration of the connection between the meanings of the words *if, and, not, or, there is, some, all,* and so forth, deserves attention.

Only the following point still requires special mention. The restriction, in § 6, to a

[3] ⟦On that point see *Frege 1879a.*⟧

single mode of inference is justified by the fact that, when the foundations for such an ideography are laid, the primitive components must be taken as simple as possible, if perspicuity and order are to be created. This does not preclude the possibility that *later* certain transitions from several judgments to a new one, transitions that this one mode of inference would not allow us to carry out except mediately, will be abbreviated into immediate ones. In fact this would be advisable in case of eventual application. In this way, then, further modes of inference would be created.

I noticed afterward that formulas (31) and (41) can be combined into a single one,

$$\vdash\!\!\!\text{——}\,(\text{┬┬}\, a \equiv a),$$

which makes some further simplifications possible.

As I remarked at the beginning, arithmetic was the point of departure for the train of thought that led me to my ideography. And that is why I intend to apply it first of all to that science, attempting to provide a more detailed analysis of the concepts of arithmetic and a deeper foundation for its theorems. For the present I have reported in the third chapter some of the developments in this direction. To proceed farther along the path indicated, to elucidate the concepts of number, magnitude, and so forth—all this will be the object of further investigations, which I shall publish immediately after this booklet.

Jena, 18 December 1878.

CONTENTS

I. Definition of the symbols

Identity of content

Functions

Generality

II. Representation and derivation of some judgments of pure thought

III. Some topics from a general theory of sequences

I. DEFINITION OF THE SYMBOLS

§ 1. The signs customarily employed in the general theory of magnitudes are of two kinds. The first consists of letters, of which each represents either a number left indeterminate or a function left indeterminate. This indeterminacy makes it possible to use letters to express the universal validity of propositions, as in

$$(a + b)c = ac + bc.$$

The other kind consists of signs such as $+$, $-$, $\surd$, 0, 1, and 2, of which each has its particular meaning.[4]

I adopt this basic idea of distinguishing two kinds of signs, which unfortunately is not strictly observed in the theory of magnitudes,[5] *in order to apply it in the more*

[4] ⟦Footnote by Jourdain (*1912*, p. 238):

Russell (*1908*) has expressed it: "A variable is a symbol which is to have one of a certain set of values, without its being decided which one. It does not have first one value of a set and then another; it has at all times *some* value of the set, where, so long as we do not replace the variable by a constant, the 'some' remains unspecified."

On the word "variable" Frege has supplied the note: "Would it not be well to omit this expression entirely, since it is hardly possible to define it properly? Russell's definition immediately raises the question what it means to say that 'a symbol has a value'. Is the relation of a sign to its significatum meant by this? In that case, however, we must insist that the sign be univocal, and the meaning (value) that the sign is to have must be determinate; then the variable would be a sign. But for him who does not subscribe to a formal theory a variable will not be a sign, any more than a number is. If, now, you write 'A variable is represented by a symbol that is to represent one of a certain set of values', the last defect is thereby removed; but what is the case then? The symbol represents, first, the variable and, second, a value taken from a certain supply without its being determined which. Accordingly, it seems better to leave the word 'symbol' out of the definition. The question as to what a variable is has to be answered independently of the question as to which symbol is to represent the variable. So we come to the definition: 'A variable is one of a certain set of values, without its being decided which one'. But the last addition does not yield any closer determination, and to belong to a certain set of values means, properly, to fall under a certain concept; for, after all, we can determine this set only by giving the properties that an object must have in order to belong to the set; that is, the set of values will be the extension of a concept. But, now, we can for every object specify a set of values to which it belongs, so that even the requirement that something is to be a value taken from a certain set does not determine anything. It is probably best to hold to the convention that Latin letters serve to confer generality of content on a theorem. And it is best not to use the expression 'variable' at all, since ultimately we cannot say either of a sign, or of what it expresses or denotes, that it is variable or that it is a variable, at least not in a sense that can be used in mathematics or logic. On the other hand, perhaps someone may insist that in '$(2 + x)(3 + x)$' the letter 'x' does not serve to confer generality of content on a proposition. But in the context of a proof such a formula will always occur as a part of a proposition, whether this proposition consists partly of words or exclusively of mathematical signs, and in such a context x will always serve to confer generality of content on a proposition. Now, it seems to me unfortunate to restrict to a particular set the values that are admissible for this letter. For we can always add the condition that a belong to this set, and then drop that condition. If an object Δ does not belong to the set, the condition is simply not satisfied and, if we replace 'a' by 'Δ' in the entire proposition, we obtain a true proposition. I would not say of a letter that it has a signification, a sense, a meaning, if it serves to confer generality of content on a proposition. We can replace the letter by the proper name 'Δ' of an object Δ; but this Δ cannot anyhow be regarded as the *meaning* of the letter; for it is not more closely allied with the letter than is any other object. Also, generality cannot be regarded as the meaning of the Latin letter; for it cannot be regarded as something independent, something that would be added to a content already complete in other respects. I would not, then, say 'terms whose meaning is indeterminate' or 'signs have variable meanings'. In this case signs have no denotations at all." [Frege, 1910.]⟧

[5] Consider 1, log, sin, lim.

comprehensive domain of pure thought in general. I therefore divide all signs that I use into *those by which we may understand different objects* and *those that have a completely determinate meaning.* The former are *letters* and they will serve chiefly to express *generality.* But, no matter how indeterminate the meaning of a letter, we must insist that throughout a given context the letter *retain* the meaning once given to it.

Judgment

§ 2. A judgment will always be expressed by means of the sign

which stands to the left of the sign, or the combination of signs, indicating the content of the judgment. If we *omit* the small vertical stroke at the left end of the horizontal one, the judgment will be transformed into a *mere combination of ideas* [*Vorstellungsverbindung*],[6] of which the writer does not state whether he acknowledges it to be true or not. For example, let

$$\vdash\!\!\!\!-\!\!\!-\!\!\!-\, A$$

stand for ⟦bedeute⟧ the judgment "Opposite magnetic poles attract each other";[7] then

$$-\!\!\!-\!\!\!-\, A$$

will not express ⟦ausdrücken⟧ this judgment;[8] it is to produce in the reader merely the idea of the mutual attraction of opposite magnetic poles, say in order to derive consequences from it and to test by means of these whether the thought is correct. When the vertical stroke is omitted, we express ourselves *paraphrastically,* using the words "the circumstance that" or "the proposition that".[9]

Not every content becomes a judgment when $\vdash\!\!\!-\!\!\!-$ is written before its sign; for

[6] ⟦Footnote by Jourdain (*1912,* p. 242):

"For this word I now simply say 'Gedanke'. The word 'Vorstellungsinhalt' is used now in a psychological, now in a logical sense. Since this creates obscurities, I think it best not to use this word at all in logic. We must be able to express a thought without affirming that it is true. If we want to characterize a thought as false, we must first express it without affirming it, then negate it, and affirm as true the thought thus obtained. We cannot correctly express a hypothetical connection between thoughts at all if we cannot express thoughts without affirming them, for in the hypothetical connection neither the thought appearing as antecedent nor that appearing as consequent is affirmed." [Frege, 1910.]⟧

[7] I use Greek letters as abbreviations, and to each of these letters the reader should attach an appropriate meaning when I do not expressly give them a definition. ⟦The "*A*" that Frege is now using is a capital alpha.⟧

[8] ⟦Jourdain had originally translated "bedeuten" by "signify", and Frege wrote (see *Jourdain 1912,* p. 242):

"Here we must notice the words 'signify' and 'express'. The former seems to correspond to 'bezeichnen' or 'bedeuten', the latter to 'ausdrücken'. According to the way of speaking I adopted I say 'A proposition expresses a thought and signifies its truth value'. Of a judgment we cannot properly say either that it signifies or that it is expressed. We do, to be sure, have a thought in the judgment, and that can be expressed; but we have more, namely, the recognition of the truth of this thought."⟧

[9] ⟦Footnote by Jourdain (*1912,* p. 243):

"Instead of 'circumstance' and 'proposition' I would simply say 'thought'. Instead of 'beurtheilbarer Inhalt' we can also say 'Gedanke'." [Frege, 1910.]⟧

example, the idea "house" does not. We therefore distinguish contents that *can become a judgment* from those that *cannot.*[10]

The horizontal stroke that is part of the sign ⊢—— *combines the signs that follow it into a totality, and the affirmation expressed by the vertical stroke at the left end of the horizontal one refers to this totality.* Let us call the horizontal stroke the *content stroke* and the vertical stroke the *judgment stroke.* The content stroke will in general serve to relate any sign to the totality of the signs that follow the stroke. *Whatever follows the content stroke must have a content that can become a judgment.*

§ 3. A distinction between *subject* and *predicate* does *not occur* in my way of representing a judgment. In order to justify this I remark that the contents of two judgments may differ in two ways: either the consequences derivable from the first, when it is combined with certain other judgments, always follow also from the second, when it is combined with these same judgments, ⟦and conversely,⟧ or this is not the case. The two propositions "The Greeks defeated the Persians at Plataea" and "The Persians were defeated by the Greeks at Plataea" differ in the first way. Even if one can detect a slight difference in meaning, the agreement outweighs it. Now I call that part of the content that is the *same* in both the *conceptual content.* Since *it alone* is of significance for our ideography, we need not introduce any distinction between propositions having the same conceptual content. If one says of the subject that it "is the concept with which the judgment is concerned", this is equally true of the object. We can therefore only say that the subject "is the concept with which the judgment is chiefly concerned". In ordinary language, the place of the subject in the sequence of words has the significance of a *distinguished* place, where we put that to which we wish especially to direct the attention of the listener (see also § 9). This may, for example, have the purpose of pointing out a certain relation of the given judgment to others and thereby making it easier for the listener to grasp the entire context. Now, all those peculiarities of ordinary language that result only from the interaction of speaker and listener—as when, for example, the speaker takes the expectations of the listener into account and seeks to put them on the right track even before the complete sentence is enunciated—have nothing that answers to them in my formula language, since in a judgment I consider only that which influences its *possible consequences.* Everything necessary for a correct inference is expressed in full, but what is not necessary is generally not indicated; *nothing is left to guesswork.* In this I faithfully follow the example of the formula language of mathematics, a language to which one would do violence if he were to distinguish between subject and predicate in it. We can imagine a language in which the proposition "Archimedes perished at the capture of Syracuse" would be expressed thus: "The violent death of Archimedes at the capture of Syracuse is a fact". To be sure, one can distinguish between subject and predicate here, too, if one wishes to do so, but the subject contains the whole content, and the predicate serves only to turn the content into a judgment. *Such a*

[10] On the other hand, the circumstance that there are houses, or that there is a house (see § 12 ⟦footnote 15⟧), is a content that can become a judgment. But the idea "house" is only a part of it. In the proposition "The house of Priam was made of wood" we could not put "circumstance that there is a house" in place of "house". For a different kind of example of a content that cannot become a judgment see the passage following formula (81).

⟦In German Frege's distinction is between "beurtheilbare" and "unbeurtheilbare" contents. Jourdain uses the words "judicable" and "nonjudicable".⟧

language would have only a single predicate for all judgments, namely, "is a fact". We see that there cannot be any question here of subject and predicate in the ordinary sense. *Our ideography is a language of this sort, and in it the sign* $\vdash$ *is the common predicate for all judgments.*

In the first draft of my formula language I allowed myself to be misled by the example of ordinary language into constructing judgments out of subject and predicate. But I soon became convinced that this was an obstacle to my specific goal and led only to useless prolixity.

§ 4. The remarks that follow are intended to explain the significance for our purposes of the distinctions that we introduce among judgments.

We distinguish between *universal* and *particular* judgments; this is really not a distinction between judgments but between contents. *We ought to say "a judgment with a universal content", "a judgment with a particular content".* For these properties hold of the content even when it is *not* advanced as a judgment but as a ⟦mere⟧ proposition (see § 2).

The same holds of negation. In an indirect proof we say, for example, "Suppose that the line segments AB and CD are not equal". Here the content, that the line segments AB and CD are not equal, contains a negation; but this content, though it can become a judgment, is nevertheless not advanced as a judgment. Hence the negation attaches to the content, whether this content becomes a judgment or not. I therefore regard it as more appropriate to consider negation as an adjunct of a *content that can become a judgment.*

The distinction between categoric, hypothetic, and disjunctive judgments seems to me to have only grammatical significance.[11]

The apodictic judgment differs from the assertory in that it suggests the existence of universal judgments from which the proposition can be inferred, while in the case of the assertory one such a suggestion is lacking. By saying that a proposition is necessary I give a hint about the grounds for my judgment. *But, since this does not affect the conceptual content of the judgment, the form of the apodictic judgment has no significance for us.*

If a proposition is advanced as possible, either the speaker is suspending judgment by suggesting that he knows no laws from which the negation of the proposition would follow or he says that the generalization of this negation is false. In the latter case we have what is usually called a *particular affirmative judgment* (see § 12). "It is possible that the earth will at some time collide with another heavenly body" is an instance of the first kind, and "A cold can result in death" of the second.

Conditionality

§ 5. If A and B stand for contents that can become judgments (§ 2), there are the following four possibilities:

(1) A is affirmed and B is affirmed;
(2) A is affirmed and B is denied;
(3) A is denied and B is affirmed;
(4) A is denied and B is denied.

[11] The reason for this will be apparent from the entire book.

Now

$$\vdash\!\!\begin{array}{l}\top\!\!-A\\ \llcorner\!\!-B\end{array}$$

stands for the judgment that *the third of these possibilities does not take place, but one of the three others does.* Accordingly, if

$$-\!\!\begin{array}{l}\top\!\!-A\\ \llcorner\!\!-B\end{array}$$

is denied, this means that the third possibility takes place, hence that A is denied and B affirmed.

Of the cases in which

$$-\!\!\begin{array}{l}\top\!\!-A\\ \llcorner\!\!-B\end{array}$$

is affirmed we single out for comment the following three:

(1) A must be affirmed. Then the content of B is completely immaterial. For example, let $\vdash\!\!-\!\!-A$ stand for $3 \times 7 = 21$ and B for the circumstance that the sun is shining. Then only the first two of the four cases mentioned are possible. There need not exist a causal connection between the two contents.

(2) B has to be denied. Then the content of A is immaterial. For example, let B stand for the circumstance that perpetual motion is possible and A for the circumstance that the world is infinite. Then only the second and fourth of the four cases are possible. There need not exist a causal connection between A and B.

(3) We can make the judgment

$$\vdash\!\!\begin{array}{l}\top\!\!-A\\ \llcorner\!\!-B\end{array}$$

without knowing whether A and B are to be affirmed or denied. For example, let B stand for the circumstance that the moon is in quadrature with the sun and A for the circumstance that the moon appears as a semicircle. In that case we can translate

$$\vdash\!\!\begin{array}{l}\top\!\!-A\\ \llcorner\!\!-B\end{array}$$

by means of the conjunction "if": "If the moon is in quadrature with the sun, the moon appears as a semicircle". The causal connection inherent in the word "if", however, is not expressed by our signs, even though only such a connection can provide the ground for a judgment of the kind under consideration. For causal connection is something general, and we have not yet come to express generality (see § 12).

Let us call the vertical stroke connecting the two horizontal ones the *condition stroke*. The part of the upper horizontal stroke to the left of the condition stroke is the content stroke for the meaning, just explained, of the combination of signs

$$-\!\!\begin{array}{l}\top\!\!-A\\ \llcorner\!\!-B;\end{array}$$

to it is affixed any sign that is intended to relate to the total content of the expression. The part of the horizontal stroke between A and the condition stroke is the content stroke of A. The horizontal stroke to the left of B is the content stroke of B. Accordingly, it is easy to see that

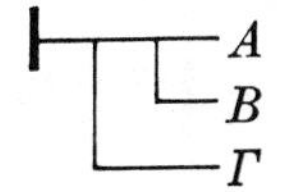

denies the case in which A is denied and B and Γ are affirmed. We must think of this as having been constructed from

and Γ in the same way as

was constructed from A and B. We therefore first have the denial of the case in which

is denied and Γ is affirmed. But the denial of

means that A is denied and B is affirmed. From this we obtain what was given above. If a causal connection is present, we can also say "A is the necessary consequence of B and Γ", or "If the circumstances B and Γ occur, then A also occurs".

It is no less easy to see that

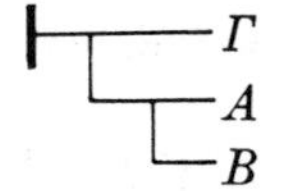

denies the case in which B is affirmed but A and Γ are denied.[12] If we assume that there exists a causal connection between A and B, we can translate the formula as "If A is a necessary consequence of B, one can infer that Γ takes place".

§ 6. The definition given in § 5 makes it apparent that from the two judgments

and $\vdash B$

the new judgment

$\vdash A$

[12] [[There is an oversight here, already pointed out by Schröder (*1880*, p. 88).]]

follows. Of the four cases enumerated above, the third is excluded by

$$\vdash\!\!\begin{array}{l} A \\ B \end{array}$$

and the second and fourth by

$$\vdash B,$$

so that only the first remains.

We could write this inference perhaps as follows:

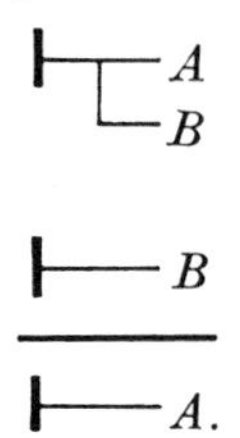

This would become awkward if long expressions were to take the places of A and B, since each of them would have to be written twice. That is why I use the following abbreviation. To every judgment occurring in the context of a proof I assign a number, which I write to the right of the judgment at its first occurrence. Now assume, for example, that the judgment

$$\vdash\!\!\begin{array}{l} A \\ B, \end{array}$$

or one containing it as a special case, has been assigned the number X. Then I write the inference as follows:

$$(\mathrm{X}):\ \frac{\vdash B}{\vdash A.}$$

Here it is left to the reader to put the judgment

$$\vdash\!\!\begin{array}{l} A \\ B \end{array}$$

together for himself from $\vdash B$ and $\vdash A$ and to see whether it is the judgment X above.

If, for example, the judgment $\vdash B$ has been assigned the number XX, I also write the same inference as follows:

$$(\mathrm{XX})::\ \frac{\vdash\!\!\begin{array}{l} A \\ B \end{array}}{\vdash A.}$$

Here the double colon indicates that $\vdash B$, which was only referred to by XX, would have to be formed, from the two judgments written down, in a way different from that above.

Furthermore if, say, the judgment $\vdash\!\!-\!\!-\; \Gamma$ had been assigned the number XXX, I would abbreviate the two judgments

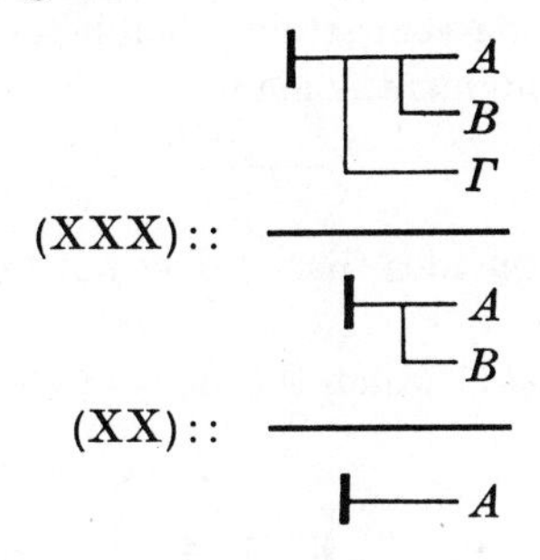

still more thus:

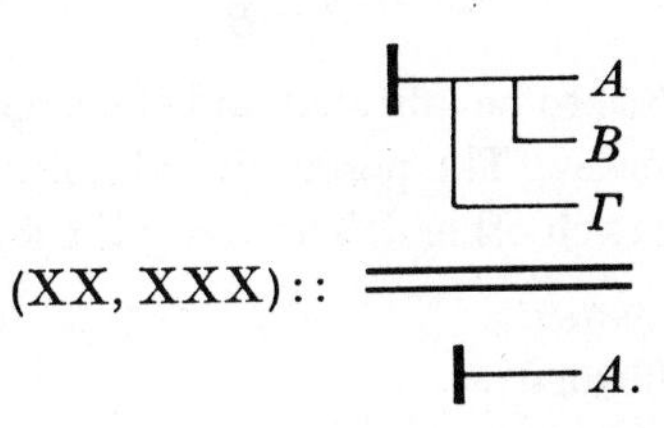

Following Aristotle, we can enumerate quite a few modes of inference in logic; I employ only this one, at least in all cases in which a new judgment is derived from more than a single one. For, the truth contained in some other kind of inference can be stated in one judgment, of the form: if M holds and if N holds, then Λ holds also, or, in signs,

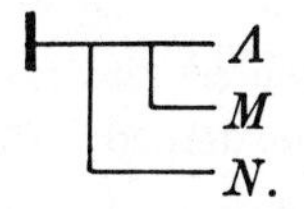

From this judgment, together with $\vdash\!\!-\!\!-\; N$ and $\vdash\!\!-\!\!-\; M$, there follows, as above, $\vdash\!\!-\!\!-\; \Lambda$. In this way an inference in accordance with any mode of inference can be reduced to our case. Since it is therefore possible to manage with a single mode of inference, it is a commandment of perspicuity to do so. Otherwise there would be no reason to stop at the Aristotelian modes of inference; instead, one could continue to add new ones indefinitely: from each of the judgments expressed in a formula in §§ 13–22 we could make a particular mode of inference. *With this restriction to a single mode of inference, however, we do not intend in any way to state a psychological proposition; we wish only to decide a question of form in the most expedient way.* Some of the judgments that take the place of Aristotelian kinds of inference will be listed in § 22 (formulas (59), (62), and 65)).

Negation

§ 7. If a short vertical stroke is attached below the content stroke, this will express the circumstance that *the content does not take place*. So, for example,

$$\vdash\!\!\!\top\!\!-\; A$$

means "A does not take place". I call this short vertical stroke the *negation stroke.*

The part of the horizontal stroke to the right of the negation stroke is the content stroke of A; the part to the left of the negation stroke is the content stroke of the negation of A. If there is no judgment stroke, then here—as in any other place where the ideography is used—no judgment is made.

```
──┬── A
```

merely calls upon us to form the idea that A does not take place, without expressing whether this idea is true.

We now consider some cases in which the signs of conditionality and negation are combined.

```
├─┬┬─ A
  └── B
```

means "The case in which B is to be affirmed and the negation of A to be denied does not take place"; in other words, "The possibility of affirming both A and B does not exist", or "A and B exclude each other". Thus only the following three cases remain:

A is affirmed and B is denied;
A is denied and B is affirmed;
A is denied and B is denied.

In view of the preceding it is easy to state what the significance of each of the three parts of the horizontal stroke to the left of A is.

```
├─┬── A
  └┬─ B
```

means "The case in which A is denied and the negation of B is affirmed does not obtain", or "A and B cannot both be denied". Only the following possibilities remain:

A is affirmed and B is affirmed;
A is affirmed and B is denied;
A is denied and B is affirmed;

A and B together exhaust all possibilities. Now the words "or" and "either—or" are used in two ways: "A or B" means, in the first place, just the same as

```
──┬── A
  └┬─ B,
```

hence it means that no possibility other than A and B is thinkable. For example, if a mass of gas is heated, its volume or its pressure increases. In the second place, the expression "A or B" combines the meanings of both

```
──┬┬─ A      and      ──┬── A
  └── B                 └┬─ B,
```

so that no third is possible besides A and B, and, moreover, that A and B exclude each other. Of the four possibilities, then, only the following two remain:

A is affirmed and B is denied;
A is denied and B is affirmed.

Of the two ways in which the expression "A or B" is used, the first, which does not exclude the coexistence of A and B, is the more important, and *we shall use the word "or" in this sense.* Perhaps it is appropriate to distinguish between "or" and "either—or" by stipulating that only the latter shall have the secondary meaning of mutual exclusion. We can then translate

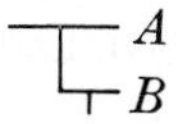

$$\begin{array}{l} A \\ B \end{array}$$

by "A or B". Similarly,

$$\begin{array}{l} A \\ B \\ \Gamma \end{array}$$

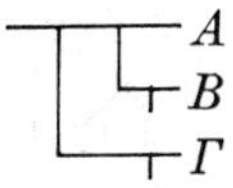

has the meaning of "A or B or Γ".

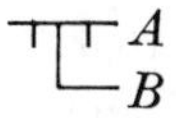

$$\begin{array}{l} A \\ B \end{array}$$

means

$$\text{“}\begin{array}{l} A \\ B \end{array}\text{ is denied”,}$$

or "The case in which both A and B are affirmed occurs". The three possibilities that remained open for

$$\begin{array}{l} A \\ B \end{array}$$

are, however, excluded. Accordingly, we can translate

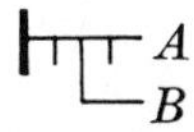

$$\begin{array}{l} A \\ B \end{array}$$

by "Both A and B are facts". It is also easy to see that

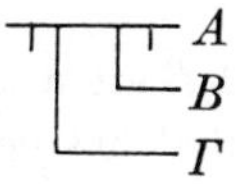

$$\begin{array}{l} A \\ B \\ \Gamma \end{array}$$

can be rendered by "A and B and Γ". If we want to represent in signs "Either A or B" with the secondary meaning of mutual exclusion, we must express

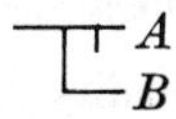

$$\text{“}\begin{array}{l} A \\ B \end{array}\quad\text{and}\quad\begin{array}{l} A\text{”} \\ B. \end{array}$$

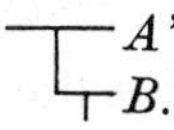

This yields

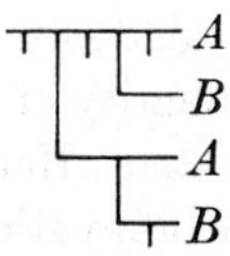

$$\begin{array}{l} A \\ B \\ A \\ B \end{array}\quad\text{or also}\quad\begin{array}{l} A \\ B \\ A \\ B. \end{array}$$

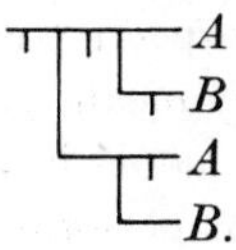

Instead of expressing the "and", as we did here, by means of the signs of conditionality and negation, we could on the other hand also represent conditionality by means of a sign for "and" and the sign of negation. We could introduce, say,

$$\left\{\begin{array}{l}\Gamma\\ \Delta\end{array}\right.$$

as a sign for the total content of Γ and Δ, and then render

$$\begin{array}{l}A\\ B\end{array}$$

by

$$\left\{\begin{array}{l}A\\ \\ B.\end{array}\right.$$

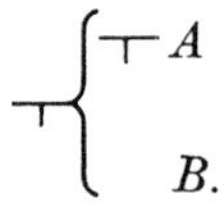

I chose the other way because I felt that it enables us to express inferences more simply. The distinction between "and" and "but" is of the kind that is not expressed in the present ideography. The speaker uses "but" when he wants to hint that what follows is different from what one might at first expect.

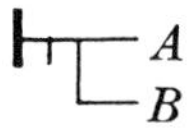

$$\begin{array}{l}A\\ B\end{array}$$

means "Of the four possibilities the third, namely, that A is denied and B is affirmed, occurs". We can therefore translate it as "B takes place and (but) A does not".

We can translate the combination of signs

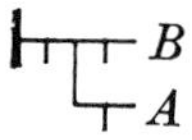

$$\begin{array}{l}B\\ A\end{array}$$

by the same words.

$$\begin{array}{l}B\\ A\end{array}$$

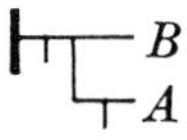

means "The case in which both A and B are denied occurs". Hence we can translate it as "Neither A nor B is a fact". What has been said here about the words "or", "and", and "neither —nor" applies, of course, only when they connect contents that *can become judgments*.

Identity of content

§ 8. Identity of content differs from conditionality and negation in that it applies to names and not to contents. Whereas in other contexts signs are merely representatives of their content, so that every combination into which they enter expresses only a relation between their respective contents, they suddenly display their own selves when they are combined by means of the sign for identity of content; for it expresses the circumstance that two names have the same content. Hence the introduction of a sign for identity of content necessarily produces a bifurcation in the meaning of all signs: they stand at times for their content, at times for themselves. At first we have the impression that what we are dealing with pertains merely to the *expression* and *not to the thought*, that we do not need different signs at all for the same content and hence no sign whatsoever for identity of content. To show that this is an empty illusion I take the following example from geometry. Assume that on the circumference of a circle there is a fixed point A about which a ray revolves. When this ray passes through the center of the circle, we call the other point at which it intersects the circle the point B associated with this position of the ray. The point of intersection, other than A, of the ray and the circumference will then be called the point B associated with the position of the ray at any time; this point is such that continuous variations in its position must always correspond to continuous variations in the position of the ray. Hence the name B denotes something indeterminate so long as the corresponding position of the ray has not been specified. We can now ask: what point is associated with the position of the ray when it is perpendicular to the diameter? The answer will be: the point A. In this case, therefore, the name B has the same content as has the name A; and yet we could not have used only one name from the beginning, since the justification for that is given only by the answer. One point is determined in two ways: (1) immediately through intuition and (2) as a point B associated with the ray perpendicular to the diameter.

To each of these ways of determining the point there corresponds a particular name. Hence the need for a sign for identity of content rests upon the following consideration: the same content can be completely determined in different ways; but that in a particular case *two ways of determining it* really yield the *same result* is the content of a *judgment*. Before this judgment can be made, two distinct names, corresponding to the two ways of determining the content, must be assigned to what these ways determine. The judgment, however, requires for its expression a sign for identity of content, a sign that connects these two names. From this it follows that the existence of different names for the same content is not always merely an irrelevant question of form; rather, that there are such names is the very heart of the matter if each is associated with a different way of determining the content. In that case the judgment that has the identity of content as its object is synthetic, in the Kantian sense. A more extrinsic reason for the introduction of a sign for identity of content is that it is at times expedient to introduce an abbreviation for a lengthy expression. Then we must express the identity of content that obtains between the abbreviation and the original form.

Now let

$$\vdash\!\!\!\!-\!\!\!-\!\!\!-(A \equiv B)$$

mean that *the sign A and the sign B have the same conceptual content, so that we can everywhere put B for A and conversely.*

Functions

§ 9. Let us assume that the circumstance that hydrogen is lighter than carbon dioxide is expressed in our formula language; we can then replace the sign for hydrogen by the sign for oxygen or that for nitrogen. This changes the meaning in such a way that "oxygen" or "nitrogen" enters into the relations in which "hydrogen" stood before. If we imagine that an expression can thus be altered, it decomposes into a stable component, representing the totality of relations, and the sign, regarded as replaceable by others, that denotes the object standing in these relations. The former component I call a function, the latter its argument. The distinction has nothing to do with the conceptual content; it comes about only because we view the expression in a particular way. According to the conception sketched above, "hydrogen" is the argument and "being lighter than carbon dioxide" the function; but we can also conceive of the same conceptual content in such a way that "carbon dioxide" becomes the argument and "being heavier than hydrogen" the function. We then need only regard "carbon dioxide" as replaceable by other ideas, such as "hydrochloric acid" or "ammonia".

"The circumstance that carbon dioxide is heavier than hydrogen" and "The circumstance that carbon dioxide is heavier than oxygen" are the same function with different arguments if we regard "hydrogen" and "oxygen" as arguments; on the other hand, they are different functions of the same argument if we regard "carbon dioxide" as the argument.

To consider another example, take "The circumstance that the center of mass of the solar system has no acceleration if internal forces alone act on the solar system". Here "solar system" occurs in two places. Hence we can consider this as a function of the argument "solar system" in various ways, according as we think of "solar system" as replaceable by something else at its first occurrence, at its second, or at both (but then in both places by the same thing). These three functions are all different. The situation is the same for the proposition that Cato killed Cato. If we here think of "Cato" as replaceable at its first occurrence, "to kill Cato" is the function; if we think of "Cato" as replaceable at its second occurrence, "to be killed by Cato" is the function; if, finally, we think of "Cato" as replaceable at both occurrences, "to kill oneself" is the function.

We now express the matter generally.

If in an expression, whose content need not be capable of becoming a judgment, a simple or a compound sign has one or more occurrences and if we regard that sign as replaceable in all or some of these occurrences by something else (but everywhere by the same thing), then we call the part that remains invariant in the expression a function, and the replaceable part the argument of the function.

Since, accordingly, something can be an argument and also occur in the function at places where it is not considered replaceable, we distinguish in the function between the argument places and the others.

Let us warn here against a false impression that is very easily occasioned by linguistic usage. If we compare the two propositions "The number 20 can be represented as the sum of four squares" and "Every positive integer can be represented as the sum of four squares", it seems to be possible to regard "being representable as the sum of four squares" as a function that in one case has the argument "the number 20" and in the other "every positive integer". We see that this view is mistaken if we observe that "the number 20" and "every positive integer" are not concepts of the same rank ⟦gleichen Ranges⟧. What is asserted of the number 20 cannot be asserted in the same sense of "every positive integer", though under certain circumstances it can be asserted of every positive integer. The expression "every positive integer" does not, as does "the number 20", by itself yield an independent idea but acquires a meaning only from the context of the sentence.

For us the fact that there are various ways in which the same conceptual content can be regarded as a function of this or that argument has no importance so long as function and argument are completely determinate. But, if the argument becomes *indeterminate*, as in the judgment "You can take as argument of 'being representable as the sum of four squares' an arbitrary positive integer, and the proposition will always be true", then the distinction between function and argument takes on a *substantive* ⟦*inhaltliche*⟧ significance. On the other hand, it may also be that the argument is determinate and the function indeterminate. In both cases, through the opposition between the *determinate* and the *indeterminate* or that between the *more* and the *less* determinate, the whole is decomposed into *function* and *argument* according to its content and not merely according to the point of view adopted.

If, given a function, we think of a sign[13] *that was hitherto regarded as not replaceable as being replaceable at some or all of its occurrences, then by adopting this conception we obtain a function that has a new argument in addition to those it had before.* This procedure yields *functions of two or more arguments.* So, for example, "The circumstance that hydrogen is lighter than carbon dioxide" can be regarded as function of the two arguments "hydrogen" and "carbon dioxide".

In the mind of the speaker the subject is ordinarily the main argument; the next in importance often appears as object. Through the choice between ⟦grammatical⟧ forms, such as active—passive, or between words, such as "heavier"—"lighter" and "give"—"receive", ordinary language is free to allow this or that component of the sentence to appear as main argument at will, a freedom that, however, is restricted by the scarcity of words.

§ 10. *In order to express an indeterminate function of the argument A, we write A, enclosed in parentheses, to the right of a letter,* for example

$$\Phi(A).$$

Likewise,

$$\Psi(A, B)$$

means a function of the two arguments A and B that is not determined any further. Here the occurrences of A and B in the parentheses represent the occurrences of A and B in the function, irrespective of whether these are single or multiple for A or for B. *Hence in general*

$$\Psi(A, B)$$

differs from

$$\Psi(B, A).$$

Indeterminate functions of more arguments are expressed in a corresponding way. We can read

$$\vdash\!\!\!\text{——}\; \Phi(A)$$

[13] We can now regard a sign that previously was considered replaceable ⟦in some places⟧ as replaceable also in those places in which up to this point it was considered fixed.

as "A has the property Φ".

$$\vdash\!\!\!-\!\!-\!\!-\, \Psi(A, B)$$

can be translated by "B stands in the relation Ψ to A" or "B is a result of an application of the procedure Ψ to the object A".

Since the sign Φ occurs in the expression $\Phi(A)$ and since we can imagine that it is replaced by other signs, Ψ or X, which would then express other functions of the argument A, *we can also regard $\Phi(A)$ as a function of the argument Φ*. This shows quite clearly that the concept of function in analysis, which in general I used as a guide, is far more restricted than the one developed here.

Generality

§ 11. In the expression of a judgment we can always regard the combination of signs to the right of $\vdash\!\!\!-\!\!-$ as a function of one of the signs occurring in it. *If we replace this argument by a German letter and if in the content stroke we introduce a concavity with this German letter in it, as in*

$$\vdash\!\!\smile^{\mathfrak{a}}\!\!-\!\!-\, \Phi(\mathfrak{a}),$$

this stands for the judgment that, whatever we may take for its argument, the function is a fact. Since a letter used as a sign for a function, such as Φ in $\Phi(A)$, can itself be regarded as the argument of a function, its place can be taken, in the manner just specified, by a German letter. The meaning of a German letter is subject only to the obvious restrictions that, if a combination of signs following a content stroke can become a judgment (§ 2), this possibility remain unaffected by such a replacement and that, if the German letter occurs as a function sign, this circumstance be taken into account. *All other conditions to be imposed on what may be put in place of a German letter are to be incorporated into the judgment.* From such a judgment, therefore, we can always derive an arbitrary number of *judgments of less general content* by substituting each time something else for the German letter and then removing the concavity in the content stroke. The horizontal stroke to the left of the concavity in

$$\vdash\!\!\smile^{\mathfrak{a}}\!\!-\!\!-\, \Phi(\mathfrak{a})$$

is the content stroke for the circumstance that, whatever we may put in place of $\mathfrak{a}$, $\Phi(\mathfrak{a})$ holds; the horizontal stroke to the right of the concavity is the content stroke of $\Phi(\mathfrak{a})$, and here we must imagine that something definite has been substituted for $\mathfrak{a}$.

According to what we said above about the significance of the judgment stroke, it is easy to see what an expression like

$$-\!\!\smile^{\mathfrak{a}}\!\!-\!\!-\, X(\mathfrak{a})$$

means. It can occur as a part of a judgment, like

$$\vdash\!\!\top\!\!\smile^{\mathfrak{a}}\!\!-\, X(\mathfrak{a}) \quad \text{or} \quad \vdash\!\!\top\!\!-\!\!-\!\!-\, A \atop \quad \llcorner\!\smile^{\mathfrak{a}}\!\!-\, X(\mathfrak{a}).$$

It is clear that from these judgments we cannot derive less general judgments by substituting something definite for $\mathfrak{a}$, as we could from

$$\vdash\!\overset{\mathfrak{a}}{\smile}\!\!-\Phi(\mathfrak{a}).$$

$\vdash\!\!\!\top\!\overset{\mathfrak{a}}{\smile}\!\!-X(\mathfrak{a})$ denies that, whatever we may put in place of $\mathfrak{a}$, $X(\mathfrak{a})$ is always a fact. This does not by any means deny that we could specify some meaning Δ for $\mathfrak{a}$ such that $X(\Delta)$ would be a fact.

$$\begin{array}{l}\vdash\!\!\!\top\!\!-\!\!-\!\!-\, A\\ \quad\llcorner\!\overset{\mathfrak{a}}{\smile}\!\!-X(\mathfrak{a})\end{array}$$

means that the case in which $-\!\overset{\mathfrak{a}}{\smile}\!\!-X(\mathfrak{a})$ is affirmed and A is denied does not occur. But this does not by any means deny that the case in which $X(\Delta)$ is affirmed and A is denied does occur; for, as we just saw, $X(\Delta)$ can be affirmed and $-\!\overset{\mathfrak{a}}{\smile}\!\!-X(\mathfrak{a})$ can still be denied. Hence we cannot put something arbitrary in place of $\mathfrak{a}$ here either without endangering the truth of the judgment. This explains why the concavity with the German letter written into it is necessary: *it delimits the scope* ⟦*Gebiet*⟧ *that the generality indicated by the letter covers. The German letter retains a fixed meaning only within its own scope*; within one judgment the same German letter can occur in different scopes, without the meaning attributed to it in one scope extending to any other. The scope of a German letter can include that of another, as is shown by the example

$$\begin{array}{l}\vdash\!\overset{\mathfrak{a}}{\smile}\!\!\top\!\!-\!\!-\!\!-\, A(\mathfrak{a})\\ \qquad\llcorner\!\overset{\mathfrak{e}}{\smile}\!\!-B(\mathfrak{a},\mathfrak{e}).\end{array}$$

In that case they must be chosen *different*; we could not put $\mathfrak{a}$ for $\mathfrak{e}$. Replacing a German letter everywhere in its scope by some other one is, of course, permitted, so long as in places where different letters initially stood different ones also stand afterward. This has no effect on the content. *Other substitutions are permitted only if the concavity immediately follows the judgment stroke*, that is, if the content of the entire judgment constitutes the scope of the German letter. Since, accordingly, that case is a distinguished one, I shall introduce the following abbreviation for it. *An* ⟦*italic*⟧ *Latin letter always is to have as its scope the content of the entire judgment*, and this fact need not be indicated by a concavity in the content stroke. If a Latin letter occurs in an expression that is not preceded by a judgment stroke, the expression is meaningless. *A Latin letter may always be replaced by a German one that does not yet occur in the judgment*; then the concavity must be introduced immediately following the judgment stroke. For example, instead of

$$\vdash\!\!-\!\!-\, X(a)$$

we can write

$$\vdash\!\overset{\mathfrak{a}}{\smile}\!\!-X(\mathfrak{a})$$

if $\mathfrak{a}$ occurs only in the argument places of $X(a)$.

It is clear also that from

$$\begin{array}{l}\vdash\!\!\top\!\!-\; \Phi(a)\\ \quad\llcorner\!-\; A\end{array}$$

we can derive

$$\begin{array}{l}\vdash\!\!\top\!\!-\!\overset{\mathfrak{a}}{\smile}\!-\; \Phi(\mathfrak{a})\\ \quad\llcorner\!-\!-\!-\!-\; A\end{array}$$

if A is an expression in which a does not occur and if a stands only in the argument places of $\Phi(a)$.[14] If $-\overset{\mathfrak{a}}{\smile}-\Phi(\mathfrak{a})$ is denied, we must be able to specify a meaning for a such that $\Phi(a)$ will be denied. If, therefore, $-\overset{\mathfrak{a}}{\smile}-\Phi(\mathfrak{a})$ were to be denied and A to be affirmed, we would have to be able to specify a meaning for a such that A would be affirmed and $\Phi(a)$ would be denied. But on account of

$$\begin{array}{l}\vdash\!\!\top\!\!-\; \Phi(\mathfrak{a})\\ \quad\llcorner\!-\; A\end{array}$$

we cannot do that; for this means that, whatever a may be, the case in which $\Phi(a)$ is denied and A is affirmed is excluded. Therefore we cannot deny $-\overset{\mathfrak{a}}{\smile}-\Phi(\mathfrak{a})$ and affirm A; that is,

$$\begin{array}{l}\vdash\!\!\top\!\!-\!\overset{\mathfrak{a}}{\smile}\!-\; \Phi(\mathfrak{a})\\ \quad\llcorner\!-\!-\!-\!-\; A.\end{array}$$

Likewise, from

$$\begin{array}{l}\vdash\!\!\top\!\!-\!\!\top\!\!-\; \Phi(a)\\ \quad|\quad\llcorner\!-\; A\\ \quad\llcorner\!-\!-\!-\!-\; B\end{array}$$

we can deduce

$$\begin{array}{l}\vdash\!\!\top\!\!-\!\!\top\!\!-\!\overset{\mathfrak{a}}{\smile}\!-\; \Phi(\mathfrak{a})\\ \quad|\quad\llcorner\!-\!-\!-\!-\; A\\ \quad\llcorner\!-\!-\!-\!-\!-\!-\; B\end{array}$$

if a does not occur in A or B and $\Phi(a)$ contains a only in the argument places. This case can be reduced to the preceding one, since

$$\begin{array}{l}\vdash\!\!\top\!\!-\!\!\top\!\!-\; \Phi(a)\\ \quad|\quad\llcorner\!-\; A\\ \quad\llcorner\!-\!-\!-\!-\; B\end{array}$$

can be written

$$\begin{array}{l}\vdash\!\!\top\!\!-\!-\!-\!-\; \Phi(a)\\ \quad\llcorner\!\!\top\!\!\top\!\!\top\!-\; A\\ \qquad\quad\llcorner\!-\; B\end{array}$$

[14] ⟦Footnote by Jourdain (*1912*, p. 248):
Frege remarked [Frege, 1910] that "it is correct that one can give up the distinguishing use of Latin, German, and perhaps also of Greek letters, but at the cost of perspicuity of formulas".⟧

and since we can transform

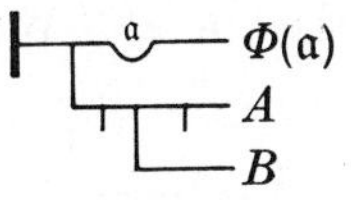

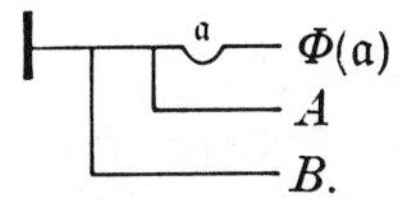

back into

$$\vdash\!\!\top\!\!\top\!\!\smile^{\mathfrak{a}}\!\!- \Phi(\mathfrak{a}) \quad A \quad B.$$

Similar considerations apply when still more condition strokes are present.

§ 12. We now consider certain combinations of signs.

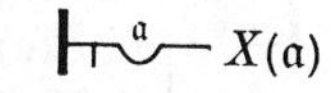

$$\vdash\!\top\!\smile^{\mathfrak{a}}\!- X(\mathfrak{a})$$

means that we could find some object, say Δ, such that $X(\Delta)$ would be denied. We can therefore translate it as "There are some objects that do not have property X".

The meaning of

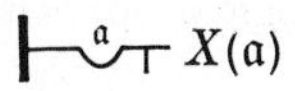

$$\vdash\!\smile^{\mathfrak{a}}\!\top X(\mathfrak{a})$$

differs from this. The formula means "Whatever $\mathfrak{a}$ may be, $X(\mathfrak{a})$ must always be denied", or "There does not exist anything having property X", or, if we call something that has property X an X, "There is no X".

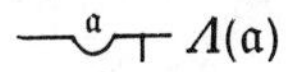

$$-\!\smile^{\mathfrak{a}}\!\top \Lambda(\mathfrak{a})$$

is denied by

$$\vdash\!\top\!\smile^{\mathfrak{a}}\!\top \Lambda(\mathfrak{a}).$$

We can therefore translate the last formula as "There are Λ".[15]

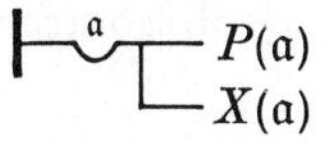

$$\vdash\!\smile^{\mathfrak{a}}\!\!\top\!- P(\mathfrak{a}) \quad X(\mathfrak{a})$$

means "Whatever we may put in place of $\mathfrak{a}$, the case in which $P(\mathfrak{a})$ would have to be denied and $X(\mathfrak{a})$ to be affirmed does not occur". Thus it is possible here that, for some meanings that can be given to $\mathfrak{a}$, $P(\mathfrak{a})$ would have to be affirmed and $X(\mathfrak{a})$ to be affirmed, for others $P(\mathfrak{a})$ would have to be affirmed and $X(\mathfrak{a})$ to be denied, and for others still $P(\mathfrak{a})$ would have to be denied and $X(\mathfrak{a})$ to be denied. We could therefore translate it as "If something has property X, it also has property P", "Every X is a P", or "All X are P".

This is the way in which causal connections are expressed.

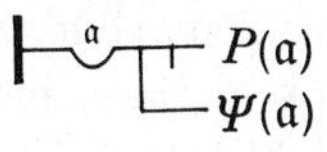

$$\vdash\!\smile^{\mathfrak{a}}\!\!\top\!\top P(\mathfrak{a}) \quad \Psi(\mathfrak{a})$$

[15] This must be understood in such a way as to include the case "There exists one Λ" as well. If, for example, $\Lambda(x)$ means the circumstance that x is a house, then

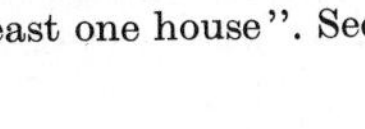

$$\vdash\!\top\!\smile^{\mathfrak{a}}\!\top \Lambda(\mathfrak{a})$$

reads "There are houses or there is at least one house". See footnote 10.

means "No meaning can be given to $\mathfrak{a}$ such that both $P(\mathfrak{a})$ and $\Psi(\mathfrak{a})$ could be affirmed". We can therefore translate it as "What has property Ψ does not have property P" or "No Ψ is a P".

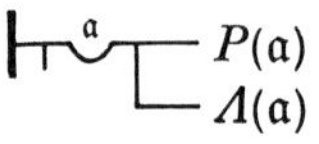

$P(\mathfrak{a})$
$\Lambda(\mathfrak{a})$

denies

$P(\mathfrak{a})$
$\Lambda(\mathfrak{a})$

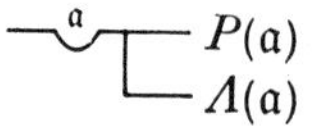

and can therefore be rendered by "Some Λ are not P".

$P(\mathfrak{a})$
$M(\mathfrak{a})$

denies that no M is a P and therefore means "Some[16] M are P", or "It is possible that a M be a P".

Thus we obtain the square of logical opposition:

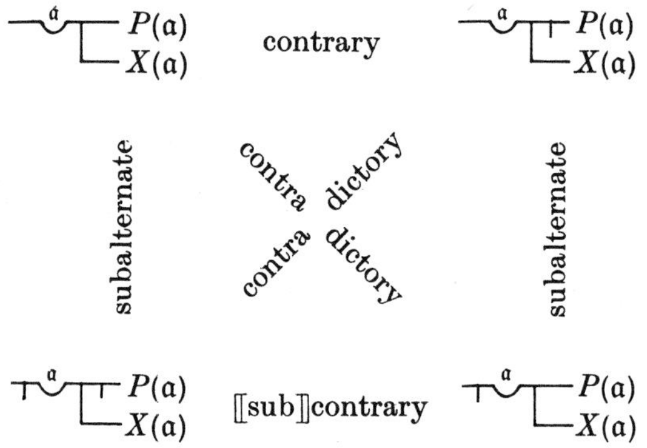

$P(\mathfrak{a})$ $X(\mathfrak{a})$ contrary $P(\mathfrak{a})$ $X(\mathfrak{a})$

subalternate contradictory contradictory subalternate

$P(\mathfrak{a})$ $X(\mathfrak{a})$ [[sub]]contrary $P(\mathfrak{a})$ $X(\mathfrak{a})$

II. REPRESENTATION AND DERIVATION OF SOME JUDGMENTS OF PURE THOUGHT

§ 13. We have already introduced a number of fundamental principles of thought in the first chapter in order to transform them into rules for the use of our signs. These rules and the laws whose transforms they are cannot be expressed in the ideography because they form its basis. Now in the present chapter a number of judgments of pure thought for which this is possible will be represented in signs. It seems natural to derive the more complex of these judgments from simpler ones, not in order to make them more certain, which would be unnecessary in most cases, but in order to make manifest the relations of the judgments to one another. Merely to know the laws is obviously not the same as to know them together with the connections that

[16] The word "some" must always be understood here in such a way as to include the case "one" as well. More explicitly we would say "some or at least one".

some have to others. In this way we arrive at a small number of laws in which, if we add those contained in the rules, the content of all the laws is included, albeit in an undeveloped state. And that the deductive mode of presentation makes us acquainted with that core is another of its advantages. Since in view of the boundless multitude of laws that can be enunciated we cannot list them all, we cannot achieve completeness except by searching out those that, *by their power*, contain all of them. Now it must be admitted, certainly, that the way followed here is not the only one in which the reduction can be done. That is why not all relations between the laws of thought are elucidated by means of the present mode of presentation. There is perhaps another set of judgments from which, when those contained in the rules are added, all laws of thought could likewise be deduced. Still, with the method of reduction presented here such a multitude of relations is exhibited that any other derivation will be much facilitated thereby.

The propositions forming the core of the presentation below are nine in number. To express three of these, formulas (1), (2), and (8), we require besides letters only the sign of conditionality; formulas (28), (31), and (41) contain in addition the sign of negation; two, formulas (52) and (54), contain that of identity of content; and in one, formula (58), the concavity in the content stroke is used.

The derivations that follow would tire the reader if he were to retrace them in every detail; they serve merely to insure that the answer to any question concerning the derivation of a law is at hand.

§ 14.

a
b
a (1)

says "The case in which a is denied, b is affirmed, and a is affirmed is excluded". This is evident, since a cannot at the same time be denied and affirmed. We can also express the judgment in words thus, "If a proposition a holds, then it also holds in case an arbitrary proposition b holds". Let a, for example, stand for the proposition that the sum of the angles of the triangle ABC is two right angles, and b for the proposition that the angle ABC is a right angle. Then we obtain the judgment "If the sum of the angles of the triangle ABC is two right angles, this also holds in case the angle ABC is a right angle".

The (1) to the right of

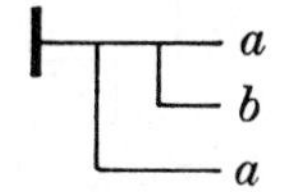

is the number of this formula.

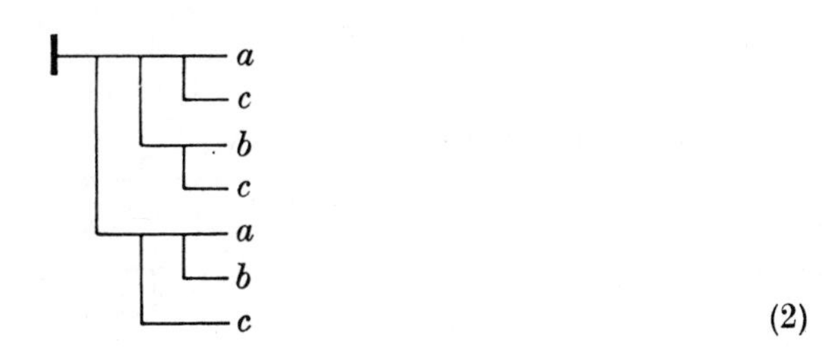

means "The case in which

$$\begin{matrix} a \\ c \\ b \\ c \end{matrix}$$

is denied and

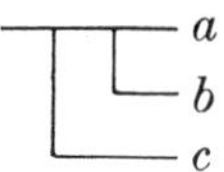

is affirmed does not take place".

But

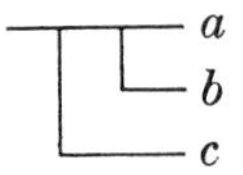

means the circumstance that the case in which a is denied, b is affirmed, and c is affirmed is excluded. The denial of

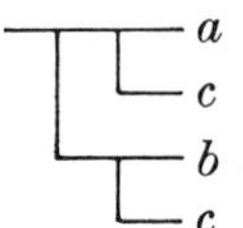

says that $\begin{matrix} a \\ c \end{matrix}$ is denied and $\begin{matrix} b \\ c \end{matrix}$ is affirmed. But the denial of $\begin{matrix} a \\ c \end{matrix}$ means that a is denied and c is affirmed. Thus the denial of

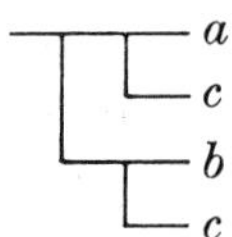

means that a is denied, c is affirmed, and $\begin{matrix} b \\ c \end{matrix}$ is affirmed. But the affirmation of $\begin{matrix} b \\ c \end{matrix}$ and that of c entails the affirmation of b. That is why the denial of

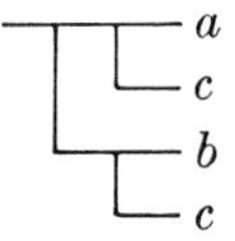

has as a consequence the denial of a and the affirmation of b and c. Precisely this case is excluded by the affirmation of

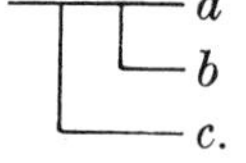

Thus the case in which

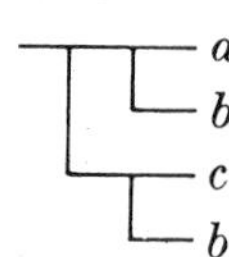

is denied and

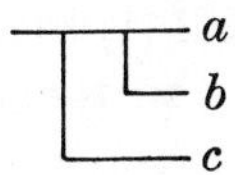

is affirmed cannot take place, and that is what the judgment

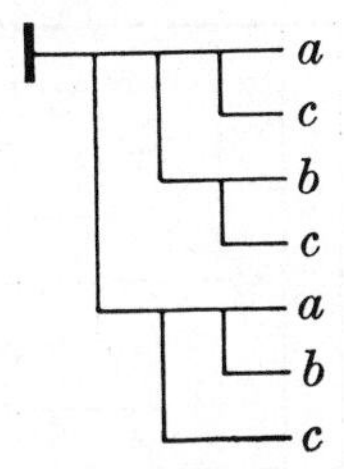

asserts. For the case in which causal connections are present, we can also express this as follows: "If a proposition a is a necessary consequence of two propositions b and c, that is, if

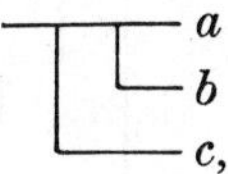

and if one of these, b, is in turn a necessary consequence of the other, c, then the proposition a is a necessary consequence of this latter one, c, alone".

For example, let c mean that in a sequence Z of numbers every successor term is greater than its predecessor, let b mean that a term M is greater than L, and let a mean that the term N is greater than L. Then we obtain the following judgment: "If from the propositions that in the number sequence Z every successor term is greater than its predecessor and that the term M is greater than L it can be inferred that the term N is greater than L, and if from the proposition that in the number sequence Z every successor term is greater than its predecessor it follows that M is greater than L, then the proposition that N is greater than L can be inferred from the proposition that every successor term in the number sequence Z is greater than its predecessor".

§ 15.

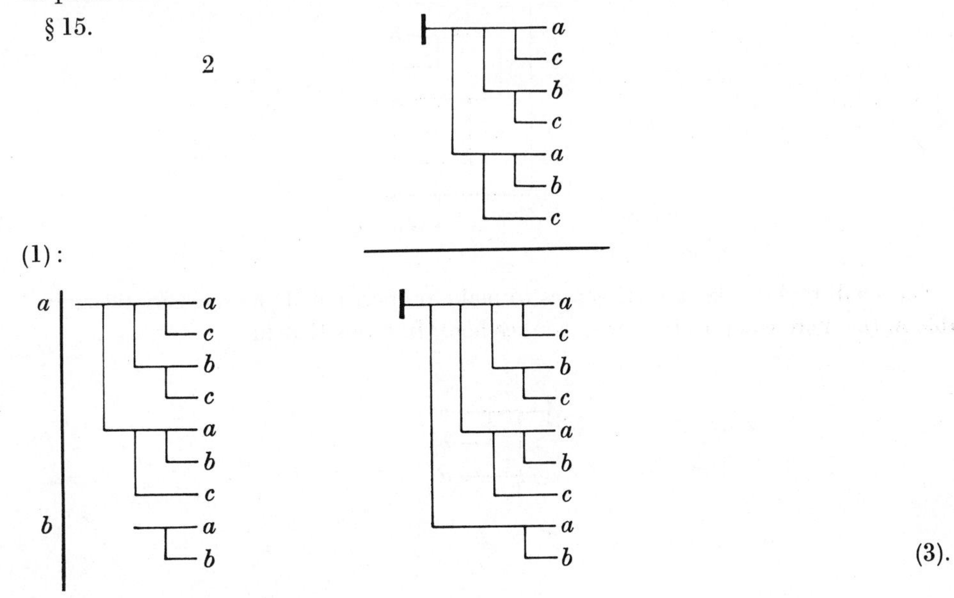

The 2 on the left indicates that formula (2) stands to its right. The inference that brings about the transition from (2) and (1) to (3) is expressed by an abbreviation in accordance with § 6. In full it would be written as follows:

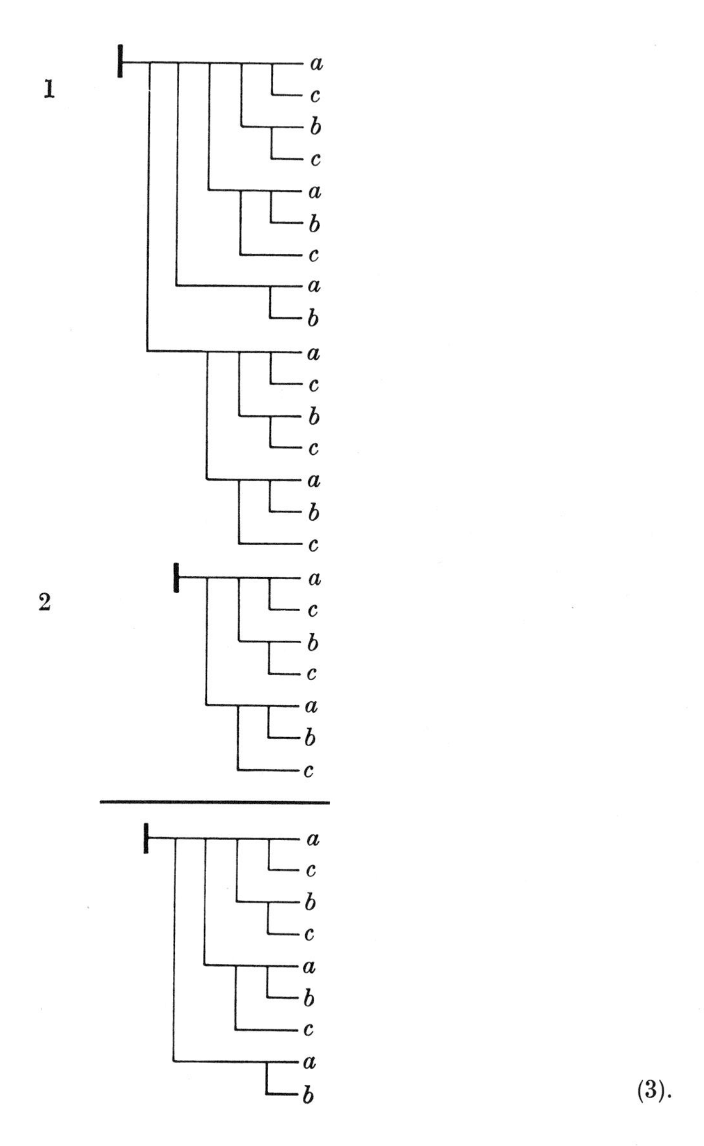

The small table under the (1) serves to make proposition (1) more easily recognizable in the more complicated form it takes here. It states that in

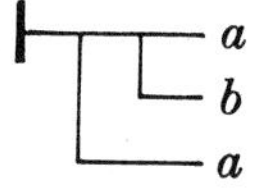

we are to put

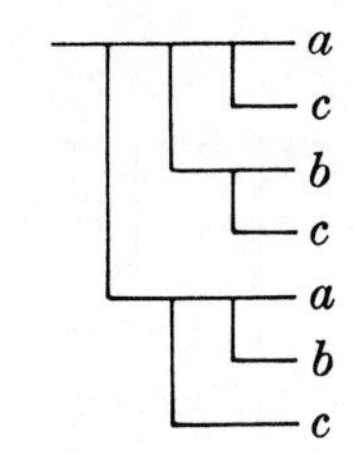

in place of a and

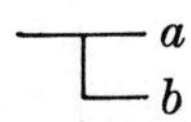

in place of b.

3

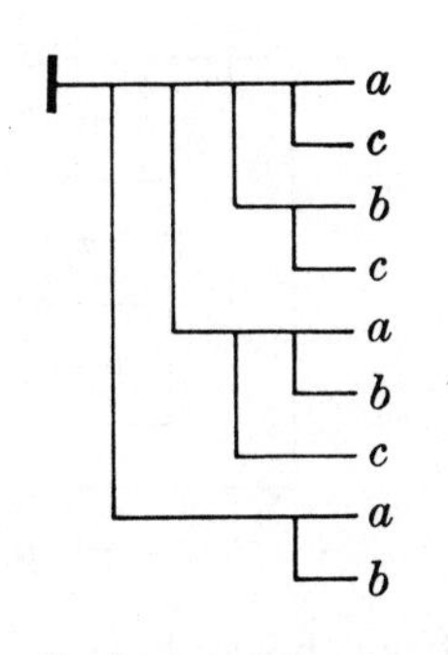

(2):

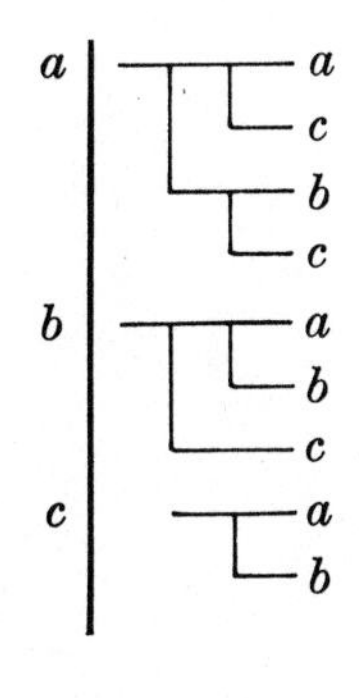

(4).

The table under the (2) means that in

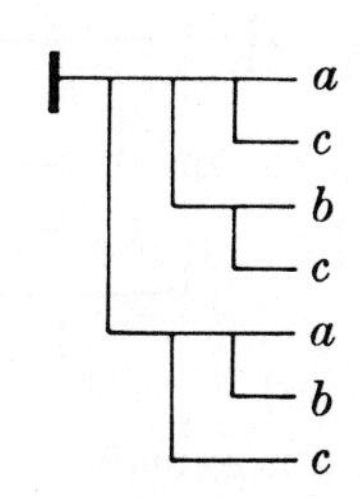

we are to put in place of a, b, and c, respectively, the expressions standing to the right of them; as a result we obtain

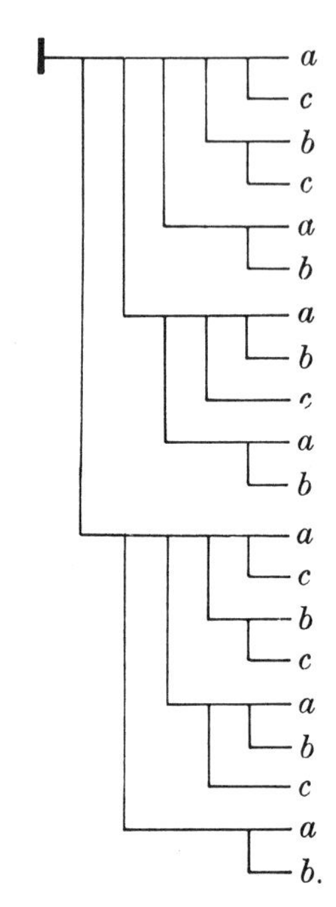

We readily see how (4) follows from this and (3).

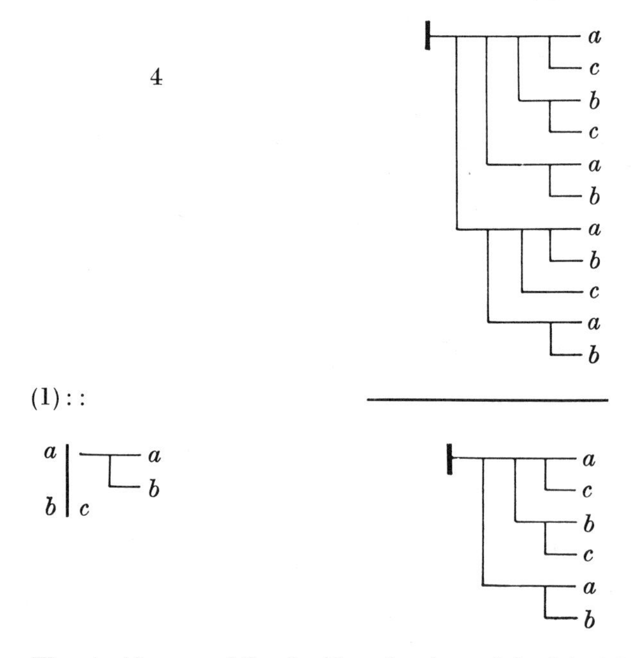

The significance of the double colon is explained in § 6.

Example for (5). Let a be the circumstance that the piece of iron E becomes magnetized, b the circumstance that a galvanic current flows through the wire D, and c the circumstance that the key T is depressed. We then obtain the judgment: "If the proposition holds that E becomes magnetized as soon as a galvanic current flows through D and if the proposition holds that a galvanic current flows through D as soon as T is depressed, then E becomes magnetized if T is depressed".

If causal connections are assumed, (5) can be expressed thus: "If b is a sufficient condition for a and if c is a sufficient condition for b, then c is a sufficient condition for a".

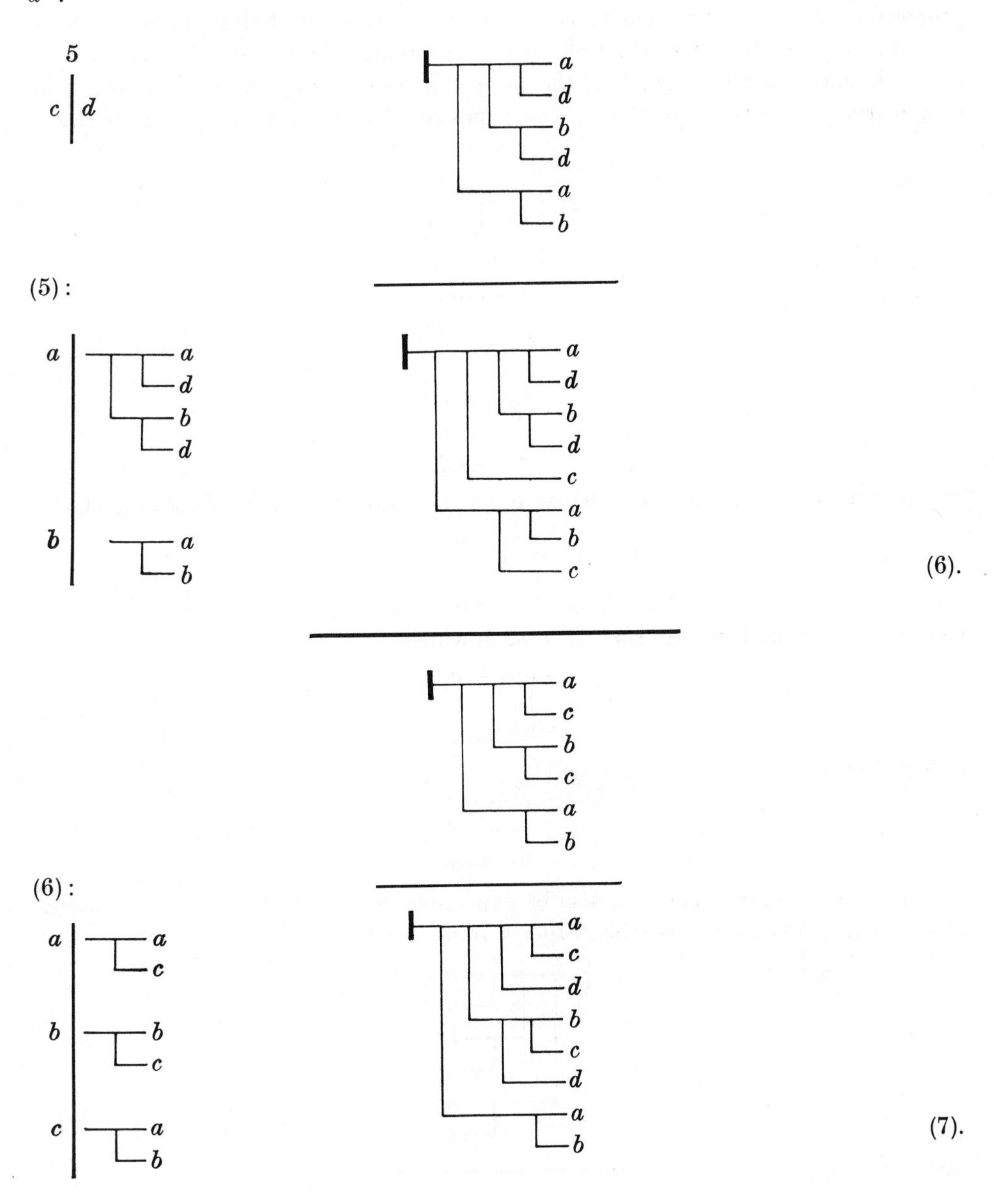

This proposition differs from (5) only in that instead of one condition, c, we now have two, c and d.

Example for (7). Let d mean the circumstance that the piston K of an air pump is moved from its leftmost position to its rightmost position, c the circumstance that the valve H is in position I, b the circumstance that the density D of the air in the cylinder of the air pump is reduced by half, and a the circumstance that the height H of a barometer connected to the inside of the cylinder decreases by half. Then we obtain the judgment: "If the proposition holds that the height H of the barometer decreases by half as soon as the density D of the air is reduced by half, and if the proposition holds that the density D of the air is reduced by half if the piston K is moved from the leftmost to the rightmost position and if the valve is in position I, then it follows that the height H of the barometer decreases by half if the piston K is moved from the leftmost to the rightmost position while the valve H is in position I".

§ 16.

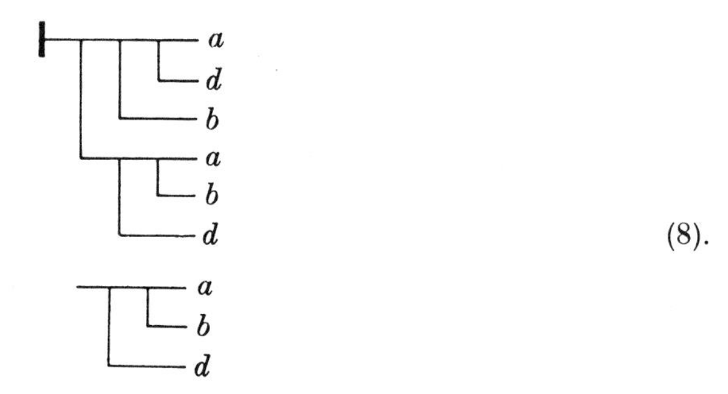

(8).

means that the case in which a is denied but b and d are affirmed does not take place;

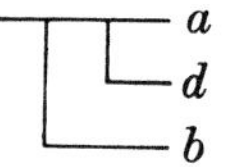

means the same, and (8) says that the case in which

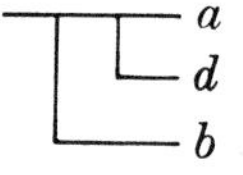

is denied and

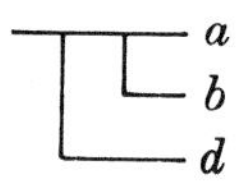

is affirmed is excluded. This can also be expressed thus: "If two conditions have a proposition as a consequence, their order is immaterial".

5

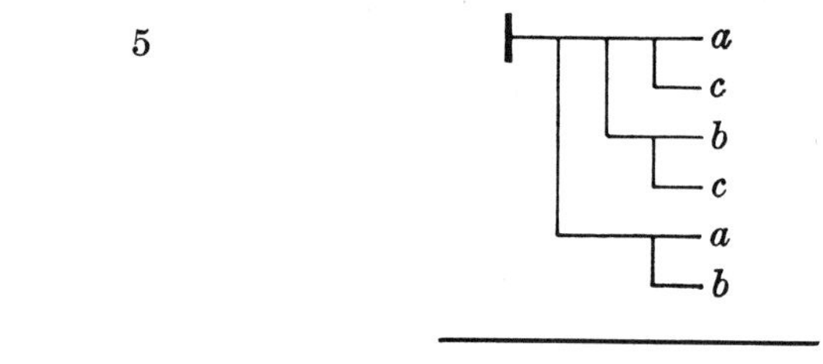

(8):

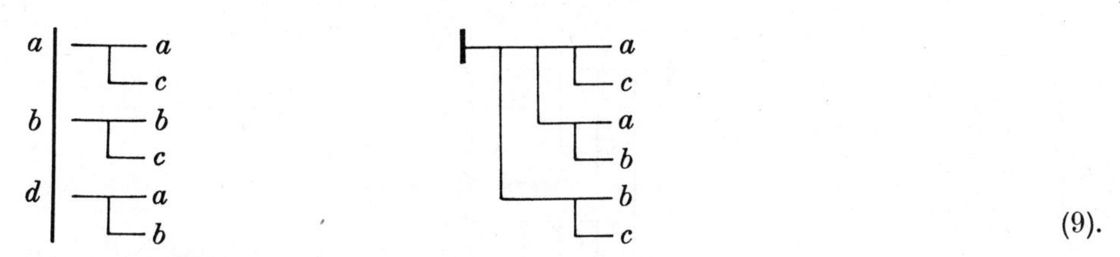

(9).

This proposition differs from (5) only in an unessential way.

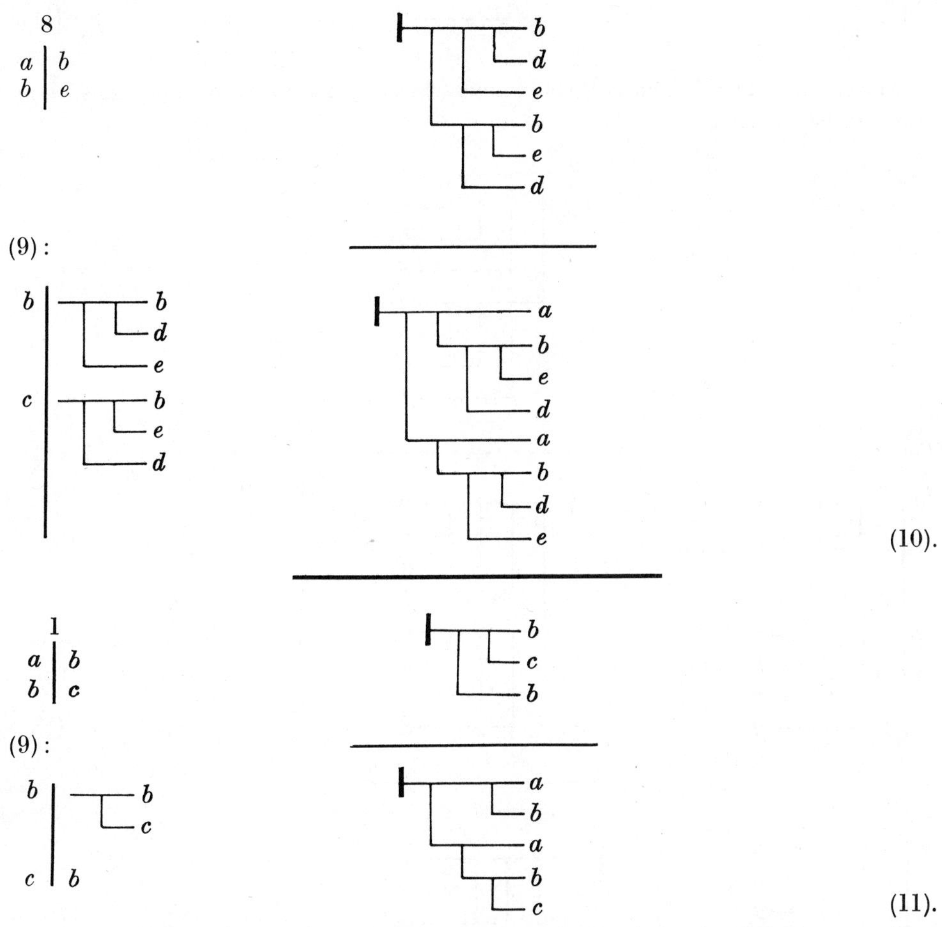

(9):

(10).

(9):

(11).

We can translate this formula thus: "If the proposition that b takes place or c does not is a sufficient condition for a, then b is by itself a sufficient condition for a".

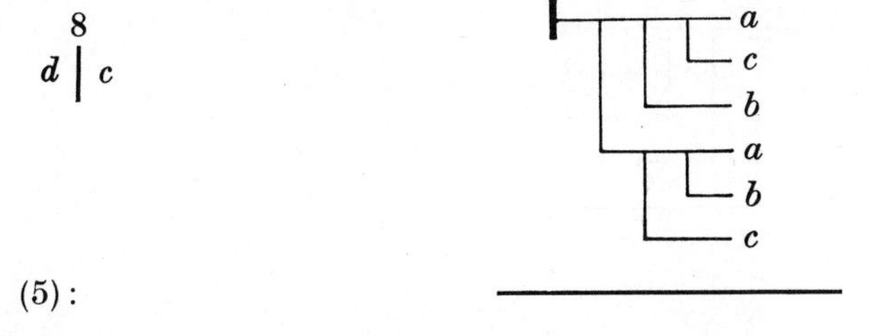

(5):

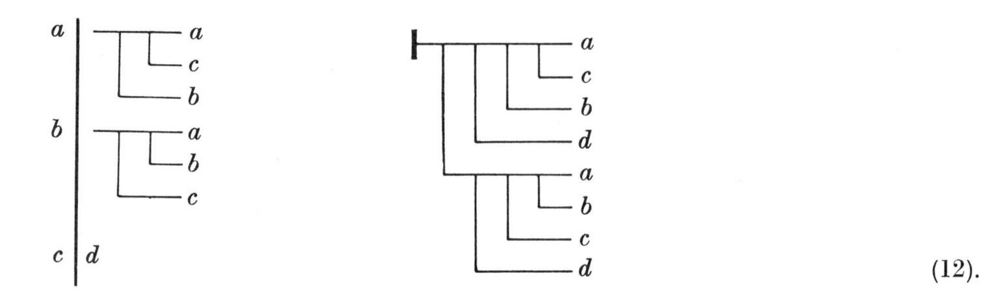

Propositions (12)–(17) and (22) show how, when there are several conditions, their order can be changed.

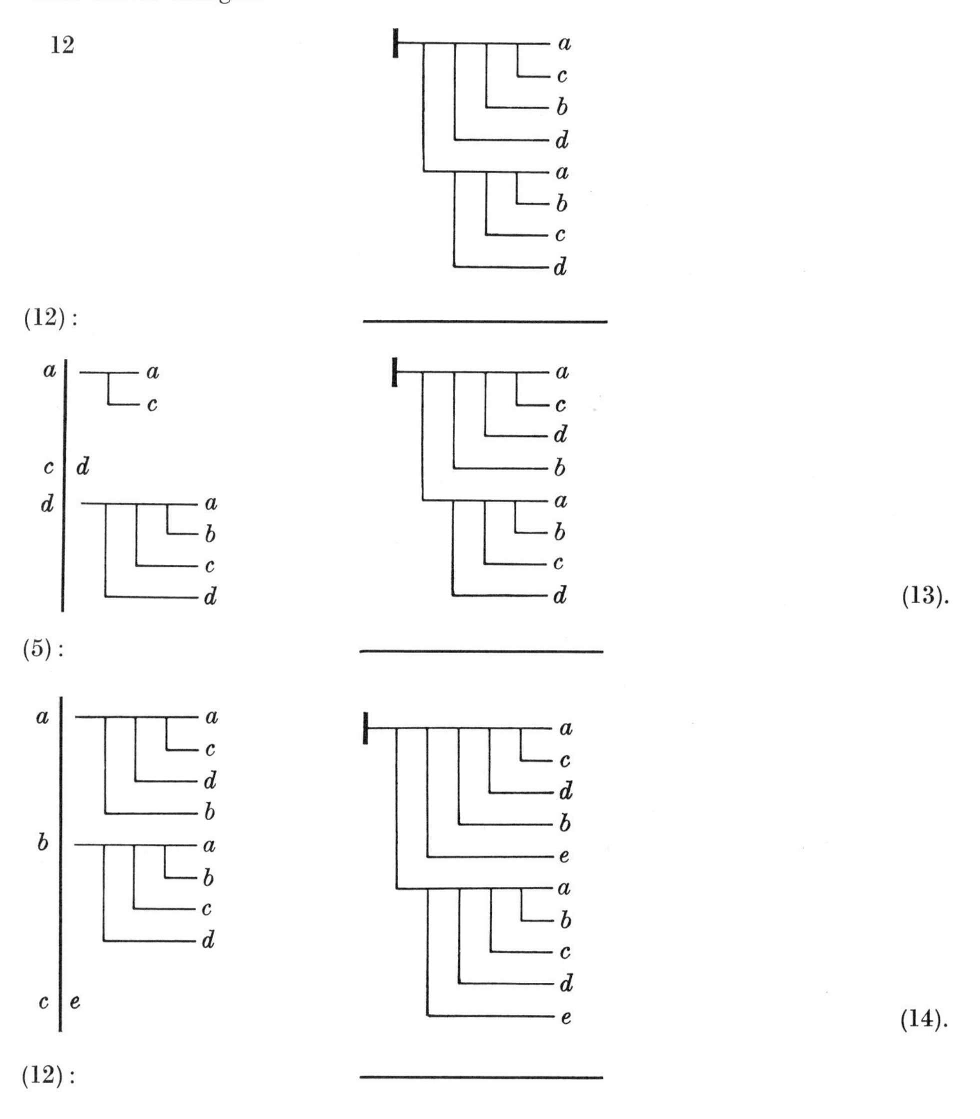

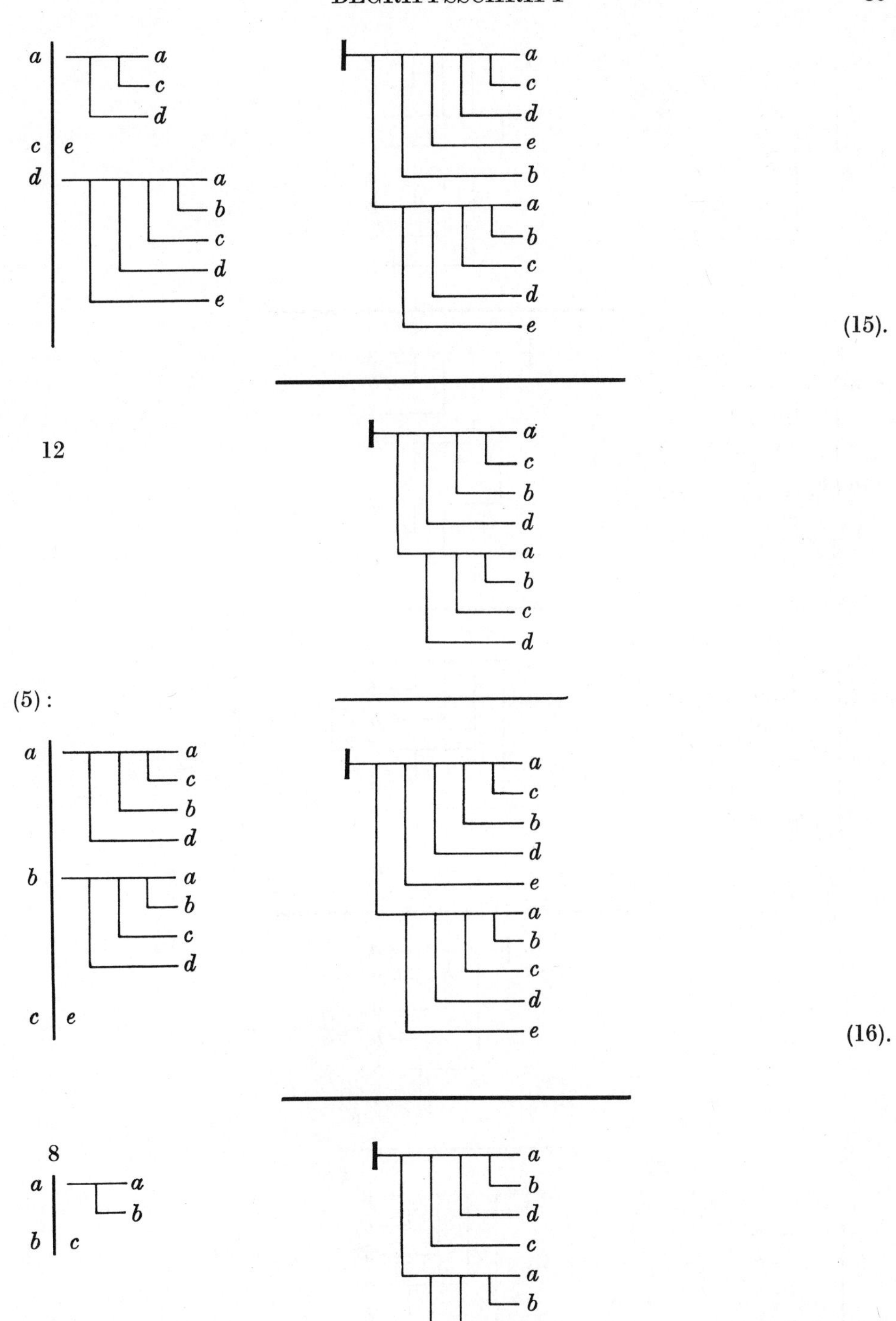
a
c
d
e
a
c
d
e
b
a
b
c
d
e
(15).
12
(5):
a
b
c
e
(16).
8
(16):

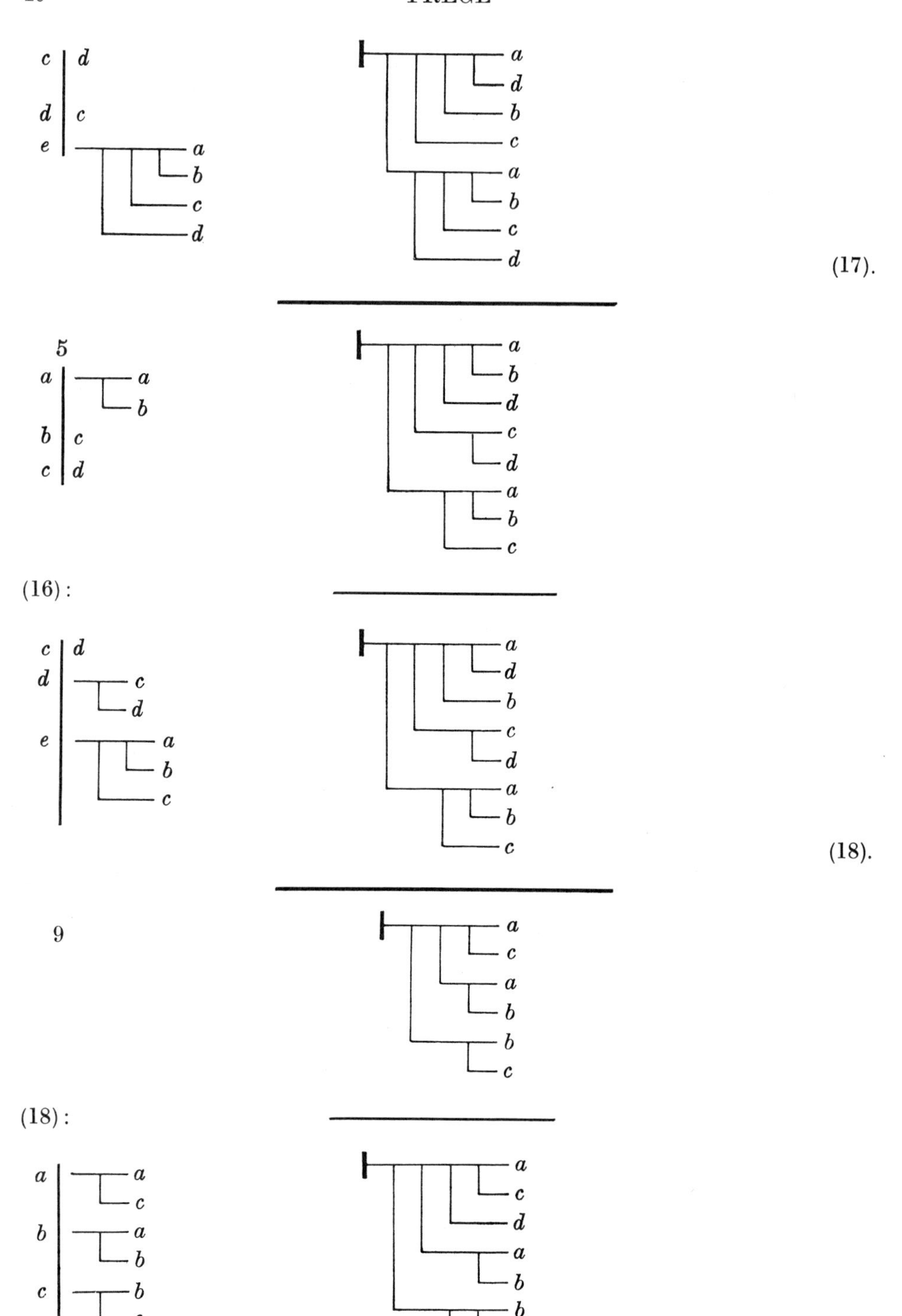
c | d
d | c
e
(17).
5
a
b | c
c | d
(16):
c | d
d
e
(18).
9
(18):
a
b
c
(19).

This proposition differs from (7) only in an unessential way.

19

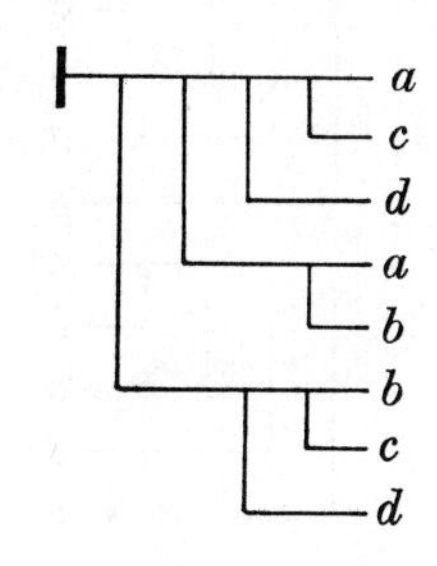

(18):

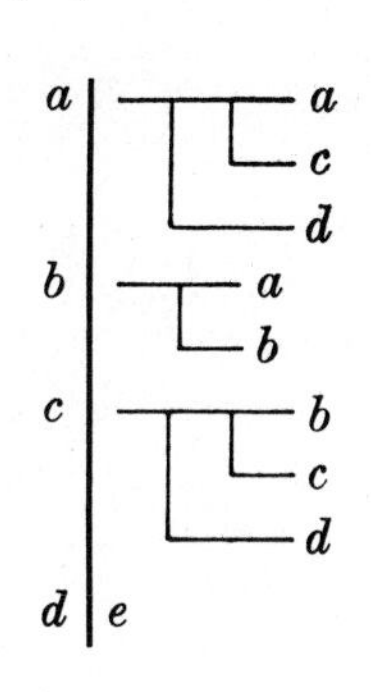

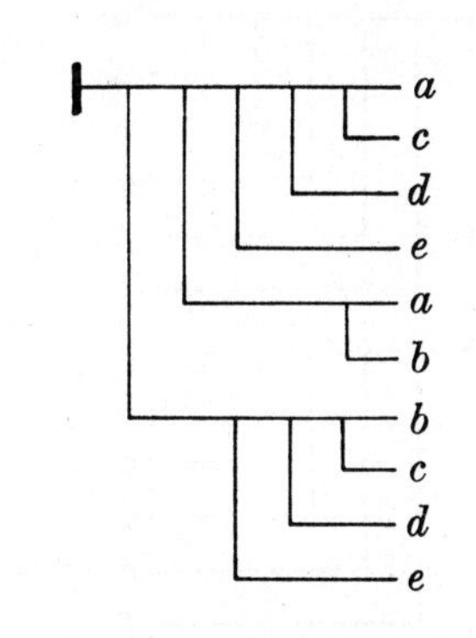

(20).

9

a	b
b	c
c	d

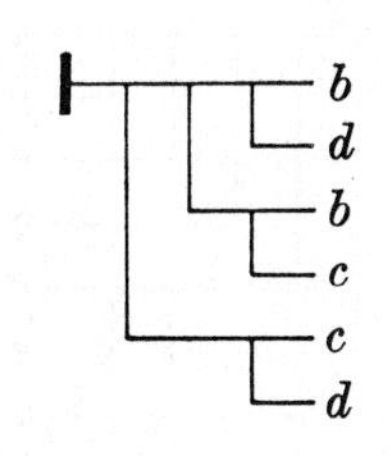

(19):

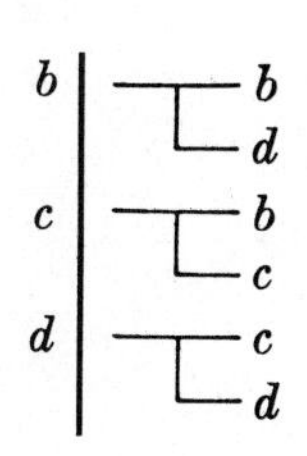

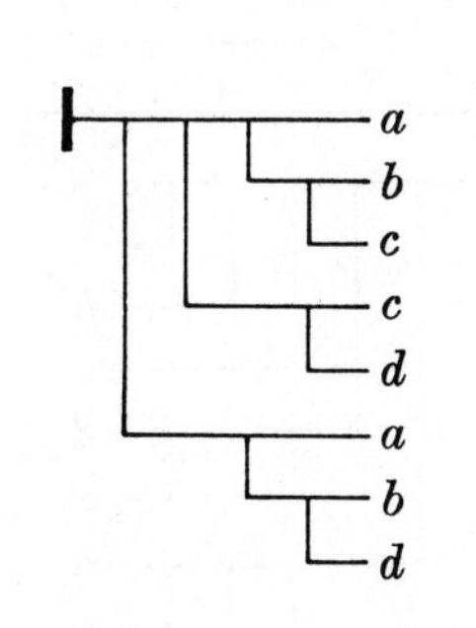

(21).

16

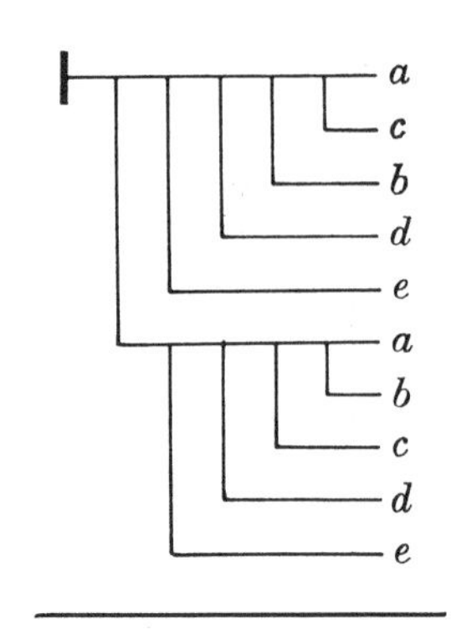

(5):

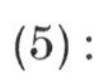

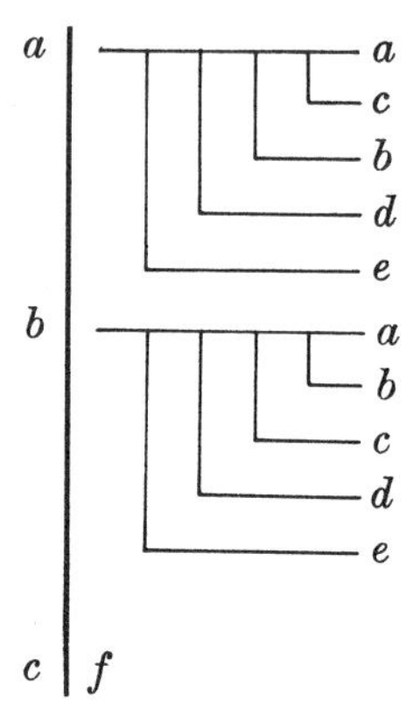

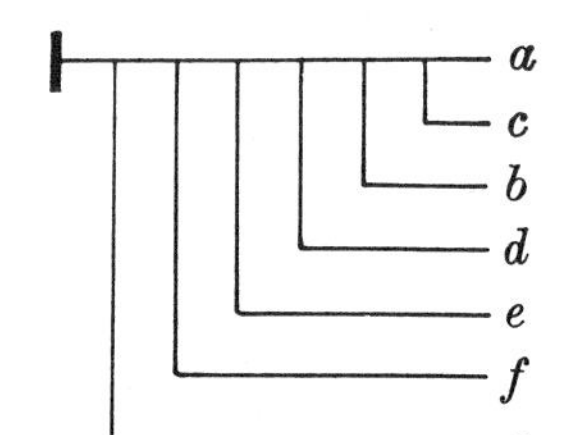

(22).

18

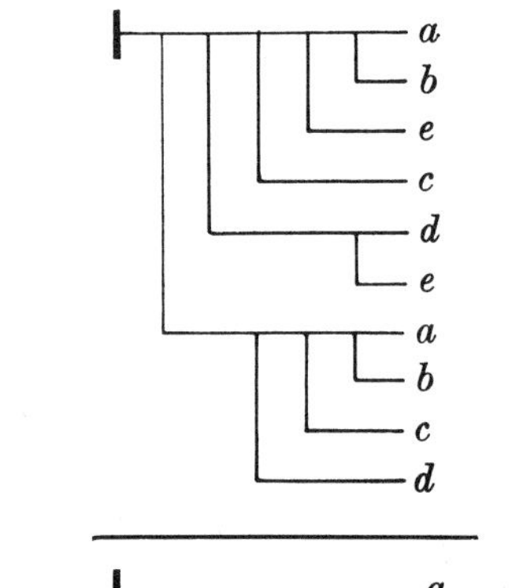

(22):

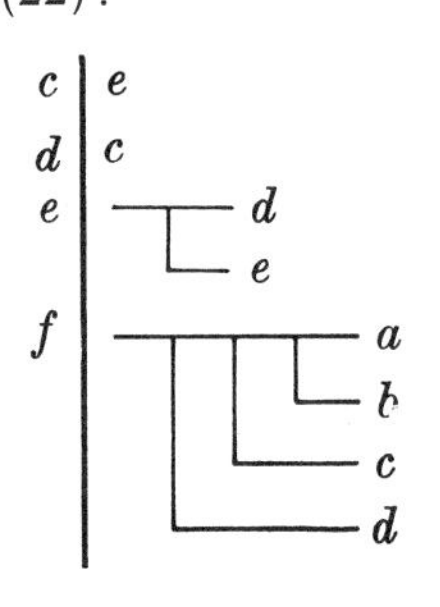

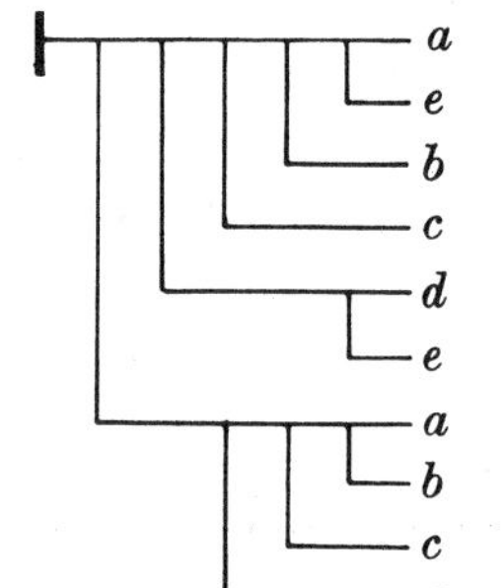

(23).

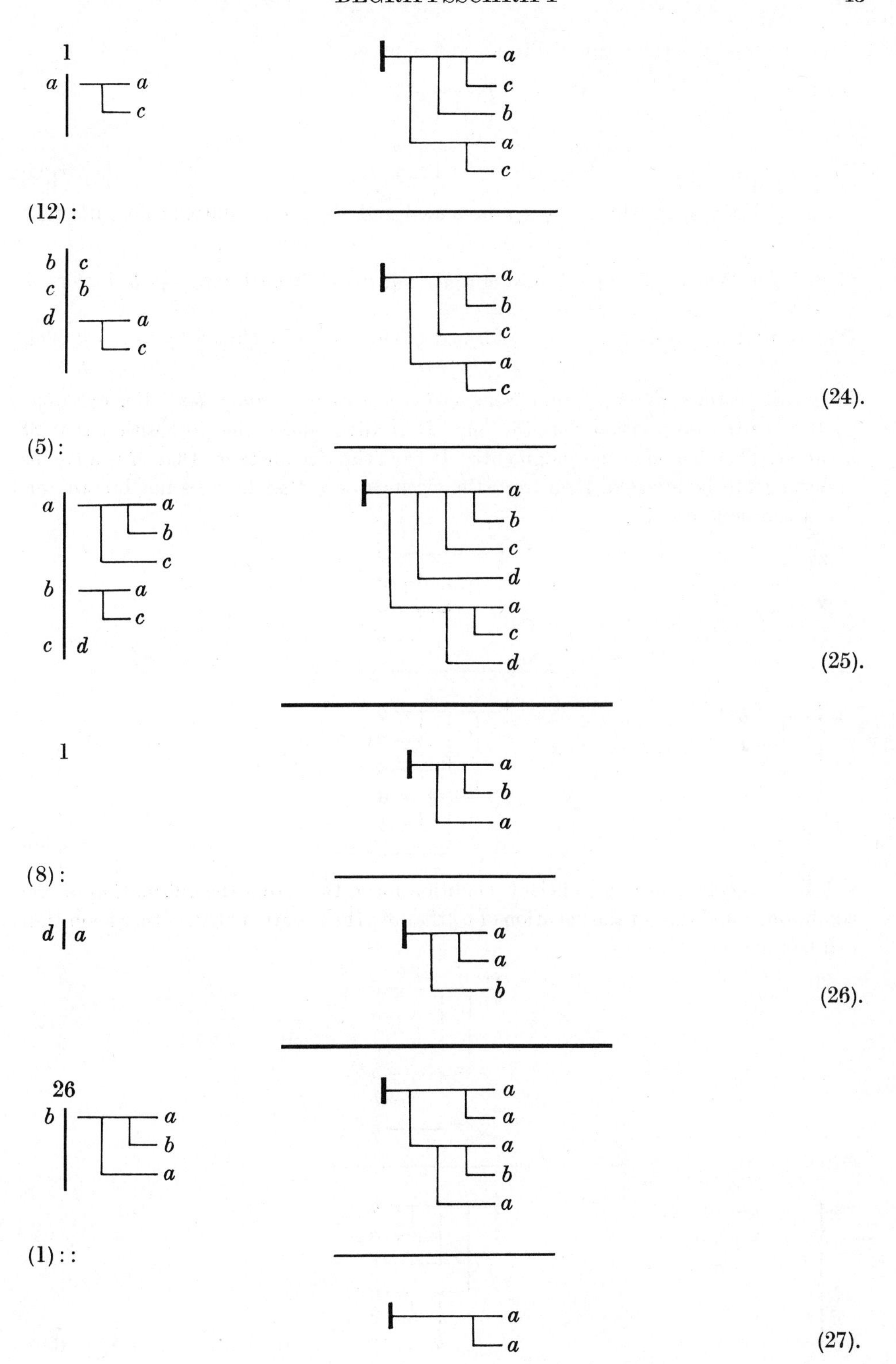
1
a
a
c
a
c
b
a
c
(12):
b c
c b
d
a
c
a
b
c
a
c
(24).
(5):
a
a
b
c
b
a
c
c d
a
b
c
d
a
c
d
(25).
1
a
b
a
(8):
d a
a
a
b
(26).
26
b
a
b
a
a
a
a
b
a
(1)::
a
a
(27).

We cannot (at the same time) affirm a and deny a.

§ 17.

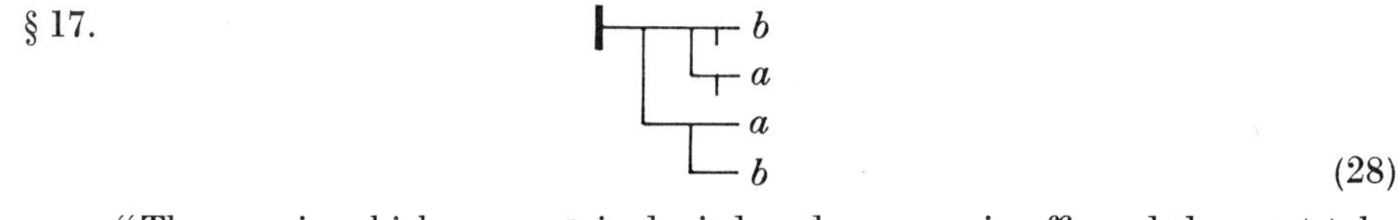

(28)

means: "The case in which b, a is denied and a, b is affirmed does not take place". The denial of b, a means that a is affirmed and b is denied, that is, that a is denied and b is affirmed. This case is excluded by a, b. This judgment justifies the transition from *modus ponens* to *modus tollens*. For example, let b mean the proposition that the man M is alive, and a the proposition that M breathes. Then we have the judgment: "If from the circumstance that M is alive his breathing can be inferred, then from the circumstance that he does not breathe his death can be inferred".

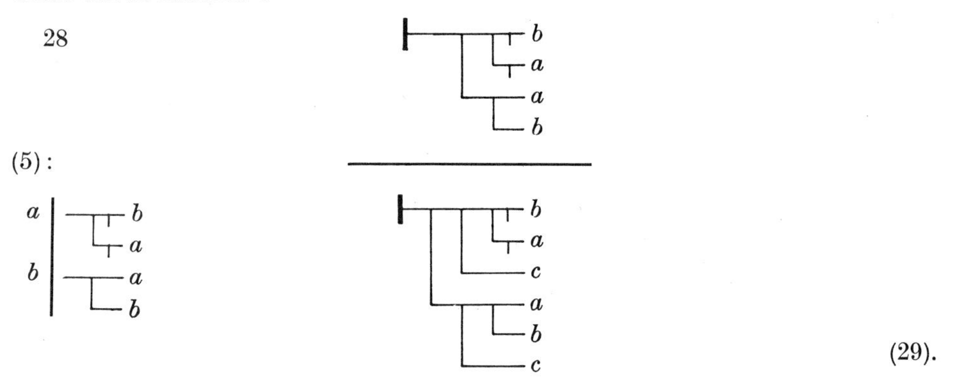

If b and c together form a sufficient condition for a, then from the affirmation of one condition, c, and that of the negation of a ⟦that of⟧ the negation of the other condition can be inferred.

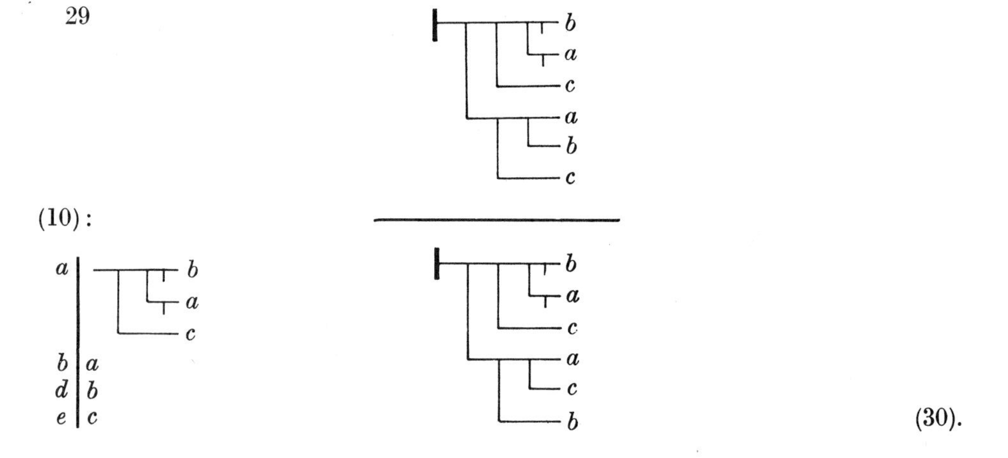

§ 18.

(31).

⊤⊤ a means the denial of the denial, hence the affirmation of a. Thus a cannot be denied and (at the same time) ⊤⊤ a affirmed. *Duplex negatio affirmat.* The denial of the denial is affirmation.

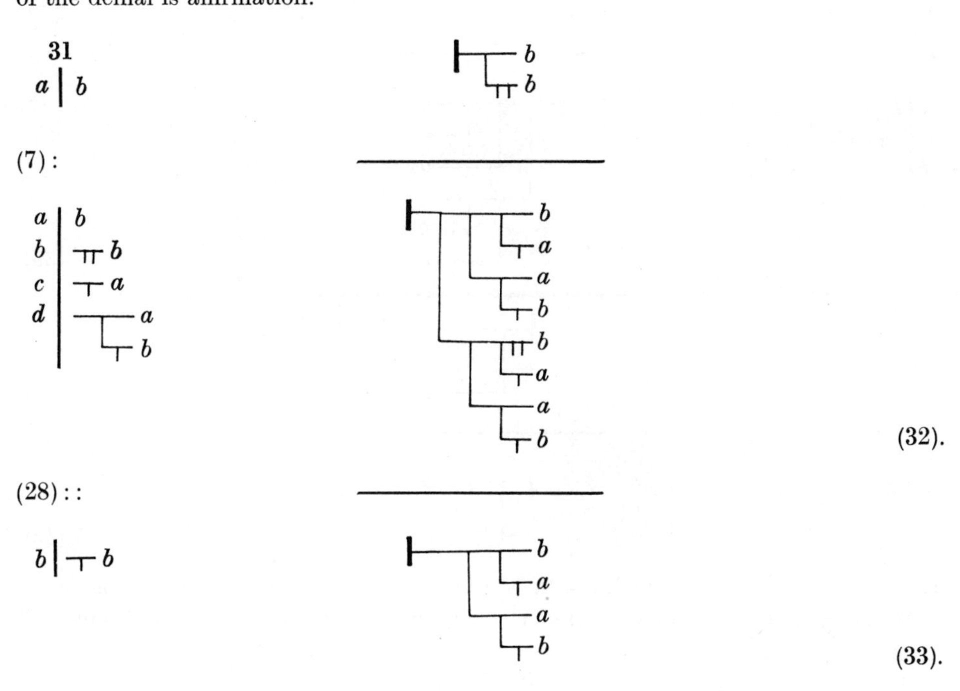

If a or b takes place, then b or a takes place.

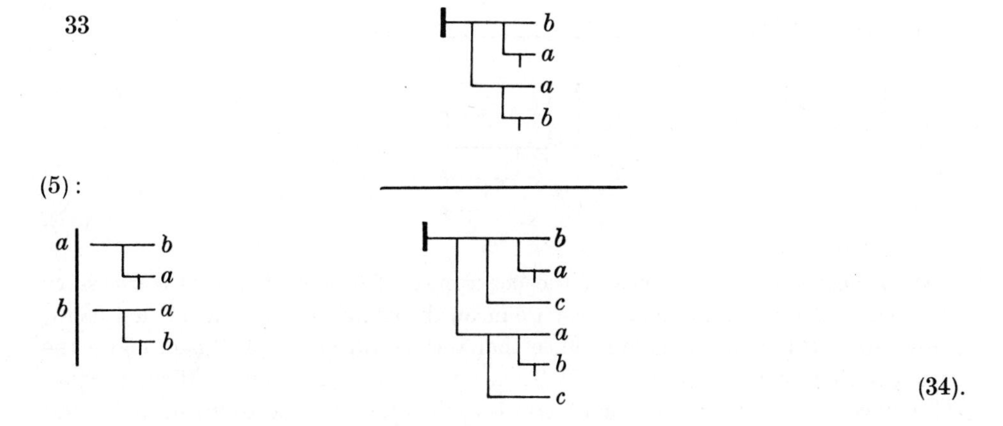

If as a consequence of the occurrence of the circumstance c, when the obstacle b is removed, a takes place, then from the circumstance that a does not take place while c occurs the occurrence of the obstacle b can be inferred.

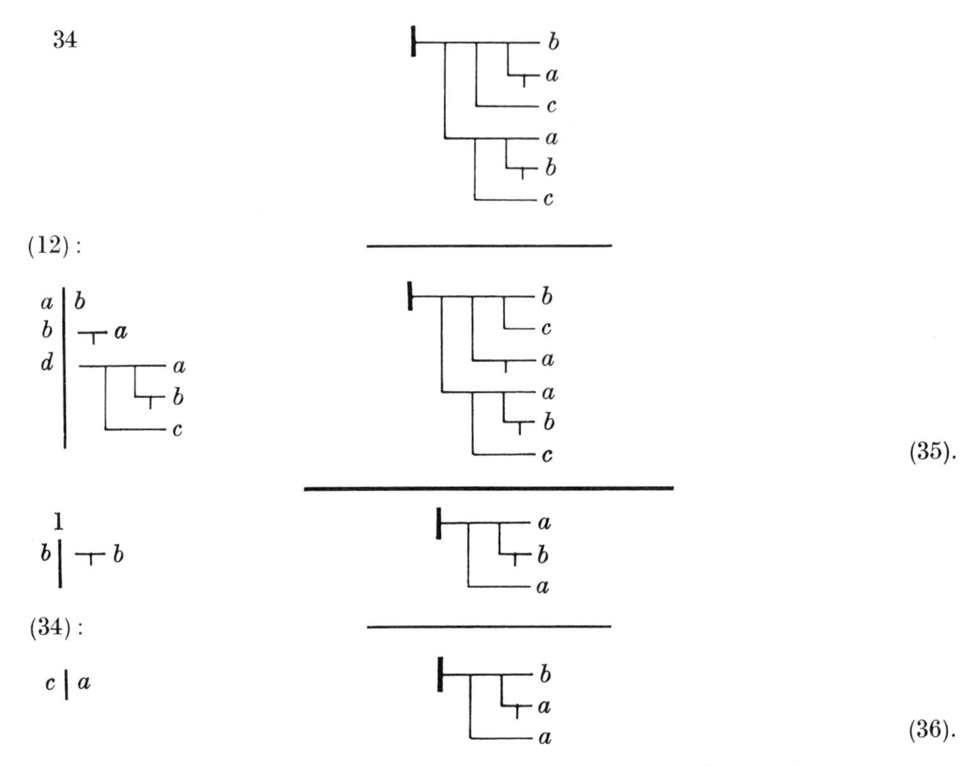

The case in which b is denied, $\top a$ is affirmed, and a is affirmed does not occur. We can express this as follows: "If a occurs, then one of the two, a or b, takes place".

36

$a \mid c$

(9) :

(37).

If a is a necessary consequence of the occurrence of b or c, then a is a necessary consequence of c alone. For example, let b mean the circumstance that the first factor of a product P is 0, c the circumstance that the second factor of P is 0, and a the circumstance that the product P is 0. Then we have the judgment: "If the product P is 0 in case the first or the second factor is 0, then from the vanishing of the second factor the vanishing of the product can be inferred".

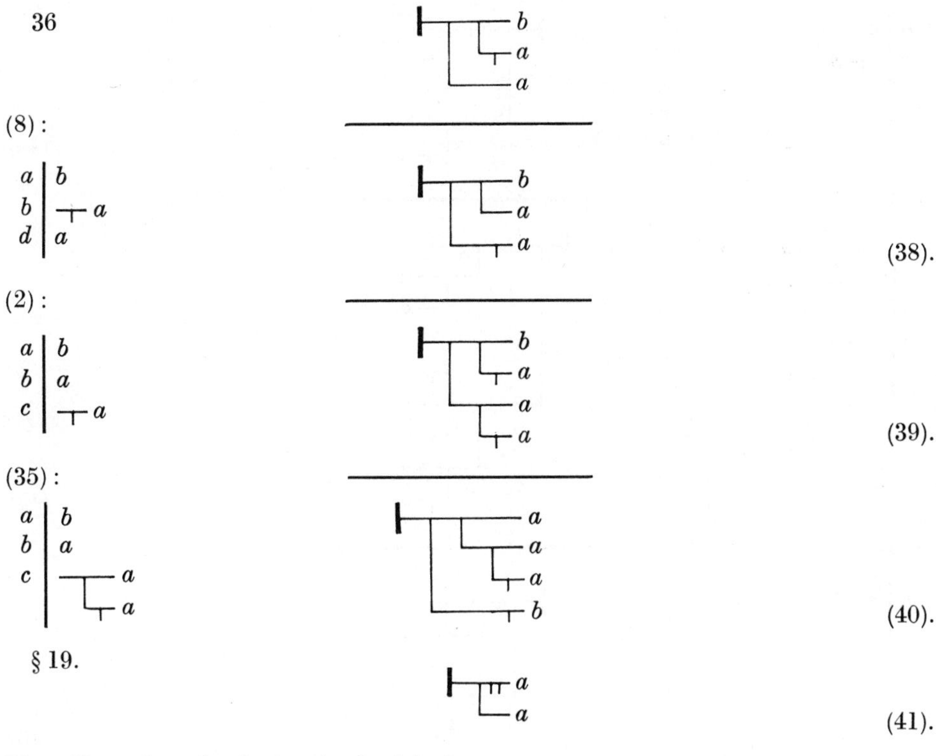

§ 19.

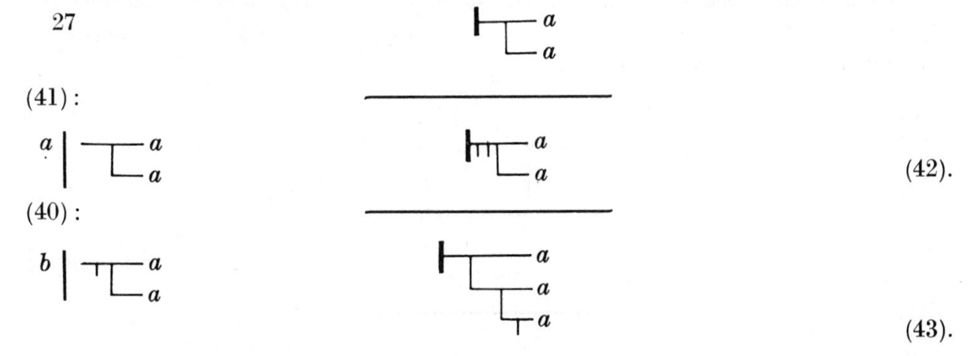

The affirmation of a denies the denial of a.

27

(41):

a

(42).

(40):

b

(43).

If there is a choice only between a and a, then a takes place. For example, we have to distinguish two cases that between them exhaust all possibilities. In following the first, we arrive at the result that a takes place; the same result holds when we follow the second. Then the proposition a holds.

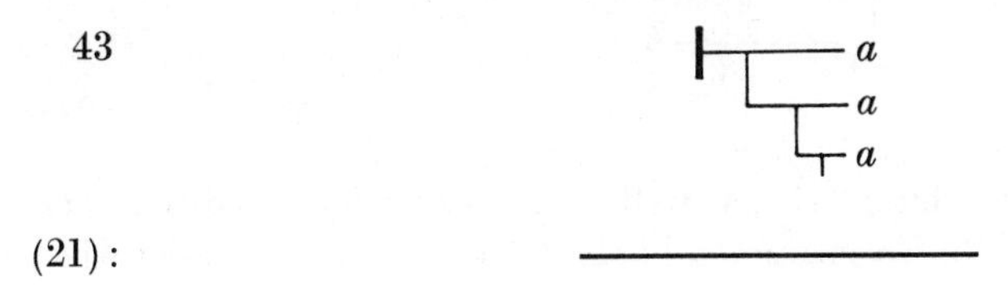

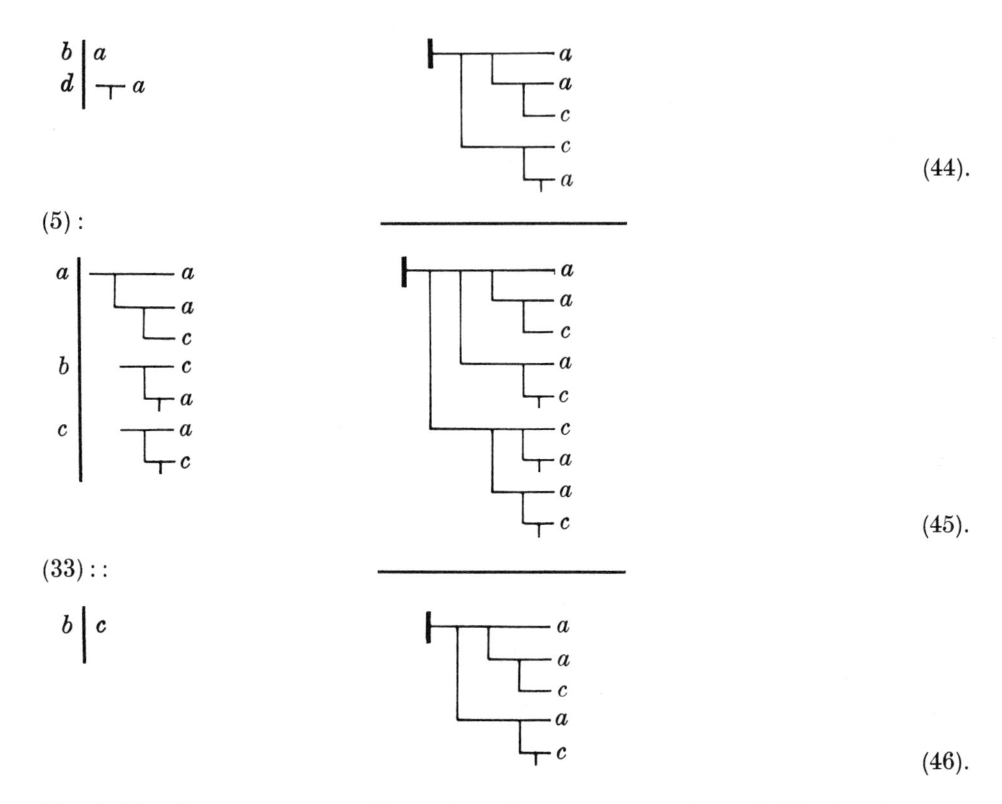

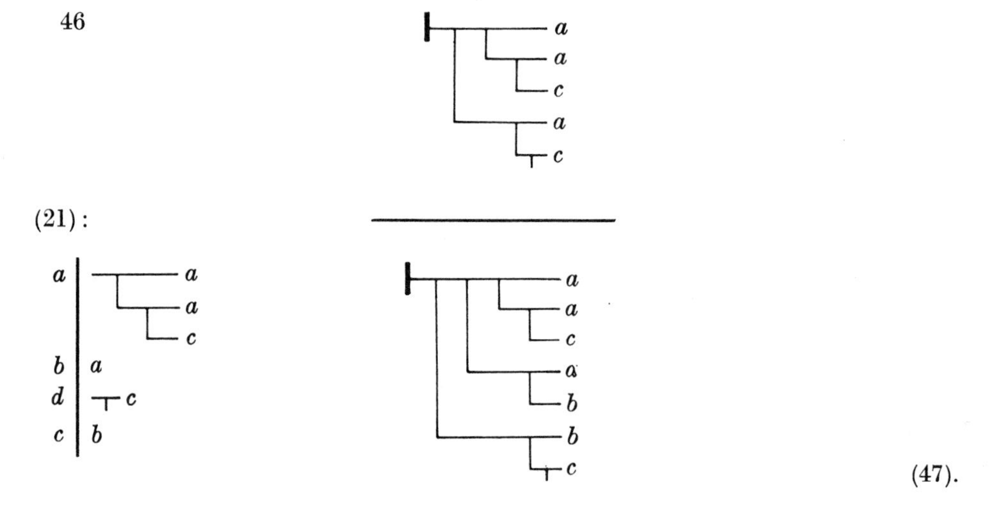

If a holds when c occurs as well as when c does not occur, then a holds. Another way of expressing it is: "If a or c occurs and if the occurrence of c has a as a necessary consequence, then a takes place".

We can express this proposition thus: "If c, as well as b, is a sufficient condition for a and if b or c takes place, then the proposition a holds". This judgment is used when two cases are to be distinguished in a proof. When more cases occur, we can always reduce them to two by taking one of the cases as the first and the totality of the others as the second. The latter can in turn be broken down into two cases, and this can be continued so long as further decomposition is possible.

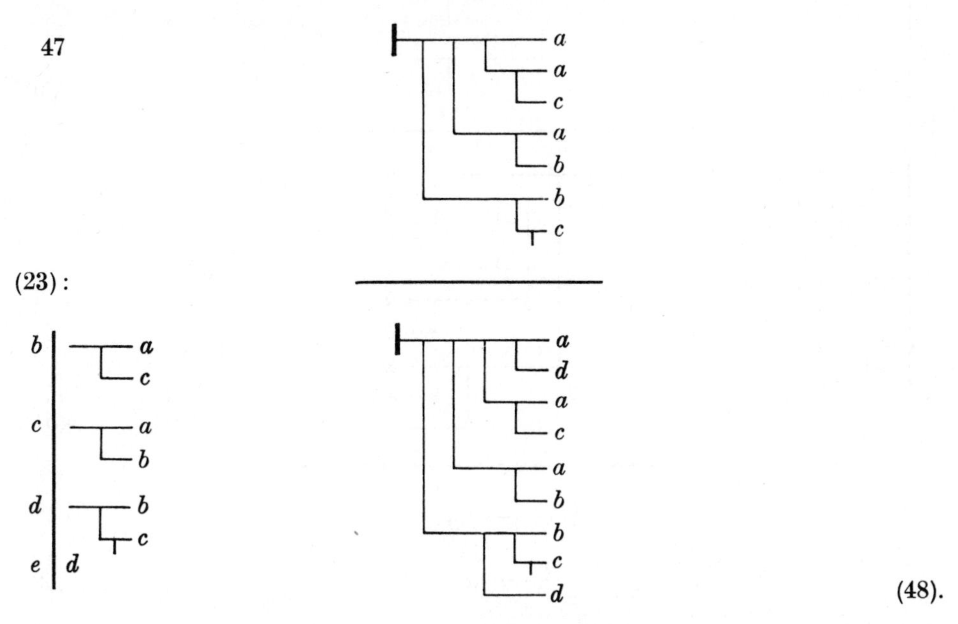

If d is a sufficient condition for the occurrence of b or c and if b, as well as c, is a sufficient condition for a, then d is a sufficient condition for a. An example of an application is furnished by the derivation of formula (101).

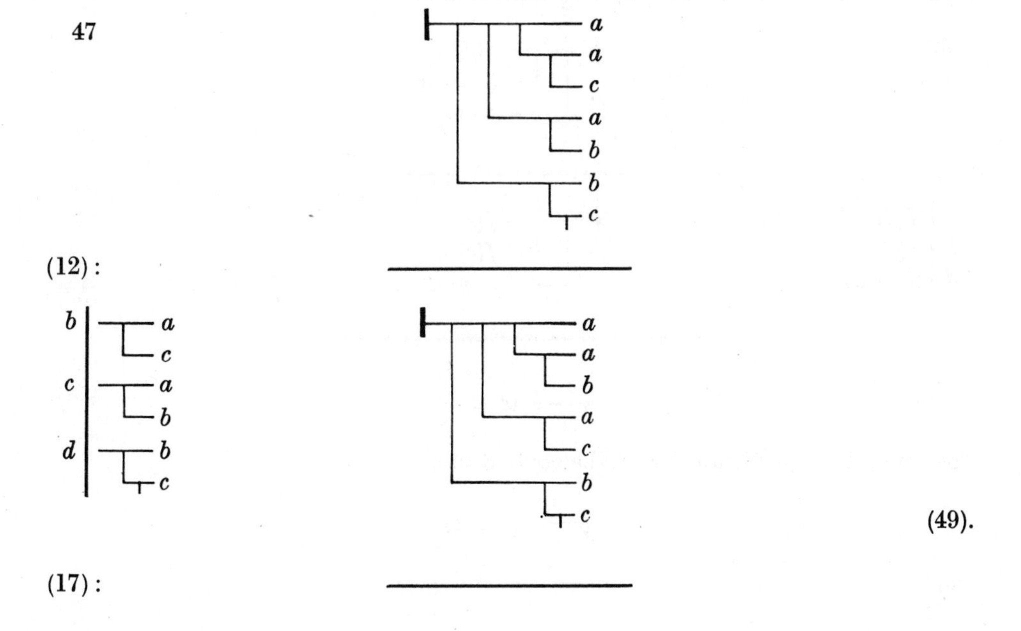

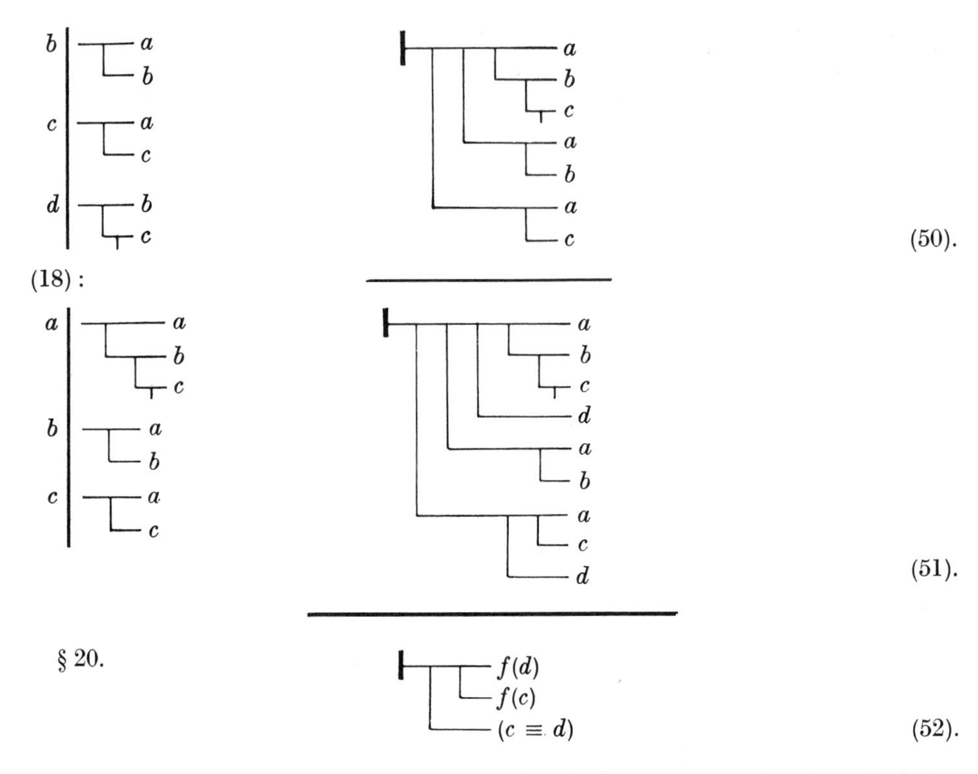

§ 20.

$f(d)$
$f(c)$
$(c \equiv d)$ (52).

The case in which the content of c is identical with the content of d and in which $f(c)$ is affirmed and $f(d)$ is denied does not take place. This proposition means that, if $c \equiv d$, we could everywhere put d for c. In $f(c)$, c can also occur in other than the argument places. Hence c may still be contained in $f(d)$.

52

$f(d)$
$(c \equiv d)$
$f(c)$

(8) :

a	$f(d)$
b	$f(c)$
d	$(c \equiv d)$

$f(d)$
$f(c)$
$(c \equiv d)$ (53).

§ 21.

$(c \equiv c)$ (54).

The content of c is identical with the content of c.

54

$(c \equiv c)$

(53) :

$f(A) \mid (A \equiv c)$

$\vdash\!\!\top\; (d \equiv c)$
$\quad\llcorner\; (c \equiv d)$ (55).

(9) :

$b \mid (d \equiv c)$
$c \mid (c \equiv d)$
$a \mid \top\; f(c)$
$\quad\;\llcorner\; f(d)$

$\vdash\; f(c)$
$f(d)$
$(c \equiv d)$
$f(c)$
$f(d)$
$(d \equiv c)$ (56).

(52) : :

$d \mid c$
$c \mid d$

$\vdash\; f(c)$
$f(d)$
$(c \equiv d)$ (57).

§ 22.

$\vdash\!\!\top\; f(c)$
$\quad\llcorner\!\smile^{\mathfrak{a}}\; f(\mathfrak{a})$ (58).

$\smile^{\mathfrak{a}}\; f(\mathfrak{a})$ means that $f(\mathfrak{a})$ takes place, whatever we may understand by $\mathfrak{a}$. If therefore $\smile^{\mathfrak{a}}\; f(\mathfrak{a})$ is affirmed, $f(c)$ cannot be denied. This is what our proposition expresses. Here $\mathfrak{a}$ can occur only in the argument places of f, since in the judgment this function also occurs outside the scope of $\mathfrak{a}$.

58

$f(A) \mid \top\; f(A)$
$\quad\;\;\llcorner\; g(A)$
$c \mid b$

$\vdash\; f(b)$
$g(b)$
$\smile^{\mathfrak{a}}\; f(\mathfrak{a})$
$g(\mathfrak{a})$

(12) :

$a \mid f(b)$
$c \mid g(b)$
$b \mid \smile^{\mathfrak{a}}\top\; f(\mathfrak{a})$
$\quad\;\;\llcorner\; g(\mathfrak{a})$

$\vdash\; \smile^{\mathfrak{a}}\; f(\mathfrak{a})$
$g(\mathfrak{a})$
$\top\; f(b)$
$g(b)$ (59).

Example. Let b mean an ostrich, that is, an individual animal belonging to the species, let $g(A)$ mean "A is a bird", and let $f(A)$ mean "A can fly". Then we have the judgment "If this ostrich is a bird and cannot fly, then it can be inferred from this that some[17] birds cannot fly".

[17] See footnote 16.

We see how this judgment replaces one mode of inference, namely, Felapton or Fesapo, between which we do not distinguish here since no subject has been singled out.

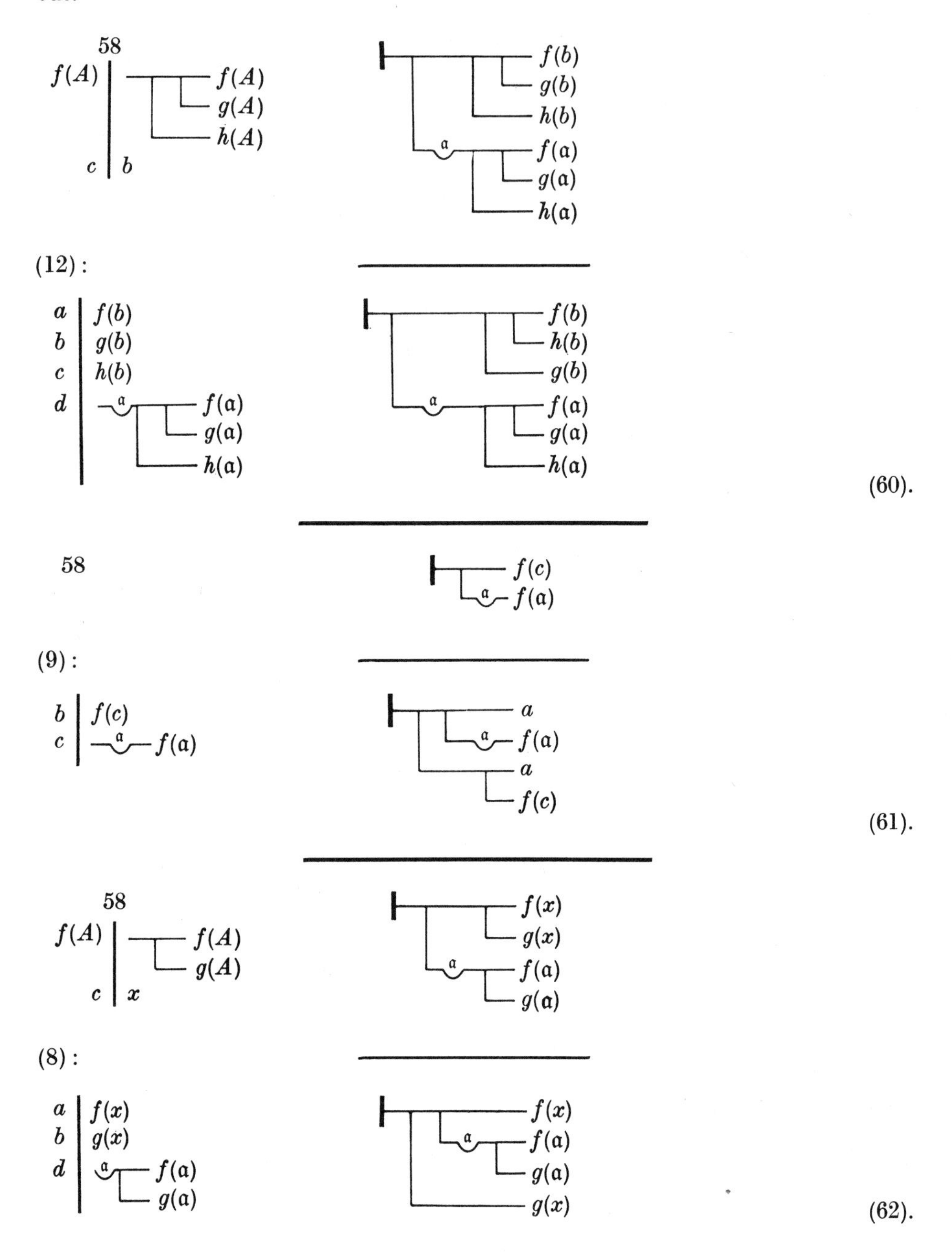

This judgment replaces the mode of inference Barbara when the minor premiss, $g(x)$, has a particular content.

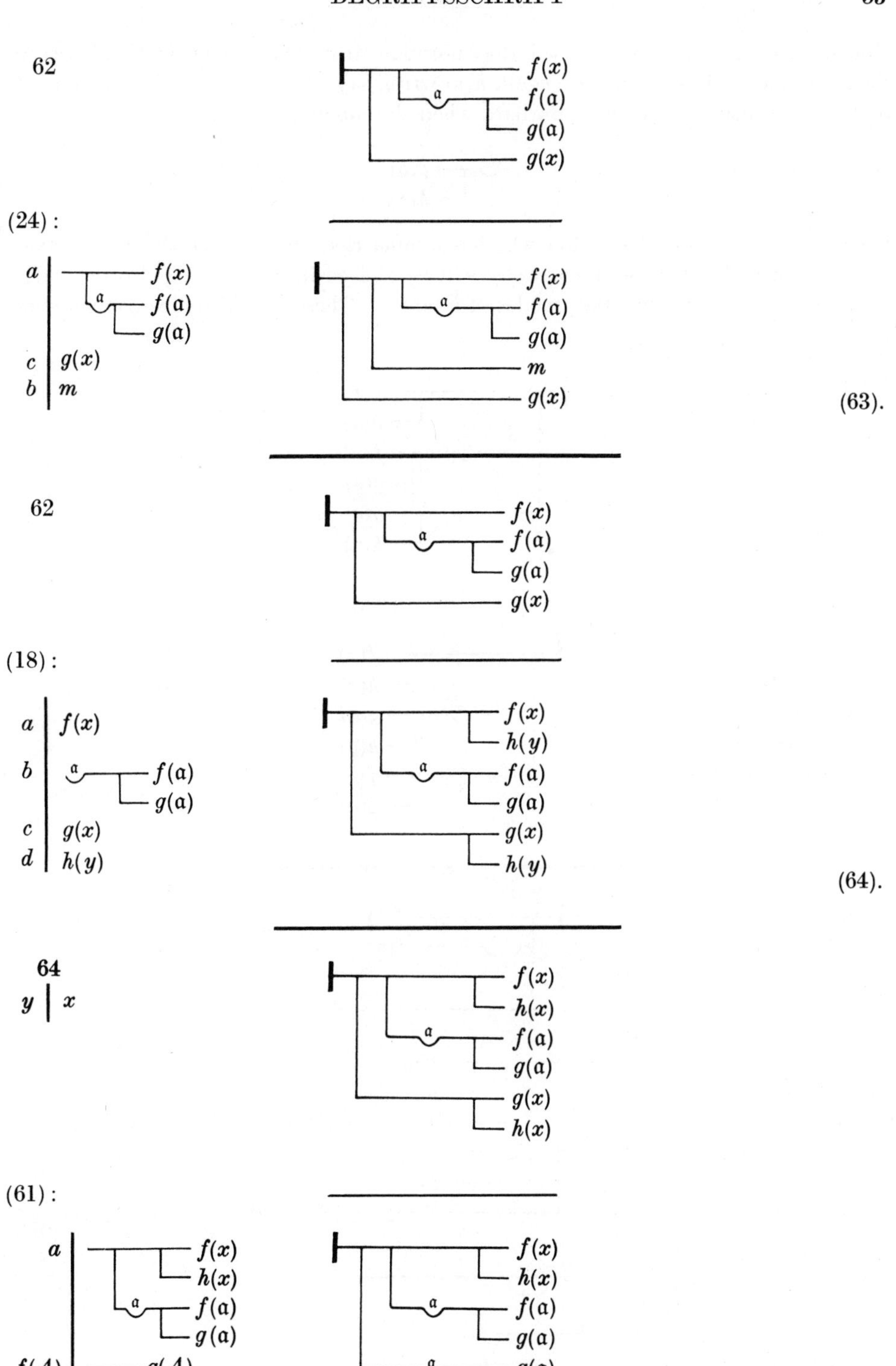
62
f(x)
f(𝔞)
g(𝔞)
g(x)
(24):
a
c g(x)
b m
(63).
62
(18):
a f(x)
b
c g(x)
d h(y)
h(y)
(64).
64
y | x
h(x)
(61):
f(A)
g(A)
h(A)
c x
g(𝔞)
h(𝔞)
(65).

Here $\mathfrak{a}$ occurs in two scopes, but this does not indicate any particular relation between them. In one of these scopes we could also write, say, $\mathfrak{e}$ instead of $\mathfrak{a}$. This judgment replaces the mode of inference Barbara when the minor premiss

$$\overset{\mathfrak{a}}{\smile}\ g(\mathfrak{a}),\ h(\mathfrak{a})$$

has a general content. The reader who has familiarized himself with the way derivations are carried out in the ideography will be in a position to derive also the judgments that answer to the other modes of inference. These should suffice as examples here.

65

$$\vdash\ f(x),\ h(x),\ \overset{\mathfrak{a}}{\smile}[f(\mathfrak{a}),\ g(\mathfrak{a})],\ \overset{\mathfrak{a}}{\smile}[g(\mathfrak{a}),\ h(\mathfrak{a})]$$

(8):

a	$f(x),\ h(x)$
b	$\overset{\mathfrak{a}}{\smile}\ f(\mathfrak{a}),\ g(\mathfrak{a})$
d	$\overset{\mathfrak{a}}{\smile}\ g(\mathfrak{a}),\ h(\mathfrak{a})$

$$\vdash\ f(x),\ h(x),\ \overset{\mathfrak{a}}{\smile}[g(\mathfrak{a}),\ h(\mathfrak{a})],\ \overset{\mathfrak{a}}{\smile}[f(\mathfrak{a}),\ g(\mathfrak{a})]$$

(66).

58

$$\vdash\ f(c),\ \overset{\mathfrak{a}}{\smile}\ f(\mathfrak{a})$$

(7):

a	$f(c)$
b	$\overset{\mathfrak{a}}{\smile}\ f(\mathfrak{a})$
c	b
d	$[(\overset{\mathfrak{a}}{\smile}\ f(\mathfrak{a})) \equiv b]$

$$\vdash\ f(c),\ b,\ [(\overset{\mathfrak{a}}{\smile}\ f(\mathfrak{a})) \equiv b],\ \overset{\mathfrak{a}}{\smile}\ f(\mathfrak{a}),\ b,\ [(\overset{\mathfrak{a}}{\smile}\ f(\mathfrak{a})) \equiv b]$$

(67).

(57)::

$f(A)$	A
c	$\overset{\mathfrak{a}}{\smile}\ f(\mathfrak{a})$
d	b

$$\vdash\ f(c),\ b,\ [(\overset{\mathfrak{a}}{\smile}\ f(\mathfrak{a})) \equiv b]$$

(68).

III. SOME TOPICS FROM A GENERAL THEORY OF SEQUENCES

§ 23. The derivations that follow are intended to give a general idea of the way in which our ideography is handled, even if they are perhaps not sufficient to demonstrate its full utility. This utility would become clear only when more involved propositions are considered. Through the present example, moreover, we see how pure thought, irrespective of any content given by the senses or even by an intuition a priori, can, solely from the content that results from its own constitution, bring forth judgments that at first sight appear to be possible only on the basis of some intuition. This can be compared with condensation, through which it is possible to transform the air that to a child's consciousness appears as nothing into a visible fluid that forms drops. The propositions about sequences developed in what follows far surpass in generality all those that can be derived from any intuition of sequences. If, therefore, one were to consider it more appropriate to use an intuitive idea of sequence as a basis, he should not forget that the propositions thus obtained, which might perhaps have the same wording as those given here, would still state far less than these, since they would hold only in the domain of precisely that intuition upon which they were based.

§ 24.

$$\Vdash \left[\left[\forall \mathfrak{d}\, \Big(F(\mathfrak{d}) \rightarrow \forall \mathfrak{a}\, \big(f(\mathfrak{d},\mathfrak{a}) \rightarrow F(\mathfrak{a}) \big) \Big) \right] \equiv \begin{matrix} \delta \\ | \\ \alpha \end{matrix} \left(\begin{matrix} F(\alpha) \\ \\ f(\delta,\alpha) \end{matrix} \right. \right] \qquad (69).$$

This proposition differs from the judgments considered up to now in that it contains signs that have not been defined before; it itself gives the definition. It does not say "The right side of the equation has the same content as the left", but "It is to have the same content". Hence this proposition is not a judgment, and consequently *not a synthetic judgment* either, to use the Kantian expression. I point this out because Kant considers all judgments of mathematics to be synthetic. If now (69) were a synthetic judgment, so would be the propositions derived from it. But we can do without the notation introduced by this proposition and hence without the proposition itself as its definition; nothing follows from the proposition that could not also be inferred without it. Our sole purpose in introducing such definitions is to bring about an extrinsic simplification by stipulating an abbreviation. They serve besides to emphasize a particular combination of signs in the multitude of possible ones, so that our faculty of representation can get a firmer grasp of it. Now, even though the simplification mentioned is hardly noticeable in the case of the small number of judgments cited here, I nevertheless included this formula for the sake of the example.

Although originally (69) is not a judgment, it is immediately transformed into one; for, once the meaning of the new signs is specified, it must remain fixed, and therefore formula (69) also holds as a judgment, but as an analytic one, since it only makes apparent again what was put into the new signs. This dual character of the formula is indicated by the use of a double judgment stroke. So far as the derivations that follow are concerned, (69) can therefore be treated like an ordinary judgment.

Lower-case Greek letters, which occur here for the first time, do not represent an independent content, as do German and Latin ones. The only thing we have to observe is whether they are identical or different; hence we can put arbitrary lower-case Greek letters for α and δ, provided only that places previously occupied by identical letters are again occupied by identical ones and that different letters are not replaced by identical ones. *Whether Greek letters are identical or different, however, is of significance only within the formula for which they were especially introduced, as they were here for*

$$\begin{array}{c}\delta\\ |\\ \alpha\end{array}\left(\begin{array}{l}F(\alpha)\\ f(\delta, \alpha).\end{array}\right.$$

Their purpose is to enable us to reconstruct unambiguously at any time from the abbreviated form

$$\begin{array}{c}\delta\\ |\\ \alpha\end{array}\left(\begin{array}{l}F(\alpha)\\ f(\delta, \alpha)\end{array}\right.$$

the full one,

$$\begin{array}{l}\overset{\mathfrak{d}}{\smile}\!\!\top\!\!\overset{\mathfrak{a}}{\smile}\!\!\top\!\!\!-\!\!-\, F(\mathfrak{a})\\ \qquad\quad\;\;\llcorner\!\!-\, f(\mathfrak{d}, \mathfrak{a})\\ \quad\;\llcorner\!\!-\!\!-\!\!-\!\!-\, F(\mathfrak{d}).\end{array}$$

For example,

$$\begin{array}{c}\alpha\\ |\\ \delta\end{array}\left(\begin{array}{l}F(\delta)\\ f(\delta, \alpha)\end{array}\right.$$

means the expression

$$\begin{array}{l}\overset{\mathfrak{d}}{\smile}\!\!\top\!\!\overset{\mathfrak{a}}{\smile}\!\!\top\!\!\!-\!\!-\, F(\mathfrak{a})\\ \qquad\quad\;\;\llcorner\!\!-\, f(\mathfrak{a}, \mathfrak{d})\\ \quad\;\llcorner\!\!-\!\!-\!\!-\!\!-\, F(\mathfrak{d}),\end{array}$$

whereas

$$\begin{array}{c}\alpha\\ |\\ \delta\end{array}\left(\begin{array}{l}F(\alpha)\\ f(\delta, \alpha)\end{array}\right.$$

has no meaning. We see that the complete expression, no matter how involved the functions F and f may be, can always be retrieved with certainty, except for the arbitrary choice of German letters.

$$\vdash\!\!-\!\!-\, f(\Gamma, \Delta)$$

can be rendered by "Δ is a result of an application of the procedure f to Γ", by "Γ is the object of an application of the procedure f, with the result Δ", by "Δ bears the relation f to Γ", or by "Γ bears the converse relation of f to Δ"; these expressions are to be taken as equivalent.

$$\begin{array}{c}\delta\\ |\\ \alpha\end{array}\left(\begin{array}{l}F(\alpha)\\ f(\delta, \alpha)\end{array}\right.$$

can be translated by "the circumstance that property F is hereditary in the f-sequence ⟦sich in der f-Reihe vererbt⟧". Perhaps the following example can make this expression acceptable. Let $\Lambda(M, N)$ mean the circumstance that N is a child of M, and $\Sigma(P)$ the circumstance that P is a human being. Then

$$\overset{\delta}{\underset{\alpha}{|}}\Big(\begin{array}{l}\Sigma(\alpha)\\ \Lambda(\delta,\alpha)\end{array} \quad \text{or} \quad \overset{\mathfrak{d}}{\smile}\!\!-\!\!\overset{\mathfrak{a}}{\smile}\begin{array}{l}\Sigma(\mathfrak{a})\\ \Lambda(\mathfrak{d},\mathfrak{a})\\ \Sigma(\mathfrak{d})\end{array}$$

is the circumstance that every child of a human being is in turn a human being, or that the property of being human is hereditary.[18] We see, incidentally, that it can become difficult and even impossible to give a rendering in words if very involved functions take the places of F and f. Proposition (69) could be expressed in words as follows:

If from the proposition that $\mathfrak{d}$ *has property* F *it can be inferred generally, whatever* $\mathfrak{d}$ *may be, that every result of an application of the procedure* f *to* $\mathfrak{d}$ *has property* F, *then I say*: "*Property* F *is hereditary in the* f-*sequence*".

§ 25.

69

$$\vdash\Bigg[\Bigg[\overset{\mathfrak{d}}{\smile}\!\!-\!\!\overset{\mathfrak{a}}{\smile}\begin{array}{l}F(\mathfrak{a})\\ f(\mathfrak{d},\mathfrak{a})\\ F(\mathfrak{d})\end{array}\Bigg] \equiv \overset{\delta}{\underset{\alpha}{|}}\Big(\begin{array}{l}F(\alpha)\\ f(\delta,\alpha)\end{array}\Bigg]$$

(68) :

$$\begin{array}{r|l} \mathfrak{a} & \mathfrak{d} \\ f(\Gamma) & \overset{\mathfrak{a}}{\smile}\begin{array}{l}F(\mathfrak{a})\\ f(\Gamma,\mathfrak{a})\\ F(\Gamma)\end{array} \\ b & \overset{\delta}{\underset{\alpha}{|}}\Big(\begin{array}{l}F(\alpha)\\ f(\delta,\alpha)\end{array} \\ c & x \end{array} \qquad \vdash\; \overset{\mathfrak{a}}{\smile}\begin{array}{l}F(\mathfrak{a})\\ f(x,\mathfrak{a})\\ F(x)\\ \overset{\delta}{\underset{\alpha}{|}}\Big(\begin{array}{l}F(\alpha)\\ f(\delta,\alpha)\end{array}\end{array} \qquad (70).$$

(19) :

$$\begin{array}{r|l} b & \overset{\mathfrak{a}}{\smile}\begin{array}{l}F(\mathfrak{a})\\ f(x,\mathfrak{a})\end{array} \\ c & F(x) \\ d & \overset{\delta}{\underset{\alpha}{|}}\Big(\begin{array}{l}F(\alpha)\\ f(\delta,\alpha)\end{array} \\ a & \begin{array}{l}F(y)\\ f(x,y)\end{array} \end{array} \qquad \vdash\; \begin{array}{l}F(y)\\ f(x,y)\\ F(x)\\ \overset{\delta}{\underset{\alpha}{|}}\Big(\begin{array}{l}F(\alpha)\\ f(\delta,\alpha)\end{array}\\ F(y)\\ f(x,y)\\ \overset{\mathfrak{a}}{\smile}\begin{array}{l}F(\mathfrak{a})\\ f(x,\mathfrak{a})\end{array}\end{array} \qquad (71).$$

(58) : :

[18] ⟦In the German text the formulas contain two misprints: at the extreme left "δ" and the "α" below it are interchanged, and, instead of "$\Lambda(\mathfrak{d},\mathfrak{a})$", the second formula contains "$\Lambda(d,\mathfrak{a})$".⟧

$$f(\Gamma) \;\Big|\; \text{—}\!\!\top\!\text{— } F(\Gamma),\ f(x, \Gamma) \qquad c \;\Big|\; y$$

$$\vdash\ F(y),\ f(x, y),\ F(x),\ \left(\begin{smallmatrix}\delta \\ | \\ \alpha\end{smallmatrix}\ \begin{smallmatrix}F(\alpha) \\ f(\delta, \alpha)\end{smallmatrix}\right) \tag{72}$$

If property F is hereditary in the f-sequence, if x has property F, and if y is a result of an application of the procedure f to x, then y has property F.

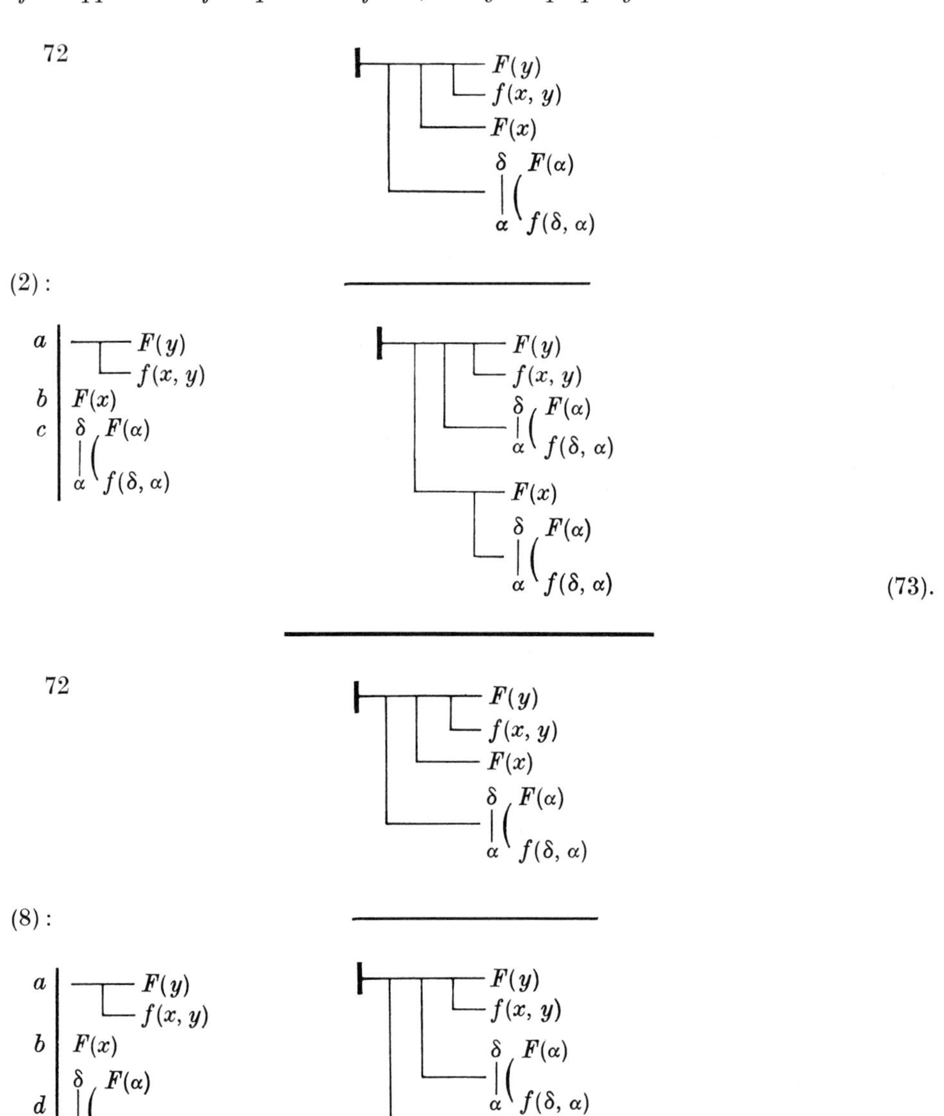

(73).

(74).

If x has a property F that is hereditary in the f-sequence, then every result of an application of the procedure f to x has property F.

69

$$\vdash\Big[\Big[\ \overset{\mathfrak{d}}{\smile}\!\!-\!\!\overset{\mathfrak{a}}{\smile}\!\!-\ F(\mathfrak{a}),\ f(\mathfrak{d},\mathfrak{a}),\ F(\mathfrak{d})\Big] \equiv \left(\begin{matrix}\delta \\ \alpha\end{matrix}\ \begin{matrix}F(\alpha)\\ f(\delta,\alpha)\end{matrix}\right)\Big]$$

(52):

$$\begin{array}{c|l} c & \overset{\mathfrak{d}}{\smile}\!\!-\!\!\overset{\mathfrak{a}}{\smile}\!\!-\ F(\mathfrak{a}),\ f(\mathfrak{d},\mathfrak{a}),\ F(\mathfrak{d}) \\ d & \left(\begin{matrix}\delta \\ \alpha\end{matrix}\ \begin{matrix}F(\alpha)\\ f(\delta,\alpha)\end{matrix}\right) \\ f(\Gamma) & \Gamma \end{array}$$

$$\vdash \left(\begin{matrix}\delta \\ \alpha\end{matrix}\ \begin{matrix}F(\alpha)\\ f(\delta,\alpha)\end{matrix}\right);\quad \overset{\mathfrak{d}}{\smile}\!\!-\!\!\overset{\mathfrak{a}}{\smile}\!\!-\ F(\mathfrak{a}),\ f(\mathfrak{d},\mathfrak{a}),\ F(\mathfrak{d}) \qquad (75).$$

If from the proposition that $\mathfrak{d}$ has property F, whatever $\mathfrak{d}$ may be, it can be inferred that every result of an application of the procedure f to $\mathfrak{d}$ has property F, then property F is hereditary in the f-sequence.

§ 26.

$$\Vdash\Big[\Big[\ \overset{\mathfrak{F}}{\smile}\!\!-\ \mathfrak{F}(y);\ \overset{\mathfrak{a}}{\smile}\!\!-\ \mathfrak{F}(\mathfrak{a}),\ f(x,\mathfrak{a});\ \left(\begin{matrix}\delta \\ \alpha\end{matrix}\ \begin{matrix}\mathfrak{F}(\alpha)\\ f(\delta,\alpha)\end{matrix}\right)\Big] \equiv \frac{\gamma}{\beta} f(x_\gamma, y_\beta)\Big] \qquad (76).$$

This is the definition of the combination of signs on the right, $\frac{\gamma}{\beta} f(x_\gamma, y_\beta)$. I refer the reader to § 24 for the use of the double judgment stroke and Greek letters. It would not do to write merely

$$\frac{x}{y} f(x, y)$$

instead of the expression above since, when a function of x and y is fully written out, these letters could still appear outside of the argument places; in that case we should not be able to tell which places were to be regarded as argument places. Hence these must be characterized as such. This is done here by means of the subscripts γ and β. These must be chosen different since it is possible that the two arguments may be identical with each other. We use Greek letters for this, so that we have a certain freedom of choice and thus can choose the symbols for the argument places of the enclosed expression different from those ⟦used for the argument places⟧ of the enclosing expression in case

$$\frac{\gamma}{\beta} f(x_\gamma, y_\beta)$$

should enclose within itself a similarly constructed expression. *Whether Greek letters are identical or different is of significance here only within the expression*

$$\overset{\gamma}{\underset{\beta}{\smile}} f(x_{\gamma}, y_{\beta});$$

outside, the same letters could be used, and this would not indicate any connection with the occurrences inside.

We translate

$$\overset{\gamma}{\underset{\beta}{\smile}} f(x_{\gamma}, y_{\beta})$$

by "y follows x in the f-sequence", a way of speaking that, to be sure, is possible only when the function f is determined. Accordingly, (76) can be rendered in words somewhat as follows:

If from the two propositions that every result of an application of the procedure f to x has property F and that property F is hereditary in the f-sequence, it can be inferred, whatever F may be, that y has property F, then I say: "y follows x in the f-sequence", or "x precedes y in the f-sequence".[19]

§ 27.

76

$$\Vdash \left[\left[\begin{array}{l} \mathfrak{F}\;\; \mathfrak{F}(y) \\ \quad \mathfrak{a}\;\; \mathfrak{F}(\mathfrak{a}) \\ \qquad f(x, \mathfrak{a}) \\ \quad \overset{\delta}{\underset{\alpha}{|}}\left(\begin{array}{l}\mathfrak{F}(\alpha) \\ f(\delta, \alpha)\end{array}\right. \end{array}\right] \equiv \overset{\gamma}{\underset{\beta}{\smile}} f(x_{\gamma}, y_{\beta})\right]$$

(68):

$$\begin{array}{c|l} \mathfrak{a} & \mathfrak{F} \\ f(\Gamma) & \begin{array}{l} \Gamma(y) \\ \quad \mathfrak{a}\;\; \Gamma(\mathfrak{a}) \\ \qquad f(x, \mathfrak{a}) \\ \overset{\delta}{\underset{\alpha}{|}}\left(\begin{array}{l}\Gamma(\alpha) \\ f(\delta, \alpha)\end{array}\right. \end{array} \\ b & \overset{\gamma}{\underset{\delta}{\smile}} f(x_{\gamma}, y_{\beta}) \\ c & F \end{array}$$

$$\vdash \begin{array}{l} F(y) \\ \quad \mathfrak{a}\;\; F(\mathfrak{a}) \\ \qquad f(x, \mathfrak{a}) \\ \overset{\delta}{\underset{\alpha}{|}}\left(\begin{array}{l}F(\alpha) \\ f(\delta, \alpha)\end{array}\right. \\ \overset{\gamma}{\underset{\beta}{\smile}} f(x_{\gamma}, y_{\beta}) \end{array} \qquad (77).$$

Here $F(y)$, $F(\mathfrak{a})$, and $F(\alpha)$ must be regarded, in accordance with § 10, as different functions of the argument F. (77) means:

If y follows x in the f-sequence, if property F is hereditary in the f-sequence, and if every result of an application of the procedure f to x has property F, then y has property F.

[19] To make clearer the generality of the concept, given hereby, of succession in a sequence, I remind the reader of a number of possibilities. Not only juxtaposition, such as pearls on a string exhibit, is subsumed here, but also branching like that of a family tree, merging of several branches, and ringlike self-linking.

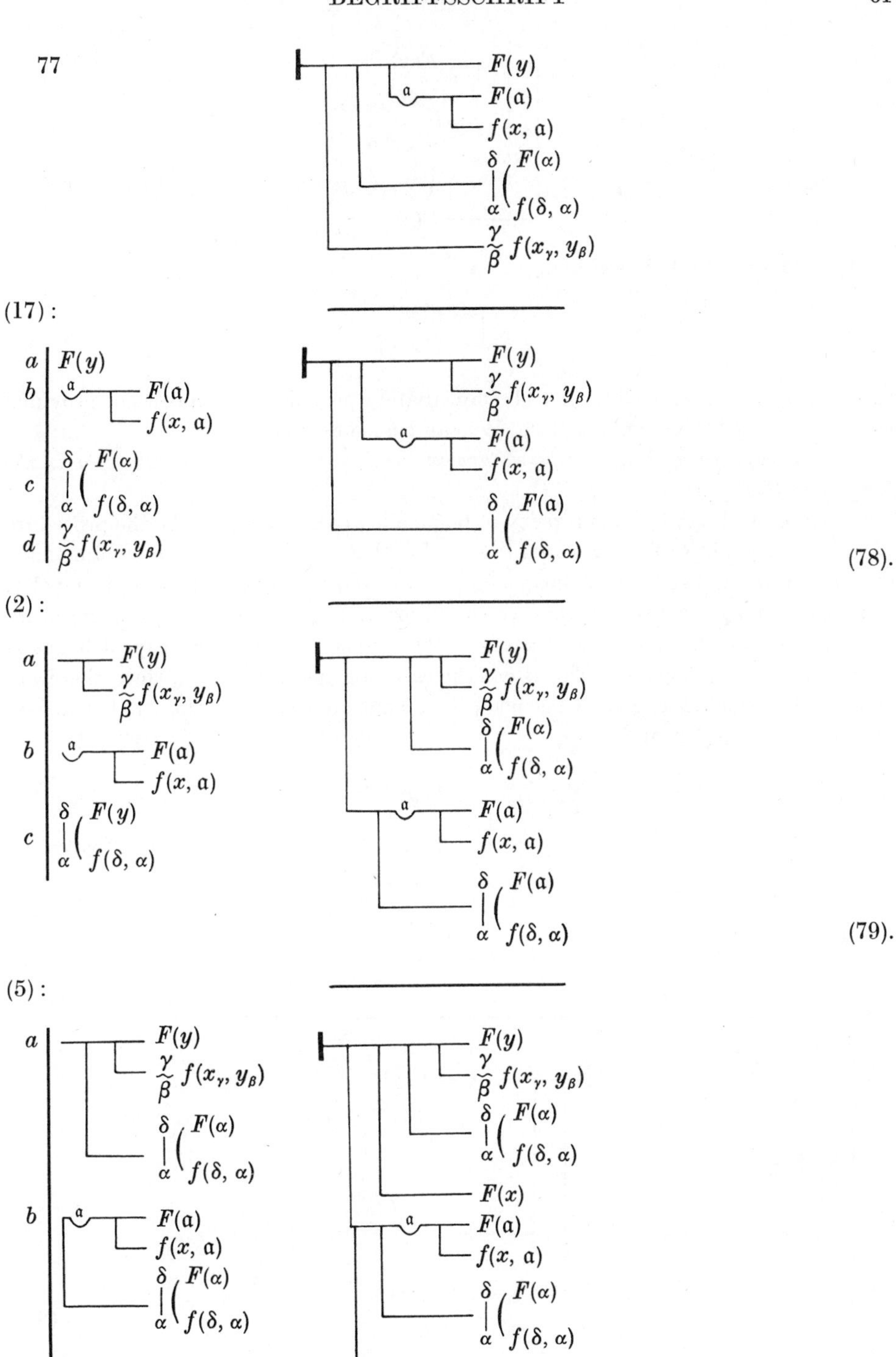
77
(17) :
(78).
(2) :
(79).
(5) :
(80).
(74) : :

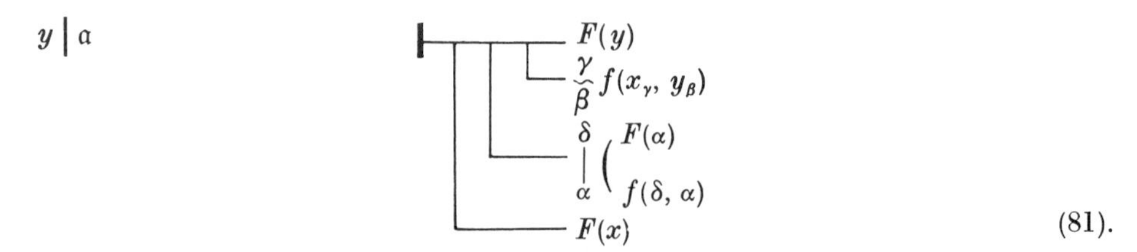

Since in (74) y occurs only in

$$\begin{array}{l} F(y) \\ f(x, y), \end{array}$$

the concavity can, according to § 11, immediately precede this expression, provided y is replaced by the German letter $\mathfrak{a}$. We can translate (81) thus:

If x has a property F that is hereditary in the f-sequence, and if y follows x in the f-sequence, then y has property F.[20]

For example, let F be the property of being a heap of beans; let f be the procedure of removing one bean from a heap of beans; so that $f(a, b)$ means the circumstance that b contains all beans of the heap a except one and does not contain anything else. Then by means of our proposition we would arrive at the result that a single bean, or even none at all, is a heap of beans if the property of being a heap of beans is hereditary in the f-sequence. This is not the case in general, however, since there are certain z for which $F(z)$ cannot become a judgment on account of the indeterminateness of the notion "heap".

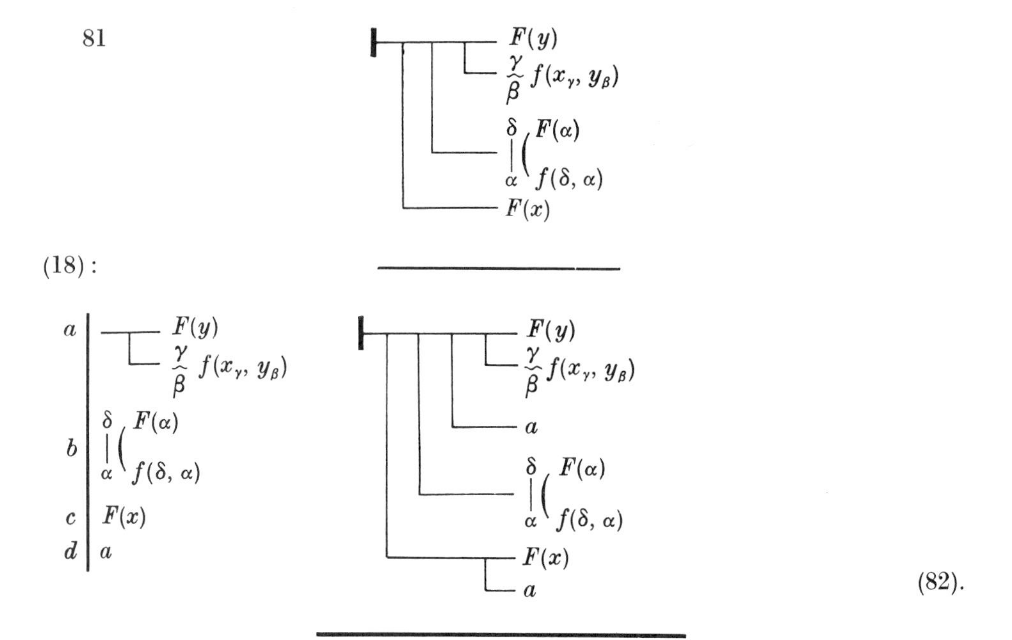

[20] Bernoulli's induction rests upon this. ⟦Jakob Bernoulli is considered one of the originators of mathematical induction, which he used from 1686 on (see *Bernoulli 1686*).⟧

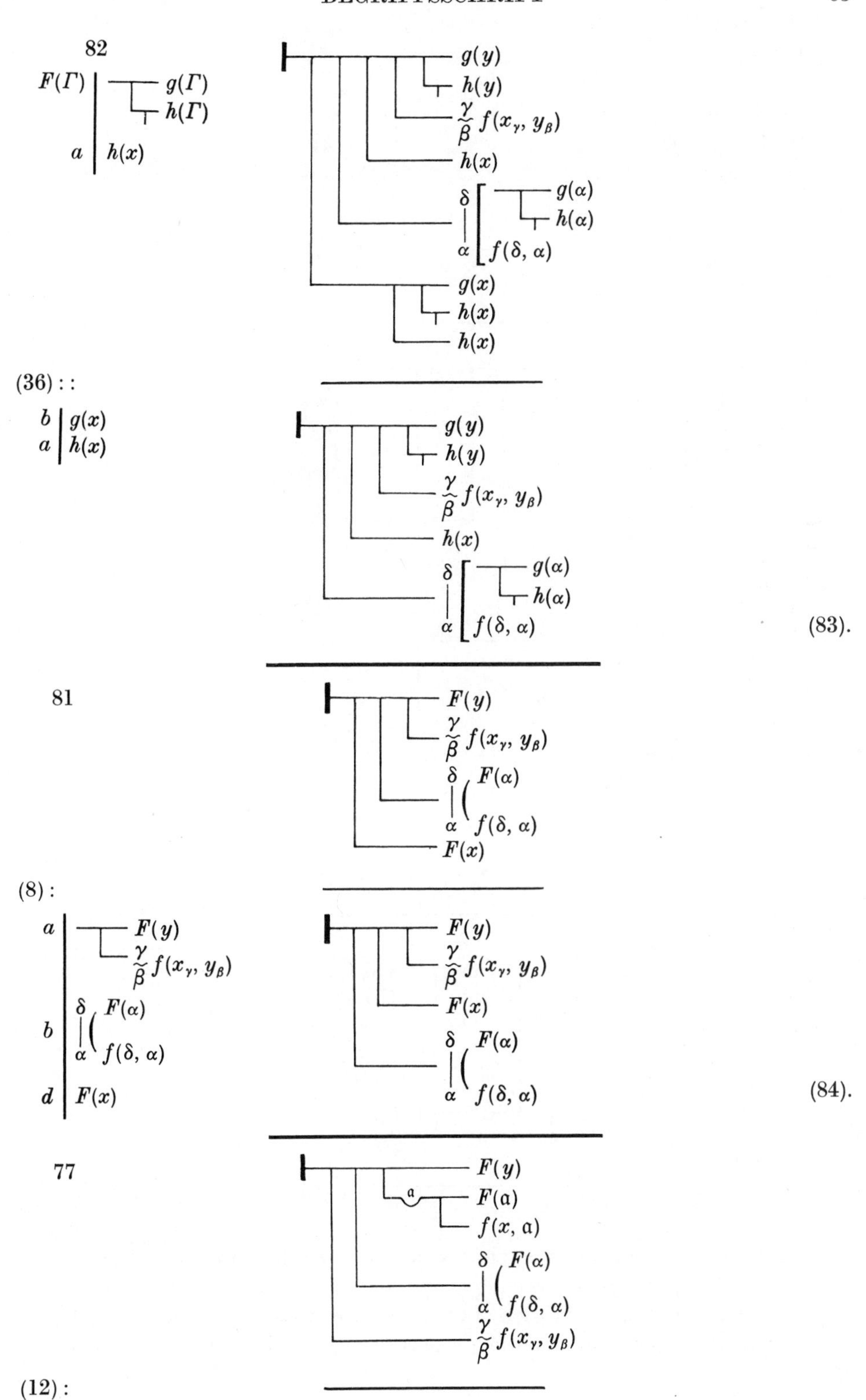
82
F(Γ)
g(Γ)
h(Γ)
a
h(x)
g(y)
h(y)
f(x_γ, y_β)
h(x)
g(α)
h(α)
f(δ, α)
g(x)
h(x)
h(x)
(36) : :
b
g(x)
a
h(x)
(83).
81
F(y)
F(α)
F(x)
(8) :
a
b
d
(84).
77
F(a)
f(x, a)
(12) :

a | $F(y)$

b | $\smile^{\mathfrak{a}}$ $F(\mathfrak{a})$, $f(x, \mathfrak{a})$

c | $\delta \,\vert\, \alpha$ $\big(F(\alpha), f(\delta, \alpha)$

d | $\frac{\gamma}{\tilde{\beta}} f(x_\gamma, y_\beta)$

$F(y)$

$\delta \,\vert\, \alpha$ $\big(F(\alpha), f(\delta, \alpha)$

$\smile^{\mathfrak{a}}$ $F(\mathfrak{a})$, $f(x, \mathfrak{a})$

$\frac{\gamma}{\tilde{\beta}} f(x_\gamma, y_\beta)$ (85).

(19):

b | $F(y)$, $\delta \,\vert\, \alpha$ $\big(F(\alpha), f(\delta, \alpha)$

c | $\smile^{\mathfrak{a}}$ $F(\mathfrak{a})$, $f(x, \mathfrak{a})$

d | $\frac{\gamma}{\tilde{\beta}} f(x_\gamma, y_\beta)$

a | $F(z)$, $f(y, z)$, $\delta \,\vert\, \alpha$ $\big(F(\alpha), f(\delta, \alpha)$

$F(z)$

$f(y, z)$

$\delta \,\vert\, \alpha$ $\big(F(\alpha), f(\delta, \alpha)$

$\smile^{\mathfrak{a}}$ $F(\mathfrak{a})$, $f(x, \mathfrak{a})$

$\frac{\gamma}{\tilde{\beta}} f(x_\gamma, y_\beta)$

$F(z)$

$f(y, z)$

$\delta \,\vert\, \alpha$ $\big(F(\alpha), f(\delta, \alpha)$

$F(y)$

$\delta \,\vert\, \alpha$ $\big(F(\alpha), f(\delta, \alpha)$ (86).

(73)::

y | z

x | y

$F(z)$

$f(y, z)$

$\delta \,\vert\, \alpha$ $\big(F(\alpha), f(\delta, \alpha)$

$\smile^{\mathfrak{a}}$ $F(\mathfrak{a})$, $f(x, \mathfrak{a})$

$\frac{\gamma}{\tilde{\beta}} f(x_\gamma, y_\beta)$ (87).

In words, the derivation of this proposition will be somewhat as follows. Assume that

(α) y follows x in the f-sequence,

(β) Every result of an application of the procedure f to x has property F, and

(γ) Property F is hereditary in the f-sequence.

From these assumptions it follows according to (85) that

(δ) y has property F.

Now,

(ε) Let z be a result of an application of the procedure f to y.

Then by (72) it follows from (γ), (δ), and (ε) that z has property F. Therefore,

If z is a result of an application of the procedure f to an object y that follows x in the f-sequence and if every result of an application of the procedure f to x has a property F that is hereditary in the f-sequence, then z has this property F.[21]

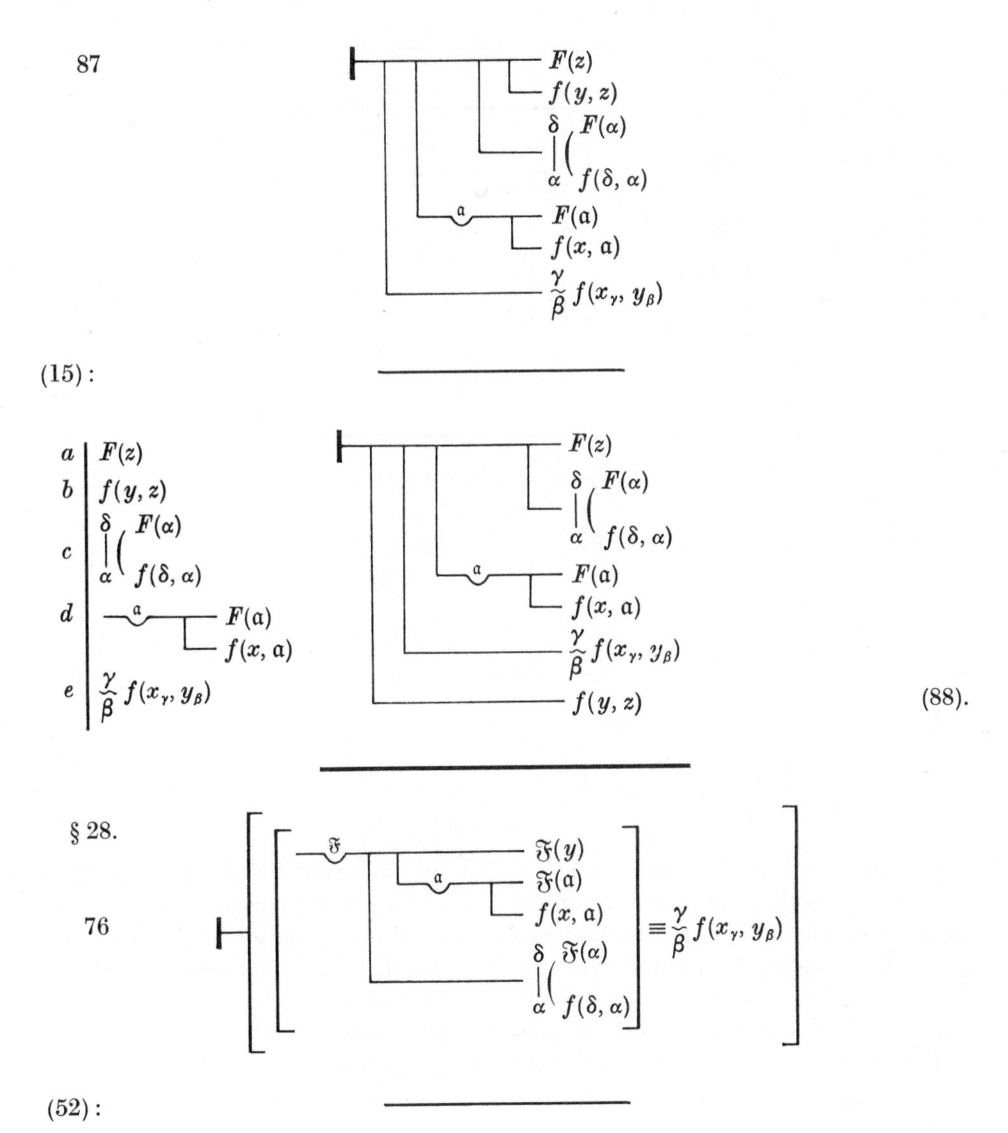

[21] [[At the place that corresponds to the last occurrence of "f" in this sentence the German text mistakenly has "F".]]

$f(\Gamma)$	Γ
c	$\mathfrak{F}$: $\mathfrak{F}(y)$; $\mathfrak{a}$: $\mathfrak{F}(\mathfrak{a})$, $f(x, \mathfrak{a})$; $\frac{\delta}{\alpha}\mathfrak{F}(\alpha) f(\delta, \alpha)$
d	$\frac{\gamma}{\beta} f(x_\gamma, y_\beta)$

$$\vdash \frac{\gamma}{\beta} f(x_\gamma, y_\beta) \quad \mathfrak{F}(y) \quad \mathfrak{F}(\mathfrak{a}) \quad f(x, \mathfrak{a}) \quad \frac{\delta}{\alpha}\mathfrak{F}(\alpha) f(\delta, \alpha)$$ (89).

(5):

a	$\frac{\gamma}{\beta} f(x_\gamma, y_\beta)$
b	$\mathfrak{F}$: $\mathfrak{F}(y)$; $\mathfrak{a}$: $\mathfrak{F}(\mathfrak{a})$, $f(x, \mathfrak{a})$; $\frac{\delta}{\alpha}\mathfrak{F}(\alpha) f(\delta, \alpha)$

$$\vdash \frac{\gamma}{\beta} f(x_\gamma, y_\beta) \quad c \quad \mathfrak{F}(y) \quad \mathfrak{F}(\mathfrak{a}) \quad f(x, \mathfrak{a}) \quad \frac{\delta}{\alpha}\mathfrak{F}(\alpha) f(\delta, \alpha) \quad c$$ (90).

63

f	$\mathfrak{F}$
x	y
$g(\Gamma)$	$f(x, \Gamma)$
m	$\frac{\delta}{\alpha}\mathfrak{F}(\alpha) f(\delta, \alpha)$

$$\vdash \mathfrak{F}(y)^{22} \quad \mathfrak{F}(\mathfrak{a}) \quad f(x, \mathfrak{a}) \quad \frac{\delta}{\alpha}\mathfrak{F}(\alpha) f(\delta, \alpha) \quad f(x, y)$$

(90):

c	$f(x, y)$

$$\vdash \frac{\gamma}{\beta} f(x_\gamma, y_\beta) \quad f(x, y)$$ (91).

Let us give here the derivation of proposition (91) in words. From the proposition (α), "Every result of an application of the procedure f to x has property $\mathfrak{F}$", it can be inferred, whatever $\mathfrak{F}$ may be, that every result of an application of the procedure f to x has property $\mathfrak{F}$. Hence it can also be inferred from proposition (α) and the proposition that property $\mathfrak{F}$ is hereditary in the f-sequence, whatever $\mathfrak{F}$ may be, that every result of an application of the procedure f to x has property $\mathfrak{F}$.

Therefore, according to (90) the following proposition holds:

Every result of an application of a procedure f to an object x follows that x in the f-sequence.

[22] Concerning the concavity with $\mathfrak{F}$ see § 11. [In fact, Frege has already used the concavity with $\mathfrak{F}$ several times, the first occurrence being in (76).]

91

$$\vdash \; \frac{\gamma}{\widetilde{\beta}} f(x_\gamma, y_\beta)$$
$$f(x, y)$$

(53):

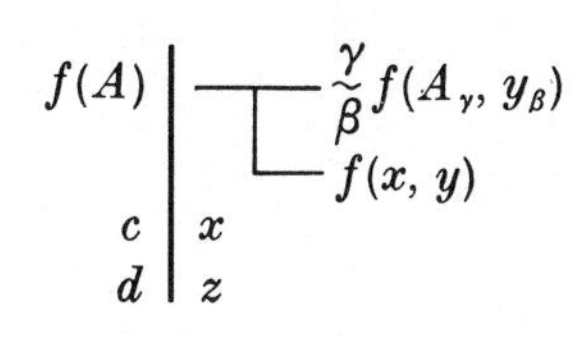

$$\vdash \; \frac{\gamma}{\widetilde{\beta}} f(z_\gamma, y_\beta)$$
$$f(x, y)$$
$$(x \equiv z)$$

(92).

60

$\mathfrak{a}$ | $\mathfrak{F}$

$f(\Gamma)$ | $\Gamma(y)$

$g(\Gamma)$ | $\delta \; \Gamma(\alpha) \quad \alpha \; f(\delta, \alpha)$

$h(\Gamma)$ | $\mathfrak{a} \; \Gamma(\mathfrak{a})$; $f(x, \mathfrak{a})$

b | $\mathfrak{F}$

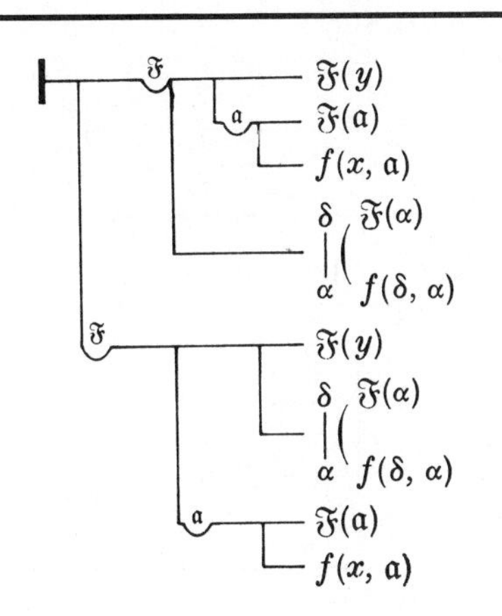

(90):

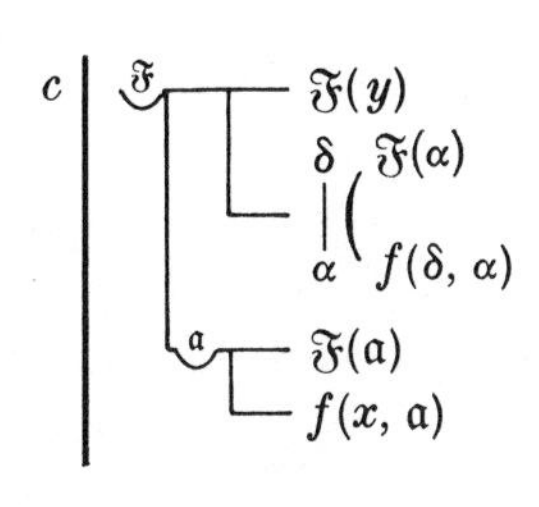

$$\vdash \; \frac{\gamma}{\widetilde{\beta}} f(x_\gamma, y_\beta)$$
$$\mathfrak{F}(y)$$
$$\delta \; \mathfrak{F}(\alpha) \quad \alpha \; f(\delta, \alpha)$$
$$\mathfrak{a} \; \mathfrak{F}(\mathfrak{a})$$
$$f(x, \mathfrak{a})$$

(93).

93

$y \mid z$

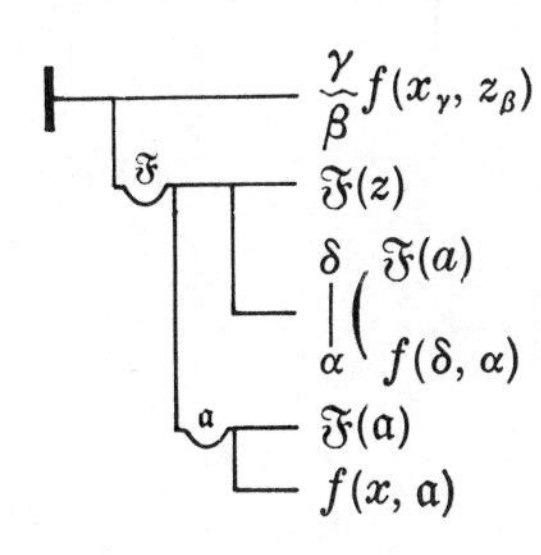

(7):

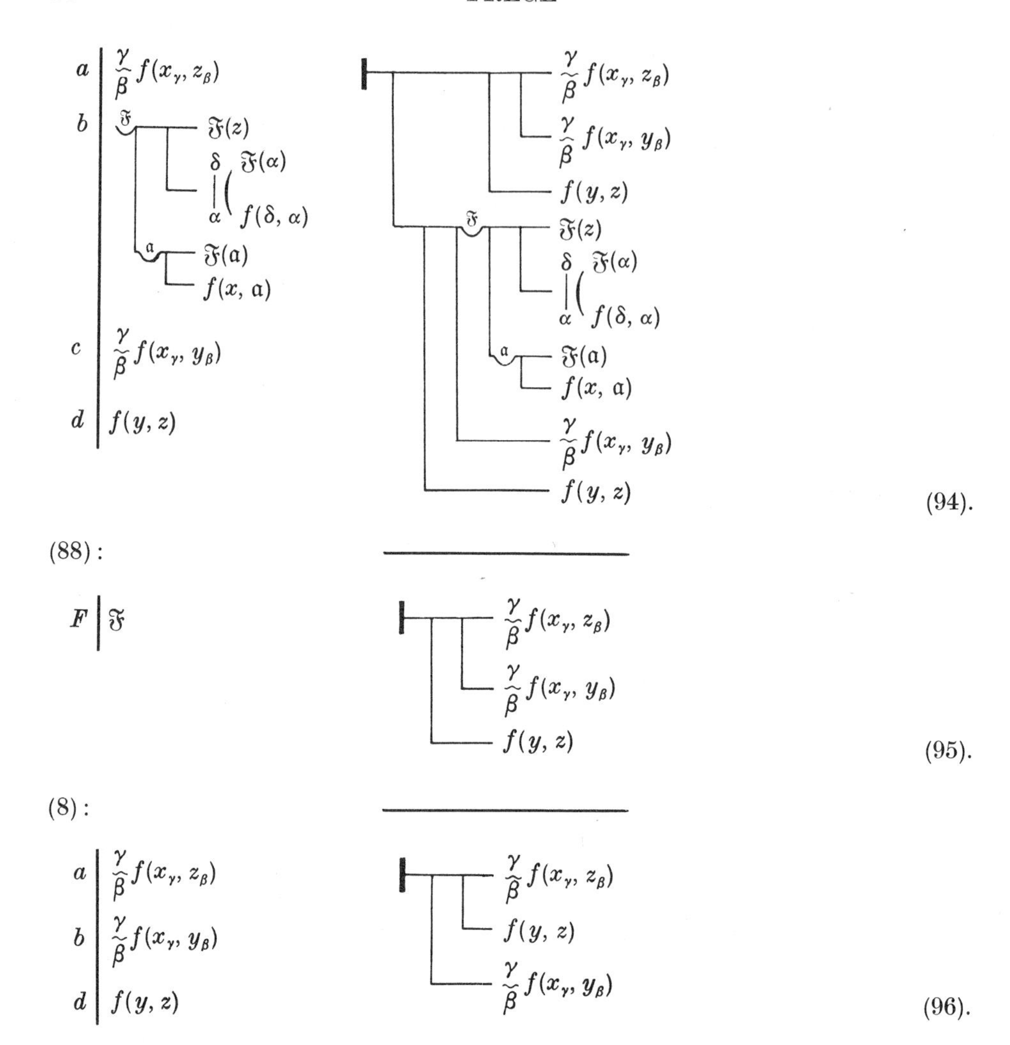

Every result of an application of the procedure f to an object that follows x in the f-sequence follows x in the f-sequence.

96

$z \mid \mathfrak{a}$

$y \mid \mathfrak{d}$

$$\vdash \overset{\mathfrak{d}}{\smile} \overset{\mathfrak{a}}{\smile} \quad \frac{\gamma}{\tilde{\beta}} f(x_\gamma, \mathfrak{a}_\beta) \quad f(\mathfrak{d}, \mathfrak{a}) \quad \frac{\gamma}{\tilde{\beta}} f(x_\gamma, \mathfrak{d}_\beta)$$

(75):

$F(\Gamma) \mid \frac{\gamma}{\tilde{\beta}} f(x_\gamma, \Gamma_\beta)$

$$\vdash \begin{array}{c}\delta \\ | \\ \alpha\end{array}\left(\begin{array}{l} \frac{\gamma}{\tilde{\beta}} f(x_\gamma, \alpha_\beta) \\ f(\delta, \alpha) \end{array}\right. \qquad (97).$$

The property of following x in the f-sequence is hereditary in the f-sequence.

97

$$\vdash\!\!\frac{\delta}{\alpha}\!\!\left(\begin{array}{l} \underset{\beta}{\overset{\gamma}{\sim}} f(x_\gamma, \alpha_\beta) \\ f(\delta, \alpha) \end{array} \right.$$

(84) :

$$\begin{array}{c|l} F(\Gamma) & \underset{\beta}{\overset{\gamma}{\sim}} f(x_\gamma, \Gamma_\beta) \\ x & y \\ y & z \end{array}$$

$$\vdash\!\!\begin{array}{l} \underset{\beta}{\overset{\gamma}{\sim}} f(x_\gamma, z_\beta) \\ \underset{\beta}{\overset{\gamma}{\sim}} f(y_\gamma, z_\beta) \\ \underset{\beta}{\overset{\gamma}{\sim}} f(x_\gamma, y_\beta) \end{array} \qquad (98).$$

If y follows x in the f-sequence and if z follows y in the f-sequence, then z follows x in the f-sequence.

§ 29.

$$\Vdash\!\!\left[\left[\begin{array}{l} (z \equiv x) \\ \underset{\beta}{\overset{\gamma}{\sim}} f(x_\gamma, z_\beta) \end{array}\right] \equiv \underset{\beta}{\overset{\gamma}{\sim}} f(x_\gamma, z_\beta)\right] \qquad (99).$$

Here I refer the reader to what was said about the introduction of new signs in connection with formulas (69) and (76). Let

$$\underset{\beta}{\overset{\gamma}{\sim}} f(x_\gamma, z_\beta)$$

be translated by "z belongs to the f-sequence beginning with x" or by "x belongs to the f-sequence ending with z". Then in words (99) reads:

If z is identical with x or follows x in the f-sequence, then I say: "z belongs to the f-sequence beginning with x" or "x belongs to the f-sequence ending with z".

99.

$$\vdash\!\!\left[\left[\begin{array}{l} z \equiv x \\ \underset{\beta}{\overset{\gamma}{\sim}} f(x_\gamma, z_\beta) \end{array}\right] \equiv \underset{\beta}{\overset{\gamma}{\sim}} f(x_\gamma, z_\beta)\right]$$

(57) :

$$\begin{array}{c|l} f(\Gamma) & \Gamma \\ c & \begin{array}{l} (z \equiv x) \\ \underset{\beta}{\overset{\gamma}{\sim}} f(x_\gamma, z_\beta) \end{array} \\ d & \underset{\beta}{\overset{\gamma}{\sim}} f(x_\gamma, z_\beta) \end{array}$$

$$\vdash\!\!\begin{array}{l} (z \equiv x) \\ \underset{\beta}{\overset{\gamma}{\sim}} f(x_\gamma, z_\beta) \\ \underset{\beta}{\overset{\gamma}{\sim}} f(x_\gamma, z_\beta) \end{array} \qquad (100).$$

(48) :

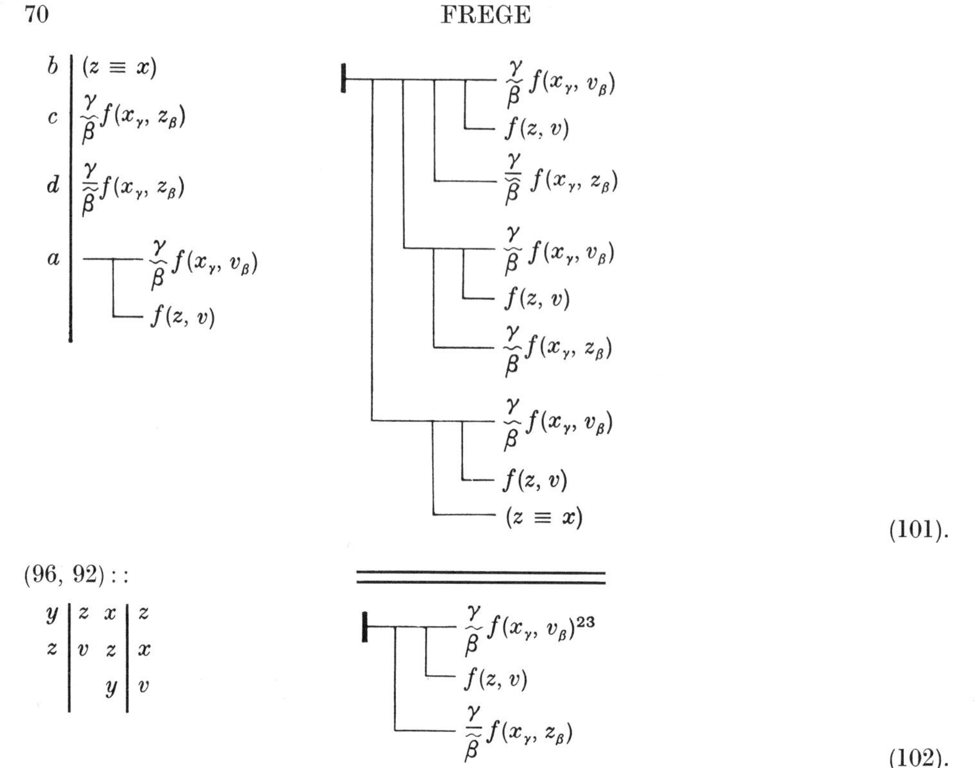

Let us here give the derivation of (102) in words.

If z is the same as x, then by (92) every result of an application of the procedure f to z follows x in the f-sequence. If z follows x in the f-sequence, then by (96) every result of an application of f to z follows x in the f-sequence.

From these two propositions it follows, according to (100), that:

If z belongs to the f-sequence beginning with x, then every result of an application of the procedure f to z follows x in the f-sequence.

100

$$(z \equiv x)$$
$$\frac{\gamma}{\tilde{\beta}} f(x_\gamma, z_\beta)$$
$$\frac{\gamma}{\tilde{\tilde{\beta}}} f(x_\gamma, z_\beta)$$

(19):

b	$(z \equiv x)$
c	$\frac{\gamma}{\tilde{\beta}} f(x_\gamma, z_\beta)$
d	$\frac{\gamma}{\tilde{\tilde{\beta}}} f(x_\gamma, z_\beta)$
a	$(x \equiv z)$

$$(x \equiv z)$$
$$\frac{\gamma}{\tilde{\beta}} f(x_\gamma, z_\beta)$$
$$\frac{\gamma}{\tilde{\tilde{\beta}}} f(x_\gamma, z_\beta)$$
$$(x \equiv z)$$
$$(z \equiv x)$$

(103).

[23] Concerning the last inference see § 6.

(55) :

d	x
c	z

$$\vdash \; (x \equiv z) \;;\; \neg\, \overset{\gamma}{\underset{\beta}{\approx}} f(x_\gamma, z_\beta) \;;\; \overset{\gamma}{\underset{\beta}{\approx}} f(x_\gamma, z_\beta)$$

(104).

§ 30.

99

$$\vdash \left[\left[\; (z \equiv x) \;;\; \neg\, \overset{\gamma}{\underset{\beta}{\sim}} f(x_\gamma, z_\beta) \right] \equiv \overset{\gamma}{\underset{\beta}{\approx}} f(x_\gamma, z_\beta)\right]$$

(52) :

$f(\Gamma)$	Γ
c	$z \equiv x \;;\; \neg\, \overset{\gamma}{\underset{\beta}{\sim}} f(x_\gamma, z_\beta)$
d	$\overset{\gamma}{\underset{\beta}{\approx}} f(x_\gamma, z_\beta)$

$$\vdash \; \overset{\gamma}{\underset{\beta}{\approx}} f(x_\gamma, z_\beta) \;;\; (z \equiv x) \;;\; \neg\, \overset{\gamma}{\underset{\beta}{\sim}} f(x_\gamma, z_\beta)$$

(105).

(37) :

a	$\overset{\gamma}{\underset{\beta}{\approx}} f(x_\gamma, z_\beta)$
b	$(z \equiv x)$
c	$\overset{\gamma}{\underset{\beta}{\sim}} f(x_\gamma, z_\beta)$

$$\vdash \; \overset{\gamma}{\underset{\beta}{\approx}} f(x_\gamma, z_\beta) \;;\; \overset{\gamma}{\underset{\beta}{\sim}} f(x_\gamma, z_\beta)$$

(106).

Whatever follows x in the f-sequence belongs to the f-sequence beginning with x.

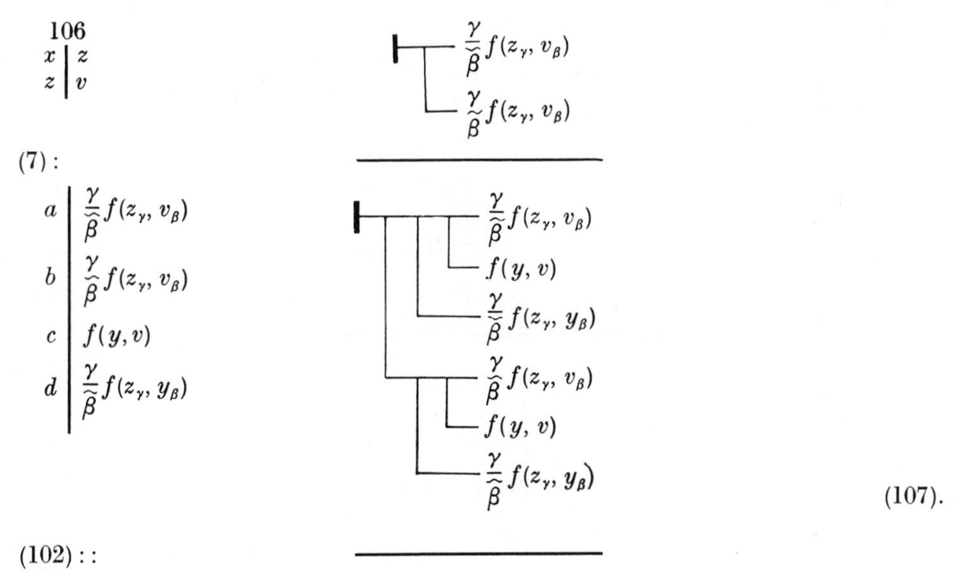

$$\begin{array}{c|c} x & z \\ z & y \end{array}$$

$$\vdash \begin{array}{l} \overset{\gamma}{\underset{\beta}{\tilde{}}} f(z_\gamma, v_\beta) \\ f(y, v) \\ \overset{\gamma}{\underset{\beta}{\tilde{}}} f(z_\gamma, y_\beta) \end{array} \qquad (108).$$

Let us here give the derivation of (108) in words.

If y belongs to the f-sequence beginning with z, then by (102) every result of an application of the procedure f to y follows z in the f-sequence. Then by (106) every result of an application of the procedure f to y belongs to the f-sequence beginning with z. Therefore,

If y belongs to the f-sequence beginning with z, then every result of an application of the procedure f to y belongs to the f-sequence beginning with z.

108

$$\begin{array}{c|c} v & \mathfrak{a} \\ z & x \\ y & \mathfrak{d} \end{array}$$

$$\vdash\!\smile^{\mathfrak{d}}\!\smile^{\mathfrak{a}} \begin{array}{l} \overset{\gamma}{\underset{\beta}{\tilde{}}} f(x_\gamma, \mathfrak{a}_\beta) \\ f(\mathfrak{d}, \mathfrak{a}) \\ \overset{\gamma}{\underset{\beta}{\tilde{}}} f(x_\gamma, \mathfrak{d}_\beta) \end{array}$$

(75):

$$\begin{array}{c|c} F(\Gamma) & \overset{\gamma}{\underset{\beta}{\tilde{}}} f(x_\gamma, \Gamma_\beta) \end{array}$$

$$\vdash \overset{\delta}{\underset{\alpha}{|}} \left(\begin{array}{l} \overset{\gamma}{\underset{\beta}{\tilde{}}} f(x_\gamma, \alpha_\beta) \\ f(\delta, \alpha) \end{array} \right. \qquad (109).$$

The property of belonging to the f-sequence beginning with x is hereditary in the f-sequence.

109

$$\vdash \overset{\delta}{\underset{\alpha}{|}} \left(\begin{array}{l} \overset{\gamma}{\underset{\beta}{\tilde{}}} f(x_\gamma, \alpha_\beta) \\ f(\delta, \alpha) \end{array} \right.$$

(78):

$$\begin{array}{c|c} F(\Gamma) & \overset{\gamma}{\underset{\beta}{\tilde{}}} f(x_\gamma, \Gamma_\beta) \\ x & y \\ y & m \end{array}$$

$$\vdash \begin{array}{l} \overset{\gamma}{\underset{\beta}{\tilde{}}} f(x_\gamma, m_\beta) \\ \overset{\gamma}{\underset{\beta}{\tilde{}}} f(y_\gamma, m_\beta) \\ \smile^{\mathfrak{a}} \begin{array}{l} \overset{\gamma}{\underset{\beta}{\tilde{}}} f(x_\gamma, \mathfrak{a}_\beta) \\ f(y, \mathfrak{a}) \end{array} \end{array} \qquad (110).$$

108

$$\vdash \begin{array}{l} \overset{\gamma}{\underset{\beta}{\tilde{}}} f(z_\gamma, v_\beta) \\ f(y, v) \\ \overset{\gamma}{\underset{\beta}{\tilde{}}} f(z_\gamma, y_\beta) \end{array}$$

(25):

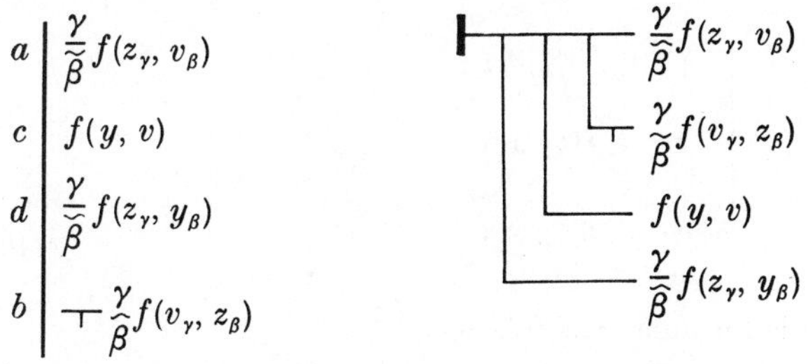

$a \mid \overset{\gamma}{\underset{\beta}{\tilde{}}} f(z_\gamma, v_\beta)$

$c \mid f(y, v)$

$d \mid \overset{\gamma}{\underset{\beta}{\tilde{}}} f(z_\gamma, y_\beta)$

$b \mid \top \overset{\gamma}{\underset{\beta}{\tilde{}}} f(v_\gamma, z_\beta)$

$\vdash \overset{\gamma}{\underset{\beta}{\tilde{}}} f(z_\gamma, v_\beta)$

$\quad \top \overset{\gamma}{\underset{\beta}{\tilde{}}} f(v_\gamma, z_\beta)$

$\quad f(y, v)$

$\quad \overset{\gamma}{\underset{\beta}{\tilde{}}} f(z_\gamma, y_\beta)$

(111).

In words the derivation of (111) is as follows:

If y belongs to the f-sequence beginning with z, then by (108) every result of an application of the procedure f to y belongs to the f-sequence beginning with z. Hence every result of an application of the procedure f to y belongs to the f-sequence beginning with z or precedes z in the f-sequence. Therefore,

If y belongs to the f-sequence beginning with z, then every result of an application of the procedure f to y belongs to the f-sequence beginning with z or precedes z in the f-sequence.

105

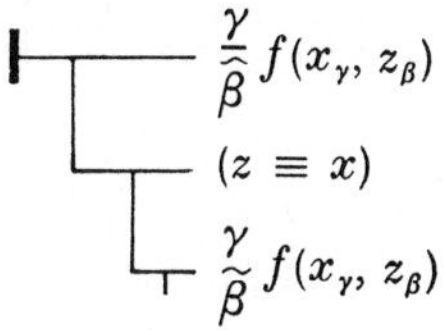

$\vdash \overset{\gamma}{\underset{\beta}{\tilde{}}} f(x_\gamma, z_\beta)$

$\quad (z \equiv x)$

$\quad \top \overset{\gamma}{\underset{\beta}{\tilde{}}} f(x_\gamma, z_\beta)$

(11):

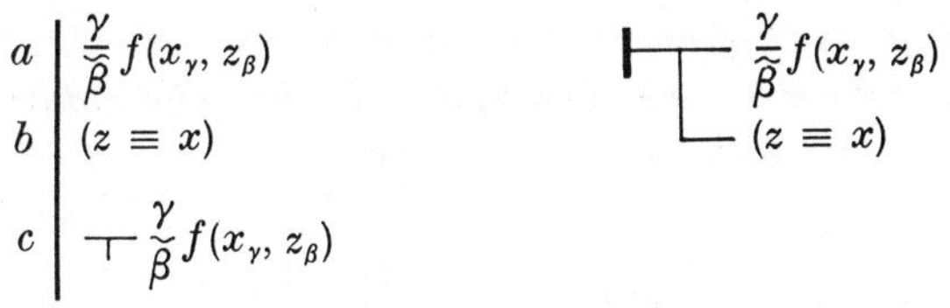

$a \mid \overset{\gamma}{\underset{\beta}{\tilde{}}} f(x_\gamma, z_\beta)$

$b \mid (z \equiv x)$

$c \mid \top \overset{\gamma}{\underset{\beta}{\tilde{}}} f(x_\gamma, z_\beta)$

$\vdash \overset{\gamma}{\underset{\beta}{\tilde{}}} f(x_\gamma, z_\beta)$

$\quad (z \equiv x)$

(112).

(7):

$a \mid \overset{\gamma}{\underset{\beta}{\tilde{}}} f(x_\gamma, z_\beta)$

$b \mid (z \equiv x)$

$c \mid \top \overset{\gamma}{\underset{\beta}{\tilde{}}} f(z_\gamma, x_\beta)$

$d \mid \overset{\gamma}{\underset{\beta}{\tilde{}}} f(z_\gamma, x_\beta)$

$\vdash \overset{\gamma}{\underset{\beta}{\tilde{}}} f(x_\gamma, z_\beta)$

$\quad \top \overset{\gamma}{\underset{\beta}{\tilde{}}} f(z_\gamma, x_\beta)$

$\quad \overset{\gamma}{\underset{\beta}{\tilde{}}} f(z_\gamma, x_\beta)$

$\quad (z \equiv x)$

$\quad \top \overset{\gamma}{\underset{\beta}{\tilde{}}} f(z_\gamma, x_\beta)$

$\quad \overset{\gamma}{\underset{\beta}{\tilde{}}} f(z_\gamma, x_\beta)$

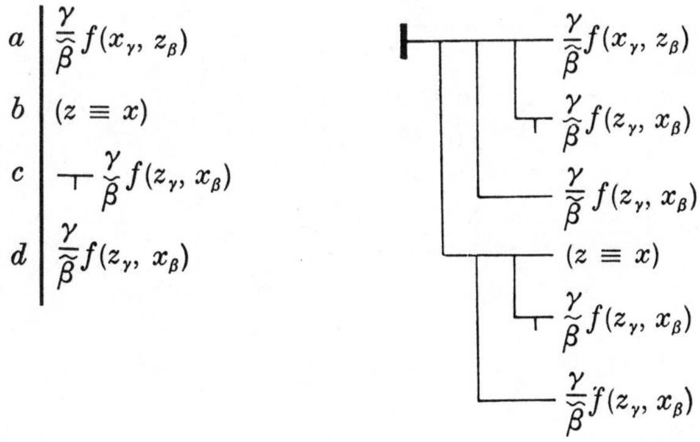

(113).

(104)::

$$\begin{array}{c|c} x & z \\ z & x \end{array}$$

$$\vdash \begin{array}{l} \underset{\beta}{\overset{\gamma}{\simeq}} f(x_\gamma, z_\beta) \\ \top\; \underset{\beta}{\overset{\gamma}{\simeq}} f(z_\gamma, x_\beta) \\ \underset{\beta}{\overset{\gamma}{\simeq}} f(z_\gamma, x_\beta) \end{array} \qquad (114).$$

In words the derivation of this formula is as follows:

Assume that x belongs to the f-sequence beginning with z. Then by (104) z is the same as x or x follows z in the f-sequence. If z is the same as x, then by (112) z belongs to the f-sequence beginning with x. From the last two propositions it follows that z belongs to the f-sequence beginning with x or x follows z in the f-sequence. Therefore,

If x belongs to the f-sequence beginning with z, then z belongs to the f-sequence beginning with x or x follows z in the f-sequence.

§ 31.

$$\Vdash \left[\left[\overset{\mathfrak{e}}{\smile}\overset{\mathfrak{d}}{\smile}\overset{\mathfrak{a}}{\smile} \begin{array}{l} (\mathfrak{a} \equiv \mathfrak{e}) \\ f(\mathfrak{d}, \mathfrak{a}) \\ f(\mathfrak{d}, \mathfrak{e}) \end{array} \right] \equiv \underset{\varepsilon}{\overset{\delta}{\mathrm{I}}} f(\delta, \varepsilon) \right]^{24} \qquad (115).$$

I translate

$$\underset{\varepsilon}{\overset{\delta}{\mathrm{I}}} f(\delta, \varepsilon)$$

by "the circumstance that the procedure f is single-valued". Then (115) can be rendered thus:

If from the circumstance that $\mathfrak{e}$ is a result of an application of the procedure f to $\mathfrak{d}$, whatever $\mathfrak{d}$ may be, it can be inferred that every result of an application of the procedure f to $\mathfrak{d}$ is the same as $\mathfrak{e}$, then I say: "The procedure f is single-valued".

115

$$\vdash \left[\left[\overset{\mathfrak{e}}{\smile}\overset{\mathfrak{d}}{\smile}\overset{\mathfrak{a}}{\smile} \begin{array}{l} (\mathfrak{a} \equiv \mathfrak{e}) \\ f(\mathfrak{d}, \mathfrak{a}) \\ f(\mathfrak{d}, \mathfrak{e}) \end{array} \right] \equiv \underset{\varepsilon}{\overset{\delta}{\mathrm{I}}} f(\delta, \varepsilon) \right]$$

(68): ______

$$\begin{array}{c|l} f(\Gamma) & \overset{\mathfrak{d}}{\smile}\overset{\mathfrak{a}}{\smile} \begin{array}{l} (\mathfrak{a} \equiv \Gamma) \\ f(\mathfrak{d}, \mathfrak{a}) \\ f(\mathfrak{d}, \Gamma) \end{array} \\ b & \underset{\varepsilon}{\overset{\delta}{\mathrm{I}}} f(\delta, \varepsilon) \\ c & x \\ \mathfrak{a} & \mathfrak{e} \end{array} \qquad \vdash \begin{array}{l} \overset{\mathfrak{d}}{\smile}\overset{\mathfrak{a}}{\smile} \begin{array}{l} (\mathfrak{a} \equiv x) \\ f(\mathfrak{d}, \mathfrak{a}) \\ f(\mathfrak{d}, x) \end{array} \\ \underset{\varepsilon}{\overset{\delta}{\mathrm{I}}} f(\delta, \varepsilon) \end{array} \qquad (116).$$

(9): ______

[24] See § 24.

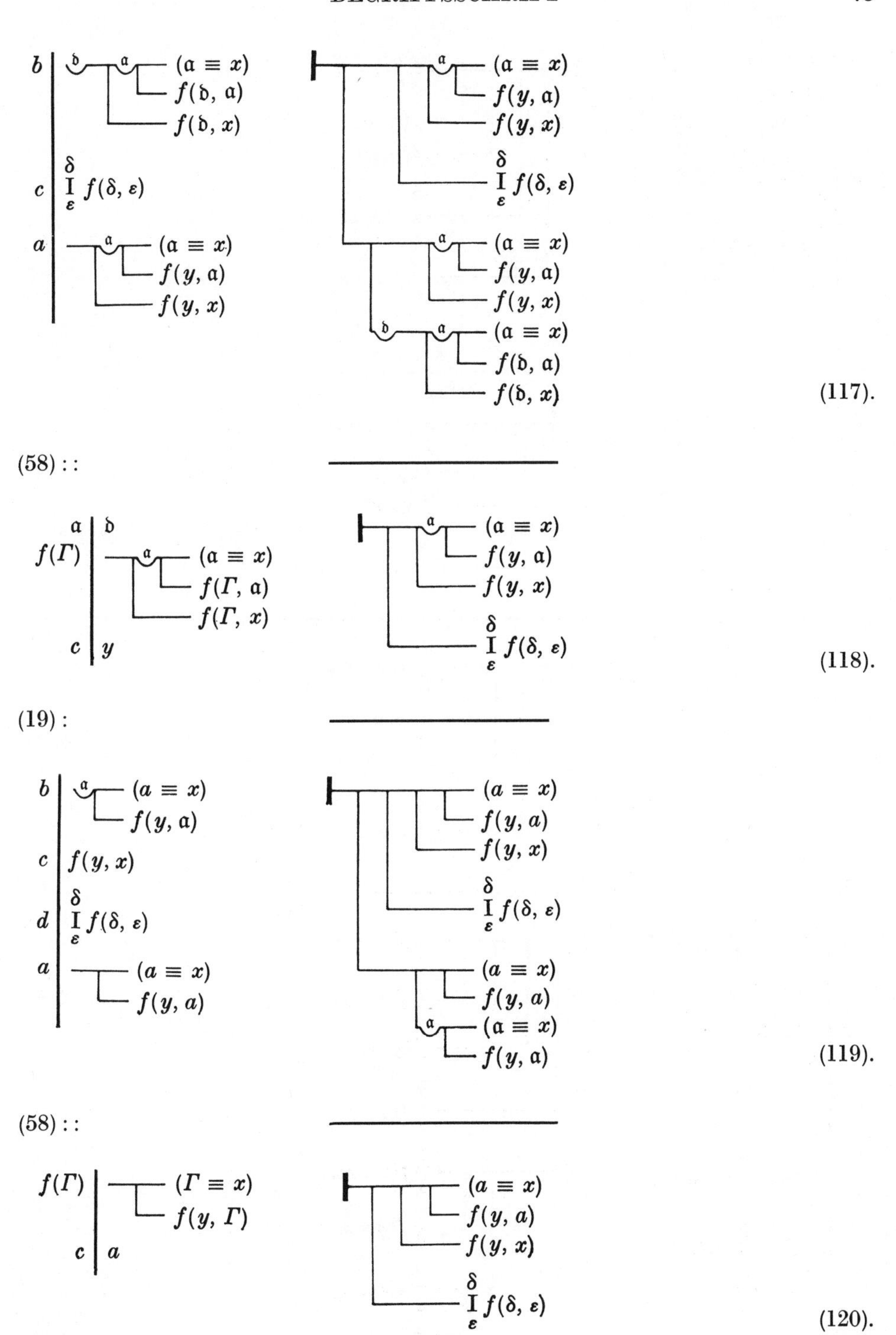

(117).
(58) : :
(118).
(19) :
(119).
(58) : :
(120).
(20) :

b	$(a \equiv x)$
c	$f(y, a)$
d	$f(y, x)$
e	$\overset{\delta}{\underset{\varepsilon}{\mathrm{I}}} f(\delta, \varepsilon)$
a	$\frac{\gamma}{\tilde{\beta}} f(x_\gamma, a_\beta)$

$\vdash$ $\frac{\gamma}{\tilde{\beta}} f(x_\gamma, a_\beta)$
$f(y, a)$
$f(y, x)$
$\overset{\delta}{\underset{\varepsilon}{\mathrm{I}}} f(\delta, \varepsilon)$
$\frac{\gamma}{\tilde{\beta}} f(x_\gamma, a_\beta)$
$(a \equiv x)$ (121).

(112) : :

z	a

$\vdash$ $\frac{\gamma}{\tilde{\beta}} f(x_\gamma, a_\beta)$
$f(y, a)$
$f(y, x)$
$\overset{\delta}{\underset{\varepsilon}{\mathrm{I}}} f(\delta, \varepsilon)$ (122).

122

a	$\mathfrak{a}$

$\vdash$ $\smile^{\mathfrak{a}}$ $\frac{\gamma}{\tilde{\beta}} f(x_\gamma, \mathfrak{a}_\beta)$
$f(y, \mathfrak{a})$
$f(y, x)$
$\overset{\delta}{\underset{\varepsilon}{\mathrm{I}}} f(\delta, \varepsilon)$

(19) :

b	$\smile^{\mathfrak{a}}$ $\frac{\gamma}{\tilde{\beta}} f(x_\gamma, \mathfrak{a}_\beta)$ / $f(y, \mathfrak{a})$
c	$f(y, x)$
d	$\overset{\delta}{\underset{\varepsilon}{\mathrm{I}}} f(\delta, \varepsilon)$
a	$\frac{\gamma}{\tilde{\beta}} f(x_\gamma, m_\beta)$ / $\frac{\gamma}{\tilde{\beta}} f(y_\gamma, m_\beta)$

$\vdash$ $\frac{\gamma}{\tilde{\beta}} f(x_\gamma, m_\beta)$
$\frac{\gamma}{\tilde{\beta}} f(y_\gamma, m_\beta)$
$f(y, x)$
$\overset{\delta}{\underset{\varepsilon}{\mathrm{I}}} f(\delta, \varepsilon)$
$\frac{\gamma}{\tilde{\beta}} f(x_\gamma, m_\beta)$
$\frac{\gamma}{\tilde{\beta}} f(y_\gamma, m_\beta)$
$\smile^{\mathfrak{a}}$ $\frac{\gamma}{\tilde{\beta}} f(x_\gamma, \mathfrak{a}_\beta)$
$f(y, \mathfrak{a})$ (123).

(110) : :

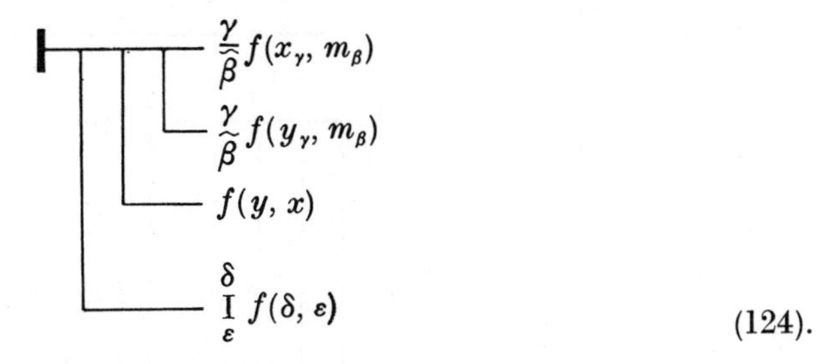

(124).

Let us give the derivation of formulas (122) and (124) in words.

Assume that x is a result of an application of the single-valued procedure f to y. Then by (120) every result of an application of the procedure f to y is the same as x. Hence by (112) every result of an application of the procedure f to y belongs to the f-sequence beginning with x. Therefore,

If x is a result of an application of the single-valued procedure f to y, then every result of an application of the procedure f to y belongs to the f-sequence beginning with x. (Formula (122).)

Assume that m follows y in the f-sequence. Then (110) yields: If every result of an application of the procedure f to y belongs to the f-sequence beginning with x, then m belongs to the f-sequence beginning with x. This, combined with (122), shows that, if x is a result of an application of the single-valued procedure f to y, then m belongs to the f-sequence beginning with x. Therefore,

If x is a result of an application of the single-valued procedure f to y and if m follows y in the f-sequence, then m belongs to the f-sequence beginning with x. (Formula (124).)

124

(20):

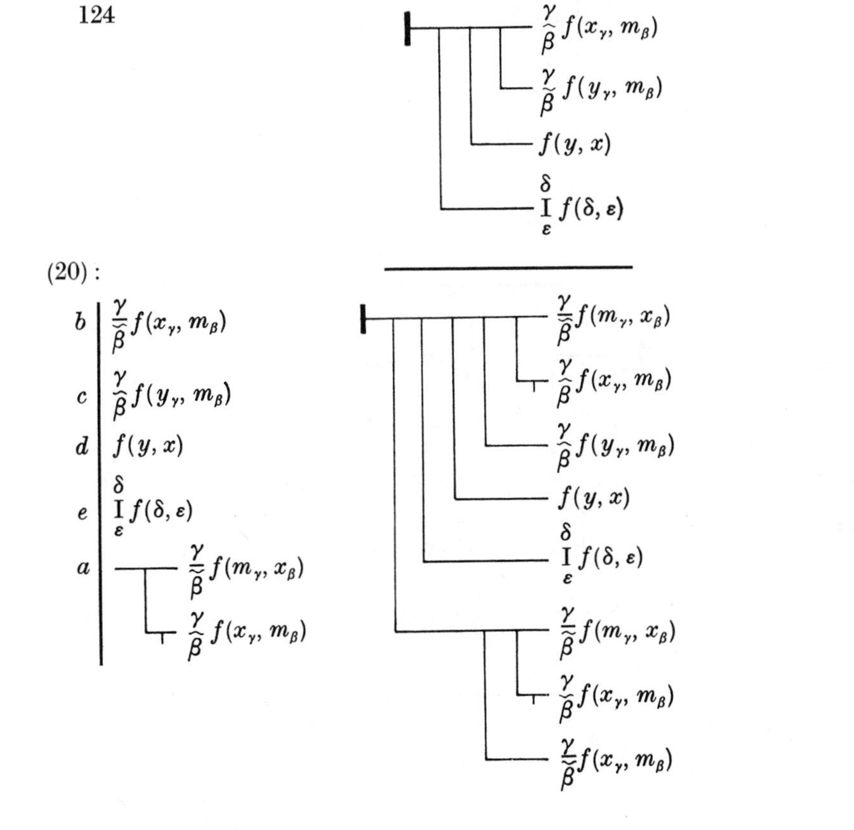

(125).

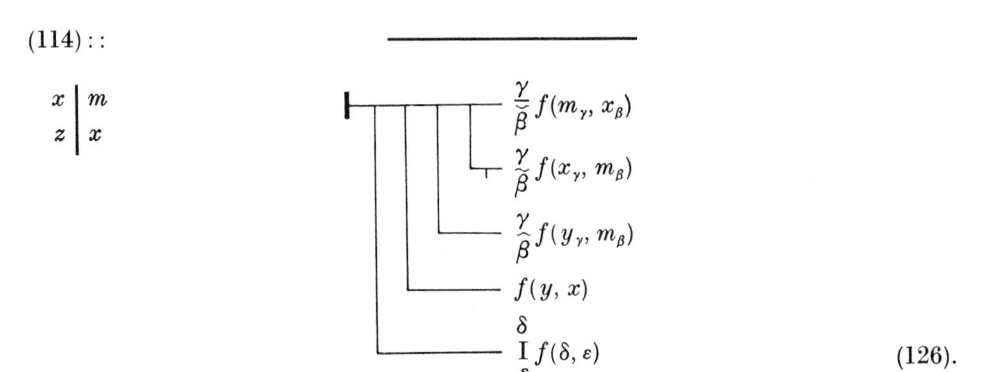

The derivation of this formula follows here in words.

Assume that x is a result of an application of the single-valued procedure f to y. Assume that m follows y in the f-sequence. Then by (124) m belongs to the f-sequence beginning with x. Consequently, by (114) x belongs to the f-sequence beginning with m or m follows x in the f-sequence. This can also be expressed as follows: x belongs to the f-sequence beginning with m or precedes m in the f-sequence. Therefore,

If m follows y in the f-sequence and if the procedure f is single-valued, then every result of an application of the procedure f to y belongs to the f-sequence beginning with m or precedes m in the f-sequence.

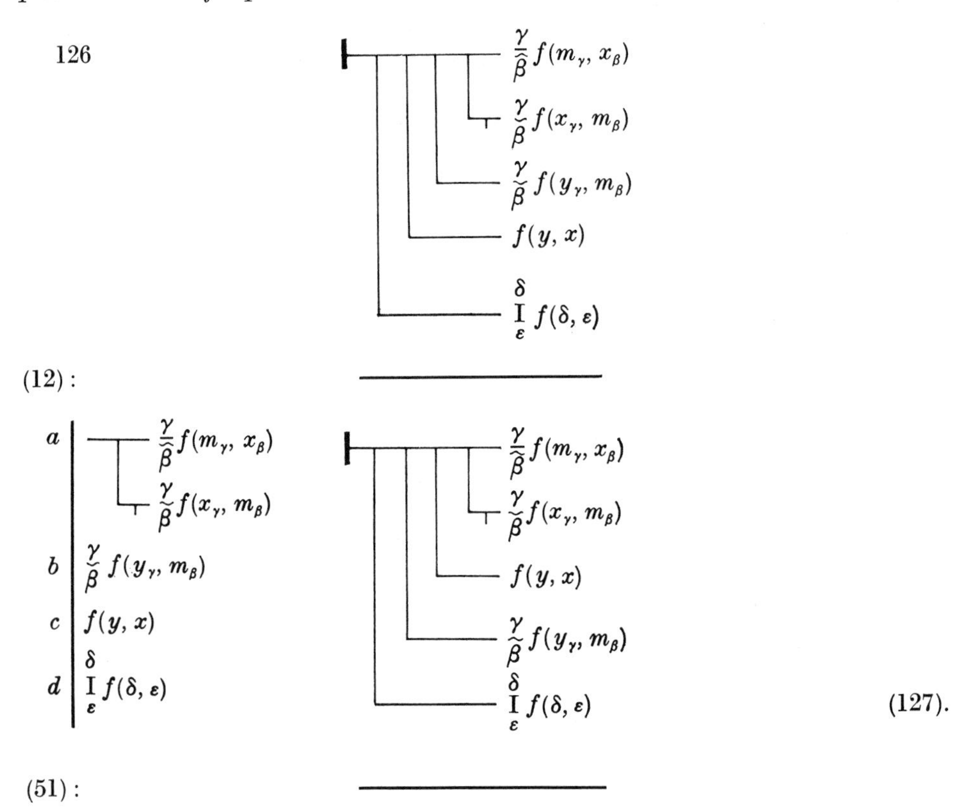

(51):

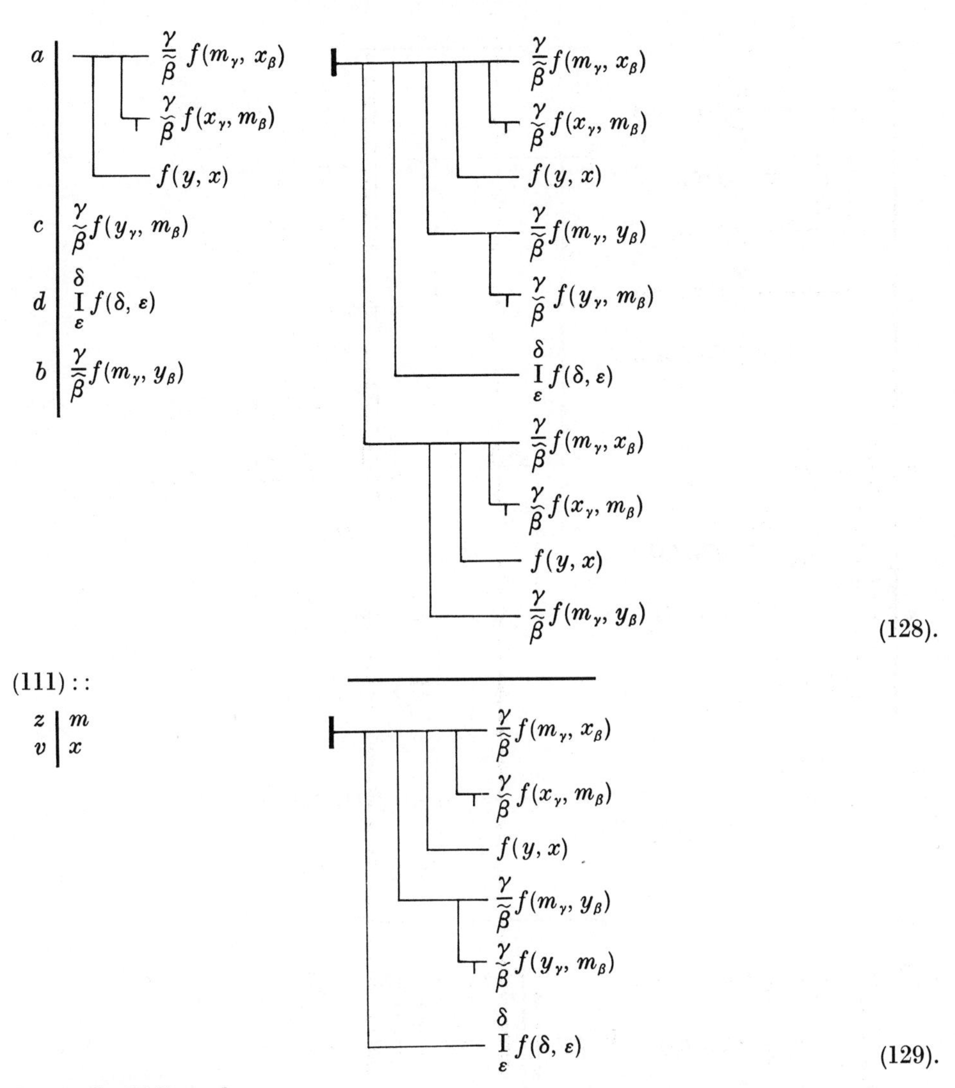

In words (129) reads:

If the procedure f is single-valued and y belongs to the f-sequence beginning with m or precedes m in the f-sequence, then every result of an application of the procedure f to y belongs to the f-sequence beginning with m or precedes m in the f-sequence.

129

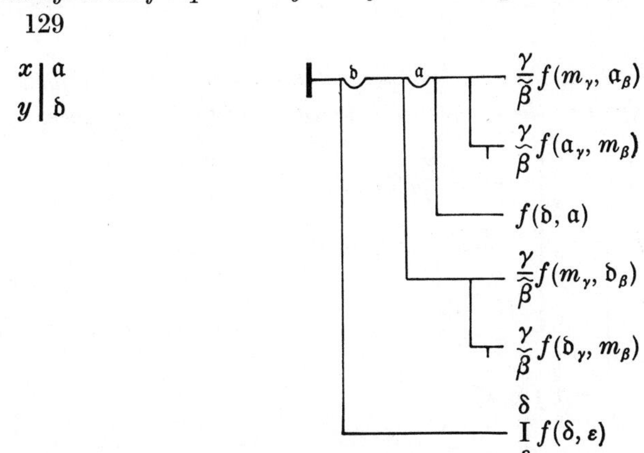

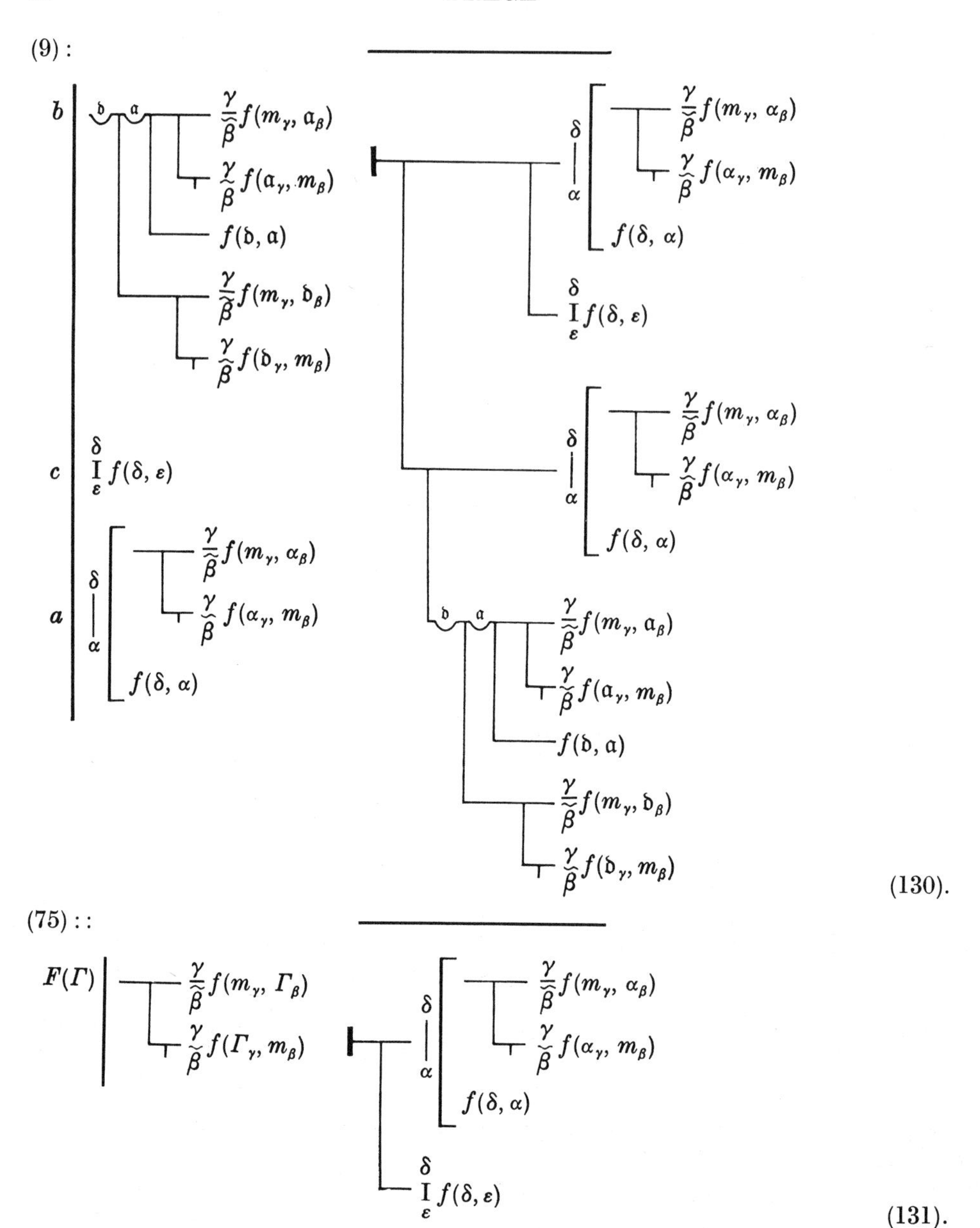

In words (131) reads:

If the procedure f is single-valued, then the property of belonging to the f-sequence beginning with m or of preceding m in the f-sequence is hereditary in the f-sequence.

131

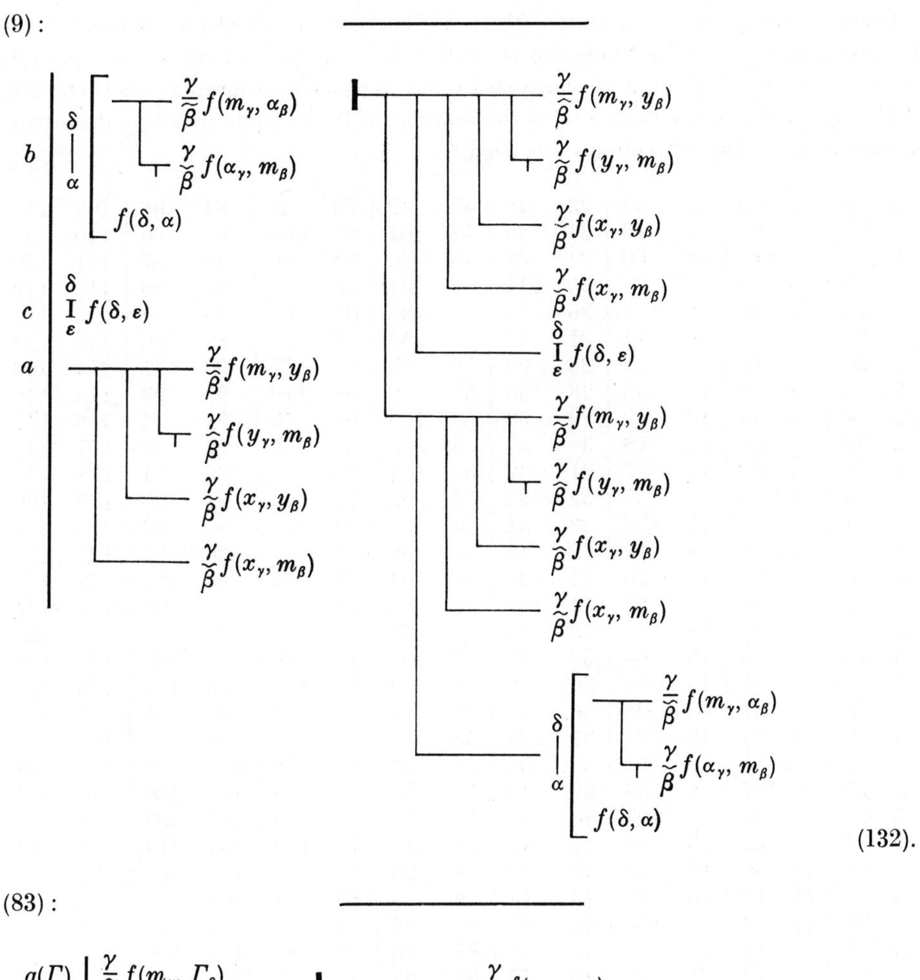

(83):

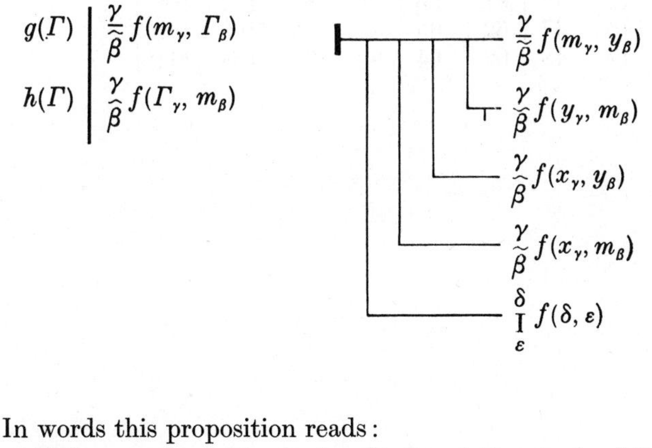

(133).

In words this proposition reads:

If the procedure f is single-valued and if m and y follow x in the f-sequence, then y belongs to the f-sequence beginning with m or precedes m in the f-sequence.

Below I give a table that shows where use has been made of one formula in the derivation of another. The table can be used to look up the ways in which a formula has been employed. From it we can also see how frequently a formula has been used.

The right column always contains the number of the formula in whose derivation the one listed in the left column was used.

1	3	7	94	12	35	23	48	47	49	63	91	84	98	109	110
1	5	7	107	12	49	24	25	48	101	64	65	85	86	110	124
1	11	7	113	12	60	24	63	49	50	65	66	86	87	111	129
1	24	8	9	12	85	25	111	50	51	66	—	87	88	112	113
1	26	8	10	12	127	26	27	51	128	67	68	88	95	112	122
1	27	8	12	13	14	27	42	52	53	68	70	89	90	113	114
1	36	8	17	14	15	28	29	52	57	68	77	90	91	114	126
2	3	8	26	15	88	28	33	52	89	68	116	90	93	115	116
2	4	8	38	16	17	29	30	52	105	69	70	91	92	116	117
2	39	8	53	16	18	30	59	52	75	69	75	92	102	117	118
2	73	8	62	16	22	31	32	53	55	70	71	93	94	118	119
2	79	8	66	17	50	32	33	53	92	71	72	94	95	119	120
3	4	8	74	17	78	33	34	54	55	72	73	95	96	120	121
4	5	8	84	18	19	33	46	55	56	72	74	96	97	121	122
5	6	8	96	18	20	34	35	55	104	73	87	96	102	122	123
5	7	9	10	18	23	34	36	56	57	74	81	97	98	123	124
5	9	9	11	18	51	35	40	57	68	75	97	98	—	124	125
5	12	9	19	18	64	36	37	57	100	75	109	99	100	125	126
5	14	9	21	18	82	36	38	58	59	75	131	99	105	126	127
5	16	9	37	19	20	36	83	58	60	76	77	100	101	127	128
5	18	9	56	19	21	37	106	58	61	76	89	100	103	128	129
5	22	9	61	19	71	38	39	58	62	77	78	101	102	129	130
5	25	9	117	19	86	39	40	58	67	77	85	102	108	130	131
5	29	9	130	19	103	40	43	58	72	78	79	103	104	131	132
5	34	9	132	19	119	41	42	58	118	78	110	104	114	132	133
5	45	10	30	19	123	42	43	58	120	79	80	105	106	133	—
5	80	11	112	20	121	43	44	59	—	80	81	105	112		
5	90	12	13	20	125	44	45	60	93	81	82	106	107		
6	7	12	15	21	44	45	46	61	65	81	84	107	108		
7	32	12	16	21	47	46	47	62	63	82	83	108	109		
7	67	12	24	22	23	47	48	62	64	83	133	108	111		

The principles of arithmetic, presented by a new method

GIUSEPPE PEANO

(*1889*)

Written in Latin, this small book was Peano's first attempt at an axiomatization of mathematics in a symbolic language. Peano had already (*1888*) used the logic of Boole and Schröder in mathematical investigations and introduced into it a number of innovations that marked a definite advance upon the work of his predecessors: for instance, the use of different signs for logical and mathematical operations, and a distinction between categorical and conditional propositions that was to lead him to quantification theory (these were innovations relatively to Boole and Schröder—not to Frege, whose work Peano did not know at that time). In the present work, after having introduced logical notions and formulas, Peano undertakes to rewrite arithmetic in symbolic notation. But he aspires to more than that: the book deals also with fractions, real numbers, even the notion of limit and definitions in point-set theory.

The initial arithmetic notions are "number", "one", "successor", and "is equal to", and nine axioms are stated concerning these notions. Today we would consider that Axioms 2, 3, 4, and 5, which deal with identity, belong to the underlying logic. This leaves the five axioms that have become universally known as "the Peano axioms". The last one, Axiom 9, is the translation of the principle of mathematical induction. It is formulated in terms of classes, and it contains a class variable, "k" (it even involves the class of all classes, K). Peano acknowledges (*1891b*, p. 93) that his axioms come from Dedekind (*1888*, art. 71, definition of a simply infinite system; see also below, pp. 100–101). As for Frege, Peano learned of his work immediately after the publication of *Arithmetices principia*.[a]

From the outset, Peano uses the notation $x + 1$ for the successor function. He then introduces addition (§ 1, 18) and multiplication (§ 4, 1 and 2) as "definitions". These definitions are recursive definitions, although Peano does not have at his disposal in his system anything like Dedekind's powerful Theorem 126 (*1888*), which justifies such definitions. Peano does not explicitly claim that these definitions are eliminable, but, just as he does for ordinary definitions (that of subtraction, for example), he puts them under the heading "Definition", although they do not satisfy his own statement on that score (p. 93), namely, that the right side of a definitional equation is "an aggregate of

[a] In the list of references appended to *Arithmetices principia* (below, p. 86, footnote 1) Frege's name does not occur; but Peano mentions and even quotes Frege in his very next paper on logic (*1891*).

signs having a known meaning". He proves for addition a theorem (§ 1, 19) stating that "for every a and b, $a, b \varepsilon N$.Ɔ. $a + b \varepsilon N$", and a similar theorem (§ 4, 3) for multiplication; but these theorems are far from having the same effect as Dedekind's Theorem 126.

The ease with which we read Peano's booklet today shows how much of his notation has found its way, either directly or in a somewhat modified form, into contemporary logic. ε is there, with the distinction between elementhood and subclasshood (except for classes of one element—see formula 56 in Part IV, p. 90 below; for such classes the distinction will appear the following year (*1890*, p. 192)). The inverted C, Ɔ, will become ⊃.

The logical part of the work presents formulas of the propositional calculus, of the calculus of classes, and a few of quantification theory. Peano's notation is quite superior to that of Boole and Schröder, and it marks an important transition toward modern logic. Some distinction is made between the calculus of propositions and that of classes (ε, for example, already introduces an asymmetry between propositions and classes); we now have two different calculi, not just two interpretations of the same calculus. The notation for the universal quantifier is new and convenient. There is, however, a grave defect. The formulas are simply listed, not derived; and they could not be derived, because no rules of inference are given. Peano introduces a notation for substitution (V 4, p. 91) but does not state any rule. What is far more important, he does not have any rule that would play the role of the rule of detachment. The result is that, for all his meticulousness in the writing of formulas, he has no logic that he can use. The point is vividly illustrated by the first proof he gives, that of

11. $2 \varepsilon N$

(below, p. 94). What is presented as a proof is actually a list of formulas that are such that, from the point of view of the working mathematician, each one is very close to the next. But, however close two successive formulas may be, the logician cannot pass from one to the next because of the absence of rules of inference. The proof does not get off the ground.

In the proof just mentioned (and it is typical of Peano's proofs), the passage from formulas (1) and (2) to formula (3) cannot be carried out by a *formal* procedure; it requires some intuitive logical argument, which the reader has to supply. The proof brings out the whole difference between an axiomatization, even written in symbols and however careful it may be, and a formalization. The absence of a rule of detachment in Peano's booklet (and other works) is apparently connected with his inadequate interpretation of the conditional. He reads "a Ɔ b" as "from a one deduces b" ("ab a deducitur b"), which remains vague; truth values are not used at all in the work below, and only marginally in Peano's subsequent writings.

In a series of papers (*1891*, *1891a*, *1891b*; see also *1890*) that form a sequel to *Arithmetices principia* and a transition toward the first volume of the *Formulaire* (*1895*), Peano undertakes to prove the logical formulas that he simply listed in the logical part of the work below. Just like his arithmetic proofs, his logical proofs suffer from the absence of rules of inference. In the proof of proposition 9 (*1891a*, p. 27), for example, he strings conditionals one after another; when ultimately he does detach, it is by a move totally unjustified in his system. For a pertinent critique of that aspect of Peano's work see *Frege 1896* and *1896a*. Some of Peano's explanations tend to suggest that his logical laws should perhaps be taken as rules of inference, not as formulas in a logical language; this, however, would not yield a coherent interpretation of his system.

In the work below and in the various editions of the *Formulaire* that were to follow, Peano intends to cover much more ground than Frege does in his *Begriffsschrift* and his subsequent works, but he does not till that ground to any depth comparable to what Frege does in his self-alloted field. Peano's writings, of minor significance for logic proper, showed how mathematical theories can be expressed in one symbolic language. These writings rapidly gained a wide influence and greatly contributed to the diffusion of the new ideas.

Arithmetices principia consists of a long explanatory preface and ten sections: § 1 Number and addition, § 2 Subtraction, § 3 Maxima and minima, § 4 Multiplication, § 5 Powers, § 6 Division, § 7 Various theorems, § 8 Ratios of numbers, § 9 Systems of rationals, irrationals, § 10 Systems of quantities. Below we print the preface and § 1 *in extenso*; from §§ 2, 4, 5, and 6 we give the "explanations" and "definitions", omitting the theorems, and we leave out the other sections entirely. The omitted parts consist almost exclusively of formulas and are readily available in Peano's collected works (*1958*).

The translation is by the editor, and it is printed here with the kind permission of the Unione matematica italiana.

Preface

Questions that pertain to the foundations of mathematics, although treated by many in recent times, still lack a satisfactory solution. The difficulty has its main source in the ambiguity of language.

That is why it is of the utmost importance to examine attentively the very words we use. My goal has been to undertake this examination, and in this paper I am presenting the results of my study, as well as some applications to arithmetic.

I have denoted by signs all ideas that occur in the principles of arithmetic, so that every proposition is stated only by means of these signs.

The signs belong either to logic or to arithmetic proper. The signs of logic that occur here are ten in number, although not all are necessary. In the first part of the present paper ⟦Logical notations⟧ the use of these signs, as well as some of their properties, is explained in ordinary language. It was not my intention to present their theory more fully there. The signs of arithmetic are explained wherever they occur.

With these notations, every proposition assumes the form and the precision that equations have in algebra; from the propositions thus written other propositions are deduced, and in fact by procedures that are similar to those used in solving equations. This is the main point of the whole paper.

Thus, having introduced the signs with which I can write the propositions of arithmetic, I have, in dealing with these propositions, used a method that, because it will have to be followed in other studies too, I shall present briefly here.

Among the signs of arithmetic, those that can be expressed by other signs of arithmetic together with the signs of logic represent the ideas that we can define. Thus, I have defined all signs except the four that are contained in the explanations of § 1. If, as I think, these cannot be reduced any further, it is not possible to define the ideas expressed by them through ideas assumed to be known previously.

Propositions that are deduced from others by the operations of logic are *theorems*; propositions that are not thus deduced I have called *axioms*. There are nine of these axioms (§ 1), and they express the fundamental properties of the signs that lack definition.

In §§ 1–6 I have proved the ordinary properties of numbers. For the sake of brevity I have omitted proofs that are similar to other proofs given before. In order to express proofs with the signs of logic, the ordinary form of these proofs has to be changed; this transformation is sometimes rather difficult, but it is by means of it that the nature of the proof reveals itself most clearly.

In subsequent sections I deal with various subjects, so that the power of the method will be more apparent.

§ 7 contains a few theorems that pertain to number theory. In §§ 8–9 the definitions of rational and irrational numbers are found.

Finally, in § 10 I present a few theorems, which, I think, are new, pertaining to the theory of those objects that Cantor has called *Punktmengen* (*ensembles de points*).

In the present paper I have made use of the studies of other writers. The logical symbols and propositions contained in parts II, III, and IV, except for a few, are to be traced to the works of many writers, especially Boole.[1]

I introduced the sign ε, which should not be confused with the sign Ɔ; I also introduced applications of inversion in logic, as well as a few other conventions, in order to be able to express any proposition whatsoever.

For proofs in arithmetic, I used *Grassmann 1861*.

The recent work of Dedekind (*1888*) was also most useful to me; in it, questions pertaining to the foundations of numbers are acutely examined.

This little book of mine is intended to give an example of the new method. With these notations we can state and prove innumerable other propositions, whether they pertain to rational or to irrational numbers. But to deal with other theories new signs denoting new objects must be introduced. However, I think that the propositions of any science can be expressed by these signs of logic alone, provided we add signs representing the objects of that science.

Logical notations

I. *Punctuation*

The letters $a, b, \ldots, x, y, \ldots, x', y', \ldots$ denote indeterminate objects. We denote well-determined objects by signs or by the letters P, K, N, ...

We shall generally write signs on a single line. To show the order in which they should be taken, we use *parentheses*, as in algebra, or *dots*, ., :, .:., ::, and so on.

To understand a formula divided by dots we first take together the signs that are not separated by any dot, next those separated by one dot, then those separated by two dots, and so on.

For example, let $a, b, c, \ldots$ be any signs. Then $ab.cd$ means $(ab)(cd)$; and $ab.cd\!:\!ef.gh\,.\!\cdot\!.\,k$ means $(((ab)(cd))((ef)(gh)))k$.

Punctuation signs may be omitted if formulas differing in punctuation have the

[1] See *Boole 1847, 1848, 1854*, and *Schröder 1877*. Schröder had already dealt with some questions relevant to logic in an earlier work (*1873*). I presented the theories of Boole and Schröder very briefly in a book of mine (*1888*). See also *Peirce 1880, 1885, Jevons 1883, MacColl 1877, 1878, 1878a*, and *1880*.

same meaning or if only one formula, which is just the one we want to write, has meaning.

To avoid the danger of ambiguity we never use . or : as signs for arithmetic operations.

The only kind of parentheses is ( ); if parentheses and dots occur in the same formula, what is contained within parentheses is taken together first.

II. *Propositions*

The sign P means *proposition.*

The sign $\cap$ is read *and.* Let a and b be propositions; then $a \cap b$ is the simultaneous affirmation of the propositions a and b. For the sake of brevity, we ordinarily write ab instead of $a \cap b$.

The sign $-$ is read *not.* Let a be a P; then $-a$ is the negation of the proposition a.

The sign $\cup$ is read *or* ⟦*vel*⟧. Let a and b be propositions; then $a \cup b$ is the same as $-:-a.-b$.

[The sign V means *the true,* or *identity*; but we never use this sign.]

The sign Λ means *the false,* or *the absurd.*

[The sign C means *is a consequence of*; thus $b \mathrel{C} a$ is read *b is a consequence of the proposition a.* But we never use this sign.]

The sign Ɔ means *one deduces* ⟦*deducitur*⟧;[2] thus a Ɔ b means the same as $b \mathrel{C} a$. If propositions a and b contain the indeterminate objects $x, y, \ldots$, that is, are conditions between these objects, then a Ɔ$_{x,y,\ldots}$ b means: whatever $x, y, \ldots$ may be, from the proposition a one deduces b. If there is no danger of any ambiguity, we write only Ɔ instead of Ɔ$_{x,y,\ldots}$.

The sign $=$ means *is equal to* ⟦*est aequalis*⟧. Let a and b be propositions; then $a = b$ means the same as a Ɔ $b.b$ Ɔ a; the proposition $a =_{x,y,\ldots} b$ means the same as a Ɔ$_{x,y,\ldots}$ $b.b$ Ɔ$_{x,y,\ldots}$ a.

III. *Propositions of logic*

Let $a, b, c, \ldots$ be propositions. Then we have:

1. a Ɔ a.
2. a Ɔ $b.b$ Ɔ c :Ɔ. a Ɔ c.
3. $a = b$.=: a Ɔ $b.b$ Ɔ a.
4. $a = a$.
5. $a = b$.=. $b = a$.
6. $a = b.b$ Ɔ c :Ɔ. a Ɔ c.
7. a Ɔ $b.b = c$:Ɔ. a Ɔ c.
8. $a = b.b = c$:Ɔ. $a = c$.
9. $a = b$.Ɔ. a Ɔ b.

[2] ⟦Peano reads a Ɔ b "ab a deducitur b". Translated word for word, this either would be awkward ("from a is deduced b") or would reverse the relative positions of a and b ("b is deduced from a"), which would lead to misinterpretations when the sign is read alone. Peano himself uses "on déduit" for "deducitur" when writing in French (for instance, *1890*, p. 184), and this led to the translation adopted here.⟧

10. $a = b .\supset. b \supset a.$

11. $ab \supset a.$
12. $ab = ba.$
13. $a(bc) = (ab)c = abc.$
14. $aa = a.$
15. $a = b .\supset. ac = bc.$
16. $a \supset b .\supset. ac \supset bc.$
17. $a \supset b . c \supset d :\supset. ac \supset bd.$
18. $a \supset b . a \supset c :=. a \supset bc.$
19. $a = b . c = d :\supset. ac = bd.$

20. $-(-a) = a.$
21. $a = b .=. -a = -b.$
22. $a \supset b .=. -b \supset -a.$

23. $a \cup b .= \therefore -: -a . -b.$
24. $-(ab) = (-a) \cup (-b).$
25. $-(a \cup b) = (-a)(-b).$
26. $a \supset . a \cup b.$
27. $a \cup b = b \cup a.$
28. $a \cup (b \cup c) = (a \cup b) \cup c = a \cup b \cup c.$
29. $a \cup a = a.$
30. $a(b \cup c) = ab \cup ac.$
31. $a = b .\supset. a \cup c = b \cup c.$
32. $a \supset b .\supset. a \cup c \supset b \cup c.$
33. $a \supset b . c \supset d :\supset: a \cup c .\supset. b \cup d.$
34. $b \supset a . c \supset a :=. b \cup c \supset a.$

35. $a - a = \Lambda.$
36. $a\Lambda = \Lambda.$
37. $a \cup \Lambda = a.$
38. $a \supset \Lambda .=. a = \Lambda.$
39. $a \supset b .=. a - b = \Lambda.$
40. $\Lambda \supset a.$
41. $a \cup b = \Lambda .=: a = \Lambda . b = \Lambda.$

42. $a \supset . b \supset c :=: ab \supset c.$
43. $a \supset . b = c :=. ab = ac.$

Let α be a relation sign (for example, $=, \supset$), so that $a \, \alpha \, b$ is a proposition. Then, instead of $-. a \, \alpha \, b$, we write $a -\alpha \, b$; that is,

$$\alpha -= b .=: -. a = b.$$

$$a -\supset b .=: -. a \supset b.$$

Thus the sign $-=$ means *is not equal to*. If proposition a contains the indeterminate

x, then $a \mathbin{-=}_x \Lambda$ means: there are x that satisfy condition a. The sign $-\supset$ means *one does not deduce*.

Similarly, if α and β are relation signs, instead of $a\,\alpha\,b . a\,\beta\,b$ and $a\,\alpha\,b\ .\!\cup.\ a\,\beta\,b$, we can write $a\ .\alpha\beta.\ b$ and $a\ .\alpha \cup \beta.\ b$, respectively. Thus, if a and b are propositions, the formula $a\ .\!\supset\!-\!=.\ b$ says: from a one deduces b, but not conversely.

$$a\ .\!\supset\!-\!=.\ b :=: a \supset b . b -\!\supset a.$$

We have the formulas:

$$a \supset b . b \supset c . a -\!\supset c := \Lambda.$$
$$a = b . b = c . a -\!= c := \Lambda.$$
$$a \supset b . b \supset\!-\!= c :\supset.\ a \supset\!-\!= c.$$
$$a \supset\!-\!= b . b \supset c :\supset.\ a \supset\!-\!= c.$$

But we shall rarely use these devices.

IV. *Classes*

The sign K means *class*, or aggregate of objects.

The sign ε means *is*. Thus $a \,\varepsilon\, b$ is read *a is a b*; $a \,\varepsilon\, \mathrm{K}$ means *a is a class*; $a \,\varepsilon\, \mathrm{P}$ means *a is a proposition*.

Instead of $-(a \,\varepsilon\, b)$ we write $a -\!\varepsilon\, b$; the sign $-\varepsilon$ means *is not*; that is,

44. $a -\!\varepsilon\, b .=: -. a \,\varepsilon\, b.$

The sign $a, b, c \,\varepsilon\, m$ means: a, b, and c are m; that is,

45. $a, b, c \,\varepsilon\, m .=: a \,\varepsilon\, m . b \,\varepsilon\, m . c \,\varepsilon\, m.$

Let a be a class; then $-a$ means the class composed of the individuals that are not a.

46. $a \,\varepsilon\, \mathrm{K} .\!\supset\!: x \,\varepsilon\, -a .=.\ x -\!\varepsilon\, a.$

Let a and b be classes; $a \cap b$, or ab, is the class composed of the individuals that are at the same time a and b; $a \cup b$ is the class composed of the individuals that are a or b.

47. $a, b \,\varepsilon\, \mathrm{K} .\!\supset\!\therefore x \,\varepsilon.\ ab :=: x \,\varepsilon\, a . x \,\varepsilon\, b.$
48. $a, b \,\varepsilon\, \mathrm{K} .\!\supset\!\therefore x \,\varepsilon.\ a \cup b :=: x \,\varepsilon\, a\ .\!\cup.\ x \,\varepsilon\, b.$

The sign Λ denotes the class that contains no individuals. Thus,

49. $a \,\varepsilon\, \mathrm{K} .\!\supset\!\therefore a = \Lambda :=: x \,\varepsilon\, a .=_x \Lambda.$

[We shall not use the sign V, which denotes the class composed of all individuals under consideration.]

The sign $\supset$ means *is contained in*. Thus $a \supset b$ means *class a is contained in class b*.

50. $a, b \,\varepsilon\, \mathrm{K} .\!\supset\!\therefore a \supset b :=: x \,\varepsilon\, a .\!\supset_x\!.\ x \,\varepsilon\, b.$

[The formula b C a could mean *class b contains class a*; but we shall not use the sign C.]

The signs Λ and $\supset$ here have a meaning that differs somewhat from the meaning given above; but no ambiguity will arise. If we are dealing with propositions, these signs are read *the absurd* and *one deduces*; but, if we are dealing with classes, they are read *nothing* and *is contained in*.

If a and b are classes, the formula $a = b$ means $a \supset b . b \supset a$. Thus,

51. $a, b \,\varepsilon\, \mathrm{K} .\supset.\because a = b :=: x \,\varepsilon\, a .=_x. x \,\varepsilon\, b$.

Propositions 1–41 still hold if $a, b, \ldots$ denote classes; in addition, we have:

52. $a \,\varepsilon\, b .\supset. b \,\varepsilon\, \mathrm{K}$.
53. $a \,\varepsilon\, b .\supset. b -= \Lambda$.
54. $a \,\varepsilon\, b . b = c :\supset. a \,\varepsilon\, c$.
55. $a \,\varepsilon\, b . b \supset c :\supset. a \,\varepsilon\, c$.

Let s be a class and k a class contained in s; then we say that k is an individual of class s if k consists of just one individual. Thus,

56. $s \,\varepsilon\, \mathrm{K} . k \supset s :\supset:: k \,\varepsilon\, s .=.\because k -= \Lambda : x, y \,\varepsilon\, k .\supset_{x,y}. x = y$.

V. *Inversion*

The sign of inversion is [ ], and we shall explain its use in Part VI. Here we simply present some special cases.

1. Let a be a proposition containing the indeterminate x; then the expression $[x \,\varepsilon]\, a$, which is read *the* x *for which* a, or *the solutions* of the condition a, or its *roots*, means the class composed of the individuals that satisfy condition a. Thus,

57. $a \,\varepsilon\, \mathrm{P} .\supset: [x \,\varepsilon]\, a .\varepsilon\, \mathrm{K}$.
58. $a \,\varepsilon\, \mathrm{K} .\supset.\because [x \,\varepsilon]. x \,\varepsilon\, a := a$.
59. $a \,\varepsilon\, \mathrm{P} .\supset.\because x \,\varepsilon. [x \,\varepsilon]\, a := a$.

Let α and β be propositions containing the indeterminate x; we have:

60. $[x \,\varepsilon]\, (\alpha\beta) = ([x \,\varepsilon]\, \alpha)([x \,\varepsilon]\, \beta)$.
61. $[x \,\varepsilon] -\alpha = -[x \,\varepsilon]\, \alpha$.
62. $[x \,\varepsilon]\, (\alpha \cup \beta) = [x \,\varepsilon]\, \alpha \cup [x \,\varepsilon]\, \beta$.
63. $\alpha \supset_x \beta .=. [x \,\varepsilon]\, \alpha \supset [x \,\varepsilon]\, \beta$.
64. $a =_x \beta .=. [x \,\varepsilon]\, \alpha = [x \,\varepsilon]\, \beta$.

2. Let x and y be any objects whatsoever; we consider as a new object the system composed of the object x and the object y, and we denote it by the sign (x, y); and similarly if we have a greater number of objects. Let α be a proposition containing the indeterminates x and y; then $[(x, y) \,\varepsilon]\, \alpha$ means the class composed of the objects (x, y) that satisfy the condition α. We have:

65. $\alpha \supset_{x,y} \beta .=. [(x, y) \,\varepsilon]\, \alpha \supset [(x, y) \,\varepsilon]\, \beta$.
66. $[(x, y) \,\varepsilon]\, \alpha -= \Lambda .=.\because [x \,\varepsilon]. [y \,\varepsilon]\, \alpha -= \Lambda :-= \Lambda$.

3. Let $x \,\alpha\, y$ be a relation between the indeterminates x and y (for example, in logic, the relations $x = y$, $x -= y$, $x \supset y$; in arithmetic, $x < y$, $x > y$, and so on). Then the sign $[\varepsilon]\, \alpha\, y$ denotes the x that satisfy the relation $x \,\alpha\, y$. For the sake of convenience, we use the sign $\ni$ instead of the sign $[\varepsilon]$. Thus, $\ni \alpha\, y .=: [x \,\varepsilon]. x \,\alpha\, y$, and the sign $\ni$ is read *the objects that*. For example, let y be a number; then $\ni < y$ denotes the class formed by the numbers x that satisfy the condition $x < y$, that is, *the objects that are smaller than* y, or simply *the objects smaller than* y. Similarly, if the sign D means *divides* or *is a*

divisor of, the formula $\ni D$ means *the objects that divide* or *the divisors*. It follows that $x \, \varepsilon \ni \alpha \, y = x \, \alpha \, y$.

4. Let α be a formula containing the indeterminate x. Then the expression $x' \, [x] \, \alpha$, which is read x' *being substituted for* x *in* α, denotes the formula obtained if, in α, we read x' for x. It follows that $x \, [x] \, \alpha = \alpha$.

5. Let α be a formula that contains the indeterminates $x, y, \ldots$. Then $(x', y', \ldots)[x, y, \ldots] \, \alpha$, which is read $x', y' \ldots$ *being substituted for* $x, y, \ldots$ *in* α, denotes the formula obtained if, in α, the letters $x_1', y', \ldots$ are written for $x, y, \ldots$. It follows that $(x, y) \, [x, y] \, \alpha = \alpha$.

VI. *Functions*

The symbols of logic introduced above suffice to express any proposition of arithmetic, and we shall use only these. We explain here briefly some other symbols that may be useful.

Let s be a class; we assume that equality is defined between the objects of the system s so as to satisfy the conditions:

$$a = a.$$
$$a = b \, .=. \, b = a.$$
$$a = b \, . \, b = c \, :\supset . \, a = c.$$

Let φ be a sign or an aggregate of signs such that, if x is an object of the class s, the expression φx denotes a new object; we assume also that equality is defined between the objects φx; further, if x and y are objects of the class s and if $x = y$, we assume it is possible to deduce $\varphi x = \varphi y$. Then the sign φ is said to be a *function presign* ⟦*praesignum*⟧ *in the class* s, and we write $\varphi \, \varepsilon \, \mathrm{F}`s$:

$$s \, \varepsilon \, \mathrm{K} \, . \supset :: \varphi \, \varepsilon \, \mathrm{F}`s \, .=.\therefore x, y \, \varepsilon \, s \, . \, x = y \, :\supset_{x,y} . \, \varphi x = \varphi y.$$

If, x being any object of the class s, the expression $x\varphi$ denotes a new object and $x\varphi = y\varphi$ follows from $x = y$, then we say that φ is a *function postsign* ⟦*postsignum*⟧ *in the class* s, and we write $\varphi \, \varepsilon \, s'\mathrm{F}$:

$$s \, \varepsilon \, \mathrm{K} \, . \supset :: \varphi \, \varepsilon \, s'\mathrm{F} \, .=.\therefore x, y \, \varepsilon \, s \, . \, x = y \, :\supset_{x,y} . \, x\varphi = y\varphi.$$

Examples. Let a be a number; then $a+$ is a function presign in the class of numbers, and $+a$ is a function postsign; for any number x, formulas $a + x$ and $x + a$ denote new numbers; $a + x = a + y$ and $x + a = y + a$ follow from $x = y$. Thus,

$$a \, \varepsilon \, \mathrm{N} \, . \supset : a+ \, .\varepsilon . \, \mathrm{F}`\mathrm{N}.$$
$$a \, \varepsilon \, \mathrm{N} \, . \supset : +a \, .\varepsilon . \, \mathrm{N}'\mathrm{F}.$$

Let φ be a function presign in the class s. Then $[\varphi] \, y$ denotes the class composed of the x that satisfy the condition $\varphi x = y$; that is,

$$\textit{Def.} \quad s \, \varepsilon \, \mathrm{K} \, . \, \varphi \, \varepsilon \, \mathrm{F}`s \, :\supset : [\varphi] \, y \, .=. \, [x \, \varepsilon] \, (\varphi x = y).$$

The class $[\varphi] \, y$ may contain one or several individuals, or none at all. We have

$$s \, \varepsilon \, \mathrm{K} \, . \, \varphi \, \varepsilon \, \mathrm{F}`s \, :\supset : y = \varphi x \, .=. \, x \, \varepsilon \, [\varphi] \, y.$$

But if φy consists of just one individual, we have $y = \varphi x \, .=. \, x = [\varphi] \, y$.

Let φ be a function postsign; we write similarly:

$$s \varepsilon \mathrm{K} . \varphi \varepsilon s\text{'}\mathrm{F} :\supset.\cdot. y [\varphi] = [x \varepsilon] (x\varphi = y).$$

The sign [ ] is called the *inversion sign*, and we have already presented some of its uses in logic. If α is a proposition containing the indeterminate x and a is a class composed of the individuals x that satisfy the condition α, we have $x \varepsilon a .= \alpha$, and then $a = [x \varepsilon] \alpha$, as in V 1.

Let α be a formula containing the indeterminate x and let φ be a function presign that yields the formula α when written before the letter x; that is, let $\alpha = \varphi x$. Then we have $\varphi = \alpha [x]$, and, if x' is a new object, we have $\varphi x' = \alpha [x] x'$; that is, if α is a formula containing the indeterminate x, then $\alpha [x] x'$ means what is obtained when, in α, we put x' for x.

Similarly, let α be a formula containing the indeterminate x and let φ be a function postsign, such that $x\varphi = \alpha$; it follows that $\varphi = [x] \alpha$. Then, if x' is a new object, we have $x'\varphi = x' [x] \alpha$; that is, $x' [x] \alpha$ again denotes what is obtained when, in α, we read x' for x, as in V 4.

The sign [ ] can have other uses in logic, which we present only briefly, since we shall not use it in these ways. Let a and b be two classes; then $[a \cap] b$ (or $b [\cap a]$) denotes the classes x that satisfy the condition $b = a \cap x$ (or the condition $b = x \cap a$). If b is not contained in a, no class satisfies this condition; if b is contained in a, the sign $b [\cap a]$ denotes all classes that contain b and are contained in $b \cup -a$.

In arithmetic, let a and b be numbers; then $b [+ a]$ (or $[a +] b$) denotes the number x that satisfies the condition $b = x + a$ (or $b = a + x$), that is, $b - a$. Similarly we have $b [\times a] = [a \times] b = b/a$. This sign can even find a use in analysis; thus,

$$y = \sin x .=. x \varepsilon [\sin] y \qquad \text{(instead of } x = \text{arc} \sin y)$$
$$d\mathrm{F}(x) = f(x)dx .=. \mathrm{F}(x) \varepsilon [d] f(x)dx \qquad \text{(instead of } \mathrm{F}(x) = \textstyle\int f(x)dx).$$

Let φ again be a function presign in a class s and let k be a class contained in s; then φk denotes the class consisting of all φx, where the x are the objects of class k; that is,

Def. $s \varepsilon \mathrm{K} . k \varepsilon \mathrm{K} . k \supset s . \varphi \varepsilon \mathrm{F}\text{'}s :\supset. \varphi k = [y \varepsilon] (k . [\varphi] y :-= \Lambda)$,

or $s \varepsilon \mathrm{K} . k \varepsilon \mathrm{K} . k \supset s . \varphi \varepsilon \mathrm{F}\text{'}s :\supset. \varphi k = [y \varepsilon] ([x \varepsilon]: x \varepsilon k . \varphi x = y .\cdot.-= \Lambda)$.

Def. $s \varepsilon \mathrm{K} . k \varepsilon \mathrm{K} . k \supset s . \varphi \varepsilon s\text{'}\mathrm{F} :\supset. k\varphi = [y \varepsilon] (k . y [\varphi] :-= \Lambda)$.

Thus, if $\varphi \varepsilon \mathrm{F}\text{'}s$, then φs denotes the class composed of all φx, where the x are objects of the class s. We have:

$$s \varepsilon \mathrm{K} . \varphi \varepsilon \mathrm{F}\text{'}s . y \varepsilon \varphi s :\supset: \varphi [\varphi] y = y.$$
$$s \varepsilon \mathrm{K} . a, b \varepsilon \mathrm{K} . a \supset s . b \supset s . \varphi \varepsilon \mathrm{F}\text{'}s :\supset. \varphi(a \cup b) = (\varphi a) \cup (\varphi b).$$
$$s \varepsilon \mathrm{K} . \varphi \varepsilon \mathrm{F}\text{'}s :\supset. \varphi\Lambda = \Lambda.$$
$$s \varepsilon \mathrm{K} . a, b \varepsilon \mathrm{K} . b \supset s . a \supset b . \varphi \varepsilon \mathrm{F}\text{'}s :\supset. \varphi a \supset \varphi b.$$
$$s \varepsilon \mathrm{K} . a, b \varepsilon \mathrm{K} . a \supset s . b \supset s . \varphi \varepsilon \mathrm{F}\text{'}s :\supset. \varphi(ab) \supset (\varphi a)(\varphi b).$$

Let a be a class; then $a \cap \mathrm{K}$ (or $\mathrm{K} \cap a$, or $\mathrm{K}a$) denotes all classes of the form $a \cap x$ (or $x \cap a$, or xa), where x is any class; that is, $\mathrm{K}a$ denotes the classes that are contained in a. The formula $x \varepsilon \mathrm{K}a$ means the same as $x \varepsilon \mathrm{K} . x \supset a$. We shall sometimes use this convention; thus KN means *a class of numbers*.

Similarly, if a is a class, $\mathrm{K} \cup a$ denotes the classes that contain a.

Let a be a number; then $a + \mathrm{N}$ (or $\mathrm{N} + a$) denotes *the numbers greater than the number* a; $a \times \mathrm{N}$ (or $\mathrm{N} \times a$ or $\mathrm{N}a$) denotes *the multiples of the number* a; a^{N} denotes *the powers of the number* a; $\mathrm{N}^2, \mathrm{N}^3, \ldots$ denote *the squares, the cubes,* and so on.

Equality, product, and powers can be defined thus for function signs:

$$\textit{Def.}\quad s \,\varepsilon\, \mathrm{K} . \varphi, \psi \,\varepsilon\, \mathrm{F}'s :\mathrm{Ɔ}.\cdot\, \varphi = \psi :=: x \,\varepsilon\, s \,.\mathrm{Ɔ}_x .\, \varphi x = \psi x.$$

$$\textit{Def.}\quad s \,\varepsilon\, \mathrm{K} . \varphi \,\varepsilon\, \mathrm{F}'s . \psi \,\varepsilon\, \mathrm{F}'\varphi s . x \,\varepsilon\, s :\mathrm{Ɔ}.\, \psi\varphi x = \psi(\varphi x).$$

Thus, if we assume this definition, we have the new function presign $\psi\varphi$; it is called the *product of the signs* ψ and φ.

Similarly if φ and ψ are function postsigns.

The following proposition holds:

$$s \,\varepsilon\, \mathrm{K} . \varphi \,\varepsilon\, \mathrm{F}'s . \varphi s \,\mathrm{Ɔ}\, s :\mathrm{Ɔ}: \varphi\varphi s \,\mathrm{Ɔ}\, s . \varphi\varphi\varphi s \,\mathrm{Ɔ}\, s . \text{and so on.}$$

The functions $\varphi\varphi, \varphi\varphi\varphi, \ldots$ are said to be *iterated* and are generally denoted by the signs $\varphi^2, \varphi^3, \ldots,$ as powers of the operation φ.

But if φ is a function postsign, we can use the following more convenient notation without ambiguity:

$$\textit{Def.}\quad s \,\varepsilon\, \mathrm{K} . \varphi \,\varepsilon\, s'\mathrm{F} . s\varphi \,\mathrm{Ɔ}\, s :\mathrm{Ɔ}: \varphi 1 = \varphi . \varphi 2 = \varphi\varphi . \varphi 3 = \varphi\varphi\varphi . \text{and so on.}$$

Assuming this definition, if $m, n \,\varepsilon\, \mathrm{N}$, we have $\varphi(m + n) = (\varphi m)(\varphi n)$; $(\varphi m)n = \varphi(mn)$.

If we use this definition in arithmetic, we obtain the following. We can denote the number that follows the number a by the more convenient sign $a+$; then $a + 1$, $a + 2, \ldots,$ and, if b is a number, $a + b$, have the meaning of $a+, a++, \ldots,$ which is clear from the definition in § 1 below. Proposition 6 in § 1 can be written $\mathrm{N} + \mathrm{Ɔ}\, \mathrm{N}$. If a, b, and c are numbers, then $a :+ b . c$ means $a + bc$, and $a :\times b . c$ means ab^c.

Function signs possess many other properties, especially if they satisfy the condition $\varphi x = \varphi y \,.\mathrm{Ɔ}.\, x = y$. A function sign that satisfies this condition is called *similar* by Dedekind (*ähnliche Abbildung*).[3]

But we lack the space to present these properties.

Remarks

A *definition*, or *Def.* for short, is a proposition of the form $x = a$ or $\alpha \,\mathrm{Ɔ}.\, x = a$, where a is an aggregate of signs having a known meaning, x is a sign or an aggregate of signs, hitherto without meaning, and α is the condition under which the definition is given.

A *theorem* (Theor. or Th.) is a proposition that is *proved.* If a theorem has the form $a \,\mathrm{Ɔ}\, \beta$, where α and β are propositions, then α is called the *hypothesis* (Hyp. or, even shorter, Hp.) and β the *thesis* (Thes. or Ts.). Hyp. and Ts. depend on the form of the theorem; in fact, if we write $-\beta \,\mathrm{Ɔ}\, -\alpha$ instead of $\alpha \,\mathrm{Ɔ}\, \beta$, then $-\beta$ is the Hp. and $-\alpha$ the Ts.; if we write $\alpha - \beta = \Lambda$, Hp. and Ts. do not exist.

In any section below, the sign P followed by a number denotes the proposition indicated by that number in the same section. Propositions of logic are indicated by the sign L and the number of the proposition.

Formulas that do not fit on one line are continued on the next line without any intervening sign.

[3] ⟦Today "similar" has another meaning and instead we would say "equivalent".⟧

§ 1. Numbers and addition

Explanations

The sign N means *number* (*positive integer*).

The sign 1 means *unity*.

The sign $a + 1$ means *the successor of a*, or *a plus* 1.

The sign $=$ means *is equal to*. We consider this sign as new, although it has the form of a sign of logic.

Axioms

1. $1 \,\varepsilon\, \mathrm{N}$.
2. $a \,\varepsilon\, \mathrm{N} \,.\!\supset\!.\, a = a$.
3. $a, b \,\varepsilon\, \mathrm{N} \,.\!\supset\!:\, a = b \,.\!=\!.\, b = a$.
4. $a, b, c \,\varepsilon\, \mathrm{N} \,.\!\supset\!.\!\!\cdot\!.\, a = b \,.\, b = c \,:\!\supset\!.\, a = c$.
5. $a = b \,.\, b \,\varepsilon\, \mathrm{N} \,:\!\supset\!.\, a \,\varepsilon\, \mathrm{N}$.
6. $a \,\varepsilon\, \mathrm{N} \,.\!\supset\!.\, a + 1 \,\varepsilon\, \mathrm{N}$.
7. $a, b \,\varepsilon\, \mathrm{N} \,.\!\supset\!:\, a = b \,.\!=\!.\, a + 1 = b + 1$.
8. $a \,\varepsilon\, \mathrm{N} \,.\!\supset\!.\, a + 1 -\!= 1$.
9. $k \,\varepsilon\, \mathrm{K} \,.\!\cdot\!.\, 1 \,\varepsilon\, k \,.\!\cdot\!.\, x \,\varepsilon\, \mathrm{N} \,.\, x \,\varepsilon\, k \,:\!\supset_x\!.\, x + 1 \,\varepsilon\, k \,::\!\supset\!.\, \mathrm{N} \supset k$.

Definitions

10. $2 = 1 + 1$; $3 = 2 + 1$; $4 = 3 + 1$; and so forth.

Theorems

11. $2 \,\varepsilon\, \mathrm{N}$.

Proof:

$\mathrm{P}\,1 \,.\!\supset\!:$	$1 \,\varepsilon\, \mathrm{N}$	(1)
$1\,[a]\,(\mathrm{P}\,6) \,.\!\supset\!:$	$1 \,\varepsilon\, \mathrm{N} \,.\!\supset\!.\, 1 + 1 \,\varepsilon\, \mathrm{N}$	(2)
$(1)\,(2) \,.\!\supset\!:$	$1 + 1 \,\varepsilon\, \mathrm{N}$	(3)
$\mathrm{P}\,10 \,.\!\supset\!:$	$2 = 1 + 1$	(4)
$(4)\,.\,(3)\,.\,(2, 1 + 1)\,[a, b]\,(\mathrm{P}\,5) \,:\!\supset\!:$	$2 \,\varepsilon\, \mathrm{N}$	(Theorem).

Note. We have written explicitly all the steps of this very easy proof. For the sake of brevity, we now write it as follows:

$$\mathrm{P}\,1 \,.\, 1\,[a]\,(\mathrm{P}\,6) \,:\!\supset\!:\, 1 + 1 \,\varepsilon\, \mathrm{N} \,.\, \mathrm{P}\,10 \,.\, (2, 1+1)\,[a, b]\,(\mathrm{P}\,5) \,:\!\supset\!:\, \mathrm{Th}.$$

or

$$\mathrm{P}\,1 \,.\, \mathrm{P}\,6 \,:\!\supset\!:\, 1 + 1 \,\varepsilon\, \mathrm{N} \,.\, \mathrm{P}\,10 \,.\, \mathrm{P}\,5 \,:\!\supset\!:\, \mathrm{Th}.$$

12. $3, 4, \ldots \,\varepsilon\, \mathrm{N}$.
13. $a, b, c, d \,\varepsilon\, \mathrm{N} \,.\, a = b \,.\, b = c \,.\, c = d \,:\!\supset\!:\, a = d$.

Proof: Hyp. $\mathrm{P}\,4 \,:\!\supset\!:\, a, c, d \,\varepsilon\, \mathrm{N} \,.\, a = c \,.\, c = d \,.\, \mathrm{P}\,4 \,:\!\supset\!:$ Thes.

14. $a, b, c \,\varepsilon\, \mathrm{N} \,.\, a = b \,.\, b = c \,.\, a -\!= c \,:\!=\, \Lambda$.

Proof: P 4 . L 39 :Ɔ. Theor.

15. $a, b, c \; \varepsilon \; \mathrm{N} . a = b . b -= c :Ɔ. a -= c.$

16. $a, b \; \varepsilon \; \mathrm{N} . a = b :Ɔ. a + 1 = b + 1.$

16′. $a, b \; \varepsilon \; \mathrm{N} . a + 1 = b + 1 :Ɔ. a = b.$

Proof: P 7 = (P 16)(P 16′).

17. $a, b \; \varepsilon \; \mathrm{N} .Ɔ: a -= b .=. a + 1 -= b + 1.$

Proof: P 7 . L 21 :Ɔ. Theor.

Definition

18. $a, b \; \varepsilon \; \mathrm{N} .Ɔ. a + (b + 1) = (a + b) + 1.$

Note. This definition has to be read as follows: if a and b are numbers, and if $(a + b) + 1$ has a meaning (that is, if $a + b$ is a number) but $a + (b + 1)$ has not yet been defined, then $a + (b + 1)$ means the number that follows $a + b$.

From this definition and also the preceding it follows that

$$a \; \varepsilon \; \mathrm{N} .Ɔ.\therefore a + 2 = a + (1 + 1) = (a + 1) + 1,$$
$$a \; \varepsilon \; \mathrm{N} .Ɔ.\therefore a + 3 = a + (2 + 1) = (a + 2) + 1,$$

and so forth.

Theorems

19. $a, b \; \varepsilon \; \mathrm{N} .Ɔ. a + b \; \varepsilon \; \mathrm{N}.$

Proof: $a \; \varepsilon \; \mathrm{N} . \mathrm{P}\,6 :Ɔ: a + 1 \; \varepsilon \; \mathrm{N} :Ɔ: 1 \; \varepsilon \, [b \; \varepsilon] \, \mathrm{Ts}.$ (1)

$a \; \varepsilon \; \mathrm{N} .Ɔ:: b \; \varepsilon \; \mathrm{N} . b \; \varepsilon \, [b \; \varepsilon] \, \mathrm{Ts} :Ɔ: a + b \; \varepsilon \; \mathrm{N} . \mathrm{P}\,6 :Ɔ: (a + b) + 1 \; \varepsilon \; \mathrm{N} . \mathrm{P}\,18 :Ɔ: a + (b + 1) \; \varepsilon \; \mathrm{N} :Ɔ: (b + 1) \; \varepsilon \, [b \; \varepsilon] \, \mathrm{Ts}.$ (2)

$a \; \varepsilon \; \mathrm{N} . (1) . (2) .Ɔ:: 1 \; \varepsilon \, [b \; \varepsilon] \, \mathrm{Ts} .\therefore b \; \varepsilon \; \mathrm{N} . b \; \varepsilon \, [b \; \varepsilon] \, \mathrm{Ts} :Ɔ: b + 1 \; \varepsilon \, [b \; \varepsilon] \, \mathrm{Ts} .\therefore ([b \; \varepsilon] \, \mathrm{Ts}) \, [k] \, \mathrm{P}\,9 ::Ɔ: \mathrm{N} \, Ɔ \, [b \; \varepsilon] \, \mathrm{Ts}. (\mathrm{L}\,50) ::Ɔ: b \; \varepsilon \; \mathrm{N} .Ɔ \, \mathrm{Ts}.$ (3)

$(3) . (\mathrm{L}\,42) :Ɔ: a, b \; \varepsilon \; \mathrm{N} .Ɔ.$ Thesis. (Theor.).

20. *Def.* $a + b + c = (a + b) + c.$

21. $a, b, c \; \varepsilon \; \mathrm{N} .Ɔ. a + b + c \; \varepsilon \; \mathrm{N}.$

22. $a, b, c \; \varepsilon \; \mathrm{N} .Ɔ: a = b .=. a + c = b + c.$

Proof: $a, b \; \varepsilon \; \mathrm{N} . \mathrm{P}\,7 :Ɔ. 1 \; \varepsilon \, [c \; \varepsilon] \, \mathrm{Ts}.$ (1)

$a, b \; \varepsilon \; \mathrm{N} .Ɔ:: c \; \varepsilon \; \mathrm{N} . c \; \varepsilon \, [c \; \varepsilon] \, \mathrm{Ts} .\therefore Ɔ.\therefore a = b .=. a + c = b + c : a + c, b + c \; \varepsilon \; \mathrm{N} : a + c = b + c .=. a + c + 1 = b + c + 1 .\therefore Ɔ.\therefore a = b .=. a + (c + 1) = b + (c + 1) .\therefore Ɔ.\therefore (c + 1) \; \varepsilon \, [c \; \varepsilon] \, \mathrm{Ts}.$ (2)

$a, b \; \varepsilon \; \mathrm{N} . (1) . (2) :Ɔ:: 1 \; \varepsilon \, [c \; \varepsilon] \, \mathrm{Ts} .\therefore c \; \varepsilon \, [c \; \varepsilon] \, \mathrm{Ts} .Ɔ. (c + 1) \; \varepsilon \, [c \; \varepsilon] \, \mathrm{Ts} ::Ɔ:: c \; \varepsilon \; \mathrm{N} .Ɔ. \mathrm{Ts}.$ (3)

(3) Ɔ Theor.

23. $a, b, c \; \varepsilon \; \mathrm{N} .Ɔ. a + (b + c) = a + b + c.$

Proof: $a, b \; \varepsilon \; \mathrm{N} . \mathrm{P}\,18 . \mathrm{P}\,20 :Ɔ. 1 \; \varepsilon \, [c \; \varepsilon] \, \mathrm{Ts}.$ (1)

$a, b \; \varepsilon \; \mathrm{N} .Ɔ.\therefore c \; \varepsilon \; \mathrm{N} . c \; \varepsilon \, [c \; \varepsilon] \, \mathrm{Ts} :Ɔ: a + (b + c) = a + b + c . \mathrm{P}\,7 :Ɔ: a + (b + c) + 1 = a + b + c + 1 . \mathrm{P}\,18 :Ɔ: a + (b + (c + 1)) = a + b + (c + 1) :Ɔ. c + 1 \; \varepsilon \, [c \; \varepsilon] \, \mathrm{Ts}.$ (2)

(1) (2) (P 9) $. \supset .$ Theor.

24. $a \,\varepsilon\, \mathrm{N} . \supset . 1 + a = a + 1.$

Proof: $\mathrm{P}\, 2 . \supset . 1 \,\varepsilon\, [a \,\varepsilon]$ Ts. (1)

$a \,\varepsilon\, \mathrm{N} . a \,\varepsilon\, [a \,\varepsilon]$ Ts $: \supset : 1 + a = a + 1 : \supset : 1 + (a + 1) = (a + 1) + 1 : \supset : (a + 1)$ $\varepsilon\, [a \,\varepsilon]$ Ts. (2)

(1) (2) $. \supset .$ Theor.

24′. $a, b \,\varepsilon\, \mathrm{N} . \supset . 1 + a + b = a + 1 + b.$

Proof: Hyp. P 24 $: \supset : 1 + a = a + 1 .$ P 22 $: \supset .$ Thesis.

25. $a, b \,\varepsilon\, \mathrm{N} . \supset . a + b = b + a.$

Proof: $a \,\varepsilon\, \mathrm{N} .$ P 24 $: \supset : 1 \,\varepsilon\, [b \,\varepsilon]$ Ts. (1)

$a \,\varepsilon\, \mathrm{N} . \supset \therefore b \,\varepsilon\, \mathrm{N} . b \,\varepsilon\, [b \,\varepsilon]$ Ts $: \supset : a + b = b + a .$ P 7 $: \supset : (a + b) + 1 = (b + a) + 1 . (a + b) + 1 = a + (b + 1) . (b + a) + 1 = 1 + (b + a) . 1 + (b + a) = (1 + b) + a . (1 + b) + a = (b + 1) + a : \supset : a + (b + 1) = (b + 1) + a : \supset : (b + 1)$ $\varepsilon\, [b \,\varepsilon]$ Ts. (2)

(1) (2) $. \supset .$ Theor.

26. $a, b, c \,\varepsilon\, \mathrm{N} . \supset : a = b . = . c + a = c + b.$

27. $a, b, c \,\varepsilon\, \mathrm{N} . \supset : a + b + c = a + c + b.$

28. $a, b, c, d \,\varepsilon\, \mathrm{N} . a = b . c = d : \supset . a + c = b + d.$

§ 2. Subtraction

Explanations

The sign $-$ is read *minus*.
The sign $<$ is read *is less than.*
The sign $>$ is read *is greater than.*

Definitions

1. $a, b \,\varepsilon\, \mathrm{N} . \supset : b - a = \mathrm{N}\, [x \,\varepsilon]\, (x + a = b).$
2. $a, b \,\varepsilon\, \mathrm{N} . \supset : a < b . = . b - a \mathrel{-\!=} \Lambda.$
3. $a, b \,\varepsilon\, \mathrm{N} . \supset : b > a . = . a < b.$

$a + b - c = (a + b) - c ; a - b + c = (a - b) + c ; a - b - c = (a - b) - c.$

§ 4. Multiplication

Definitions

1. $a \,\varepsilon\, \mathrm{N} . \supset . a \times 1 = a.$
2. $a, b \,\varepsilon\, \mathrm{N} . \supset . a \times (b + 1) = a \times b + a.$

$ab = a \times b ; ab + c = (ab) + c ; abc = (ab)\, c.$

§ 5. Powers

Definitions

1. $a \,\varepsilon\, \mathrm{N} . \supset . a^1 = a.$
2. $a, b \,\varepsilon\, \mathrm{N} . \supset . a^{b+1} = a^b\, a.$

§ 6. Division

Explanations

The sign / is read *divided by*.
The sign D is read *divides*, or *is a divisor of*.
The sign ᗡ is read *is a multiple of*.
The sign Np is read *prime number*.
The sign π is read *is prime to*.

Definitions

1. $a, b \,\varepsilon\, \mathrm{N} \,.\text{Ɔ}.\, b/a = \mathrm{N}\,[x\,\varepsilon]\,(xa = b).$
2. $a, b \,\varepsilon\, \mathrm{N} \,.\text{Ɔ}\!:\, a \,\mathrm{D}\, b \,.\!=\!.\, b/a \mathrel{-\!=} \Lambda.$
3. $a, b \,\varepsilon\, \mathrm{N} \,.\text{Ɔ}\!:\, b \,\text{ᗡ}\, a \,.\!=\!.\, a \,\mathrm{D}\, b.$
4. $\mathrm{Np} = \mathrm{N}\,[x\,\varepsilon]\,(\text{ɜ}\,\mathrm{D}\,x \,.\, \text{ɜ} > 1 \,.\, \varepsilon < x := \Lambda).$
5. $a, b \,\varepsilon\, \mathrm{N} \,.\text{Ɔ}::\, a \,\pi\, b \;.\!\cdot\!.\!=\!.\!\cdot\!.\; \text{ɜ}\,\mathrm{D}\,a \,.\, \text{ɜ}\,\mathrm{D}\,b \,.\, \text{ɜ} > 1 := \Lambda.$
6. $a, b \,\varepsilon\, \mathrm{N} \,.\text{Ɔ}.\!\cdot\!.\; \text{ɜ}\,\mathrm{D}\,(a, b) :=: \text{ɜ}\,\mathrm{D}\,a \,.\!\cap\!.\, \text{ɜ}\,\mathrm{D}\,b.$
7. $a, b \,\varepsilon\, \mathrm{N} \,.\text{Ɔ}.\!\cdot\!.\; \text{ɜ}\,\text{ᗡ}\,(a, b) :=: \text{ɜ}\,\text{ᗡ}\,a \,.\!\cap\!.\, \text{ɜ}\,\text{ᗡ}\,b.$

$ab/c = (ab)/c;\ a/b/c = (a/b)/c;\ a/b \times c = (a/b)\,c.$

Letter to Keferstein

RICHARD DEDEKIND

(*1890a*)

Hans Keferstein, *Oberlehrer* in Hamburg, published a paper (*1890*) on the notion of number in which he commented on Frege's (*1884*) and Dedekind's (*1888*) books on the subject. His comments on Dedekind's work, although not entirely negative, included a number of suggestions for amending the text that revealed his lack of real understanding of some fundamental points, for example, the equivalence of two sets. Dedekind felt obliged to answer with an essay (*1890*) in which he showed how pointless the "corrections" were. He sent it to Keferstein on 9 February 1890, with a letter in which he suggested that the Hamburg Mathematical Society, in whose yearly *Mitteilungen* Keferstein's paper had appeared, publish either the essay or, should Keferstein realize that his suggestions were based upon misunderstandings, a declaration to that effect.

Dedekind's essay dealt with three points. The first was an objection of Keferstein's to Dedekind's proof that there exists an infinite set. This proof has often been criticized (see, for instance, below, p. 131); but Keferstein's objection rested upon a wrong argument, a plain confusion of the equivalence relation between sets with their identity, and Dedekind had no difficulty in answering him. The second point was Keferstein's claim that he had found two conflicting definitions of infinite sets in the book; Dedekind pointed out that one was in fact merely a stylistic variant of the other. The third point was the substitution by Keferstein of a new definition of simply infinite sets for that given by Dedekind (*1888*, art. 71). Keferstein's purpose was to avoid the notion of chain, and his proposal amounted in effect to the abandonment of mathematical induction; Dedekind showed that this proposal would bar the possibility of providing an adequate foundation for the theory of natural numbers.

On 14 February 1890 Keferstein acknowledged receipt of the essay, announcing that at the next meeting of the Society he would propose its publication, that he was confident that the proposition would be accepted, that, moreover, he did not consider his criticisms, especially the third one, as mere misunderstandings on his part and would return to them in case the essay should be published.

On 27 February 1890 Dedekind sent to Keferstein a long letter that is a brilliant presentation of the development of his ideas on the notion of natural number. In it he tried to show that his assumptions had not been haphazardly chosen and that each one of them had a profound justification. This is especially true, Dedekind insisted, of the notion of chain, which Keferstein wanted to eliminate. Professor Hao Wang published (*1957*) an English translation of a major part of the letter, with commentaries. The text below is a translation of the whole letter.

On 19 March 1890 Keferstein thanked

Dedekind for the letter and asked his permission to use it in a lecture before the Hamburg Mathematical Society. Dedekind granted this permission in his next letter, dated 1 April 1890, adding a few lines of explanation on the notion of chain.

On 17 November 1890, as publication of the yearly volume of the *Mitteilungen* of the Society was drawing near, Keferstein wrote an "Erwiderung" (*1890b*) that was to follow Dedekind's essay (*1890*) in the volume. But on 19 December 1890 he had to inform Dedekind that the editorial board of the Society had declined to publish Dedekind's essay, as well as Keferstein's rejoinder, the reason invoked being the lack of space and the fact that Dedekind's reply was longer than Keferstein's original criticism of Dedekind's book. Keferstein also announced his intention of publishing in the Society's coming yearly volume a note withdrawing his proposed "corrections" to Dedekind's work. The note appeared in volume 3 of the *Mitteilungen* (p. 31), published in February 1891; it consisted of a few lines incorporated in a report of the 11 October 1890 meeting of the Society.

On 23 December 1890 Dedekind wrote his last letter to Keferstein, acknowledging receipt of his returned manuscript as well as of a copy of Keferstein's "Erwiderung". He expressed his regrets that, although the polemic and the correspondence had taken so much of his time, Keferstein's reply still contained many misunderstandings.

Dedekind's time and efforts were, however, not wasted at all. The controversy produced the letter below, which remains a masterly presentation of his ideas.

Dedekind's essay and the Dedekind–Keferstein correspondence are preserved in the Niedersächsische Staats- und Universitätsbibliothek in Göttingen. They come from the Dedekind estate. Keferstein's letters are the originals, as received by Dedekind; Dedekind's letters and his essay, as well as Keferstein's "Erwiderung", are clean copies in Dedekind's hand. Dedekind's letter of 27 February 1890 is reproduced below with the kind permission of the Library (where it has the classmark: Göttingen, UB, Cod. Ms. Nachlass Dedekind, 13).

Stefan Bauer-Mengelberg translated the parts of the letter omitted from Professor Wang's paper and introduced some changes into the text of Professor Wang's translation. Permission to make use of that translation was granted by Professor Wang and *The journal of symbolic logic*.

My dear Doctor,

I should like to express my sincerest thanks for your kind letter of the 14th of this month and for your willingness to publish my reply. But I would ask you not to rush anything in this matter and to come to a decision only after you have once more carefully read and thoroughly considered the most important definitions and proofs in my essay on numbers, if you have the time. For I think that most probably you will then be converted on all points to my conception and to my treatment of the subject; and this is just what I should value most, since I am convinced that you really have a deep interest in the matter.

In order to further this rapprochement wherever possible, I should like to ask you to lend your attention to the following train of thought, which constitutes the genesis of my essay. How did my essay come to be written? Certainly not in one day; rather, it is a synthesis constructed after protracted labor, based upon a prior analysis of the sequence of natural numbers just as it presents itself, in experience, so to speak, for our consideration. What are the mutually independent fundamental properties of the sequence N, that is, those properties that are not derivable from one another but from which all others follow? And how should we divest these properties of their specifically arithmetic character so that they are subsumed under more general notions and under activities of the understanding *without* which no thinking is possible at all but *with* which a foundation is provided for the reliability and completeness of proofs and for the construction of consistent notions and definitions?

When the problem is posed in this way, one is, I believe, forced to accept the following facts:

(1) The number sequence N is a *system* of individuals, or elements, called numbers. This leads to the general consideration of systems as such (§ 1 of my essay).

(2) The elements of the system N stand in a certain relation to one another; a certain order obtains, which consists, to begin with, in the fact that to each definite number n there corresponds a definite number n', the succeeding, or next greater, number. This leads to the consideration of the general notion of a *mapping* φ of a system (§ 2), and since the image $\varphi(n)$ of every number n is again a *number*, n', and therefore $\varphi(N)$ is a part of N, we are here concerned with the mapping φ of a system N *into itself*, of which we must therefore make a general investigation (§ 4).

(3) Distinct numbers a and b are succeeded by distinct numbers a' and b'; the mapping φ, therefore, has the property of distinctness, or *similarity*[1] (§ 3).

(4) Not every number is a successor n'; in other words, $\varphi(N)$ is a proper part of N. This (together with the preceding) is what makes the number sequence N infinite (§ 5).

(5) And, in particular, the number 1 is the *only* number that does not lie in $\varphi(N)$. Thus we have listed the facts that you (p. 124, ll. 8–14) regard as the complete characterization of an ordered, simply infinite system N.

(6) I have shown in my reply (III),[2] however, that these facts are still far from being adequate for completely characterizing the nature of the number sequence N. All these facts would hold also for every system S that, besides the number sequence N, contained a system T, of arbitrary additional elements t, to which the mapping φ could always be extended while remaining similar and satisfying $\varphi(T) = T$. But such a system S is obviously something quite different from our number sequence N, and I could so choose it that scarcely a single theorem of arithmetic would be preserved in it. What, then, must we add to the facts above in order to cleanse our system S again of such alien intruders t as disturb every vestige of order and to restrict it to N? This was one of the most difficult points of my analysis and its mastery required lengthy reflection. If one presupposes knowledge of the sequence N of natural numbers and, accordingly, allows himself the use of the language of arithmetic, then, of course, he has an easy time of it. He need only say: an element n belongs to the sequence N if and only if, starting with the element 1 and counting on and on steadfastly, that is, going through a finite number of iterations of the mapping φ (see the end of article 131 in my essay), I actually reach the element n at some time; by this procedure, however, I shall never reach an element t outside of the sequence N. But this way of characterizing the distinction between those elements t that are to be ejected from S and those elements n that alone are to remain is surely quite useless for our purpose; it would, after all, contain the most pernicious and obvious kind of vicious

[1] [[See footnote 3, p. 93 above.]]

[2] [[This refers to sec. III in *Dedekind 1890*; see introductory note.]]

circle. The mere words "finally get there at some time", of course, will not do either; they would be of no more use than, say, the words "karam sipo tatura", which I invent at this instant without giving them any clearly defined meaning. Thus, how can I, without presupposing any arithmetic knowledge, give an unambiguous conceptual foundation to the distinction between the elements n and the elements t? Merely through consideration of the *chains* (articles 37 and 44 of my essay), and yet, by means of these, completely! If I wanted to avoid my technical expression "chain" I would say: an element n of S belongs to the sequence N if and only if n is an element of *every* part K of S that possesses the following two properties: (i) the element 1 belongs to K and (ii) the image $\varphi(K)$ is a part of K. In my technical language: N is the intersection [[Gemeinheit]] 1_0, or $\varphi_0(1)$, of all those chains K (in S) to which the element 1 belongs. Only now is the sequence N characterized completely. In passing I would like to make the following remark on this point. Frege's *Begriffsschrift* and *Grundlagen der Arithmetik* came into my possession for the first time for a brief period last summer (1889), and I noted with pleasure that his way of defining the nonimmediate succession of an element upon another in a sequence agrees in *essence* with my notion of chain (articles 37 and 44); only, one must not be put off by his somewhat inconvenient terminology.

(7) After the essential nature of the simply infinite system, whose abstract type is the number sequence N, had been recognized in my analysis (articles 71 and 73), the question arose: does such a system *exist* at all in the realm of our ideas? Without a logical proof of existence it would always remain doubtful whether the notion of such a system might not perhaps contain internal contradictions. Hence the need for such proofs (articles 66 and 72 of my essay).

(8) After this, too, had been settled, there was the question: does what has been said so far also contain a *method of proof* sufficient to establish, in full generality, propositions that are supposed to hold for *all* numbers n? Yes! The famous method of proof by induction rests upon the secure foundation of the notion of chain (articles 59, 60, and 80 of my essay).

(9) Finally, is it possible also to set up the *definitions* of numerical operations consistently for *all* numbers n? Yes! This is in fact accomplished by the theorem of article 126 of my essay.

Thus the analysis was completed and the synthesis could begin; but this still caused me trouble enough! Indeed the reader of my essay does not have an easy task either; apart from sound common sense, it requires very strong determination to work everything through completely.

I shall now turn to some parts of your paper that I did not mention in my recent reply [[*1890*]] because they are not as important; but perhaps my remarks about them will contribute something more to the clarification of the issue.

(*a*) P. 121, l. 19.[3] Why the mention of a *part* here? I later (article 161 of my essay) ascribe a *number* [[*Anzahl*]] to each *finite* system and to no other.

(*b*) P. 122, l. 8.[4] Here we have a confusion between *mapping* and *map*; instead of

[3] [[Keferstein had written: "In fact he [[Dedekind]] later ascribes each number to a certain part [[Teil]] of such a system...".]]

[4] [[Keferstein had written: "... to the mapping φ of S we can match an inverse mapping $\bar{\varphi}(S')$ ...".]]

"mapping $\bar{\varphi}(S')$" it should be "mapping $\bar{\varphi}$ of the system S'". Not $\bar{\varphi}(S')$ but $\bar{\varphi}$ is a *mapping* (the mapping cartographer) ⟦*Abbildung* (der abbildende Maler)⟧, which generates the *map* $\bar{\varphi}(S') = S$ from the *system* S' (the original). Such confusions can become quite dangerous in our investigations.

(*c*) P. 123, ll. 1–2.[5] These words might perhaps apply to Frege, but they certainly do not apply to me. I define the *number* ⟦*Zahl*⟧ 1 as the basic element of the number sequence without any ambiguity in articles 71 and 73, and, just as unambiguously, I arrive at the *number* ⟦*Anzahl*⟧ 1 in the theorem of article 164 as a consequence of the general definition in article 161. Nothing further *may* be added to this at all if the matter is not to be muddled.

(*d*) P. 123, ll. 29–31.[6] The preceding remark, (*c*), has already taken care of this. And how would the greater reliability and the lesser prolixity shape up in *actual fact*?

(*e*) P. 124, ll. 21–24.[7] The meaning of these lines (as well as of the preceding and subsequent ones) is not quite clear to me. Do they perhaps express the desire that my definition of the number sequence N and of the way in which the element n' follows the element n be propped up, if possible, by an *intuitive* sequence? If so, I would resist that with the utmost determination, since the danger would immediately arise that from such an intuition we might perhaps unconsciously also take as self-evident theorems that must rather be derived quite abstractly from the logical definition of N. If I *call* (article 73) n' the element *following* n, that is only a new *technical expression* by means of which I merely bring some variety into my *language*; this language would sound even more monotonous and repelling if I had to deny myself this variety and were always to call n' only the *map* $\varphi(n)$ of n. But one expression is to *mean* exactly the same as the other.

(*f*) P. 124, l. 33—p. 125, l. 7.[8] The word "merely" ⟦"lediglich"⟧, taken from the third line of my definition in article 73, is obviously meant to indicate the sole *restriction* to which the word "entirely" ⟦"gänzlich"⟧, which occurs just before, is subject;[9]

[5] ⟦Keferstein had written: "In our opinion, both Frege (*1884*, pp. 89–90) and Dedekind, who incidentally derives the notion of cardinal number only from the previously defined notion of ordinal number (*1888*, pp. 21 ⟦article 73⟧ and 54 ⟦article 161⟧), have, when all is said and done, introduced the notion of the number 1 without an adequate definition".⟧

[6] ⟦Keferstein had written: "... especially since, by the previous introduction of the number 1, the latter ⟦Dedekind⟧ seems not only to gain in reliability but also to lose in prolixity".⟧

[7] ⟦Keferstein had written: "Since Dedekind does not emphasize this fact ⟦that N can be regarded as a sequence in which $\varphi(n) = n'$ immediately follows n⟧, the notions of sequence and of succession in a sequence turn up in an *apparently* abrupt way in the definition of ordinal numbers that comes at that point".⟧

[8] ⟦Keferstein had written: "When the above comments are properly taken into account, there remains in these propositions at most one point that could give offense, namely, the demand that we *entirely* disregard the particular character of the elements and retain merely their distinguishability, since objects remain distinguishable, after all, only if they still exhibit differences. If we strike out the words 'ihre Unterscheidbarkeit festhält und nur' ⟦see footnote 9⟧, however, the difficulty vanishes, since the relations in which the elements are put with one another by the ordering mapping φ are conceived by precisely a pure mental activity that remains completely independent of the particular character of the objects toward which it is directed".⟧

[9] ⟦The German text to which Dedekind refers reads: "Wenn man bei der Betrachtung eines einfach unendlichen, durch eine Abbildung φ geordneten Systems N von der besonderen Beschaffenheit der Elemente gänzlich absieht, lediglich ihre Unterscheidbarkeit festhält und nur die Beziehungen auffaßt, in die sie durch die ordnende Abbildung φ zueinander gesetzt sind, so heißen diese Elemente *natürliche Zahlen* oder *Ordinalzahlen* oder auch schlechthin *Zahlen*, und das Grundelement 1 heißt die *Grundzahl* der *Zahlenreihe* N". (*1888*, art. 73).⟧

if one were to remove this restriction—if, in other words, the word "entirely" were to assume its full meaning—then we would lose the distinguishability of the elements, which, after all, is indispensable for the notion of the simply infinite system. This "merely", therefore, does not seem at all superfluous to me, but necessary. I do not understand how it could arouse any objection.

Repeating the wish I expressed at the beginning and begging you to excuse the thoroughness of my discussion, I remain with kindest regards

Yours very truly,

R. Dedekind

27 February 1890
Petrithorpromenade 24

A question on transfinite numbers

CESARE BURALI-FORTI

(*1897*)

Dated "Turin, February 1897", Burali-Forti's paper was communicated at the 28 March 1897 meeting of the Circolo matematico di Palermo and published in its *Rendiconti*.

The paper is the first published statement of a modern paradox. It immediately aroused the interest of the mathematical world, and it provoked lively discussions in the years that followed its publication. Dozens of papers dealt with it, and it gave a strong impulse to a reexamination of the foundations of set theory.

The statement of the paradox is quite simple. On naive set-theoretic assumptions, the set of ordinals is well-ordered, hence has an ordinal; this ordinal is at once an element of the set of ordinals and greater than any ordinal in the set.

Burali-Forti himself considered the contradiction as establishing, by *reductio ad absurdum*, the result that the natural ordering of ordinals is just a partial ordering (thus contradicting the result that Cantor was to publish a few months later (*1897*), that the set of ordinals is linearly ordered). Burali-Forti was not followed on that road. Russell's first explanation (*1903*, p. 43) was that the set of ordinals, although linearly ordered, is not well-ordered; but this could not be maintained, in the face of the result that the initial segment up to any given ordinal is well-ordered. Jourdain saw a way out in the distinction between consistent and inconsistent multiplicities, a distinction already privately used by Cantor for several years (see below, p. 114). Soon Russell (*1905a*) questioned the existence of the set of ordinals; Zermelo took the same path (see below, p. 199), and so did subsequent workers in the field (although for some the ordinals form a class, in a technical sense of that word).

The paper contains an erroneous conception of well-ordered sets. This notion, introduced by Cantor in 1883, was somehow slow in finding its way; in August 1897, hence after the writing of Burali-Forti's paper, at the First International Congress of Mathematicians, Hadamard (*1897*, p. 201) still gave a wrong definition of well-ordered set. Since what Burali-Forti conceived to be Cantor's notion of well-ordered set was much too weak, he had to introduce his own "perfectly ordered" sets. However, the two notions do not coincide; every well-ordered set is perfectly ordered, but not conversely. Burali-Forti soon realized his oversight and wrote a short note, "On well-ordered classes", dated "Turin, October 1897" and reproduced below, in which he indicates the difference between the two notions, but does not recast his proof. In fact, in a letter to Couturat written sometime in 1905 or at the beginning of 1906 (see *Couturat 1906*), he seems to think that the contradiction revealed in his paper vanishes once the distinction between "well-ordered" and "perfectly ordered" sets is recognized. (The discussion is befogged by a number of editorial errors in the key passages of Couturat's paper (see *Poincaré 1906*, p. 304).) The conclusion, however, is quite clear and simple: Burali-Forti's argument is readily transferred to Cantor's ordinal numbers, and the antinomy stands in its full force. The crux of the argument is given by Cantor in his letter to Dedekind in 1899 (below, p. 115). On Burali-Forti's oversight one can consult *Couturat 1906*, *Poincaré 1906*, and *Young 1929*.

The reader should not have any difficulty with the notation used in the present paper, especially if he has read *Peano 1889*, reproduced above, pp. 85–97. It should perhaps be mentioned that "$\overline{h\,\varepsilon}\{\cdots\}$" denotes the class of the h's such that $\{\cdots\}$ (see *Peano 1894*, p. 20).

The translation is by the editor, and it is printed here with the kind permission of the Circolo matematico di Palermo.

The principal object of this note is to prove that there actually exist *transfinite numbers*[1] (or *order types*) a and b such that a is not equal to b, not smaller than b, and not larger than b.

On the basis of results already known, we could prove in a few words what we have just stated; but to remove any doubt from the reader's mind as to the validity of our proof we have deemed it necessary to establish exactly (in §§ 1–8) the meaning of the terms we use,[2] repeating—perhaps in a somewhat different form—matters already presented by us in these *Rendiconti* (*1894*).

§ 1. *Order of the elements of a class.* Let u be a class. We say that h is an "order of the u's" if:[3]

(a) h is a correspondence between the elements of u and classes formed of elements of u;

(b) The correspondence h is *transitive*; that is, if x, y, and z are elements of u, if x is an element of the class hy, and if y is an hz, then x is an hz;

(c) If x and y are elements of u, then it cannot be that, at the same time, x is an hy and y an hx;

(d) If x and y are arbitrary elements of u, then always x is identical with y, or x is an hy, or y is an hx.

In symbols, writing Ord u for "order of the u's", we have

1. $u \mathbin{\varepsilon} \mathrm{K} .\supset. \mathrm{Ord}\, u = (\mathrm{K}u)\mathrm{f}u \cap \overline{h\,\varepsilon}\, \{x, y, z \mathbin{\varepsilon} u . x \mathbin{\varepsilon} hy . y \mathbin{\varepsilon} hz .\supset_{x,y,z}. x \mathbin{\varepsilon} hz .\therefore. x, y \mathbin{\varepsilon} u . x \mathbin{\varepsilon} hy . y \mathbin{\varepsilon} hx .=_{x,y} \Lambda .\therefore. x, y \mathbin{\varepsilon} u .\supset_{x,y}: x = y .\cup. x \mathbin{\varepsilon} hy .\cup. y \mathbin{\varepsilon} hx\}$ (Def).

If, x being an element of u, we call the elements of the class hx "successors of x in u with respect to the order h", we easily see that the properties (a)–(d) attributed to h are those ordinarily attached to the idea of *order*.

We invert the relation $y \mathbin{\varepsilon} hx$ (y is a successor of x) by writing $x \mathbin{\varepsilon} h|y$, and from now on $h|y$ can be used for "predecessor of y in u with respect to the order h". We call the element x of u that has no predecessors ($h|x = \Lambda$) "the *first* u with respect

[1] *Cantor 1895*; Italian translation, *1895a*. For the previous works of Cantor on this subject consult the *Lista bibliografica* of Vivanti appended to Part VI of the *Formulaire* published by the *Rivista di matematica* ⟦*Vivanti 1893*⟧.

[2] For the meaning of the terms *class, correspondence, finite class,* . . . we refer the reader to *Burali-Forti 1896*.

[3] For further explanations regarding the matters contained in §§ 1–8 see *Burali-Forti 1894*.

to the order h", and we call the element y of u that has no successors ($hy = \Lambda$) "the *last* u with respect to the order h". If x is an element of u, we use the term "*immediate successor* of x" for the element y of u that is a successor of x and is such that there exists no u that is at the same time a successor of x and a predecessor of y ($y \mathrel{\varepsilon} hx . u \cap hx \cap h|y = \Lambda$). The first u or the last u or the immediate successor of an arbitrary element of u need not exist; but if it exists, it is uniquely determined (which readily follows from conditions (a)–(d)).

§ 2. *Ordered class.*[4] Let u and v be arbitrary classes; let h be an order of the u's and k an order of the v's. We write (u, h) for "class u whose elements are arranged in the order h", or for "class u ordered by criterion h".

It does not seem possible at present to define (u, h) by putting (u, h) equal (*identical*) to a complex of signs having a known meaning. We define (u, h) as an *abstract object*. stating the condition under which two ordered classes have to be considered identical. We write

$$2. \quad u, v \mathrel{\varepsilon} \mathrm{K} . h \mathrel{\varepsilon} \mathrm{Ord}\, u . k \mathrel{\varepsilon} \mathrm{Ord}\, v \mathrel{.\supset.\!\cdot} (u, h) = (v, k) \mathrel{.=:} u = v . h = k \qquad \text{(Def)};$$

that is, we assume that (u, h) is identical with (v, k) if u is identical with v and the correspondence h is identical with the correspondence k.

If u is a class containing just one element x ($u \mathrel{\varepsilon} \mathrm{Un}$), then every order h of the u's is such that $hx = \Lambda$. We can henceforth agree to indicate by the sign Λ every order of u, and consequently (u, Λ) will represent the class u ordered.

Writing Ko for "ordered class", we put

$$3. \quad \mathrm{Ko} = \overline{(u, h)} \mathrel{\varepsilon} \{u \mathrel{\varepsilon} \mathrm{K} . h \mathrel{\varepsilon} \mathrm{Ord}\, u\} \qquad \text{(Def)};$$

that is, we say that (u, h) is an ordered class if u is a class and h is an order of the u's.

§ 3. *Perfectly ordered class.* Let u be a nonempty class (one that actually contains some elements) and let h be an order of the u's.

We say that (u, h) is a "*perfectly ordered* class" if:

(a) There exists an element of u that occupies the first place with respect to the order h;

(b) Every element of u that has successors has an immediate successor;

(c) For any arbitrary element x of u either x has no immediate predecessor, or there exists a predecessor y of x that has no immediate predecessor and is such that the u's that are at the same time successors of y and predecessors of x form a *finite class*.

In symbols, writing Kpo for "perfectly ordered class", we have

$$4. \quad \mathrm{Kpo} = \mathrm{Ko} \cap \overline{(u, h)} \mathrel{\varepsilon} \{u \mathrel{-\!=} \Lambda :\cdot\!: x \mathrel{\varepsilon} u . h|x = \Lambda \mathrel{-\!=_x} \Lambda :\cdot\!: x \mathrel{\varepsilon} u . hx \mathrel{-\!=} \Lambda \mathrel{.\supset_x}: y \mathrel{\varepsilon} hx . u \cap hx \cap h|y = \Lambda \mathrel{.-\!=_y} \Lambda :\cdot\!: x \mathrel{\varepsilon} u \mathrel{.\supset_x ::} y \mathrel{\varepsilon} h|x . u \cap hy \cap h|x = \Lambda \mathrel{.=_y} \Lambda \mathrel{.\!\cdot\!.\cup.\!\cdot\!.} y \mathrel{\varepsilon} h|x : m \mathrel{\varepsilon} h|y . u \cap hm \cap h|y = \Lambda \mathrel{.=_m} \Lambda : (u \cap hy \cap h|x) \mathrel{\varepsilon} \mathrm{Kfin} \mathrel{:-\!=_y} \Lambda\}^{5} \qquad \text{(Def)}.$$

[4] Cantor (*1895*, p. 496 [[or *1932*, p. 296]]) states: "We say that a set M is *simply ordered* if there exists between its elements a definite *order by rank* (*Rangordnung*) according to which, for any two elements m_1 and m_2, one occupies the *lower* rank and the other the *higher* rank, and this in such a way that, if of three elements, m_1, m_2, and m_3, m_1 is lower in rank than m_2 and m_2 lower than m_3, then m_1 is also lower in rank than m_3."

[5] The definition just given appears in a somewhat complicated symbolic form because we did not want to introduce signs for *immediate predecessor* or *immediate successor* of x, for *first* or *last* u, and so on. It would be convenient to introduce such signs if we wanted to give a complete treatment of the theory of ordered classes.

G. Cantor calls an ordered class satisfying conditions (*a*) and (*b*) a "well-ordered class". There is a marked difference between a well-ordered class and a perfectly ordered class. Let, for example, $a_1, a_2, a_3, \ldots$ and $b_1, b_2, b_3, \ldots$ be the elements of two denumerable classes; if we consider the ordered class

$$a_1, a_2, a_3, \ldots, b_3, b_2, b_1,$$

we easily see that it is a well-ordered, but not a perfectly ordered, class, while

$$a_1, a_2, a_3, \ldots, b_1, b_2, b_3, \ldots$$

is a perfectly ordered, hence a well-ordered, class.

§ 4. *Relations between ordered classes.* Let (u, h) and (v, k) be ordered classes. We write $(v, k)\mathrm{f}(u, h)$ for "ordered correspondence between the u's and the v's ordered by criteria h and k". We say that f is such a correspondence if f is a *single-valued and reciprocal* correspondence between the u's and the v's and is such that, if x and y are elements of u and x is a successor of y with respect to h, fx is a successor of fy with respect to k. In symbols we have:

5. $(u, h), (v, k) \,\varepsilon\, \mathrm{Ko} \,.\!\supset\!.\, (v, k)\mathrm{f}(u, h) = (vfu)\mathrm{rcp} \cap \overline{f \,\varepsilon}\, \{x, y \,\varepsilon\, u \,.\, x \,\varepsilon\, hy \,.\!\supset_{x,y}\!.\, fx \,\varepsilon\, k(fy)\}$ (Def).

We say that (u, h) is *equivalent*[6] to (v, k), and we write $(u, h) \sim (v, k)$, if an ordered correspondence can be established between (u, h) and (v, k). We say that (u, h) is *smaller* than (v, k), and we write $(u, h) < (v, k)$, if there exists a class w contained in v such that $(u, h) \sim (w, k)$.[7] In symbols we have:

6. $(u, h), (v, k) \,\varepsilon\, \mathrm{Ko} \,.\!\supset\!:\, (u, h) \sim (v, k) \,.\!=\!.\, (v, k)\mathrm{f}(u, h) \,-\!\!=\, \Lambda$ (Def);
7. $\ldots \,.\!\supset\!.\!\cdot\!.\, (u, h) < (v, k) \,.\!=\!:\, w \,\varepsilon\, \mathrm{K}v \,.\, (u, h) \sim (w, k) \,.\!-\!\!=_w \Lambda$ (Def).

From propositions 6 and 7 the following are easily deduced:

$$(u, h), (v, k), (w, l) \,\varepsilon\, \mathrm{Ko} \,.\!\supset\!.\!\cdot\!.$$

8. $(u, h) \sim (u, h)$,
9. $(u, h) \sim (v, k) \,.\!=\!.\, (v, k) \sim (u, h)$,
10. $(u, h) \sim (v, k) \,.\, (v, k) \sim (w, l) \,.\!\supset\!.\, (u, h) \sim (w, l)$,
11. $(u, h) \sim (v, k) \,.\!\supset\!.\, (u, h) < (v, k)$,
12. $(u, h) \sim (v, k) \,.\, (v, k) < (w, l) \,.\!\supset\!.\, (u, h) < (w, l)$,
13. $(u, h) < (v, k) \,.\, (v, k) < (w, l) \,.\!\supset\!.\, (u, h) < (w, l)$,
14. $(u, h) \sim (v, k) \,.\!\supset\!.\, u \sim v$,
15. $(u, h) < (v, k) \,.\!\supset\!.\, u < v$.[8]

§ 5. *Operation* S. Let (u, h) and (v, k) be ordered classes such that u and v have no common elements. By the symbol $(u, h)\mathrm{S}(v, k)$, which we read "ordered class (u, h) *followed* by ordered class (v, k)", we shall denote the class $u \cup v$ (logical sum of u and v) when ordered by a criterion l such that, x and y being elements of $u \cup v$, the statement

[6] Cantor says that the two classes are *similar* (ähnlich). We keep the term *equivalent* because this term, having been defined for two classes (see *Burali-Forti 1896*), is now without meaning for two ordered classes. The same can be said about the sign < introduced by us in the same paper.

[7] We observe that every order of the v's is also an order of every class contained in v (§ 1); hence, if w is a part of v, then (w, k) is also an ordered class.

[8] See *Burali-Forti 1896*.

$x \; \varepsilon \; ly$ amounts to saying: x and y are elements of u and $x \; \varepsilon \; hy$, or x and y are elements of v and $x \; \varepsilon \; ky$, or x is a v and y is a u. In symbols we have

16. $(u, h), (v, k) \; \varepsilon \; \mathrm{Ko} . u \cap v = \Lambda . \supset . (u, h)\mathrm{S}(v, k) = \iota[\overline{(u \cup v, l) \; \varepsilon} \{l \; \varepsilon \; \mathrm{Ord} \; (u \cup v) :\cdot : x, y \; \varepsilon \; (u \cup v) . \supset_{x,y} :: x \; \varepsilon \; ly \; .=.\!\cdot . \; x, y \; \varepsilon \; u . x \; \varepsilon \; hy \; : \cup : \; x, y \; \varepsilon \; v . x \; \varepsilon \; ky \; : \cup : \; x \; \varepsilon \; v . y \; \varepsilon \; u\}]$ (Def).

It is rather easy to prove that $(u, h)\mathrm{S}(v, k)$ is a uniquely determined ordered class. In an analogous way we have

17. $(u, h), (v, k) \; \varepsilon \; \mathrm{Kpo} . u \cap v = \Lambda . \supset . (u, h)\mathrm{S}(v, k) \; \varepsilon \; \mathrm{Kpo}$.

§ 6. *Order type*. If (u, h) is an ordered class, we write $\mathrm{T}'(u, h)$ for "order type of the u's ordered by criterion h". We regard $\mathrm{T}'(u, h)$ as an *abstract object* that is a function of (u, h) and that (u, h) has in common with all ordered classes equivalent to itself; that is, we put

18. $(u, h), (v, k) \; \varepsilon \; \mathrm{Ko} . \supset : \mathrm{T}'(u, h) = \mathrm{T}'(v, k) \; .=. \; (u, h) \sim (v, k)$ (Def).

Writing T for "order type", we put

19. $\mathrm{T} = \overline{x \; \varepsilon} \{(u, h) \; \varepsilon \; \mathrm{Ko} . x = \mathrm{T}'(u, h) \; .-=_{(u,h)} \Lambda\}$ (Def).

From propositions 8, 9, and 10 it immediately follows that for order types the equality defined by proposition 18 has the properties of being *reflexive*, *symmetric*, and *transitive*; that is, if a and b are arbitrary order types, we have

$$a = a; \quad a = b \; .=. \; b = a; \quad a = b . b = c . \supset . a = c.$$

§ 7. *Greater and smaller order types*. Let (u, h) and (v, k) be ordered classes. We say that $\mathrm{T}'(u, h)$ is *smaller* than $\mathrm{T}'(v, k)$, or that $\mathrm{T}'(v, k)$ is *greater* than $\mathrm{T}'(u, h)$, and we write

$$\mathrm{T}'(u, h) < \mathrm{T}'(v, k), \qquad \text{or} \qquad \mathrm{T}'(v, k) > \mathrm{T}'(u, h),$$

if there exists a part v_1 of v such that $(u, h) \sim (v_1, k)$ but no part u_1 of u such that $(u_1, h) \sim (v, k)$. By virtue of proposition 7 we have

20. $(u, h), (v, k) \; \varepsilon \; \mathrm{Ko} . \supset .\!\cdot . \; \mathrm{T}'(u, h) < \mathrm{T}'(v, k) :=: \mathrm{T}'(v, k) > \mathrm{T}'(u, h) :=: (u, h) < (v, k) . (v, k) -\!< (u, h)$. (Def).

We have the following propositions:

$$a, b, c \; \varepsilon \; \mathrm{T} . \supset :$$

21. $a = b . a < b \; .=. \; \Lambda$,
22. $a < b . a > b \; .=. \; \Lambda$,
23. $a = b . b < c . \supset . a < c$,
24. $a < b . b < c . \supset . a < c$.

Propositions 21 and 22 express the fact that, of the three cases $a = b$, $a > b$, $a < b$, two cannot occur at the same time; they are immediate consequences of the propositions of § 4. Proposition 23 is also an immediate consequence of these propositions. We now prove proposition 24. Let (u, h), (v, k), and (w, l) be ordered classes. From proposition 13 we obtain

$$(u, h) < (v, k) . (v, k) < (w, l) . (w, l) < (u, h) . \supset : (v, k) < (u, h) . \cup . (w, l) < (v, k);$$[9]

transferring the two terms of the thesis into the hypothesis and the third factor of the hypothesis into the thesis, we have

$$(u, h) < (v, k) . (v, k) -\!< (u, h) . (v, k) < (w, l) . (w, l) -\!< (v, k) . \supset . (w, l) -\!< (u, h);$$

[9] Because, if a, b, and c are propositions, we have $b . \supset . b \cup c$ and therefore $a \supset b . \supset : a . \supset . b \cup c$.

affixing to the thesis the factor $(u, h) < (w, l)$, which is a consequence of the hypothesis, and recalling proposition 20, we obtain proposition 24.

However, we cannot prove that

$A.\ a, b \,\varepsilon\, \mathrm{T} .\supset: a = b .\cup. a < b .\cup. a > b,$

that is, that for two arbitrary types one of the three cases $a = b$, $a < b$, $a > b$ must always occur. Proceeding as in our *1896a*, we reduce proposition A to the logical product of the following two:

I. $(u, h), (v, k) \,\varepsilon\, \mathrm{Ko} .\supset: (u, h) < (v, k) .\cup. (v, k) < (u, h),$

II. $(u, h), (v, k) \,\varepsilon\, \mathrm{Ko}.(u, h) < (v, k).(v, k) < (u, h) .\supset. (u, h) \sim (v, k).$

If we assume for a moment that proposition A, or at least proposition II, is true, it follows from proposition 20 after some simple logical transformations that

$B.\ (u, h), (v, k) \,\varepsilon\, \mathrm{Ko} .\supset\therefore \mathrm{T}‘(u, h) < \mathrm{T}‘(v, k) .=: (u, h) < (v, k).(u, h) -\!\sim (v, k);$

that is, $\mathrm{T}‘(u, h)$ is smaller than $\mathrm{T}‘(v, k)$[10] if they are not equal and there exists a part v_1 of v such that $(u, h) \sim (v_1, k)$. It is in this form that Cantor (*1887*) gives the definition of the relation containing the sign $<$. Our proposition 20 and proposition B are equivalent if proposition A, or at least proposition II, holds. If we assumed proposition B as a definition, we would not know how to prove proposition 24 unless we assumed proposition II.

From proposition 20 and proposition B it follows quite readily that

$C.\ (u, h), (v, k) \,\varepsilon\, \mathrm{Ko}.(u, h) < (v, k) .\supset. \mathrm{T}‘(u, h) \leqq \mathrm{T}‘(v, k).$

§ 8. *Ordinal numbers and sum.* We say "ordinal number",[11] and we write No, for "order type of a perfectly ordered class";

25. $\mathrm{No} = \mathrm{T}‘\mathrm{Kpo}.$ (Def).

We put

26. $1 = \iota\, \mathrm{T}‘\{\mathrm{Ko} \cap \overline{(u, h) \,\varepsilon}\, (u \,\varepsilon\, \mathrm{Un})\}$ (Def),

that is, we say that *one* is the order type of the ordered classes (u, h) such that u contains just one element. It immediately follows that

27. $1 \,\varepsilon\, \mathrm{No}.$

We define the sum of two order types, and therefore also that of two ordinal numbers, by writing

28. $(u, h), (v, k) \,\varepsilon\, \mathrm{Ko}.u \cap v = \Lambda .\supset. \mathrm{T}‘(u, h) + \mathrm{T}‘(v, k) = \mathrm{T}‘\{(u, h)\mathrm{S}(v, k)\}$ (Def).

If we observe that two equivalent ordered classes followed by the same class are still equivalent, we have the result that the sum now defined is independent of u, v, h, k; moreover, from proposition 17 it follows that

29. $a, b \,\varepsilon\, \mathrm{No} .\supset. a + b \,\varepsilon\, \mathrm{No},$

and this proposition still holds if we put T for No.[12]

[10] [Here the text reads, "il $\mathrm{T}‘(u, h)$ è minore del $T‘(u, h)$", which is a misprint.]

[11] Cantor in fact says *ordinal number* for "order type of a well-ordered class". However, the properties of our ordinal numbers seem to us to agree with those of Cantor's ordinal numbers. Besides, this does not affect our conclusions, for which it suffices to prove that our T are precisely Cantor's order types.

[12] It is understood that we have to assume the existence of at least one infinite class (see *Burali-Forti 1896*). Besides, if there exist no infinite classes, the *order types* are the *integers*.

§ 9. *Consequences of proposition A*. Let us assume that proposition A is true and see what propositions logically follow from it.

30. $a \,\varepsilon\, \text{No} \,.\!\supset.\, a + 1 > a.$ (A)

Proof. Let (u, h) be a perfectly ordered class having a for its ordinal number and let v be a class containing just one element. If we write

$$P = (u, h) \quad \text{and} \quad Q = (u, h)\text{S}(v, \Lambda),$$

we have $P < Q$. Proposition 30 will be proved if, using proposition B, we can prove that $P \mathrel{-\!\!\sim} Q$. If P has no last element, then $P \mathrel{-\!\!\sim} Q$, because Q has a last element, namely $\bar{\iota}v$. If P has a last element, x, then there exists (§ 3) an element y of u that has no immediate predecessor and is such that the elements $(u \cap hy) \cup \iota y$ form a finite class of n elements; then, if $P \sim Q$, the classes P_1 and Q_1 that we obtain when we remove the last n elements from P and Q will also be equivalent; but this is absurd, because P_1 has no last element and Q_1 has y as last element.[13] Hence $P \mathrel{-\!\!\sim} Q$, and consequently proposition 30 holds. If, following Cantor, we call the order type of a well-ordered class an ordinal number, we observe that proposition 30 is not true in general, as immediately follows from the example given at the end of § 3.

31. $a \,\varepsilon\, \text{No} \,.\!\supset\!:\, x \,\varepsilon\, \text{No} . a < x . x < a + 1 \,.\!=_x \Lambda$ (A).

This proposition—which says that there exists no ordinal number between a and $a + 1$—readily follows from proposition C.

If a is an ordinal number, we can denote by $\bar{\varepsilon} > a$ the ordinal numbers greater than a, because the relation $x > a$ is equivalent to the relation $x \,\varepsilon\, (\bar{\varepsilon} >)\, a$. Consequently, if $\bar{\varepsilon} >$ is an order criterion for ordinal numbers, $(\text{No}, \bar{\varepsilon} >)$ denotes the class of ordinal numbers thus ordered.

We now want to prove that

32. $(\text{No}, \bar{\varepsilon} >) \,\varepsilon\, \text{Kpo}$ (A).

Proof. Since $(\text{No}, \bar{\varepsilon} >)$ is an ordered class because conditions (b) and (c) of § 1 hold by virtue of propositions 24, 21, and 22, condition (d) is equivalent to proposition A. Propositions 27, 30, and 31 prove that conditions (a) and (b) of § 3 are satisfied by the ordered class $(\text{No}, \bar{\varepsilon} >)$. Let now (u, h) be a perfectly ordered class whose ordinal number is a; since (u, h) satisfies condition (c) of § 3, either a has no immediate predecessor (that is, by the criterion $\bar{\varepsilon} >$), or there exists an ordinal number x smaller than a that has no immediate predecessor and is such that we obtain a if we repeat the operation $+1$ a finite number of times on x. Hence $(\text{No}, \bar{\varepsilon} >)$ also satisfies condition (c) of § 3.

Since $(\text{No}, \bar{\varepsilon} >)$ is a perfectly ordered class, we can put

33. $\Omega = \text{T}\text{‘}(\text{No}, \bar{\varepsilon} >)$ (Def),

and we have

34. $\Omega \,\varepsilon\, \text{No}.$ (A).

We can now prove that

35. $a \,\varepsilon\, \text{No} \,.\!\supset.\, a \lesseqgtr \Omega$ (A).

[13] It is understood that we make use here of the *principle of mathematical induction*, proved in *Burali-Forti 1896*.

Proof. Let (u, h) be a perfectly ordered class having ordinal number a. Let (v, k) be the ordered class obtained by the following specifications: for any element x of u, let the class $u \cap -hx$ be an element of v, and let v have only such elements (hence v turns out to be a class of classes); if x and y are elements of u and if $x \varepsilon hy$, let us put $(u \cap -hx) \varepsilon k(u \cap -hy)$.

It immediately follows that $(u, h) \sim (v, k)$ and therefore (v, k) is a perfectly ordered class having ordinal number a. From the preceding propositions we also readily obtain the result that the class of the ordinal numbers of the elements of v, this class being ordered by the criterion k, is a class of ordinal numbers smaller than a, perfectly ordered in increasing order, and having ordinal number a; but such a class is smaller than $(\text{No}, \bar{\varepsilon} >)$ and thus proposition 35 follows from proposition C.

§ 10. *Conclusion.* If we write Ω for a in proposition 30 and $\Omega + 1$ for a in proposition 35, we have, by virtue of propositions 34, 26, and 29,

$$\Omega + 1 > \Omega \qquad \text{and} \qquad \Omega + 1 \leqq \Omega,$$

and these, by propositions 21 and 22, turn out to be contradictory.

Hence if we assume proposition A, we are led to an absurdity, and therefore it has been rigorously proved that there exist at least two order types a and b (and there certainly exist some among the ordinal numbers) such that a is not equal to b, is not larger than b, and is not smaller than b.

It is therefore impossible to order the order types in general, or even the ordinal numbers in particular; that is to say, the order types cannot provide a *standard* class for the ordered classes, as the class of integers, ordered according to magnitude, does for the finite classes and the denumerable class ⟦that is, the class of order type ω⟧.[14] It seems that the order types thus necessarily fall short of one of their most important objectives.

ON WELL-ORDERED CLASSES

(1897a)

In my note *A question on transfinite numbers* I said that, according to Cantor, a class u is *well-ordered* if it satisfies the following two conditions:

(a) There is in u a first element;

(b) Every element of u that has successors has an immediate successor.

In a recent paper (*1897*) and also, as I was able to verify, in volume 21 of *Mathematische Annalen* (*1883*, p. 548) Cantor, to define well-ordered classes, adds to conditions (a) and (b) the following:

(c) If u_1 is a class included in u and such that there exist in u elements greater than any element of u_1, then there exists an element a of u such that there exists no element smaller than a and greater than every element of u_1.

[14] Nor can the class of order types that are not ordinal numbers form the standard ordered class for the ordered classes that are not perfectly ordered. It can in fact be proved rather easily that, for the T there are —No, either proposition A is not true, or the following proposition is not true: "Given a Ko—Kpo there exist two T—No, x and y, such that the order types not smaller than x and not greater than y, ordered in increasing order, form an ordered class equivalent to the given class". Naturally this proposition must be verified together with A if the class of the T—No is to be a standard class for the Ko—Kpo.

I thought it advisable to indicate explicitly this involuntary omission of mine, although I made use neither of the well-ordered classes defined by Cantor nor of the ordered classes that satisfy conditions (*a*) and (*b*), but only of the classes that I called *perfectly ordered* classes and that, besides satisfying conditions (*a*) and (*b*), are also such that:

(c') For any arbitrary element x of u, either x has no immediate predecessor, or there exists a predecessor y of x that has no immediate predecessor and is such that there is a finite number of u's that are at the same time successors of y and predecessors of x.

It readily follows that every well-ordered class is also perfectly ordered, but not conversely. The reader can check which propositions in the note of mine mentioned above are verified also by the well-ordered classes.

Letter to Dedekind

GEORG CANTOR

(*1899*)

In his edition of Cantor's collected papers (*1932*), Zermelo included a few letters exchanged between Cantor and Dedekind and left unpublished until then. The one reproduced below in translation is dated "Halle, 28 July 1899".

In the first part of the letter Cantor deals with the same contradiction as Burali-Forti considered in the paper printed immediately above, that engendered by the ordinal of the multiplicity of all ordinals. The contradiction leads Cantor to abandon, not the well-ordering of the multiplicity, but its sethood. Multiplicities become divided into "consistent" and "inconsistent" multiplicities, the former alone being called sets. This partition prefigures the distinction between sets and classes, introduced (under different names) by von Neumann in 1925, a quarter of a century later (see below, pp. 393–394). However, Cantor's criterion for the distinction remains imprecise: a multiplicity is a set if we can consider it, without contradiction, as *one* object. The idea will become sharply defined when it will be specified that a multiplicity is a set whenever it is an element of another multiplicity. The distinction between consistent and inconsistent multiplicities had already been introduced by Schröder (*1890*, p. 213), a multiplicity being consistent if its elements are compatible ("verträglich") with each other, and inconsistent if they are not. It is remarkable that Schröder introduced this distinction independently of the paradoxes, still unknown then in their modern form.

One of Cantor's statements about multiplicities ("Two equivalent multiplicities either are both 'sets' or are both inconsistent") can be considered an early formulation of the axiom of replacement, introduced by Fraenkel and Skolem in 1922 (see below, p. 291).

Cantor and Dedekind exchanged letters over a long period (1872–1899), and their correspondence was published by Emmy Noether and Jean Cavaillès (*1937*).

In the text below interpolations in square brackets are by Zermelo.

Cantor's letter was translated by Stefan Bauer-Mengelberg and the editor, and the translation is printed here with the kind permission of Springer Verlag.

. . . As you know, many years ago I had already arrived at a well-ordered sequence of cardinalities ⟦Mächtigkeiten⟧, or transfinite cardinal numbers, which I call "alephs":

$$\aleph_0, \aleph_1, \aleph_2, \ldots, \aleph_{\omega_0}, \ldots.$$

$\aleph_0$ means the cardinality of the sets "denumerable" in the usual sense, $\aleph_1$ is the next greater cardinal number, $\aleph_2$ is the next greater still, and so on; $\aleph_{\omega_0}$ is the one next following (that is, next greater than) all the $\aleph_\nu$ and equals

$$\lim_{\nu \to \omega_0} \aleph_\nu,$$

and so on.

The big question was whether, besides the alephs, there were still other cardinalities of sets; for two years now I have been in possession of a proof that there are no others, so that, for example, the arithmetic linear continuum (the totality of all real numbers) has a definite aleph as its cardinal number.

If we start from the notion of a definite multiplicity [[Vielheit]] (a system, a totality) of things, it is necessary, as I discovered, to distinguish two kinds of multiplicities (by this I always mean *definite* multiplicities).

For a multiplicity can be such that the assumption that *all* of its elements "are together" leads to a contradiction, so that it is impossible to conceive of the multiplicity as a unity, as "one finished thing". Such multiplicities I call *absolutely infinite* or *inconsistent multiplicities.*

As we can readily see, the "totality of everything thinkable", for example, is such a multiplicity; later still other examples will turn up.

If on the other hand the totality of the elements of a multiplicity can be thought of without contradiction as "being together", so that they can be gathered together into "*one* thing", I call it a *consistent multiplicity* or a "set". (In French and in Italian this notion is aptly expressed by the words "ensemble" and "insieme".)

Two equivalent multiplicities either are both "sets" or are both inconsistent.

Every submultiplicity [[Teilvielheit]] of a set is a set.

Whenever we have a set of sets, the elements of these sets again form a set.

If a set M is given, I call the general notion that applies to it and to all sets equivalent to it, and to these alone, its *cardinal number* or also its *cardinality*, and I denote it by $\mathfrak{m}$. I then arrive at the system of all cardinalities—which will later turn out to be an *inconsistent* multiplicity—in the following way.

A multiplicity is said to be "simply ordered" if between its elements there exists a rank order such that, for any two of its elements, one is the earlier and the other the later, and that, for any three of its elements, one is the earliest, another is the middle one, and the remaining one is the last by rank among them.

If a simply ordered multiplicity is a *set*, then by its *type* μ I understand the general notion that applies to it and to all ordered sets *similar* to it, and to these alone. (I use the notion of *similarity* in a more restricted sense than you do;[1] I say that two simply ordered multiplicities are *similar* if they can be brought into a one-to-one relation such that the rank order of corresponding elements is the same in both.)

A multiplicity is said to be *well-ordered* if it satisfies the condition that every *submultiplicity* of it has a *first* element; I call such a multiplicity a "sequence" for short.

Every part [[Teil]] of a sequence is a sequence.

If now a sequence F is a set, I call the type of F its *ordinal number* or, more briefly, its *number*; thus, when in what follows I speak simply of numbers, I shall have in mind only ordinal numbers, that is, types of well-ordered sets.

I now consider the system of *all numbers* and denote it by Ω.

[1] [[Dedekind uses "ähnlich" in the sense of "equivalent".]]

I proved (*1897*, p. 216) that, of two distinct numbers α and β, one is always the smaller, the other the greater, and that, if for three numbers we have $\alpha < \beta$ and $\beta < \gamma$, we also have $\alpha < \gamma$.

Ω is therefore a simply ordered system.

But it also follows easily from the theorems on well-ordered sets proved in § 13 that every multiplicity of numbers, that is, every part of Ω, contains a *least* number.

Hence the system Ω, when naturally ordered according to magnitude, forms a sequence.

If we then add 0 to this sequence as an element—putting it first, of course—we obtain a sequence Ω',

$$0, 1, 2, 3, \ldots, \omega_0, \omega_0 + 1, \ldots, \gamma, \ldots,$$

in which, as we can readily see, *every* number is the *type* of the *sequence of all elements preceding it* (including 0). (The sequence Ω has this property only from $\omega_0 + 1$ on ⟦in fact, from ω_0 on⟧.)

Ω' *cannot* be a *consistent* multiplicity (and therefore neither can Ω); if Ω' were consistent, then, since it is a well-ordered set, there would correspond to it a number δ greater than all numbers of the system Ω; but the number δ also occurs in the system Ω, because this system contains *all* numbers; δ would thus be greater than δ, which is a contradiction. Therefore

A. *The system Ω of all numbers is an inconsistent, absolutely infinite multiplicity.*

Since the *similarity* of well-ordered sets establishes at the same time their *equivalence*, to every number γ there corresponds a definite cardinal number $\aleph(\gamma) = \bar{\gamma}$, namely, the cardinal number of any well-ordered set whose type is γ.

The cardinal numbers that correspond in this sense to the *transfinite* numbers of the system Ω I call "alephs", and the *system of all alephs* is denoted by ת (*tav*, the last letter of the Hebrew alphabet).

I call the system of all numbers γ corresponding to one and the same cardinal number $\mathfrak{c}$ a "number class", and, more specifically, the number class $Z(\mathfrak{c})$. We readily see that in every number class there occurs a least number γ_0 and that there is a number γ_1 falling outside of $Z(\mathfrak{c})$ such that the condition

$$\gamma_0 \leqq \gamma < \gamma_1$$

is equivalent to the fact that the number γ belongs to the number class $Z(\mathfrak{c})$. Every number class is therefore a definite "segment" of the sequence Ω.[2]

Certain numbers of the system Ω form, each one by itself, a number class; they are the *finite* numbers, $1, 2, 3, \ldots, \nu, \ldots$, to which correspond the various "finite" cardinal numbers, $\bar{1}, \bar{2}, \bar{3}, \ldots, \bar{\nu}, \ldots$.

Let ω_0 be the least transfinite number; I call the aleph corresponding to it $\aleph_0$, so that

$$\aleph_0 = \bar{\omega}_0;$$

$\aleph_0$ is the *least* aleph and determines the number class

$$Z(\aleph_0) = \Omega_0.$$

[2] Here we constantly use the theorem mentioned a few paragraphs above according to which *every* totality of numbers, hence *every* submultiplicity of Ω, has a *minimum*, a *least number*.

The numbers α of $Z(\aleph_0)$ satisfy the condition

$$\omega_0 \leqq \alpha < \omega_1$$

and are characterized by it; here ω_1 is the least transfinite number whose cardinal number is not equal to $\aleph_0$. If we put

$$\bar{\omega}_1 = \aleph_1,$$

then $\aleph_1$ is not only distinct from $\aleph_0$, but it is also the next greater aleph, for we can prove that there is no cardinal number at all that would lie between $\aleph_0$ and $\aleph_1$. We thus obtain the number class $\Omega_1 = Z(\aleph_1)$, which immediately follows Ω_0. It contains all numbers β that satisfy the condition

$$\omega_1 \leqq \beta < \omega_2;$$

here ω_2 is the least transfinite number whose cardinal number differs from $\aleph_0$ and $\aleph_1$.

$\aleph_2$ is the next greater aleph after $\aleph_1$; it determines the number class $\Omega_2 = Z(\aleph_2)$ that immediately follows Ω_1 and consists of all numbers γ that are $\geqq \omega_2$ and $< \omega_3$, where ω_3 is the least transfinite number whose cardinal number differs from $\aleph_0$, $\aleph_1$, and $\aleph_2$; and so on.

I would still like to stress the following:

$$\bar{\bar{\Omega}}_0 = \aleph_1, \bar{\bar{\Omega}}_1 = \aleph_2, \ldots, \bar{\bar{\Omega}}_\nu = \aleph_{\nu+1},$$

$$\sum_{\nu' = 0,1,2,\ldots,\nu} \aleph_{\nu'} = \aleph_\nu;$$

all this is easy to prove.

Among the transfinite numbers of the system Ω to which no $\aleph_\nu$ [with finite ν] corresponds as a cardinal number, there is again a least, which we call ω_{ω_0}, and with it we obtain a new aleph,

$$\aleph_{\omega_0} = \bar{\omega}_{\omega_0},$$

which is also definable by means of the equation

$$\aleph_{\omega_0} = \sum_{\nu = 1,2,3,\ldots} \aleph_\nu$$

and which we recognize as the next greater cardinal number after all the $\aleph_\nu$.

We see that this process of formation of the alephs and of the number classes of the system Ω that correspond to them is *absolutely* limitless.

B. *The system* ת *of all alephs, when ordered according to magnitude,*

$$\aleph_0, \aleph_1, \ldots, \aleph_{\omega_0}, \aleph_{\omega_0+1}, \ldots, \aleph_{\omega_1}, \ldots,$$

forms a sequence that is similar to the system Ω *and therefore likewise inconsistent, or absolutely infinite.*

The question now arises whether *all transfinite cardinal numbers* are contained in the system ת. In other words, is there a *set* whose cardinality is *not an aleph*?

This question is to be answered *negatively*, and the reason for this lies in the *inconsistency* that we discerned in the systems Ω and ת.

Proof. If we take a definite multiplicity V and assume that *no aleph* corresponds to it *as its cardinal number*, we conclude that V must be *inconsistent*.

For we readily see that, on the assumption made, the whole system Ω is projectible into the multiplicity V, that is, there must exist a submultiplicity V' of V that is equivalent to the system Ω.[3]

V' is *inconsistent* because Ω is, and the same must therefore be asserted of V. [See page 444 ⟦of *Cantor 1932*, or 115 of the present volume⟧.]

Accordingly, every transfinite *consistent multiplicity*, that is, every transfinite set, must have a *definite aleph* as its cardinal number. Hence

C. *The system* ת *of all alephs is nothing but the system of all transfinite cardinal numbers.*

All sets, and in particular all "*continua*", are therefore "*denumerable*" in an *extended sense.*

Furthermore C makes it clear that I was right when I stated (*1895*, p. 484) the theorem:

"If $\mathfrak{a}$ and $\mathfrak{b}$ are arbitrary cardinal numbers, then $\mathfrak{a} = \mathfrak{b}$ or $\mathfrak{a} < \mathfrak{b}$ or $\mathfrak{a} > \mathfrak{b}$."

For, as we have seen, these relations of magnitude obtain between the alephs.

[3] [It is precisely at this point that the weakness of the proof sketched here lies. It has *not* been proved that the entire number sequence Ω would necessarily be "projectible" into every multiplicity V that has no aleph as its cardinal number. Cantor apparently thinks that successive and arbitrary elements of V are assigned to the numbers of Ω in such a way that every element of V is used only *once*. *Either* this procedure would of necessity come to an end once all elements of V had been exhausted, and then V would be mapped onto a *segment* of the number sequence and its cardinality would be an aleph, contrary to the assumption, *or* V would remain inexhaustible, hence contain a constituent part that is equivalent to all of Ω and therefore inconsistent. Thus the intuition of time is applied here to a process that goes beyond all intuition, and a fictitious entity is posited of which it is assumed that it could make *successive* arbitrary choices and thereby define a subset V' of V that, by the conditions imposed, is precisely *not* definable. Only through the use of the "axiom of choice", which postulates the possibility of a *simultaneous* choice and which Cantor uses unconsciously and instinctively everywhere but does not formulate explicitly anywhere, could V' be defined as a subset of V. But even then there would still remain a doubt: perhaps the proof involves "inconsistent" multiplicities, indeed possibly contradictory notions, and is logically inadmissible already because of that. It is precisely doubts of this kind that impelled the editor a few years later to base his own proof of the well-ordering theorem (*1904*) purely upon the axiom of choice without using inconsistent multiplicities.]

Logical introduction to any deductive theory

ALESSANDRO PADOA

(*1900*)

The text given here is part of an "Essai d'une théorie algébrique des nombres entiers, précédé d'une introduction logique à une théorie déductive quelconque", which was Padoa's contribution to the Third International Congress of Philosophy, held in Paris on 1–5 August 1900, a congress at which were present, among others, Poincaré, Peano, and Russell. (In the days that immediately followed the congress some of the participants were again together at the Second International Congress of Mathematicians (6–12 August), at which Hilbert delivered his famous address on unsolved mathematical problems (*1900a*); Padoa again presented his method at this congress (*1900a*).)

Padoa read his paper, dated "Rome, 1 July 1900", on 3 August 1900. His "logical introduction" deals with the relation between a system and its interpretations. In particular, it shows how to solve two symmetric problems concerning a deductive system: (1) Is a term definable by means of other terms? (2) Is a proposition derivable from other propositions? Peano already knew at that time how to deal with the second problem and had elaborated a method that was to become standard in independence proofs. (On this point see footnote 539 in *Church 1956*.) Padoa was the first to present a method for establishing the undefinability, in a given system, of a term by means of other terms.

The method is merely stated, and Padoa seems to consider it intuitively evident.[a] He does not offer a proof of its correctness, and such a proof could hardly have been undertaken then, because, first, Padoa's system (which is Peano's) is ill-defined and, second, many results requisite for such a proof were still unknown at the time.

Padoa's method somehow remained in the twilight for years. Tarski and Lindenbaum (*1926*) drew attention to it, announcing a theorem of Tarski's that would justify it. Tarski's results were subsequently published (*1934*, *1935*; see also *1956*, chap. X), and they establish that the method can be applied to any system formalized in a theory of types. McKinsey, apparently not knowing Tarski's work, gave (*1935*) a partial justification of the method and some examples of its application. Beth (*1953*, but see also *1959*, pp. 288–293) dealt with the case of first-order logic. In the case of this weaker logic the proof that Padoa's method is correct is essentially more involved because the predicate whose independence has to be ascertained is not in the range of any quantifier.

Using his previously proved interpolation lemma (*1957*), Craig gave (*1957a*; see also *1956*) a simpler proof of Beth's result. A. Robinson (*1955*) had obtained a theorem that is equivalent to the interpolation lemma and had derived Beth's

---

[a] He also seems to have overestimated its power (see *Padoa 1903*).

result from it. Schütte (*1962*) extended the interpolation lemma and Beth's theorem to intuitionistic logic; for further results on Padoa's method see Svenonius (*1959*), Craig (*1963*), Makkai (*1964*), and Chang (*1964*); de Bouvère's work (*1959*) is closely connected with Padoa's method.

The part of the "Essai" not reproduced here contains an axiomatization of the signed integers using the primitive notions "integer", "successor", and "opposite".

One point should perhaps be mentioned concerning the translation. Padoa, as a member of the Peano school, uses the symbols "P" and "Df" for "proposition" and "definition" respectively, even in the plural; he speaks, for instance, of "all particular P". In the translation the words "proposition" and "definition" have been restored.

The translation is by the editor, and it is reproduced here with the kind permission of the Société française de philosophie and the Librairie Armand Colin.

If x and y are individuals,[1] then $x = y$ or $x \neq y$. For us, these are the *only relations* that we can consider between *individuals* without transgressing the boundaries that separate *general logic* from *particular deductive theories*.[2]

As is customarily done, we assume at the beginning of any deductive theory that the symbols and the propositions of *general logic* used in the theory are known; and all *other* symbols considered in it and all *other* propositions stated in it we call *particular symbols* and *particular propositions* of the theory, respectively.

If to *define* a symbol means to express it by means of other symbols already considered and if to *prove* a proposition means to deduce it from other propositions already stated, then it is obvious, for any deductive theory, that

1. When a particular symbol is said to be or not to be *definable*, we must add: by means of the other particular symbols $a, b, c, \ldots$; and when a particular proposition is said to be or not to be *provable*, we must add: by means of the other particular propositions $a, b, c, \ldots$;[3]

2. It is not possible to define all particular symbols and to prove all particular propositions.[4]

[1] Whatever x and y may be, they are *individuals* of the class "[equal to x] or [equal to y]".

[2] We say *particular deductive theories* (but in the title and in what follows we omit the word *particular*, leaving it implied) because we could first frame a *general deductive theory* by considering only the symbols and the propositions of *general logic*.

Concerning this we read in *Peano 1899*, p. 11: "The proof of a proposition of logic has as its purpose, in general, not to assure us of its truth, but rather to reduce this mode of reasoning to simpler ones, which cannot be decomposed any further and which will be called *primitive* propositions. Leibniz has expressed several proofs in ordinary language; many can be found in *Schröder 1890*. Systems of *primitive* propositions and of symbolic proofs, as well as remarks about them, are contained in *Peano 1895, 1897, Revue de mathématiques, Burali-Forti 1894a, Padoa 1898*, and *Couturat 1899*. It would be interesting to coordinate these theories". However, we do not intend to undertake that here.

[3] Whereas, according to Blaise Pascal (*De l'esprit géométrique*, I), "there are words that cannot be defined ... because these terms designate so naturally the things they mean ... that to try to clarify them would bring obscurity rather than enlightenment", and there are propositions that are not provable because there is "nothing clearer to prove them".

[4] In the same passage Blaise Pascal, before stating "the way to prove truth and to present it to men ... as geometry does", gives "the idea of a still superior and more accomplished method, but one that men can never attain.... This true method ... would consist of two main things ... to define all terms and prove all propositions". Having arrived at what is our statement 2, he

Thus, in any deductive theory there necessarily are particular symbols that are not defined and that, accordingly, we call *undefined symbols* of the theory.

Since there is some arbitrariness in the choice of the *system of undefined symbols* of a deductive theory (because it is almost always possible to interchange the role of *defined* and *undefined* symbols for some of the particular symbols of a given theory), it seems to us necessary and natural to begin any deductive theory by explicitly and completely stating what this *system* of symbols is; it seems, however, that few authors bother about that.

(Should we remark that the undefined symbols of a theory do not necessarily represent its *simplest ideas*? Besides, could we imagine a *rule* enabling us to choose infallibly *the simpler of two ideas*?)

In any deductive theory a *symbolic definition* is nothing but the convention of replacing a sequence a of already considered symbols (that is, *logical, undefined,* or *already defined* symbols) by a new symbol x (new so far as the theory is concerned), namely, the convention of taking $x = a$ from now on.

Therefore, although symbolic definitions are very useful for the conciseness of the propositions that follow them, they are not necessary. In fact, by a simple *symbolic translation* (that is, replacing each defined symbol by the sequence of symbols that defines it) we can reduce the system of particular symbols of the theory to the system of undefined symbols.

It also follows from what we have said that, among the particular propositions of any deductive theory that are not symbolic definitions, there necessarily are propositions that are not proved; accordingly we call them *unproved propositions*.

For these propositions we can repeat the remarks made about undefined symbols, replacing the words "symbol", "defined", "idea", and "simple" by the words "proposition", "proved", "fact", and "obvious", respectively.

We have already seen that the necessary point of departure for any *deductive theory* whatsoever is a *system* of *undefined symbols* and a *system* of *unproved propositions*. Besides, it is obvious that a deductive theory has no *practical significance* if its undefined symbols and its unproved propositions do not represent (or cannot represent) *ideas* and *facts*, respectively. Thus the *psychological* origin of a deductive theory is *empirical*; nevertheless, its *logical* point of departure can be considered to be a *matter of convention*.

Indeed, during the period of *elaboration* of any deductive theory we choose the *ideas* to be represented by the undefined symbols and the *facts* to be stated by the unproved propositions; but, when we begin to *formulate* the theory, we can imagine that the undefined symbols are *completely devoid of meaning* and that the unproved propositions (instead of stating *facts*, that is, *relations* between the *ideas* represented by the undefined symbols) are simply *conditions* imposed upon the undefined symbols.

Then, the *system* of *ideas* that we have initially chosen is simply *one interpretation* of the *system* of *undefined symbols*; but from the deductive point of view this interpreta-

concludes, "Hence it seems that men are naturally and immutably incapable of treating any science whatsoever in an absolutely perfect order".

This pessimism seems unjustified to us, for, if *men can never attain* the *method* that Pascal calls *true*, it is not because they are *naturally incapable*, and so on, but because there is a *contradiction* between this *method* and the *meaning* of the words "to define" and "to prove".

tion can be ignored by the reader, who is free to replace it in his mind by *another interpretation* that satisfies the conditions stated by the *unproved propositions*. And since these propositions, from the deductive point of view, do not state *facts*, but *conditions*, we cannot consider them genuine *postulates*.

Logical questions thus become completely independent of *empirical* or *psychological* questions (and, in particular, of the *problem of knowledge*), and every question concerning the *simplicity of ideas* and the *obviousness of facts* disappears.

Still, if a *script* is adopted by *convention*, it is appropriate to state the *system* of *ideas* intended to be represented by the system of undefined symbols; or, what amounts to the same, it is appropriate to propose a *reading* of these *symbols* by words or sentences of the ordinary language (adding also *explanations* in order to make more precise the meaning of these words or sentences), provided all that is nothing but a *commentary* on the theory, very useful in facilitating the reading and the understanding of the text but completely useless from the deductive point of view; for what is necessary to the logical development of a deductive theory is not *the empirical knowledge of properties of things*,[5] but *the formal knowledge of relations between symbols*.

From now on we shall consider only deductive theories in which the meaning of undefined symbols is stated as mere *commentary* and that we may call *generic theories*.

It may be that there are several (or even infinitely many) *interpretations* of the system of *undefined* symbols that verify the system of *unproved* propositions and, therefore, all the propositions of a theory. The system of undefined symbols can then be regarded as the *abstraction* obtained from all these interpretations, and the *generic theory* can be regarded as the *abstraction* obtained from the *specialized theories* that result when in the generic theory the system of undefined symbols is successively replaced by each of the interpretations of this theory.

Thus, by means of just one argument that proves a proposition of the generic theory we prove implicitly a proposition in each of the specialized theories.[6]

We may add that a *symbolic definition* does not *individualize* the meaning of the *symbol* it defines; it simply expresses a *relation* between this symbol and the symbols already considered, a relation sufficient to *individualize* its *meaning* as soon as we choose an *interpretation* of the system of *undefined* symbols (or only of the undefined symbols that explicitly constitute the definition considered).

We can now settle completely (and, we believe, for the first time) a question of the greatest logical importance.

Let us consider an arbitrary generic theory and assume (to facilitate our study without decreasing its generality) that in the unproved propositions no particular symbols are used other than the undefined symbols.[7]

We say that *the system of undefined symbols is* IRREDUCIBLE *with respect to the system of unproved propositions* when *no symbolic definition of any undefined symbol can be*

[5] Sometimes this empirical knowledge is even dangerous, because it may fill in and hide gaps in the arguments.

[6] The *principle of duality* in *projective geometry* offers a confirmation of what we assert and provides one of the most interesting examples for it.

[7] If we do not want to give up *symbolic definitions*, it is sufficient to state them *after* the *unproved* propositions.

deduced from the system of unproved propositions, that is, when we cannot deduce from the system a relation of the form $x = a$, where x is one of the undefined symbols and a is a sequence of other such symbols (and logical symbols).

Obviously, the failure of attempts to deduce some relation of the form mentioned from the system of unproved propositions is not sufficient to *prove* the irreducibility of which we are speaking; we have to find a *method of proving* this *irreducibility*.[8]

Let us assume that, after an *interpretation* of the system of undefined symbols that verifies the system of unproved propositions has been determined, all these propositions are still verified if we suitably change the meaning of the undefined symbol x *only*. Then, since the meaning of x is not *individualized* once we have chosen an *interpretation* of the *other* undefined symbols, we can assert that it is impossible to deduce a relation of the form $x = a$, where a is a sequence of other undefined symbols, from the unproved propositions.

Conversely, in order to be able to assert that it is impossible to deduce, from the unproved propositions, a relation of the form mentioned we must show that the meaning of x is not individualized once we have chosen an interpretation of the other undefined symbols; and this we do by establishing an interpretation of the system of undefined symbols that verifies the system of unproved propositions and that still does so if we suitably change the meaning of x only.

Consequently, *to prove that the system of undefined symbols is irreducible with respect to the system of unproved propositions it is necessary and sufficient to find, for each undefined symbol, an interpretation of the system of undefined symbols that verifies the system of unproved propositions and that continues to do so if we suitably change the meaning of only the symbol considered.*

In our essay on *algebra*,[9] we *prove* that the system of *undefined symbols* is irreducible with respect to the system of unproved propositions; and that, we believe, has never been done before, not even for other theories.

Let us again consider a theory satisfying the conditions stated above.

We say that *the system of unproved propositions is* IRREDUCIBLE (or that these propositions are *absolutely independent*) when *it is not possible to deduce any unproved proposition from other such propositions* (and the logical propositions).

Just as above in the case of the irreducibility of the system of primitive symbols, we could not accept as a *proof* of the irreducibility now considered the *failure* of attempts to deduce any unproved proposition from other such propositions. But a *method of proving* this *irreducibility* has been known for a long time.[10]

Let us assume that we have established an *interpretation* of the system of undefined symbols that verifies the system of unproved propositions, except for *one* of these propositions. Then this proposition is not a logical consequence of the other propositions; that is, it is not possible to deduce the proposition in question from the other unproved propositions.

Conversely, in order to deny the possibility of deducing the proposition in question from the other unproved propositions, we must show that it could be false even if all

[8] In case it obtains, of course.

[9] [[This refers to the part of the paper that follows the text reproduced here.]]

[10] In case it obtains, of course. For example, *Peano 1899*, p. 30, contains a proof of the absolute independence of the *unproved propositions* of *arithmetic*.

the others were true; and this we do by establishing an interpretation of the system of undefined symbols that verifies all the other unproved propositions.

Consequently, *to prove that the system of unproved propositions is irreducible it is necessary and sufficient to find, for each of these propositions, an interpretation of the system of undefined symbols that verifies all the other unproved propositions but not that one.*

In our essay on *algebra* we prove that the system of *unproved propositions* is *irreducible.*

Another remark has to be made about *symbolic definitions.*

It is sometimes convenient to represent the unique individual belonging to a class (already considered in the theory), for example a, by a *new symbol,* for example x. But before stating the proposition "$x =$ the a" as a symbolic definition we have to ascertain that, from the preceding propositions of the theory or from the hypothesis of the definition itself, we can deduce the *existence* and the *uniqueness* of the individual to be defined, that is, whether from them we can deduce the propositions "there is some a" and "for every y and z, if y and z are a, then $y = z$" (that is, there are no two distinct a). For, if we could not deduce them, then, after the proposition "$x =$ the a" ⟦had been introduced⟧, we would have the logical right to assert the *existence* and the *uniqueness* of x, whereas we did not have that right before ⟦introducing the proposition⟧; this would make the proposition "$x =$ the a" lose the character of a symbolic definition ⟦and become an additional axiom⟧.[11]

[11] Otherwise we would justify those who used to say that *God exists by definition.* (This is not intended to be a theological criticism, but merely a logical remark.)

Nevertheless, some authors state as symbolic definitions propositions of the form "$x =$ the a" without taking the trouble of first deducing from the preceding propositions of the theory the existence and the uniqueness of the individual to be defined or of restricting the definition by a suitable hypothesis to the case in which this existence and this uniqueness obtain. If they almost never make mistakes, it is simply because from their empirical knowledge they draw a proof that does not always follow from the propositions stated; and sometimes, by doing this, they seem to decrease the number of unproved propositions.

Letter to Frege

BERTRAND RUSSELL

(*1902*)

Bertrand Russell discovered what became known as the Russell paradox in June 1901 (see *1944*, p. 13). In the letter below, written more than a year later and hitherto unpublished, he communicates the paradox to Frege. The paradox shook the logicians' world, and the rumbles are still felt today.

The Burali-Forti paradox, discovered a few years earlier, involves the notion of ordinal number; it seemed to be intimately connected with Cantor's set theory, hence to be the mathematicians' concern rather than the logicians'. Russell's paradox, which makes use of the bare notions of set and element, falls squarely in the field of logic. The paradox was first published by Russell in *The principles of mathematics* (*1903*) and is discussed there in great detail (see especially pp. 101–107). After various attempts, Russell considered the paradox solved by the theory of types (*1908a*). Zermelo (below, p. 191, footnote 9) states that he had discovered the paradox independently of Russell and communicated it to Hilbert, among others, prior to its publication by Russell.

In addition to the statement of the paradox, the letter offers a vivid picture of Russell's attitude toward Frege and his work at the time.

The formula in Peano's notation at the end of the letter can be read more easily if one compares it with formula 450 in *Peano 1898a*, p. VII (or *1897*, p. 15).

Russell wrote the letter in German, and it was translated by Beverly Woodward. Lord Russell read the translation and gave permission to print it here.

Friday's Hill, Haslemere, 16 June 1902

Dear colleague,

For a year and a half I have been acquainted with your *Grundgesetze der Arithmetik*, but it is only now that I have been able to find the time for the thorough study I intended to make of your work. I find myself in complete agreement with you in all essentials, particularly when you reject any psychological element ⟦Moment⟧ in logic and when you place a high value upon an ideography ⟦Begriffsschrift⟧ for the foundations of mathematics and of formal logic, which, incidentally, can hardly be distinguished. With regard to many particular questions, I find in your work discussions, distinctions, and definitions that one seeks in vain in the works of other logicians. Especially so far as function is concerned (§ 9 of your *Begriffsschrift*), I have been led on my own to views that are the same even in the details. There is just one point where I have encountered a difficulty. You state (p. 17 ⟦p. 23 above⟧) that a function, too, can act as the indeterminate element. This I formerly believed, but now this view seems doubtful to me because of the following contradiction. Let w be the predicate: to be a predicate that cannot be predicated of itself. Can w be predicated of itself? From each answer its opposite follows. Therefore we must conclude that w is not a predicate. Likewise there is no class (as a totality) of those classes which, each taken as a totality, do not belong to themselves. From this I conclude that under certain circumstances a definable collection ⟦Menge⟧ does not form a totality.

I am on the point of finishing a book on the principles of mathematics and in it I should like to discuss your work very thoroughly.[1] I already have your books or shall buy them soon, but I would be very grateful to you if you could send me reprints of your articles in various periodicals. In case this should be impossible, however, I will obtain them from a library.

The exact treatment of logic in fundamental questions, where symbols fail, has remained very much behind; in your works I find the best I know of our time, and therefore I have permitted myself to express my deep respect to you. It is very regrettable that you have not come to publish the second volume of your *Grundgesetze*; I hope that this will still be done.

Very respectfully yours,

BERTRAND RUSSELL

The above contradiction, when expressed in Peano's ideography, reads as follows:

$$w = \mathrm{cls} \cap x \ni (x \sim\varepsilon\, x) . \supset : w \,\varepsilon\, w \,.=.\, w \sim\varepsilon\, w.$$

I have written to Peano about this, but he still owes me an answer.

[1] ⟦This was done in *Russell 1903*, Appendix A, "The logical and arithmetical doctrines of Frege".⟧

Letter to Russell

GOTTLOB FREGE

(*1902*)

This is Frege's prompt answer to Russell's letter published above. Frege first calls Russell's attention to an error in *Begriffsschrift*; it is a mere oversight, without any consequence (see above, p. 15, footnote 12). He then describes his reaction to the paradox that Russell has just communicated to him, and he begins to look for the source of the predicament. He incriminates the "transformation of the generalization of an equality into an equality of courses-of-values". For Frege a function is something incomplete, "unsaturated". When it is written $f(x)$, x is something extraneous that merely serves to indicate the kind of supplementation that is needed; we might just as well write $f(\ )$. Consider now two functions that, for the same argument, always have the same value: $(x)(f(x) = g(x))$. (This is not Frege's notation, but its modern equivalent.) Since f and g, or rather $f(\ )$ and $g(\ )$, are something incomplete, we cannot simply write $f = g$. Functions are not objects, and in order to treat them, in some respect, as objects Frege introduces their *Werthverlauf*. The *Werthverlauf* of a function $f(x)$ is denoted by $\grave{\varepsilon}f(\varepsilon)$ (where ε is a dummy; we can also write $\grave{\alpha}f(\alpha), \ldots$). The expression "the function $f(x)$ has the same *Werthverlauf* as the function $g(x)$" is taken to mean "for the same argument the function $f(x)$ always has the same value as the function $g(x)$", and we can write (in modern notation)

$$(*)\quad (x)(f(x) = g(x)) \equiv (\grave{\varepsilon}f(\varepsilon) = \grave{\alpha}g(\alpha)).$$

This is the "transformation of the generalization of an equality into an equality of courses-of-values". Whereas the function is unsaturated and is not an object, its *Werthverlauf* is "something complete in itself", an object, in particular so far as substitution is concerned. There Frege sees the origin of the paradox.

Frege soon made his point more specific. He received Russell's letter while the second volume of his *Grundgesetze der Arithmetik* was at the printshop, and he barely had the time to add an appendix in which he shows how the schema (*) above (or rather half of it, the implication from right to left) allows the derivation of the paradox; he also proposed a restriction in the schema to prevent that. Russell, whose *Principles of mathematics* was at the printshop when he received Frege's volume, added to his book an appendix in which he endorsed Frege's emendation. But soon thereafter he tried out various other solutions (*1905a*); he finally proposed his theory of types (*1908a*).

Russell's paradox has been leaven in modern logic, and countless works have dealt with it. For a late and thorough study of Frege's "way out", see *Quine 1955*.

When Lord Russell was asked whether he would consent to the publication of his letter to Frege (*1902*), he replied with the following letter, in which the reader will find a stirring tribute to Frege.

Penrhyndeudraeth, 23 November 1962

Dear Professor van Heijenoort,

I should be most pleased if you would publish the correspondence between Frege and myself, and I am grateful to you for suggesting this. As I think about acts of integrity and grace, I realise that there is nothing in my knowledge to compare with Frege's dedication to truth. His entire life's work was on the verge of completion, much of his work had been ignored to the benefit of men infinitely less capable, his second volume was about to be published, and upon finding that his fundamental assumption was in error, he responded with intellectual pleasure clearly submerging any feelings of personal disappointment. It was almost superhuman and a telling indication of that of which men are capable if their dedication is to creative work and knowledge instead of cruder efforts to dominate and be known.

Yours sincerely,

Bertrand Russell

The translation of Frege's letter is by Beverly Woodward, and it is printed here with the kind permission of Verlag Felix Meiner and the Institut für mathematische Logik und Grundlagenforschung in Münster, who are preparing an edition of Frege's scientific correspondence and hitherto unpublished writings; this edition will include the German text of the letter.

Jena, 22 June 1902

Dear colleague,

Many thanks for your interesting letter of 16 June. I am pleased that you agree with me on many points and that you intend to discuss my work thoroughly. In response to your request I am sending you the following publications:

1. "Kritische Beleuchtung" ⟦*1895*⟧,
2. "Ueber die Begriffsschrift des Herrn Peano" ⟦*1896*⟧,
3. "Ueber Begriff und Gegenstand" ⟦*1892*⟧,
4. "Über Sinn und Bedeutung" ⟦*1892a*⟧,
5. "Ueber formale Theorien der Arithmetik" ⟦*1885*⟧.

I received an empty envelope that seems to be addressed by your hand. I surmise that you meant to send me something that has been lost by accident. If this is the case, I thank you for your kind intention. I am enclosing the front of the envelope.

When I now read my *Begriffsschrift* again, I find that I have changed my views on many points, as you will see if you compare it with my *Grundgesetze der Arithmetik*. I ask you to delete the paragraph beginning "Nicht minder erkennt man" on page 7 of my *Begriffsschrift* ⟦"It is no less easy to see", p. 15 above⟧, since it is incorrect; incidentally, this had no detrimental effects on the rest of the booklet's contents.

Your discovery of the contradiction caused me the greatest surprise and, I would almost say, consternation, since it has shaken the basis on which I intended to build arithmetic. It seems, then, that transforming the generalization of an equality into an equality of courses-of-values ⟦die Umwandlung der Allgemeinheit einer Gleichheit in eine Werthverlaufsgleichheit⟧ (§ 9 of my *Grundgesetze*) is not always permitted, that my Rule V (§ 20, p. 36) is false, and that my explanations in § 31 are not sufficient to ensure that my combinations of signs have a meaning in all cases. I must reflect further on the matter. It is all the more serious since, with the loss of my Rule V, not only the foundations of my arithmetic, but also the sole possible foundations of arithmetic, seem to vanish. Yet, I should think, it must be possible to set up conditions for the transformation of the generalization of an equality into an equality of courses-of-values such that the essentials of my proofs remain intact. In any case your discovery is very remarkable and will perhaps result in a great advance in logic, unwelcome as it may seem at first glance.

Incidentally, it seems to me that the expression "a predicate is predicated of itself" is not exact. A predicate is as a rule a first-level function, and this function requires an object as argument and cannot have itself as argument (subject). Therefore I would prefer to say "a notion is predicated of its own extension". If the function $\Phi(\xi)$ is a concept, I denote its extension (or the corresponding class) by "$\grave{\varepsilon}\Phi(\varepsilon)$" (to be sure, the justification for this has now become questionable to me). In "$\Phi(\grave{\varepsilon}\Phi(\varepsilon))$" or "$\grave{\varepsilon}\Phi(\varepsilon) \cap \grave{\varepsilon}\Phi(\varepsilon)$"[1] we then have a case in which the concept $\Phi(\xi)$ is predicated of its own extension.

The second volume of my *Grundgesetze* is to appear shortly. I shall no doubt have to add an appendix in which your discovery is taken into account. If only I already had the right point of view for that!

Very respectfully yours,

G. Frege

[1] [["$\cap$" is a sign used by Frege for reducing second-level functions to first-level functions. See *Frege 1893*, § 34.]]

On the foundations of logic and arithmetic

DAVID HILBERT

(*1904*)

This is the text of an address delivered by Hilbert on 12 August 1904 at the Third International Congress of Mathematicians, held in Heidelberg on 8–13 August 1904.

In the last years of the nineteenth century Hilbert provided a satisfactory axiomatization of geometry (*1899*). He then (*1900*) offered a set of axioms for the real numbers and indicated that the question of the consistency of geometry comes down to that of the real-number system. At the Paris International Congress of Mathematicians in 1900, as a natural continuation of this work, he placed the consistency of the real-number system on a list of problems challenging the mathematical world (*1900a*, pp. 264–266). He did not outline any approach, simply stressing that a relative consistency proof seemed out of the question and that, therefore, the problem presented a fundamental difficulty.

Meanwhile the Russell paradox became known, and the question of consistency became more pressing. In 1904, in the paper below, Hilbert presents a first attempt at proving the consistency of arithmetic. In fact, his plan—to show that all the formulas of a certain class possess a certain property (that of being "homogeneous") by showing that the initial formulas have it and the rules transmit it—is the prototype of a device now current in investigations of that nature. Besides the search for a consistency proof the paper offers a critique of the various points of view held at that time on the foundations of arithmetic and introduces the themes that Hilbert is going to develop, modify, or make more precise in his further work in the foundations of mathematics: the reduction of mathematics to a collection of formulas, the extralogical existence of basic objects, like 1, and their combinations, and the construction of logic in parallel with the study of these combinations.

The presentation remains tentative and sketchy. Only many years later (*1917*) will Hilbert come back to the problems of the foundations of mathematics and then present the mature and enriched papers of the twenties (*1922*, *1922a*, *1925*, *1927*). The 1904 paper provides a helpful landmark in the development of Hilbert's conceptions.

The paper was commented upon by Poincaré (*1905*, pp. 17–27; *1908*, pp. 179–191) and Pieri (*1906*). Later commentaries can be found in Bernays (*1935*, pp. 199–200) and Blumenthal (*1935*, p. 422). The paper greatly influenced Julius König's book (*1914*), which in turn inspired von Neumann in his search for a consistency proof of arithmetic (*1927*, footnote 8, p. 22).

An English translation of Hilbert's paper was published (*1905*) in *The monist*, but we have not found it possible to use it. The present translation is by Beverly Woodward, and it is printed here with the kind permission of B. G. Teubner Verlagsgesellschaft, Stuttgart.

While we are essentially in agreement today as to the paths to be taken and the goals to be sought when we are engaged in research into the foundations of geometry, the situation is quite different with regard to the inquiry into the foundations of arithmetic; here investigators still hold a wide variety of sharply conflicting opinions.

In fact, some of the difficulties in the foundations of arithmetic are different in nature from those that had to be overcome when the foundations of geometry were established. In examining the foundations of geometry it was possible for us to leave aside certain difficulties of a purely arithmetic nature; but recourse to another fundamental discipline does not seem to be allowed when the foundations of arithmetic are at issue. The principal difficulties that we encounter when providing a foundation for arithmetic will be brought out most clearly if I submit the points of view of several investigators to a brief critical discussion.

L. Kronecker, as is well known, saw in the notion of the integer the real foundation of arithmetic; he came up with the idea that the integer—and, in fact, the integer as a general notion (parameter value)—is directly and immediately given; this prevented him from recognizing that the notion of integer must and can have a foundation. I would call him a *dogmatist,* to the extent that he accepts the integer with its essential properties as a dogma and does not look further back.

H. Helmholtz represents the standpoint of the *empiricist*; the standpoint of pure experience, however, seems to me to be refuted by the objection that the existence, possible or actual, of an arbitrarily large number can never be derived from experience, that is, through experiment. For even though the number of things that are objects of our experience is large, it still lies below a finite bound.

E. B. Christoffel and all those opponents of Kronecker's who, guided by the correct feeling that without the notion of irrational number the whole of analysis would be condemned to sterility, attempt to save the existence of the irrational number by discovering "positive" properties of this notion or by similar means, I would call *opportunists.* In my opinion they have not succeeded in giving a pertinent refutation of Kronecker's conception.

Among the scholars who have probed more deeply into the essence of the integer I mention the following.

G. Frege sets himself the task of founding the laws of arithmetic by the devices of *logic,* taken in the traditional sense. He has the merit of having correctly recognized the essential properties of the notion of integer as well as the significance of inference by mathematical induction. But, true to his plan, he accepts among other things the fundamental principle that a concept (a set) is defined and immediately usable if only it is determined for every object whether the object is subsumed under the concept or not, and here he imposes no restriction on the notion "every"; he thus exposes himself to precisely the set-theoretic paradoxes that are contained, for example, in the notion of the set of all sets and that show, it seems to me, that the conceptions and means of investigation prevalent in logic, taken in the traditional sense, do not measure up to the rigorous demands that set theory imposes. *Rather, from the very beginning a major goal of the investigations into the notion of number should be to avoid such contradictions and to clarify these paradoxes.*

R. Dedekind clearly recognized the mathematical difficulties encountered when a foundation is sought for the notion of number; for the first time he offered a construction of the theory of integers, and in fact an extremely sagacious one. However, I would call his method *transcendental* insofar as in proving the existence of the infinite he follows a method that, though its fundamental idea is used in a similar way by philosophers, I cannot recognize as practicable or secure because it employs the notion of the totality of all objects, which involves an unavoidable contradiction.

G. Cantor sensed the contradictions just mentioned and expressed this awareness by differentiating between "consistent" and "inconsistent" sets. But, since in my opinion he does not provide a precise criterion for this distinction, I must characterize his conception on this point as one that still leaves latitude for *subjective* judgment and therefore affords no objective certainty.

It is my opinion that all the difficulties touched upon can be overcome and that we can provide a rigorous and completely satisfying foundation for the notion of number, and in fact by a method that I would call *axiomatic* and whose fundamental idea I wish to develop briefly in what follows.

Arithmetic is often considered to be a part of logic, and the traditional fundamental logical notions are usually presupposed when it is a question of establishing a foundation for arithmetic. If we observe attentively, however, we realize that in the traditional exposition of the laws of logic certain fundamental arithmetic notions are already used, for example, the notion of set and, to some extent, also that of number. Thus we find ourselves turning in a circle, and that is why a partly simultaneous development of the laws of logic and of arithmetic is required if paradoxes are to be avoided.

In the brief space of an address I can merely indicate how I conceive of this common construction. I beg to be excused, therefore, if I succeed only in giving you an approximate idea of the direction my researches are taking. In addition, to make myself more easily understood, I shall make more use of ordinary language "in words" and of the laws of logic indirectly expressed in it than would be desirable in an exact construction.

Let an object of our thought be called a *thought-object* ⟦*Gedankending*⟧ or, briefly, an *object* ⟦*Ding*⟧ and let it be denoted by a sign.

We take as a basis of our considerations a first thought-object, 1 (one). We call what we obtain by putting together two, three, or more occurrences of this object, for example,

$$11, \ 111, \ 1111,$$

combinations ⟦*Kombinationen*⟧ of the object 1 with itself; also, any combinations of these combinations, such as

$$(1)(11), \ (11)(11)(11), \ ((11)(11))(11), \ ((111)(1))(1),$$

are again called combinations of the object 1 with itself. The combinations likewise are just called objects and then, to distinguish it, the basic thought-object 1 is called a *simple* object.

We now add a second simple thought-object and denote it by the sign = (equals). Then we form combinations of these two thought-objects, for example,

$$1{=}, \ 11{=}, \ldots, \ (1)({=}1)({=}{=}{=}), \quad ((11)(1)({=}))({=}{=}), \quad 1{=}1, \quad (11){=}(1)(1).$$

We say that the combination a of the simple objects 1 and = *differs* from the combination b of these objects if the combinations deviate in any way from each other with regard to the mode and order of succession in the combinations or the choice and place of the objects 1 and = themselves, that is, if a and b are not *identical* with each other.

Now we think of the combinations of these two simple objects as falling into two classes, the *class of entities* ⟦die *Klasse der Seienden*⟧ and that *of nonentities* ⟦die *der Nichtseienden*⟧: each object belonging to the class of entities differs from each object belonging to the class of nonentities. Every combination of the two simple objects 1 and = belongs to one of these two classes.

If a is a combination of the two objects 1 and = taken as primitive, then we denote also by a the *proposition* that a belongs to the class of entities and by $\bar{a}$ the *proposition* that a belongs to the class of nonentities. We call a a *true* proposition if a belongs to the class of entities; on the other hand, let $\bar{a}$ be called a *true* proposition if a belongs to the class of nonentities. The propositions a and $\bar{a}$ form a *contradiction.*

The composite ⟦Inbegriff⟧ of two propositions A and B, expressed in signs by

$$A \mid B,$$

and in words by "from A, B follows" or "if A is true, so is B", is also called a proposition; here A is called the *supposition* ⟦*Voraussetzung*⟧ and B the *assertion* ⟦*Behauptung*⟧. Supposition and assertion may themselves in turn consist of several propositions A_1, A_2, or B_1, B_2, B_3, and so forth, and we have in signs

$$A_1 \text{ a. } A_2 \mid B_1 \text{ o. } B_2 \text{ o. } B_3,$$

in words "from A_1 and A_2, B_1 or B_2 or B_3 follows", and so forth.

With the sign o. (or) at our disposal it would be possible to avoid the sign |, since negation has already been introduced; I use it in this address merely in order to follow ordinary language as closely as possible.

We shall understand by $A_1, A_2, \ldots$ the propositions that, briefly stated, result from a proposition $A(x)$ if we take the thought-objects 1 and = and their combinations in place of the "*arbitrary object*" ⟦der "*Willkürlichen*"⟧ x; then we write the propositions $A_1 \text{ o. } A_2 \text{ o. } A_3 \ldots$ and $A_1 \text{ a. } A_2 \text{ a. } A_3 \ldots$ also as follows: $A(x^{(o)})$, in words "for at least one x", and $A(x^{(a)})$, in words "for every x", respectively; we regard this merely as an abbreviated way of writing.

From the two objects 1 and = taken as primitive we now form the following propositions:

1. $x = x$,
2. $\{x = y \text{ a. } w(x)\} \mid w(y)$.

Here x (in the sense of $x^{(a)}$) means each of the two thought-objects taken as primitive and every combination of them; in 2, y (in the sense of $y^{(a)}$) likewise can be each of these objects and every combination; further, $w(x)$ is an "arbitrary" combination containing the "arbitrary object" x (in the sense of $x^{(a)}$). Proposition 2 reads in words "from $x = y$ and $w(x)$, $w(y)$ follows".

Propositions 1 and 2 form the *definition of the notion* = (equals) and accordingly are also called *axioms.*

If we put the simple objects 1 and = or particular combinations of them in place of the arbitrary objects x and y in Axioms 1 and 2, particular propositions result, which may be called *consequences* ⟦*Folgerungen*⟧ of these axioms. We consider a sequence of certain consequences such that the suppositions of the last consequence of the sequence are identical with the assertions of the preceding consequences. If we then take the suppositions of the preceding consequences as supposition, and the assertion of the last consequence as assertion, a new proposition results, which can in turn be called a *consequence* of the axioms. By continuing this deduction process we can obtain further consequences.

We now select from these consequences those that have the simple form of the proposition a (assertion without supposition), and we gather the objects a thus obtained into the class of entities, whereas the objects that differ from these are to belong to the class of nonentities. We recognize that only consequences of the form $\alpha = \alpha$ result from 1 and 2, where α is a combination of the objects 1 and $=$. Axioms 1 and 2 for their part, too, are satisfied with regard to this partition of the objects into the two classes, that is, they are true propositions, and because of this property of Axioms 1 and 2 we say that the notion $=$ (equals) defined by them is a *consistent* notion.

I would like to call attention to the fact that Axioms 1 and 2 do not contain any proposition of the form $\bar{a}$ at all, that is, a proposition according to which a combination is to be found in the class of nonentities. Therefore, we could also satisfy the axioms by including all the combinations of the two simple objects in the class of entities and leaving the class of nonentities empty. But the partition, chosen above, into two classes shows better how we must proceed in subsequent, more difficult, cases.

We now carry the construction of the logical foundations of mathematical thought further by adjoining to the two thought-objects 1 and $=$ the three additional thought-objects $\mathfrak{u}$ (infinite set, infinite ⟦Unendlich⟧), $\mathfrak{f}$ (following ⟦Folgendes⟧), $\mathfrak{f}'$ (accompanying operation ⟦begleitende Operation⟧) and stipulating for them the following axioms:

3. $\mathfrak{f}(\mathfrak{u}x) = \mathfrak{u}(\mathfrak{f}'x)$,
4. $\mathfrak{f}(\mathfrak{u}x) = \mathfrak{f}(\mathfrak{u}y) \mid \mathfrak{u}x = \mathfrak{u}y$,
5. $\overline{\mathfrak{f}(\mathfrak{u}x) = \mathfrak{u}1}$.

Here the arbitrary object x (in the sense of $x^{(a)}$) stands for each of the five thought-objects now taken as primitive and every combination of them. The thought-object $\mathfrak{u}$ will be called, simply, *infinite set* and the combination $\mathfrak{u}x$ (for example, $\mathfrak{u}1$, $\mathfrak{u}(11)$, $\mathfrak{u}\mathfrak{f}$) an *element* of this infinite set $\mathfrak{u}$. Axiom 3 then states that each element $\mathfrak{u}x$ is followed by a definite thought-object $\mathfrak{f}(\mathfrak{u}x)$, which is equal to an element of the set $\mathfrak{u}$, namely, the element $\mathfrak{u}(\mathfrak{f}'x)$, that is, which likewise belongs to the set $\mathfrak{u}$. Axiom 4 expresses the fact that, if the same element follows two elements of the set $\mathfrak{u}$, these two elements themselves are equal. According to Axiom 5 there is no element in $\mathfrak{u}$ that is followed by the element $\mathfrak{u}1$; the element $\mathfrak{u}1$ may therefore be called the first element in $\mathfrak{u}$.

We now have to subject Axioms 1–5 to an investigation corresponding to that previously carried out for Axioms 1 and 2; in doing so we must observe that Axioms 1 and 2 now apply to a larger class of objects inasmuch as the arbitrary objects x and y now denote any arbitrary combination of the five simple objects taken as primitive.

We ask again whether certain consequences from Axioms 1–5 form a contradiction or whether, on the contrary, the five thought-objects taken as primitive, 1, $=$, $\mathfrak{u}$, $\mathfrak{f}$, $\mathfrak{f}'$, and their combinations can be so distributed into the class of entities and the class of nonentities that Axioms 1–5 are satisfied with regard to this partition into two classes, that is, that every consequence of these axioms becomes a true proposition with regard to this partition. To answer this question we note that Axiom 5 is the only one giving rise to propositions of the form $\bar{a}$, that is, to propositions asserting that a combination a of the five thought-objects taken as primitive is to belong to the class of nonentities. Accordingly, propositions that form a contradiction with 5 must certainly be of the form

6. $\mathfrak{f}(\mathfrak{u}x^{(o)}) = \mathfrak{u}1$;

but such a consequence cannot result from Axioms 1–4 in any way.

To see this, we call the equation (that is, the thought-object) $a = b$ a homogeneous equation if a and b are combinations of two simple objects each, likewise if a and b are combinations of three simple objects each or of four, and so forth; for example,

$$(11) = (\mathfrak{f}\mathfrak{u}), \quad (\mathfrak{f}\mathfrak{f} = (\mathfrak{u}\mathfrak{f}'), \quad (\mathfrak{f}11) = (\mathfrak{u}1=),$$
$$(\mathfrak{f}1)(\mathfrak{f}1) = (1111), \quad (\mathfrak{f}(\mathfrak{f}\mathfrak{f}'\mathfrak{u})) = (1\mathfrak{u}\mathfrak{u}1),$$
$$((\mathfrak{f}\mathfrak{f})(111)) = ((1)(11)(11)), \quad (\mathfrak{f}\mathfrak{u}111=) = (\mathfrak{u}\mathfrak{u}111\mathfrak{u})$$

are called homogeneous equations. From Axioms 1 and 2 alone follow, as we have seen previously, only homogeneous equations, namely, the equations of the form $\alpha = \alpha$. In the same way Axiom 3 yields only homogeneous equations if in it we take any thought-object for x. Likewise Axiom 4 certainly always exhibits a homogeneous equation in the assertion if the assumption is a homogeneous equation, and consequently from Axioms 1–4 only homogeneous equations can appear as consequences. But now equation 6, which after all was the one to be proved, is certainly not a homogeneous equation, since in it we are to take a combination in place of $x^{(o)}$ and the left side thereby becomes a combination of three or more simple objects, while the right side remains a combination of the two simple objects $\mathfrak{u}$ and 1.

The fundamental idea needed for recognizing the truth of my assertion has thus, I believe, been presented. In order to carry out the proof completely we need the notion of finite ordinal number, as well as certain theorems concerning the notion of equinumerousness, which we could in fact easily state and derive at this point; to develop completely the fundamental idea presented here, we must still consider those points of view to which I shall refer briefly at the close of my address (see V below).

Thus we obtain the desired partition if we put all objects a, where a is a consequence of Axioms 1–4, into the class of entities and all objects that differ from these —in particular, the objects $\mathfrak{f}(\mathfrak{u}x) = \mathfrak{u}1$—into the class of nonentities. Having thus established a certain property for the axioms adopted here, we recognize that they never lead to any contradiction at all, and therefore we speak of the thought-objects defined by means of them, $\mathfrak{u}$, $\mathfrak{f}$, and $\mathfrak{f}'$, as *consistent* notions or operations, or as *consistently existing*. So far as the notion of the infinite $\mathfrak{u}$, in particular, is concerned, the assertion of the *existence of the infinite* $\mathfrak{u}$ appears to be justified by the argument outlined above; for it now receives a definite meaning and a content that henceforth is always to be employed.

The considerations just sketched constitute the first case in which a direct proof of consistency has been successfully carried out for axioms, whereas the method of a suitable specialization, or of the construction of examples, which is otherwise customary for such proofs—in geometry in particular—necessarily fails here.

We see that the success of this direct proof here is essentially due to the fact that a proposition of the form $\bar{a}$, that is, a proposition according to which a certain combination is to belong to the class of nonentities, occurs as assertion in only one place, namely, in Axiom 5.

If we translate the well-known axioms for mathematical induction into the language I have chosen, we arrive in a similar way at the consistency of this larger number of axioms, that is, at the proof of the consistent *existence of what we call the smallest infinite*[1] (that is, *of the ordinal type* 1, 2, 3, . . .).

It is not difficult to provide a foundation for the notion of finite ordinal according to the principles adopted above; this can be done on the basis of an axiom stating that every set containing the first element of the ordinal and, whenever any element belongs to it, containing the succeeding one also, must certainly always contain the last element. Here we very easily obtain a proof of the consistency of the axioms by adducing an example, for instance, the number two. The point then is to show that the elements of the finite ordinal can be so ordered that every subset of it has a first and a last element—a fact that we prove by defining a thought-object $<$ through the axiom

$$(x < y \text{ a. } y < z) \,|\, x < z$$

and then recognizing the consistency of the axioms obtained when this new axiom is added, provided x, y, and z denote arbitrary elements of the finite ordinal. If we also make use of the fact that the smallest infinite exists, it then follows that for every finite ordinal a still greater one can be found.

The principles that must constitute the standard for the construction and further elaboration of the laws of mathematical thought in the way envisaged here are, briefly, the following.

I. Once arrived at a certain stage in the development of the theory, I may say that a further proposition is true as soon as we recognize that no contradiction results if it is added as an axiom to the propositions previously found true, that is, that it leads to consequences that all are true propositions with regard to a certain partition of the objects into the class of entities and that of nonentities.

II. In the axioms the arbitrary objects—taking the place of the notion "every" or "all" in ordinary logic—represent only those thought-objects and their mutual combinations that at this stage are taken as primitive or are to be newly defined. In the derivation of consequences from the axioms the arbitrary objects that occur in the axioms may therefore be replaced only by such thought-objects and their combinations. We must also duly note that, when a new thought-object is added and taken as primitive, the axioms previously assumed apply to a larger class of objects or must be suitably modified.

III. A set is generally defined as a thought-object m, and the combinations mx are called the elements of the set m, so that—contrary to the usual conception—the

[1] See *Hilbert 1900a*, sec. 2, "Die Widerspruchslosigkeit der arithmetischen Axiome".

notion of element of a set appears only as a subsequent product of the notion of set itself.

Just like the notion "set", the notions "mapping", "transformation", "relation", and "function" are also thought-objects, for which, exactly as was the case above with the notion "infinite", we have to consider suitable axioms and which can then be recognized as consistently existing if the combinations in question can be distributed into the class of entities and that of nonentities.

Point I expresses the creative principle that, in its freest use, justifies us in forming ever new notions, with the sole restriction that we avoid a contradiction. The paradoxes mentioned at the beginning of this address become impossible by virtue of II and III; this holds in particular for the paradox of the set of all sets that do not contain themselves as elements.

To show that the notion of set defined in III agrees to a large extent in content with the ordinary notion of set, I will prove the following theorem:

Let $1, \ldots, \alpha, \ldots, \mathfrak{k}$ be the thought-objects taken as primitive at a certain stage in the development and let $a(\xi)$ be a combination of them containing the arbitrary object ξ; further let $a(\alpha)$ be a true proposition (that is, let $a(\alpha)$ be in the class of entities). Then there certainly exists a thought-object m such that $a(mx)$ represents only true propositions for the arbitrary object x (that is, that $a(mx)$ is always in the class of entities), and conversely, also, every object ξ for which $a(\xi)$ represents a true proposition is equated to a combination $mx^{(o)}$, so that the proposition

$$\xi = mx^{(o)}$$

is true, that is, the objects ξ for which $a(\xi)$ becomes a true proposition form the elements of a set m in the sense of the above definition.

To prove this, we take the following axiom: let m be a thought-object for which the propositions

7. $a(\xi) \mid m\xi = \xi,$
8. $\overline{a(\xi)} \mid m\xi = \alpha,$

are true; that is, if ξ is an object such that $a(\xi)$ belongs to the class of entities, then $m\xi = \xi$ is to hold, otherwise $m\xi = \alpha$. We add this axiom to the axioms that hold for the objects $1, \ldots, \alpha, \ldots, \mathfrak{k}$, and we then assume that a contradiction thus appears, that is, that for the objects $1, \ldots, \alpha, \ldots, \mathfrak{k}, m$, the propositions

$$p(m) \quad \text{and} \quad \overline{p(m)},$$

say, are at the same time consequences, $p(m)$ being a certain combination of the objects $1, \ldots, \mathfrak{k}, m$. Here 8 means in words the stipulation: $m\xi = \alpha$ if $a(\xi)$ belongs to the class of nonentities. Wherever in $p(m)$ the object m appears in the combination $m\xi$, let us, in accordance with Axioms 7 and 8 and taking 2 into account, replace the combination $m\xi$ by ξ or α; let $q(m)$ (where $q(m)$ now no longer contains the object m in a combination mx) be obtained in this way from $p(m)$; then $q(m)$ would have to be a consequence of the axioms[2] originally posited for $1, \ldots, \alpha, \ldots, \mathfrak{k}$, and therefore would have to remain true even if we took for m any one of these objects, say, the

[2] ⟦The German text has "dem...Axiome", but the argument seems to call for the plural; the version in *Hilbert 1930a* has the plural.⟧

object 1. Since the same considerations also hold for the proposition $\overline{p(m)}$, the contradiction

$$q(1) \quad \text{and} \quad \overline{q(1)}$$

would therefore also exist at the original stage at which the objects $1, \ldots, \alpha, \ldots, \mathfrak{k}$ were taken as primitive; but this cannot be, if we assume that the objects $1, \ldots, \mathfrak{k}$ exist consistently. We must therefore reject our assumption that a contradiction occurs; that is, m exists consistently, which was to be proved.

IV. If we want to investigate a given system of axioms according to the principles above, we must distribute the combinations of the objects taken as primitive into two classes, that of entities and that of nonentities, with the axioms playing the role of prescriptions that the partition must satisfy. The main difficulty will consist in recognizing the possibility of distributing all objects into the two classes, that of entities and that of nonentities. The question whether this distribution is possible is essentially equivalent to the question whether the consequences we can obtain from the axioms by specialization and combination in the sense explained earlier lead to a contradiction or not, *if we still add the familiar modes of logical inference such as*

$$\{(a \mid b) \text{ a. } (\bar{a} \mid b)\} \mid b,$$
$$\{(a \text{ o. } b) \text{ a. } (a \text{ o. } c)\} \mid \{a \text{ o. } (b \text{ a. } c)\}.$$

We can then recognize the consistency of the axioms either by showing how a possible contradiction would already have to have occurred at an earlier stage in the development of the theory or by making the assumption that there is a proof leading from the axioms to a certain contradiction and then showing that such a proof is not possible, because it would itself contain a contradiction. Thus the proof sketched above for the consistent existence of the infinite came down to the recognition that a proof of equation 6 from Axioms 1–4 is not possible.

V. Whenever in the preceding we spoke of *several* thought-objects, of *several* combinations, of *various* kinds of combinations, or of *several* arbitrary objects, a bounded number of such objects was to be understood. Now that we have established the definition of the finite number we are in a position to comprehend the general meaning of this way of speaking. The meaning of the "arbitrary" consequence and of the "differing" of one proposition from all propositions of a certain kind is also now, on the basis of the definition of the finite number (corresponding to the idea of mathematical induction) susceptible of an exact description by means of a recursive procedure. It is in this way that we can carry out completely the proof, sketched above, that the proposition $\mathfrak{f}(\mathfrak{u}x^{(o)}) = \mathfrak{u}1$ differs from every proposition obtained as a consequence of Axioms 1–4 by a finite number of steps; we need only consider the proof itself to be a mathematical object, namely, a finite set whose elements are connected by propositions stating that the proof leads from 1–4 to 6, and we must then show that such a proof contains a contradiction and therefore does not exist consistently in the sense defined by us.

The existence of the totality ⟦Inbegriff⟧ of real numbers can be demonstrated in a way similar to that in which the existence of the smallest infinite can be proved; in fact, the axioms for real numbers as I have set them up (*1903*, pp. 24–26) can be expressed by precisely such formulas as the axioms hitherto assumed. In particular, so far as the axiom I called the completeness axiom [[Vollständigkeitsaxiom]] is concerned, it expresses the fact that the totality of real numbers contains, in the sense of a one-to-one correspondence between elements, any other set whose elements satisfy also the axioms that precede; thus considered, the completeness axiom, too, becomes a stipulation expressible by formulas constructed like those above, and the axioms for the totality of real numbers do not differ qualitatively in any respect from, say, the axioms necessary for the definition of the integers. In the recognition of this fact lies, I believe, the real refutation of the conception of the foundations of arithmetic associated with L. Kronecker and characterized at the beginning of my lecture as dogmatic.

In the same way we can show that the fundamental notions of Cantor's set theory, in particular Cantor's alephs, have a consistent existence.

Proof that every set can be well-ordered

ERNST ZERMELO

(*1904*)

Zermelo's very short paper is a part of a letter to Hilbert dated "Münden in Hannover, 24 September 1904". Cantor (*1883*, p. 550, or *1932*, p. 169) had conjectured that "every well-defined set can be brought into the form of a well-ordered set" and had seen in that a "fundamental law of thought [[Denkgesetz]] of great consequence, particularly noteworthy for its universal validity". Several times he announced a proof of the conjecture. At the International Congress of Mathematicians in Paris Hilbert (*1900a*, p. 264) listed the discovery of such a proof as one of the most important tasks confronting the mathematical world. At the Third International Congress of Mathematicians in Heidelberg, J. König presented on 10 August 1904 what he considered a proof that there is no well-ordering of the continuum, but he had to withdraw it shortly afterward (see below, p. 192, footnote 10). Less than two months later, Zermelo completed his proof of the existence of such a well-ordering.

The proof remains relatively simple because it makes use of a powerful new mathematical tool, the axiom of choice. What Zermelo proves is, in fact, that the axiom of choice implies the well-ordering theorem. Tacitly used for years in mathematical arguments, incidentally stated and rejected by Peano (*1890*, p. 210), recognized as a new mathematical principle by Beppo Levi (*1902*, p. 864), suggested to Zermelo by Erhard Schmidt, the axiom was immediately put in the limelight by the importance of the theorem that could now be proved.

The proof provoked sharp reactions among mathematicians, so sharp that a few years later Zermelo had to discuss them at length, after presenting a second version of the proof (see below, pp. 183–198).

The translation is by Stefan Bauer-Mengelberg, and it is printed here with the kind permission of Springer Verlag.

... The proof in question grew out of conversations that I had last week with Mr. Erhard Schmidt, and it is as follows.

(1) Let M be an arbitrary set of cardinality $\mathfrak{m}$, let m denote an arbitrary element of it, let M', of cardinality $\mathfrak{m}'$, be a subset of M that contains at least one element m and may even contain all elements of M, and let $M - M'$ be the subset "complementary" to M'. Two subsets are regarded as distinct if one of them contains some element that does not occur in the other. Let the set of all subsets M' be denoted by M.

(2) *Imagine that with every subset M' there is associated an arbitrary element m_1' that*

occurs in M' itself; let m_1' be called the "distinguished" element of M'. This yields a "covering" γ of the set M by certain elements of the set M. The number of these coverings γ is equal to the product $\Pi\mathfrak{m}'$ taken over all subsets M' and is therefore certainly different from 0. In what follows we take an arbitrary covering γ and derive from it a definite well-ordering of the elements of M.

(3) *Definition.* Let us apply the term "γ-set" to any well-ordered set M_γ that consists entirely of elements of M and has the following property: whenever a is an arbitrary element of M_γ and A is the "associated" segment, which consists of the elements x of M such that $x \prec a$, a is the distinguished element of $M - A$.

(4) *There are γ-sets included in M.* Thus, for example, [[the set containing just]] m_1, the distinguished element of M' when $M' = M$, is itself a γ-set; so is the (ordered) set $M_2 = (m_1, m_2)$, where m_2 is the distinguished element of $M - m_1$.

(5) *Whenever M_γ' and M_γ'' are any two distinct γ-sets* (associated, however, with the same covering γ chosen once for all!), *one of the two is identical with a segment of the other.*

For, of the two well-ordered sets, let M_γ' be the one for which there exists a similar mapping onto the other, M_γ'', or onto one of its segments. Then any two elements corresponding to each other under this mapping must be identical. For the first element of *every* γ-set is m_1, since the associated segment A contains no element and therefore $M - A = M$. If now m' were the *first* element of M_γ' that differs from the corresponding element m'', the associated segments A' and A'' would still have to be identical, consequently also the complements $M - A'$ and $M - A''$, and thus their distinguished elements m' and m'' themselves, contrary to assumption.

(6) *Consequences.* If two γ-sets have an element a in common, they also have the segment A of the preceding elements in common. If they have *two* elements a and b in common, then either in *both* sets $a \prec b$ or in *both* sets $b \prec a$.

(7) If we call any element of M that occurs in some γ-set a "γ-element", the following theorem holds: *The totality L_γ of all γ-elements can be so ordered that it will itself be a γ-set, and it contains all elements of the original set M. M itself is thereby well-ordered.*

(I) If a and b are two arbitrary γ-elements and if M_γ' and M_γ'' are any two γ-sets to which they respectively belong, then according to (5) the larger of the two γ-sets contains both elements and determines whether the order relation is $a \prec b$ or $b \prec a$. According to (6) this order relation is independent of the γ-sets selected.

(II) If a, b, and c are three arbitrary γ-elements and if $a \prec b$ and $b \prec c$, then always $a \prec c$. For according to (6) every γ-set containing c also contains b, hence also a, and then, since it is simply ordered, within the set, $a \prec c$ indeed follows from $a \prec b$ and $b \prec c$. The set L_γ is therefore *simply ordered.*

(III) If L_γ' is an arbitrary subset of L and a is one of its elements, belonging, say, to the γ-set M_γ, then according to (6) M_γ contains all elements preceding a, hence includes the subset L_γ'' that is obtained from L_γ' when all elements following a are removed; L_γ'', being a subset of the well-ordered set M_γ, possesses a *first* element, which is also the first element of L_γ'. L_γ is therefore also *well-ordered.*

(IV) If a is an arbitrary γ-element and A the totality of *all* preceding elements $x \prec a$, then according to (6), in every set M_γ containing a, A is the segment associated with a; according to (3), consequently, a is the distinguished element of $M - A$. *Therefore L_γ is itself a γ-set.*

(V) If there existed an element of M that belonged to *no* γ-set, that consequently was an element of $M - L_\gamma$, there would also exist a distinguished element m_1' of $M - L_\gamma$, and the ordered set (L_γ, m_1'), in which every γ-element precedes the element m_1', would itself according to (3) be a γ-set. Then m_1' too would be a γ-element, contrary to assumption; so really $L_\gamma = M$, and thus M is itself a *well-ordered set.*

Accordingly, to every covering γ there corresponds a definite well-ordering of the set M, even if the well-orderings that correspond to two distinct coverings are not always themselves distinct. There must at any rate exist *at least one* such well-ordering, and every set for which the totality of subsets, and so on, is meaningful may be regarded as well-ordered and its cardinality as an "aleph". It therefore follows that, for every transfinite cardinality,

$$\mathfrak{m} = 2\mathfrak{m} = \aleph_0\mathfrak{m} = \mathfrak{m}^2, \text{ and so forth;}$$

and any two sets are "comparable"; that is, one of them can always be mapped one-to-one onto the other or one of its parts.

The present proof rests upon the assumption that coverings γ actually do exist, hence upon the principle that even for an infinite totality of sets there are always mappings that associate with every set one of its elements, or, expressed formally, that the product of an infinite totality of sets, each containing at least one element, itself differs from zero. This logical principle cannot, to be sure, be reduced to a still simpler one, but it is applied without hesitation everywhere in mathematical deduction. For example, the validity of the proposition that the number of parts into which a set decomposes is less than or equal to the number of all of its elements cannot be proved except by associating with each of the parts in question one of its elements.

I owe to Mr. Erhard Schmidt the idea that, by invoking this principle, we can take an *arbitrary* covering γ as a basis for the well-ordering; the proof, as I carried it through, then rests upon the fusion of the various possible "γ-sets", that is, of the well-ordered segments resulting from the ordering principle.

The principles of mathematics and the problem of sets

JULES RICHARD

(*1905*)

Richard's paper was written as a letter to the editor of the *Revue générale des sciences pures et appliquées* and was published in the issue of 30 June 1905, with comments by the editor (Louis Olivier); it was reprinted in *Acta mathematica* (*1906*).

Richard, then a mathematics teacher at the *lycée* in Dijon, was prompted to write his letter by some editorial comments in the issue of 30 March 1905 of the *Revue*. There it was stated that at the Heidelberg Congress of Mathematicians (8–15 August 1904) J. König had established that the continuum cannot be well-ordered, while shortly thereafter (24 September 1904) Zermelo had given a proof that any set can be well-ordered. No indication of König's argument was given. What Richard learned from the 30 March issue of the *Revue* was simply, as he states at the beginning of his letter, that there are "certain contradictions" in set theory and these contradictions are somehow connected with the notions of well-ordering and ordinal number. Meanwhile König abandoned the argument that he had presented at the 1904 congress (see *König 1904, 1905*, and footnote 10, p. 192 below). The new argument that he advanced for the impossibility of the well-ordering of the continuum took the form of a paradox, which bears a resemblance to Richard's. König communicated this paradox to the Hungarian Academy of Sciences on 20 June 1905. Hence the letter presenting Richard's paradox and the paper presenting König's were written during the same weeks, perhaps even the same days, and independently of each other.

Richard's paradox became known when the stir aroused by Russell's, published two years before, had not yet abated. It bears some similarity to Russell's in that both rest upon the most elementary set-theoretic notions, namely, set, subset, and element, and are independent of more involved notions like well-ordering, used in Burali-Forti's and König's paradoxes. In his letter Richard himself offered a solution of the paradox, and this solution was soon endorsed by Poincaré (*1906*). Yet perhaps the most penetrating commentaries on the paradox were made by Peano (*1906a*), who rejected Richard's explanation and proposed his own: "Richard's example does not belong to mathematics, but to linguistics; an element that is fundamental in the definition of ⟦the diagonal number⟧ N cannot be defined in an exact way (according to the rules of mathematics)". Today the paradox is generally considered solved by the distinction of language levels. Gödel's informal presentation of his famous proof (*1931*) bears some similarity to Richard's argument but skirts the paradox (see p. 598 below).

The translation is by the editor, and it is printed here with the kind permission of the Librairie Armand Colin.

In its issue of 30 March 1905 the *Revue* draws attention to certain contradictions that are encountered in general set theory.

It is not necessary to go so far as the theory of ordinal numbers to find such contradictions. Here is one that presents itself the moment we study the continuum and to which some others could probably be reduced.

I am going to define a certain set of numbers, which I shall call the set E, through the following considerations.

Let us write all permutations of the twenty-six letters of the French alphabet taken two at a time, putting these permutations in alphabetical order; then, after them, all permutations taken three at a time, in alphabetical order; then, after them, all permutations taken four at a time, and so forth. These permutations may contain the same letter repeated several times; they are permutations with repetitions.

For any integer p, any permutation of the twenty-six letters taken p at a time will be in the table; and, since everything that can be written with finitely many words is a permutation of letters, everything that can be written will be in the table formed as we have just indicated.

The definition of a number being made up of words, and these words of letters, some of these permutations will be definitions of numbers. Let us cross out from our permutations all those that are not definitions of numbers.

Let u_1 be the first number defined by a permutation, u_2 the second, u_3 the third, and so on.

We thus have, written in a definite order, *all numbers that are defined by finitely many words*.

Therefore, the numbers that can be defined by finitely many words form a denumerably infinite set.

Now, here comes the contradiction. We can form a number not belonging to this set. "Let p be the digit in the nth decimal place of the nth number of the set E; let us form a number having 0 for its integral part and, in its nth decimal place, $p + 1$ if p is not 8 or 9, and 1 otherwise." This number N does not belong to the set E. If it were the nth number of the set E, the digit in its nth decimal place would be the same as the one in the nth decimal place of that number, which is not the case.

I denote by G the collection of letters between quotation marks.

The number N is defined by the words of the collection G, that is, by finitely many words; hence it should belong to the set E. But we have seen that it does not.

Such is the contradiction.

Let us show that this contradiction is only apparent. We come back to our permutations. The collection G of letters is one of these permutations; it will appear in my table. But, at the place it occupies, it has no meaning. It mentions the set E, which has not yet been defined. Hence I have to cross it out. The collection G has meaning only if the set E is totally defined, and this is not done except by infinitely many words. *Therefore there is no contradiction.*

We can make a further remark. The set containing ⟦the elements of⟧ the set E and the number N represents a new set. This new set is denumerably infinite. The number N can be inserted into the set E at a certain rank k if we increase by 1 the rank of each number of rank [[equal to or]] greater than k. Let us still denote by E the thus modified set. Then the collection of words G will define a number N′ *distinct from* N, since the number N now occupies rank k and the digit in the kth decimal place of N′ is not equal to the digit in the kth decimal place of the kth number of the set E.

On the foundations of set theory and the continuum problem

JULIUS KÖNIG

(*1905a*)

König's paper was written at about the same time as Richard's (see above, p. 142). Like Richard, König deals with the set of finitely definable numbers, but, instead of using a diagonal construction, he considers the complement of that set. If the set of real numbers could be well-ordered, the set of not finitely definable real numbers would have a first element, which, after all, would be finitely definable. König's conclusion is that the set of real numbers cannot be well-ordered. (This would imply the rejection of the continuum hypothesis; at the 1904 International Congress of Mathematicians König had presented another argument, subsequently withdrawn, for that rejection; see footnote 10, below, p. 192.)

After the discovery of the Richard paradox, in which the well-ordering problem does not play any role, König's conclusion can hardly be accepted. Rather, taking for granted Zermelo's result that the set of real numbers can be well-ordered, we infer the existence of a first element in the complement of the set of finitely definable real numbers, and we thus have a paradox (sometimes called the Zermelo-König paradox) bearing on the notion of definability.

A third paradox, related to both Richard's and König's, is Berry's paradox, first published by Bertrand Russell (see below, p. 153). The paradox is formulated in terms of integers, rather than real numbers, but this is unessential. The notion "finitely definable" is replaced by (say) "definable in fewer than nineteen syllables". The set of numbers thus definable being finite, the argument is perhaps still simpler.

The translation is by Stefan Bauer-Mengelberg, and it is printed here with the kind permission of Springer Verlag.

It is only after overcoming serious misgivings that I have decided to publish what follows. But no matter how the conception advanced here may be received, I believe that the questions raised cannot be ignored in the further development of set theory.

That the word "set" is being used indiscriminately for completely different notions and that this is the source of the apparent paradoxes of this young branch of science, that, moreover, set theory itself can no more dispense with axiomatic assumptions than can any other exact science and that these assumptions, just as in other disciplines, are subject to a certain arbitrariness, even if they lie much deeper here—I do not want to represent any of this as something new. But I believe that even in this fragmentary preliminary communication I offer some new points of view concerning these questions. In particular, it is presumably impossible to regard the *special* theory of well-ordered sets as having been provided with a complete foundation so long as the questions treated in Section 4 below are not clarified.

1. If $a_1, a_2, \ldots, a_k, \ldots$ is a denumerably infinite sequence (a sequence of type ω) of positive integers, the objects

$$(a_1, a_2, \ldots, a_k, \ldots)$$

shall define the set called the "continuum". Or, if some other definition of the continuum is taken as a point of departure, the objects $(a_1, a_2, \ldots, a_k, \ldots)$ are symbols that on the one hand univocally determine the elements of the continuum and on the other clearly distinguish them from one another.

An element of the continuum will be said to be "finitely defined" if, by means of a language capable of giving a definite form to our scientific thinking, we can in a finite span of time specify a procedure (law) that conceptually distinguishes that element of the continuum from any other one, or—to put it differently—that for an arbitrarily chosen k leads to the existence of one and only one associated number a_k.

At the same time it must be expressly emphasized that the finite conceptual differentiation demanded here is not to be confused with the demand for a well-defined, or even finite, procedure by which the a_k can be determined.

It is easy to show that *the finitely defined elements of the continuum determine a subset of the continuum that has the cardinality* $\aleph_0$, a subset that we shall henceforth denote by E.

For such a finite definition must be completely given by means of a finite number of letters and punctuation marks, and of these only a definite, finite number is available. Furthermore, we can then order these various finite definitions in such a way that to any definition there corresponds one and only one definite positive integer as a (finite) ordinal.

Since, now, for every finitely defined element of the continuum several finite definitions can and actually do exist, every such element determines a sequence of positive integers, among which, then, the least is uniquely determined; conversely, this integer, by means of the associated combination of signs, uniquely determines the corresponding finitely defined element of the continuum.

E is therefore equivalent to a subset of the set of positive integers. But since

$$(a, a, \ldots, a, \ldots),$$

where a is an arbitrary positive integer, is a finitely defined element of the continuum, it in turn follows that

$$\mathfrak{e} = \aleph_0,$$

where $\mathfrak{e}$ stands for the cardinality of E.

However, since the continuum, in consequence of its definition, is not denumerable, there must exist elements of the continuum that cannot be finitely defined.

2. Even if at present I must still forgo giving an exact and systematic presentation, it is nevertheless absolutely necessary to make precise the axiomatic assumptions contained in the train of thought up to this point.

(*a*) Above all, the "fact" has been assumed that there are processes taking place in our consciousness that satisfy the formal laws of logic and are called "scientific thinking" and that among these there are some that stand in a one-to-one relation to other processes of that kind, namely, the generation of the previously described sequences of signs.

The question how such a relation comes to be, or even how far these relations can be extended, is not touched upon at all here. (*Metalogical axiom.*)

(*b*) All "arbitrary" sequences of type ω formed from the positive integers—and the "totality of all such sequences", which we call the "continuum"—are "possible concepts", that is, they do not in themselves lead to a logical contradiction. (*Continuum axiom.*)

A further, thorough analysis of this assertion is, I believe, contained in the considerations presented by Hilbert (*1904*) at the Third International Congress of Mathematicians in Heidelberg.

In particular, this definition of the continuum entails the assertion that its cardinality is $\aleph_0^{\aleph_0}$.

That, furthermore, $\aleph_0^{\aleph_0} > \aleph_0$ can be proved, for example, by a method that I gave in my *1905*.

I am aware that I am thus adopting a position in conflict with the assumption that it is *not* permissible to go beyond "finite laws". To make this assumption is, in my opinion, to deny the existence of the continuum and of the continuum problem. The assumption I employ here is, on the contrary, that there exist elements of the continuum that we cannot "think through to the end" [["zu Ende denken"]] and that are nevertheless consistent; thus they are "ideal" elements, if I may be permitted to use this expression in what is, to be sure, an entirely new sense.

(*c*) If, with the assumptions we have made so far, we may now speak of an "arbitrary" element of the continuum, it is because we are employing, lastly, the *logical antithesis* "given an arbitrary element of the continuum, either it is finitely defined or this is not the case". In view of (*a*) and (*b*) it is presumably impossible to reject this antithesis, especially since without changing our inferences we can also formulate it subjectively: "For an arbitrary element of the continuum we surely have a finite definition or this is not the case".

3. The assumptions developed thus far lead in a remarkably simple way to the conclusion that *the continuum cannot be well-ordered.*

If we think of the elements of the continuum as a well-ordered set, then those elements that cannot be finitely defined form a subset of that well-ordered set, and this subset certainly contains elements of the continuum. But then this subset, too, is well-ordered and contains one and only one first element. Let us observe, further, that according to the assumptions now in force the continuum, like any well-ordered set, defines an unbroken sequence of specific ordinals; it does so in such a way, moreover, that to every element of the continuum there corresponds one and only one of these ordinals, and conversely. Accordingly, "the ordinal corresponding to a finitely defined element of the continuum", as well as "the element of the continuum corresponding to a finitely defined ordinal of that kind", is finitely defined. Therefore in that sequence there would have to occur a first ordinal that was not finitely definable. But this is impossible.

For there exists a definite (well-ordered) set of finitely defined ordinals that follow each other in an unbroken sequence beginning with the first. But "the first ordinal that exceeds all of these in magnitude" would in fact be finitely defined by the very words in quotation marks, whereas according to the assumption it cannot be finitely defined.

The assumption that the continuum can be well-ordered has therefore led to a contradiction.

4. Militating against the considerations presented thus far is the objection, which almost immediately suggests itself, that they can be applied word for word to every well-ordered nondenumerable set, that therefore such sets could not exist at all. But since Cantor's second number class $Z(\aleph_0)$, "the totality of all order types of well-ordered sets of cardinality $\aleph_0$", defines such a "set" consistently, an error must have occurred in our reasoning up to this point. That this seemingly paradoxical result need not *necessarily* imply such an error is yet to be explained in greater detail.

The word "set" is used for two completely different notions in the two cases.

When the notion of the continuum is formed, it is the "arbitrary" sequence $(a_1, a_2, \ldots, a_k, \ldots)$ that is primary, or fundamental. Through the stipulation that $a_1, a_2, \ldots$ are to be replaced by definite positive integers, it becomes a "definite" sequence, an element of the continuum, which cannot become an object of our thought without being conceptually distinct from any other element. The further stipulation that we consider the *totality* ⟦*Inbegriff*⟧ of these "well-distinguished" objects then leads to the continuum.

The situation is quite different in the case of the number class $Z(\aleph_0)$. Its "elements" are determined by the "property" of being order types of well-ordered sets of cardinality $\aleph_0$. To be sure, we know such elements: ω, $\omega + 1, \ldots$; but this property is only an abstraction, at best a means of distinguishing between objects belonging and objects not belonging to the class; however, it is certainly not a rule according to which *every* element of $Z(\aleph_0)$ can be formed. What is primary, or fundamental, here is the collective notion ⟦Kollektivbegriff⟧, which for this very reason, following Cantor's nomenclature, I would not call a "set" but a "class"; it is only *afterward* that elements belonging to the class are constructed that exemplify this notion.

That the second number class $Z(\aleph_0)$ is definable as a "completed" set of well-distinguished elements, that is, of elements conceptually altogether distinct, cannot be considered probable even at the present stage of our knowledge of set theory. To the extent that the conclusions derived earlier are correct, the present discussion might even be said to contain a proof that the second number class cannot be considered to be a completed set, that is, a totality of well-distinguished elements that are altogether conceptually distinct.[1]

[1] In this fact, I believe, lies the origin of those paradoxes that appear in the theory of ordinal numbers and to which Burali-Forti first called attention.

I should like to add here a few short remarks that may perhaps facilitate understanding of the considerations in Section 4.

Even the totality of the positive integers is originally given only as a "class". Hilbert, too, defines (*1904*) the "smallest infinite" in this way. But, it seems, the stipulation that *this* class be considered a completed set is possible, that is, consistent in itself.

On the other hand, the continuum would be conceivable only as a "completed set", and the

In closing these fragmentary remarks I take satisfaction in being able to note that, insofar as they are correct, they only throw a new light on the great value of what Cantor's genius created, despite their partially oppositionist character. The opposition is directed only against certain of Cantor's conjectures; the content of the theorems that he proved remains completely intact. I remark, finally, that the distinction here drawn between "set" and "class" completely resolves the paradoxes cited ("set of all sets", and so forth).

In substance the above is a lecture given before the Hungarian Academy of Sciences on 20 June 1905.

second number class only as a class, or (if the expression be permitted) as a "set in the process of becoming".

I should like, still, to point out a very elementary collective notion that certainly does *not* permit being considered a "completed set".

We take as a point of departure the totality of all terminating decimal fractions, which, however, we write as nonterminating decimal fractions by putting the digit 0 in all places in which there is no digit.

In these objects every place may be arbitrarily filled; that is, for any digit we can put any other digit without going out of the totality of the defined objects.

It would nevertheless be wholly inadmissible to speak of the totality of all places as places that can be arbitrarily filled; obviously this would eliminate the "principle of inhibition" ⟦"Hemmungsprinzip"⟧ contained in the definition of the places that can be arbitrarily filled. This principle of inhibition can be formulated as follows: "The kth place can be arbitrarily filled, but there must exist a positive integer $l > k$ such that from the lth place on only the digit 0 is used".

The question "How many places are there that can simultaneously be arbitrarily filled?" cannot be answered with any cardinal number (in Cantor's sense) and requires the creation of a new "concept of numeration" ⟦eines neuen "Zählbegriffs"⟧.

Mathematical logic as based on the theory of types

BERTRAND RUSSELL

(*1908a*)

⟦It was in June 1901 that Russell discovered the paradox of the class of all classes that do not contain themselves as elements. He communicated it to Frege on 16 June 1902 (see pp. 124–125 above). Discussing "the Contradiction", as he calls it, in *The principles of mathematics* (*1903*) in a passage probably written in 1901, he mentions, without much elaboration, that "the class as many is of a different type from the terms of the class" and that "it is the distinction of logical types that is the key to the whole mystery". The solution to the problem is presented in less than thirty lines (§ 104). He also examines other solutions, finds them less satisfactory, and concludes that "no peculiar philosophy is involved in the above contradiction, which springs directly from common sense, and can only be solved by abandoning some common-sense assumption". But before the volume came out (the preface is dated December 1902 and the volume 1903), Russell felt that the subject deserved more attention. He wrote Appendix B, of almost six pages, where the doctrine of types is put forward "tentatively", since "it requires, in all probability, to be transformed into some subtler shape before it can answer all difficulties". At that time Russell knew, of course, of other paradoxes, for instance the Burali-Forti paradox and that of the greatest cardinal.

By December 1905 Russell had abandoned the theory of types. To overcome the difficulties raised by the paradoxes he then presented (*1905a*) three theories: (1) the zigzag theory ("propositional functions determine classes when they are fairly simple, and only fail to do so when they are complicated and recondite"), (2) the theory of limitation of size ("there is not such a thing as the class of all entities"), and (3) the no-classes theory ("classes and relations are banished altogether"). The theory of types is not even mentioned in the paper. In a note added at the end of the paper, on 5 February 1906, Russell wrote: "From further investigation I now feel hardly any doubt that the no-classes theory affords the complete solution of all the difficulties stated in the first section of this paper", that is, the paradoxes.⟧

The central idea of the no-classes theory was that, instead of speaking of the class of all the objects that fulfill some given sentence, one might speak of the sentence itself and of substitutions within it. Now discourse about specified classes lends itself well enough to paraphrase in terms thus of sentences and substitution, but when we talk rather of classes in general, as values of quantifiable variables, it is not evident how to continue such paraphrase. Russell (*1905a*) had already acknowledged that the no-classes theory might prove inadequate to much of classical mathematics; and his more sanguine postscript of February 1906 was the expression only of a renewed hope that he was shortly to abandon. For he soon turned back to his theory of types and proceeded to develop it in detail. The result, published in July 1908, is the paper reproduced below.

Russell sees the universe as dividing into levels, or *types*. We can speak of all the things fulfilling a given condition only if they are all of the same type. The members of a class, then, must all be of one type. So must the values of the variable of any one quantification. Russell is thus led to propound a distinction between "all" and "any": "all" is expressed by the bound ("apparent") variable of universal quantification, which ranges over a type, and "any" is expressed by the free ("real") variable, which refers schematically to any unspecified thing irrespective of type. Because the free variable is in this way unhampered, Russell likes to suppress a universal quantifier when it has the whole of a theorem as its scope. This contrast between asserting a universal quantification and asserting a "propositional function" carries over into the first edition of Whitehead and Russell's *Principia mathematica* and puzzles readers who do not perceive that it is pointless apart from a certain aspect of the theory of types.

Russell sees the key to the paradoxes in what he calls the vicious-circle principle: "No totality can contain members defined in terms of itself". He implements it by declaring "whatever contains an apparent variable" to be of higher type than the things over which that variable ranges. The formulation is troublesome. Variables, in the easiest sense, are letters; and what contain them are notational expressions. Is Russell then assigning types to his objects or to his notations? The confusion persists as he proceeds to define "*n*th-order propositions". His lowest type comprises individuals; his next comprises what he calls first-order propositions; and so on up. These propositions, unlike the individuals, are evidently notation; at any rate they can contain variables. Yet, like the individuals, they have type and they figure as values of quantified variables.

From his hierarchy of types of propositions Russell derives a hierarchy of types of propositional functions. He speaks here of substitution, in a way that suggests that his functions also are notational in character; they seem simply to be open sentences, sentences with free variables. Still, he assigns them types and lets them be values of quantified variables. Insofar, they should be viewed not as open sentences but as *attributes*, or, when they are functions of two or more arguments, *relations*. Failure to distinguish thus between open sentences on the one hand and attributes and relations on the other had grave consequences for this paper and equally for *Principia mathematica*, for which this paper sets the style.

This theory of types of propositional functions is what has come to be called the *ramified* theory of types. It has been so called because the type of a function depends both on the types of its arguments and on the types of the apparent variables contained in it (or in its expression), in case these exceed the types of the arguments.

This second limitation on types proved onerous. It would make for a constructive set theory like Weyl's or that of intuitionism, both inadequate as a foundation for classical analysis. So Russell canceled the second limitation, the one concerning apparent variables, by propounding his *axiom of reducibility*. It says that any function is coextensive with what he calls a *predicative* function: a function in which the types of apparent variables run no higher than the types of the arguments. Granted this axiom, the predicative functions will suffice for the work of propositional functions generally. Restrictions occasioned by apparent variables that out-type the arguments are summarily abolished.

The method is indeed oddly devious.

If for every propositional function there is a coextensive predicative one, then the symbols for propositional functions could have been construed from the start as referring outright just to the corresponding predicative ones. In short, the types of propositional functions could have been described in the first place as depending simply on the types of the arguments. The axiom of reducibility is self-effacing: if it is true, the ramification that it is meant to cope with was pointless to begin with. Russell's failure to appreciate this point is due to his failure to distinguish between propositional functions as notations and propositional functions as attributes and relations.

The wisdom of dropping the ramification and the axiom of reducibility was urged by Ramsey (*1931*, pp. 20–29), though he likewise failed to grasp the point in its full simplicity. Ramsey was prompted rather by Peano's observation (see above, p. 142) that the paradoxes divide into two kinds: those of pure set theory and those which, like (1), (4), (5), and (6) of Russell's list, hinge rather on semantic concepts such as falsity and specifiability. The only use Russell made of the ramification in his theory of types was, Ramsey rightly observed, in his effort to solve the semantic paradoxes; and these paradoxes, Ramsey pursued, might better have been left out of account because they hinge on concepts extraneous to logic and mathematics. I would add that because of the confusion of propositions with sentences, and of attributes with their expressions, Russell's purported solution of the semantic paradoxes was enigmatic anyway.

Proceeding to the formal presentation of his system, Russell begins with a list of axioms and rules of inference for the logic of truth functions and quantifiers. His axiom (5) was later shown by Bernays (*1926*) to be derivable from the others.

The difference between classes and attributes is extensionality: classes with entirely the same members are identical, whereas attributes of entirely the same objects can be distinct. Accordingly Russell introduces discourse of classes as a transcription of discourse of attributes (or propositional functions), the transcription being so contrived as to assure extensionality of classes. This treatment of classes, like the theory of descriptions dealt with in Whitehead and Russell (*1910*, pp. 216–223 below), illustrates the stratagem of incomplete symbols. There is perhaps some trace in it, indeed, of Russell's no-classes theory. He finally found that theory untenable, but this trace is innocent enough.

The treatment of relations (or relations-in-extension, as against propositional functions) is parallel to that of classes. And the theory of descriptions is stated here too, consonantly with the later selection but more briefly.

Numerous useful concepts from the theory of relations are defined. They go back largely to Peano, Peirce, and Schröder. There is a needless proliferation of notation; thus Russell's $\overrightarrow{R}$‘x, $\overleftarrow{R}$‘x, D‘R, and $\mathrm{C\!\!\!I}$‘R could be rendered quite briefly enough, using others of his notations, as R“ι‘x, $\breve{R}$“ι‘x, R“V, and $\breve{R}$“V. In *Principia mathematica*, where the ideas of this essay are reproduced and developed in great detail, the price of this notational excess is evident; scores of theorems serve merely to link up the various ways of writing things.

W. V. Quine

The paper is printed here with the kind permission of Lord Russell and the Johns Hopkins Press.

The following theory of symbolic logic recommended itself to me in the first instance by its ability to solve certain contradictions, of which the one best known to mathematicians is Burali-Forti's concerning the greatest ordinal.[1] But the theory in question seems not wholly dependent on this indirect recommendation; it has also, if I am not mistaken, a certain consonance with common sense which makes it inherently credible. This, however, is not a merit upon which much stress should be laid; for common sense is far more fallible than it likes to believe. I shall therefore begin by stating some of the contradictions to be solved and shall then show how the theory of logical types effects their solution.

I. The contradictions

(1) The oldest contradiction of the kind in question is the *Epimenides*. Epimenides the Cretan said that all Cretans were liars, and all other statements made by Cretans were certainly lies. Was this a lie? The simplest form of this contradiction is afforded by the man who says "I am lying"; if he is lying, he is speaking the truth, and vice versa.

(2) Let w be the class of all those classes which are not members of themselves. Then, whatever class x may be, "x is a w" is equivalent[2] to "x is not an x". Hence, giving to x the value w, "w is a w" is equivalent to "w is not a w".

(3) Let T be the relation which subsists between two relations R and S whenever R does not have the relation R to S. Then, whatever relations R and S may be, "R has the relation T to S" is equivalent to "R does not have the relation R to S". Hence, giving the value T to both R and S, "T has the relation T to T" is equivalent to "T does not have the relation T to T".

(4) The number of syllables in the English names of finite integers tends to increase as the integers grow larger and must gradually increase indefinitely, since only a finite number of names can be made with a given finite number of syllables. Hence the names of some integers must consist of at least nineteen syllables, and among these there must be a least. Hence "the least integer not nameable in fewer than nineteen syllables" must denote a definite integer; in fact, it denotes 111,777. But "the least integer not nameable in fewer than nineteen syllables" is itself a name consisting of eighteen syllables; hence the least integer not nameable in fewer than nineteen syllables can be named in eighteen syllables, which is a contradiction.[3]

(5) Among transfinite ordinals some can be defined, while others cannot; for the total number of possible definitions is $\aleph_0$, while the number of transfinite ordinals exceeds $\aleph_0$. Hence there must be indefinable ordinals, and among these there must be a least. But this is defined as "the least indefinable ordinal", which is a contradiction.[4]

(6) Richard's paradox[5] is akin to that of the least indefinable ordinal. It is as follows: Consider all decimals that can be defined by means of a finite number of words; let E be the class of such decimals. Then E has $\aleph_0$ terms; hence its members can be ordered as the 1st, 2nd, 3rd, ... Let N be a number defined as follows: If the

[1] See below.

[2] Two propositions are called *equivalent* when both are true or both are false.

[3] This contradiction was suggested to me by Mr. G. G. Berry of the Bodleian Library.

[4] See *König 1905a, Dixon 1906*, and *Hobson 1906*. The solution offered in the last of these papers does not seem to me adequate.

[5] See *Poincaré 1906*, especially sections VII and IX, and *Peano 1906a*, pp. 149–157.

nth figure in the nth decimal is p, let the nth figure in N be $p + 1$ (or 0, if $p = 9$). Then N is different from all the members of E, since, whatever finite value n may have, the nth figure in N is different from the nth figure in the nth of the decimals composing E, and therefore N is different from the nth decimal. Nevertheless we have defined N in a finite number of words, and therefore N ought to be a member of E. Thus N both is and is not a member of E.

(7) Burali-Forti's contradiction (*1897*) may be stated as follows: It can be shown that every well-ordered series has an ordinal number, that the series of ordinals up to and including any given ordinal exceeds the given ordinal by one, and (on certain very natural assumptions) that the series of all ordinals (in order of magnitude) is well-ordered. It follows that the series of all ordinals has an ordinal number, Ω say. But in that case the series of all ordinals including Ω has the ordinal number $\Omega + 1$, which must be greater than Ω. Hence Ω is not the ordinal number of all ordinals.

In all the above contradictions (which are merely selections from an indefinite number) there is a common characteristic, which we may describe as self-reference or reflexiveness. The remark of Epimenides must include itself in its own scope. If *all* classes, provided they are not members of themselves, are members of w, this must also apply to w; and similarly for the analogous relational contradiction. In the cases of names and definitions, the paradoxes result from considering nonnameability and indefinability as elements in names and definitions. In the case of Burali-Forti's paradox, the series whose ordinal number causes the difficulty is the series of all ordinal numbers. In each contradiction something is said about *all* cases of some kind, and from what is said a new case seems to be generated, which both is and is not of the same kind as the cases of which *all* were concerned in what was said. Let us go through the contradictions one by one and see how this occurs.

(1) When a man says "I am lying", we may interpret his statement as: "There is a proposition which I am affirming and which is false". All statements that "there is" so-and-so may be regarded as denying that the opposite is always true; thus "I am lying" becomes "It is not true of all propositions that either I am not affirming them or they are true"; in other words, "It is not true for all propositions p that if I affirm p, p is true". The paradox results from regarding this statement as affirming a proposition, which must therefore come within the scope of the statement. This, however, makes it evident that the notion of "all propositions" is illegitimate; for otherwise, there must be propositions (such as the above) which are about all propositions and yet cannot, without contradiction, be included among the propositions they are about. Whatever we suppose to be the totality of propositions, statements about this totality generate new propositions which, on pain of contradiction, must lie outside the totality. It is useless to enlarge the totality, for that equally enlarges the scope of statements about the totality. Hence there must be no totality of propositions, and "all propositions" must be a meaningless phrase.

(2) In this case, the class w is defined by reference to "all classes" and then turns out to be one among classes. If we seek help by deciding that no class is a member of itself, then w becomes the class of all classes, and we have to decide that this is not a member of itself, that is, is not a class. This is only possible if there is no such thing as the class of all classes in the sense required by the paradox. That there is no such

class results from the fact that, if we suppose there is, the supposition immediately gives rise (as in the above contradiction) to new classes lying outside the supposed total of all classes.

(3) This case is exactly analogous to (2) and shows that we cannot legitimately speak of "all relations".

(4) "The least integer not nameable in fewer than nineteen syllables" involves the totality of names, for it is "the least integer such that all names either do not apply to it or have more than nineteen syllables". Here we assume, in obtaining the contradiction, that a phrase containing "all names" is itself a name, though it appears from the contradiction that it cannot be one of the names which were supposed to be all the names there are. Hence "all names" is an illegitimate notion.

(5) This case, similarly, shows that "all definitions" is an illegitimate notion.

(6) This is solved, like (5), by remarking that "all definitions" is an illegitimate notion. Thus the number E is *not* defined in a finite number of words, being in fact not defined at all.[6]

(7) Burali-Forti's contradiction shows that "all ordinals" is an illegitimate notion; for if not, all ordinals in order of magnitude form a well-ordered series, which must have an ordinal number greater than all ordinals.

Thus all our contradictions have in common the assumption of a totality such that, if it were legitimate, it would at once be enlarged by new members defined in terms of itself.

This leads us to the rule: "Whatever involves *all* of a collection must not be one of the collection", or, conversely: "If, provided a certain collection had a total, it would have members only definable in terms of that total, then the said collection has no total".[7]

The above principle is, however, purely negative in its scope. It suffices to show that many theories are wrong, but it does not show how the errors are to be rectified. We cannot say: "When I speak of *all* propositions, I mean all except those in which 'all propositions' are mentioned"; for in this explanation we have mentioned the propositions in which all propositions are mentioned, which we cannot do significantly. It is impossible to avoid mentioning a thing by mentioning that we won't mention it. One might as well, in talking to a man with a long nose, say: "When I speak of noses, I except such as are inordinately long", which would not be a very successful effort to avoid a painful topic. Thus it is necessary, if we are not to sin against the above negative principle, to construct our logic without mentioning such things as "all propositions" or "all properties" and without even having to say that we are excluding such things. The exclusion must result naturally and inevitably from our positive doctrines, which must make it plain that "all propositions" and "all properties" are meaningless phrases.

The first difficulty that confronts us is as to the fundamental principles of logic known under the quaint name of "laws of thought". "All propositions are either true or false", for example, has become meaningless. If it were significant, it would be a

[6] See *Russell 1906a*.

[7] When I say that a collection has no total, I mean that statements about *all* its members are nonsense. Furthermore, it will be found that the use of this principle requires the distinction of *all* and *any* considered in Section II.

proposition and would come under its own scope. Nevertheless, some substitute must be found, or all general accounts of deduction become impossible.

Another more special difficulty is illustrated by the particular case of mathematical induction. We want to be able to say: "If n is a finite integer, n has all properties possessed by 0 and by the successors of all numbers possessing them". But here "all properties" must be replaced by some other phrase not open to the same objections. It might be thought that "all properties possessed by 0 and by the successors of all numbers possessing them" might be legitimate even if "all properties" were not. But in fact this is not so. We shall find that phrases of the form "all properties which etc." involve *all* properties of which the "etc." can be significantly either affirmed or denied, and not only those which in fact have whatever characteristic is in question; for, in the absence of a catalogue of properties having this characteristic, a statement about all those that have the characteristic must be hypothetical and of the form: "It is always true that, if a property has the said characteristic, then etc.". Thus mathematical induction is prima facie incapable of being significantly enunciated if "all properties" is a phrase destitute of meaning. This difficulty, as we shall see later, can be avoided; for the present we must consider the laws of logic, since these are far more fundamental.

II. All and any

Given a statement containing a variable x, say "$x = x$", we may affirm that this holds in all instances, or we may affirm any one of the instances without deciding as to which instance we are affirming. The distinction is roughly the same as that between the general and particular enunciation in Euclid. The general enunciation tells us something about (say) all triangles, while the particular enunciation takes one triangle and asserts the same thing of this one triangle. But the triangle taken is *any* triangle, not some one special triangle; and thus, although, throughout the proof, only one triangle is dealt with, yet the proof retains its generality. If we say: "Let ABC be a triangle, then the sides AB and AC are together greater than the side BC", we are saying something about *one* triangle, not about *all* triangles; but the one triangle concerned is absolutely ambiguous, and our statement consequently is also absolutely ambiguous. We do not affirm any one definite proposition, but an undetermined one of all the propositions resulting from supposing ABC to be this or that triangle. This notion of ambiguous assertion is very important, and it is vital not to confound an ambiguous assertion with the definite assertion that the same thing holds in *all* cases.

The distinction between (1) asserting any value of a propositional function and (2) asserting that the function is always true is present throughout mathematics, as it is in Euclid's distinction of general and particular enunciations. In any chain of mathematical reasoning, the objects whose properties are being investigated are the arguments to *any* value of some propositional function. Take as an illustration the following definition:

"We call $f(x)$ continuous for $x = a$ if, for every positive number σ, different from 0, there exists a positive number ε, different from 0, such that, for all values of δ which are numerically less than ε, the difference $f(a + \delta) - f(a)$ is numerically less than σ."

Here the function f is *any* function for which the above statement has a meaning; the statement is *about* f, and varies as f varies. But the statement is not *about* σ or ε or δ, because *all* possible values of these are concerned, not one undetermined value. (In regard to ε, the statement "there exists a positive number ε such that etc." is the denial that the denial of "etc." is true of *all* positive numbers.) For this reason, when *any* value of a propositional function is asserted, the argument (for example, f in the above) is called a *real* variable, whereas, when a function is said to be *always* true, or to be not always true, the argument is called an *apparent* variable.[8] Thus in the above definition, f is a real variable, and σ, ε, δ are apparent variables.

When we assert *any* value of a propositional function, we shall say simply that we assert the *propositional function*. Thus if we enunciate the law of identity in the form "$x = x$", we are asserting the function "$x = x$"; that is, we are asserting any value of this function. Similarly we may be said to deny a propositional function when we deny any instance of it. We can only truly assert a propositional function if, whatever value we choose, that value is true; similarly we can only truly deny it if, whatever value we choose, that value is false. Hence in the general case, in which some values are true and some false, we can neither assert nor deny a propositional function.[9]

If φx is a propositional function, we will denote by "$(x).\varphi x$" the proposition "φx is always true". Similarly "$(x, y).\varphi(x, y)$" will mean "$\varphi(x, y)$ is always true", and so on. Then the distinction between the assertion of all values and the assertion of any is the distinction between (1) asserting $(x).\varphi x$ and (2) asserting φx where x is undetermined. The latter differs from the former in that it cannot be treated as one determinate proposition.

The distinction between asserting φx and asserting $(x).\varphi x$ was, I believe, first emphasized by Frege (*1893*, p. 31). His reason for introducing the distinction explicitly was the same which had caused it to be present in the practice of mathematicians, namely, that deduction can only be effected with *real* variables, not with apparent variables. In the case of Euclid's proofs, this is evident: we need (say) some one triangle ABC to reason about, though it does not matter what triangle it is. The triangle ABC is a *real* variable; and although it is *any* triangle, it remains the *same* triangle throughout the argument. But in the general enunciation the triangle is an apparent variable. If we adhere to the apparent variable, we cannot perform any deductions, and this is why in all proofs real variables have to be used. Suppose, to take the simplest case, that we know "φx is always true", that is, "$(x).\varphi x$", and we know "φx always implies ψx", that is, "$(x).\{\varphi x$ implies $\psi x\}$". How shall we infer "ψx is always true", that is, "$(x).\psi x$"? We know it is always true that, if φx is true and if φx implies ψx, then ψx is true. But we have no premisses to the effect that φx is true and φx implies ψx; what we have is: φx is *always* true, and φx *always* implies ψx. In order to make our inference, we must go from "φx is always true" to φx, and from "φx always implies ψx" to "φx implies ψx", where the x, while remaining any possible argument, is to be the same in both. Then, from "φx" and "φx implies ψx",

[8] These two terms are due to Peano (*1903a*), who uses them approximately in the above sense.

[9] Mr. MacColl speaks of "propositions" as divided into the three classes of certain, variable, and impossible. We may accept this division as applying to propositional functions. A function which can be asserted is certain, one which can be denied is impossible, and all others are (in Mr. MacColl's sense) variable.

we infer "ψx"; thus ψx is true for any possible argument and therefore is always true. Thus, in order to infer "$(x).\psi x$" from "$(x).\varphi x$" and "$(x).\{\varphi x$ implies $\psi x\}$", we have to pass from the apparent to the real variable and then back again to the apparent variable. This process is required in all mathematical reasoning which proceeds from the assertion of all values of one or more propositional functions to the assertion of all values of some other propositional function, as, for example, from "all isosceles triangles have equal angles at the base" to "all triangles having equal angles at the base are isosceles". In particular, this process is required in proving *Barbara* and the other moods of the syllogism. In a word, *all deduction operates with real variables* (or with constants).

It might be supposed that we could dispense with apparent variables altogether, contenting ourselves with *any* as a substitute for *all*. This, however, is not the case. Take, for example, the definition of a continuous function quoted above: in this definition σ, ε, and δ must be apparent variables. Apparent variables are constantly required for definitions. Take, for example, the following: "An integer is called a prime when it has no integral factors except 1 and itself". This definition unavoidably involves an apparent variable in the form: "If n is an integer other than 1 or the given integer, n is not a factor of the given integer, for all possible values of n".

The distinction between *all* and *any* is, therefore, necessary to deductive reasoning and occurs throughout mathematics, though, so far as I know, its importance remained unnoticed until Frege pointed it out.

For our purposes it has a different utility, which is very great. In the case of such variables as propositions or properties, "any value" is legitimate, though "all values" is not. Thus we may say: "p is true or false, where p is any proposition", though we cannot say "all propositions are true or false". The reason is that, in the former, we merely affirm an undetermined one of the propositions of the form "p is true or false", whereas in the latter we affirm (if anything) a new proposition, different from all the propositions of the form "p is true or false". Thus we may admit "any value" of a variable in cases where "all values" would lead to reflexive fallacies; for the admission of "any value" does not in the same way create new values. Hence the fundamental laws of logic can be stated concerning *any* proposition, though we cannot significantly say that they hold of *all* propositions. These laws have, so to speak, a particular enunciation but no general enunciation. There is no one proposition which *is* the law of contradiction (say); there are only the various instances of the law. Of any proposition p, we can say: "p and not-p cannot both be true"; but there is no such proposition as: "Every proposition p is such that p and not-p cannot both be true".

A similar explanation applies to properties. We can speak of *any* property of x, but not of *all* properties, because new properties would be thereby generated. Thus we can say: "If n is a finite integer, and if 0 has the property φ and $m + 1$ has the property φ provided m has it, it follows that n has the property φ". Here we need not specify φ; φ stands for "any property". But we cannot say: "A finite integer is defined as one which has *every* property φ possessed by 0 and by the successors of possessors". For here it is essential to consider *every* property,[10] not *any* property; and in using such a definition we assume that it embodies a *property* distinctive of

[10] This is indistinguishable from "all properties".

finite integers, which is just the kind of assumption from which, as we saw, the reflexive contradictions spring.

In the above instance, it is necessary to avoid the suggestions of ordinary language, which is not suitable for expressing the distinction required. The point may be illustrated further as follows: If induction is to be used for defining finite integers, induction must state a definite property of finite integers, not an ambiguous property. But, if φ is a real variable, the statement "n has the property φ provided this property is possessed by 0 and by the successors of possessors" assigns to n a property which varies as φ varies, and such a property cannot be used to define the class of finite integers. We wish to say: "'n is a finite integer' means: 'Whatever property φ may be, n has the property φ provided φ is possessed by 0 and by the successors of possessors'". But here φ has become an *apparent* variable. To keep it a real variable, we should have to say: "Whatever property φ may be, 'n is a finite integer' means: 'n has the property φ provided φ is possessed by 0 and by the successors of possessors'". But here the meaning of 'n is a finite integer' varies as φ varies, and thus such a definition is impossible. This case illustrates an important point, namely the following: "The scope[11] of a real variable can never be less than the whole propositional function in the assertion of which the said variable occurs". That is, if our propositional function is (say) "φx implies p", the assertion of this function will mean "any value of 'φx implies p' is true", *not* "'any value of φx is true' implies p". In the latter, we have really "*all* values of φx are true", and the x is an *apparent* variable.

III. The meaning and range of generalized propositions

In this section we have to consider first the meaning of propositions in which the word *all* occurs and then the kind of collections which admit of propositions about all their members.

It is convenient to give the name *generalized propositions* not only to such as contain *all*, but also to such as contain *some* (undefined). The proposition "φx is sometimes true" is equivalent to the denial of "not-φx is always true"; "some A is B" is equivalent to the denial of "all A is not B", that is, of "no A is B". Whether it is possible to find interpretations which distinguish "φx is sometimes true" from the denial of "not-φx is always true", it is unnecessary to inquire; for our purposes we may *define* "φx is sometimes true" as the denial of "not-φx is always true". In any case, the two kinds of propositions require the same kind of interpretation and are subject to the same limitations. In each there is an apparent variable; and it is the presence of an apparent variable which constitutes what I mean by a generalized proposition. (Note that there cannot be a *real* variable in any proposition; for what contains a real variable is a propositional function, not a proposition.)

The first question we have to ask in this section is: How are we to interpret the word *all* in such propositions as "all men are mortal"? At first sight, it might be thought that there could be no difficulty, that "all men" is a perfectly clear idea, and that we say of all men that they are mortal. But to this view there are many objections.

[11] The *scope* of a real variable is the whole function of which "any value" is in question. Thus in "φx implies p" the scope of x is not φx, but "φx implies p".

(1) If this view were right, it would seem that "all men are mortal" could not be true if there were no men. Yet, as Mr. Bradley has urged (*1883*, p. 47), "Trespassers will be prosecuted" may be perfectly true even if no one trespasses; and hence, as he further argues, we are driven to interpret such propositions as hypotheticals, meaning "if anyone trespasses, he will be prosecuted", that is, "if x trespasses, x will be prosecuted", where the range of values which x may have, whatever it is, is certainly not confined to those who really trespass. Similarly "all men are mortal" will mean "if x is a man, x is mortal, where x may have any value within a certain range". What this range is, remains to be determined; but in any case it is wider than "men", for the above hypothetical is certainly often true when x is not a man.

(2) "All men" is a denoting phrase; and it would appear, for reasons which I have set forth elsewhere (*1905*), that denoting phrases never have any meaning in isolation, but only enter as constituents into the verbal expression of propositions which contain no constituent corresponding to the denoting phrases in question. That is to say, a denoting phrase is defined by means of the propositions in whose verbal expression it occurs. Hence it is impossible that these propositions should acquire their meaning through the denoting phrases; we must find an independent interpretation of the propositions containing such phrases and must not use these phrases in explaining what such propositions mean. Hence we cannot regard "all men are mortal" as a statement about "all men".

(3) Even if there were such an object as "all men", it is plain that it is not this object to which we attribute mortality when we say "all men are mortal". If we were attributing mortality to this object, we should have to say "*all men* is mortal". Thus the supposition that there is such an object as "all men" will not help us to interpret "all men are mortal".

(4) It seems obvious that, if we meet something which may be a man or may be an angel in disguise, it comes within the scope of "all men are mortal" to assert "if this is a man, it is mortal". Thus again, as in the case of the trespassers, it seems plain that we are really saying "if anything is a man, it is mortal", and that the question whether this or that is a man does not fall within the scope of our assertion, as it would do if the *all* really referred to "all men".

(5) We thus arrive at the view that what is meant by "all men are mortal" may be more explicitly stated in some such form as "it is always true that if x is a man, x is mortal". Here we have to inquire as to the scope of the word *always*.

(6) It is obvious that *always* includes some cases in which x is not a man, as we saw in the case of the disguised angel. If x were limited to the case when x is a man, we could infer that x is a mortal, since if x is a man, x is a mortal. Hence, with the same meaning of *always*, we should find "it is always true that x is mortal". But it is plain that, without altering the meaning of *always*, this new proposition is false, though the other was true.

(7) One might hope that "always" would mean "for all values of x". But "all values of x", if legitimate, would include as parts "all propositions" and "all functions", and such illegitimate totalities. Hence the values of x must be somehow restricted within some legitimate totality. This seems to lead us to the traditional doctrine of a "universe of discourse" within which x must be supposed to lie.

(8) Yet it is quite essential that we should have some meaning of *always* which

does not have to be expressed in a restrictive hypothesis as to x. For suppose "always" means "whenever x belongs to the class i". Then "all men are mortal" becomes "whenever x belongs to the class i, if x is a man, x is mortal", that is, "it is always true that if x belongs to the class i, then, if x is a man, x is mortal". But what is our new *always* to mean? There seems no more reason for restricting x, in this new proposition, to the class i, than there was before for restricting it to the class *man*. Thus we shall be led on to a new wider universe, and so on ad infinitum, unless we can discover some natural restriction upon the possible values of (that is, some restriction given with) the function "if x is a man, x is mortal", and not needing to be imposed from without.

(9) It seems obvious that, since all men are mortal, there cannot be any *false* proposition which is a value of the function "if x is a man, x is mortal". For if this is a proposition at all, the hypothesis "x is a man" must be a proposition, and so must the conclusion "x is mortal". But if the hypothesis is false, the hypothetical is true; and if the hypothesis is true, the hypothetical is true. Hence there can be no false propositions of the form "if x is a man, x is mortal".

(10) It follows that, if any values of x are to be excluded, they can only be values for which there is no proposition of the form "if x is a man, x is mortal", that is, for which this phrase is meaningless. Since, as we saw in (7), there must be excluded values of x, it follows that the function "if x is a man, x is mortal" must have a certain *range of significance*,[12] which falls short of all imaginable values of x, though it exceeds the values which are men. The restriction on x is therefore a restriction to the range of significance of the function "if x is a man, x is mortal".

(11) We thus reach the conclusion that "all men are mortal" means "if x is a man, x is mortal, always", where *always* means "for all values of the function 'if x is a man, x is mortal'". This is an *internal* limitation upon x, given by the nature of the function; and it is a limitation which does not require explicit statement, since it is impossible for a function to be true more generally than for all its values. Moreover, if the range of significance of the function is i, the function "if x is an i, then if x is a man, x is mortal" has the same range of significance, since it cannot be significant unless its constituent "if x is a man, x is mortal" is significant. But here the range of significance is again implicit, as it was in "if x is a man, x is mortal"; thus we cannot make ranges of significance explicit, since the attempt to do so only gives rise to a new proposition in which the same range of significance is implicit.

Thus generally: "$(x) . \varphi x$" is to mean "φx always". This may be interpreted, though with less exactitude, as "φx is always true", or, more explicitly: "All propositions of the form φx are true", or "All values of the function φx are true".[13] Thus the fundamental *all* is "all values of a propositional function", and every other *all* is derivative from this. And every propositional function has a certain *range of significance*, within which lie the arguments for which the function has values.

[12] A function is said to be significant for the argument x if it has a value for this argument. Thus we may say shortly "φx is significant", meaning "the function φ has a value for the argument x". The range of significance of a function consists of all the arguments for which the function is true, together with all the arguments for which it is false.

[13] A linguistically convenient expression for this idea is "φx is true for all *possible* values of x", a possible value being understood to be one for which φx is significant.

Within this range of arguments, the function is true or false; outside this range, it is nonsense.

The above argumentation may be summed up as follows:

The difficulty which besets attempts to restrict the variable is that restrictions naturally express themselves as hypotheses that the variable is of such or such a kind, and that, when so expressed, the resulting hypothetical is free from the intended restriction. For example, let us attempt to restrict the variable to *men*, and assert that, subject to this restriction, "x is mortal" is always true. Then what is always true is that if x is a man, x is mortal; and this hypothetical is true even when x is not a man. Thus a variable can never be restricted within a certain range if the propositional function in which the variable occurs remains significant when the variable is outside that range. But if the function ceases to be significant when the variable goes outside a certain range, then the variable is *ipso facto* confined to that range, without the need of any explicit statement to that effect. This principle is to be borne in mind in the development of logical types, to which we shall shortly proceed.

We can now begin to see how it comes that "all so-and-so's" is sometimes a legitimate phrase and sometimes not. Suppose we say "all terms which have the property φ have the property ψ". That means, according to the above interpretation, "φx always implies ψx". Provided the range of significance of φx is the same as that of ψx, this statement is significant; thus, given any definite function φx, there are propositions about "all the terms satisfying φx". But it sometimes happens (as we shall see more fully later on) that what appears verbally as one function is really many analogous functions with different ranges of significance. This applies, for example, to "p is true", which, we shall find, is not really one function of p, but is different functions according to the kind of proposition that p is. In such a case, the *phrase* expressing the ambiguous function may, owing to the ambiguity, be significant throughout a set of values of the argument exceeding the range of significance of any one function. In such a case, *all* is not legitimate. Thus if we try to say "all true propositions have the property φ", that is, "'p is true' always implies φp", the possible arguments to "p is true" necessarily exceed the possible arguments to φ, and therefore the attempted general statement is impossible. For this reason, genuine general statements about all true propositions cannot be made. It may happen, however, that the supposed function φ is really ambiguous like "p is true"; and, if it happens to have an ambiguity precisely of the same kind as that of "p is true", we may be able always to give an interpretation to the proposition "'p is true' implies φp". This will occur, for example, if φp is "not-p is false". Thus we get an appearance, in such cases, of a general proposition concerning *all* propositions; but this appearance is due to a systematic ambiguity about such words as *true* and *false*. (This systematic ambiguity results from the hierarchy of propositions which will be explained later on.) We may, in all such cases, make our statement about *any* proposition, since the meaning of the ambiguous words will adapt itself to any proposition. But if we turn our proposition into an apparent variable and say something about *all*, we must suppose the ambiguous words fixed to this or that possible meaning, though it may be quite irrelevant which of their possible meanings they are to have. This is how it happens both that *all* has limitations which exclude "all propositions" and that there nevertheless *seem* to be true statements about "all propositions". Both these points will become plainer when the theory of types has been explained.

It has often been suggested[14] that what is required in order that it may be legitimate to speak of *all* of a collection is that the collection should be finite. Thus "all men are mortal" will be legitimate because men form a finite class. But that is not really the reason why we can speak of "all men". What is essential, as appears from the above discussion, is not finitude, but what may be called *logical homogeneity*. This property is to belong to any collection whose terms are all contained within the range of significance of some one function. It would always be obvious at a glance whether a collection possessed this property or not, if it were not for the concealed ambiguity in common logical terms such as *true* and *false*, which gives an appearance of being a single function to what is really a conglomeration of many functions with different ranges of significance.

The conclusions of this section are as follows: Every proposition containing *all* asserts that some propositional function is always true; and this means that all values of the said function are true, not that the function is true for all arguments, since there are arguments for which any given function is meaningless, that is, has no value. Hence we can speak of *all* of a collection when and only when the collection forms part or the whole of the *range of significance* of some propositional function, the range of significance being defined as the collection of those arguments for which the function in question is significant, that is, has a value.

IV. The hierarchy of types

A *type* is defined as the range of significance of a propositional function, that is, as the collection of arguments for which the said function has values. Whenever an apparent variable occurs in a proposition, the range of values of the apparent variable is a type, the type being fixed by the function of which "all values" are concerned. The division of objects into types is necessitated by the reflexive fallacies which otherwise arise. These fallacies, as we saw, are to be avoided by what may be called the "vicious-circle principle", that is, "no totality can contain members defined in terms of itself". This principle, in our technical language, becomes: "Whatever contains an apparent variable must not be a possible value of that variable". Thus whatever contains an apparent variable must be of a different type from the possible values of that variable; we will say that it is of a *higher* type. Thus the apparent variables contained in an expression are what determines its type. This is the guiding principle in what follows.

Propositions which contain apparent variables are generated from such as do not contain these apparent variables by processes of which one is always the process of *generalization*, that is, the substitution of a variable for one of the terms of a proposition and the assertion of the resulting function for all possible values of the variable. Hence a proposition is called a *generalized* proposition when it contains an apparent variable. A proposition containing no apparent variable we will call an *elementary* proposition. It is plain that a proposition containing an apparent variable presupposes others from which it can be obtained by generalization; hence all generalized

[14] For example, by Poincaré (*1906*).

propositions presuppose elementary propositions. In an elementary proposition we can distinguish one or more *terms* from one or more *concepts*; the *terms* are whatever can be regarded as the *subject* of the proposition, while the concepts are the predicates or relations asserted of these terms.[15] The terms of elementary propositions we will call *individuals*; these form the first or lowest type.

It is unnecessary, in practice, to know what objects belong to the lowest type, or even whether the lowest type of variable occurring in a given context is that of individuals or some other. For in practice only the *relative* types of variables are relevant; thus the lowest type occurring in a given context may be called that of individuals, so far as that context is concerned. It follows that the above account of individuals is not essential to the truth of what follows; all that is essential is the way in which other types are generated from individuals, however the type of individuals may be constituted.

By applying the process of generalization to individuals occurring in elementary propositions, we obtain new propositions. The legitimacy of this process requires only that no individuals should be propositions. That this is so, is to be secured by the meaning we give to the word *individual*. We may define an individual as something destitute of complexity; it is then obviously not a proposition, since propositions are essentially complex. Hence in applying the process of generalization to individuals we run no risk of incurring reflexive fallacies.

Elementary propositions together with such as contain only individuals as apparent variables we will call *first-order propositions*. These form the second logical type.

We have thus a new totality, that of *first-order propositions*. We can thus form new propositions in which first-order propositions occur as apparent variables. These we will call *second-order propositions*; these form the third logical type. Thus, for example, if Epimenides asserts "all first-order propositions affirmed by me are false", he asserts a second-order proposition; he may assert this truly, without asserting truly any first-order proposition, and thus no contradiction arises.

The above process can be continued indefinitely. The $(n + 1)$th logical type will consist of propositions of order n, which will be such as contain propositions of order $n - 1$, but of no higher order, as apparent variables. The types so obtained are mutually exclusive, and thus no reflexive fallacies are possible so long as we remember that an apparent variable must always be confined within some one type.

In practice, a hierarchy of *functions* is more convenient than one of propositions. Functions of various orders may be obtained from propositions of various orders by the method of *substitution*. If p is a proposition and a a constituent of p, let "$p/a'x$" denote the proposition which results from substituting x for a wherever a occurs in p. Then p/a, which we will call a *matrix*, may take the place of a function; its value for the argument x is $p/a'x$, and its value for the argument a is p. Similarly, if "$p/(a, b)'(x, y)$" denotes the result of first substituting x for a and then substituting y for b, we may use the double matrix $p/(a, b)$ to represent a double function. In this way we can avoid apparent variables other than individuals and propositions of various orders. The *order* of a matrix will be defined as being the order of the proposition in which the substitution is effected, which proposition we will call the *prototype*. The order

[15] See *Russell 1903*, § 48.

of a matrix does not determine its type: in the first place because it does not determine the number of arguments for which others are to be substituted (that is, whether the matrix is of the form p/a or $p/(a, b)$ or $p/(a, b, c)$, etc.); in the second place because, if the prototype is of more than the first order, the arguments may be either propositions or individuals. But it is plain that the type of a matrix is definable always by means of the hierarchy of propositions.

Although it is *possible* to replace functions by matrices, and although this procedure introduces a certain simplicity into the explanation of types, it is technically inconvenient. Technically, it is convenient to replace the prototype p by φa, and to replace $p/a{}^{\cdot}x$ by φx; thus where, if matrices were being employed, p and a would appear as apparent variables, we now have φ as our apparent variable. In order that φ may be legitimate as an apparent variable, it is necessary that its values should be confined to propositions of some one type. Hence we proceed as follows.

A function whose argument is an individual and whose value is always a first-order proposition will be called a first-order function. A function involving a first-order function or proposition as apparent variable will be called a second-order function, and so on. A function of one variable which is of the order next above that of its argument will be called a *predicative* function; the same name will be given to a function of several variables if there is one among these variables in respect of which the function becomes predicative when values are assigned to all the other variables. Then the type of a function is determined by the type of its values and the number and type of its arguments.

The hierarchy of functions may be further explained as follows. A first-order function of an individual x will be denoted by $\varphi!x$ (the letters ψ, χ, θ, f, g, F, G will also be used for functions). No first-order function contains a function as apparent variable; hence such functions form a well-defined totality, and the φ in $\varphi!x$ can be turned into an apparent variable. Any proposition in which φ appears as apparent variable and there is no apparent variable of higher type than φ is a second-order proposition. If such a proposition contains an individual x, it is not a predicative function of x; but, if it contains a first-order function φ, it is a predicative function of φ and will be written $f!(\psi!\hat{z})$. Then f is a *second-order predicative function*; the possible values of f again form a well-defined totality, and we can turn f into an apparent variable. We can thus define *third-order predicative functions*, which will be such as have third-order propositions for their values and second-order predicative functions for their arguments. And in this way we can proceed indefinitely. A precisely similar development applies to functions of several variables.

We will adopt the following conventions. Variables of the lowest type occurring in any context will be denoted by small Latin letters (excluding f and g, which are reserved for functions); a predicative function of an argument x (where x may be of any type) will be denoted by $\varphi!x$ (where ψ, χ, θ, f, g, F or G may replace φ); similarly a predicative function of two arguments x and y will be denoted by $\varphi!(x, y)$; a general function of x will be denoted by φx, and a general function of x and y by $\varphi(x, y)$. In φx, φ cannot be made into an apparent variable, since its type is indeterminate; but in $\varphi!x$, where φ is a *predicative* function whose argument is of some given type, φ *can* be made into an apparent variable.

It is important to observe that since there are various types of propositions and functions, and since generalization can only be applied within some one type, all phrases containing the words "all propositions" or "all functions" are prima facie meaningless, though in certain cases they are capable of an unobjectionable interpretation. The contradictions arise from the use of such phrases in cases where no innocent meaning can be found.

If we now revert to the contradictions, we see at once that some of them are solved by the theory of types. Wherever "all propositions" are mentioned, we must substitute "all propositions of order n", where it is indifferent what value we give to n but it is essential that n should have *some* value. Thus when a man says "I am lying", we must interpret him as meaning: "There is a proposition of order n which I affirm and which is false". This is a proposition of order $n + 1$; hence the man is not affirming any proposition of order n; hence his statement is false, and yet its falsehood does not imply, as that of "I am lying" appeared to do, that he is making a true statement. This solves the Liar.

Consider next "the least integer not nameable in fewer than nineteen syllables". It is to be observed, in the first place, that *nameable* must mean "nameable by means of such-and-such assigned names", and that the number of assigned names must be finite. For if it is not finite, there is no reason why there should be any integer not nameable in fewer than nineteen syllables, and the paradox collapses. We may next suppose that "nameable in terms of names of the class N" means "being the only term satisfying some function composed wholly of names of the class N". The solution of this paradox lies, I think, in the simple observation that "nameable in terms of names of the class N" is never itself nameable in terms of names of that class. If we enlarge N by adding the name "nameable in terms of names of the class N", our fundamental apparatus of names is enlarged; calling the new apparatus N', "nameable in terms of names of the class N'" remains not nameable in terms of names of the class N'. If we try to enlarge N till it embraces *all* names, "nameable" becomes (by what was said above) "being the only term satisfying some function composed wholly of names". But here there is a function as apparent variable; hence we are confined to predicative functions of some one type (for nonpredicative functions cannot be apparent variables). Hence we have only to observe that nameability in terms of such functions is nonpredicative in order to escape the paradox.

The case of "the least indefinable ordinal" is closely analogous to the case we have just discussed. Here, as before, "definable" must be relative to some given apparatus of fundamental ideas; and there is reason to suppose that "definable in terms of ideas of the class N" is not definable in terms of ideas of the class N. It will be true that there is some definite segment of the series of ordinals consisting wholly of definable ordinals and having the least indefinable ordinal as its limit. This least indefinable ordinal will be definable by a slight enlargement of our fundamental apparatus; but there will then be a new ordinal which will be the least that is indefinable with the new apparatus. If we enlarge our apparatus so as to include all possible ideas, there is no longer any reason to believe that there is any indefinable ordinal. The apparent force of the paradox lies largely, I think, in the supposition that if all the ordinals of a certain class are definable, the class must be definable, in which case its successor is of course also definable; but there is no reason for accepting this supposition.

The other contradictions, that of Burali-Forti in particular, require some further developments for their solution.

V. The axiom of reducibility

A propositional function of x may, as we have seen, be of any order; hence any statement about "all properties of x" is meaningless. (A "property of x" is the same thing as a "propositional function which holds of x".) But it is absolutely necessary, if mathematics is to be possible, that we should have some method of making statements which will usually be equivalent to what we have in mind when we (inaccurately) speak of "all properties of x". This necessity appears in many cases, but especially in connection with mathematical induction. We can say, by the use of *any* instead of *all*, "Any property possessed by 0, and by the successors of all numbers possessing it, is possessed by all finite numbers". But we cannot go on to: "A finite number is one which possesses *all* properties possessed by 0 and by the successors of all numbers possessing them". If we confine this statement to all first-order properties of numbers, we cannot infer that it holds of second-order properties. For example, we shall be unable to prove that, if m and n are finite numbers, then $m + n$ is a finite number. For, with the above definition, "m is a finite number" is a second-order property of m; hence the fact that $m + 0$ is a finite number, and that, if $m + n$ is a finite number, so is $m + n + 1$, does not allow us to conclude by induction that $m + n$ is a finite number. It is obvious that such a state of things renders much of elementary mathematics impossible.

The other definition of finitude, by the nonsimilarity of whole and part, fares no better. For this definition is: "A class is said to be finite when every one-one relation whose domain is the class and whose converse domain is contained in the class has the whole class for its converse domain". Here a variable relation appears, that is, a variable function of two variables; we have to take *all* values of this function, which requires that it should be of some assigned order; but any assigned order will not enable us to deduce many of the propositions of elementary mathematics.

Hence we must find, if possible, some method of reducing the order of a propositional function without affecting the truth or falsehood of its values. This seems to be what common sense effects by the admission of *classes*. Given any propositional function, φx, of whatever order, this is assumed to be equivalent, for all values of x, to a statement of the form "x belongs to the class α". Now this statement is of the first order, since it makes no allusion to "all functions of such-and-such a type". Indeed its only practical advantage over the original statement φx is that it is of the first order. There is no advantage in assuming that there really are such things as classes, and the contradiction about the classes which are not members of themselves shows that, if there are classes, they must be something radically different from individuals. I believe the chief purpose which classes serve, and the chief reason which makes them linguistically convenient, is that they provide a method of reducing the order of a propositional function. I shall, therefore, not assume anything of what may seem to be involved in the common-sense admission of classes, except this: that every propositional function is equivalent, for all its values, to some predicative function.

This assumption with regard to functions is to be made whatever may be the type of their arguments. Let φx be a function, of any order, of an argument x, which may itself be either an individual or a function of any order. If φ is of the order next above x, we write the function in the form $\varphi ! x$; in such a case we will call φ a *predicative* function. Thus a predicative function of an individual is a first-order function; and for higher types of arguments, predicative functions take the place that first-order functions take in respect of individuals. We assume, then, that every function is equivalent, for all its values, to some predicative function of the same argument. This assumption seems to be the essence of the usual assumption of classes; at any rate, it retains as much of classes as we have any use for, and little enough to avoid the contradictions which a less grudging admission of classes is apt to entail. We will call this assumption the *axiom of classes*, or the *axiom of reducibility*.

We shall assume similarly that every function of two variables is equivalent, for all its values, to a predicative function of those variables, where a predicative function of two variables is one such that there is one of the variables in respect of which the function becomes predicative (in our previous sense) when a value is assigned to the other variable. This assumption is what seems to be meant by saying that any statement about two variables defines a relation between them. We will call this assumption the *axiom of relations*, or the *axiom of reducibility*.

In dealing with relations between more than two terms, similar assumptions would be needed for three, four, ... variables. But these assumptions are not indispensable for our purpose and are therefore not made in this paper.

By the help of the axiom of reducibility, statements about "all first-order functions of x" or "all predicative functions of α" yield most of the results which otherwise would require "all functions". The essential point is that such results are obtained in all cases where only the truth or falsehoold of values of the functions concerned are relevant, as is invariably the case in mathematics. Thus mathematical induction, for example, need now only be stated for all predicative functions of numbers; it then follows from the axiom of classes that it holds of *any* function of whatever order. It might be thought that the paradoxes for the sake of which we invented the hierarchy of types would now reappear. But this is not the case, because, in such paradoxes, either something beyond the truth or falsehood of values of functions is relevant, or expressions occur which are unmeaning even after the introduction of the axiom of reducibility. For example, such a statement as "Epimenides asserts ψx" is not equivalent to "Epimenides asserts $\varphi ! x$", even though ψx and $\varphi ! x$ are equivalent. Thus "I am lying" remains unmeaning if we attempt to include *all* propositions among those which I may be falsely affirming and is unaffected by the axiom of classes if we confine it to propositions of order n. The hierarchy of propositions and functions, therefore, remains relevant in just those cases in which there is a paradox to be avoided.

VI. Primitive ideas and propositions of symbolic logic

The primitive ideas required in symbolic logic appear to be the following seven:

(1) Any propositional function of a variable x or of several variables $x, y, z \dots$. This will be denoted by φx or $\varphi(x, y, z, \dots)$.

(2) The negation of a proposition. If p is the proposition, its negation will be denoted by $\sim p$.

(3) The disjunction or logical sum of two propositions, that is, "this or that". If p and q are the two propositions, their disjunction will be denoted by $p \vee q$.[16]

(4) The truth of *any* value of a propositional function, that is, of φx where x is not specified.

(5) The truth of *all* values of a propositional function. This is denoted by $(x).\varphi x$ or $(x):\varphi x$ or whatever larger number of dots may be necessary to bracket off the proposition.[17] In $(x).\varphi x$, x is called an *apparent variable*, whereas when φx is asserted, where x is not specified, x is called a *real variable*.

(6) Any predicative function of an argument of any type; this will be represented by $\varphi!x$ or $\varphi!\alpha$ or $\varphi!R$, according to circumstances. A predicative function of x is one whose values are propositions of the type next above that of x, if x is an individual or a proposition, or that of values of x if x is a function. It may be described as one in which the apparent variables, if any, are all of the same type as x or of lower type; and a variable is of lower type than x if it can significantly occur as argument to x, or as argument to an argument to x, and so forth.

(7) Assertion, that is, the assertion that some proposition is true, or that any value of some propositional function is true. This is required to distinguish a proposition actually asserted from one merely considered, or from one adduced as hypothesis to some other. It will be indicated by the sign "$\vdash$" prefixed to what is asserted, with enough dots to bracket off what is asserted.[18]

Before proceeding to the primitive propositions, we need certain definitions. In the following definitions, as well as in the primitive propositions, the letters p, q, and r are used to denote propositions.

$$p \supset q \,.=.\, \sim p \vee q \quad \text{Df.}$$

This definition states that "$p \supset q$" (which is read "p implies q") is to mean "p is false or q is true". I do not mean to affirm that "implies" cannot have any other meaning, but only that this meaning is the one which it is most convenient to give to "implies" in symbolic logic. In a definition, the sign of equality and the letters "Df." are to be regarded as one symbol, meaning jointly "is defined to mean". The sign of equality without the letters "Df." has a different meaning, to be defined shortly.

$$p.q \,.=.\, \sim(\sim p \vee \sim q) \quad \text{Df.}$$

This defines the logical product of two propositions p and q, that is, "p and q are both true". The above definition states that this is to mean: "It is false that either p is false or q is false". Here again, the definition does not give the only meaning which can be given to "p and q are both true" but gives the meaning which is most convenient for our purposes.

$$p \equiv q \,.=.\, p \supset q \,.\, q \supset p \quad \text{Df.}$$

[16] In a previous article (*1906*), I took implication as indefinable, instead of disjunction. The choice between the two is a matter of taste; I now choose disjunction, because it enables us to diminish the number of primitive propositions.

[17] The use of dots follows Peano's usage. It is fully explained by Whitehead (*1902*; *1906*, p. 472).

[18] This sign, as well as the introduction of the idea which it expresses, is due to Frege. See *Frege 1879*, p. 1, and *1893*, p. 9.

That is, "$p \equiv q$", which is read "p is equivalent to q", means "p implies q and q implies p"; whence, of course, it follows that p and q are both true or both false.

$$(\exists x) . \varphi x .=. \sim\{(x) . \sim\varphi x\} \quad \text{Df.}$$

This defines "there is at least one value of x for which φx is true". We define it as meaning "it is false that φx is always false".

$$x = y .=: (\varphi):\varphi ! x .\supset. \varphi ! y \quad \text{Df.}$$

This is the definition of identity. It states that x and y are to be called identical when every predicative function satisfied by x is satisfied by y. It follows from the axiom of reducibility that if x satisfies ψx, where ψ is any function, predicative or nonpredicative, then y satisfies ψy.

The following definitions are less important and are introduced solely for the purpose of abbreviation.

$$(x, y) . \varphi(x, y) .=: (x):(y). \varphi(x, y) \quad \text{Df.}$$

$$(\exists x, y) . \varphi(x, y) .=: (\exists x):(\exists y). \varphi(x, y) \quad \text{Df.}$$

$$\varphi x .\supset_x. \psi x :=: (x):\varphi x \supset \psi x \quad \text{Df.}$$

$$\varphi x .\equiv_x. \psi x :=: (x):\varphi x .\equiv. \psi x \quad \text{Df.}$$

$$\varphi(x, y) .\supset_{x,y}. \psi(x, y) :=: (x, y):\varphi(x, y) .\supset. \psi(x, y) \quad \text{Df.},$$

and so on for any number of variables.

The primitive propositions required are as follows. (In 2, 3, 4, 5, 6, and 10, p, q, and r stand for propositions.)

(1) A proposition implied by a true premiss is true.

(2) $\vdash: p \vee p . \supset . p$.

(3) $\vdash: q . \supset . p \vee q$.

(4) $\vdash: p \vee q . \supset . q \vee p$.

(5) $\vdash: p \vee (q \vee r) . \supset . q \vee (p \vee r)$.

(6) $\vdash:. q \supset r . \supset : p \vee q . \supset . p \vee r$.

(7) $\vdash: (x) . \varphi x . \supset . \varphi y$;

that is, "if all values of $\varphi \hat{x}$ are true, then φy is true, where φy is any value".[19]

(8) If φy is true, where φy is any value of $\varphi \hat{x}$, then $(x) . \varphi x$ is true. This cannot be expressed in our symbols; for if we write "$\varphi y . \supset . (x) . \varphi x$," that means "$\varphi y$ implies that all values of $\varphi \hat{x}$ are true, where y may have any value of the appropriate type", which is not in general the case. What we mean to assert is: "If, however y is chosen, φy is true, then $(x) . \varphi x$ is true", whereas what is expressed by "$\varphi y . \supset . (x) . \varphi x$" is: "However y is chosen, if φy is true, then $(x) . \varphi x$ is true", which is quite a different statement, and in general a false one.

(9) $\vdash: (x) . \varphi x . \supset . \varphi a$, where a is any definite constant.

This principle is really as many different principles as there are possible values of a. That is, it states that, for example, whatever holds of all individuals holds of Socrates; also that it holds of Plato; and so on. It is the principle that a general rule

[19] It is convenient to use the notation $\varphi \hat{x}$ to denote the function itself, as opposed to this or that value of the function.

may be applied to particular cases; but in order to give it scope, it is necessary to mention the particular cases, since otherwise we need the principle itself to assure us that the general rule that general rules may be applied to particular cases may be applied (say) to the particular case of Socrates. It is thus that this principle differs from (7); our present principle makes a statement about Socrates, or about Plato, or some other definite constant, whereas (7) made a statement about a variable.

The above principle is never used in symbolic logic or in pure mathematics, since all our propositions are general, and even when (as in "one is a number") we seem to have a strictly particular case, this turns out not to be so when closely examined. In fact, the use of the above principle is the distinguishing mark of *applied* mathematics. Thus, strictly speaking, we might have omitted it from our list.

(10) $\vdash :. (x) . p \vee \varphi x . \supset : p . \vee . (x) . \varphi x$;

that is, "if 'p or φx' is always true, then either p is true, or φx is always true".

(11) When $f(\varphi x)$ is true whatever argument x may be, and $F(\varphi y)$ is true whatever possible argument y may be, then $\{f(\varphi x) . F(\varphi x)\}$ is true whatever possible argument x may be.

This is the axiom of the "identification of variables". It is needed when two separate propositional functions are each known to be always true and we wish to infer that their logical product is always true. This inference is only legitimate if the two functions take arguments of the same type, for otherwise their logical product is meaningless. In the above axiom, x and y must be of the same type, because both occur as arguments to φ.

(12) If $\varphi x . \varphi x \supset \psi x$ is true for any possible x, then ψx is true for any possible x.

This axiom is required in order to assure us that the range of significance of ψx, in the case supposed, is the same as that of $\varphi x . \varphi x \supset \psi x . \supset . \psi x$; both are in fact the same as that of φx. We know, in the case supposed, that ψx is true whenever $\varphi x . \varphi x \supset \psi x$ and $\varphi x . \varphi x \supset \psi x . \supset . \psi x$ are both significant, but we do not know, without an axiom, that ψx is true whenever ψx is significant. Hence the need of the axiom.

Axioms (11) and (12) are required, for example, in proving

$$(x) . \varphi x : (x) . \varphi x \supset \psi x :\supset . (x) . \psi x .$$

By (7) and (11),

$$\vdash :. (x) . \varphi x : (x) . \varphi x \supset \psi x :\supset : \varphi y . \varphi y \supset \psi y ,$$

whence by (12)

$$\vdash :. (x) . \varphi x : (x) . \varphi x \supset \psi x :\supset : \psi y ,$$

whence the result follows by (8) and (10).

(13) $\vdash :. (\exists f) :. (x) : \varphi x . \equiv . f ! x$.

This is the axiom of reducibility. It states that, given any function $\varphi \hat{x}$, there is a predicative function $f ! \hat{x}$ such that $f ! x$ is always equivalent to φx. Note that, since a proposition beginning with "$(\exists f)$" is, by definition, the negation of one beginning with "(f)", the above axiom involves the possibility of considering "all predicative functions of x". If φx is *any* function of x, we cannot make propositions beginning with "(φ)" or "$(\exists \varphi)$", since we cannot consider "all functions", but only "*any* function" or "all *predicative* functions".

(14) $\vdash :. (\exists f) :. (x, y) : \varphi(x, y) . \equiv . f ! (x, y)$.

This is the axiom of reducibility for double functions.

In the above propositions, our x and y may be of any type whatever. The only way in which the theory of types is relevant is that (11) only allows us to identify real variables occurring in different contents when they are shown to be of the same type by both occurring as arguments to the same function, and that, in (7) and (9), y and a must respectively be of the appropriate type for arguments to $\varphi\hat{z}$. Thus, for example, suppose we have a proposition of the form $(\varphi).f!(\varphi!\hat{z}, x)$, which is a second-order function of x. Then by (7),

$$\vdash :(\varphi).f!(\varphi!\hat{z}, x).\supset.f!(\psi!\hat{z}, x),$$

where $\psi!\hat{z}$ is any *first*-order function. But it will not do to treat $(\varphi)\ f!(\varphi!\hat{z}, x)$ as if it were a first-order function of x, and take this function as a possible value of $\psi!\hat{z}$ in the above. It is such confusions of types that give rise to the paradox of the Liar.

Again, consider the classes which are not members of themselves. It is plain that, since we have identified classes with functions,[20] no class can be significantly said to be or not to be a member of itself; for the members of a class are arguments to it, and arguments to a function are always of lower type than the function. And if we ask: "But how about the class of all classes? Is not that a class, and so a member of itself?", the answer is twofold. First, if "the class of all classes" means "the class of all classes of whatever type", then there is no such notion. Secondly, if "the class of all classes" means "the class of all classes of type t", then this is a class of the next type above t and is therefore again not a member of itself.

Thus although the above primitive propositions apply equally to all types, they do not enable us to elicit contradictions. Hence in the course of any deduction it is never necessary to consider the absolute type of a variable; it is only necessary to see that the different variables occurring in one proposition are of the proper relative types. This excludes such functions as that from which our fourth contradiction was obtained, namely: "The relation R holds between R and S". For a relation between R and S is necessarily of higher type than either of them, so that the proposed function is meaningless.

VII. Elementary theory of classes and relations

Propositions in which a function φ occurs may depend, for their truth-value, upon the particular function φ, or they may depend only upon the *extension* of φ, that is, upon the arguments which satisfy φ. A function of the latter sort we will call *extensional*. Thus, for example, "I believe that all men are mortal" may not be equivalent to "I believe that all featherless bipeds are mortal", even if men are coextensive with featherless bipeds; for I may not know that they are coextensive. But "all men are mortal" must be equivalent to "all featherless bipeds are mortal" if men are coextensive with featherless bipeds. Thus "all men are mortal" is an extensional function of the function "x is a man", while "I believe all men are mortal" is a function which is not extensional; we will call functions *intensional* when they are not extensional. The functions of functions with which mathematics is specially concerned are all extensional. The mark of an extensional function f of a function $\varphi!\hat{z}$ is

$$\varphi!x.\equiv_x.\psi!x:\supset_{\phi,\psi}:f(\varphi!\hat{z}).\equiv.f(\psi!\hat{z}).$$

[20] This identification is subject to a modification to be explained shortly.

From any function f of a function $\varphi!\hat{z}$ we can derive an associated extensional function as follows. Put

$$f\{\hat{z}(\psi z)\} .=: (\exists\varphi):\varphi!x .\equiv_x. \psi x : f\{\varphi!\hat{z}\} \quad \text{Df.}$$

The function $f\{\hat{z}(\psi z)\}$ is in reality a function of $\psi\hat{z}$, though not the same function as $f(\psi\hat{z})$, supposing this latter to be significant. But it is convenient to treat $f\{\hat{z}(\psi z)\}$ technically as though it had an argument $\hat{z}(\psi z)$, which we call "the class defined by ψ". We have

$$\vdash :. \varphi x .\equiv_x. \psi x :\supset: f\{\hat{z}(\varphi z)\} .\equiv. f\{\hat{z}(\psi z)\},$$

whence, applying to the fictitious objects $\hat{z}(\varphi z)$ and $\hat{z}(\psi z)$ the definition of identity given above, we find

$$\vdash :. \varphi x .\equiv_x. \psi x :\supset. \hat{z}(\varphi z) = \hat{z}(\psi z).$$

This, with its converse (which can also be proved), is the distinctive property of classes. Hence we are justified in treating $\hat{z}(\varphi z)$ as the class defined by φ. In the same way we put

$$f\{\hat{x}\hat{y}\psi(x, y)\} .=: (\exists\varphi): \varphi!(x, y) .\equiv_{x,y}. \psi(x, y) : f\{\varphi!(\hat{x}, \hat{y})\} \quad \text{Df.}$$

A few words are necessary here as to the distinction between $\varphi!(\hat{x}, \hat{y})$ and $\varphi!(\hat{y}, \hat{x})$. We will adopt the following convention: When a function (as opposed to its values) is represented in a form involving $\hat{x}$ and $\hat{y}$, or any other two letters of the alphabet, the value of this function for the arguments a and b is to be found by substituting a for $\hat{x}$ and b for $\hat{y}$; that is, the argument mentioned first is to be substituted for the letter which comes earlier in the alphabet, and the argument mentioned second for the later letter. This sufficiently distinguishes between $\varphi!(\hat{x}, \hat{y})$ and $\varphi!(\hat{y}, \hat{x})$; for example:

The value of $\varphi!(\hat{x}, \hat{y})$ for arguments a and b is $\varphi!(a, b)$.
The value of $\varphi!(\hat{x}, \hat{y})$ for arguments b and a is $\varphi!(b, a)$.
The value of $\varphi!(\hat{y}, \hat{x})$ for arguments a and b is $\varphi!(b, a)$.
The value of $\varphi!(\hat{y}, \hat{x})$ for arguments b and a is $\varphi!(a, b)$.

We put

$$x \,\varepsilon\, \varphi!\hat{z} .=. \varphi!x \quad \text{Df.},$$

whence

$$\vdash :. x \,\varepsilon\, \hat{z}(\psi z) .=: (\exists\varphi): \varphi!y .\equiv_y. \psi y : \varphi!x.$$

Also by the reducibility axiom we have

$$(\exists\varphi): \varphi!y .\equiv_y. \psi y,$$

whence

$$\vdash : x \,\varepsilon\, \hat{z}(\psi z) .\equiv. \psi x.$$

This holds whatever x may be. Suppose now we want to consider $\hat{z}(\psi z) \,\varepsilon\, \hat{\varphi} f\{\hat{z}(\varphi!z)\}$. We have, by the above,

$$\vdash :. \hat{z}(\psi z) \,\varepsilon\, \hat{\varphi} f\{\hat{z}(\varphi!z)\} .\equiv: f\{\hat{z}(\psi z)\} :\equiv: (\exists\varphi):\varphi!y .\equiv_y. \psi y : f\{\varphi!z\},$$

whence

$$\vdash :. \hat{z}(\psi z) = \hat{z}(\chi z) . \supset : \hat{z}(\psi z) \, \varepsilon \, \kappa . \equiv_{\kappa} . \hat{z}(\chi z) \, \varepsilon \, \kappa,$$

where κ is written for any expression of the form $\hat{\varphi} f\{\hat{z}(\varphi ! z)\}$.

We put

$$cls = \hat{\alpha}\{(\exists \varphi) . \alpha = \hat{z}(\varphi ! z)\} \quad \text{Df.}$$

Here *cls* has a meaning which depends upon the type of the apparent variable φ. Thus, for example, the proposition "$cls \,\varepsilon\, cls$", which is a consequence of the above definition, requires that "*cls*" should have a different meaning in the two places where it occurs. The symbol "*cls*" can only be used where it is unnecessary to know the type; it has an ambiguity which adjusts itself to circumstances. If we introduce as an indefinable the function "Indiv!x", meaning "x is an individual", we may put

$$Kl = \hat{\alpha}\{(\exists \varphi) . \alpha = \hat{z}(\varphi ! z . \text{Indiv} ! z)\} \quad \text{Df.}$$

Then Kl is an unambiguous symbol meaning "classes of individuals".

We will use lower-case Greek letters (other than ε, φ, ψ, χ, θ) to represent classes of whatever type, that is, to stand for symbols of the form $\hat{z}(\varphi ! z)$ or $\hat{z}(\varphi z)$.

The theory of classes proceeds, from this point on, much as in Peano's system; $\hat{z}(\varphi z)$ replaces $z \ni (\varphi z)$. Also I put

$$\alpha \subset \beta . = : x \,\varepsilon\, \alpha . \supset_x . x \,\varepsilon\, \beta \quad \text{Df.}$$

$$\exists ! \alpha . = . (\exists x) . x \,\varepsilon\, \alpha \quad \text{Df.}$$

$$V = \hat{x}(x = x) \quad \text{Df.}$$

$$\Lambda = \hat{x}\{\sim (x = x)\} \quad \text{Df.}$$

where Λ, as with Peano, is the null class. The symbols $\exists$, Λ, V, like *cls* and ε, are ambiguous and acquire a definite meaning only when the type concerned is otherwise indicated.

We treat relations in exactly the same way, putting

$$a\{\varphi ! (\hat{x}, \hat{y})\} b . = . \varphi ! (a, b) \quad \text{Df.}$$

(the order being determined by the alphabetical order of x and y and the typographical order of a and b); whence

$$\vdash :. a\{\hat{x}\hat{y}\psi(x, y)\} b . \equiv : (\exists \varphi) : \psi(x, y) . \equiv_{x, y} . \varphi ! (x, y) : \varphi ! (a, b),$$

whence, by the reducibility axiom,

$$\vdash : a\{\hat{x}\hat{y}\psi(x, y)\} b . \equiv . \psi(a, b).$$

We use Latin capital letters as abbreviations for such symbols as $\hat{x}\hat{y}\psi(x, y)$, and we find

$$\vdash :. R = S . \equiv . xRy . \equiv_{x, y} . xSy,$$

where

$$R = S . = : f ! R . \supset_f . f ! S \quad \text{Df.}$$

We put

$$\text{Rel} = \hat{R}\{(\exists \varphi) . R = \hat{x}\hat{y}\varphi ! (x, y)\} \quad \text{Df.},$$

and we find that everything proved for classes has its analogue for dual relations. Following Peano, we put

$$\alpha \cap \beta = \hat{x}(x \, \varepsilon \, \alpha \, . \, x \, \varepsilon \, \beta) \quad \text{Df.},$$

defining the product, or common part, of two classes;

$$\alpha \cup \beta = \hat{x}(x \, \varepsilon \, \alpha \, . \vee . \, x \, \varepsilon \, \beta) \quad \text{Df.},$$

defining the sum of two classes; and

$$-\alpha = \hat{x}\{\sim(x \, \varepsilon \, \alpha)\} \quad \text{Df.},$$

defining the negation of a class. Similarly for relations we put

$$R \mathbin{\dot{\cap}} S = \hat{x}\hat{y}(xRy \, . \, xSy) \quad \text{Df.}$$

$$R \mathbin{\dot{\cup}} S = \hat{x}\hat{y}(xRy \, . \vee . \, xSy) \quad \text{Df.}$$

$$\dot{-} R = \hat{x}\hat{y}\{\sim(xRy)\} \quad \text{Df.}$$

VIII. Descriptive functions

The functions hitherto considered have been propositional functions, with the exception of a few particular functions such as $R \mathbin{\dot{\cap}} S$. But the ordinary functions of mathematics, such as x^2, $\sin x$, $\log x$, are not propositional. Functions of this kind always mean "the term having such-and-such a relation to x". For this reason they may be called *descriptive* functions, because they *describe* a certain term by means of its relation to their argument. Thus "$\sin \pi/2$" describes the number 1; yet propositions in which $\sin \pi/2$ occurs are not the same as they would be if 1 were substituted. This appears, for example, from the proposition "$\sin \pi/2 = 1$", which conveys valuable information, whereas "$1 = 1$" is trivial. Descriptive functions have no meaning by themselves, but only as constituents of propositions; and this applies generally to phrases of the form "the term having such-and-such a property". Hence in dealing with such phrases, we must define any proposition in which they occur, not the phrases themselves.[21] We are thus led to the following definition, in which "$(\imath x)(\varphi x)$" is to be read "*the* term x which satisfies φx".

$$\psi\{(\imath x)(\varphi x)\} \, . = : (\exists b) : \varphi x \, . \equiv_x . \, x = b : \psi b \quad \text{Df.}$$

This definition states that "the term which satisfies φ satisfies ψ" is to mean: "There is a term b such that φx is true when and only when x is b, and ψb is true". Thus all propositions about "*the* so-and-so" will be false if there are no so-and-so's or several so-and-so's.

The general definition of a descriptive function is

$$R\text{‘}y = (\imath x)(xRy) \quad \text{Df.};$$

that is, "$R\text{‘}y$" is to mean "the term which has the relation R to y". If there are several terms or none having the relation R to y, all propositions about $R\text{‘}y$ will be false. We put

$$E!(\imath x)(\varphi x) \, . = : (\exists b) : \varphi x \, . \equiv_x . \, x = b \quad \text{Df.}$$

[21] See *Russell 1905*, where the reasons for this view are given at length.

Here "$E!(\imath x)(\varphi x)$" may be read "there is such a term as the x which satisfies φx", or "the x which satisfies φx exists". We have

$$\vdash :. E! R^{\backprime} y . \equiv : (\exists b) : xRy . \equiv_x . x = b.$$

The inverted comma in $R^{\backprime}y$ may be read *of*. Thus if R is the relation of father to son, "$R^{\backprime}y$" is "the father of y". If R is the relation of son to father, all propositions about $R^{\backprime}y$ will be false unless y has one son and no more.

From the above it appears that descriptive functions are obtained from relations. The relations now to be defined are chiefly important on account of the descriptive functions to which they give rise.

$$\mathrm{Cnv} = \hat{Q}\hat{P}\{xQy . \equiv_{x,y} . yPx\} \quad \mathrm{Df.}$$

Here "Cnv" is short for "converse". It is the relation of a relation to its converse, for example, of *greater* to *less*, of parentage to sonship, of preceding to following, and so on. We have

$$\vdash . \mathrm{Cnv}^{\backprime}P = (\imath Q)\{xQy . \equiv_{x,y} . yPx\}.$$

For a shorter notation, often more convenient, we put

$$\breve{P} = \mathrm{Cnv}^{\backprime}P \quad \mathrm{Df.}$$

We want next a notation for the class of terms which have the relation R to y. For this purpose, we put

$$\overrightarrow{R} = \hat{\alpha}\hat{y}\{\alpha = \hat{x}(xRy)\} \quad \mathrm{Df.},$$

whence

$$\vdash . \overrightarrow{R}^{\backprime}y = \hat{x}(xRy).$$

Similarly we put

$$\overleftarrow{R} = \hat{\beta}\hat{x}\{\beta = \hat{y}(xRy)\} \quad \mathrm{Df.},$$

whence

$$\vdash . \overleftarrow{R}^{\backprime}x = \hat{y}(xRy).$$

We want next the *domain* of R (that is, the class of terms which have the relation R to something), the *converse domain* of R (that is, the class of terms to which something has the relation R), and the *field* of R, which is the sum of the domain and the converse domain. For this purpose we define the relations of the domain, converse domain, and field, to R. The definitions are:

$$D = \hat{\alpha}\hat{R}\{\alpha = \hat{x}((\exists y) . xRy)\} \quad \mathrm{Df.}$$

$$\text{Ɑ} = \hat{\beta}\hat{R}\{\beta = \hat{y}((\exists x) . xRy)\} \quad \mathrm{Df.}$$

$$C = \hat{\gamma}\hat{R}\{\gamma = \hat{x}((\exists y) : xRy . \vee . yRx)\} \quad \mathrm{Df.}$$

Note that the third of these definitions is only significant when R is what we may call a *homogeneous* relation, that is, one in which, if xRy holds, x and y are of the same type. For otherwise, however we may choose x and y, either xRy or yRx will be meaningless. This observation is important in connection with Burali-Forti's contradiction.

We have, in virtue of the above definitions,

$$\vdash . D'R = \hat{x}\{(\exists y).xRy\},$$
$$\vdash . \text{Ɑ}'R = \hat{y}\{(\exists x).xRy\},$$
$$\vdash . C'R = \hat{x}\{(\exists y):xRy . \vee . yRx\},$$

the last of these being significant only when R is homogeneous. "$D'R$" is read "the domain of R"; "Ɑ$'R$" is read "the converse domain of R", and "$C'R$" is read "the field of R". The letter C is chosen as the initial of the word "campus".

We want next a notation for the relation, to a class α contained in the domain of R, of the class of terms to which some member of α has the relation R, and also for the relation, to a class β contained in the converse domain of R, of the class of terms which have the relation R to some member of β. For the second of these we put

$$R_\varepsilon = \hat{\alpha}\hat{\beta}\{\alpha = \hat{x}((\exists y).y \,\varepsilon\, \beta . xRy)\} \quad \text{Df.}$$

So that

$$\vdash . R_\varepsilon{}'\beta = \hat{x}\{(\exists y).y \,\varepsilon\, \beta . xRy\}.$$

Thus if R is the relation of father to son, and β is the class of Etonians, $R_\varepsilon{}'\beta$ will be the class "fathers of Etonians"; if R is the relation "less than" and β is the class of proper fractions of the form $1 - 2^{-n}$ for integral values of n, $R_\varepsilon{}'\beta$ will be the class of fractions less than some fraction of the form $1 - 2^{-n}$; that is, $R_\varepsilon{}'\beta$ will be the class of proper fractions. The other relation mentioned above is $(\breve{R})_\varepsilon$.

We put, as an alternative notation often more convenient,

$$R''\beta = R_\varepsilon{}'\beta \quad \text{Df.}$$

The *relative product* of two relations R and S is the relation which holds between x and z whenever there is a term y such that xRy and yRz both hold. The relative product is denoted by $R \mid S$. Thus

$$R \mid S = \hat{x}\hat{z}\{(\exists y).xRy.yRz\} \quad \text{Df.}$$

We put also

$$R^2 = R \mid R \quad \text{Df.}$$

The product and sum of a class of classes are often required. They are defined as follows:

$$s'\kappa = \hat{x}\{(\exists \alpha).\alpha \,\varepsilon\, \kappa . x \,\varepsilon\, \alpha) \quad \text{Df.}$$
$$p'\kappa = \hat{x}\{\alpha \,\varepsilon\, \kappa . \supset_\alpha . x \,\varepsilon\, \alpha\} \quad \text{Df.}$$

Similarly for relations we put

$$\dot{s}'\lambda = \hat{x}\hat{y}\{(\exists R).R \,\varepsilon\, \lambda . xRy\} \quad \text{Df.}$$
$$\dot{p}'\lambda = \hat{x}\hat{y}\{R \,\varepsilon\, \lambda . \supset_R . xRy\} \quad \text{Df.}$$

We need a notation for the class whose only member is x. Peano uses ιx, hence we shall use $\iota' x$. Peano showed (what Frege also had emphasized) that this class cannot be identified with x. With the usual view of classes, the need for such a distinction remains a mystery; but with the view set forth above, it becomes obvious.

We put

$$\iota = \hat{\alpha}\hat{x}\{\alpha = \hat{y}(y = x)\} \quad \text{Df.},$$

whence

$$\vdash . \iota ` x = \hat{y}(y = x),$$

and

$$\vdash : E ! \breve{\iota} ` \alpha . \supset . \breve{\iota} ` \alpha = (\imath x)(x \,\varepsilon\, \alpha) ;$$

that is, if α is a class which has only one member, then $\breve{\iota} ` \alpha$ is that one member.[22]

For the class of classes contained in a given class, we put

$$\text{Cl} ` \alpha = \hat{\beta}(\beta \subset \alpha) \quad \text{Df.}$$

We can now proceed to the consideration of cardinal and ordinal numbers, and of how they are affected by the doctrine of types.

IX. Cardinal numbers

The cardinal number of a class α is defined as the class of all classes *similar* to α, two classes being similar when there is a one-one relation between them. The class of one-one relations is denoted by $1 \rightarrow 1$ and defined as follows:

$$1 \rightarrow 1 = \hat{R}\{xRy . x'Ry . xRy' . \supset_{x, y, x', y'} . x = x' . y = y'\} \quad \text{Df.}$$

Similarity is denoted by "Sim"; its definition is

$$\text{Sim} = \hat{\alpha}\hat{\beta}\{(\exists R) . R \,\varepsilon\, 1 \rightarrow 1 . D ` R = \alpha . \text{Ꭰ} ` R = \beta\} \quad \text{Df.}$$

Then $\overrightarrow{\text{Sim}} ` \alpha$ is, by definition, the cardinal number of $\alpha$; this we will denote by $Nc ` \alpha$; hence we put

$$Nc = \overrightarrow{\text{Sim}} \quad \text{Df.},$$

whence

$$\vdash . Nc ` \alpha = \overrightarrow{\text{Sim}} ` \alpha .$$

The class of cardinals we will denote by NC; thus

$$NC = Nc `` cls \quad \text{Df.}$$

0 is defined as the class whose only member is the null class, Λ, so that

$$0 = \iota ` \Lambda \quad \text{Df.}$$

The definition of 1 is:

$$1 = \hat{\alpha}\{(\exists c) : x \,\varepsilon\, \alpha . \equiv_x . x = c\} \quad \text{Df.}$$

It is easy to prove that 0 and 1 are cardinals according to the definition.

It is to be observed, however, that 0 and 1 and all the other cardinals, according to the above definitions, are ambiguous symbols, liks *cls*, and have as many meanings as there are types. To begin with 0: the meaning of 0 depends upon that of Λ, and the meaning of Λ is different according to the type of which it is the null class. Thus

[22] Thus $\breve{\iota} ` \alpha$ is what Peano calls $\imath \alpha$.

there are as many 0's as there are types; and the same applies to all the other cardinals. Nevertheless, if two classes α and β are of different types, we can speak of them as having the same cardinal, or of one as having a greater cardinal than the other, because a one-one relation may hold between the members of α and the members of β, even when α and β are of different types. For example, let β be ι"α, that is, the class whose members are the classes consisting of single members of α. Then ι"α is of higher type than α, but similar to α, being correlated with α by the one-one relation ι.

The hierarchy of types has important results in regard to addition. Suppose we have a class of α terms and a class of β terms, where α and β are cardinals; it may be quite impossible to add them together to get a class of α and β terms, since, if the classes are not of the same type, their logical sum is meaningless. Where only a finite number of classes are concerned, we can obviate the practical consequences of this, owing to the fact that we can always apply operations to a class which raise its type to any required extent without altering its cardinal number. For example, given any class α, the class ι"α has the same cardinal number, but is of the next type above α. Hence, given any finite number of classes of different types, we can raise all of them to the type which is what we may call the lowest common multiple of all the types in question; and it can be shown that this can be done in such a way that the resulting classes shall have no common members. We may then form the logical sum of all the classes so obtained, and its cardinal number will be the arithmetical sum of the cardinal numbers of the original classes. But where we have an infinite series of classes of ascending types, this method cannot be applied. For this reason, we cannot now prove that there must be infinite classes. For suppose there were only n individuals altogether in the universe, where n is finite. There would then be 2^n classes of individuals, and 2^{2^n} classes of classes of individuals, and so on. Thus the cardinal number of terms in each type would be finite; and though these numbers would grow beyond any assigned finite number, there would be no way of adding them so as to get an infinite number. Hence we need an axiom, so it would seem, to the effect that no finite class of individuals contains all individuals; but if any one chooses to assume that the total number of individuals in the universe is (say) 10,367, there seems no a priori way of refuting his opinion.

From the above mode of reasoning, it is plain that the doctrine of types avoids all difficulties as to the greatest cardinal. There is a greatest cardinal in each type, namely, the cardinal number of the whole of the type; but this is always surpassed by the cardinal number of the next type, since, if α is the cardinal number of one type, that of the next type is 2^α, which, as Cantor has shown, is always greater than α. Since there is no way of adding different types, we cannot speak of "the cardinal number of all objects, of whatever type", and thus there is no absolutely greatest cardinal.

If it is admitted that no finite class of individuals contains all individuals, it follows that there are classes of individuals having any finite number. Hence all finite cardinals exist as individual-cardinals, that is, as the cardinal numbers of classes of individuals. It follows that there is a class of $\aleph_0$ cardinals, namely, the class of finite cardinals. Hence $\aleph_0$ exists as the cardinal of a class of classes of classes of individuals. By forming all classes of finite cardinals, we find that $2^{\aleph_0}$ exists as the cardinal of a class of classes of classes of classes of individuals; and so we can proceed indefinitely. The existence of $\aleph_n$ for every finite value of n can also be proved; but this requires the consideration of ordinals.

If, in addition to assuming that no finite class contains all individuals, we assume the multiplicative axiom (that is, the axiom that, given a set of mutually exclusive classes, none of which are null, there is at least one class consisting of one member from each class in the set), then we can prove that there is a class of individuals containing $\aleph_0$ members, so that $\aleph_0$ will exist as an individual-cardinal. This somewhat reduces the type to which we have to go in order to prove the existence theorem for any given cardinal, but it does not give us any existence theorem which cannot be got otherwise sooner or later.

Many elementary theorems concerning cardinals require the multiplicative axiom.[23] It is to be observed that this axiom is equivalent to Zermelo's,[24] and therefore to the assumption that every class can be well-ordered.[25] These equivalent assumptions are, apparently, all incapable of proof, though the multiplicative axiom, at least, appears highly self-evident. In the absence of proof, it seems best not to assume the multiplicative axiom, but to state it as a hypothesis on every occasion on which it is used.

X. *Ordinal numbers*

An ordinal number is a class of ordinally similar well-ordered series, that is, of relations generating such series. Ordinal similarity or *likeness* is defined as follows:

$$\mathrm{Smor} = \hat{P}\hat{Q}\{\exists S).S \,\varepsilon\, 1 \rightarrow 1.\text{ᗡ}\text{‘}S = C\text{‘}Q.P = S\,|\,Q\,|\,\breve{S}\} \quad \mathrm{Df.},$$

where "Smor" is short for "similar ordinally".

The class of serial relations, which we will call "Ser", is defined as follows:

$$\mathrm{Ser} = \hat{P}\{xPy\,.\supset_{x,y}.\, \sim(x = y):xPy.yPz\,.\supset_{x,y,z}.\, xPz: \\ x \,\varepsilon\, C\text{‘}P\,.\supset_x.\, \overrightarrow{P}\text{‘}x \cup \iota\text{‘}x \cup \overleftarrow{P}\text{‘}x = C\text{‘}P\} \quad \mathrm{Df.}$$

That is, reading P as "precedes", a relation is serial if (1) no term precedes itself, (2) a predecessor of a predecessor is a predecessor, (3) if x is any term in the field of the relation, then the predecessors of x together with x together with the successors of x constitute the whole field of the relation.

Well-ordered serial relations, which we will call Ω, are defined as follows:

$$\Omega = \hat{P}\{P \,\varepsilon\, \mathrm{Ser}:\alpha \subset C\text{‘}P.\exists!\alpha.\supset_\alpha.\exists!(\alpha - \breve{P}\text{‘‘}\alpha)\} \quad \mathrm{Df.};$$

that is, P generates a well-ordered series if P is serial and any class α contained in

[23] See *Russell 1905a*, Part III.

[24] See *Russell 1905a* for a statement of Zermelo's axiom and for the proof that this axiom implies the multiplicative axiom. The converse implication results as follows: Putting Prod‘k for the multiplicative class of k, consider

$$Z\text{‘}\beta = \hat{R}\{(Ex).x \,\varepsilon\, \beta.D\text{‘}R = \iota\text{‘}\beta.\text{ᗡ}\text{‘}R = \iota\text{‘}x\} \quad \mathrm{Df.},$$

and assume

$$\gamma \,\varepsilon\, \mathrm{Prod}\text{‘}Z\text{‘‘}cl\text{‘}\alpha.R = \hat{\xi}\hat{x}\{(S).\exists S \,\varepsilon\, \gamma.\xi Sx\}.$$

Then R is a Zermelo correlation. Hence if Prod’Z‘‘cl‘α is not null, at least one Zermelo correlation for α exists.

[25] See *Zermelo 1904*.

the field of P and not null has a first term. (Note that $\breve{P}``\alpha$ are the terms coming after some term of α.)

If we denote by $No`P$ the ordinal number of a well-ordered relation P and by NO the class of ordinal numbers, we shall have

$$No = \hat{\alpha}\hat{P}\{P \,\varepsilon\, \Omega \,.\, \alpha = \overrightarrow{\text{Smor}}`P\} \quad \text{Df.}$$

$$NO = No``\Omega.$$

From the definition of No we have

$$\vdash : P \,\varepsilon\, \Omega \,.\supset.\, No`P = \overrightarrow{\text{Smor}}`P$$

$$\vdash : \sim(P \,\varepsilon\, \Omega) \,.\supset.\, \sim E!\, No`P.$$

If we now examine our definitions with a view to their connection with the theory of types, we see, to begin with, that the definitions of "Ser" and Ω involve the *fields* of serial relations. Now the field is only significant when the relation is homogeneous; hence relations which are not homogeneous do not generate series. For example, the relation ι might be thought to generate series of ordinal number ω, such as

$$x,\ \iota`x,\ \iota`\iota`x,\ \ldots,\ \iota^{n}`x,\ \ldots,$$

and we might attempt to prove in this way the existence of ω and $\aleph_0$. But x and $\iota`x$ are of different types, and therefore there is no such series according to the definition.

The ordinal number of a series of individuals is, by the above definition of No, a class of relations of individuals. It is therefore of a different type from any individual and cannot form part of any series in which individuals occur. Again, suppose all the finite ordinals exist as individual-ordinals, that is, as the ordinals of series of individuals. Then the finite ordinals themselves form a series whose ordinal number is ω; thus ω exists as an ordinal-ordinal, that is, as the ordinal of a series of ordinals. But the type of an ordinal-ordinal is that of classes of relations of classes of relations of individuals. Thus the existence of ω has been proved in a higher type than that of the finite ordinals. Again, the cardinal number of ordinal numbers of well-ordered series that can be made out of finite ordinals is $\aleph_1$; hence $\aleph_1$ exists in the type of classes of classes of classes of relations of classes of relations of individuals. Also the ordinal numbers of well-ordered series composed of finite ordinals can be arranged in order of magnitude, and the result is a well-ordered series whose ordinal number is ω_1. Hence ω_1 exists as an ordinal-ordinal-ordinal. This process can be repeated any finite number of times, and thus we can establish the existence, in appropriate types, of $\aleph_n$ and ω_n for any finite value of n.

But the above process of generation no longer leads to any totality of *all* ordinals, because, if we take all the ordinals of any given type, there are always greater ordinals in higher types; and we cannot add together a set of ordinals of which the type rises above any finite limit. Thus all the ordinals in any type can be arranged by order of magnitude in a well-ordered series, which has an ordinal number of higher type than that of the ordinals composing the series. In the new type, this new ordinal is not the greatest. In fact, there is no greatest ordinal in any type, but in every type all ordinals are less than some ordinals of higher type. It is impossible to complete the series of ordinals, since it rises to types above every assignable finite limit; thus although every segment of the series of ordinals is well-ordered, we cannot say that the whole series is well-ordered, because the "whole series" is a fiction. Hence Burali-Forti's contradiction disappears.

From the last two sections it appears that, if it is allowed that the number of individuals is not finite, the existence of all Cantor's cardinal and ordinal numbers can be proved, short of $\aleph_\omega$ and ω_ω. (It is quite possible that the existence of these may also be demonstrable.) The existence of all *finite* cardinals and ordinals can be proved without assuming the existence of anything. For, if the cardinal number of terms in any type is n, that of terms in the next type is 2^n. Thus if there are no individuals, there will be one class (namely, the null class), two classes of classes (namely, that containing no class and that containing the null class), four classes of classes of classes, and generally 2^{n-1} classes of the nth order. But we cannot add together terms of different types, and thus we cannot in this way prove the existence of any infinite class.

We can now sum up our whole discussion. After stating some of the paradoxes of logic, we found that all of them arise from the fact that an expression referring to *all* of some collection may itself appear to denote one of the collection; as, for example, "all propositions are either true or false" appears to be itself a proposition. We decided that, where this appears to occur, we are dealing with a false totality and that in fact nothing whatever can significantly be said about *all* of the supposed collection. In order to give effect to this decision, we explained a doctrine of *types* of variables, proceeding upon the principle that any expression which refers to *all* of some type must, if it denotes anything, denote something of a higher type than that to all of which it refers. Where *all* of some type is referred to, there is an *apparent variable* belonging to that type. Thus *any expression containing an apparent variable is of higher type than that variable.* This is the fundamental principle of the doctrine of types. A change in the manner in which the types are constructed, should it prove necessary, would leave the solution of contradictions untouched so long as this fundamental principle is observed. The method of constructing types explained above was shown to enable us to state all the fundamental definitions of mathematics and at the same time to avoid all known contradictions. And it appeared that in practice the doctrine of types is never relevant except where existence theorems are concerned or where applications are to be made to some particular case.

The theory of types raises a number of difficult philosophical questions concerning its interpretation. Such questions are, however, essentially separable from the mathematical development of the theory and, like all philosophical questions, introduce elements of uncertainty which do not belong to the theory itself. It seemed better, therefore, to state the theory without reference to philosophical questions, leaving these to be dealt with independently.

A new proof of the possibility of a well-ordering

ERNST ZERMELO

(*1908*)

The present paper consists of two parts. In the first, Zermelo offers a new proof of the well-ordering theorem. Like the original (*1904*), the new proof makes use of the axiom of choice, and with the same strength: the choice set contains an element of every nonempty subset of any given set. The difference between the two proofs lies in the remaining set-theoretic assumptions. The second proof assumes much less, especially with respect to well-ordering and ordinals, and what is needed for the proof is derived anew, through the use of Θ-chains, which are a generalization of Dedekind's "chains".

The second part of the paper is a discussion of the objections raised against the first proof. These objections had been numerous, and they sprang from three main sources: the old mistrust, still lingering, of Cantor's set theory; a wariness of the new device, the principle of choice; and a suspicion of any argument reminiscent of those leading to the paradoxes. Zermelo's answer to the objections is lively and on the whole cogent, and it has been upheld by subsequent developments. After Zermelo's second proof the discussion subsided, and his result, at least in the sense that the principle of choice implies the well-ordering theorem, was generally accepted by the mathematical world.

The translation is by Stefan Bauer-Mengelberg, and it is printed here with the kind permission of Springer Verlag.

Although I still fully uphold my "Proof that every set can be well-ordered", published in 1904, in the face of the various objections that will be thoroughly discussed in § 2, the new proof that I give below of the same theorem may yet be of interest, since, on the one hand, it presupposes no specific theorems of set theory and, on the other, it brings out, more clearly than the first proof did, the purely formal character of the well-ordering, which has nothing at all to do with spatiotemporal arrangement.

§ 1. The new proof

The assumptions and forms of inference that I use in the proof of the theorem below can be reduced to the following postulates.

I. All elements of a set M that have a property $\mathfrak{E}$ well-defined for every single element are the elements of another set, $M_{\mathfrak{E}}$, a "subset" of M.

Thus to every subset M_1 of M there corresponds a "complementary subset",

$M - M_1$, that contains all elements not occurring in M_1 and, when $M_1 = M$, reduces to the (empty) "null set".

II. All subsets of a set M, that is, all sets M_1 whose elements are also elements of M, are the elements of a set $\mathfrak{U}(M)$ determined by M.

Postulate I easily yields the following proposition:

III. All elements that are common to all of the sets $A, B, C, \ldots$, these being elements of a higher set T, are the elements of a set $Q = \mathfrak{D}(T)$, which will be called the "intersection" or the "common component" of the sets $A, B, C, \ldots$.

THEOREM. *If with every nonempty subset of a set M an element of that subset is associated by some law as "distinguished element", then $\mathfrak{U}(M)$, the set of all subsets of M, possesses one and only one subset M such that to every arbitrary subset P of M there always corresponds one and only one element P_0 of M that includes P as a subset and contains an element of P as its distinguished element. The set M is well-ordered by M.*

Proof. If A is any nonempty subset of M and hence an element of $\mathfrak{U}(M)$, and if $a = \varphi(A)$ is its distinguished element, let $A' = A - \{a\}$ be the part of A that results when the distinguished element is removed. Now $\mathfrak{U}(M)$, the set of all subsets of M, possesses the following three properties:

(1) It contains the element M;

(2) Along with each of its elements A it also contains the corresponding A';

(3) Along with each of its subsets $\mathsf{A} = \{A, B, C, \ldots\}$ it also contains the corresponding intersection $Q = \mathfrak{D}(\mathsf{A})$ as an element.

If now a subset Θ of $\mathfrak{U}(M)$ that also has these three properties is called a "Θ-*chain*", it immediately follows that the intersection of several Θ-chains is itself always a Θ-chain, and the intersection M of all existing Θ-chains, which according to I and II are the elements of a well-defined subset of $\mathfrak{U}\mathfrak{U}(M)$, is therefore the smallest possible Θ-chain; therefore no proper subset of M can be a Θ-chain any longer.

Now let A be an element of M such that all other elements X of M fall into two classes with respect to A: (1) elements U_A that are parts of A, and (2) elements V_A that include the set A as a part, as for instance M itself does. Then, as we shall now show, every U_A always has the property of being a W_A; that is to say, it is a subset of $A' = A - \{\varphi(A)\}$. In fact, every V'_A, since it cannot be a U_A and yet must be an element of M, is either A itself or a V_A, and every intersection of several V_A is again a V_A or A. On the other hand, A', as well as every W'_A, is again a W_A; and likewise every intersection of several W_A, as well as the intersection of some W_A and some V_A or A, is again a W_A. Thus the W_A together with the V_A and A already form a Θ-chain; they therefore exhaust the smallest Θ-chain M, and every U_A is actually a W_A, that is, a subset of A'. But from this it immediately follows that A', too, has the same property as A, namely, that all other elements of M are either parts of A' or include A' as a part. If, finally, Q is the intersection of several $A, B, C, \ldots$ that have the property just assumed of A and if X is any other element of M, then only two cases are possible: either X includes one of the sets $A, B, C, \ldots$, and therewith also Q, as a part, or X is included in all of the sets $A, B, C, \ldots$, and therewith also in Q, as a subset, that is, Q too possesses the above-mentioned property of A. Since, finally, M includes all elements of M as subsets and therefore is itself an A, the elements of M that are constituted like A again form a Θ-chain, namely, M itself, and for two arbitrary [[distinct]] elements A and B of M the alternative holds that either B must be a subset of A' or A a subset of B'.

Now let P be an arbitrary subset of M, and let P_0 be the intersection of all elements of M that include P as a subset and to which at least the element M belongs. Then P_0 also is an element of M, and the distinguished element p_0 of P_0 must be an element of P, since otherwise $P_0' = P_0 - \{p_0\}$ also would contain all elements of P and would still only be a part of P_0. Every other element P_1 of M that includes P as a subset must then include P_0 as a part; that is, P_0, according to what has just been proved, is a subset of P_1', and the distinguished element p_1 of P_1 cannot be an element of P, since it does not occur in P_1' and hence not in P_0 either. So there really exists only a single element P_0 of M that includes P as a subset and contains an element of P as its distinguished element.

If we here choose for P a set of the form $\{a\}$, where a is any element of M, it follows in particular that to every element a of M there corresponds a single element A of M in which a is the distinguished element; let this element A be denoted by $\mathfrak{R}(a)$. If a and b are any two distinct elements of M, then either $\mathfrak{R}(a)$ or $\mathfrak{R}(b)$ is the element P_0 of M corresponding to the set $P = \{a, b\}$, that is, either $\mathfrak{R}(a)$ contains the element b or $\mathfrak{R}(b)$ the element a, but never both. If, finally, a, b, and c are any three [[distinct]] elements of M and if, say, b is an element of $\mathfrak{R}(a)$ and c an element of $\mathfrak{R}(b)$, then only $\mathfrak{R}(a)$ can be the element P_0 corresponding to the set $P = \{a, b, c\}$, that is, c also is an element of $\mathfrak{R}(a)$. Therefore, if we write $a \prec b$ when b is an element of $\mathfrak{R}(a)$ and $a \neq b$ (we then say that the element a "precedes the element b") the trichotomy

$$a \prec b, \quad a = b, \quad \text{or} \quad b \prec a$$

obtains for any two elements a and b, and from

$$a \prec b \quad \text{and} \quad b \prec c$$

it always follows that $a \prec c$.

The set M, therefore, is "simply ordered" by means of the set M, and, moreover, it is "well-ordered" in the sense of Cantor; for to every subset P of M there corresponds a "first element", namely, the distinguished element p_0 of $P_0 = \mathfrak{R}(p_0)$, which precedes all other elements p of P, since all these p are elements of P_0.

If, conversely, the set M is well-ordered in any way, then to every element a of M there corresponds a certain subset $\mathfrak{R}(a)$ of M that contains, besides a, all elements "following a"; let us call it the "remainder" associated with a. If from such a remainder $\mathfrak{R}(a)$ we remove the first element a, what is left is the remainder of the "next" element a'. Likewise, the common component, or intersection, of several remainders is always again a remainder, and, finally, the entire set M is the remainder $\mathfrak{R}(e)$ of its first element. Thus the totality of all remainders, in the sense specified above, forms a Θ-chain, in which for every remainder the first element is the distinguished element. If now $\mathfrak{U}(M)$ were to include, besides M, a second subset M_1 constituted as required by the theorem, then M_1 also would determine a well-ordering of M with the same distinguished elements and would therefore, as a Θ-chain, include the intersection M of all Θ-chains as a component. If z_0 then denotes the distinguished element of an element Z of $\mathsf{M}_1 - \mathsf{M}$, z_0 would be the distinguished element of two elements of M_1—namely, of Z as well as of the $\mathfrak{R}(z_0)$ determined by M—and this would contradict the property assumed of M_1. Therefore, the well-ordering M is indeed uniquely determined by the choice of distinguished elements, and the theorem asserted is proved in its entirety.

Now in order to apply our theorem to arbitrary sets, we require only ⟦!⟧ the additional assumption that *a simultaneous choice of distinguished elements is in principle always possible for an arbitrary set of sets*, or, to be more precise, that the same consequences always hold as if such a choice were possible. In this formulation, to be sure, the principle taken as fundamental still appears to be somewhat tainted with subjectivity and liable to misinterpretation. But since, as I shall show in more detail elsewhere ⟦below, p. 209⟧, we can, by means of elementary and indispensable set-theoretic principles, always replace an arbitrary set T' of sets A', B', C', ... by a set T of mutually disjoint sets A, B, C, ... that are equivalent to the sets A', B', C', ..., respectively, the general principle of choice can be reduced to the following axiom, whose purely objective character is immediately evident.

IV. Axiom. *A set S that can be decomposed into a set of disjoint parts A, B, C, ..., each containing at least one element, possesses at least one subset S_1 having exactly one element in common with each of the parts A, B, C, ... considered.*

Then the application of this axiom, just as in my note of 1904, yields the general theorem that *every set can be well-ordered.*

The definition of well-ordering that has already appeared in the formulation of the Theorem and forms the basis of our new proof has the advantage that it rests exclusively upon the elementary notions of set theory, whereas experience shows that, with the usual presentation, the uninformed are only too prone to look for some mystical meaning behind Cantor's relation $a \prec b$, which is suddenly introduced. Let us now once more formulate our definition explicitly, as follows.

Definition. *A set M is said to be well-ordered if to any element a of M there corresponds a unique subset $\mathfrak{R}(a)$ of M, the remainder of a, and every nonempty subset P of M contains one and only one first element, that is, an element p_0 such that its remainder $\mathfrak{R}(p_0)$ includes the set P as a subset.*

§ 2. Discussion of the objections to the earlier proof

Since 1904, the date of my "Proof that every set can be well-ordered", a number of objections have been made to it and various critiques of it have been published. Let me take this opportunity to discuss them together.

a. *Objections to the principle of choice*

In first place we here consider the objections that are directed against the "postulate of choice" formulated above and therefore strike at both of my proofs in the same way. I concede that they are to some extent justified, since I just cannot *prove* this postulate, as I expressly emphasized at the end of my note,[1] and therefore cannot compel anyone to accept it apodictically. Hence if Borel (*1905*, but see also *1905a*) and Peano (*1906a*, pp. 145–148) in their critiques note the lack of a proof, they

[1] "This logical principle cannot, to be sure, be reduced to a still simpler one..." (*1904*, p. 516 ⟦above, p. 141⟧).

have merely adopted my own point of view. They would even have put me in their debt had they now for their part established the unprovability I asserted—namely, that this postulate is logically independent of the others—thereby corroborating my conviction.

Now even in mathematics *unprovability*, as is well known, is in no way equivalent to *nonvalidity*, since, after all, not everything can be proved, but every proof in turn presupposes unproved principles. Thus, in order to reject such a fundamental principle, one would have had to ascertain that in some particular case it did not hold or to derive contradictory consequences from it; but none of my opponents has made any attempt to do this.

Even Peano's *Formulaire* (*1897*), which is an attempt to reduce all of mathematics to "syllogisms" (in the Aristotelian-Scholastic sense),[2] rests upon quite a number of unprovable principles; one of these is equivalent to the principle of choice for a single set and can then be extended syllogistically to an arbitrary finite number of sets.[3] But the general axiom that, following other researchers, I permitted myself to apply to arbitrary sets in this new case just is not to be found among Peano's principles, and Peano himself assures us that he could not derive it from them either. He is content to note this fact, and that finishes the principle for him. The idea that possibly his *Formulaire* might be incomplete in precisely this point does, after all, suggest itself, and, since there are no infallible authorities in mathematics, we must also take that possibility into account and not reject it without objective examination.

First, how does Peano arrive at his own fundamental principles and how does he justify their inclusion in the *Formulaire*, since, after all, he cannot prove them either? Evidently by analyzing the modes of inference that in the course of history have come to be recognized as valid and by pointing out that the principles are intuitively evident and necessary for science—considerations that can all be urged equally well in favor of the disputed principle. That this axiom, even though it was never formulated in textbook style, has frequently been used, and successfully at that, in the most diverse fields of mathematics, especially in set theory, by Dedekind, Cantor, F. Bernstein, Schoenflies, J. König, and others is an indisputable fact, which is only corroborated by the opposition that, at one time or another, some logical purists directed against it. Such an extensive use of a principle can be explained only by its *self-evidence*, which, of course, must not be confused with its provability. No matter if this self-evidence is to a certain degree subjective—it is surely a necessary source of mathematical principles, even if it is not a tool of mathematical proofs, and Peano's assertion (*1906a*, p. 147) that it has nothing to do with mathematics fails to do justice to manifest facts. But the question that can be objectively decided, whether the principle is *necessary for science*, I should now like to submit to judgment by presenting

[2] See *Peano 1906a*, p. 147.

[3] See *Peano 1906a*, pp. 145–147. This proof, incidentally, can be carried out only by mathematical induction; hence it is binding only if we define the finite numbers in Peano's way by means of their order type. If on the other hand we take as a basis Dedekind's definition of a finite set as one that is not equivalent to any of its parts, no proof is possible even for finite sets, since the reduction of the two definitions to each other, as we shall show below (example 4), again requires the principle of choice. In this sense, therefore, Poincaré's remark in *1906*, p. 313 [[see below, p. 190, footnote 6]], is justified.

a number of elementary and fundamental theorems and problems that, in my opinion, could not be dealt with at all without the principle of choice.

(1) If a set M can be decomposed into disjoint parts, A, B, C, . . . , the set of these parts is equivalent to a subset of M, or, in other words, the set of summands always has a cardinality lower than, or the same as, that of the sum.

To prove this we must mentally associate with each of these parts one of its elements.[4]

(2) The sums of equivalent sets are again equivalent, provided all terms are mutually disjoint, a theorem upon which the entire calculus of cardinalities rests.

Here it is necessary to consider a system of mappings that *simultaneously* correlate any two equivalent summands with each other; thus from all the possible mappings associated with each pair of equivalent summands we must choose a single one, and we must make this choice for every pair.

(3) The product of several cardinalities can vanish only if one factor vanishes, that is, Cantor's "connection set" ⟦see below, p. 204⟧ of several sets A, B, C, . . . , each containing at least one element, must likewise contain at least one element. But since every such element is a set having exactly one element in common with each of the sets A, B, C, . . . , the theorem is merely another expression of the postulate of choice for disjoint sets (the axiom in IV, end of § 1 above).

(4) A set that is not equivalent to any of its parts can always be ordered in such a way that every subset possesses a first as well as a last element.

This theorem, upon which the theory of *finite* sets rests, is most simply proved by means of my well-ordering theorem. Dedekind (*1888*, art. 159) proved a logically equivalent theorem—that a set not equivalent to any segment of his "number sequence" must have a component equivalent to the entire number sequence—by simultaneously mapping a system of equivalent pairs of sets, hence, as in (2) above, also by using the principle of choice.[5] I do not know of any other proof.

(5) A denumerable set of finite or denumerable sets always possesses a denumerable sum.

Upon this theorem rests the theory of denumerable sets and of the "second number class"; but it can be proved only if we *simultaneously* order all the finite or denumerable sets in question like the natural numbers ⟦nach dem Normaltypus⟧.

[4] That a particular principle of inference is used here was probably first stated by Beppo Levi (*1902*) in connection with a proof by F. Bernstein. According to Bernstein (*1905*, p. 193), however, the "hypothesis" that a choice is possible is said to be "dispensable" in all similar cases, for instance also in my proof, if one employs the notion "multivalued equivalence" that he introduces. According to him (*1904*) two sets M and N are said to stand in the relation of multivalued equivalence if an entire set A of one-to-one mappings φ, χ, ψ, . . . , "among which none is distinguished", is given for them instead of just a single one. Hence a pure relational notion such as "distinguished" is used here, without supplementary determination or definition, as an absolute characteristic, and the attempt to differentiate between multivalued and ordinary equivalence is logically not realizable. In the examples considered, however, we are not at all concerned with the "multiplicity", that is, the *cardinality* of the set A of mappings, but merely with the question whether *at least one* such mapping φ exists, a question that cannot be evaded here by any definition and can be settled only by means of an axiom.

[5] This fact, which has frequently been overlooked, is expressly acknowledged also by Hessenberg (*1906*, preface).

(6) Does there exist a "base" for all real numbers, that is, a system of real numbers such that no linear relation with a finite number of integral coefficients holds between them and that any other real number can be obtained from them by a linear relation with a finite number of integral coefficients?

(7) Are there discontinuous solutions of the functional equation

$$f(x + y) = f(x) + f(y)?$$

The last two questions were answered affirmatively by Hamel (*1905*) on the assumption that the continuum can be well-ordered.

Cantor's theory of cardinalities, therefore, certainly requires our postulate, and so does Dedekind's theory of finite sets, which forms the foundation of arithmetic. That in the theory of functions we can usually circumvent its use is to be explained simply by the fact that there as a rule we deal with "completed" ⟦"abgeschlossenen"⟧ sets, for which distinguished elements can be unambiguously defined without any difficulty. Where this is not the case, hence especially in the theory of everywhere discontinuous functions, the principle is often indispensable, as our last example shows.

Now so long as the relatively simple problems mentioned here remain inaccessible to Peano's expedients, and so long as, on the other hand, the principle of choice cannot be definitely refuted, no one has the right to prevent the representatives of productive science from continuing to use this "hypothesis"—as one may call it for all I care—and developing its consequences to the greatest extent, especially since any possible contradiction inherent in a given point of view can be discovered only in that way. We need merely separate the theorems that necessarily require the axiom from those that can be proved without it in order to delimit the whole of Peano's mathematics as a special branch, as an artificially mutilated science, so to speak. Banishing fundamental facts or problems from science merely because they cannot be dealt with by means of certain prescribed principles would be like forbidding the further extension of the theory of parallels in geometry because the axiom upon which this theory rests has been shown to be unprovable. Actually, principles must be judged from the point of view of science, and not science from the point of view of principles fixed once and for all. Geometry existed before Euclid's *Elements*, just as arithmetic and set theory did before Peano's *Formulaire*, and both of them will no doubt survive all further attempts to systematize them in such a textbook manner.

Of course, there remains to Peano a simple way of proving the theorems in question, as well as many others, from his own principles. He need only use Russell's antinomy, lately much discussed, since, as is well known, everything can be proved from contradictory premisses. Indeed the principles of the *Formulaire*, which make no distinction between "set" and "class", do not exclude this contradiction. On the other hand, as I shall soon show elsewhere ⟦*1908a*⟧, those who champion set theory as a purely mathematical discipline that is not confined to the basic notions of traditional logic are certainly in a position to avoid, by suitably restricting their axioms, all antinomies discovered until now. Thus while the domain of Peano's principles, as we have just shown, is too narrow to permit the development of our science in its full beauty, it is on the other hand too wide to exclude internal contradictions; and, so long as the antinomies of this system are not eliminated, one is hardly justified in seeking in it the definitive foundation for the science of mathematics.

b. *Objection concerning nonpredicative definition*

The point of view maintained here, that we are dealing with a productive science resting ultimately upon intuition, was recently urged, in opposition to Peano's "logistic", by Poincaré, too, in a series of essays (*1905*, *1906*, *1906a*); in these he also does full justice to the principle of choice, which he considers an unprovable but indispensable axiom.[6] However, he presses his attack so far—since his opponents made use chiefly of set theory—as to identify all of Cantor's theory, this original creation of specifically mathematical thought and the intuition of a genius, with the logistic that he combats and to deny it any right to exist, without regard for its positive achievements, solely on the ground of antinomies that have not yet been resolved.[7] If he was concerned only to show that in the foundations of arithmetic there are "synthetic judgments a priori", among which, he believed, he could reckon the principle of mathematical induction above all, it would have sufficed, so far as the set-theoretic proofs of this principle are concerned, to ascribe a synthetic character to the fundamental propositions upon which these proofs rest; even the champions of set theory could have accepted that, since the distinction between "synthetic" and "analytic" would then be a purely philosophical one and not touch mathematics as such. Instead, he undertook to combat mathematical proofs with the weapons of formal logic, thus venturing upon a territory in which his logistic opponents are his betters.

To make Poincaré's conception clearer, the simplest thing would probably be to choose an example from the proof given in § 1 of the present paper. There I defined a special class of sets, which I called Θ-chains, and then proved that the common component M of all these Θ-chains is itself a Θ-chain. This procedure is modeled upon the theory of "chains" on which Dedekind (*1888*, § 4) bases his theory of finite numbers, and it is also customary elsewhere in set theory. But according to Poincaré (*1906*, p. 307) a definition is "predicative" and logically admissible only if it *excludes* all objects that are "dependent" upon the notion defined, that is, that can in any way be determined by it. Accordingly, in the example cited here the set M, which itself is determined only by the totality of the Θ-chains, would have had to be excluded from the definition of these chains, and my definition, which counts M itself as a Θ-chain, would be "nonpredicative" and contain a vicious circle. In two passages that are quite similar (*1906*, pp. 314–315), the latter referring to the "γ-sets" of my 1904 proof, this is expressly elaborated as a criticism of my proof procedure.

Now, on the one hand, proofs that have this logical form are by no means confined to set theory; exactly the same kind can be found in analysis wherever the maximum or the minimum of a previously defined "completed" set of numbers Z is used for

[6] See *Poincaré 1906*, pp. 311–313, especially p. 313: "Hence this is a synthetic judgment a priori; without it the theory of cardinals would be impossible, for finite numbers as well as for infinite ones".

[7] See *Poincaré 1906*, p. 316: "There is no actual infinite; the Cantorians forgot that, and they fell into contradiction".

further inferences. This happens, for example, in the well-known Cauchy proof of the fundamental theorem of algebra, and up to now it has not occurred to anyone to regard this as something illogical. On the other hand, it is precisely the form of definition said to be predicative that contains something circular; for unless we already have the notion, we cannot know at all what objects might at some time be determined by it and would therefore have to be excluded. In truth, of course, the question whether an arbitrary given object is subsumed under a definition must be decidable independently of the notion still to be defined, by means of an *objective* criterion. But once such a criterion is given, as is in fact the case everywhere in the examples drawn from my proofs, nothing can prevent some of the objects subsumed under the definition from having in addition a special relation to the same notion and thus being determined by, or distinguished from, the remaining ones—say, as common component or minimum. After all, an object is not created through such a "determination"; rather, every object can be determined in a wide variety of ways, and these different determinations do not yield identical but merely equivalent notions, that is, notions having the same extension. In fact, the existence of equivalent notions seems to be what Poincaré has overlooked in his critique, as Peano[8] especially emphasizes in this connection. A definition may very well rely upon notions that are equivalent to the one to be defined; indeed, in every definition *definiens* and *definiendum* are equivalent notions, and the strict observance of Poincaré's demand would make every definition, hence all of science, impossible.

c. *Objections based upon the set W*

The criticisms discussed so far, which are directed against the principles and proof methods of set theory in general, naturally met with little approval among the mathematicians who, like J. König, Jourdain, and F. Bernstein, already have themselves been productively active in this field and thus had the opportunity of convincing themselves that the devices mentioned are indispensable. On the other hand, the "Burali-Forti antinomy", recently again so much the subject of discussion, concerning the "set W of all Cantor ordinals" seems to have instilled in some of them an all too pervasive skepticism toward the theory of well-ordering. And yet, even the elementary form that Russell[9] gave to the set-theoretic antinomies could have

[8] *1906a*, p. 152. At issue here is the new proof of the Schröder-Bernstein theorem on the equivalence of sets that I had communicated to Poincaré by letter in January 1906. He had reproduced this proof, correctly as to content, on pages 314–315 of the May number of the *Revue de métaphysique et de morale* and had made it the object of a critique, to which the cited passage from Peano refers (in part quoting literally). But Peano, without mentioning this state of affairs, links his presentation to a previous version of his own proof, which essentially coincides with mine, except that it is formulated in his ideography; Poincaré apparently had not yet seen this first form (*Peano 1906*) of the proof when he wrote his article. Why does Peano avoid mentioning my name here, where he agrees with me, and then direct his opposition to the principle of choice immediately afterward so expressly at me? It would seem obvious, after all, that not mathematical principles, which are common property, but only the proofs based upon them can be the possession of an individual mathematician. I had expressly stated at the end of my note, incidentally, that I owe to a suggestion by Erhard Schmidt (now in Bonn) the idea of invoking the principle of choice in order to form a "γ-covering".

[9] *1903*, pp. 366–368. I had, however, discovered this antinomy myself, independently of Russell, and had communicated it prior to 1903 to Professor Hilbert among others.

persuaded them that the solution of these difficulties is not to be sought in the surrender of well-ordering but only in a suitable restriction of the notion of set. Already in my 1904 proof, having such reservations in mind, I avoided not only all notions that were in any way dubious but also the use of ordinals in general; I clearly restricted myself to principles and devices that have not yet by themselves given rise to any antinomy. If some critics nevertheless deploy this ominous "set W" against my proof, they must first project it into that proof artificially, and all arguments drawn from the inconsistent character of this "set" turn back upon their authors. Now I succeeded in completing my new proof without even the device of rank-ordering, and I hope thereby to have definitively cut off every possibility of introducing W.

J. König, it seems, is at least not far from this W point of view. For although he himself still relied upon the principle of choice in his Heidelberg lecture,[10] thus accepting the most essential assumption of my theorem, he treated the question whether the continuum can be well-ordered as an unsolved problem in later publications,[11] without regard for my already published note; but up to now he has abstained from giving public expression to his reservations about any specific step in my proof.

Jourdain claims to have proved the well-ordering theorem before I did,[12] and,

[10] *1904*. [This is the published version of König's address at the Third International Congress of Mathematicians, in Heidelberg. On what happened at the congress itself we have the testimony of Schoenflies (*1922a*, p. 100): "For Cantor to claim that every set can be well-ordered and, in particular, that the continuum has the second cardinality was a kind of dogma that was part and parcel of what he knew and believed in set theory. Consequently König's address, which culminated in the proposition that the continuum could not be an aleph (hence could not be well-ordered either), had a stunning effect, especially since its presentation was extremely elaborate and precise. It was based upon a relation between alephs that had been established by F. Bernstein; since this relation soon turned out not to be universally valid, König's result, as is well known, had to be restricted to alephs with transfinite subscripts and thus was robbed of its initial significance." (See also *Schoenflies 1928*, pp. 560–561.) The published version of König's address (*1904*) contains an indication of the loophole in the proof.]

[11] *1905*, *1905a*. [Save for an emendation, the first paper is identical with *1904*.] Concerning the Richard antinomy, which König tries to adduce in the second of these papers, see *Peano 1906a*, pp. 148–157, as well as, especially, *Hessenberg 1906*, chap. 23, "Die Paradoxie der endlichen Bezeichnung", where the present fallacy is in my opinion appropriately exposed. "Finitely definable" is not an absolute notion but only a relative one, and it is always related to the chosen language, or notation. The conclusion that [the set of] all finitely definable objects must be denumerable holds only if one and the same system of signs is to be used for all of them, and the question whether a single individual can or cannot have a finite designation is in and of itself meaningless, since to any object we could, if necessary, arbitrarily assign any designation whatever. That the continuum can be well-ordered, incidentally, does not have much more to do with this antinomy, basically, than does any other proposition; all can equally well be proved and refuted by the use of a contradiction. In fact, it is precisely by means of finite definability that F. Bernstein once (*1905a*) wanted to prove that the continuum is equivalent to the second number class, hence to a well-ordered set; thus, starting from the same notion as König, he arrived at the opposite conclusion. To be sure, the promised realization of this "proof" was never published.

[12] *Jourdain 1905a*. In the earlier papers cited by him (*1904*, *1904a*, *1905*), upon which he rests his claims to priority, there is, however, no mention at all of the possibility of a well-ordering. Rather, in the first of these articles his "proof that every cardinal number is an aleph" confines itself merely to an attempt to exclude the possibility of cardinalities greater than all alephs by reference to the Burali-Forti antinomy. Hence it is assumed there *without proof* that a set whose cardinal number is not itself an aleph must have a constituent part similar to the totality of all alephs; a mere reference to the methods and results of Cantor and Hardy, which concern the first two cardinalities, cannot, after all, possibly replace a proof.

moreover, in a simpler and more complete way; but, having W in his mind, he interprets the theorem in such a way that, as he expressly states (*1905a*, p. 469), not even the continuum need be an aleph. According to him, only "consistent" sets, that is, those that have no component similar to W, are to possess order types and cardinal numbers, and precisely in thus allowing "inconsistent" sets he sees the greater "completeness" of his result. But now, since in Cantor's theory "order types" and "cardinal numbers" are nothing but convenient *means of expression* for the comparison of sets with respect to the similarity or equivalence of their parts, I cannot extract any intelligible meaning from the proposition that a well-ordered set possesses no order type or cardinal number, and this attempt to resolve the antinomy while retaining W seems to me to amount to a mere word game. One can, to be sure, arbitrarily ignore all higher order types, from, say, the second or third number class on; one can refuse to recognize them as such any longer and thus obtain a W of type ω, or Ω, respectively, which is certainly consistent under the assumptions made, insofar as one disregards its "order type". But then this W remains *completely indeterminate*, and, what is more, it is just not the *same* W as is at issue in the antinomy, namely, a set so constituted that to every arbitrary well-ordered set an element of W corresponds as order type. A similar reservation seems afterward to have occurred to Jourdain himself and to have led him, if not to renounce his W, in any event to introduce a *second*, likewise well-ordered set $\mathfrak{W}$, which being "absolutely infinite" like Bernstein's W, to be discussed below, is claimed to be no longer capable of any "continuation".

Now, finally, concerning the *proof* that Jourdain, in the note cited (*1905a*, p. 468), opposes to mine as being simpler, the procedure he proposes for well-ordering an arbitrary set M is the following. Take an arbitrary element as first, then another, and so forth; after an arbitrary finite or infinite number of elements take an arbitrary element of the remainder as the next one; and continue in this way until the entire set is exhausted. This idea of a stepwise construction is not new; it was communicated to me orally quite some time ago by F. Bernstein, and it probably goes back to Cantor, who, however, apparently had reservations about accepting it as a *proof.* Borel (*1905*), too, recommends the same construction, only to reject it immediately without further explanation, thereby, as he believes, reducing the principle of choice to absurdity. But in this he does not succeed at all; for it is not the infinitely repeated *choice* that vitiates this "proof", but simply the fact that the proof does not lead to the goal. For, once we accept the above-mentioned principle of Peano, which permits a choice from a single set, there is no longer any limit to its repeated application. But what, then, do all these considerations *prove*? Evidently no more than that every well-ordered proper subset M' of M can still be extended by the addition of an arbitrary element m' taken from the remainder; or, rather, this is the *assumption* that—in strict contrast to the conception of Bernstein and Schoenflies—forms the basis of the entire procedure. If it can now be proved that among the well-ordered components of M there *exists a largest one*, L, in the same way as, for example, I *prove* the existence of a largest γ-set L_γ, then necessarily $L = M$, and M is well-ordered. Jourdain, too, wants to make precisely the same inference; only he lacks the essential premiss, the proof of the existence of L. Rather, he presupposes it *without proof*, by assuming that his procedure, insofar as it does not exhaust the set

M, would have to terminate in a well-ordered part similar to $\mathfrak{W}$. Thus this proof is so "simple" that it reduces to a single inference, albeit a fallacious one.[13]

While for all that Jourdain, as we saw, looks upon the "inconsistent" set W with some doubts, F. Bernstein (*1905*) already makes it the object of a dogmatic theory. Since, as is well known, the contradictory character of this set of all ordinals becomes manifest when we add to it a further element, e, that follows all elements of W, Bernstein believes that he can overcome all difficulties by declaring that to append such a new element is inadmissible on the grounds that this would contradict the definition of W. The set W is to contain only the order types of all "continuable" well-ordered sets, or of all "segments of well-ordered sets", but W itself is to be "noncontinuable". From this point of view he then criticizes the transition (7 V) from the well-ordered set L to (L, m'_1) in my proof of 1904, since, after all, L might possibly be similar to the set W,[14] and by means of W he constructs a set Z that is said not to be capable of any well-ordering.

This, of course, is love's labor lost. For, as Bernstein expressly admits, once the set W exists, it is well-ordered, with a definite order type β, and any other well-ordered set is similar either to W itself or to a segment of W. According to Cantor's definition, which is fixed once and for all, of the relations "greater than" and "less than" for ordinals, β would hence be greater than any other ordinal α, and, in the set (W, β), when it is ordered according to magnitude, β would rank *behind* all elements of W, that is, W would in fact be "continuable", contrary to its definition and in spite of all prohibitions. Bernstein does not know what to say in reply to this undesired consequence other than "that the contradiction comes about only because β is *assumed* to follow upon all elements of W. If only the union $(W; \beta)$ is formed and *no order relation is stipulated* between β and the elements of W, no contradiction is generated".[15] As if it were only a matter of the term "order relation" or the *notation* $\alpha \prec \beta$, and as if we could eliminate an objective mathematical *fact* by avoiding a *term*! Mathematics would not be an international science if its propositions did not have an objective content independent of the language in which we express them. In the examination of a contradiction the issue, after all, is not whether a questionable consequence is *actually realized* and officially *accepted*, but merely whether it is formally *possible* at all; and to deny this possibility merely *because* it leads to a contradiction would obviously be begging the question or falling into a vicious circle. In

[13] Hardy's alleged "construction of a part of the continuum having the second cardinality" (*1904*), which was expressly accepted also by Schoenflies (*1905*) among others, suffers from the same indefiniteness as the procedure of Cantor and Jourdain. He gives a rule for deriving, from a previously constructed denumerable and well-ordered part A, a new element of the continuum that is distinct from all preceding ones. Since this rule is *not univocal*, however, but depends upon the representation [[of an element of the continuum]] by means of fundamental sequences, which is arbitrary within wide limits, there is no reason to prefer his procedure to Cantor's very much simpler diagonal procedure, which merely requires a reordering according to the type ω; no more than Cantor's does Hardy's procedure yield a *definite* part having the second cardinality; rather, at best it yields a new proof of the fact that the cardinality of the continuum is $\geqq \aleph_1$.

[14] Therewith Bernstein rejects not only a "part", as he says, but the entire content of Jourdain's proof, even though his own W corresponds exactly to Jourdain's $\mathfrak{W}$. For Jourdain, L, precisely because it is continuable, cannot be similar to $\mathfrak{W}$, whereas Bernstein, on the contrary, infers that it is not continuable because it is similar to W. Hence here, too, opposite consequences are drawn from a common assumption.

[15] *1905*, p. 189. I have merely replaced e by β and italicized a few words.

fact, however, the procedure followed in the justification of W amounts to this: the contradiction inherent in its definition is not resolved but *ignored*. If according to universal principles an assumption A yields two contrary consequences B and B', then A must be rejected as untenable. But in the present case, we are told, it is permissible to opt for one of these consequences, B, while prohibiting the other, B', by means of some special decree or veiling it by a change of name, lest it yield a contradiction with B. Since this procedure would obviously be applicable to every arbitrary hypothesis A, there would never be any contradiction at all; we could assert everything but prove nothing, since along with the possibility of a contradiction that of a proof would also be eliminated, and no science of mathematics could exist.[16]

That it is in fact always possible to add a further element u as a last element to an *arbitrary* well-ordered set M can, incidentally, be *proved* in an elementary way from the general principles of set theory, provided we adopt a purely formal definition of well-ordering, such as that given here at the end of § 1. For if the well-ordering of M is given by the system of remainders $\mathfrak{R}(x)$ associated with the elements x of M, we need merely add the new element to each of these remainders (independently of its ordering) by means of a simple union with the set $\{u\}$ to obtain then, together with the set $\{u\}$, a new system of remainders that furnishes the desired well-ordering of $M_1 = M + \{u\}$. In fact it is then very easy to see that every subset of M_1 again possesses a first element in the sense defined above and that all elements of M precede the element u. This proof, which, merely on account of its trivial simplicity, I did not find it necessary to include in my note of 1904, also assures, according to what was said above, the nonexistence of W, and all consequences drawn from W come to nought. Since now, on the other hand, "order type of a well-ordered set" is certainly a logically admissible *notion*, it follows further—as already appears in a much simpler way from Russell's antinomy, to be sure—that it is not permissible to treat the extension of every arbitrary notion as a set and that therefore the customary definition of set is too wide. But if in set theory we confine ourselves to a number of established principles such as those that constitute the basis of our proof—principles that enable us to form initial sets and to derive new sets from given ones—then all such contradictions can be avoided.

d. *Objection concerning particular generating principles and the set Z*

The critique of Schoenflies (*1905*, p. 181), likewise directed against the last step (7 V) of my proof, is inspired by the same belief in W and is therefore taken care of by the preceding discussion simultaneously with Bernstein's. His paper, however, contains further errors and misunderstandings that cannot be ignored here.

To begin with, Schoenflies distinguishes in the theory of well-ordered sets between "a general and a special part" and claims that only the theorems of the general part rest upon Cantor's familiar *definition* of well-ordering (*1897*), while the others rest

[16] Concerning the various attempts to "save W", see also *Hessenberg 1906*, chap. 24, "Ultrafinite Paradoxieen", where at the end of § 98 we read: "The set W itself is, incidentally, ungrateful in the highest degree for all attempts to redeem its honor. Thus in volume 60 of *Mathematische Annalen* both Bernstein and Jourdain exert themselves on behalf of its consistency, the former proving on the basis of the properties of W that there are sets that cannot be well-ordered, while the latter succeeds in proving the opposite."

upon "generating principles". In truth, however, *all* theorems concerning a notion must be provable from its definition; otherwise they would not be proved at all. If there are two definitions at hand for a notion, we must opt definitely for one of them or prove the equivalence of the two; to vacillate in any way between two definitions or to complement one by means of the other is logically completely inadmissible. Schoenflies then continues: "If we want to prove the theorem in question on this general basis, we must show that no infinite sequence of sets M_1 with decreasing cardinalities can exist; every decreasing sequence of cardinalities, $\mathfrak{m}_1 > \mathfrak{m}_2 > \mathfrak{m}_3 > \ldots$, would have to terminate after a finite number of terms. This is the necessary *and sufficient* condition for the theorem." Necessary but *not* sufficient, and for precisely this reason not usable for a proof of the theorem. The criterion cited applies merely to the well-ordering of the *cardinalities* ordered according to magnitude, and not at all to the well-ordering of the *sets* themselves, which is the subject of my theorem. If all cardinalities are alephs, then, to be sure, they are also well-ordered according to magnitude, but one *cannot* infer the converse, whereas Schoenflies's proposal apparently requires that. This method does not even allow one to prove that arbitrary sets *can be compared* with respect to their cardinality; rather, the method already presupposes this as a basis. "Zermelo's proof does not follow this path." Indeed not, since a proof just cannot be carried out in that way. "He operates with devices that concern the special part of the theory of well-ordered sets, namely, their generation." Quite on the contrary, my proof rests exclusively upon Cantor's classical definition and has nothing to do with generating principles, in the sense assumed, as independent sources of proof.

"Generating principles can at first only be *postulated axiomatically*; then one must demonstrate that they are justified. Even the introduction of the numbers of the second number class and of the principle of passage from n to ω was originally possible only by means of such an axiom. The proof that this axiom is admissible was carried out, as we must recognize, by the detailed theory that Cantor gave of these numbers, whose regularity admits of no exception." Rather, Cantor (*1897*, p. 221) defines the numbers of the second number class as the *order types* that can be associated with well-ordered denumerable sets, and he *proves* everything else from this definition. Only the existence of denumerable sets, or the type ω by which they are defined, is assumed; there is no need of a further axiom. But if the existence of such order types were in any way doubtful, not even the most beautiful formalism would be of help. In an analogous way every higher number class is defined by means of the preceding one, *without* "any need of a new creation, or of a new axiom and the proof of its justification". In my proof, too, Schoenflies discovered a new "postulate", "which concerns the generation of well-ordered sets. For it asserts that, if L is any well-ordered set, so is (L, m)". Actually not a postulate at all, but, as we saw above, a *provable proposition*. "This assumption, and in particular the use that Zermelo makes of it, so to speak *includes within itself all possible generating principles*. It *contains even more, however*, and for precisely this reason it is inadmissible." Here, too, everything is just the opposite of what Schoenflies asserts. Cantor does *not* use the generating principles for the production of well-ordered sets but for the systematic discovery of all order types of a given number class. For him the addition of a single element at the end of a set of given order type ξ, hence the operation $\xi + 1$, is the first generating principle, which yields new order types within *each* number class and constitutes the *presupposition* of all other generating principles. But already in the second number class it *fails to suffice*, since, beginning with ω, it could not even generate $\omega \cdot 2$ and hence requires supplementation by the second generating principle, which relates to the "fundamental sequences" of type ω. In the higher number classes one would require still *further* principles *besides* these two. But the first generating principle $\xi + 1$ never leads us out of any number class, since the *cardinality* of a transfinite set, which by definition forms the distinguishing characteristic of the various number classes, is, as is well known, not changed by the addition of just a single element. Accordingly, the well-ordering of the entire set is, of course, *not* "produced" by means of *this* operation in my proof either, but, as I expressly noted at the end, by the "*fusion* of the various possible γ-sets". It is only that each single γ-set can be "continued" in accordance with this principle, a fact upon which I then base the inference that necessarily $L_\gamma = M$. Very strange, furthermore, is the assertion that "*nowhere* else is an assumption required such as Zermelo's proof employs" —when in fact the entire theory of ordinals and generating principles is based upon this operation $\xi + 1$.

But now we come to the main point. "The notion of the *totality* of all possible generating principles for well-ordered sets is in my opinion a *well-defined set-theoretic notion*, as is the notion of the well-ordered sets that can be produced with them . . . in the same sense . . . as the totality of all integers." Thus the dogma of W is proclaimed. Schoenflies does not even venture to justify this inconsistent notion, as F. Bernstein after all had at least *attempted* to do; his "opinion" is to suffice for us here. The distinction that he then makes between well-ordered sets that "can be produced" and those that "cannot be produced" is not really intelligible, especially since already in the next paragraph a set that cannot be produced is also characterized without further ado as "logically inconsistent". "In any case we arrive at a *totality* of well-ordered sets that is likewise *well-defined*. . . . Let Z be the well-ordered set thus determined. According to its definition, it provides the limit beyond which we never proceed when we actually produce a well-ordered set." Apparently this Z is to be the totality of the possible *order types* of such sets, but the notions of an ordered "set" and its "order type", so carefully differentiated by Cantor, are not distinguished at all in Schoenflies's paper, an imprecision that, as we shall see immediately, has not remained without fatal consequences either.

"Let us assume for the time being that Z is the second number class, that is, that no well-ordered set that we . . . can form ever takes us beyond the second number class. . . . In that case the set (Z, m) represents a notion that is inconsistent in itself." Thus according to this passage Schoenflies seems to regard the order type of (Z, m) as the *first* that no longer belongs to the second number class. But now Cantor proved that *the totality of the numbers of the second number class is itself no longer denumerable*, hence that its order type already belongs to the next higher number class, whereas when we employ the first generating principle $\xi + 1$ we always remain within the same number class. Accordingly, on the basis of the assumption made, not only (Z, m) but already Z would *itself be a contradictory notion*. And in fact Schoenflies's Z, just like Bernstein's W, is afflicted with internal contradictions in any case, precisely because, as we saw above, it cannot contain its own order type as an element.

"On the preceding basis" a new method for the examination of the theorem in question is then proposed, a method that, as Schoenflies strains to show, also leads to *no result*. There are always various methods of *not* proving a given theorem, to be sure, and in particular it would seem that vague and inconsistent notions are especially appropriate when one wants to avoid giving a proof. But then, of course, one cannot refute genuine proofs by means of such devices. The proposed procedure, however, consists in comparing the well-ordered component L of the set M under consideration with the sets W and Z characterized above. This, then, yields several cases, all of which are said to be logically equally possible, whereas only one of them corresponds to the assumptions made in my proof. This cannot surprise us, of course, after Schoenflies has already incorporated into his assumptions all those contradictions that lead to the exclusion of the other cases.

Thus, when Schoenflies remarks at the end of his paper that "in my opinion we should, in set theory, treat assumptions that lead to inconsistent notions or results just as we are accustomed to treating them elsewhere", we can surely concur. For this means that such assumptions are to be excluded and that no consequences should be derived from inconsistent notions. In my proof this procedure has been strictly observed, too, but not in Schoenflies's critique.

e. *Summary*

The preceding discussion of the opposition to my 1904 proof can perhaps be summarized most simply by the following statements. Except for Poincaré, whose critique, based on formal logic—a critique that would threaten the existence of all of mathematics—has hitherto not met with any assent whatsoever, all opponents can be divided into two classes. Those who have no objection at all to my deductions protest the use of an unprovable general principle, without reflecting that such axioms constitute the basis of every mathematical theory and that precisely the one I adduced is indispensable for the extension of the science in other respects, too. The other critics, however, who have been able to convince themselves of this indispensability by a deeper involvement with set theory, base their objections upon the Burali-Forti antinomy, which in fact *is without significance* for my point of view, since the principles I employed *exclude* the existence of a set W.

The relatively large number of criticisms directed against my short note testifies to the fact that, apparently, strong prejudices stand in the way of the theorem that any arbitrary set can be well-ordered. But the fact that in spite of a searching examination, for which I am indebted to all the critics, no *mathematical error* could be demonstrated in my proof and the objections raised against my *principles* are mutually contradictory and thus in a sense cancel each other allows me to hope that in time all of this resistance can be overcome through adequate clarification.

Investigations in the foundations of set theory I

ERNST ZERMELO

(*1908a*)

This paper presents the first axiomatic set theory. Cantor's definition of set[a] had hardly more to do with the development of set theory than Euclid's definition of point with that of geometry. Dedekind, whom Zermelo considers one of the two creators of set theory, had explicitly stated (*1888*, § 1) a number of principles about sets (which he called "systems"), but his attempt had remained fragmentary and had been somewhat discredited by the nonmathematical way in which he justified the existence of an infinite set (*1888*, art. 66). In spite of the great advances that set theory was making, the very notion of set remained vague. The situation became critical after the appearance of the Burali-Forti paradox and intolerable after that of the Russell paradox, the latter involving the bare notions of set and element. One response to the challenge was Russell's theory of types (above, pp. 150–182). Another, coming at almost the same time, was Zermelo's axiomatization of set theory. The two responses are extremely different; the former is a far-reaching theory of great significance for logic and even ontology, while the latter is an immediate answer to the pressing needs of the working mathematician.

Zermelo's basic idea resembles, if anything, Russell's "theory of limitation of size" (*1905a*); both refuse to take as sets collections that are too "big", that of all "things" or that of all ordinals, for example.[b] Sets are not simply collections; they are objects satisfying certain axiomatic conditions. Zermelo's axioms are surprisingly few in number. The most original is perhaps Axiom III, the axiom of separation. In Peano and Russell sets were part of logic, being intimately connected with "conditions", or propositional functions. To express the relation, generally felt to exist, between a set and a stipulation asserted by a statement, Zermelo introduces the notion "definite property" and, together with it, the axiom of separation: a definite property separates a subset from an already given set. His definition of "definite property", however, invokes "the universally valid laws of logic"; since Zermelo pays no attention at all to the underlying logic, these laws are left unspecified, and the notion of definite property remains hazy. The flaw will be removed, in different manners, by Weyl, Fraenkel, Skolem, and von Neumann (see below, p. 285).

Zermelo was perhaps the first to see clearly that the existence of infinite sets has to be insured by a special axiom (Axiom VII, of infinity). The powerful new tool that he used in his proofs of the well-ordering theorem appears in two forms, as axiom of choice (Axiom VI) and as general principle of choice (Article 29). The latter, which does not assume that

---

[a] See below, p. 200, footnote 1.

[b] Cantor already had a similar idea; see above, p. 114. See also *König 1905a*, above, p. 148, and *von Neumann 1925*, below, pp. 396–398.

the sets from which the choice is made are disjoint, is here derived from Axiom VI.

Zermelo does not have the Cartesian product; he makes do with the "connection set", the set of unordered pairs, and that renders his treatment of equivalence of sets (§ 2), for example, somewhat cumbersome.

Zermelo states his axioms, declares that he has been unable to prove their consistency, and shows that the usual derivations of a number of known paradoxes cannot be obtained from them. He then proves theorems about sets. The development goes as far as Cantor's theorem,[c] König's theorem, and the theorem of Article 36, which connects two notions of infinitude. A second paper, dealing with a theory of well-ordering and a set-theoretic definition of natural numbers, is announced at the end of the introduction but was never published. However, in a paper (*1908b*) prepared shortly after the one below, Zermelo briefly shows how the natural numbers can be defined in a theory of finite sets.

The paper is dated "Chesières, 30 July 1907". The translation is by Stefan Bauer-Mengelberg, and it is printed here with the kind permission of Springer Verlag.

[c] "Every set is of lower cardinality than the set of its subsets."

Set theory is that branch of mathematics whose task is to investigate mathematically the fundamental notions "number", "order", and "function", taking them in their pristine, simple form, and to develop thereby the logical foundations of all of arithmetic and analysis; thus it constitutes an indispensable component of the science of mathematics. At present, however, the very existence of this discipline seems to be threatened by certain contradictions, or "antinomies", that can be derived from its principles—principles necessarily governing our thinking, it seems—and to which no entirely satisfactory solution has yet been found. In particular, in view of the "Russell antinomy" (*1903*, pp. 101–107 and 366–368) of the set of all sets that do not contain themselves as elements, it no longer seems admissible today to assign to an arbitrary logically definable notion a set, or class, as its extension. Cantor's original definition of a set (*1895*) as "a collection, gathered into a whole, of certain well-distinguished objects of our perception or our thought"[1] therefore certainly requires some restriction; it has not, however, been successfully replaced by one that is just as simple and does not give rise to such reservations. Under these circumstances there is at this point nothing left for us to do but to proceed in the opposite direction and, starting from set theory as it is historically given, to seek out the principles required for establishing the foundations of this mathematical discipline. In solving the problem we must, on the one hand, restrict these principles sufficiently to exclude all contradictions and, on the other, take them sufficiently wide to retain all that is valuable in this theory.

Now in the present paper I intend to show how the entire theory created by Cantor and Dedekind can be reduced to a few definitions and seven principles, or axioms, which appear to be mutually independent. The further, more philosophical, question about the origin of these principles and the extent to which they are valid will not be discussed here. I have not yet even been able to prove rigorously that my axioms

[1] ⟦Cantor's full definition reads: "Unter einer 'Menge' verstehen wir jede Zusammenfassung M von bestimmten wohlunterschiedenen Objekten m unsrer Anschauung oder unseres Denkens (welche die 'Elemente' von M genannt werden) zu einem Ganzen" (*1895*, p. 481, or *1932*, p. 282).⟧

are consistent, though this is certainly very essential; instead I have had to confine myself to pointing out now and then that the antinomies discovered so far vanish one and all if the principles here proposed are adopted as a basis. But I hope to have done at least some useful spadework hereby for subsequent investigations in such deeper problems.

The present paper contains the axioms and their most immediate consequences, as well as a theory of equivalence based upon these principles that avoids the formal use of cardinal numbers. A second paper, which will develop the theory of well-ordering together with its application to finite sets and the principles of arithmetic, is in preparation.[2]

§ 1. Fundamental definitions and axioms

1. Set theory is concerned with a *domain* $\mathfrak{B}$ of individuals, which we shall call simply *objects* and among which are the *sets*. If two symbols, a and b, denote the same object, we write $a = b$, otherwise $a \neq b$. We say of an object a that it "exists" if it belongs to the domain $\mathfrak{B}$; likewise we say of a class $\mathfrak{K}$ of objects that "there exist objects of the class $\mathfrak{K}$" if $\mathfrak{B}$ contains at least one individual of this class.

2. Certain *fundamental relations* of the form $a \varepsilon b$ obtain between the objects of the domain $\mathfrak{B}$. If for two objects a and b the relation $a \varepsilon b$ holds, we say "a is an *element* of the set b", "b contains a as an element", or "b possesses the element a". An object b may be called a *set* if and—with a single exception (Axiom II)—only if it contains another object, a, as an element.

3. If every element x of a set M is also an element of the set N, so that from $x \varepsilon M$ it always follows that $x \varepsilon N$, we say that M is a *subset* of N and we write $M \Subset N$.[3] We always have $M \Subset M$, and from $M \Subset N$ and $N \Subset R$ it always follows that $M \Subset R$. Two sets M and N are said to be *disjoint* if they possess no common element, or if no element of M is an element of N.

4. A question or assertion $\mathfrak{E}$ is said to be *definite* if the fundamental relations of the domain, by means of the axioms and the universally valid laws of logic, determine without arbitrariness whether it holds or not. Likewise a "propositional function" ⟦"Klassenaussage"⟧ $\mathfrak{E}(x)$, in which the variable term x ranges over all individuals of a class $\mathfrak{K}$, is said to be definite if it is definite for *each single* individual x of the class $\mathfrak{K}$. Thus the question whether $a \varepsilon b$ or not is always definite, as is the question whether $M \Subset N$ or not.

The fundamental relations of our domain $\mathfrak{B}$, now, are subject to the following *axioms*, or *postulates*.

Axiom I. (Axiom of extensionality ⟦Axiom der Bestimmtheit⟧.) If every element of a set M is also an element of N and vice versa, if, therefore, both $M \Subset N$ and $N \Subset M$, then always $M = N$; or, more briefly: Every set is determined by its elements.

The set that contains only the elements $a, b, c, \ldots, r$ will often be denoted briefly by $\{a, b, c, \ldots, r\}$.

[2] ⟦This paper is apparently *1909*; see also *1908b*.⟧

[3] This sign of inclusion was introduced by Schröder (*1890*). Peano and, following him, Russell, Whitehead, and others use the sign $\supset$ instead.

Axiom II. (Axiom of elementary sets ⟦Axiom der Elementarmengen⟧.) There exists a (fictitious) set, the *null set*, 0, that contains no element at all. If a is any object of the domain, there exists a set $\{a\}$ containing a and only a as element; if a and b are any two objects of the domain, there always exists a set $\{a, b\}$ containing as elements a and b but no object x distinct from both.

5. According to Axiom I, the elementary sets $\{a\}$ and $\{a, b\}$ are always uniquely determined and there is only a single null set. The question whether $a = b$ or not is always definite (No. 4), since it is equivalent to the question whether or not $a \varepsilon \{b\}$.

6. The null set is a subset of every set M: $0 \in M$; a subset of M that differs from both 0 and M is called a *part* ⟦*Teil*⟧[4] of M. The sets 0 and $\{a\}$ do not have parts.

Axiom III. (Axiom of separation ⟦Axiom der Aussonderung⟧.) Whenever the propositional function $\mathfrak{E}(x)$ is definite for all elements of a set M, M possesses a subset $M_{\mathfrak{E}}$ containing as elements precisely those elements x of M for which $\mathfrak{E}(x)$ is true.

By giving us a large measure of freedom in defining new sets, Axiom III in a sense furnishes a substitute for the general definition of set that was cited in the introduction and rejected as untenable. It differs from that definition in that it contains the following restrictions. In the first place, sets may never be *independently defined* by means of this axiom but must always be *separated* as subsets from sets already given; thus contradictory notions such as "the set of all sets" or "the set of all ordinal numbers", and with them the "ultrafinite paradoxes", to use Hessenberg's expression (*1906*, chap. 24), are excluded. In the second place, moreover, the defining criterion must always be definite in the sense of our definition in No. 4 (that is, for each single element x of M the fundamental relations of the domain must determine whether it holds or not), with the result that, from our point of view, all criteria such as "definable by means of a finite number of words", hence the "Richard antinomy" and the "paradox of finite denotation",[5] vanish. But it also follows that we must, prior to each application of our Axiom III, prove the criterion $\mathfrak{E}(x)$ in question to be definite, if we wish to be rigorous; in the considerations developed below this will indeed be proved whenever it is not altogether evident.

7. If $M_1 \in M$, then M always possesses another subset, $M - M_1$, the *complement* of M_1, which contains all those elements of M that are *not* elements of M_1. The complement of $M - M_1$ is M_1 again. If $M_1 = M$, its complement is the null set, 0; the complement of any part (No. 6) M_1 of M is again a part of M.

8. If M and N are any two sets, then according to Axiom III all those elements of M that are also elements of N are the elements of a subset D of M; D is also a subset of N and contains all elements *common* to M and N. This set D is called the *common component*, or *intersection*, of the sets M and N and is denoted by $[M, N]$. If $M = N$, then $[M, N] = M$; if $N = 0$ or if M and N are disjoint (No. 3), then $[M, N] = 0$.

9. Likewise, for several sets $M, N, R, \ldots$ there always exists an intersection $D = [M, N, R, \ldots]$. For, if T is any set whose elements are themselves sets, then according to Axiom III there corresponds to every object a a certain subset T_a of T that contains all those elements of T that contain a as an element. Thus it is definite

[4] ⟦Below, Zermelo also uses the expression "echter Teil" in the same sense. In the translation "part" has been used throughout for "nonempty proper subset".⟧

[5] See *Hessenberg 1906*, chap. 23; on the other hand, see *König 1905a*.

for every a whether $T_a = T$, that is, whether a is a common element of all elements of T; if A is an arbitrary element of T, all elements a of A for which $T_a = T$ are the elements of a subset D of A that contains all these common elements. This set D is called the intersection associated with T and is denoted by $\mathfrak{D}T$. If the elements of T do not possess a common element, $\mathfrak{D}T = 0$, and this is always the case if, for example, an element of T is not a set or if it is the null set.

10. THEOREM. Every set M possesses at least one subset M_0 that is not an element of M.

Proof. It is definite for every element x of M whether $x \varepsilon x$ or not; the possibility that $x \varepsilon x$ is not in itself excluded by our axioms. If now M_0 is the subset of M that, in accordance with Axiom III, contains all those elements of M for which it is not the case that $x \varepsilon x$, then M_0 cannot be an element of M. For either $M_0 \varepsilon M_0$ or not. In the first case, M_0 would contain an element $x = M_0$ for which $x \varepsilon x$, and this would contradict the definition of M_0. Thus M_0 is surely not an element of M_0, and in consequence M_0, if it were an element of M, would also have to be an element of M_0, which was just excluded.

It follows from the theorem that not all objects x of the domain $\mathfrak{B}$ can be elements of one and the same set; that is, *the domain* $\mathfrak{B}$ *is not itself a set*, and this disposes of the Russell antinomy so far as we are concerned.

AXIOM IV. (Axiom of the power set [[Axiom der Potenzmenge]].) To every set T there corresponds another set $\mathfrak{U}T$, the *power set* of T, that contains as elements precisely all subsets of T.

AXIOM V. (Axiom of the union [[Axiom der Vereinigung]].) To every set T there corresponds a set $\mathfrak{S}T$, the *union* of T, that contains as elements precisely all elements of the elements of T.

11. If no element of T is a set different from 0, then, of course, $\mathfrak{S}T = 0$. If $T = \{M, N, R, \ldots\}$, where $M, N, R, \ldots$ all are sets, we also write $\mathfrak{S}T = M + N + R + \ldots$ and call $\mathfrak{S}T$ the *sum* of the sets $M, N, R, \ldots$, whether some of these sets $M, N, R, \ldots$ contain common elements or not. Always $M = M + 0 = M + M = M + M + \ldots$.

12. For the "addition" of sets that we have just defined, the commutative and associative laws hold:

$$M + N = N + M, \quad M + (N + R) = (M + N) + R.$$

Finally, for sums and intersections (No. 8) the distributive law also holds, in the two forms:

$$[M + N, R] = [M, R] + [N, R]$$

and

$$[M, N] + R = [M + R, N + R].$$

The proof is carried out by means of Axiom I and consists in a demonstration that every element of the set on the left is also an element of the set on the right, and conversely.[6]

[6] The complete theory of this logical addition and multiplication can be found in *Schröder 1890*.

13. *Introduction of the product.* If M is a set different from 0 and a is any one of its elements, then according to No. 5 it is definite whether $M = \{a\}$ or not. *It is therefore always definite whether a given set consists of a single element or not.*

Now let T be a set whose elements, $M, N, R, \ldots$, are various (mutually disjoint) sets, and let S_1 be any subset of its union $\mathfrak{S}T$. Then it is definite for every element M of T whether the intersection $[M, S_1]$ consists of a single element or not. Thus all those elements of T that have exactly one element in common with S_1 are the elements of a certain subset T_1 of T, and it is again definite whether $T_1 = T$ or not. All subsets S_1 of $\mathfrak{S}T$ that have exactly one element in common with each element of T then are, according to Axiom III, the elements of a set $P = \mathfrak{P}T$, which, according to Axioms III and IV, is a subset of $\mathfrak{U}\mathfrak{S}T$ and will be called the *connection set* ⟦*Verbindungsmenge*⟧ associated with T or the *product* of the sets $M, N, R, \ldots$. If $T = \{M, N\}$, or $T = \{M, N, R\}$, we write $\mathfrak{P}T = MN$, or $\mathfrak{P}T = MNR$, respectively, for short.

In order, now, to obtain the theorem that *the product of several sets can vanish* (that is, be equal to the null set) *only if a factor vanishes* we need a further axiom.

Axiom VI. (Axiom of choice ⟦Axiom der Auswahl⟧.) If T is a set whose elements all are sets that are different from 0 and mutually disjoint, its union $\mathfrak{S}T$ includes at least one subset S_1 having one and only one element in common with each element of T.

We can also express this axiom by saying that it is always possible to *choose* a single element from each element $M, N, R, \ldots$ of T and to combine all the chosen elements, $m, n, r, \ldots$, into a set S_1.[7]

The preceding axioms suffice, as we shall see, for the derivation of all essential theorems of general set theory. But in order to secure the existence of infinite sets we still require the following axiom, which is essentially due to Dedekind.[8]

Axiom VII. (Axiom of infinity ⟦Axiom des Unendlichen⟧.) There exists in the domain at least one set Z that contains the null set as an element and is so constituted that to each of its elements a there corresponds a further element of the form $\{a\}$, in other words, that with each of its elements a it also contains the corresponding set $\{a\}$ as an element.

14_{VII}.[9] If Z is an arbitrary set constituted as required by Axiom VII, it is definite for each of its subsets Z_1 whether it possesses the same property. For, if a is any element of Z_1, it is definite whether $\{a\}$, too, is an element of Z_1, and all elements a of Z_1 that satisfy this condition are the elements of a subset Z_1' for which it is definite whether $Z_1' = Z_1$ or not. Thus all subsets Z_1 having the property in question are the elements of a subset T of $\mathfrak{U}Z$, and the intersection (No. 9) $Z_0 = \mathfrak{D}T$ that corresponds to them is a set constituted in the same way. For, on the one hand, 0 is a common element of all elements Z_1 of T, and, on the other, if a is a common element of all of these Z_1, then $\{a\}$ is also common to all of them and is thus likewise an element of Z_0.

[7] For the justification of this axiom see my *1908*, where in § 2, pp. 111–128 ⟦above, pp. 186–198⟧, the relevant literature is discussed.

[8] *1888*, art. 66. The "proof" that Dedekind there attempts to give of this principle cannot be satisfactory, since it takes its departure from "the set of everything thinkable", whereas from our point of view the domain $\mathfrak{B}$ itself, according to No. 10, does *not* form a set.

[9] The subscript VI, or VII, on the number of a section indicates that explicit or implicit use has been made of Axiom VI, or VII, respectively, in establishing the theorem of that section.

Now if Z' is any other set constituted as required by the axiom, there corresponds to it a smallest subset Z'_0 having the same property, exactly as Z_0 corresponds to Z. But now the intersection $[Z_0, Z'_0]$, which is a common subset of Z and Z', must be constituted in the same way as Z and Z'; and just as, being a subset of Z, it must contain the component Z_0, so, as a subset of Z', it must contain the component Z'_0. According to Axiom I it then necessarily follows that $[Z_0, Z'_0] = Z_0 = Z'_0$ and that Z_0 thus is the *common component of all possible sets constituted like* Z, even though these need not be elements of a set. The set Z_0 contains the elements 0, $\{0\}$, $\{\{0\}\}$, and so forth, and it may be called the *number sequence*, because its elements can take the place of the numerals. It is the simplest example of a denumerably infinite set (below, No. 36).

§ 2. Theory of equivalence

From our point of view, the equivalence of two sets (*Cantor 1895*, p. 483) cannot be defined at first except for the case in which the sets are disjoint (No. 3); it is only afterward that the definition can be extended to the general case.

15. *Definition* A. Two disjoint sets M and N are said to be *immediately equivalent*, $M \sim N$, if their product MN (No. 13) possesses at least one subset Φ such that each element of $M + N$ occurs as an element in one and only one element $\{m, n\}$ of Φ. A subset Φ of MN thus constituted is called a *mapping of M onto N*; two elements m and n that occur together in one element of Φ are said to be "mapped onto each other"; they "correspond to each other", or one "is the image of the other".

16. If Φ is any subset of MN and therefore an element of $\mathfrak{U}(MN)$ and if x is any element of $M + N$, it is always definite (No. 4) whether the elements of Φ that contain x form a set consisting of a single element (No. 13). Thus it is also definite whether *all* elements x of $M + N$ possess this property, that is, whether Φ represents a mapping of M onto N or not. According to Axiom III, all of the mappings Φ therefore are the elements of a certain subset Ω of $\mathfrak{U}(MN)$, and it is definite whether Ω differs from 0 or not. *It is therefore always definite for two disjoint sets M and N whether they are equivalent or not.*

17. If two equivalent disjoint sets M and N are mapped onto each other by Φ, there also corresponds to each subset M_1 of M an equivalent subset N_1 of N under a mapping Φ_1, that is a subset of Φ.

For it is definite for every element $\{m, n\}$ of Φ whether $m \varepsilon M_1$ or not, and therefore all elements of Φ thus associated with M_1 are the elements of a subset Φ_1 of Φ. If we now denote by N_1 the intersection (No. 8) of $\mathfrak{S}\Phi_1$ with N, each element of $M_1 + N_1$ occurs as an element in only a single element of Φ_1, since otherwise it would occur more than once in Φ as well; and, according to No. 15, we in fact have $M_1 \sim N_1$.

18. If two disjoint sets M and N are disjoint from and equivalent to one and the same third set, R, or if $M \sim R$, $R \sim R'$, and $R' \sim N$, where each of these pairs of equivalent sets is assumed to be disjoint, then always also $M \sim N$.

Let the subset Φ of MR, the subset X of RR', and the subset Ψ of $R'N$ be three mappings (No. 15) that map M onto R, R onto R', and R' onto N, respectively. If then $\{m, n\}$ is any element of MN, it is definite whether there exist an element r of R and an element r' of R' such that $\{m, r\} \varepsilon \Phi$, $\{r, r'\} \varepsilon \mathrm{X}$, and $\{r', n\} \varepsilon \Psi$. All elements

$\{m, n\}$ for which this is the case therefore are the elements of a subset Ω of MN, which represents a mapping of M onto N. For, if m is any element of M, there always correspond to it a single element r of R, a single element r' of R', and therefore also a single element n of N that satisfy the required condition; an analogous statement holds for each element n of N. Therefore to each element of $M + N$ there actually corresponds a single element $\{m, n\}$ of Ω, and we in fact have $M \sim N$.

19. THEOREM. If M and N are any two sets, there always exists a set M' that is equivalent to one, M, and disjoint from the other, N.

Proof. Let $S = \mathfrak{S}\{M + N\}$, in accordance with Axiom V, be the set that contains the elements of $M + N$, and let r, in accordance with No. 10, be an object that is not an element of $M + S$. Then the sets M and $R = \{r\}$ are disjoint, and the product $M' = MR$ possesses the property required by the theorem. Indeed, every element of M' is then, according to No. 13, a set m' of the form $\{m, r\}$ (where $m \varepsilon M$) but never an element of $M + N$, since otherwise r would be an element of an element of $M + N$, hence, by Axiom V, an element of S, contrary to the assumption. Thus M' is disjoint from both sets, M and N.

Further, there corresponds to each element m of M one and only one element $m' = \{m, r\}$, and conversely each m' contains as an element only a single element m of M, since r was assumed not to be an element of M. To each element of $M + M'$, therefore, there corresponds a single element $\{m, m'\}$ of MM' for which $m' = \{m, r\}$, and if all pairs $\{m, m'\}$ that are so constituted are assumed to form a subset Φ of MM', then, according to No. 15, Φ is a mapping of M onto M', and $M \sim M'$.

It follows from our theorem that *it is not possible for all sets equivalent to a nonempty set M to be the elements of a set T*; for if T is an arbitrary set, there always exists a set M', equivalent to M, that is disjoint from the union $\mathfrak{S}T$ and therefore *not* an element of T.

20. If M and N are any two sets, it is always definite whether there is a set R that is simultaneously disjoint from and equivalent to both sets, M and N.

For let M', in accordance with No. 19, be a set that is equivalent to M and disjoint from $M + N$. Then, according to No. 16, it is definite whether $M' \sim N$ or not. If $M' \sim N$, then $R = M'$ is a set constituted as required; otherwise, such a set R cannot exist at all, since, according to No. 18, it would always necessarily follow from $M' \sim M$, $M \sim R$, and $R \sim N$ that $M' \sim N$, contrary to the assumption.

The preceding theorem, in combination with No. 18, now justifies the following extension of our Definition A:

21. *Definition* B. Two arbitrary (not disjoint) sets M and N are said to be *mediately equivalent*, $M \sim N$, if there exists a third set, R, that is disjoint from both and immediately equivalent to both in the sense of Definition A.

Such an equivalence, "mediated" by R, of two sets M and N is given by means of *two simultaneous* mappings, a subset Φ of MR and a subset Ψ of NR, and two elements, m of M and n of N, are said to "correspond" or "be mapped onto each other" if they correspond to one and the same element r of R, so that both $\{m, r\} \varepsilon \Phi$ and $\{n, r\} \varepsilon \Psi$. In the case of such a mediated mapping, too, there corresponds to each subset M_1 of M, as in No. 17, an equivalent subset R_1 of R, and consequently again an equivalent subset N_1 of N.

On account of No. 18, Definition B may also be applied to disjoint sets M and N, and according to No. 20 *it is always definite whether two arbitrary sets are equivalent or not* in the sense of this definition.

22. Every set is equivalent to itself. If two sets, M and N, are equivalent to a third, R, they are equivalent to each other.

For if, in accordance with No. 19, M' is a set that is disjoint from and equivalent to M, both $M \sim M'$ and $M' \sim M$; therefore, according to No. 21, we in fact have $M \sim M$.

If, furthermore, the equivalence of the sets M and R is mediated by M', and that of R and N by N', where M' is assumed to be disjoint from M and R, and N' to be disjoint from N and R, then we choose, in accordance with No. 19, a sixth set, R', equivalent to R and disjoint from the sum $M + N + R$, and we now have, on account of No. 18, $M \sim M' \sim R \sim R'$, therefore $M \sim R'$, and $N \sim N' \sim R \sim R'$, therefore $N \sim R'$, so that according to No. 21 the equivalence of M and N is mediated by R'.

23. The null set is equivalent only to itself. Every set of the form $\{a\}$ is equivalent to all other sets $\{b\}$ of the same form, and to no other set.

For, since the product $0 \cdot M$ is always equal to 0,[10] no set $M \neq 0$ can be (immediately) equivalent to the null set in the sense of No. 15, and therefore no set M' can be mediately equivalent to it in the sense of No. 21.

If, furthermore, $\{a\}$ is disjoint from M, that is, if a is not an element of M, then all elements of the product $\{a\}M$ are of the form $\{a, m\}$, and, if M were to contain, besides m, another element, p, then $\{a, m\}$ and $\{a, p\}$ would not be disjoint, and this, according to No. 15, would prevent any subset Φ of $\{a\}M$ from being a mapping. On the other hand, $\{a\} \cdot \{b\} = \{a, b\}$ is always a mapping of $\{a\}$ onto $\{b\}$.

24. Theorem. If $M \sim M'$ and $N \sim N'$, while M and N on the one hand, and M' and N' on the other, are mutually disjoint, then always $M + N \sim M' + N'$.

Proof. We first consider the case in which $M + N$ and $M' + N'$ are disjoint. Then Definition A (No. 15) is applicable to both of the equivalences $M \sim M'$ and $N \sim N'$, and there are two mappings, the subset Φ of MM' and the subset Ψ of NN', whose sum $\Phi + \Psi$ represents the required mapping of $M + N$ onto $M' + N'$. For, if $p \in (M + N)$, either $p \in M$ or $p \in N$, but, since $[M, N] = 0$, not both; and Φ in one case, and Ψ in the other, contains a single element of the form $\{p, q\}$. Likewise there corresponds to each element q of $M' + N'$ one and only one element $\{p, q\}$ of $\Phi + \Psi$.

If $M + N$ and $M' + N'$ are not themselves disjoint, there exists, according to No. 19, a set S'' equivalent to $M' + N'$ and disjoint from the sum $M + N + M' + N'$; and, given a mapping X of $M' + N'$ onto S'', equivalent and disjoint parts M'' and N'' of S'' will, according to No. 17, correspond to the two parts M' and N'. Then $M \sim M' \sim M''$, as well as $N \sim N' \sim N''$, and, since now $M + N$ and $M'' + N''$ are disjoint,

$$M + N \sim M'' + N'' = S'' \sim M' + N'$$

according to what has just been proved; therefore again

$$M + N \sim M' + N'.$$

[10] [[Up to this point Zermelo has used only juxtaposition for the product; from here on he occasionally uses a dot.]]

25. Theorem. If a set M is equivalent to one of its parts, M', it is also equivalent to any other part M_1, that includes M' as component.

Proof. Let

$$M \sim M' \in M_1 \in M \quad \text{and} \quad Q = M_1 - M'.$$

Because of the equivalence $M \sim M'$ that has been assumed, there exists, according to No. 21, a mapping $\{\Phi, \Psi\}$ of M onto M', mediated by, say, M''. If now A is an arbitrary subset of M, a certain subset A' of M' will correspond to it under the mapping in question, and it is definite whether $A' \in A$ or not. Thus all elements A of $\mathfrak{U}M$ for which we have both $Q \in A$ and $A' \in A$ are, according to Axiom III, the elements of a certain subset T of $\mathfrak{U}M$, and, in particular, M is itself an element of T. The common component $A_0 = \mathfrak{D}T$ (No. 9) of all elements of T now possesses the following properties: (1) $Q \in A_0$, since Q is a common subset of all elements A of T; (2) $A_0' \in A_0$, because every element x of A_0 is a common element of all elements A of T and its map $x' \varepsilon A' \in A$ is thus also a common element of all A. On account of (1) and (2), therefore, also $A_0 \varepsilon T$. Finally we have (3) $A_0 = Q + A_0'$. For, since $A_0' \in A_0$ and also $A_0' \in M' \in M - Q$, on the one hand $A_0' \in A_0 - Q$. On the other hand, however, every element r of $A_0 - Q$ is also an element of A_0', and therefore $A_0 - Q \in A_0'$. Indeed, if r were not an element of A_0', then $A_1 = A_0 - \{r\}$ would still have A_0', and a fortiori A_1', as a component, and, since it still includes Q, it would itself be an element of T, whereas it is in fact only a *part* of $A_0 = \mathfrak{D}T$. Therefore

$$M_1 = Q + M' = (Q + A_0') + (M' - A_0') = A_0 + (M' - A_0'),$$

where the two summands on the right have no element in common, since Q and M' are disjoint. But now, since A_0 is equivalent to A_0' and $M' - A_0'$ is equivalent to itself, it follows according to No. 24 that

$$M_1 \sim A_0' + (M' - A_0') = M' \sim M;$$

that is, $M_1 \sim M$ as asserted.

26. *Corollary.* If a set M is equivalent to one of its parts, M', it is also equivalent to any set M_1 that is obtained from M when a single element is removed or added.

Let $M \sim M' = M - R$ and $M_1 = M - \{r\}$, where r is some element of R. Then

$$M' = M - \{r\} - (R - \{r\}) \in M - \{r\} = M_1,$$

and, according to the previous theorem, $M \sim M_1$.

If, furthermore, $M_2 = M - \{a\}$, where $a \varepsilon M' = M - R$, let $M_0 = M - \{a, r\}$, and we have, according to Nos. 23 and 24,

$$M_2 = M_0 + \{r\} \sim M_0 + \{a\} = M_1 \sim M;$$

therefore also $M_2 \sim M$.

If, finally, $M_3 = M + \{c\}$, where c is not an element of M, it follows from $M \sim M'$, again according to No. 24, that

$$M_3 = M + \{c\} \sim M' + \{c\} = M - R + \{c\} = M_3 - R,$$

and furthermore, according to what was proved previously,

$$M = M_3 - \{c\} \sim M_3;$$

thus the theorem is proved in its entirety.

27. Equivalence theorem. If each of the two sets M and N is equivalent to a subset of the other, M and N are themselves equivalent.

Let $M \sim M' \in N$ and $N \sim N' \in M$. Then on account of No. 21 there corresponds to the subset M' of N an equivalent set M'' such that $M'' \in N' \in M$, and we have $M \sim M' \sim M''$; therefore, according to the theorem of No. 25, also $M \sim N' \sim N$, q.e.d.[11]

28. Theorem. If all the sets $M, N, R, \ldots$ are elements of an arbitrary set T, they can all be simultaneously mapped onto [[respectively]] equivalent sets $M', N', R', \ldots$ that are the elements of a new set, T', and are disjoint from one another as well as from a given set Z.

Proof. Let $S = \mathfrak{S}T = M + N + R + \cdots$, in accordance with Axiom V, be the sum of all elements of T, and, in accordance with No. 19, let T'' be a set equivalent to T and disjoint from the sum $T + S + \mathfrak{S}(S + Z)$, so that under a mapping Ω to each element of T there corresponds a certain element of T'': $M'', N'', R'', \ldots$ to $M, N, R, \ldots$, respectively. An arbitrary element of the product ST'' (No. 13) will then have the form $\{s, M''\}$, where $s \varepsilon S$ and $M'' \varepsilon T''$, and for every such element it is definite (No. 4) whether $s \varepsilon M$, where M is assumed to be the element of T corresponding to M'' under Ω, that is, according to No. 15, to be such that $\{M, M''\} \varepsilon \Omega$. All elements of the product that are so constituted then are, on account of Axiom III, the elements of a subset S' of ST'', and this set S' is disjoint from $S + Z$, since otherwise an element M'' of T'', being an element of $\{s, M''\}$, would be an element of an element of $S + Z$ and thus, on account of Axiom V, also an element of $\mathfrak{S}(S + Z)$, contrary to the assumption made about T''. If, furthermore, M is an arbitrary element of T and M'' the corresponding one of T'', those elements $\{s, M''\}$ of S' that contain M'' as an element form, according to Axiom III, a certain subset M' of S', and $M' \sim M$ by virtue of a mapping M ($\mathsf{M} \in MM' \in SS'$) under which to each element m of M there corresponds an element $m' = \{m, M''\}$ of M' and conversely. Likewise, with any other element N of T there are associated an equivalent subset N' of S' and a mapping N ($\mathsf{N} \in NN' \in SS'$) under which to each element n of N there corresponds an element $\{n, N''\}$ of N'. The two subsets M' and N', which correspond to two different elements M and N of T, are, however, always disjoint, for, if, say, $\{m, M''\} = \{n, N''\}$ were a common element of M' and N', then M'', being an element of $\{n, N''\}$, would have to be equal either to N'' or to n, and in the first case M would also be equal to N, and in the second T'' and S would not be disjoint, contrary to the assumption. The subsets $M', N', R', \ldots$ of S', which by virtue of the mappings M $\mathsf{N}, \mathsf{P}, \ldots$ are equivalent to the elements $M, N, R, \ldots$ of T, are thus indeed disjoint from one another, and they are also disjoint from the set Z since S' is. Finally, it is always definite for every subset S_1' of S' containing an element $\{s, M''\}$ whether that subset is identical with the corresponding set M', and all of those among $M', N', R', \ldots$ for which this is the case are, according to Axioms III and IV,

[11] The proof of the equivalence theorem given here in Nos. 25 and 27 (first published by Poincaré (*1906*, pp. 314–315) on the basis of a letter that I wrote in January 1906) rests solely upon Dedekind's chain theory (*1888*, § 4) and, unlike the older proofs by Schröder and F. Bernstein as well as the latest proof by J. König (*1906*), avoids any reference to ordered sequences of order ω or to the principle of mathematical induction. At approximately the same time Peano (*1906*) published a proof that was quite similar; the paper containing that proof also contains a discussion of the objection directed by Poincaré against my proof. See § 2 b in my *1908* [[above, pp. 190–191]].

the elements of a certain subset T' of $\mathfrak{U}S'$. The theorem is therefore proved in its entirety.

29_{VI}. General principle of choice. If T is a set whose elements $M, N, R, \ldots$ all are sets different from the null set, there always exist sets P that, according to a certain rule, uniquely correlate with each element M of T one element m of that M.

Proof. Apply to T the procedure specified in No. 28 above, letting $Z = 0$.[12] This yields simultaneous mappings of all sets $M, N, R, \ldots$ onto the equivalent sets $M', N', R', \ldots$, which are mutually disjoint and form the elements of a set T'. If now P, in accordance with Axiom VI, is a subset of $\mathfrak{S}T'$ that has exactly one element in common with each element of T', then P provides the desired correlation. For, if M is any element of T and M' the corresponding element of T', P contains only a single element m' of M', and to this there again corresponds a well-determined element m of M.

30_{VI}. Theorem. Let T and T' be two equivalent sets containing as elements the mutually disjoint sets $M, N, R, \ldots$ and the mutually disjoint sets $M', N', R', \ldots$, respectively. If T and T' are mapped onto each other in such a way that to each element M of one set there corresponds an equivalent set M' as an element of the other, the associated sums $\mathfrak{S}T$ and $\mathfrak{S}T'$, as well as the corresponding products $\mathfrak{P}T$ and $\mathfrak{P}T'$, are also equivalent.

Proof. We first prove the theorem on the assumption that $S = \mathfrak{S}T$ and $S' = \mathfrak{S}T'$ are mutually disjoint, in which case each element of T would have to be disjoint from each element of T', too. It then follows from $M \sim M'$, according to No. 15, that $\mathfrak{U}(MM')$ possesses a subset A_M, different from 0, that contains as elements all possible mappings $\mathsf{M}, \mathsf{M}', \mathsf{M}'', \ldots$ of M onto M'. Likewise, to any other element N of T there corresponds a subset A_N of $\mathfrak{U}(NN')$ that contains all mappings of N onto N', and A_N is also different from 0. All these mapping-sets $\mathsf{A}_M, \mathsf{A}_N, \mathsf{A}_R, \ldots$ are subsets of $\mathfrak{U}(SS')$ and therefore are, according to Axioms III and IV, the elements of a certain subset T of $\mathfrak{U}\mathfrak{U}(SS')$. Since, now, all elements of T are sets that are different from 0 and mutually disjoint (because from the fact that MM' and NN' are disjoint it follows that their subsets are disjoint), according to Axiom VI the product $\mathfrak{P}\mathsf{T}$ is also different from 0, and an arbitrary element Θ of $\mathfrak{P}\mathsf{T}$ is a set of the form $\{\mathsf{M}, \mathsf{N}, \mathsf{P}, \ldots\}$ that contains exactly one element of each of the sets $\mathsf{A}_M, \mathsf{A}_N, \mathsf{A}_R, \ldots$. We could also have inferred the existence of such a "combined" mapping from the theorem of No. 29, and more quickly at that. If we now form, in accordance with Axiom V, the union

$$\Omega = \mathfrak{S}\Theta = \mathsf{M} + \mathsf{N} + \mathsf{P} + \cdots \in SS',$$

then Ω provides the required mapping of S onto S'. For each element s of S must belong as an element to one and only one element of T, say M, and must therefore occur as an element in a single element of the corresponding mapping M, while no element of M occurs in any of the remaining summands $\mathsf{N}, \mathsf{P}, \ldots$. An analogous statement holds for every element of s' of S', and according to Definition A (No. 15) we thus have $S \sim S'$.

By means of the same Ω and its subsets, each subset p of S is also mapped, on account of No. 17, onto an equivalent subset p' of S', and if in particular $p = \{m, n, \ldots\}$ is, in accordance with No. 13, an element of $P = \mathfrak{P}T$, the subset $p' = \{m', n',$

[12] [[Instead of "T" the German text has "T'", which is a misprint.]]

$\ldots\}$ of S' corresponding to it is an element of $\mathfrak{P}T'$. For, if M' is an arbitrary element of T' and M the corresponding element of T, p contains one and only one element m of M and p' contains the corresponding element m' of M' but no other element of M', since such an element would also have to correspond to a second element of M in p. Likewise, there corresponds to each element p' of $\mathfrak{P}T'$ one and only one element p of $\mathfrak{P}T$, and we indeed obtain a certain subset Π of $\mathfrak{P}T \cdot \mathfrak{P}T'$ as a mapping of $\mathfrak{P}T$ onto $\mathfrak{P}T'$, so that these two products are mutually equivalent.

But if now S and S' are no longer assumed to be disjoint, we can, according to No. 19, introduce a third set, S'', that is equivalent to S' and disjoint from $S + S'$. Then, on account of No. 17, there corresponds to a subset M' of S' an equivalent subset M'' of S'', and, since $M', N', R', \ldots$ are mutually disjoint, the same holds of the corresponding $M'', N'', R'', \ldots$. Since, furthermore, every element s'' of S'' corresponds to an element s' of S' belonging to one of the sets $M', N', R', \ldots, S''$ is the *sum* of all these $M'', N'', R'', \ldots$, which are the elements of a certain subset T'' of $\mathfrak{U}S''$. But now we have $M \sim M' \sim M''$, $N \sim N' \sim N'', \ldots$; thus every element M of T is equivalent to the corresponding element M'' of T'', and, since now $S'' = \mathfrak{S}T''$ is disjoint from both of the sums $S' = \mathfrak{S}T'$ and $S = \mathfrak{S}T$, it follows according to what has been proved above that

$$\mathfrak{S}T \sim \mathfrak{S}T'' \sim \mathfrak{S}T' \quad \text{and} \quad \mathfrak{P}T \sim \mathfrak{P}T'' \sim \mathfrak{P}T';$$

thus the theorem is proved in full generality.

31. *Definition.* If a set M is equivalent to a subset of the set N, but N is not equivalent to a subset of M, we say that M is *of lower cardinality than* N, and we write $M < N$ for short.

Corollaries. (*a*) Since according to No. 21 it is definite for any two sets whether they are equivalent or not, it is also definite whether M is equivalent to at least one element of $\mathfrak{U}N$, as well as whether N is equivalent to some element of $\mathfrak{U}M$. *It is therefore always definite whether $M < N$ or not.*

(*b*) The three relations $M < N$, $M \sim N$, and $M > N$ are mutually exclusive.

(*c*) Whenever we have $M < N$ and either $N < R$ or $N \sim R$, we have $M < R$.

(*d*) If M is equivalent to a subset of N, then either $M \sim N$ or $M < N$. This is a consequence of the equivalence theorem (No. 27).

(*e*) The null set is of lower cardinality than any other set; likewise, every set $\{a\}$ consisting of a single element is of lower cardinality than any set M that has parts (see No. 23).

32. Cantor's theorem. If M is an arbitrary set, then always $M < \mathfrak{U}M$. Every set is of lower cardinality than the set of its subsets.

Proof. To each element m of M there corresponds a subset $\{m\}$ of M. Now since it is definite for each subset M_1 of M whether it contains only a single element (No. 13), all subsets of the form $\{m\}$ are the elements of a subset U_0 of $\mathfrak{U}M$, and $M \sim U_0$.

If on the other hand $U = \mathfrak{U}M$ were equivalent to a subset M_0 of M, then under a mapping Φ of U onto M_0 there would correspond to each subset M_1 of M a certain element m_1 of M_0, so that $\{M_1, m_1\}$ would be an element of Φ, and it would always be definite whether $m_1 \varepsilon M_1$ or not. All those elements m_1 of M_0 for which m_1 is not an element of M_1 would then be the elements of a set M' ($M' \Subset M_0 \Subset M$), which likewise would be an element of U. But no element m' of M_0 could correspond to this subset

M' of M. For if m' were an element of M', this would contradict the definition of M'. But if m' were not an element of M', then, according to the same definition, M' would also have to contain this element m', contrary to the assumption. Thus it follows that U cannot be equivalent to any subset of M, and, in combination with what was proved first, $M < \mathfrak{U}M$.

The theorem holds for *all* sets M, even, for instance, for $M = 0$, and indeed

$$0 < \{0\} = \mathfrak{U}0.$$

Likewise, for every a,

$$\{a\} < \{0, \{a\}\} = \mathfrak{U}\{a\}.$$

Finally, it follows from the theorem that for every arbitrary set T of sets M, N, $R, \ldots$ there always exist further sets of higher cardinality; for example, the set

$$P = \mathfrak{U}\mathfrak{S}T > \mathfrak{S}T \gtrsim M, N, R, \ldots$$

possesses this property.

33_{VI}. Theorem. Let T and T' be two equivalent sets containing as elements the mutually disjoint sets M, N, $R, \ldots$ and the mutually disjoint sets M', N', $R', \ldots$, respectively. If T and T' are mapped onto each other in such a way that each element M of T is of lower cardinality than the corresponding element M' of T', the sum $S = \mathfrak{S}T$ of all elements of T is also of lower cardinalty than the product $P' = \mathfrak{P}T'$ of all elements of T'.

Proof. It suffices to prove the theorem for the case in which the two sums $S = \mathfrak{S}T$ and $S' = \mathfrak{S}T'$ are disjoint. The extension to the general case is then accomplished by a method analogous to that used for the theorem in No. 30, through interposition of a third set, S'', equivalent to S and disjoint from S', and by means of that theorem.

First it must be shown that S is equivalent to a subset of P'. Because $M < M'$, there exists a subset A_M of $\mathfrak{U}(MM')$, different from 0, all of whose elements M, M', $\mathsf{M}'', \ldots$ are mappings that map M onto subsets M_1', $M_2' \ldots$ of M'. Such mapping-sets A_M, A_N, $\mathsf{A}_R, \ldots$ exist for any two corresponding elements $\{M, M'\}$, $\{N, N'\}$, $\{R, R'\}, \ldots$ of T and T', and each element $\Theta = \{\mathsf{M}, \mathsf{N}, \mathsf{P}, \ldots\}$ of their product $\mathfrak{P}\mathsf{T} = \mathsf{A}_M \cdot \mathsf{A}_N \cdot \mathsf{A}_R \ldots$ furnishes, just as in No. 30, a simultaneous mapping of all elements M, N, $R, \ldots$ of T onto equivalent subsets M_1', N_1', $R_1', \ldots$ of the corresponding elements of T'. By means of $\Omega = \mathfrak{S}\Theta \in SS'$, therefore, every element s of S is mapped onto an element s' of S', even though not every element of S' is mapped onto one of S.

But now the complements $M' - M_1'$, $N' - N_1'$, $R' - R_1', \ldots$, which are the elements of a subset T_1' of $\mathfrak{U}S'$, all are different from 0, since, because $M < M'$, the case $M \sim M_1' = M'$ is always excluded. Thus also the product $\mathfrak{P}T_1' \neq 0$, and there exists at least one set $q \in \mathfrak{P}T_1'$ of the form $\{m_0', n_0', r_0', \ldots\} \in S'$ that has exactly one element in common with each of the sets $M' - M_1'$, $N' - N_1', \ldots$ and is therefore also an element of P'.

If now s is any element of S, and s' is the element of S' corresponding to it under Ω, there corresponds to both of them yet another element s_0 of the subset q of S', such that s' and s_0' always belong to one and the same element of T', and consequently for $s \varepsilon M$ always $s'_0 = m'_0$, and so forth.[13] But, since $s'_0 \varepsilon (M' - M'_1)$ whenever $s' \varepsilon M'_1$, s' and s'_0 are always distinct. If we now form the set $q_s = q - \{s'_0\} + \{s'\}$, which we obtain from q when we replace one of these elements, s'_0, by the other, s', we again obtain an element of P', namely, a subset of S' that has exactly one element in common with each of the sets M', N', R', ... But the elements of P' such as q_s, which are the elements of a subset P'_0 of P', all are distinct. For if, say, m_1 and m_2 are two distinct elements of the *same* element M of T, the corresponding elements m'_1 and m'_2 of M'_1, which take the place of s', are also distinct, and thus

$$q_{m_1} = q - \{m'_0\} + \{m'_1\} \neq q - \{m'_0\} + \{m'_2\} = q_{m_2},$$

since, except for m'_0, q has no element in common with M'. But if m and n are two elements of S that belong to *distinct* sets M and N, then $q_m = q - \{m'_0\} + \{m'\}$ has one element m' of M'_1 in common with M', while $q_n = q - \{n'_0\} + \{n'\}$ has only the element m'_0 of $(M' - M'_1)$ in common with M', and the two sets are likewise distinct. Thus the pairs $\{s, q_s\}$ are the elements of a subset Φ of SP'_0, which, according to No. 15, possesses the character of a mapping, and we in fact have $S \sim P'_0 \in P'$.

On the other hand, P' cannot be equivalent to any subset S_0 of S. For if this were the case, there would have to correspond to every element s of S_0 an element p_s of P' under a mapping Ψ ($\Psi \in S_0P' \in SP'$). Let us consider in particular those elements p_m that correspond to elements m of the intersection $M_0 = [M, S_0]$. Each of these p_m contains, then, an element m'' of M', namely, the one that p_m, as an element of P', has in common with M'; but the m'' belonging to distinct m are not necessarily always distinct. In any case, all m'' belonging to the elements m of M_0 are the elements of a subset M'_2 of M' that is distinct from M' itself, since otherwise M' would be equivalent to a subset of M_0, which in turn is a subset of M, contrary to the assumption that $M < M'$.[14] Similarly, there answer to *all* elements M, N, R, ... of T certain parts M'_2, N'_2, R'_2, ... of the corresponding elements M', N', R', ... of T'. The respective complements $M' - M'_2$, $N' - N'_2$, $R' - R'_2$, ..., therefore, all are different from 0 and are the elements of a subset T'_2 of $\mathfrak{U}S'$. If, now, p'_0 is any element of $\mathfrak{P}T'_2 \neq 0$, it is also an element of P', but under the mapping Ψ whose existence was assumed it cannot correspond to any element s of S_0. For if, say, p'_0 were equal to p_m, if, therefore, p'_0 were to correspond to an element of M_0, then according to the assumption made it would necessarily have an element m'' of M'_2 in common with M', while actually p'_0 cannot have any element in common with M' other than one of $M' - M'_2$. Nor can p'_0 correspond to any element of N_0, R_0, ... ; thus it corresponds to no element of the subset S_0 of S at all, and the assumption that $P' \sim S_0$ leads to a contradiction, which completes the proof of the assertion that $S < P'$.

This theorem (communicated by me to the Göttingen Mathematical Society at the end of 1904) is the most general theorem now known concerning the comparison of cardinalities, one from which all the others can be derived. The proof rests upon a generalization of a procedure applied by J. König in a special case (see below).

34_{VI}. *Corollary.* (*J. König's theorem.*) If a set T whose elements all are sets that are mutually disjoint is mapped onto a subset T' of T in such a way that to each element

13 ⟦For the first occurrence in this sentence of "s'_0" the German text has "s_0", which is a misprint.⟧

14 Here, too, use is made of Axiom VI (axiom of choice).

M of T there corresponds an element M' of T' of higher cardinality ($M < M'$), then $\mathfrak{S}T < \mathfrak{P}T$ whenever $\mathfrak{P}T \neq 0$.[15]

According to the theorem of No. 33, we always have $\mathfrak{S}T < \mathfrak{P}T'$ in the case under consideration; it therefore remains to be shown only that $\mathfrak{P}T'$ is here equivalent to a subset of $\mathfrak{P}T$. When $T' = T$ this is trivial; but in the other case we have $\mathfrak{P}(T - T') \neq 0$, else, on account of Axiom VI, the null set would be any element of $T - T'$, and, contrary to the assumption, $\mathfrak{P}T$ would equal 0. But if q is any element of $\mathfrak{P}(T - T')$ and p' any element of $\mathfrak{P}T'$, then $p' + q$ is an element of $\mathfrak{P}T$, namely, a subset of $\mathfrak{S}T' + \mathfrak{S}(T - T') = \mathfrak{S}T$ that has exactly one element in common with each element of T' as well as of $T - T'$. Thus for a fixed q there corresponds to each element p' of $\mathfrak{P}T'$ a certain element $p' + q$ of $\mathfrak{P}T$, and all these $p' + q$ are the elements of a certain subset P_q of $\mathfrak{P}T$ that is equivalent to $\mathfrak{P}T'$.

35. Cantor's theorem (No. 32) can also be obtained as a special case of the general theorem of No. 33.

Let M be an arbitrary set; let M' be a set—and according to No. 19 such a set does exist—equivalent to and disjoint from M, and let the subset Φ of MM' be an arbitrary mapping of M onto M'. Then to each element m of M there corresponds a definite element $\{m, m'\}$ of Φ and, according to No. 31 (*e*), always

$$\{m\} < \{m, m'\}.$$

These sets $\{m\}$ obviously are the elements of a new set, T, that is equivalent to M, and, according to the theorem of No. 33,

$$M = \mathfrak{S}T < \mathfrak{P}\Phi.$$

It thus remains to be shown only that $\mathfrak{P}\Phi \sim \mathfrak{U}M$. Now every element of $\mathfrak{P}\Phi$ is a set of the form $M_1 + (M' - M_1')$, where M_1 is a subset of M and M_1' the corresponding subset of M'. Then there indeed corresponds to each element M_1 of $\mathfrak{U}M$ one and only one element of $\mathfrak{P}\Phi$ and conversely; and, as asserted,

$$M < \mathfrak{P}\Phi \sim \mathfrak{U}M.$$

36_{VII}. THEOREM. The number sequence Z_0 (No. 14) is an *infinite* set, that is, one that is equivalent to one of its parts. Conversely, also, every infinite set M contains a *denumerably infinite* component M_0, that is, one that is equivalent to the number sequence.

Proof. Let Z be an arbitrary set that, in accordance with Axiom VII, contains the element 0 and, for each of its elements a, also the corresponding element $\{a\}$, and let this set Z be mapped by means of the subset Ω of ZZ' onto a set Z' equivalent to and disjoint from it, which, according to No. 19, is possible. Now, whenever $\{z, x'\}$ is an arbitrary element of ZZ' and $\{x, x'\}$ an element of Ω for the same x', it is definite whether $z = \{x\}$ or not. All elements of ZZ' that in fact have the form $\{\{x\}, x'\}$ thus are, according to Axiom III, the elements of a certain subset Φ of ZZ', and Φ is a mapping of Z' onto the subset Z_1 of Z that contains all elements z of the form $\{x\}$. In fact, to every element x' of Z' there corresponds a certain element $\{x\}$ of Z_1 and conversely; that is, each element of $Z_1 + Z'$ occurs in one and only one

[15] See *König 1905* for the special case in which the elements of T, when ordered according to cardinality, form a sequence of type ω.

element of Φ. Thus, according to No. 21, $Z \sim Z' \sim Z_1$, where Z_1, since it does not contain the element 0, is only a *part* of Z; and every set constituted like Z, hence also Z_0, is infinite.

To prove now the second half of the theorem as well, we consider an arbitrary infinite set M, which, however, we may in view of No. 19 assume without loss of generality to be disjoint from Z_0. Thus let $M \sim M' = M - R$, let r be an arbitrary element of $R \neq 0$, and let $\{\Phi, \Psi\}$ be a mapping, whose existence is possible according to No. 21, under which there corresponds to each element m of M an element m' of M' and conversely. Furthermore let A be a subset of MZ_0 that possesses the following properties: (1) it contains the element $\{r, 0\}$; and (2) if $\{m, z\}$ is any element of A, then A also contains the further element $\{m', z'\}$, where m' is the element of M' corresponding to m, and $z' = \{z\}$ is likewise an element of Z_0 on account of No. 14. If now $\mathsf{A}_0 = \mathfrak{D}\mathsf{T}$ is the common component of all subsets of MZ_0 constituted like A, which, on account of Axioms III and IV, are the elements of a certain subset T of $\mathfrak{U}(MZ_0)$, then A_0 also possesses properties (1) and (2), as we see immediately, and is thus likewise an element of T. Furthermore, with the sole exception of $\{r, 0\}$, every element of A_0, too, has the form $\{m', z'\}$; for in the contrary case we could remove it, and the remainder of A_0 would still possess properties (1) and (2), without, however, including the component A_0, which all elements of T do include. From this it follows first that the element $\{r, 0\}$ is disjoint from all other elements of A_0, since neither $r = m' \varepsilon M'$ nor $0 = \{z\} = z'$ is possible, and therefore no further element $\{m', z'\}$ can contain one of the elements r or 0. Furthermore, if an element $\{m, z\}$ of A_0 is disjoint from all the remaining ones, the same also holds for the corresponding element $\{m', z'\}$, since to each element of the form $\{m', z_1'\}$ or $\{m_1', z'\}$ there would have to correspond a further element, $\{m, z_1\}$ or $\{m_1, z\}$. All those elements of A_0 that are disjoint from all the others, therefore, are the elements of a subset A_0' of A_0 possessing properties (1) and (2); hence, being an element of T, A_0' now includes A_0 as a subset, that is, is identical with A_0. Every element of

$$\mathfrak{S}\mathsf{A}_0 = M_0 + Z_{00} \Subset M + Z_0,$$

where M_0 is the common component of $\mathfrak{S}\mathsf{A}_0$ and M, and Z_{00} that of $\mathfrak{S}\mathsf{A}_0$ and Z_0, can therefore figure as an element in only a single element of A_0, and, on account of No. 15, $M_0 \sim Z_{00}$. But now Z_{00} is a subset of Z_0 containing the element 0 and, for every one of its elements z, also the associated $z' = \{z\}$; Z_{00} must therefore on account of No. 14 contain the entire number sequence Z_0 as a component; that is, we have $Z_{00} = Z_0$ and, as asserted, $Z_0 \sim M_0 \Subset M$.

Incomplete symbols: Descriptions

ALFRED NORTH WHITEHEAD
AND BERTRAND RUSSELL

(*1910*)

Whitehead published the first thick volume of *A treatise of universal algebra* in 1898. His pupil Russell published the first thick volume of *The principles of mathematics* in 1903. Then Whitehead was persuaded to abandon plans for his second volume and rather to collaborate with Russell on the second volume of *The principles*. This volume never materialized either. Instead, the collaboration bore as its fruit a decade later a work independent of *The principles*, the three-volume *Principia mathematica*, from which the present selection is drawn. This monumental work is a detailed study of logic and set theory, and a construction, on that basis, of classical mathematics. It came as a fulfillment of what Frege and Peano, in their different ways, had projected and partly carried through.

The present selection from the Introduction to *Principia* is concerned with the logical device of *description*, "$(\imath x)(\varphi x)$", "the object x such that φx". In Peano's notation (*1899*, pp. 22–23) the class whose sole member is x was referred to as ιx, and conversely the sole member x of a class y was $\imath y$; so Whitehead and Russell's "$(\imath x)(\varphi x)$" was an adaptation of Peano's "$\imath x \ni \varphi x$", "the member of the class $x \ni \varphi x$". Frege had a notation for the same purpose earlier (*1893*, § 11). But there were differences in the handling of the waste cases—cases in which there is not one and only one object x such that φx. Peano left his "$\imath$" unexplained in the waste cases. Frege arbitrarily equated description to class abstraction in the waste cases; "the" object x fulfilling a given condition became for him the class of all such objects x if there were many or none.

Whitehead and Russell contrive to avoid both the incompleteness of Peano's line and the artificiality of Frege's by defining the notation of description for all cases and yet giving it no object in the waste cases. They use contextual definition: they indicate how any sentence containing a description can as a whole be expanded into primitive notation, but they equate the description by itself to no term in primitive notation.

The general idea of contextual definition goes back to Jeremy Bentham (*1843*, vol. 8, pp. 246–247), who called it *paraphrasis* and saw in it a method of accounting for fictions by explaining various purported terms away. A mathematical application of the method became familiar in the differential operators; and it is with this illustration that the selection from Whitehead and Russell begins.

The theory of descriptions that occupies these pages was expounded earlier by Russell (*1905*) in a way that showed motives not unlike Bentham's. Russell was dissatisfied with such theories as "that of Meinong, [which] regards any grammatically correct denoting phrase as standing for an *object*. Thus 'the present King of France', 'the round square', etc., are supposed to be genuine objects" (*1905*, p. 483). At Whitehead and Russell's hands the method of paraphrasis receives sharper formulation than it had had from Bentham.

A disadvantage of Whitehead and Russell's way of handling descriptions, as compared with Frege's or Peano's, is that it raises the question of *scope*, to which a good part of the appended selection is given over. An advantage, we saw, is completeness without artificiality. Another advantage is that this way of introducing description does not use class abstraction, and hence leaves us free to define the notation "$x \ni \varphi x$" or "$\hat{x}\varphi x$" of class abstraction in terms of description as $(\imath y)(x)(x \varepsilon y . \equiv \varphi x)$. But this is an advantage that, in the continuing pages of *Principia mathematica*, Whitehead and Russell do not use. Instead they assume, not quite explicitly, a primitive notation for the abstraction of attributes (or propositional functions), and then with its help they introduce class abstraction by contextual definition (see above, p. 152).

More generally, Whitehead and Russell's way of eliminating descriptions enables us to make a clean sweep not only of class abstracts but of all other sorts of singular terms as well, except variables. Instead of assuming a constant singular term or proper name, say "a", we can always assume or define a predicate "A" that is true only of the object a, and then take "a" itself as short for the eliminable description "$(\imath x)Ax$". Similarly, instead of assuming functors, say the two-place functor "$+$", we can assume or define a three-place predicate "Σ" such that $\Sigma zxy \equiv . z = x + y$, and then take "$x + y$" itself as short for the eliminable description "$(\imath z)\Sigma zxy$". An important thing about thus absorbing all singular terms other than variables is that the logic of quantification and identity no longer needs to be conceived otherwise than in its simplest form, involving just predicate letters, variables, quantifiers, truth functions, and "$=$".

W. V. Quine

The text is reproduced here by arrangement with Cambridge University Press.

By an "incomplete" symbol we mean a symbol which is not supposed to have any meaning in isolation, but is only defined in certain contexts. In ordinary mathematics, for example, d/dx and $\int_a^b$ are incomplete symbols: something has to be supplied before we have anything significant. Such symbols have what may be called a "definition in use". Thus, if we put

$$\nabla^2 = \frac{\partial^2}{\partial x^2} + \frac{\partial^2}{\partial y^2} + \frac{\partial^2}{\partial z^2} \quad \text{Df},$$

we define the *use* of ∇^2, but ∇^2 by itself remains without meaning. This distinguishes such symbols from what (in a generalized sense) we may call *proper names*: "Socrates", for example, stands for a certain man and therefore has a meaning by itself, without the need of any context. If we supply a context, as in "Socrates is mortal", these words express a fact of which Socrates himself is a constituent: there is a certain object, namely Socrates, which does have the property of mortality, and this object is a constituent of the complex fact which we assert when we say "Socrates is mortal". But in other cases this simple analysis fails us. Suppose we say: "The round square does not exist." It seems plain that this is a true proposition, yet we cannot regard it as denying the existence of a certain object called "the round square". For if there were such an object, it would exist: we cannot first assume that there is a certain object and then proceed to deny that there is such an object. Whenever the grammatical subject of a proposition can be supposed not to exist, without rendering the proposition meaningless, it is plain that the grammatical subject is not a proper name, that is, not a name directly representing some object. Thus, in all such cases, the proposition must be capable of being so analyzed that what was the grammatical subject shall have disappeared. Thus, when we say "the round square does not exist", we may, as a first attempt at such analysis, substitute "it is false that there is an object x which is both round and square". Generally, when "the so-and-so" is said not to exist, we have a proposition of the form[1]

$$\text{``} \sim E!(\imath x)(\varphi x) \text{''},$$

that is,

$$\sim\{(\exists c) : \varphi x \,.\!\equiv_x .\, x = c\},$$

or some equivalent. Here the apparent grammatical subject $(\imath x)(\varphi x)$ has completely disappeared; thus, in "$\sim E!(\imath x)(\varphi x)$", $(\imath x)(\varphi x)$ is an *incomplete* symbol.

By an extension of the above argument, it can easily be shown that $(\imath x)(\varphi x)$ is *always* an incomplete symbol. Take, for example, the following proposition: "Scott is the author of Waverley". (Here "the author of Waverley" is $(\imath x)(x$ wrote Waverley).) This proposition expresses an identity; thus, if "the author of Waverley" could be taken as a proper name and supposed to stand for some object c, the proposition would be "Scott is c". But if c is any one except Scott, this proposition is false; while if c *is* Scott, the proposition is "Scott is Scott", which is trivial and plainly different from "Scott is the author of Waverley". Generalizing, we see that the proposition

$$a = (\imath x)(\varphi x)$$

is one which may be true or may be false, but is never merely trivial, like $a = a$; whereas, if $(\imath x)(\varphi x)$ were a proper name, $a = (\imath x)(\varphi x)$ would necessarily be either false or the same as the trivial proposition $a = a$. We may express this by saying that $a = (\imath x)(\varphi x)$ is not a value of the propositional function $a = y$, from which it follows that $(\imath x)(\varphi x)$ is not a value of y. But, since y may be anything, it follows that $(\imath x)(\varphi x)$ is nothing. Hence, since in use it has meaning, it must be an incomplete symbol.

It might be suggested that "Scott is the author of Waverley" asserts that "Scott" and "the author of Waverley" are two names for the same object. But a little reflection will show that this would be a mistake. For if that were the meaning of "Scott is the author of Waverley", what would be required for its truth would be that Scott should have been *called* the author of Waverley: if he had been so called, the proposition would be true, even if some one else had written Waverley; while, if no one called him so, the proposition would be false, even if he had written Waverley. But in fact he was the author of Waverley at a time when no one called him so, and he would not have been the author if every one had called him so but some one else had written Waverley. Thus the proposition "Scott is the author of Waverley" is not a proposition about names, like "Napoleon is Bonaparte"; and this illustrates the sense in which "the author of Waverley" differs from a true proper name.

[1] See *Whitehead and Russell 1910*, pp. 31–32 ⟦or *1925*, pp. 30–31⟧.

Thus all phrases (other than propositions) containing the word *the* (in the singular) are incomplete symbols: they have a meaning in use, but not in isolation. For "the author of Waverley" cannot mean the same as "Scott", or "Scott is the author of Waverley" would mean the same as "Scott is Scott", which it plainly does not; nor can "the author of Waverley" mean anything other than "Scott", or "Scott is the author of Waverley" would be false. Hence "the author of Waverley" means nothing.

It follows from the above that we must not attempt to define "$(\iota x)(\varphi x)$", but must define the *uses* of this symbol, that is, the propositions in whose symbolic expression it occurs. Now in seeking to define the uses of this symbol, it is important to observe the import of propositions in which it occurs. Take as an illustration: "The author of Waverley was a poet". This implies (1) that Waverley was written, (2) that it was written by one man, and not in collaboration, (3) that the one man who wrote it was a poet. If any one of these fails, the proposition is false. Thus "the author of 'Slawkenburgius on Noses' was a poet" is false, because no such book was ever written; "the author of 'The Maid's Tragedy' was a poet" is false, because this play was written by Beaumont and Fletcher jointly. These two possibilities of falsehood do not arise if we say "Scott was a poet". Thus our interpretation of the uses of $(\iota x)(\varphi x)$ must be such as to allow for them. Now taking φx to replace "x wrote Waverley", it is plain that any statement apparently about $(\iota x)(\varphi x)$ requires (1) $(\exists x).(\varphi x)$ and (2) $\varphi x . \varphi y . \supset_{x,y} . x = y$; here (1) states that *at least* one object satisfies φx, while (2) states that *at most* one object satisfies φx. The two together are equivalent to

$$(\exists c) : \varphi x . \equiv_x . x = c,$$

which we defined as

$$E!(\iota x)(\varphi x).$$

Thus "$E!(\iota x)(\varphi x)$" must be part of what is affirmed by any proposition about $(\iota x)(\varphi x)$. If our proposition is $f\{(\iota x)(\varphi x)\}$, what is further affirmed is fc, if $\varphi x . \equiv_x . x = c$. Thus we have

$$f\{(\iota x)(\varphi x)\} . = : (\exists c) : \varphi x . \equiv_x . x = c : fc \quad \text{Df},$$

that is, "the x satisfying φx satisfies fx" is to mean: "There is an object c such that φx is true when, and only when, x is c, and fc is true", or, more exactly: "There is a c such that 'φx' is always equivalent to 'x is c' and fc". In this, "$(\iota x)(\varphi x)$" has completely disappeared; thus "$(\iota x)(\varphi x)$" is merely symbolic, and does not directly represent an object, as single small Latin letters are assumed to do.[2]

The proposition "$a = (\iota x)(\varphi x)$" is easily shown to be equivalent to "$\varphi x . \equiv_x . x = a$". For, by the definition, it is

$$(\exists c) : \varphi x . \equiv_x . x = c : a = c,$$

that is, "there is a c for which $\varphi x . \equiv_x . x = c$, and this c is a", which is equivalent to "$\varphi x . \equiv_x . x = a$". Thus "Scott is the author of Waverley" is equivalent to

"'x wrote Waverley' is always equivalent to 'x is Scott'",

that is, "x wrote Waverley" is true when x is Scott and false when x is not Scott.

[2] We shall generally write "$f(\iota x)(\varphi x)$" rather than "$f\{(\iota x)(\varphi x)\}$" in future.

Thus, although "$(\imath x)(\varphi x)$" has no meaning by itself, it may be substituted for y in any propositional function fy, and we get a significant proposition, though not a value of fy.

When $f\{(\imath x)(\varphi x)\}$, as above defined, forms part of some other proposition, we shall say that $(\imath x)(\varphi x)$ has a *secondary* occurrence. When $(\imath x)(\varphi x)$ has a secondary occurrence, a proposition in which it occurs may be true even when $(\imath x)(\varphi x)$ does not exist. This applies, for example, to the proposition: "There is no such person as the King of France". We may interpret this as

$$\sim\{E\,!\,(\imath x)(\varphi x)\},$$

or as

$$\sim\{(\exists c)\,.\,c = (\imath x)(\varphi x)\},$$

if "φx" stands for "x is King of France". In either case, what is asserted is that a proposition p in which $(\imath x)(\varphi x)$ occurs is false, and this proposition p is thus part of a larger proposition. The same applies to such a proposition as the following: "If France were a monarchy, the King of France would be of the House of Orleans".

It should be observed that such a proposition as

$$\sim f\{(\imath x)(\varphi x)\}$$

is ambiguous; it may deny $f\{(\imath x)(\varphi x)\}$, in which case it will be true if $(\imath x)(\varphi x)$ does not exist, or it may mean

$$(\exists c): \varphi x \,.\!\equiv_x\!.\, x = c : \sim fc,$$

in which case it can be true only if $(\imath x)(\varphi x)$ exists. In ordinary language, the latter interpretation would usually be adopted. For example, the proposition "the King of France is not bald" would usually be rejected as false, being held to mean "the King of France exists and is not bald", rather than "it is false that the King of France exists and is bald". When $(\imath x)(\varphi x)$ exists, the two interpretations of the ambiguity give equivalent results; but, when $(\imath x)(\varphi x)$ does not exist, one interpretation is true and one is false. It is necessary to be able to distinguish these in our notation; and generally, if we have such propositions as

$$\psi(\imath x)(\varphi x) \,.\!\supset\!.\, p,$$

$$p \,.\!\supset\!.\, \psi(\imath x)(\varphi x),$$

$$\psi(\imath x)(\varphi x) \,.\!\supset\!.\, \chi(\imath x)(\varphi x),$$

and so on, we must be able by our notation to distinguish whether the whole or only part of the proposition concerned is to be treated as the "$f(\imath x)(\varphi x)$" of our definition. For this purpose, we will put "$[(\imath x)(\varphi x)]$" followed by dots at the beginning of the part (or whole) which is to be taken as $f(\imath x)(\varphi x)$, the dots being sufficiently numerous to bracket off the $f(\imath x)(\varphi x)$; that is, $f(\imath x)(\varphi x)$ is to be everything following the dots until we reach an equal number of dots not signifying a logical product, or a greater number signifying a logical product, or the end of the sentence, or the end of a bracket enclosing "$[(\imath x)(\varphi x)]$". Thus

$$[(\imath x)(\varphi x)]\,.\,\psi(\imath x)(\varphi x) \,.\!\supset\!.\, p$$

will mean

$$(\exists c): \varphi x \,.\!\equiv_x\!.\, x = c : \psi c \,:\!\supset\!.\, p,$$

but

$$[(\imath x)(\varphi x)]:\psi(\imath x)(\varphi x) \,.\supset.\, p$$

will mean

$$(\exists c):\varphi x \,.\equiv_x.\, x = c:\psi c \,.\supset.\, p.$$

It is important to distinguish these two, for, if $(\imath x)(\varphi x)$ does not exist, the first is true and the second false. Again

$$[(\imath x)\varphi x)].\sim\psi(\imath x)(\varphi x)$$

will mean

$$(\exists c):\varphi x \,.\equiv_x.\, x = c:\sim\psi c,$$

while

$$\sim\{[(\imath x)(\varphi x)].\psi(\imath x)(\varphi x)\}$$

will mean

$$\sim\{(\exists c):\varphi x \,.\equiv_x.\, x = c:\psi c\}.$$

Here again, when $(\imath x)(\varphi x)$ does not exist, the first is false and the second true.

In order to avoid this ambiguity in propositions containing $(\imath x)(\varphi x)$, we amend our definition, or rather our notation, putting

$$[(\imath x)(\varphi x)]\ f(\imath x)(\varphi x) \,.=:\, (\exists c):\varphi x \,.\equiv_x.\, x = c:fc \quad \text{Df.}$$

By means of this definition, we avoid any doubt as to the portion of our whole asserted proposition which is to be treated as the "$f(\imath x)(\varphi x)$" of the definition. This portion will be called the *scope* of $(\imath x)(\varphi x)$. Thus in

$$[(\imath x)(\varphi x].f(\imath x)(\varphi x) \,.\supset.\, p$$

the scope of $(\imath x)(\varphi x)$ is $f(\imath x)(\varphi x)$; but in

$$[(\imath x)(\varphi x)]:f(\imath x)(\varphi x) \,.\supset.\, p$$

the scope is

$$f(\imath x)(\varphi x) \,.\supset.\, p;$$

in

$$\sim\{[(\imath x)(\varphi x)].f(\imath x)(\varphi x)\}$$

the scope is $f(\imath x)(\varphi x)$; but in

$$[(\imath x)(\varphi x)].\sim f(\imath x)(\varphi x)$$

the scope is

$$\sim f(\imath x)(\varphi x).$$

It will be seen that, when $(\imath x)(\varphi x)$ has the whole of the proposition concerned for its scope, the proposition concerned cannot be true unless $E!(\imath x)(\varphi x)$; but, when $(\imath x)(\varphi x)$ has only part of the proposition concerned for its scope, it may often be true even when $(\imath x)(\varphi x)$ does not exist. It will be seen further that, when $E!(\imath x)(\varphi x)$, we may enlarge or diminish the scope of $(\imath x)(\varphi x)$ as much as we please without altering the truth value of any proposition in which it occurs.

If a proposition contains two descriptions, say $(\imath x)(\varphi x)$ and $(\imath x)(\psi x)$, we have to distinguish which of them has the larger scope, that is, we have to distinguish

(1) $$[(\imath x)(\varphi x)]:[(\imath x)(\psi x)].f\{(\imath x)(\varphi x),\ (\imath x)(\psi x)\},$$

(2) $$[(\imath x)(\psi x)]:[(\imath x)(\varphi x)].f\{(\imath x)(\varphi x),\ (\imath x)(\psi x)\}.$$

The first of these, eliminating $(\imath x)(\varphi x)$, becomes

(3) $$(\exists c):\varphi x .\equiv_x. x = c:[(\imath x)(\psi x)].f\{c, (\imath x)(\psi x)\},$$

which, eliminating $(\imath x)(\psi x)$, becomes

(4) $$(\exists c):.\varphi x .\equiv_x. x = c:.(\exists d):\psi x .\equiv_x. x = c:f(c, d),$$

and the same proposition results if, in (1), we eliminate first $(\imath x)(\psi x)$ and then $(\imath x)(\varphi x)$. Similarly (2) becomes, when $(\imath x)(\varphi x)$ and $(\iota x)(\psi x)$ are eliminated,

(5) $$(\exists d):.\psi x .\equiv_x. x = d:.(\exists c):\varphi x .=_x. x = c:f(c, d).$$

(4) and (5) are equivalent, so that the truth value of a proposition containing two descriptions is independent of the question which has the larger scope.

It will be found that, in most cases in which descriptions occur, their scope is, in practice, the smallest proposition enclosed in dots or other brackets in which they are contained. Thus, for example,

$$[(\imath x)(\varphi x)].\psi(\imath x)(\varphi x) .\supset. [(\imath x)(\varphi x)].\chi(\imath x)(\varphi x)$$

will occur much more frequently than

$$[(\imath x)(\varphi x)]:\psi(\imath x)(\varphi x) .\supset. \chi(\imath x)(\varphi x).$$

For this reason it is convenient to decide that, when the scope of an occurrence of $(\imath x)(\varphi x)$ is the smallest proposition, enclosed in dots or other brackets, in which the occurrence in question is contained, the scope need not be indicated by "$[(\imath x)(\varphi x)]$". Thus, for example,

$$p .\supset. a = (\imath x)(\varphi x)$$

will mean

$$p .\supset. [(\imath x)(\varphi x)].a = (\imath x)(\varphi x);$$

and

$$p .\supset. (\exists a).a = (\imath x)(\varphi x)$$

will mean

$$p .\supset. (\exists a).[(\imath x)(\varphi x)].a = (\imath x)(\varphi x);$$

and

$$p .\supset. a \neq (\imath x)(\varphi x)$$

will mean

$$p .\supset. [(\imath x)(\varphi x)].\sim\{a = (\imath x)(\varphi x)\};$$

but

$$p .\supset. \sim\{a = (\imath x)(\varphi x)\}$$

will mean

$$p .\supset. \sim\{[(\imath x)(\varphi x)].a = (\imath x)(\varphi x)\}.$$

This convention enables us, in the vast majority of cases that actually occur, to dispense with the explicit indication of the scope of a descriptive symbol; and it will be found that the convention agrees very closely with the tacit conventions of ordinary language on this subject. Thus, for example, if "$(\imath x)(\varphi x)$" is "the so-and-so", "$a \neq (\imath x)(\varphi x)$" is to be read "$a$ is not the so-and-so", which would ordinarily be regarded as implying that "the so-and-so" exists; but "$\sim\{a = (\imath x)(\varphi x)\}$" is to be read "it is not true that a is the so-and-so", which would generally be allowed to hold if "the so-and-so" does not exist. Ordinary language is, of course, rather loose and fluctuating in its implications on this matter; but, subject to the requirement of definiteness, our convention seems to keep as near to ordinary language as possible.

In the case when the smallest proposition enclosed in dots or other brackets contains two or more descriptions, we shall assume, in the absence of any indication to the contrary, that one which typographically occurs earlier has a larger scope than one which typographically occurs later. Thus

$$(\imath x)(\varphi x) = (\imath x)(\psi x)$$

will mean

$$(\exists c) : \varphi x . \equiv_x . x = c : [(\imath x)(\psi x)] . c = (\imath x)(\psi x),$$

while

$$(\imath x)(\psi x) = (\imath x)(\varphi x)$$

will mean

$$(\exists d) : \psi x . \equiv_x . x = d : [(\imath x)(\varphi x)] . (\imath x)(\varphi x) = d.$$

These two propositions are easily shown to be equivalent.

A simplification of the logic of relations

NORBERT WIENER

(*1914*)

Wiener obtained his doctorate at Harvard University in June 1913, with a thesis (*1913*) in which Schröder's system was compared with that of Whitehead and Russell. Harvard University then granted him a traveling fellowship, and from June 1913 till April 1914 he was in Cambridge, England, where he attended two courses by Bertrand Russell, one being a reading course on *Principia mathematica*. It is in these circumstances that he came to write the note below. This piece of work, as Wiener recalls (*1953*, p. 191), excited no particular approval on the part of Russell.

Wiener's note was a communication to the Cambridge Philosophical Society; it was presented on 23 February 1914 by G. H. Hardy, with whom Wiener had taken a mathematics course and who came to greatly influence his mathematical training (*1953*, p. 190). By giving a definition of the ordered pair of two elements in terms of class operations, the note reduced the theory of relations to that of classes.

Wiener introduces the dissymmetry between the two elements of an ordered pair by using the null set. At approximately the same time, Hausdorff (*1914*, p. 32) gave the definition of the ordered pair (a, b) as $\{\{a, 1\}, \{b, 2\}\}$, where 1 and 2 are two distinct objects different from a and b. A few years later Kuratowski (*1921*) offered a definition that has been widely used ever since, namely $\{\{a, b\}, \{a\}\}$.

The Wiener-Hausdorff-Kuratowski definition has the effect that the ordered pair has a type higher by 2 than that of the elements (when these are of the same type). Under certain set-theoretic assumptions this difference in type can be reduced to 1 or even to 0 (see *Goodman 1941*, *Quine 1945*, and *Schwabhäuser 1954*).

The paper is reproduced here with the kind permission of Professor Wiener and the Cambridge Philosophical Society.

Two axioms, known as the axioms of reducibility, are stated on page 174 of the first volume of *Principia mathematica* of Whitehead and Russell. One of these, ∗12·1, is essential to the treatment of identity, descriptions, classes, and relations; the other, ∗12·11, is involved only in the theory of relations. ∗12·11 is applied directly only in

∗20·701·702·703 and ∗21·12·13·151·3·701·702·703.

It states that, given any propositional function φ of two variable individuals, there is another propositional function of two variable individuals, involving no apparent variables, and having the same truth value as φ for the same arguments, or in symbols:

$$\vdash :(\exists f):\varphi(x, y) .\equiv. f!(x, y).$$

In ∗20 and ∗21·701·702·703 all that is done with ∗12·11 is to extend it to cases where the arguments of φ and f are classes and relations; ∗12·11 is essential to the development of the calculus of relations only owing to its application in ∗21·12·13·151·3. Here it is needed to make the transition between the definition of a binary relation and its uses. This is due to the fact that a binary relation itself is not defined, but only propositions about it, and ∗12·11 is needed to assure us that these propositions about it behave as if there were a real object with which they concern themselves. The authors of *Principia* wish to treat a binary relation as the extension of a propositional function of two variables; that is, when they speak about the relation between x and y when $\varphi(x, y)$, they mean to speak of any propositional function which holds of those values of x and y, and only those values, of which φ holds. Now, as it leads one into vicious-circle paradoxes to speak directly of "any propositional function which holds of those values of x and y, and those only, of which φ holds", they first define a proposition concerning the relation between x and y when $\varphi(x, y)$ as a proposition concerning *a propositional function involving no apparent variables* which holds of x and y when and only when $\varphi(x, y)$. Then they need to use ∗12·11 to assure us that, whatever φ may be, there always is some such propositional function. Now, if we can discover a propositional function ψ of one variable so correlated with φ that its extension is determined uniquely by that of φ, and vice versa —if, to put it in symbols, when ψ' bears to φ' the same relation that ψ bears to φ, $\vdash :. \varphi'(x, y) .\equiv_{x,y}. \varphi(x, y) :\equiv: \psi'\alpha .\equiv_{\alpha}. \psi\alpha$—we can entirely avoid the use of ∗12·11 and interpret any proposition concerning the extension of φ as if it concerned the extension of ψ; for the existence of the extension of a propositional function of one variable is assured to us by ∗12·1, quite as that of one of two variables is by ∗12·11. Now, the propositional function

$$(\exists x, y) . \varphi(x, y) . \alpha = \iota`(\iota`\iota`x \cup \iota`\Lambda) \cup \iota`\iota`\iota`y$$

is such a ψ. For it is clear that for each ordered pair of values of x and y there is one and only one value of α, and vice versa. On the one hand, as $\iota`(\iota`\iota`x \cup \iota`\Lambda)$ is determined uniquely by x, and $\iota`\iota`\iota`y$ is determined uniquely by $y$, $\iota`(\iota`\iota`x \cup \iota`\Lambda) \cup \iota`\iota`\iota`y$ is determined uniquely by x and y. On the other hand, if

$$\iota`(\iota`\iota`x \cup \iota`\Lambda) \cup \iota`\iota`\iota`y = \iota`(\iota`\iota`z \cup \iota`\Lambda) \cup \iota`\iota`\iota`w,$$

either $\iota`\iota`y = \iota`\iota`z \cup \iota`\Lambda$ or $\iota`\iota`y = \iota`\iota`w$. The former supposition is clearly impossible, for, as $\iota`z \neq \Lambda$, $\iota`\iota`z \cup \iota`\Lambda$ is not a unit class. From the latter alternative we conclude immediately that $y = w$. Similarly, $x = z$.

Therefore, when x and y are of the same type, we can make the following definition:

$$\hat{x}\hat{y}\varphi(x, y) = \hat{\alpha}\{(\exists x, y) . \varphi(x, y) . \alpha = \iota`(\iota`\iota`x \cup \iota`\Lambda) \cup \iota`\iota`\iota`y\} \quad \text{Df.}$$ [1]

It will be seen that in this definition of $\hat{x}\hat{y}\varphi(x, y)$ it is essential that the x and the y should be of the same type, for if they are not $\iota`(\iota`\iota`x \cup \iota`\Lambda)$ and $\iota`\iota`\iota`y$ will not be, and $\iota`(\iota`\iota`x \cup \iota`\Lambda) \cup \iota`\iota`\iota`y$ will be meaningless. To overcome this limitation and secure typical ambiguity for domain and converse domain of $\hat{x}\hat{y}\varphi(x, y)$ separately, we make the following definitions:

$$\hat{\alpha}\hat{y}\varphi(\alpha, y) = \hat{\kappa}\{(\exists \alpha, y) . \varphi(\alpha, y) . \kappa = \iota`(\iota`\iota`\alpha \cup \iota`\Lambda) \cup \iota`\iota`(\iota`\iota`y \cup \iota`\Lambda)\} \quad \text{Df.}$$

[1] This may seem circular as ι is a relation, defined in *Principia* as $\overrightarrow{I}$, but it really is not circular, for $\iota`x$ may be defined directly as the class $\hat{y}(y = x)$.

$$\hat{\kappa}\hat{y}\varphi(\kappa, y) = \hat{\mu}\{(\exists\kappa, y) . \varphi(\kappa, y) . \mu = \iota\text{‘}(\iota\text{‘}\iota\text{‘}\kappa \cup \iota\text{‘}\Lambda) \cup \iota\text{‘}\iota\text{‘}[\iota\text{‘}(\iota\text{‘}\iota\text{‘}y \cup \iota\text{‘}\Lambda) \cup \iota\text{‘}\Lambda]\} \quad \text{Df.}$$

and so forth,

$$\hat{x}\hat{\beta}\varphi(x, \beta) = \hat{\kappa}\{(\exists x, \beta) . \varphi(x, \beta) . \kappa = \iota\text{‘}[\iota\text{‘}(\iota\text{‘}\iota\text{‘}x \cup \iota\text{‘}\Lambda) \cup \iota\text{‘}\Lambda] \cup \iota\text{‘}\iota\text{‘}\iota\text{‘}\beta\} \quad \text{Df.}$$

$$\hat{x}\hat{\lambda}\varphi(x, \lambda) = \hat{\mu}\{(\exists x, \lambda) . \mu = \iota\text{‘}\{\iota\text{‘}[\iota\text{‘}(\iota\text{‘}\iota\text{‘}x \cup \iota\text{‘}\Lambda) \cup \iota\text{‘}\Lambda] \cup \iota\text{‘}\Lambda\} \cup \iota\text{‘}\iota\text{‘}\iota\text{‘}\lambda\} \quad \text{Df.,}$$

and so forth.

Though these definitions may seem to conflict with one another, they really do not conflict, for where one of them is applicable, the others are meaningless, since they define relations between objects of different types. Moreover, it is easy to see that our definitions are so chosen that

$$\vdash : \hat{\mu}\hat{\nu}\varphi(\mu, \nu) = \hat{\omega}\hat{\rho}\psi(\omega, \rho) . \supset . t\text{‘}\text{D}\text{‘}\hat{\mu}\hat{\nu}\varphi(\mu, \nu)$$
$$= t\text{‘}\text{D}\text{‘}\hat{\omega}\hat{\rho}\psi(\hat{\omega}, \rho) . t\text{‘}\text{ᗡ}\text{‘}\hat{\mu}\hat{\nu}\varphi(\mu, \nu) = t\text{‘}\text{ᗡ}\text{‘}\hat{\omega}\hat{\rho}\psi(\omega, \rho).$$

This is important, as we might easily have defined relations so that they might have several domains or converse domains of different types. This is why we did not define $\hat{\alpha}\hat{y}\varphi(\alpha, y)$ simply as

$$\hat{\kappa}\{(\exists\alpha, y) \ \varphi(\alpha, y) . \kappa = \iota\text{‘}(\iota\text{‘}\iota\text{‘}\alpha \cup \iota\text{‘}\Lambda) \cup \iota\text{‘}\iota\text{‘}\iota\text{‘}\iota\text{‘}y\},$$

for this would also represent

$$\hat{\alpha}\hat{\beta}\{(\exists y) \ \varphi(\alpha, y) \ \beta = \iota\text{‘}y\}.$$

It will be seen that what we have done is practically to revert to Schröder's treatment of a relation as a class of ordered couples. The complicated apparatus of ι‘s and Λ‘s of which we have made use is simply and solely devised for the purpose of constructing a class which shall depend only on an ordered pair of values of x and y, and which shall correspond to only one such pair. The particular method selected of doing this is largely a matter of choice; for example, I might have substituted V, or any other constant class not a unit class, and existing in every type of classes, in every place I have written Λ.

Our changed definition of $\hat{x}\hat{y}\varphi(x, y)$ renders it necessary to give new definitions of several other symbols fundamental to the theory of relations. I give the following table of such definitions:

$$\text{Rel} = \hat{\kappa}\{\kappa \subset \hat{x}\hat{y}(x = x . y = y)\} \quad \text{Df.}$$
$$xRy . = . \hat{z}\hat{w}\{z = x . w = y\} \subset R . R \,\varepsilon\, \text{Rel} \quad \text{Df.}$$
$$\varphi R . = . (\exists\alpha) . \alpha = R . \alpha \,\varepsilon\, \text{Rel} . \varphi\alpha \quad \text{Df.}^2$$
$$(R) . \varphi R : = : \alpha \,\varepsilon\, \text{Rel} . \supset_\alpha . \varphi\alpha \quad \text{Df.}$$
$$(\exists R) . \varphi R : = . (\exists\alpha) . \alpha \,\varepsilon\, \text{Rel} . \varphi\alpha \quad \text{Df.}$$

[2] We shall understand in this way any propositional functions containing capital letters in the positions proper to their arguments. Thus $\sim\varphi R$ shall be understood as

$$(\exists\alpha) . \alpha = R . \alpha \,\varepsilon\, \text{Rel} . \sim\varphi\alpha,$$

and not as

$$\alpha = R . \alpha \,\varepsilon\, \text{Rel} . \supset_\alpha . \sim\varphi\alpha.$$

We make this definition as well as the two following ones because a propositional function of a class of the sort we have defined as a relation may significantly take as arguments classes of the same type which are not relations, and we wish to define propositional functions of relations in such a manner as to require that their arguments be relations.

The first two and the last two of these definitions replace ∗21·03·02 and ∗21·07·071 respectively. From these definitions and the laws of the calculus of classes it is an exceedingly simple matter to deduce any of the propositions of ∗21 which are not explicitly used for the purpose of deriving the properties of relations from the particular definition of relations given there, and from this it is easy to prove that the formal properties of the objects I call relations are essentially the same as those of the relations of *Principia.*

But it is obvious that since they are also classes, our relations will possess some formal properties not possessed by those of *Principia.* I give in conclusion a table of some of the more interesting of these:

$$\vdash . R \cup S = R \mathbin{\dot\cup} S$$

$$\vdash . R \cap S = R \mathbin{\dot\cap} S$$

$$\vdash : R \subset S . \equiv . R \mathbin{\dot\subset} S$$

$$\vdash . R - S . \equiv . R \mathbin{\dot-} S$$

$$\vdash . \dot{\mathrm{V}} \subset \mathrm{V}$$

$$\vdash . \Lambda = \dot\Lambda$$

$$\vdash . \mathrm{Rel} \subset \mathrm{Cls}$$

$$\vdash : Rp\kappa . \equiv . R\dot{p}\kappa$$

$$\vdash : Rs\kappa . \equiv . R\dot{s}\kappa$$

$$\vdash . \alpha + \beta \ \mathrm{sm}\ s`\alpha \uparrow \beta$$

$$\vdash . \alpha \times \beta \ \mathrm{sm}\ \alpha \uparrow \beta.$$

On possibilities in the calculus of relatives

LEOPOLD LÖWENHEIM

(*1915*)

Löwenheim's paper deals with problems connected with the validity, in different domains, of formulas of the first-order predicate calculus and with various aspects of the reduction and the decision problems. All these topics had remained alien to the trend that had by then become dominant in logic, that of Frege–Peano–Russell. In the following decades, however, these problems were to come more and more into the foreground, and the paper is now rightly considered a pioneer work in logic.

Löwenheim's work links up with that of Peirce and Schröder. The latter's treatise (*1890*, *1891*, *1895*), although often prolix, followed lines that led to some of the problems treated here. Löwenheim's paper uses Schröder's notation, today fallen into disuse (for a late defense of Schröderian see *Löwenheim 1940* and Bernays's comments (*1940a*)), and some explanations are perhaps in order.

The logic of relatives was introduced by Peirce (*1870*, *1880*, *1882*, *1883*) and codified, more or less in agreement with what Peirce had done, by Schröder. A basic domain of individuals, or universe of discourse, called *Denkbereich* and denoted by 1^1, contains at least one individual (for Löwenheim; two for Schröder —see footnote 2 below, p. 234);[a] it may be finite or infinite and, if infinite, denumerable or not. Subscripts $i, j, k, l, \ldots$ range over these individuals. The domains $1^2, 1^3, \ldots$ are the Cartesian products $1^1 \times 1^1, 1^1 \times 1^1 \times 1^1, \ldots$.

A binary relative coefficient, a_{ij}, denotes the proposition stating that a certain property holds of the ordered pair (i, j). (The calculus suffers from the ambiguities inherent in the "algebra of logic" and due to the confusion between the class interpretation and the propositional interpretation; in the paper below the binary relative coefficients in fact play the role of predicate variables or constants provided with individual variables, like $F(x, y)$ or $x = y$ in contemporary notation.) We write $a_{ij} = b_{ij}$ if and only if both a_{ij} and b_{ij} hold or neither does. (The sign "=" is also used for identity of individuals, as in $i = j$.) The *negation* of a_{ij}, $\bar{a}_{ij}$, holds if and only if a_{ij} does not. The *converse* of a_{ij}, $\breve{a}_{ij}$, is such that $\breve{a}_{ij} = a_{ji}$. The sum $a_{ij} + b_{ij}$ is the alternation $a_{ij} \vee b_{ij}$. The *product* $a_{ij} \cdot b_{ij}$, or $a_{ij}b_{ij}$, is the conjunction $a_{ij} \wedge b_{ij}$. What in German is called "Subsumption", $a_{ij} \Subset b_{ij}$, is the conditional $a_{ij} \supset b_{ij}$.

The universal quantifier is denoted by Π and the existential by Σ. Several Π, or Σ, can be replaced by one. $\Pi_i \Pi_j$ is replaced by $\Pi_{i,j}$, and so on. $\Pi_{i,j}$ is a *multiple* Π; Π_i is a *simple* Π.

A *binary relative* a is the set of ordered

[a] Occasionally Löwenheim speaks of the domain of individuals as the domain of *the first degree* (*erster Ordnung*).

pairs for which a_{ij} holds. $a = b$ if and only if $a_{ij} = b_{ij}$ for all ordered pairs (i, j). $\bar{a}$ is a relative such that $(\bar{a})_{ij} = \bar{a}_{ij}$. $\breve{a}$ is such that $(\breve{a})_{ij} = \breve{a}_{ij} = a_{ji}$. The *identical sum*, $a + b$, is such that $(a + b)_{ij} = a_{ij} + b_{ij}$. The *identical product*, $a \cdot b$, or ab, is such that $(a \cdot b)_{ij} = a_{ij} \cdot b_{ij}$. The *relative product*, $a;b$, is such that $(a;b)_{ij} = \Sigma_h a_{ih} b_{hj}$. The *relative sum*, $a \dotplus b$, is such that $(a \dotplus b)_{ij} = \Pi_h (a_{ih} + b_{hj})$.

We now define four relatives, 1, 0, 1′, 0′, called *modules* (*Moduln*): 1 is such that 1_{ij} holds of all ordered pairs (i, j), and $0 = \bar{1}$; 1′ is such that $1'_{ij}$ holds if and only if $i = j$, and $0' = \overline{1'}$.

We can define n-ary relative coefficients and relatives in a completely analogous manner.

The quantifiers Π and Σ can range over relatives also, but then letters other than i, j, k, l are used as subscripts. A formula that contains no such quantifier is what Löwenheim calls a *Zählausdruck*. In the paper below (as well as in some other papers in the present volume) this term has been translated as *first-order expression*, and since the relatives 1′ and 0′ are used, it should be understood as *well-formed formula of the first-order predicate calculus with identity*.

In § 1 Löwenheim introduces a number of definitions and symbols. In particular, an "identical equation" is a valid well-formed formula; a "fleeing equation" is a formula that is not valid but is n-valid for every finite n; a "halting equation" is a formula that, for some finite n, is not n-valid.

At the beginning of § 2 Löwenheim deals with a calculus of relatives that contains no individual variables, hence no relative coefficients and no quantifiers. The relative product and relative sum permit this elimination of the quantifiers. The calculus thus obtained contains variables and constants for binary relations, and the operations are the Boolean operations, together with the converse, the relative product, and the relative sum (some of these being definable in terms of others). The singulary predicate calculus of first order can, as is well known, be translated into a Boolean algebra of classes. The calculus of relations considered by Löwenheim can be regarded as an analogous algebra, and the question arises whether the binary predicate calculus of first order can be translated into such a calculus of relations. Following Schröder (*1895*, p. 550), Löwenheim calls a formula that is thus translatable a *condensable* formula, and he presents Korselt's result (communicated to him in a letter) that not every formula is condensable. This result was subsequently sharpened by Tarski (*1941*, p. 89) and by Lyndon (*1950*). A more general calculus of relations (not necessarily binary) was presented by Bernays (*1957*).

Theorem 2, in § 2, is the now famous Löwenheim theorem. It states that, if a finitely valid well-formed formula is not valid, it is not $\aleph_0$-valid. By contraposition, we obtain the more customary formulation of the theorem (for the first-order predicate calculus with identity): If a well-formed formula is $\aleph_0$-valid, it is valid, provided there is not some finite domain in which it is invalid.

A preliminary part of the proof consists in showing that, given a first-order formula, one can obtain a formula that has a certain normal form and is equivalent, so far as satisfiability is concerned, to the given formula (there is a shift from validity to satisfiability because the formula constitutes the left side of an equation whose right side is 0, and to write that a formula is equal to 0 amounts to writing a negation sign in front of the formula). The formula in normal form contains existential quantifiers, universal quantifiers, and a quantifier-free expression, in that order. It is the left side of (3) below, on page 237. Outwardly, it resembles a formula in Skolem normal form. There is an essential difference, however. In Löwenheim's normal form the number of existential quantifiers depends on the number of individuals in the *Denkbereich*. For domains of different cardinalities, the original formula is replaced by different formulas; for infinite domains there are infinitely many quantifiers, and the formula is replaced by what now is sometimes called an infinitely long expression, which no longer satisfies Löwenheim's own definition of *Zählausdruck* ("Σ and Π occur only a finite number of times", p. 232 below). Löwenheim obtains his normal form by commuting a universal and an existential quantifier through a procedure that he borrows from Schröder (*1895*, pp. 513–517). This yields, on page 236 below, the equivalence

$$\text{(I)} \qquad \underset{i}{\Pi}\underset{k}{\Sigma} A_{ik} = \underset{k_\lambda}{\textstyle\sum^{\lambda}} \Pi A_{ik_i},$$

in which A_{ik} is quantifier-free, Π is a universal quantifier, k_i is a "fleeing subscript" (in fact, a function of i), and the twofold sigma represents a (not necessarily finite, not even necessarily denumerably infinite) "string" of existential quantifiers; λ ranges over the individuals of the domain 1^1, and $\underset{k_\lambda}{\textstyle\sum^{\lambda}}$ stands for

$$\underset{k_1}{\Sigma}\,\underset{k_2}{\Sigma}\,\underset{k_3}{\Sigma}\ldots,$$

with as many sigmas as there are individuals in the domain 1^1. For example, for a domain of two individuals the right side of (I) becomes

$$\underset{k_1}{\Sigma}\,\underset{k_2}{\Sigma} A_{1k_1}A_{2k_2},$$

and for a domain of $\aleph_0$ individuals it becomes

$$\underset{k_1}{\Sigma}\,\underset{k_2}{\Sigma}\,\underset{k_3}{\Sigma}\cdots A_{1k_1}A_{2k_2}A_{3k_3}\cdots.$$

The example presented by Löwenheim on page 237 should help the reader to follow the method.

The twofold sigma is to be regarded as a "string" of α existential quantifiers (α being the cardinal number of 1^1), hence as expressing the existence of a value of k_i for each i. But, now, we can adopt a convention that is more natural than Löwenheim's and regard the twofold sigma as ranging over functions from individuals of 1^1 to individuals of 1^1. If we change "i" to "x", "k" to "y", the fleeing subscript "k_i" to "$f(x)$", "$=$" to "$\equiv$", and use the notation now current for quantifiers, (I) becomes

$$\text{(II)} \qquad (x)(Ey)A(x, y) \equiv (Ef)(x)A(x, f(x)).$$

The right side of (II), just like that of (I) above, is not a *Zählausdruck*; it is a well-formed formula of a second-order predicate calculus. But, since Löwenheim is dealing with the satisfiability of the formula, the initial existential quantifier can be discarded, and the problem is reduced to that of the satisfiability of the formula

$$\text{(III)} \qquad (x)A(x, f(x)),$$

which is again a *Zählausdruck*. This shows that Löwenheim's normal form does not correspond to Skolem's normal form, but to the universal closure of the functional form for satisfiability (see below, p. 508). The fleeing subscripts, just like the "Skolem functions", allow the elimination of the existential quantifiers. Since the formula is again of finite length, Löwenheim has taken, according to Skolem (below, p. 293), a detour through the transfinite. During this detour an argument that justifies the equivalence (I), or (II), in the case of a finite domain is—unwarrantedly—extended to the case of an infinite domain. But this could easily have been avoided, for instance by the use of the axiom of choice. (A few paragraphs below we shall see that Löwenheim, when he wants to establish the law of infinite conjunction, again unjustifiably extends an argument from the finite to the infinite case.)

At this point Löwenheim considers that he has reduced the question of the validity of an arbitrary first-order equation to that of the satisfiability of a formula

ΠF, where F is quantifier-free. The proof now proceeds by arguments that were subsequently used by Skolem, Herbrand, and Gödel. If ΠF contains individual constants or free variables, an initial domain (here called C_1, a notation not used by Löwenheim and in fact borrowed from Herbrand) is assumed to consist of individuals corresponding to them; otherwise, C_1 consists of an arbitrary individual. For every assignment of individuals of C_1 to the productation subscripts of F in ΠF, that is, the subscripts of Π, individuals are introduced for the fleeing subscripts of F. These individuals together with those of C_1 form C_2. When the productation subscripts of F range over C_2, new individuals are again introduced for the fleeing subscripts of F, and we obtain C_i. And so forth. The n_i possible assignments of individuals of C_1 to the productation subscripts of F are ordered according to some rule, with the condition that, for $1 \leqq \alpha < i$, no assignment entirely in $C_1 \cup C_2 \cup \ldots \cup C_\alpha$ follows any assignment that contains an element of $C_{\alpha+1} \cup \ldots \cup C_i$. Let F_i^ν be the result of replacing the productation subscripts of F by the νth such assignment, and let P_i be defined as follows:

$$P_1 = F_1^1 \wedge F_1^2 \wedge \cdots \wedge F_1^{n_1},$$

$$P_2 = P_1 \wedge F_2^1 \wedge \cdots \wedge F_2^{n_2},$$

$$\cdots\cdots\cdots\cdots,$$

$$P_i = P_{i-1} \wedge F_i^1 \wedge F_i^2 \wedge \cdots \wedge F_i^{n_i}.$$

To take identity in account,[b] we now assign to each equation $a_\alpha = a_\beta$, where a_α and a_β are elements of C_{i+1}, the value "true" or the value "false". For any such system of equations, the relatives $1'$ and $0'$ can now be evaluated in P_i, and we call the result P'_i. Performing the operation for all possible such systems of equations, we obtain $P'_i, P''_i, \ldots, P_i^{(s_i)}$.

We now assign truth values to the atomic parts of the F_i^ν. If, for some i and every such assignment, the alternation $P'_i \vee P''_i \vee \ldots \vee P_i^{(s_i)}$ is truth-functionally false, $\Pi F = 0$ is valid. Otherwise, there exists, for each i, a $P_i^{(j)}$, to be called Q_i, and an assignment of truth values such that Q_i is truth-functionally true under the assignment. At this point of his argument Löwenheim concludes, from the existence of a Q_i for each i, to the satisfiability of ΠF (hence to the nonvalidity of $\Pi F = 0$) in the (denumerable) union of the C_i. His line of reasoning is that, since for each i some assignment makes Q_i equal to 1 (which, in the Boole-Schröder notation, means that it is true), there is an assignment such that the infinite product $Q_1Q_2Q_3\ldots$ is equal to 1. Löwenheim writes: "Now, for those values of the summation subscripts whose substitution yielded $Q_1, Q_2, Q_3, \ldots, \Pi F$ is $= Q_1Q_2Q_3\ldots$, hence $= 1$." What has to be proved is that, from the assignments thus obtained for all i, there can be formed one assignment such that ΠF is true, that is, $\Pi F = 0$ is false. This Löwenheim does not do. The missing step is what Quine (*1955a*) calls the law of infinite conjunction, which, incidentally, is graphically represented by Löwenheim's (unproved) formula

$$1 = Q_1Q_2Q_3\ldots \text{ ad infinitum.}$$

The first application that Löwenheim makes of his theorem is to the Boolean calculus of classes. Schröder's "Gebietekalkul" had been codified by Eugen Müller in his edition of some of Schröder's manuscripts (*Schröder 1909, 1910*), and the calculus, under the name of "algebra of logic", had been studied by Huntington (*1904*) and Couturat (*1905*). (The precise formulation and interpretation of the calculus vary from author to author; for Löwenheim's see *Löwenheim 1910*.) It is interesting to note that, although this calculus is a very small fragment of set theory, Löwenheim's application of

[b] Löwenheim deals directly with the predicate calculus with identity. Compare his method with that of Gödel (*1930a*), who first settles the case of the pure calculus and then deals with identity by means of an entirely new argument.

his theorem to it somewhat foreshadows Skolem's paradoxical result in *1922* concerning Zermelo's set theory (see below, p. 295). (Löwenheim's criticism of Huntington's work (below, p. 242) is hardly justified; this work was important on account of the methods used as well as the results obtained.) The end of § 2 shows Löwenheim wrestling with the startling consequences of his theorem for set theory; these difficulties were settled much later, after the work of Skolem, Gödel, and others.

§ 3 contains the (positive) solution of the decision problem for the singulary predicate calculus of first order with identity. The proof of this result was greatly simplified by Skolem (*1919*) and Behmann (*1922*). § 4 presents the reduction of the decision problem for the first-order predicate calculus to that of its binary fragment. Herbrand (*1931*) and Kalmár (*1932*, *1934*, *1936*) sharpened Löwenheim's result and used simpler methods.

The translation is by Stefan Bauer-Mengelberg, and it is printed here with the kind permission of Springer Verlag.

§ 1. Definitions

We put

$$1'_{ijk} = 1'_{ij} + 1'_{ik} + 1'_{jk},$$
$$0'_{ijk} = 0'_{ij}0'_{ik}0'_{jk},$$

and, in general,

$$1'_{ijkl\cdots} = 1'_{ij} + 1'_{ik} + 1'_{jk} + 1'_{il} + 1'_{jl} + 1'_{kl} + \cdots,$$
$$0'_{ijkl\cdots} = 0'_{ij}0'_{ik}0'_{jk}0'_{il}0'_{jl}0'_{kl}\cdots.$$

We could, moreover, also put

$$1''_{ijkl\cdots} = 1'_{ij}1'_{ik} + 1'_{ij}1'_{il} + 1'_{ij}1'_{kl} + 1'_{ik}1'_{jk} + \cdots,$$
$$1'''_{ijkl\cdots} = 1'_{ij}1'_{ik}1'_{il} + \cdots,$$
$$\cdot \quad \cdot \quad \cdot \quad \cdot \quad \cdot \quad \cdot \quad \cdot \quad \cdot,$$

and similarly, by duality, for $0''$, $0'''$, ...

Then we would have, for example,

$$1'''_{ijk} = 1'_{ij}1'_{ik}1'_{jk} = (i = j = k) = 1''_{ijk},$$
$$1''''_{ijk} = 0.$$

$1'''_{ijk}$ and $0'''_{ijk}$ can perhaps be important, but they will not be used in the present paper.

Let us note in passing that the definition of a' follows from the expansion theorem:

$$a' = a \cdot 1' + \bar{a} \cdot 0'.$$

In what follows we shall always understand by *relative expression* [[*Relativausdruck*]] an expression that is formed from relatives or (not necessarily binary) relative coefficients and in which Σ and Π occur only a finite number of times, each Σ or Π ranging either over the subscripts—that is, over all individuals of the domain of the first degree, which, following Schröder, we call 1^1—or over all relatives that can be formed by means of this domain. All sums and products that do not range over the individuals of 1^1 or over all the relatives are assumed to be finite and will always be denoted by $+$ or by $\cdot$ (or the juxtaposition of factors) and never by Σ or by Π.

Equating two relative expressions yields a *relative equation* [[*Relativgleichung*]], which we shall always think of as brought into zero form. All important problems of mathematics and of the calculus of logic can, it seems, be reduced to [[questions about]] such relative equations.

A relative expression in which every Σ and Π ranges over the subscripts, that is, over the individuals of 1^1 (in other words, none ranges over the relatives), will be called a *first-order expression* [[*Zählausdruck*]]. By equating two such expressions we obtain a *first-order equation* [[*Zählgleichung*]]. An example of such an equation is

$$\Sigma_{l} \Pi_{i,j,h} (\bar{z}_{hi} + \bar{z}_{hj} + 1'_{ij})\bar{z}_{li}\Sigma_{k}\bar{z}_{ki} = 0,$$

or, *condensed*, that is, transformed into an equation between relatives, not relative coefficients,

$$0 \dotplus \{[1' + (\breve{\bar{z}} \dotplus \bar{z})](0 \dotplus \breve{\bar{z}})\cdot\breve{z}\,;1\} \dotplus 0 = 0.$$

To *condense* a relative expression means to transform it so that no Σ or Π occurs any longer. For example, $\Sigma_{h} a_{ih}b_{hj}$, when condensed, yields $(a\,;b)_{ij}$.

A relative equation can be

(*a*) An identical equation;

(*b*) A *fleeing equation* [[*Fluchtgleichung*]], that is, an equation that is not satisfied in every 1^1 but is satisfied in every finite 1^1 (or, more explicitly, an equation that is not identically satisfied but is satisfied whenever the summation or productation subscripts run through a finite 1^1);

(*c*) A *halting equation* [[*Haltgleichung*]], that is, one that is not even satisfied in every finite 1^1 for arbitrary values of the subscripts.

§ 2. First-order equations

THEOREM 1. *There are uncondensable equations, for example,*

$$\Sigma_{h,i,j,k} 0'_{hijk} = 0 \text{ or } 1,$$

$$\Sigma_{h,i,j,k,l} 0'_{hijkl} = 0 \text{ or } 1.$$

$$\cdot \quad \cdot \quad \cdot \quad \cdot \quad \cdot \quad \cdot \quad \cdot \quad \cdot,$$

hence there certainly are uncondensable first-order expressions.

By means of the examples above, Korselt proved the theorem (except for unessential gaps) in a letter to me, and he noted in addition that the following can be condensed:

$$\Sigma_{i,j} 0'_{ij} = 0 \text{ into } 0' = 0,$$

$$\Sigma_{i,j,k} 0'_{ijk} = 0 \text{ into } 0'(0'\,;0') = 0.$$

Also, we have[1]

$$\Sigma_{i,j} 0'_{ij} = (1\,;0'\,;1),$$

$$\Sigma_{i,j,k} 0'_{ijk} = [1\,;0'(0'\,;0')\,;1].$$

[1] [[In the original the right side of each of the two equations that follow has i and j as subscripts, which is an error.]]

So far as the equation

$$\underset{h,i,j,k}{\Sigma} 0'_{hijk} = 0$$

is concerned, however, it is easy to see that it is a halting equation that is satisfied if and only if 1^1 contains at most three elements. The equation

$$\underset{h,i,j,k}{\Sigma} 0'_{hijk} = 1$$

therefore asserts in words "1^1 contains at least four elements". Similarly,

$$\underset{h,i,j,k,l}{\Sigma} 0'_{hijkl} = 0$$

asserts "1^1 contains at most four elements", and so on.

If now the equations in Theorem 1 could be condensed, then, after they were condensed and brought into zero form, on the left side there would occur no relative coefficient but merely the "modules", 1, 0, 1', and 0', connected by means of some of the six logical operations $+$, $\cdot$, $\dotplus$, ;, $\bar{\ }$, and $\breve{\ }$. Negation and conversion can always be performed on module expressions; thus we can see to it that only the connectives $+$, $\cdot$, $\dotplus$, and ; occur on the left side.

The expression on the left side could then be computed by means of Schröder's "abacus of relatives" (*1895*, pp. 122–123, (13)–(19)).

But now we have (see (19)):

$$0' ; 0' = \begin{cases} 0 = 0' & \text{if } 1^1 \text{ consists of one element,} \\ \quad 1' & \text{if } 1^1 \text{ consists of two elements,} \\ \quad 1 & \text{if } 1^1 \text{ consists of more than two elements.} \end{cases}$$

Also, and here Schröder errs,[2] we have

$$1 ; 0' = 0' ; 1 = \begin{cases} 0 & \text{if } 1^1 \text{ consists of one element,} \\ 1 & \text{if } 1^1 \text{ consists of more than one element.} \end{cases}$$

Similarly, by duality, for $1' \dotplus 1'$ and $0 \dotplus 1'$, which is $= 1' \dotplus 0$.

The remaining module connectives are independent of the number of elements in 1^1. Therefore, whenever a connective is applied to two modules in a 1^1 of three elements, the result is exactly the same as in one of four elements. Hence, when the left side is computed, the final result must be the same for a 1^1 of three elements as for one of four elements; if, therefore, the equation obtained by condensation is satified in a 1^1 of three elements, it is also satisfied in one of four elements. Thus it cannot have resulted by mere transformation (condensation) of the equations in Theorem 1, since for these the opposite is the case.

Schröder (*1895*, p. 551) declares that condensation can always be performed; but to carry it out he employs the formula $a_{\kappa\lambda} = (\breve{\kappa} ; a ; \lambda)_{ij}$, in which the elements of 1^1

[2] ⟦It is true that Schröder (*1895*, p. 123) gives $1 ; 0' = 0' ; 1 = 1$ without any condition. But on page 5 of the same volume he states that the domains he will consider have more than one element and that "this assumption is requisite for the validity of almost all the theorems of the theory".⟧

are interpreted as relatives.[3] If we consider this admissible, however, condensation becomes such a trivial matter that it does not deserve the name and is utterly worthless.

I should like to take this opportunity to remark that the condition that the system a contain at most three elements can be written more clearly than in Schröder if we use the form

$$\Sigma_{h,i,j,k} a_h a_i a_j a_k 0'_{hijk} = 0.$$

That Schröder's attempted condensation of the condition is impossible follows from the preceding if we set $a = 1$. We can similarly express the condition that the system a possess at most 4, 5, ... elements. "The system a possesses at least three elements" is expressed by

$$\Pi_{h,i,j} (\bar{a}_h + \bar{a}_i + \bar{a}_j + 1'_{hij}) = 0,$$

and the conjunction of the last two equations asserts that the system a contains exactly three elements.

THEOREM 2. *If the domain is at least denumerably infinite, it is no longer the case that a first-order fleeing equation is satisfied for arbitrary values of the relative coefficients.*

For the proof we think of the equation as brought into zero form. We prove first that every first-order equation can be brought into a certain normal form, which is given on page 237 under (3). *First we try to see to it that no Π or Σ ever occurs under a* (simple or multiple) Π. If we assume that a productand contains at least one Π or Σ ⟦which may be $\cdot$ or $+$, respectively, as a special case⟧, we can distinguish four cases:

1. The productand is a $\cdot$ product. We can eliminate such a product by using the formula

$$\Pi_i A_i B_i = \Pi_i A_i \Pi_i B_i.$$

(In particular, for example, we have $\Pi_{i,j} A_i B_j C_{ij} = \Pi_i A_i \Pi_j B_j \Pi_{i,j} C_{ij}$.)

2. The productand is a Π product (say $\Pi_k A_{ik}$, where the A_{ik} are functions of relative coefficients). Then Π can be extricated from the productand by means of the formula

$$\Pi_i (\Pi_k A_{ik}) = \Pi_{i,k} A_{ik}.$$

3. The productand is a $+$ sum, hence something like, say,

$$A + B + C + \cdots \text{ not ad infinitum.}$$

Here we distinguish two subcases:

(*a*) One or several of the A, B, ... are ($\cdot$ or Π) products. This case can be reduced

[3] ⟦Following Peirce, Schröder (*1895*, p. 25) treats elements as relatives by introducing the definition

$$i_{kj} = 1'_{ik}.$$

With this definition condensation can always be carried out, but the interesting point, that the value of a condensable expression is independent of the cardinality (beyond 2) of the domain, is no longer correct.⟧

to Case 1 or Case 2 by means of the formula $a + bc = (a + b)(a + c)$, that is, by what we may call "adding out".

(*b*) None of the $A, B, \ldots$ is a product. Indeed, a + sum cannot be one if the A, $B, \ldots$ are really the *last* summands into which the productand can be decomposed without the use of Σ. Therefore each of the $A, B, \ldots$ is either a (negated or unnegated) relative coefficient or a Σ. If all of these summands are relative coefficients, we have already reached our goal; but if, for example,

$$A = \Sigma_k A_{ik}, \quad B = \Sigma_k B_{ik},$$

then the productand can be written in the form

$$\Sigma_k (A_{ik} + B_{ik} + C + \cdots),$$

which reduces this case to Case 4.

4. The productand is a Σ sum. Then our task consists in transforming the $\Pi\Sigma$ into a $\Sigma\Pi$, that is, in multiplying out the product. This is done by means of the formula

$$\Pi_i \Sigma_k A_{ik} = \sum_{k_\lambda}{}^{\lambda} \Pi_i A_{ik_i}.$$

Here the k_λ under the $\sum$ means that k_λ is to run through *all* subscripts, that is, through all elements of 1^1, and the λ on the right of the $\sum$ means that *each* of the k_λ is to run through these subscripts, hence that we have an n-fold sum if 1^1 possesses n elements (where n can also denote a transfinite cardinality). The A are functions of (not necessarily binary) relative coefficients.

In order to make the formula above more intelligible, I want once to expand in part the Σ and Π that occur in it, that is, I want to use the symbol + and juxtaposition (or dots) for these in one exceptional case here (contrary to the stipulation on page 232). I shall denote the subscripts by $1, 2, 3, \ldots$. Then the formula reads

$$\begin{aligned}\Pi_i (A_{i1} + A_{i2} + A_{i3} + \cdots &= \Sigma_{k_1, k_2, k_3, \ldots} A_{1k_1} A_{2k_2} A_{3k_3} \cdots \\ &= \Sigma_{k_1 = 1,2,3,\ldots} \Sigma_{k_2 = 1,2,3,\ldots} \Sigma_{k_3 = 1,2,3,\ldots} \cdots A_{1k_1} A_{2k_2} A_{3k_3} \cdots.\end{aligned}$$

In case a multiple Σ is preceded by a multiple Π our formula must be generalized as follows:

$$\Pi_{i,i',\ldots} \Pi_{k,k',\ldots} A_{ii'\ldots kk'\ldots} = \sum_{k_{ii'\ldots} k'_{ii'\ldots} \ldots} A_{ii'\ldots k_{ii'\ldots} k'_{ii'\ldots} \ldots}.$$

By means of the procedure described in Cases 1–4 every Σ and Π can be removed from the factor step by step. In this process it may very well happen that the transformation given in Case 4 must be applied several times in succession, and we still want to indicate how this is done:

$$\Pi_h \Sigma_i \Pi_k \Sigma_l A_{hikl} = \Pi_h \Sigma_i \sum_{l_k}{}^{k} \Pi A_{hikl_k} = \sum_{i_h} \sum_{l_{kh}}{}^{k,h} \Pi_{h,k} A_{hi_h kl_{kh}}.$$

That we do not have to write $l_{(k_h)}$ instead of l_{kh} is, to be sure, unessential for the course of the proof; but I want to make it clear nevertheless by expanding the formula for a 1^1 with only two elements, 1 and 2, again exceptionally using the +

and juxtaposition (contrary to the stipulation on page 232). Also I shall write $(hikl)$ for short in place of A_{hikl}.

$$\underset{h}{\Pi}\underset{i}{\Sigma}\underset{k}{\Pi}\underset{l}{\Sigma}(hikl) = \underset{h}{\Pi}\underset{i}{\Sigma}[(hi11)(hi21) + (hi11)(hi22) + (hi12)(hi21) + (hi12)(hi22)]$$

$$= \underset{h}{\Pi}[(h111)(h121) + (h111)(h122) + (h112)(h121) + (h112)(h122)$$

$$+ (h211)(h221) + (h211)(h222) + (h212)(h221) + (h212)(h222)]$$

$$= \sum_{p,q,r,s,t,u=1,2} (1p1r)(1p2s)(2q1t)(2q2u).$$

For $1^1 = (1, 2, 3)$ we obtain

$$\sum_{i_1,i_2,i_3,l_{11},l_{12},\ldots,l_{33}=1,2,3} (1i_11l_{11})(1i_12l_{12})(1i_13l_{13})(2i_21l_{21})(2i_22l_{22})(2i_23l_{23})$$

$$(3i_31l_{31})(3i_32l_{32})(3i_33l_{33}).$$

After the productands have all been freed from Σ and Π, it remains only to remove all parentheses that do not immediately follow a Σ or Π and to multiply out the products of the Σ. This is done by means of the formula

$$\underset{i}{\Sigma}A_i\underset{i}{\Sigma}B_i = \underset{i,k}{\Sigma}A_iB_k,$$

and similarly in the case of multiple sums.

In exactly the same way a $\sum$ can also be multiplied by another, or by a Σ, and likewise for multiple sums.

The final result is an equation of the form

$$\begin{aligned} (1) \qquad & C + \Sigma D_1 + \Sigma D_2 + \cdots \text{ not ad infinitum} \\ & + \Pi E_1 + \Pi E_2 + \cdots \text{ not ad infinitum} \\ & + \Sigma\Pi F_1 + \Sigma\Pi F_2 + \cdots \text{ not ad infinitum} = 0, \end{aligned}$$

where the sums and products will in general be multiple ones and the C, D_ν, E_ν, and F_ν are identical functions of relative coefficients without Σ or Π. In our example of page 233 we obtain, by means of the transformations sketched,

$$\underset{l}{\Sigma}\underset{k_\lambda}{\sum}{}^{\lambda}\underset{i,j,h}{\Pi}(\bar{z}_{hi} + \bar{z}_{hj} + 1'_{ij})\bar{z}_{li}z_{k_ii} = 0.$$

Now in (1), to begin with, we can see to it that exactly the same summation subscripts occur under each of the Σ by merely adding any missing ones (since $\underset{i}{\Sigma}a_i = \underset{i,j}{\Sigma}a_i$). To the terms without Σ we can simply add a Σ (since $a = \underset{i}{\Sigma}a$). Thereupon we can combine all Σ into a single Σ (since $\underset{i}{\Sigma}a_i + \underset{i}{\Sigma}b_i = \underset{i}{\Sigma}(a_i + b_i)$). We obtain an equation of the following form:

$$(2) \qquad \Sigma(F_0 + \Pi F_1 + \Pi F_2 + \cdots \text{ not ad infinitum}) = 0$$

or, added out according to the formula $\underset{i}{\Pi}a_i + \underset{i}{\Pi}b_i = \underset{i,j}{\Pi}(a_i + b_j)$,

$$\Sigma\Pi(F_0 + F_1 + F_2 + \cdots \text{ not ad infinitum}) = 0,$$

or, for short,

$$(3) \qquad \Sigma\Pi F = 0.$$

If we now want to decide whether or not (3) is identically satisfied in some domain, then in our discussion we can omit the Σ and examine the equation

$$\Pi F = 0, \tag{4}$$

or, in our example,

$$\Pi_{h,i,j} (\bar{z}_{hi} + \bar{z}_{hj} + 1'_{ij})\bar{z}_{li}z_{k_i i} = 0.$$

For, after all, that this equation be *identically* satisfied means nothing but that it be satisfied for *arbitrary* values of (z and) l, as well as of the k_i (that is, of $k_1, k_2, \ldots$). But the omitted Σ did not assert anything else, either, and was therefore superfluous, at least for us. (It would have been better to omit all Σ already in (1) and, likewise, in our example of page 233 to omit the Σ_l *before* transforming the formula into normal form, but not the Σ_k in that example and in general no Σ that occurs under a Π, since for such a Σ the considerations above do not hold.)

F can contain three kinds of subscripts:

(i) *Constant* subscripts, that is, such as must always be the same in every factor of the Π (l in our example); let us denote them, in some order, by the numbers $1, 2, \ldots, n$; hence we put $l = 1$ in our example;

(ii) *Productation* subscripts (i, j, and k in our example); they run through all elements of the domain independently of one another, so that to *every* system of values for these subscripts there corresponds a factor of the Π, and conversely;

(iii) *Fleeing* subscripts (like k_i in our example, as well as i_h and l_{kh} on page 236); their *subsubscripts* (i, h, or k and h, respectively) are productation subscripts, and the fleeing subscripts are (not necessarily one-to-one) functions of their subsubscripts; that is, l_{kh}, for example, denotes one and the same element in all those factors of Π in which the productation subscripts k and h have the same values (but l_{kh} does not necessarily denote different elements in other factors).

Now of the factors of the Π in (4) we first write down only all those in which no productation subscript has any value other than the values $1, 2, \ldots, n$ defined above under (i), or, if constant subscripts are lacking, we take any element of the domain, denote it by 1, and write down the factor in which all productation subscripts have the value 1. In this case we put $n = 1$. But in F there will also occur fleeing subscripts, say

$$i_j, k_{lm}, \ldots.$$

In each of the factors written down so far, $j, l, m, \ldots$, being productation subscripts, have as their values some of the numbers $1, 2, \ldots, n$; hence in these factors we shall have as fleeing subscripts

$$i_1, i_2, \ldots, i_n, k_{11}, k_{12}, k_{21}, \ldots, k_{nn}, \ldots.$$

These are no longer functions of subscripts but denote quite specific elements, which we shall denote, again in some order, by the numbers $n + 1, n + 2, \ldots, n_1$. (Let us remark expressly that two elements denoted by distinct numbers taken among $1, \ldots, n_1$ are not assumed to be either equal *or distinct*.)

The product written down thus far we call P_1. Hence in our example we would have

$$P_1 = \bar{z}_{11}(\bar{z}_{11} + \bar{z}_{11} + 1'_{11})z_{21} = \bar{z}_{11}z_{21}.$$

(Here we were permitted to put $1'_{11} = 1$; but, if $1'_{12}$ had occurred, we would not have been permitted to set it equal to 0, since it is not being assumed, after all, that 2 denotes an element different from that denoted by 1. Rather, we would have had to let $1'_{12}$ stand.)

ΠF will certainly vanish identically in every domain if P_1 does, that is, if P_1 vanishes when all elements $1, 2, \ldots, n_1$ are mutually distinct as well as when arbitrarily many equations hold among them. In order to see whether this is the case, we go through all these possibilities; thus we form from P_1 all those (finitely many) specializations $P'_1, P''_1, P'''_1, \ldots$ that we obtain when we introduce arbitrarily many or few equations among the elements $1, 2, \ldots, n_1$ (and then in the course of this the relative coefficients of $1'$ and $0'$ are evaluated, too).

Therefore, if all $P_1^{(\nu)}$ vanish identically, (4) is identically satisfied. If not, we now write down, in addition to the factors of ΠF already included in P_1, all those that are not yet included in P_1 and in which no productation subscript has a value other than a number from 1 to n_1. The resulting product (which, therefore, will also contain the old factors of P_1) we call P_2. In P_2 the fleeing subscripts $i_j, k_m, \ldots$ will have the values

$$i_1, i_2, \ldots, i_{n_1}, k_{11}, k_{12}, k_{21}, \ldots, k_{n_1 n_1}, \ldots;$$

of these we denote by the numbers $n_1 + 1, n_1 + 2, \ldots, n_2$ those that are not already denoted by some number. (We do not assume of these, either, that they represent mutually distinct elements or elements that differ from the old ones.)

If in our example we write for short $\kappa\lambda$ instead of $z_{\kappa\lambda}$, we have

$$\begin{aligned} P_2 &= P_1(\overline{11} + \overline{12} + 1'_{12})(\overline{12} + \overline{11} + 1'_{21})(12 + \overline{12} + 1'_{22})(\overline{21} + \overline{21} + 1'_{11}) \\ &\qquad (\overline{21} + \overline{22} + 1'_{12})(\overline{22} + \overline{21} + 1'_{21})(\overline{22} + \overline{22} + 1'_{22})\overline{12}\cdot 32 \\ &= P_1(\overline{21} + \overline{22} + 1'_{12})\overline{12}\cdot 32 = (\overline{22} + 1'_{12})\cdot \overline{11}\cdot \overline{12}\cdot 21\cdot 32. \end{aligned}$$

Now in P_2 (as before in P_1) the subscripts used are taken as either equal to or distinct from each other in every conceivable way. The products thus resulting from P_2 we call

$$P'_2, P''_2, P'''_2, \ldots.$$

If they all vanish, equation (4) is identically satisfied. If not, we form P_3 by writing down all the factors of ΠF in which the productation subscripts lie between 1 and n_2. We call the new fleeing subscripts $n_2 + 1, n_2 + 2, \ldots, n_3$.

In our example we have

$$\begin{aligned} P_3 = (\overline{22} + 1'_{12})(\overline{23} + 1'_{13})(\overline{22} + \overline{23} + 1'_{23})(\overline{31} + 1'_{12})(\overline{31} + \overline{33} + 1'_{13})(\overline{33} + 1'_{23}) \\ \cdot \overline{11}\cdot \overline{12}\cdot \overline{13}\cdot 21\cdot 32\cdot 43. \end{aligned}$$

By taking the subscripts as either equal or distinct we now form

$$P'_3, P''_3, P'''_3, \ldots.$$

And so forth. Since at this point it is easy to describe how, starting from P_n, we form P_{n+1}, as well as $P'_{n+1}, P''_{n+1}, P'''_{n+1}, \ldots$, the denumerably infinite sequence of the P_κ is to be regarded as defined herewith, and likewise the $P_\kappa^{(\nu)}$.

If for some κ (hence also for all succeeding ones) all $P_\kappa^{(\nu)}$ vanish, the equation is identically satisfied. If they do not all vanish, then the equation is no longer satisfied in the denumerable domain of the first degree just constructed. For then among P_1', P_1'', P_1''', ... there is at least one Q_1 that occurs in infinitely many of the nonvanishing $P_\kappa^{(\nu)}$ as a factor (since, after all, each of the infinitely many nonvanishing $P_\kappa^{(\nu)}$ contains one of the finitely many $P_1^{(\nu)}$ as a factor). Furthermore, among P_2', P_2'', P_2''', ... there is at least one Q_2 that contains Q_1 as a factor and occurs in infinitely many of the nonvanishing $P_\kappa^{(\nu)}$ as a factor (since each of the infinitely many nonvanishing $P_\kappa^{(\nu)}$ that contain Q_1 as a factor contains one of the finitely many $P_2^{(\nu)}$ as a factor). Likewise, among P_3', P_3'', P_3''', ... there is at least one Q_3 that contains Q_2 as a factor and occurs in infinitely many of the nonvanishing $P_\kappa^{(\nu)}$ as a factor. And so forth.

Every Q_ν is $=1$; therefore we also have

$$1 = Q_1 Q_2 Q_3 \ldots \text{ ad infinitum.}$$

But now, for those values of the summation[4] subscripts whose substitution yielded $Q_1, Q_2, Q_3, \ldots, \Pi F$ is $= Q_1 Q_2 Q_3 \ldots$, hence $=1$. Therefore ΠF does not vanish identically. Hence equation (4) is no longer satisfied even in a denumerable domain. Q.e.d.

Application: *All questions concerning the dependence or independence of Schröder's, Müller's, or Huntington's class axioms* ⟦*Gebietsaxiome*⟧ *are decidable (if at all) already in a denumerable domain.*

The axiom systems for the class calculus ⟦Gebictekalkül⟧ of Schröder, Müller, and others can be written as relative equations if we let the domain of the first degree 1^1 consist of all classes and denote the relation "sub" between two classes by a relative, s. (Thus we must carefully distinguish here between the relations of inclusion, addition, and so forth for the *classes*, on the one hand, and the relations for the *relative coefficients* on the other. The former inclusion, "a sub b", will be denoted by $s_{ab} = 1$, the latter, say "s_{ab} sub s_{cd}", by $s_{ab} \mathrel{\overline{\in}} s_{cd}$.)

Müller's axioms (I)–(V), namely,

(I)	$a \mathrel{\overline{\in}} a,$
(II)	$(a \mathrel{\overline{\in}} b)(b \mathrel{\overline{\in}} c) \mathrel{\overline{\in}} (a \mathrel{\overline{\in}} c),$
(III)	$(a \mathrel{\overline{\in}} b)(b \mathrel{\overline{\in}} a) = (a = b),$
(IV$_\times$)	$0 \mathrel{\overline{\in}} a,$
(IV$_+$)	$a \mathrel{\overline{\in}} 1,$
(V)	$1 \mathrel{\overline{\notin}} 0,$

yield, if in them we write n for 0 and e for 1 to avoid confusion,

(I)	$s_{aa} = 1,$
(II)	$s_{ab}s_{bc} \mathrel{\overline{\in}} s_{ac},$
(III)	$s_{ab}s_{ba} = 1'_{ab},$
(IV$_\times$)	$s_{na} = 1,$
(IV$_+$)	$s_{ae} = 1,$
(V)	$s_{en} = 0.$

[4] ⟦The "summation subscripts" are those that should be attached to Σ in (3); they are in fact the free variables of ΠF.⟧

But, since the axioms are to hold for arbitrary a, b, and c, we should still prefix $\underset{a}{\Pi}$, $\underset{a,b}{\Pi}$, or $\underset{a,b,c}{\Pi}$ to them and, in addition, $\underset{n}{\Sigma}$ and $\underset{e}{\Sigma}$ to $(IV_{\times})$ and (IV_{+}), respectively, in accordance with the meaning of these axioms.

(VI) could be expressed by means of ternary relations π and σ if we put $(\pi_{abc} = 1) = (ab = c)$ and $(\sigma_{abc} = 1) = (a + b = c)$. But we can also express (VI) without recourse to these ternary relatives. $(VI_{\times})$ stipulates that for every two classes a and b there exists a largest subclass c (that is, one that ranks above all other subclasses), called the product of a and b, a class c, therefore, for which in the Schröder-Müller notation

(1) $$(c ⋲ a)(c ⋲ b)$$

and

(2) $$\underset{x}{\Pi}(x ⋲ a)(x ⋲ b) ⋲ (x ⋲ c),$$

that is, in the new notation to be used here,

(1) $$s_{ca}s_{cb} = 1$$

and

(2) $$\underset{x}{\Pi}(s_{xa}s_{xb} ⋲ s_{xc}).$$

Accordingly we have

$$(\pi_{abc} = 1) = (s_{ca}s_{cb} = 1)\underset{x}{\Pi}(s_{xa}s_{xb} ⋲ s_{xc}),$$

or

$$\pi_{abc} = s_{ca}s_{cb}\underset{x}{\Pi}(\bar{s}_{xa} + \bar{s}_{xb} + s_{xc}),$$

and similarly, by duality, for σ_{abc}.

We have

$(VI_{\times})$ $$\underset{a,b}{\Pi}\underset{c}{\Sigma}\pi_{abc} = 1$$

and

(VI_{+}) $$\underset{a,b}{\Pi}\underset{c}{\Sigma}\sigma_{abc} = 1,$$

where the expressions above, which contain only binary coefficients, are to be substituted for π_{abc} and σ_{abc}, that is,

$(VI_{\times})$ $$\underset{a,b}{\Pi}\underset{c}{\Sigma}s_{ca}s_{cb}\underset{d}{\Pi}(\bar{s}_{da} + \bar{s}_{db} + s_{dc}) = 1,$$

and similarly, by duality, for (VI_{+}).

In Müller, (VII) reads $(a + z)(\bar{a} + z) = z = az + \bar{a}z$ and stipulates the existence of a "complement" of a that satisfies precisely this equation. If we then write b for $\bar{a}$, we must still put $\underset{b}{\Sigma}$ before the entire equation and then $\underset{a,z}{\Pi}$ before the entire expression. In order to be able to replace the notation for products and sums of classes, which is not admissible here, by π and σ, we decompose VII as follows:

$$\underset{a,z}{\Pi}\underset{b}{\Sigma}\underset{c,d,f,g,h,i}{\Pi}[(a + z = c)(b + z = d)(cd = f)(az = g)(bz = h)(g + h = i) = (f = z = i)].$$

Thus, through the use of π and σ, VII becomes

$$\mathop{\Pi}_{a,z}\mathop{\Sigma}_{b}\mathop{\Pi}_{c,d,f,g,h,i}(\sigma_{azc}\sigma_{bzd}\pi_{cdf}\pi_{azg}\pi_{bzh}\sigma_{ghi} \Subset 1''_{fzi}),$$

where the values given on the preceding page are still to be substituted for the relative coefficients of π and σ.

Now in investigations concerning independence it must be decided whether from certain axioms another follows. That this is the case can, however, be expressed by a relative equation that can be made primary.[5] Thus the problem in such investigations is to decide whether a certain relative equation, and in particular, in view of the preceding, a first-order equation, is identically satisfied or not. But, if this is not the case, then according to Theorem 2 it is possible to give a counterexample already in a finite or denumerable domain. Q.e.d.

On the basis of these reflections I have investigated the independence of the individual axioms, and I am thinking of publishing the results on another occasion.[6] I regard the work of Huntington on this topic as mistaken, since, when axioms turn out to be inconvenient, he simply adds to their formulation "if the preceding axioms are satisfied", a cheap way of evading difficulties at will!

Theorem 3. *The evaluation of a product or a sum over relatives is not always possible.*

For by means of Schröder's devices it is at once possible to write down an equation that asserts that the domain is finite or denumerable, that is, that every system $a;1$ is finite or ~ 1 (equivalent to the domain):

$$(a;1 \text{ finite}) + (a;1 \sim 1).$$

That $a;1$ is finite means that no mapping z maps $a;1$ one-to-one onto a proper part b of itself. But that z maps a system a one-to-one onto b is expressed, according to Schröder (*1895*, p. 605, (10)), by

$$(1) \qquad (z;\breve{z} + \breve{z};z \Subset 1')(b \Subset z;a)(a \Subset \breve{z};b),$$

that b is a proper part of a by

$$(2) \qquad (b \Subset a)(b \neq a)(b \neq 0),$$

and $a;1 \sim 1$, according to (1), by

$$\mathop{\Sigma}_{z}(z;\breve{z} + \breve{z};z \Subset 1')(1 \Subset z;a;1)(a;1 \Subset \breve{z};1).$$

Thus, that 1^1 is finite or denumerable is expressed by

$$\{0 = \mathop{\Sigma}_{z}\mathop{\Sigma}_{b}(z;\breve{z} + \breve{z};z \Subset 1')(b \Subset z;a)(a \Subset \breve{z};b)(b \Subset a)(b \neq a)(b \neq 0)\}$$
$$+ \mathop{\Sigma}_{z}(z;\breve{z} + \breve{z};z \Subset 1')(1 \Subset z;a;1)(a;1 \Subset \breve{z};1).$$

This, like any secondary equation, is easily transformed into a primary equation (see *Schröder 1895*, pp. 150–151), and that in turn into a relation between coefficients. If now $\mathop{\Pi}_{z}$ and $\mathop{\Sigma}_{z}$ could be evaluated, the equation would, according to Theorem 2, either have to be identically satisfied or, already in a finite or denumerable domain, fail to be satisfied, none of which is the case.

[5] ⟦A relative equation is *primary* if its left and right sides do not contain any conditional or equivalence.⟧

[6] ⟦This was never done.⟧

The reader will ask why the proof of Theorem 2 cannot be carried over word for word to the equation above, too, since this equation certainly is not satisfied in a finite or denumerable 1^1. To be sure, by the method of that proof a domain in which the equation is not identically satisfied can be constructed; but this domain will turn out to be nondenumerable. For, since we are taking products also over *relatives* z, there will also occur fleeing subscripts of the form i_z, where z is not a "subsubscript" but a "subrelative", so that a subscript is associated with every z. But if the number of subscripts approaches infinity, the number of possible z, and hence also the number of fleeing subscripts required (that is, the cardinality of the required domain), approaches the continuum. (Therewith, however, the set of possible z in turn becomes one of yet higher cardinality than that of the continuum, and consequently so does the set of required fleeing subscripts, and so on ad infinitum.)

§ 3. Singulary equations

THEOREM 4. *There are no fleeing equations between singulary relative coefficients, not even when the relative coefficients of* $1'$ *and* $0'$ *are included as the only binary ones.*

If no binary coefficient occurs, it is very easy to show that we can do without fleeing subscripts altogether and that consequently in the construction of the domain as on pages 238–240 we have no occasion at all to make it infinite.

If coefficients of $1'$ and $0'$ occur, we construct a certain normal form. The left side of the equation in zero form is symmetric with respect to all subscripts, that is, all the elements of 1^1, except for a *finite* number of quite specific ones, which I shall call the *distinguished subscripts*. The relative coefficients corresponding to these subscripts, as well as the classes without subscript that perhaps occur, I shall call the *distinguished classes*. They may, however, also be lacking. All subscripts written down in the equation are summation, productation, or distinguished subscripts.

We now consider the Boolean expansion, in terms of the distinguished classes, of the left side of the equation. I first wish to describe the normal form in question, whose constructibility I asserted above. It is characterized

(1) By the fact that the coefficients of the expansion are functions that are symmetric with respect to *all* (even the distinguished) subscripts;

(2) By a property of precisely these symmetric functions that is now to be discussed.

We may assume, first, that these symmetric functions (which, after all, were coefficients in the Boolean expansion in terms of the distinguished classes) no longer contain classes without subscript (since these also belong to the distinguished classes in terms of which the expansion was carried out). The symmetric functions, we claim, contain no coefficients of $1'$ or $0'$. Following Boole, we can expand them in such a way that only the coefficients 0 and 1 occur. Although we shall think only of the terms with the coefficient 1 as written down, there may yet be infinitely many terms, so that we must consider an abbreviation.

If, for example, the term

$$a_1 a_2 a_3 \bar{a}_4 \bar{a}_5 \bar{a}_6 \bar{a}_7 \ldots, \text{ which is } = a_1 a_2 a_3 \Pi_i (\bar{a}_i + 1'_{123i}),$$

occurs in the expansion, then there also occur all the terms that we obtain from this one by interchanging subscripts. I denote their sum by $(3, \infty)_a$; hence

$$(3, \infty)_a = \sum_{h,i,j} a_h a_i a_j 0'_{hij} \prod_k (\bar{a}_k + 1'_{hijk}),$$

and the sum of the terms that result from

$$\bar{a}_1 \bar{a}_2 \bar{a}_3 a_4 a_5 a_6 a_7 \ldots$$

by interchange of subscripts I denote by $(\infty, 3)_a$. In general, we put

$$(n, \infty)_a = \sum_{i_1, i_2, \ldots, i_n} a_{i_1} a_{i_2} \ldots a_{i_n} 0'_{i_1 i_2 \cdots i_n} \prod_k (\bar{a}_k + 1'_{i_1 i_2 \cdots i_n k}).$$

Similarly, we obtain $(\infty, n)_a$ from this by interchange of a_λ and $\bar{a}_\lambda$ for every λ.

If several relatives occur, say a and b, then, for example, we denote by

$$(2, 3, 1, \infty)_{a,b}$$

the sum of the terms that result from

$$(a_1 b_1 a_2 b_2)(a_3 \bar{b}_3 a_4 \bar{b}_4 a_5 \bar{b}_5)(\bar{a}_6 b_6)(\bar{a}_7 \bar{b}_7 \bar{a}_8 \bar{b}_8 \ldots)$$

by interchange of subscripts. Likewise, for example, we denote by

$$(i, \infty, k, l)_{a,b}$$

the sum of the terms[7] resulting from

$$(a_1 b_1 a_2 b_2 \cdots a_i b_i)(\prod_\rho a_\rho \bar{b}_\rho + 1'_{1,2,\ldots,i+k+l,\rho})(\bar{a}_{i+1} b_{i+1} \bar{a}_{i+2} b_{i+2} \cdots a_{i+k} b_{i+k})$$
$$(\bar{a}_{i+k+1} \bar{b}_{i+k+1} \bar{a}_{i+k+2} \bar{b}_{i+k+2} \cdots \bar{a}_{i+k+l} \bar{b}_{i+k+l})$$

by interchange of subscripts. At this point, the general law of formation should be clear. That indeed there are always only finite numbers in all places except one will have to be taken as one characteristic of precisely these symmetric functions.

But even in this abbreviated notation the expansion can still contain infinitely many terms. As an example, we now denote by

$$(\leqq 2, \leqq 5)_a$$

the sum of all those terms in which there occur at least two unnegated and at least five negated a; that is, we have

$$(\geqq 2, \geqq 5)_a = (2, \infty)_a + (3, \infty)_a + (4, \infty)_a + \cdots + (\infty, 5)_a + (\infty, 6)_a + (\infty, 7)_a + \cdots$$
$$= \sum_{\kappa, \lambda, \mu, \ldots, \sigma} a_\kappa a_\lambda \bar{a}_\mu \bar{a}_\nu \bar{a}_\pi \bar{a}_\rho \bar{a}_\sigma 0_{\kappa\lambda\mu\nu\pi\rho\sigma}.$$

In general,

$$(\geqq p, \geqq q)_a = \sum_{\substack{i_1, i_2, \ldots, i_p \\ j_1, j_2, \ldots, j_q}} a_{i_1} a_{i_2} \cdots a_{i_p} \bar{a}_{j_1} \bar{a}_{j_2} \cdots \bar{a}_{j_q} 0'_{i_1 i_2 \cdots i_p j_1 j_2 \cdots j_q}.$$

That with the use of this notation the entire symmetric function can now be represented as a sum of *finitely* many terms is a second characteristic of precisely these symmetric functions, and this is really the essential point of the entire proof.

[7] [[The German text has "Summe" for what should apparently be "Glieder".]]

I shall only sketch the method of obtaining such a normal form. Every relative expression is constructed from relative expressions that do not contain any Σ or Π, and therefore already possess the desired normal form, by means of repeated use of the four operations $+$, $\cdot$, Σ, and Π. Thus we can bring any relative expression into normal form if we can bring into normal form an expression that results from a *single* application of any of those four operations to expressions already in normal form. The elaboration of a method for each of these four operations is not difficult and is left to the reader.

Thus the normal form can be constructed. It is a Boolean expansion. If during its construction everything drops out, the equation is identically satisfied in every domain whatsoever. But, if even a single term remains, we can immediately specify a *finite* domain in which this term, and consequently the entire left side of the equation, does not vanish identically. If, for example, a term

$$p(\geqq 2, \geqq 3, 1, \infty)_{a,b}$$

remains, where p is a product of distinguished classes, then the term does not vanish in a domain of six elements in which

$$\begin{array}{ll} p = 1, & \\ a_1 = a_2 = 1, & b_1 = b_2 = 1, \\ a_3 = a_4 = a_5 = 1, & b_3 = b_4 = b_5 = 0, \\ a_6 = 0, & b_6 = 1 \end{array}$$

Corollary. *Fleeing equations contain, besides the modules, other relatives.*

THEOREM 5. *The evaluation of a Π or Σ over all relatives is always possible for an expression that contains only singulary relative coefficients* (*or, in addition, at most the coefficients of* $1'$ *or* $0'$) *and finitely many Σ or Π over all subscripts.*

Assume that we have to evaluate a Σ over relatives (for a Π we would proceed similarly, by duality).

Expand into the normal form of the previous theorem. Then the terms can be evaluated individually. For example, we have

$$\Sigma_b(5, 7, 3, \infty)_{a,b} = (5 + 7, 3 + \infty)_a = (12, \infty)_a,$$

$$\Sigma_a(5, 7, \geqq 3, 4)_{a,b} = (\geqq 5 + 3, 7 + 4)_a = (\geqq 8, 11)_a.$$

If under a Σ there occurs a distinguished coefficient a_i, then in the evaluation it must be struck out (that is, set $= 1$); $\bar{a}_i$ would have to be struck out likewise.

There remains only the task of transforming the result back from the normal form into an ordinary relative expression. The following example will indicate how this is done:

$$(2, 1, 3, \infty)_{a,b} = \Sigma_{h,i,j,k,l,m}(a_h b_h a_i b_i)(a_j \bar{b}_j)(\bar{a}_k b_k \bar{a}_l b_l \bar{a}_m b_m)0'_{hijklm}\Pi_n(\bar{a}_n \bar{b}_n + 1'_{hijklmn}).$$

§ 4. REDUCTION OF THE HIGHER CALCULUS OF RELATIVES TO THE BINARY

THEOREM 6. *Every relative equation* (*or first-order equation*) *is equivalent to a binary one*; that is, if an arbitrary relative equation (or first-order equation) $f = 0$ is given, then we can specify a binary relative equation (or binary first-order equation) $F = 0$

that is satisfied identically, not identically, or never (that is, for no system of values of the parameters) if and only if the corresponding statement holds for $f = 0$; likewise, we can specify a binary equation (or binary first-order equation) $F' = 0$ that is satisfied for some systems of values if and only if $f = 0$ is. The latter statement can, after all, be reduced to the former. But F' does not coincide with F.

In the calculus of relatives this theorem has the same significance as does in algebra the theorem of Weierstrass [[*1884*]] according to which we can solve by means of the complex numbers with two fundamental units all the problems that can be solved by means of complex numbers that have more than two fundamental units and for which the same formulas hold as for our numbers. Likewise, all problems that can be solved by means of a calculus containing ternary or higher relatives can already be settled in the binary calculus. (Under certain circumstances, to be sure, recourse to ternary and higher relatives may be simpler.)

We can gauge the significance of our theorem by reflecting upon the fact that every theorem of mathematics, or of any calculus that can be invented, can be written as a relative equation; the mathematical theorem then stands or falls according as the equation is satisfied or not. This transformation of arbitrary mathematical theorems into relative equations can be carried out, I believe, by anyone who knows the work of Whitehead and Russell. Since, now, according to our theorem the whole relative calculus can be reduced to the binary relative calculus, it follows that we can decide whether an arbitrary mathematical proposition is true provided we can decide whether a binary relative equation is identically satisfied or not.

Now, first, let a quaternary relative equation be given; the ternary can be considered a special case of this. For, if ternary relative coefficients a_{ijk} are given, we can define quaternary ones by means of the equations $a_{ijkl} = a_{ijk}$.

We consider a new domain, whose elements are the *element-pairs* of the old 1^1, and which we must therefore call 1^2. Thus if

$$1^1 = (i, j, k, \ldots),$$

then

$$1^2 = ((i, i), (i, j), (j, i), (j, j), (i, k), \ldots).$$

The new domain $\mathfrak{E}$, which we shall take as basic in what follows, now results from 1^2 by replacement of (i, i) by i, (j, j) by j, and so on. Thus

$$\mathfrak{E} = (i, (i, j), (j, i), j, (i, k), \ldots).$$

It is convenient to denote the elements by single letters, say $I, K, L, \ldots$. If $I = (i, j)$, we call i the *first element* and j the *last element* of I. We shall regard i as its own first and last element.

Now, with every quaternary relative a whose subscripts must run through 1^1 we associate a binary relative A whose subscripts are to run through $\mathfrak{E}$, and we do this in such a way that for $I = (i, j)$ and $K = (k, l)$

$$a_{ijkl} = A_{IK}, \quad a_{iikl} = A_{iK}, \quad a_{ijkk} = A_{Ik}, \quad \text{and} \quad a_{iikk} = A_{ik}.$$

Thus, if a is regarded as given, A is completely defined as a binary relative with $\mathfrak{E}$ as a domain of the first degree.

On the basis of this correspondence we now replace all quaternary relatives in

$f = 0$ by the associated binary ones. Since the relative coefficients are no longer simply elements of 1^1 but elements of $\mathfrak{E}$ also, the summation and productation subscripts, too, must be modified so that they range over the entire $\mathfrak{E}$. If, for example,

$$(1) \qquad \Sigma_i a_{ijkl} b_{jkli}$$

occurs in f, and if a_{ijkl} is replaced by A_{IK} and b_{jkli} by B_{JL}, then the summation subscript i occurs in I $(= (i, j))$ and in L $(=(l, i))$; therefore we first replace the summation over i by one over I and L, and we write

$$(2) \qquad \Sigma_I \Sigma_L A_{IK} B_{JL}.$$

But now we must realize that this new sum contains more summands than the old (since I and L now range over the entire $\mathfrak{E}$ and not only over 1^1). We must remedy this inconvenience, for, after all, we want to construct an equation that contains exactly as many summands and factors as the old one, that, in fact, does not differ from the old one in anything but the notation (with one exception, to be sure). We achieve our purpose by multiplying each $A_{IK} B_{JL}$ by a factor equal to 1 for those summands that also occur in the old sum and equal to 0 for the others. We define two relatives V and H by

$$(3) \qquad \begin{cases} (V_{IJ} = 1) = (I \text{ is the first element of } J), \\ (H_{IJ} = 1) = (I \text{ is the last element of } J). \end{cases}$$

These definitional properties of V and H will still have to be incorporated into the equation. In the meantime we regard V and H as given relatives whose subscripts denote elements of $\mathfrak{E}$; for, the first and the last elements also are elements of $\mathfrak{E}$. We have

$$(4) \qquad \begin{cases} ((\breve{V}; V)_{IJ} = 1) = (\text{first element of } I = \text{first element of } J), \\ ((\breve{V}; H)_{IJ} = 1) = (\text{first element of } I = \text{last element of } J), \\ ((\breve{H}; H)_{IJ} = 1) = (\text{last element of } I = \text{last element of } J). \end{cases}$$

The sum

$$(5) \qquad \Sigma_I \Sigma_L A_{IK} B_{JL} H_{jI} V_{lL}$$

is a summation over $\mathfrak{E}$ in form only;[8] in reality it is just a summation over 1^1, since the factors added to (2) cancel all terms but those in which the summation subscripts have j as their last element and l as their first. Nevertheless, even this sum still contains more summands than (1); and only those summands are to be kept for which the first element of I coincides with the last element of L; that is, the factor $(\breve{V}; H)_{IL}$ must still be added.

$$(6) \qquad \Sigma_I \Sigma_L A_{IK} B_{JL} H_{jI} V_{lL} (\breve{V}; H)_{IL}$$

now contains exactly as many terms as the given formula (1), but (6) is more general

[8] ⟦In the original the second subscript on "H" in formula (5) is a "J", which is a misprint.⟧

than (1), since the coincidence of the two subscripts j, of the two k, and of the two l, which is visible in (1), is not expressed in (6). This must still be done by means of (4).

$$\sum_{I}\sum_{L} A_{IK} B_{JL} H_{jI} V_{lL} (\check{V}; H)_{IL} (\check{V}; H)_{JI} (\check{V}; H)_{KJ} (\check{V}; H)_{LK} \tag{7}$$

must take the place of (1), and it is of the same generality as (1). By duality, we can proceed for productation in a way similar to that for summation.

Just as in this example we expressed the equality of certain subscripts of the given equation (an equality that for a time was no longer expressed when the new relative coefficients were introduced) by means of the last factors, so in $f = 0$ we must look for all places where identical subscripts occur or where the equality (or inequality) of subscripts is expressed by coefficients such as $1'_{ij}$ (or $0'_{ij}$); and where, then, through the introduction of the new relative coefficients this equality (or inequality) is no longer expressed, we must see to it that it is expressed by introducing suitable summands or factors, as in the last example. Only then can we be sure that the new equation permits exactly the transformations corresponding to those of $f = 0$ and consequently is identically satisfied exactly in case $f = 0$ is, since both equations then contain exactly the same functions of relative coefficients, except that the arguments, that is, the relative coefficients, are named differently.

Now we must still incorporate the properties of V and H in the new equation. In what follows I shall always write relatives in the form of a table; that is, if, for example, $\alpha, \beta, \gamma, \ldots$ are the elements of a domain of the first degree, I shall write the relative a in the following way:

$$a = \left\{ \begin{array}{c|cccc} & \alpha & \beta & \gamma & \cdots \\ \hline \alpha & a_{\alpha\alpha} & a_{\alpha\beta} & a_{\alpha\gamma} & \cdots \\ \beta & a_{\beta\alpha} & a_{\beta\beta} & a_{\beta\gamma} & \cdots \\ \gamma & a_{\gamma\alpha} & a_{\gamma\beta} & a_{\gamma\gamma} & \cdots \\ . & . & . & . & . \; . \; . \end{array} \right.$$

To every subset of the domain of the first degree there then corresponds a "system" in the sense of Schröder; for example, to the set

$$\mathfrak{A} = (\alpha, \gamma, \delta)$$

there corresponds the system

$$\mathfrak{A} = \left\{ \begin{array}{c|cccccc} & \alpha & \beta & \gamma & \delta & \varepsilon & \cdots \\ \hline \alpha & 1 & 1 & 1 & 1 & 1 & \cdots \\ \beta & 0 & 0 & 0 & 0 & 0 & \cdots \\ \gamma & 1 & 1 & 1 & 1 & 1 & \cdots \\ \delta & 1 & 1 & 1 & 1 & 1 & \cdots \\ \varepsilon & 0 & 0 & 0 & 0 & 0 & \cdots \\ . & . & . & . & . & . & . \end{array} \right.$$

in which the lines associated with the elements α, γ, δ of $\mathfrak{A}$ contain only ones and the remaining lines only zeros.

Now to the subset 1^1 of $\mathfrak{C}$ there corresponds a system q, namely,

$$q = \left\{ \begin{array}{c|cccccc} & i & (i,j) & (j,i) & j & (i,k) & \cdots \\ \hline i & 1 & 1 & 1 & 1 & 1 & \cdots \\ (i,j) & 0 & 0 & 0 & 0 & 0 & \cdots \\ (j,i) & 0 & 0 & 0 & 0 & 0 & \cdots \\ j & 1 & 1 & 1 & 1 & 1 & \cdots \\ (i,k) & 0 & 0 & 0 & 0 & 0 & \cdots \\ \cdot & \cdot & \cdot & \cdot & \cdot & \cdot & \cdots \end{array} \right.$$

Further we have

$$V = \left\{ \begin{array}{c|cccccc} & i & (i,j) & (j,i) & j & (i,k) & \cdots \\ \hline i & 1 & 1 & 0 & 0 & 1 & \cdots \\ (i,j) & 0 & 0 & 0 & 0 & 0 & \cdots \\ (j,i) & 0 & 0 & 0 & 0 & 0 & \cdots \\ j & 0 & 0 & 1 & 1 & 0 & \cdots \\ (i,k) & 0 & 0 & 0 & 0 & 0 & \cdots \\ \cdot & \cdot & \cdot & \cdot & \cdot & \cdot & \cdots \end{array} \right.$$

and

$$H = \left\{ \begin{array}{c|cccccc} & i & (i,j) & (j,i) & j & (i,k) & \cdots \\ \hline i & 1 & 0 & 1 & 0 & 0 & \cdots \\ (i,j) & 0 & 0 & 0 & 0 & 0 & \cdots \\ (j,i) & 0 & 0 & 0 & 0 & 0 & \cdots \\ j & 0 & 1 & 0 & 1 & 0 & \cdots \\ (i,k) & 0 & 0 & 0 & 0 & 0 & \cdots \\ \cdot & \cdot & \cdot & \cdot & \cdot & \cdot & \cdots \end{array} \right.$$

Now we want to express the properties of V and H by means purely of relative equations. What these properties amount to is, briefly stated, that the elements of $\mathfrak{C}$ are arranged by them into a square schema. In what follows, this property must now be logically analyzed and translated into relative equations.

First of all, V and H are what Schröder calls functions in the strict sense; that is, V and H associate with each element I of $\mathfrak{C}$ one and only one element (namely, the first element, or the last element, respectively, of I). This property is expressed by

$$(1) \qquad V;\breve{V} \in 1' \in \breve{V};V,$$

$$(2) \qquad H;\breve{H} \in 1' \in \breve{H};H$$

(see *Schröder 1895*, p. 587, (17)).

The set of all first elements is the former 1^1. Thus the system of all first elements is q; that is, we have $q = V$. But, since also $q = H;1$, we have

$$(3) \qquad V;1 = H;1.$$

A further property of V and H is that every element of 1^1 is its own first element, that is,

$$q \cdot 1' \in V \cdot 1',$$

or

$$(4) \qquad V; 1 \cdot 1' \mathrel{-\!\!\!\!\in} V \cdot 1';$$

likewise

$$(5) \qquad H; 1 \cdot 1' \mathrel{-\!\!\!\!\in} H \cdot 1'.$$

On account of $V \mathrel{-\!\!\!\!\in} V; 1$ we may also write $=$ instead of $\mathrel{-\!\!\!\!\in}$ in (4) and (5).

Let us now imagine that we have constructed a square schema of the following kind:

	i	j	k	.	.
i					
j					
k					
.					
.					

where $i, j, k, \ldots$ are all the elements of 1^1, that is, all the first and last elements. Now we write the element of $\mathfrak{E}$ that possesses the first element m and the last element n in the square that is at the intersection of the mth row and the nth column of our schema. (1), (2), and (3) express the fact that every element of $\mathfrak{E}$ has exactly one first element and one last element, that is, that according to our rule it is written in exactly one square of our schema. We must now express the fact that every square contains one and only one element.

That every square contains *at most* one element means the following: if two elements I and J have the same first element and the same last element, if therefore

$$(\breve{V}; V)_{IJ} = 1 \quad \text{and} \quad (\breve{H}; H)_{IJ} = 1,$$

then I and J coincide, that is, $1'_{IJ} = 1$; hence we have

$$(\breve{V}; V)_{IJ}(\breve{H}; H)_{IJ} \mathrel{-\!\!\!\!\in} 1'_{IJ},$$

and, since this holds for arbitrary I and J, we have

$$(6) \qquad \breve{V}; V \cdot \breve{H}; H \mathrel{-\!\!\!\!\in} 1'.$$

Finally, that every square of the schema contains *at least* one element of $\mathfrak{E}$ means the following: if I and J belong to 1^1, if therefore

$$q_{IJ} = 1 \quad \text{and} \quad q_{JI} \equiv \breve{q}_{IJ} = 1,$$

then there must exist an element K having I as its first element and J as its last element; hence we must have[9]

$$1 = \sum_K V_{IK} H_{JK} \equiv \sum_K V_{IK} \breve{H}_{KJ} \equiv (V; \breve{H})_{IJ}.$$

[9] ⟦In the formula that follows, the penultimate subscript is, in the German text, "K", which is a misprint.⟧

Therefore $q_{IJ}\breve{q}_{IJ} \in (V; \breve{H})_{IJ}$ for arbitrary I and J; that is,

$$q\breve{q} \in V; \breve{H},$$

or

$$V; 1 \cdot 1; \breve{H} \in V; \breve{H}. \tag{7}$$

Therefore we must still write (1)–(7) as a premiss before the equation $F' = 0$:

$$(1)\text{–}(7) \in (F' = 0). \tag{8}$$

This, when transformed into a primary equation and brought into zero form, yields the desired equation $F = 0$.

If in fact $f = 0$ is not identically satisfied, then there are a domain 1^1 and a system of values of the relative coefficients of the relatives $a, b, \ldots$ occurring in f for which the equation is not satisfied. If we then construct the domain $\mathfrak{E}$ from 1^1 by the formation of pairs and define V and H as on page 247, the premiss (1)–(7) is satisfied for these V and H and can therefore be omitted, and $F = 0$ can be replaced by $F' = 0$. But to every coefficient occurring in $f = 0$ there corresponds a coefficient occurring in $F' = 0$; and for any system of values that does not satisfy $f = 0$ there is also one that does not satisfy $F' = 0$, hence not $F = 0$ either. Q.e.d.

If, conversely, a domain $\mathfrak{E}$ and a system of values for the relative coefficients of the relatives $A, B, \ldots, H$, and V occurring in $F = 0$ are given such that $F = 0$ is not satisfied, then to this system of values there also belong values of V and H, and, in particular, such values as satisfy (1)–(7); for $F = 0$ is, after all, identically satisfied for other systems of values, according to (8). Therefore, by (8), $F' = 0$ cannot be satisfied by the given system of values. We can now find a domain 1^1 and, associated with it, a system of values of the relative coefficients occurring in $f = 0$ for which $f = 0$ is not satisfied either. For, if we put $V; 1 = q$, then to this system there corresponds a set that we shall take as a domain 1^1 for $f = 0$. If in $F' = 0$ there occurs a coefficient A_{IK}, then by (1)–(7) there exist exactly one element i, one j, one k, and one l, such that

$$V_{iI} = 1, \quad H_{jI} = 1, \quad V_{kK} = 1, \quad \text{and} \quad H_{lK} = 1.$$

When these values i, j, k, and l have been determined, we let a_{ijkl} correspond to A_{IK} and assign to a_{ijkl} the value that is given for A_{IK}. Thus we arrive at a system of values that does not satisfy $f = 0$, since it does not satisfy $F' = 0$. Consequently, $F = 0$ is identically satisfied along with $f = 0$, and conversely.

If now a higher relative equation is given, say a senary one with coefficients a_{ijklmn}, then we can transform it by the same method into a ternary one (whose transformation into a binary one we already know) if we put

$$(i, j) = I, \quad (k, l) = K, \quad \text{and} \quad (m, n) = M,$$

or, with an unessential modification of the method, we can also transform it into a binary one immediately by putting

$$(i, j, k) = I \quad \text{and} \quad (l, m, n) = L.$$

Equations that are quinary, septenary, and so forth can be reduced to equations that are senary, octonary, and so forth.

Logico-combinatorial investigations in the satisfiability or provability of mathematical propositions: A simplified proof of a theorem by L. Löwenheim and generalizations of the theorem

THORALF SKOLEM

(*1920*)

The text below is § 1 of Skolem's paper (the other sections deal with decision procedures for Boolean algebras and elementary geometry, as well as with dense sets). The first result presented in the paper is that every well-formed formula of the first-order predicate calculus has what is now known as a Skolem normal form for satisfiability. Such a normal form consists of a string of universal quantifiers, followed by a string of existential quantifiers and then a quantifier-free matrix. Skolem proves that a well-formed formula is satisfiable in a given domain if and only if its Skolem normal form for satisfiability is satisfiable in the domain. Skolem normal forms have since become one of the logician's standard tools. These forms were used, in particular, by Gödel in his proof of the completeness of quantification theory (*1930a*) and by Bernays in his arithmetization of that proof (*Hilbert and Bernays 1939*, pp. 234–253). Skolem's result can be regarded as establishing that the well-formed formulas in Skolem normal form constitute a reduction class for quantification theory. More generally, formulas in prenex form came to play a major role in the study of the reduction problem and of special cases of the decision problem. The number and order of quantifiers in the prefix of a prenex formula came to yield a convenient standard for measuring the complexity of a formula of quantification theory. All decision and reduction problems of quantification theory, so far as classes of formulas are determined by this standard, were eventually solved, the last gap being filled by the result of Kahr, Moore, and Wang (*1961*).

After having introduced the normal form of a formula, Skolem offers (Theorem 2) a new proof of Löwenheim's theorem. With the normal form at his disposal, he can dispense with the infinite strings of quantifiers considered by Löwenheim (see above, pp. 229–230), and, thanks to considerations suggested by Dedekind's notion of chain, he is able to fill a serious gap left open by Löwenheim in his proof (see above, p. 231). Skolem notes the use of the axiom of choice in his proof. In *1929*, p. 24, Skolem gives a simplified version of his 1920 proof, likewise using the axiom of choice.

In Skolem's formulation of Löwenheim's theorem and its generalizations the phrases "is a contradiction" and "is contradictory" have to be understood in the semantic sense, that is, as "is not satisfiable"; this is apparent from the assumption that Skolem actually uses in the proofs. The same remark applies to Skolem's subsequent papers (*1922*, *1928*, and *1929*), where he also uses "consistent" ("widerspruchsfrei", "widerspruchslos") in the sense of "satisfiable".

Skolem generalizes Löwenheim's theorem to denumerably infinite sets of formulas (actually, he deals with infinite conjunctions of formulas, but this is unessential). The remainder of the part of the paper that is given below presents a number of theorems about expressions of infinite length. The first are about infinite conjunctions and alternations, and the results can readily be translated into terms of infinite sets of formulas; then Theorems 7 and 8 introduce formulas that contain an infinite sequence of individual variables bound by an infinite string of existential quantifiers. Skolem never again worked with expressions of infinite length; rather, soon afterward he came to stress the finite character of formal systems.

Some years later (*1938*, pp. 25–30) Skolem compares various proofs of the Löwenheim-Skolem theorem (for a single formula, but everything we are going to say applies to the case of a set of formulas) and remarks that these proofs yield, in fact, two different versions of the theorem. Let us consider a formula F of quantification theory and, not to burden the discussion with unessential details, assume that F contains no individual constant and no free individual variable. Skolem's two versions of the theorem can be stated as:

(I) If F is satisfiable, it is satisfiable in the domain of natural numbers, some suitable assignment being made to the predicate letters of F;

(II) If F is satisfiable, that is, if there is a domain D in which F is satisfiable, F is satisfiable in a denumerable subdomain D' of D, the same meaning being assigned to the predicate letters of the formula ("si on conserve la signification" of the predicate letters, *1938*, p. 25; see also *1929*, p. 23, last sentence of first paragraph of § 4).

In order to state (II) in a somewhat more precise form, let us first recall a few definitions. An *interpretation* $\langle D, \alpha\rangle$ of F is a nonempty domain D together with an assignment α that with each n-ary predicate letter of F associates an n-ary relation defined in D. By the usual rules, an interpretation assigns the letter t or the letter f to F. If the letter is t, the interpretation is a *model* of F. The cardinality of the model is the cardinality of D. A model $\langle D', \alpha'\rangle$ of F is a *submodel* of a model $\langle D, \alpha\rangle$ of F if and only if D' is a subdomain of D and, for each predicate letter φ of F, $\alpha'(\varphi)$ is the restriction of $\alpha(\varphi)$ to D'.

We can now state (II) as

(III) If F has a model $\langle D, \alpha\rangle$, F has a denumerable model $\langle D', \alpha'\rangle$ such that $\langle D', \alpha'\rangle$ is a submodel of $\langle D, \alpha\rangle$.

Skolem's 1920 proof of the Löwenheim-Skolem theorem, which makes uses of the axiom of choice, establishes (II), or (III), while his 1922 proof (see below, pp. 293–294), which does not make use of the axiom of choice, establishes the weaker statement (I). Löwenheim's proof contains several gaps, which can be filled in various ways; but it seems to be closer to Skolem's 1922 proof than to that of 1920.

In Skolem's 1920 proof the denumerable submodel is obtained from the original model by a process of thinning out, a process performed with the help of the axiom of choice. In the 1922 proof the denumerable model is built up by a process of successive "approximations", and it may be unrelated to the original model. In particular, if version (I) of the Löwenheim-Skolem theorem is used to show that a certain axiomatic set theory T has a denumerable model $\langle D, \alpha\rangle$ and if the elements of D are individuals of a standard model of T, then $\alpha(\varepsilon)$, that is, the relation that in D corresponds to the ε that enters in the axioms of T, is not necessarily the standard ε. If version (II)–(III) of the theorem is used, $\alpha(\varepsilon)$ is the standard ε. (Compare the notion "absolute" in *Gödel 1940*, p. 42.) In Maltsev's (*1936*, § 6, pp. 332–335) and Tarski and Vaught's (*1956*, p. 92, Theorem 2.1) extensions of the Löwenheim-Skolem theorem the distinction is made between the two versions. In Tarski's extension of the Löwenheim-Skolem theorem mentioned in the last paragraph of *Bemerkung der Redaktion* in *Skolem 1934*, p. 161, the distinction is not made.

The translation is by Stefan Bauer-Mengelberg, and it is printed here with the kind permission of Professor Skolem and the Norske Videnskaps-Akademi i Oslo.

In volume 76 of *Mathematische Annalen* Löwenheim proved an interesting and very remarkable theorem on what are called "first-order expressions". The theorem states that every first-order expression is either contradictory or already satisfiable in a denumerably infinite domain. By a first-order expression Löwenheim understands an expression constructed from relative coefficients by means of the five fundamental logical operations, namely, in Schröder's terminology, identical multiplication and addition, negation, productation, and summation, with productations and summations ranging over individuals only. The five operations mentioned are denoted by a dot (or simply juxtaposition), the sign $+$, the bar $^{-}$, and the signs Π and Σ, respectively. Löwenheim proves his theorem by means of Schröder's "development" of products and sums, a procedure that takes a Π sign across and to the left of a Σ sign, or vice versa. But this procedure is somewhat involved and makes it necessary to introduce for individuals symbols that are subsubscripts on the relative coefficients. In what follows I want to give a simpler proof, in which such subsubscripts are avoided, and, besides, prove some lemmas that are of interest in themselves; finally I shall also establish some generalizations of Löwenheim's theorem.

Instead of speaking of first-order expressions I would rather speak of first-order propositions.

Definition 1. *A first-order proposition is a proposition constructed from relative coefficients in the sense of Schröder by means of the five operations mentioned above, with productations and summations ranging over individuals only.*

Examples. (1) $\Pi_x\Sigma_y R_{xy}$. In words this reads: For every x there exists a y such that the relation R obtains between x and y.

(2) $\Sigma_x\Pi_y\Sigma_z(R_{xy} + T_{xyz})$. In words this reads: There exists an x such that for every y it is possible to determine a z such that either the binary relation R holds between x and y or the ternary relation T holds between x, y, and z.

(3) $\Sigma_x A_x . \Sigma_y\Pi_z R_{yz}S_{zx}$. In words this reads: There exists an x that belongs to the class A, and moreover there exists a y such that for every z the relation R obtains between y and z, and the relation S between z and x.

Definition 2. *A first-order proposition will be said to be in normal form if it is written in such a way that it begins with Π signs, these being followed by Σ signs and then by an expression that is free of Π and Σ signs. Every first-order proposition that contains only Π or only Σ signs will also be said to be in normal form, provided these signs stand at the beginning and follow each other.*

Examples. $\Pi_x\Sigma_y R_{xy}$ is in normal form, likewise $\Pi_x\Sigma_y\Sigma_z(R_{xy} + S_{yz})$ and $\Pi_x\Pi_y\Sigma_z$

$(R_{xy}S_{xz} + \bar{A}_x)$. On the other hand, $\Sigma_x \Pi_y R_{xy}$ is not in normal form. But $\Pi_x(\bar{A}_x + B_x)$ and $\Sigma_x A_x$ again are in normal form.

THEOREM 1. *If U is an arbitrary first-order proposition, there exists a first-order proposition U' in normal form with the property that U is satisfiable in a given domain whenever U' is, and conversely.*

Proof. We can consider first a proposition of the form

$$\Pi_{x_1}\Pi_{x_2}\cdots\Pi_{x_m}\Sigma_{y_1}\Sigma_{y_2}\cdots\Sigma_{y_n}\Pi_{z_1}\cdots\Pi_{z_p}U_{x_1x_2\cdots x_my_1y_2\cdots y_nz_1\cdots z_p}, \tag{1}$$

where $U_{x_1\cdots z_p}$ is a proposition constructed from relative coefficients by exclusive use of the three operations of identical multiplication, identical addition, and negation. We can introduce a relative coefficient $R_{x_1\cdots x_my_1\cdots y_n}$ here, setting

$$R_{x_1\cdots x_my_1\cdots y_n} = \Pi_{z_1}\cdots\Pi_{z_p}U_{x_1\cdots x_my_1\cdots y_nz_1\cdots z_p}$$

for all values of $x_1,\ldots, x_m, y_1,\ldots, y_n$; hence, if we wish, we can write

$$\Pi_{x_1}\cdots\Pi_{x_m}\Pi_{y_1}\cdots\Pi_{y_n}(R_{x_1\cdots x_my_1\cdots y_n} = \Pi_{z_1}\cdots\Pi_{z_p}U_{x_1\cdots x_my_1\cdots y_nz_1\cdots z_p}). \tag{2}$$

It is clear that, no matter how the relatives whose coefficients occur in U might be chosen, there exists in the domain (or, more accurately, in the domain 1^{m+n}, as it would be written in Schröder's notation) an $(m + n)$-ary relative R such that (2) is satisfied. In fact, (2) is nothing other than an ordinary definition of a relation R. Now proposition (2) can be transformed into

$$\Pi_{x_1}\cdots\Pi_{x_m}\Pi_{y_1}\cdots\Pi_{y_n}(\bar{R}_{x_1\cdots y_n} + \Pi_{z_1}\cdots\Pi_{z_p}U_{x_1\cdots z_p}) \\ (\Sigma_{u_1}\cdots\Sigma_{u_p}\bar{U}_{x_1\cdots y_nu_1\cdots u_p} + R_{x_1\cdots y_n}),$$

and this in turn into

$$\Pi_{x_1}\cdots\Pi_{x_m}\Pi_{y_1}\cdots\Pi_{y_n}\Pi_{z_1}\cdots\Pi_{z_p}\Sigma_{u_1}\cdots\Sigma_{u_p} \\ (\bar{R}_{x_1\cdots y_n} + U_{x_1\cdots y_nz_1\cdots z_p})(R_{x_1\cdots y_n} + \bar{U}_{x_1\cdots y_nu_1\cdots u_p}), \tag{3}$$

and (3) is in normal form. But in consequence of (2) we can also write (1) as

$$\Pi_{x_1}\cdots\Pi_{x_m}\Sigma_{y_1}\cdots\Sigma_{y_n}R_{x_1\cdots x_my_1\cdots y_n}, \tag{4}$$

which is also in normal form. The logical conjunction (propositional product) of (3) and (4), finally, can be written thus:

$$\Pi_{\xi_1}\cdots\Pi_{\xi_m}\Pi_{x_1}\cdots\Pi_{x_m}\Pi_{y_1}\cdots\Pi_{y_n}\Pi_{z_1}\cdots\Pi_{z_p}\Sigma_{\eta_1}\cdots\Sigma_{\eta_n}\Sigma_{u_1}\cdots\Sigma_{u_p} \\ R_{\xi_1\cdots\xi_m\eta_1\cdots\eta_n}(\bar{R}_{x_1\cdots y_n} + U_{x_1\cdots y_nz_1\cdots z_p})(R_{x_1\cdots y_n} + \bar{U}_{x_1\cdots y_nu_1\cdots u_p}). \tag{5}$$

This proposition, too, is in normal form.

Now, if there exist in the given domain values of the relative symbols occurring in (1) such that (1) is satisfied, it will also be possible to specify in the domain values for the relative symbols occurring in (5) such that (5) will be satisfied. In fact, apart from R, the relative symbols occurring in (5) are the same as those in (1), and R can always be found by virtue of the "definition" (2) if the relatives occurring in (1) are known. It is immediately clear that, conversely, (1) is satisfiable if (5) is, since (5) is the logical conjunction of (2) and (4), from which (1) follows.

This proves that Theorem 1 is true in the case of a formula in which we encounter only two changes from a Π to a Σ sign as we read from left to right. But it is now easy to see how the general proof is to be carried out.

For let the proposition

$$(6)\quad \Pi_{x_1 \cdots x_{n_1}} \Sigma_{x_{n_1+1} \cdots x_{n_1+n_2}} \Pi_{x_{n_1+n_2+1} \cdots x_{n_1+n_2+n_3}} \cdots \Sigma_{x_{n_1+\cdots+n_{\gamma-1}+1} \cdots x_{n_1+\cdots+n_\gamma}} U_{x_1 \cdots x_{n_1+n_2+\cdots+n_\gamma}}$$

be given. (Thus I assume here that the rightmost quantifiers are Σ signs, whereas in the case just considered we had Π signs as rightmost quantifiers; but the procedure is always the same.) I then define a number of auxiliary relations by means of the following equations:

$$(7)\quad \begin{cases} \Sigma_{x_{n_1+\cdots+n_{\gamma-1}+1}} \Sigma_{x_{n_1+\cdots+n_{\gamma-1}+2}} \cdots \Sigma_{x_{n_1+n_2+\cdots+n_\gamma}} U_{x_1 \cdots x_{n_1+\cdots+n_\gamma}} \\ \qquad = R^{(1)}_{x_1 \cdots x_{n_1+\cdots+n_{\gamma-1}}}, \\ \Pi_{x_{n_1+\cdots+n_{\gamma-2}+1}} \Pi_{x_{n_1+\cdots+n_{\gamma-2}+2}} \cdots \Pi_{x_{n_1+\cdots+n_{\gamma-1}}} R^{(1)}_{x_1 \cdots x_{n_1+\cdots+n_{\gamma-1}}} \\ \qquad = R^{(2)}_{x_1 \cdots x_{n_1+\cdots+n_{\gamma-2}}}, \\ \cdots\cdots\cdots\cdots\cdots\cdots \\ \Pi_{x_{n_1+n_2+1}} \Pi_{x_{n_1+n_2+2}} \cdots \Pi_{x_{n_1+n_2+n_3}} R^{(\gamma-3)}_{x_{n_1} \cdots x_{n_1+n_2+n_3}} = R^{(\gamma-2)}_{x_{n_1} \cdots x_{n_1+n_2}}. \end{cases}$$

Strictly speaking, every equation here should be prefixed by Π signs, namely, those corresponding to all variables x that occur free. We can then easily convince ourselves that each of these propositions, hence the definitions of $R^{(1)}$, $R^{(2)}, \ldots$, can be brought into normal form in the same way as we brought (2) into the normal form (3) above. But then, of course, the logical conjunction of these definitions of $R^{(1)}$, $R^{(2)}, \ldots$, can also be written in normal form. By virtue of (7), however, proposition (6) takes the form

$$(8)\quad \Pi_{x_1 \cdots x_{n_1}} \Sigma_{x_{n_1+1} \cdots x_{n_1+n_2}} R^{(\gamma-2)}_{x_1 \cdots x_{n_1+n_2}},$$

which is also normal. Hence the logical conjunction of (7) and (8) can also be written as a proposition U' in normal form. If it is possible now to find, for the relative symbols occurring in (6), values such that (6) is satisfied, we can by (7) successively find $R^{(1)}$, $R^{(2)}, \ldots$, and thus U' will be satisfied. If, conversely, U' is satisfied for certain values of the relative symbols occurring in U', then (7) and (8) are satisfied and therefore also (6). This proves in full generality that Theorem 1 is true.

The question when a first-order proposition is satisfiable can therefore be replaced by the simpler one: when is a proposition in normal form satisfiable? On that point the following theorem holds:

THEOREM 2. *Every proposition in normal form either is a contradiction or is already satisfiable in a finite or denumerably infinite domain.*

It would now be extremely easy to prove this theorem in Löwenheim's way, but *without the use of symbols for individuals as subsubscripts.* However, I would rather carry out the proof in another way, one in which we do not reason with successive choices but proceed more directly according to the customary methods of mathematical logic.[1]

[1] For the more general theorems that appear later in this section I shall carry out the proofs in a way closer to Löwenheim's.

First we can consider the simplest possible case of a first-order proposition in normal form containing both Π and Σ signs, namely,

$$\Pi_x \Sigma_y U_{xy},$$

where U_{xy} is a proposition constructed by means of conjunction, disjunction, and negation from relative coefficients having only x and y as subscripts. Let us now assume that this proposition is satisfied in a given domain for certain values of the relatives. Then, *by virtue of the principle of choice*, we can imagine that for every x a uniquely determined y is chosen in such a way that U_{xy} comes out true. For this is a choice of one element from each of the classes that we obtain if for every x we gather the y for which U_{xy} is true. This defines a single-valued mapping of the domain into itself. Assume that, for every x, x' is the image of x. Then, for the values assigned to the relative symbols, the proposition $U_{xx'}$ is true for every x. We can write that as follows:

$$\overset{O}{\Pi}_x U_{xx'},$$

where O is the domain. Let a be a particular individual of O. Then there exist certain classes X included in O that, first, contain a as element (X_a is true) and, second, contain x' whenever they contain x ($X_{x'}$ is true whenever X_x is true, or, in other words, $\overline{X}_x + X_{x'}$ is true for every x). Now let X^0 be the logical (identical) product (intersection) of all these classes. Then, as is well known, X^0 is either a finite or a denumerably infinite class (see Dedekind's theory of chains in *1888*). But it is clear, further, that

$$\overset{X^0}{\Pi}_x U_{xx'}$$

must hold. Hence this proves that, if $\Pi_x \Sigma_y U_{xy}$ is satisfied in a domain O, it is also satisfiable in a finite or denumerably infinite domain.

In order to prove Theorem 2 in full generality it is most convenient to prove first two simple lemmas.

Lemma 1. Let $R_{x_1 \cdots x_m y_1 \cdots y_n}$ be an $(m + n)$-ary relation with the property that, if $x_1, \ldots, x_m$ are arbitrarily given (in the given domain O), there is one and only one y_1, one and only one y_2, and so forth up to one and only one y_n such that $R_{x_1 \cdots x_m y_1 \cdots y_n}$ holds. Let K be an arbitrary finite class, and let K_1 be the class of all values of $y_1, \ldots, y_n$ that correspond to the various possible choices of $x_1, \ldots, x_m$ in K. Further, let $K' = K + K_1$ be the sum of the two classes K and K_1. Then K', too, is a finite class.

It is indeed quite clear that this proposition is true. If K consists of k objects, there exist altogether k^m possible choices of $x_1, \ldots, x_m$, hence also k^m corresponding sequences $y_1, \ldots, y_n$. Thus the number of elements of K_1 is at most $k^m n$, and K' contains at most $k^m n + k$ objects.

Lemma 2. Let R be an $(m + n)$-ary relative with the property mentioned in Lemma 1, and let Ξ be the logical product of all classes X that possess the following two properties:

(1) a is an element of X;

(2) If $x_1, \ldots, x_m$ are arbitrarily chosen in X, then X also contains the objects $y_1, \ldots, y_n$ for which $R_{x_1 \cdots x_m y_1 \cdots y_n}$ holds.

Then Ξ is either a finite or a denumerably infinite class.

Proof. We can denote by K_1 the class of all objects y occurring in the sequences $y_1, \ldots, y_n$ that correspond to the various possible choices of $x_1, \ldots, x_m$ in K, and in general we denote by K' the class $K + K_1$. The passage from K to K' therefore yields a single-valued mapping of the totality of all classes into itself. I shall denote by $\{a\}$ the class that contains a as its only element. I then first consider those classes of classes having the two properties that (1) they contain $\{a\}$ as element and (2) whenever they contain a class K, they contain K' as element. Let A be the intersection of these classes of classes. Then A is an ordinary Dedekind chain and, as is well known, consists of the classes $\{a\}$, $\{a\}'$, $\{a\}''$, and so forth ad infinitum. The classes that are elements of A need not, however, all be mutually distinct; in any case A is a finite or a denumerably infinite class of classes.

In consequence of Lemma 1 it is clear, moreover, that every element of A must be a finite class. For, first, $\{a\}$ is finite, and, second, K' is finite whenever K is. The class of all finite classes must therefore, according to the definition of A, contain all of A; that is, every element of A is a finite class.

By well-known set-theoretic theorems it then follows that the sum of all classes that are elements of A—call it SA—must be a finite or a denumerably infinite class.

Finally, it can be shown that $\Xi = SA$. For, first, $\{a\}$ is included in Ξ as a subclass, and it is clear that, if K is a subclass of Ξ, K' too is a subclass of Ξ. From this it follows that every element of A is a subclass of Ξ, and thence it follows that SA is included in Ξ as a subclass. But, conversely, Ξ must also be a subclass of SA. For a is an element of SA, and, if $x_1, \ldots, x_m$ are arbitrarily chosen in SA, then—since $\{a\}$ is a subclass of $\{a\}'$, $\{a\}'$ in turn a subclass of $\{a\}''$, and so forth—there exists an element K of A that contains all of $x_1, \ldots, x_m$; then every y of the corresponding sequence $y_1, \ldots, y_n$ belongs to K', which is the successor element of K in A, and consequently all of these y belong to SA. According to the definition of Ξ, Ξ must then be a subclass of SA. Hence $SA = \Xi$, from which it follows that Ξ is either finite or denumerably infinite.

It is now easy to carry out the proof of Theorem 2. Let a first-order proposition in normal form,

$$\Pi_{x_1} \cdots \Pi_{x_m} \Sigma_{y_1} \cdots \Sigma_{y_n} U_{x_1 \cdots x_m y_1 \cdots y_n},$$

be given, and let us assume that it is satisfied in a given domain for certain values of the relative symbols occurring in U. By the principle of choice we can then, for every choice of $x_1, \ldots, x_m$, imagine a definite sequence chosen from among the corresponding sequences $y_1, \ldots, y_n$, that is, the sequences $y_1, \ldots, y_n$ for which $U_{x_1 \cdots x_m y_1 \cdots y_n}$ is true. We can appropriately denote by $y_1(x_1, \ldots, x_m)$, $y_2(x_1, \ldots, x_m), \ldots, y_n(x_1, \ldots, x_m)$, or, more briefly, by $y_1(x), y_2(x), \ldots, y_n(x)$ the sequence of the y that correspond to the sequence $x_1, \ldots, x_m$. Then the proposition

$$U_{x_1 \cdots x_m y_1(x) y_2(x) \cdots y_n(x)}$$

holds for every choice of $x_1, \ldots, x_m$. Moreover, $U_{x_1 \cdots x_m y_1(x) \cdots y_n(x)}$ is a relation with the property considered in the lemmas. Hence, if we assume that a is a particular individual and that Ξ is the intersection of all classes X that contain a as element and contain the y of the sequence $y_1(x_1, \ldots, x_m)$, $y_2(x_1, \ldots, x_m), \ldots, y_n(x_1, \ldots, x_m)$ whenever they contain $x_1, \ldots, x_m$, then Ξ is either finite or denumerably infinite, and, moreover,

$$U_{x_1 \cdots x_m y_1(x) \cdots y_n(x)}$$

of course holds for all possible choices of $x_1, \ldots, x_m$ in Ξ, with $y_1(x), \ldots, y_n(x)$ belonging to Ξ whenever $x_1, \ldots, x_m$ do.

Theorem 2 admits generalizations of high order. Thus, it is not difficult to prove the following: Either it is contradictory to suppose that a simply infinite sequence of first-order propositions in normal form is simultaneously satisfiable or the sequence is already simultaneously satisfiable in a denumerably infinite domain. By means of the reduction, presented above, of arbitrary first-order propositions to normal form we recognize that this must also hold for every product of a simply infinite sequence of arbitrary first-order propositions. It will be proved below that such a theorem will continue to hold even if an infinite number of logical disjunctions occur. To make the presentation simpler, I shall here confine myself to giving a proof for the case in which we have a logical product of a simply infinite sequence of propositions of the form $\Pi_x \Sigma_y U_{xy}$. The extension to the general case of an infinite sequence of arbitrary propositional factors in normal form presents no difficulty in principle.

Thus, let the propositional product

$$(9) \qquad \Pi_{x_1}\Sigma_{y_1} U^1_{x_1 y_1} \Pi_{x_2}\Sigma_{y_2} U^2_{x_2 y_2} \cdots \text{ad infinitum}$$

be given, where the U are propositions constructed from relative coefficients by means of a finite number of applications of conjunction, disjunction, and negation only. Of course, we need not consider in greater detail the case in which no two arbitrary propositional factors of the product (9) have any common relative symbol at all; for, if the entire propositional product is then not a contradiction, it is immediately clear that none of the propositional factors is a contradiction, and then each of these by itself can be satisfied in a denumerably infinite domain. But, since the propositional factors are in this case completely independent of one another, that will also make the entire product a true proposition. Incidentally, if no two of the individual propositional factors ever have relative symbols in common, it is clear that it is not necessary to assume that the set of factors is denumerable. Hence we shall from now on consider only the case in which the factors have relative symbols in common.

Let us now assume that proposition (9) is satisfied in a domain O. Then, a fortiori, the proposition

$$\Pi_{x_r}\Sigma_{y_r} U^r_{x_r y_r}$$

is satisfied for every r. By virtue of the principle of choice we can then choose, for every x_r, a uniquely determined y_r from among those for which $U^r_{x_r y_r}$ is true; let us denote it by $y_r(x_r)$. Furthermore we choose an arbitrary individual in O and call it 1. Then, further, $y_1(1)$ will be called 2, $y_1(2)$ will be called 3, $y_2(1)$ will be called 4, and so forth. The law governing these successive assignments of names will be clear from the following table:

$$y_1(1) = 2$$

$$y_1(2) = 3 \qquad y_2(1) = 4$$

$$y_1(3) = 5 \qquad y_1(4) = 6 \qquad y_2(2) = 7 \qquad y_3(1) = 8$$

$$y_1(5) = 9 \qquad y_1(6) = 10 \qquad y_1(7) = 11 \qquad y_1(8) = 12 \qquad y_2(3) = 13 \qquad y_2(4) = 14$$

$$y_3(2) = 15 \qquad y_4(1) = 16$$

. . . .

Now, according to the assumption, all the propositions $U^r_{x_r y_r(x)_r}$ are true, and consequently the propositional product

$$U^1_{1,2}U^1_{2,3}U^1_{3,5}\cdots U^2_{1,4}U^2_{2,7}\cdots U^3_{1,8}U^3_{2,15}\cdots U^4_{1,16}\cdots$$

must also be true. But from this we see that the proposition

$$\Pi_{x_1}\Sigma_{y_1}U^1_{x_1y_1}\Pi_{x_2}\Sigma_{y_2}U^2_{x_2y_2}\cdots$$

is satisfied, even if the productations and summations range only over the values 1, 2, 3, ... of the subscripts; that is, proposition (9) is also satisfiable in a denumerable domain.

Hence the following theorem holds:

THEOREM 3. *If a proposition can be represented as a product of a denumerable set of first-order propositions, it either is a contradiction or is already satisfiable in a denumerable domain.*

Remark. One easily obtains propositions of that form when dealing with the formation of chains. If, for example, $\overset{*}{R}$ denotes the chain corresponding to the binary relation R (the "ancestral relation" of Russell and Whitehead (*1910*, p. 569)), then we have

$$\overset{*}{R}_{xy} = 1'_{xy} + R_{xy} + \Sigma_z R_{xz}R_{zy} + \Sigma_u\Sigma_v R_{xu}R_{uv}R_{vy} + \cdots,$$

where $1'$ is Schröder's notation for the relation "identical with". The negation of $\overset{*}{R}_{xy}$ will then be the product of an infinite sequence of first-order propositions.

In order to be able to answer the question what will happen when an infinite number of disjunctions also occur in the given proposition, it is necessary to take into consideration above all the following self-evident fact:

A propositional sum is satisfied in a certain domain for certain values of the relative symbols that occur in it if and only if at least one of the propositional summands is satisfied for precisely these values of the symbols.

The following theorem can then be proved:

THEOREM 4. *A (denumerable or nondenumerable) sum of infinitely many first-order propositions either is not satisfiable at all or is already satisfied in a denumerably infinite domain for certain values of the relative symbols that occur in the propositions.*

Proof. If the given propositional sum U is satisfied in a domain for certain values of the relative symbols that occur in the propositions, then (according to the remark that we have just made) at least one of the summands, say A, must be satisfied for precisely these values of the symbols. But, since all summands are first-order propositions, A must, according to Theorem 2, already be satisfied in a denumerable domain for certain values of the symbols occurring in A. In this way, now, the entire propositional sum U will also be satisfied in the denumerable domain if in addition arbitrary values are chosen in this very domain for the relative symbols, if any, that occur in U but not in A.

It is now easy to see also that *in general all propositions constructed from first-order propositions by means of a finite or denumerably infinite number of applications of conjunction and disjunction can already be satisfied in the denumerable infinite if they can be satisfied at all.*

In the first place it is clear that *an infinite sum* $U_1 + U_2 + \cdots$, *where every* U_r *is a product of infinitely many first-order propositions* U_r^s $(s = 1, 2, \ldots)$, *is either never satisfied or already satisfied in the denumerable infinite.* The proof of this can, in fact, be carried out in exactly the same way as the proof of Theorem 4; it makes use of Theorem 3.

We further have the following theorem:

THEOREM 5. *Assume that the proposition* U *is a product of infinitely many propositions* U_r, *each of which in turn is a sum of infinitely many first-order propositions, namely,*

$$U_r = \overset{\infty}{\underset{1}{\Sigma_s}} U_r^s,$$

where each U_r^s *is a first-order proposition. Then* U *is either never satisfied or already satisfied in a denumerable domain for a suitable choice of values for the relative symbols.*

Proof. The propositional multiplication can be carried out so that U becomes a sum of products of first-order propositions, though, to be sure, the set of these products is not denumerable. If U is satisfied for certain values of the relative symbols that occur in it, at least one summand must be satisfied for these values of the symbols. But such a summand is the product of a simply infinite sequence of first-order propositions and, by Theorem 3, is therefore already satisfied in a denumerable domain. Then U itself is also satisfied in the denumerable domain if in addition arbitrary values are chosen in this denumerable domain for the relative symbols, if any, that occur in U but not in the summand in question.

The same method of proof can, of course, also be employed if the structure of the given proposition is still more complex. We could, for example, prove a very similar theorem about all the propositions U that are a product of infinitely many propositions U_r, each of which is a sum of propositions U_r^s, where each U_r^s is a product of propositions $U_{r,r'}^s$, each of which in turn is a sum of first-order propositions $U_{r,r'}^{s,s'}$, and so forth.

From these theorems there follows, among others, the following interesting result:

THEOREM 6. *Every proposition constructed from relative coefficients by means of a finite number of applications of conjunction, disjunction, negation, and productation and summation over symbols for individuals, as well as by chain formation, is either never satisfied or already satisfied in a denumerable domain for certain values of the relative symbols.*

It is indeed quite clear that the theorem is true, since every proposition generated by chain formation can be written in the form of a simply infinite sum of first-order propositions. See the remark on page 260.

Examples. A relative equation like the one below (only binary relatives are to occur, and the notation is Schröder's except for the symbol $\overset{*}{R}$ for the chain of R and the sign † for relative addition (see the remark on page 260)):

$$(x\breve{x} + \overline{x}\breve{\overline{x}}) \dagger \overset{*}{x} ; \overline{x} \dagger \overset{*}{\overline{x}} ; x = 0$$

is solvable either in no domain or already in a denumerably infinite domain.

Likewise, the relative equation

$$(\overset{*}{x} + \overset{*}{\bar{y}})(\overset{*}{\breve{x}} + \bar{\overset{*}{y}});(\bar{x} \dagger y) \dagger x;y = 0$$

is also either not solvable at all or already solvable in a denumerable domain.

Precisely the same, of course, holds for the following equation, too:

$$(x, y) + (\overset{*}{x}, y);x\bar{y} = 0,$$

where (u, v) is to mean the relative $1' + u;v + u;u;v;v + u;u;u;v;v;v + \cdots$ ad infinitum for arbitrary values of the binary relatives u and v.

And so forth.

But I also want to present other generalizations of Theorem 2 as well. The following theorem, too, holds:

THEOREM 7. *A proposition of the form*

$$\Pi_{x_1}\Pi_{x_2}\cdots\Pi_{x_m}\Sigma_{y_1}\Sigma_{y_2}\cdots \text{ad inf. } U_{x_1\cdots x_m y_1 y_2 \cdots \text{ad inf.}}$$

either is a contradiction or is satisfied in a denumerable domain for certain values of the relative symbols occurring in U.

Proof. We assume that a proposition of this form is satisfied in a domain O for certain values of the relative symbols. We then imagine that for every sequence of values $x_1, \ldots, x_m$ we have chosen, with the help of the principle of choice, a definite sequence from among the sequences of values $y_1, y_2, \ldots$ for which $U_{x_1 \cdots x_m y_1 y_2 \ldots}$ is true. The y that occur in this sequence I denote by $y_r(x_1, \ldots, x_m)$, or, more briefly, by $y_r(R)$, where R denotes the sequence of the x. Let a be a particular individual in O. The objects $y_r(a, a, \ldots, a)$ then form an at most denumerably infinite set M_1'. (Indeed, it may be a finite set since the y need not all be distinct.) Let M_1 be the set containing the elements of M_1' and, besides, also a; hence $M_1 = M_1' + \{a\}$. Then M_1 is at most denumerably infinite. The sequences of values $x_1, \ldots, x_m$ that can be chosen within M_1 and are distinct from the sequence $a, a, \ldots, a$ already considered again form an at most denumerably infinite set. The sequences of the y corresponding to these sequences R of values therefore also form an at most denumerably infinite set, and the same holds for the objects $y_r(R)$ that occur in these sequences. Let M_2' be the set of these y and let $M_2 = M_1 + M_2'$. The set of all sequences of values $x_1, \ldots, x_m$ that can be chosen in M_2 but not yet in M_1 is again at most denumerable. Consequently the set of the corresponding sequences $y_r(x_1, \ldots, x_m)$, just like the set of objects $y_r(x_1, \ldots, x_m)$ occurring in them, is itself at most denumerable. Let M_3' be the set of these y and let $M_3 = M_2 + M_3'$. We continue in this way indefinitely. We obtain an infinite sequence $M_1, M_2, M_3, \ldots$ of ever more comprehensive sets. The limit set $M_\omega = M_1 + M_2 + \cdots$ is then also at most denumerably infinite. If $x_1, \ldots, x_m$ are arbitrarily chosen within M_ω, then $x_1, \ldots, x_m$ must already occur in the sets $M_{\gamma_1}, \ldots, M_{\gamma_m}$, respectively, where $\gamma_1, \ldots, \gamma_m$ are finite numbers, and among these numbers there exists a largest, γ. Then $x_1, \ldots, x_m$ all occur in M_γ, and consequently every $y_r(x_1, \ldots, x_m)$ $(r = 1, 2, \ldots)$ occurs in M_γ', hence in $M_{\gamma+1}$, and therefore also in M_ω. Moreover, the proposition

$$U_{x_1 \cdots x_m y_1(x_1, \ldots, x_m) y_2(x_1, \ldots, x_m) \ldots}$$

holds for every choice of $x_1, \ldots, x_m$ in M_ω. This proves the theorem.

In a similar way we can also prove the following still more general theorem:

THEOREM 8. *If a proposition U is the product of a denumerable set of propositions of the form mentioned in Theorem 7, then U is already satisfied in a denumerably infinite domain for certain values of the relative symbols occurring in U, provided U can be satisfied at all.*

Proof. If the proposition $U = \overset{\infty}{\underset{1}{\Pi}}{}_r U_r$, where, for every r, U_r is a proposition of the form

$$\Pi_{x_1^r}\Pi_{x_2^r}\cdots\Pi_{x_{m_r}^r}\Sigma_{y_1^r}\Sigma_{y_2^r}\cdots U^r_{x_1^r\cdots x_{m_r}^r y_1^r y_2^r\cdots},$$

is satisfied for certain values of the relatives that occur in it, then we imagine, for every sequence $x_1^r, \ldots, x_{m_r}^r$, a definite sequence chosen from among the corresponding sequences $y_1^r, y_2^r, \ldots$ for which U^r is true. For every choice of the number r and of the objects $x_1^r, \ldots, x_{m_r}^r$, then, the $y_1^r, y_2^r, \ldots$ are uniquely determined. Let a be a particular individual of the domain. Then the objects y corresponding to the choice $x_1^1 = a$, $x_2^1 = a, \ldots, x_{m_1}^1 = a$ form an at most denumerable set M_1'. Let $M_1 = M_1' + \{a\}$. The various new sequences of values $x_1^1, \ldots, x_{m_1}^1$, together with all possible sequences of values $x_1^2, \ldots, x_{m_2}^2$ that are taken within M_1, again form an at most denumerable set, and the objects y occurring in the sequences of the y that correspond to these sequences therefore also form an at most denumerably infinite set M_2'. Let $M_2 = M_1 + M_2'$. Then the sequences of values $x_1^1, \ldots, x_{m_1}^1, x_1^2, \ldots, x_{m_2}^2, x_1^3, \ldots, x_{m_3}^3$ that can be chosen in M_2 but were not already chosen in M_1 form an at most denumerable set, and so do the objects y that occur in the various corresponding sequences. Let M_3' be this latter set and let $M_3 = M_2 + M_3'$. We continue in this way indefinitely and ultimately obtain a limit set M_ω, the sum of the increasing sets $M_1, M_2, \ldots$, which is also at most denumerably infinite, whereas, as is easily seen, each of the propositions U_r, and consequently also U, is satisfied in M_ω.

Introduction to a general theory of elementary propositions

EMIL LEON POST

(*1921*)

The present paper is Post's doctoral dissertation at Columbia University, completed in 1920 and published the following year. The propositional calculus, carved out of the system of *Principia mathematica*, is systematically studied in itself, as a well-defined fragment of logic; moreover, a sharp line is drawn between the formal system and considerations about the system. Both the truth-table and the axiomatic approaches are clearly presented. The calculus is proved to be complete, in the sense that the set of provable well-formed formulas coincides with the set of truth-functionally valid formulas. The paper also establishes another kind of completeness, sometimes called completeness in the sense of Post; a system is complete in that sense if every well-formed formula becomes provable once we adjoin to the axioms any well-formed formula that is not provable. What Post himself calls a "complete" system is one in which every truth function can be written in terms of the primitive truth functions, and he shows that the calculus under study, in which the connectives are $\sim$ and $\vee$, is complete in that sense. A consistency proof of the calculus is given. A new definition of consistency, sometimes called consistency in the sense of Post, is presented; a calculus that contains propositional variables is consistent in that sense if no well-formed formula consisting of a single propositional variable is provable. Consistency in that sense, too, is established.

Post then generalizes the propositional fragment of *Principia* in two directions: by considering, first, an arbitrary number of (not necessarily binary) primitive truth functions, and, second, an arbitrary finite number of axioms and rules of inference. (On problems connected with the first generalization see *Post 1920a* and *1941*, and on problems connected with the second see *Post 1941a*, pp. 346–402.) This part of Post's study of the propositional calculus forms the point of departure of his investigations of formal systems (*1941a, 1943*).

The last part of the paper contains a generalization of another character: m-valued truth tables are introduced. At the same time Łukasiewicz (*1920*) was studying three-valued logic, but while he had a philosophical interest in the new logics, Post's was almost exclusively mathematical.

Two points rather naturally connected with Post's investigations are not discussed in the present paper. One is the question of the independence of the axioms that Post took over from *Principia* for the propositional calculus. The problem had been solved by Bernays in 1918 (though Bernays's paper was not published till 1926), and it turns out that the second formula on the first line of

IV, on page 267, can be proved when the other four formulas of IV are taken as axioms. The second point left untouched by Post here is the problem of finding the minimal bases of the propositional calculus, a problem subsequently investigated by Żyliński (*1925*), Post himself (*1941*), and Wernick (*1942*).

The paper is reprinted here with the kind permission of the Johns Hopkins Press.

Introduction

In the general theory of logic built up by Whitehead and Russell (*1910*, *1912*, *1913*) to furnish a basis for all mathematics there is a certain subtheory[1] which is unique in its simplicity and precision, and, though all other portions of the work have their roots in this subtheory, it itself is completely independent of them. Whereas the complete theory requires for the enunciation of its propositions real and apparent variables, which represent both individuals and propositional functions of different kinds, and as a result necessitates the introduction of the cumbersome theory of types, this subtheory uses only real variables, and these real variables represent but one kind of entity, which the authors have chosen to call elementary propositions. The most general statements are formed by merely combining these variables by means of the two primitive propositional functions of propositions, negation and disjunction; and the entire theory is concerned with the process of asserting those combinations which it regards as true propositions, employing for this purpose a few general rules, which tell how to assert new combinations from old, and a certain number of primitive assertions from which to begin.

This theory in a somewhat different form has long been the subject matter of symbolic logic.[2] However, although it had reached a high state of development as a theory of classes, it had this incurable defect as a logic of propositions that it used informally in its proofs the very propositions whose formal statements it tried to prove. This defect appears to be entirely overcome in the development of *Principia*. But owing to the particular purpose the authors had in view they decided not to burden their work with more than was absolutely necessary for its achievement, and so gave up the generality of outlook which characterized symbolic logic.

It is with the recovery of this generality that the first portion of our paper deals. We here wish to emphasize that the theorems of this paper are *about* the logic of propositions but are *not included* therein. More particularly, whereas the propositions of *Principia* are *particular* assertions introduced for their interest and usefulness in later portions of the work, those of the present paper are about the set of *all* such possible assertions. Our most important theorem gives a uniform method for testing the truth of any proposition of the system; and by means of this theorem it becomes possible to exhibit certain general relations which exist between these propositions. These relations definitely show that the postulates of *Principia* are capable of developing the complete system of the logic of propositions without ever introducing results extraneous to that system—a conclusion that could hardly have been arrived at by the particular processes used in that work.

Further development suggests itself in two directions. On the one hand this general

[1] *Whitehead and Russell 1910*, part I, sec. A.

[2] See *Lewis 1918*. An extensive bibliography is given there.

procedure might be extended to other portions of *Principia*, and we hope at some future time to present the beginning of such an attempt. On the other hand we might take cognizance of the fact that the system of *Principia* is but one particular development of the theory—particular in the primitive functions it employs and in the postulates it imposes on those functions—and so might construct a general theory of such developments. This we have tried to do in the other portions of the paper. Our first generalization leads to systems which are essentially equivalent to that of *Principia* and connects up with the work of Sheffer (*1913*) and Nicod (*1916*) in reducing the number of primitive functions and of primitive propositions respectively. The second generalization, on the other hand, while including the first, also seems to introduce essentially new systems. One class of such systems, and we study these in detail, seems to have the same relation to ordinary logic that geometry in a space of an arbitrary number of dimensions has to the geometry of Euclid. Whether these "non-Aristotelian" logics and the general development which includes them will have a direct application we do not know; but we believe that, inasmuch as the theory of elementary propositions is at the base of the complete system of *Principia*, this broadened outlook upon the theory will serve to prepare us for a similar analysis of that complete system, and so ultimately of mathematics.

Finally a word must be said about the viewpoint that is adopted in this paper and the method that is used. We have consistently regarded the system of *Principia* and the generalizations thereof as purely *formal developments*,[3] and so have used whatever instruments of logic or mathematics we found useful for a study of these developments. The fact that one of the interpretations of the system of *Principia* is part of the informal logic we have used in this study makes the full significance of this *interpretation*, at least with regard to proofs of consistency, uncertain, but it in no way affects the actual content of the paper, which is in connection with the *formal systems*.

The system of *Principia mathematica*

1. Description of the system

Let $p, p_1, p_2, \ldots, q, q_1, q_2, \ldots, r, r_1, r_2, \ldots$ arbitrarily represent the variable elementary propositions mentioned in the introduction. Then by means of the two primitive functions $\sim p$ (read not p, the function of negation) and $p \vee q$ (p or q, the function of disjunction) with the aid of the primitive propositions

I. If p is an elementary proposition, $\sim p$ is an elementary proposition,

If p and q are elementary propositions, $p \vee q$ is an elementary proposition,

we combine these variables to form the various propositions or, rather, ambiguous values of propositional functions of the system. It is desirable in what follows to have before us the vision of the totality of these functions streaming out from the unmodified variable p through forms of ever-growing complexity to form the infinite triangular array

$$
\begin{array}{c}
p \\
p \vee p, \quad p_1 \vee p_2, \quad \sim p \\
p \vee \sim p, \quad \ldots, \quad \sim p_1 \vee \sim p_2, \quad \ldots, \quad (p_1 \vee p_2) \vee (p_3 \vee p_4), \\
\sim(p_1 \vee p_2), \quad \sim(p \vee p), \quad \sim\sim p \\
\cdots\cdots\cdots\cdots\cdots\cdots\cdots\cdots
\end{array}
$$

[3] For a general statement of this viewpoint see *Lewis 1918*, chap. VI, sec. III.

and to note and remember that this array of functions formed merely through combining p's by $\sim$'s and $\vee$'s constitutes the entire set of enunications it is possible to make in the theory of elementary propositions of *Principia*.

But the actual theory is concerned with the assertion of a certain subset of these functions. We denote the assertion of a function by writing $\vdash$ before it. Then the motive power for the resulting process of deduction is furnished by the two rules of operation:

II. The assertion of a function involving a variable p produces the assertion of any function found from the given one by substituting for p any other variable q, or $\sim q$, or $(q \vee r)$;[4]

III. "$\vdash P$" and "$\vdash : \sim P . \vee . Q$" produce "$\vdash Q$".

These enable us to assert new functions from old or rather, in the form in which we have put them, generate new assertions from old. And the complete set of assertions is produced by applying II and III both to the following assertions, which give us the start, and to all derived assertions that may result:

IV. $\vdash : \sim(p \vee p) . \vee . p$, $\vdash : \sim[p \vee (q \vee r)] . \vee . q \vee (p \vee r)$,
$\vdash : \sim q . \vee . p \vee q$, $\vdash :. \sim(\sim q \vee r) . \vee : \sim(p \vee q) . \vee . p \vee r$,
$\vdash : \sim(p \vee q) . \vee . q \vee p$.

We here again point out what was emphasized in the introduction that this theory concerns itself exclusively with the production of particular assertions through the detailed use of the rules of operation upon the primitive assertions, and as a consequence the set of theorems of this portion of *Principia* consists of the assertions of a certain number of particular functions of the above infinite set.[5]

2. *Truth table development*[6]

Let us denote the truth value of any proposition p by $+$ if it is true and by $-$ if it is false. This meaning of $+$ and $-$ is convenient to bear in mind as a guide to thought, but in the actual development that follows they are to be considered merely as symbols which we manipulate in a certain way. Then if we attach these two primitive truth tables to $\sim$ and $\vee$

p	$\sim p$
$+$	$-$
$-$	$+$

p, q	$p \vee q$
$+ \; +$	$+$
$+ \; -$	$+$
$- \; +$	$+$
$- \; -$	$-$

[4] This operation is not explicitly stated in *Principia* but is pointed out to be necessary by Russell (*1919*, p. 151). Its particular form was suggested to us by the first portion of the operation of "substitution" given by Lewis (*1918*, p. 295). It will be noticed that the effect of II is to enable us to substitute any function of the system for a variable of an asserted function.

[5] We have consistently ignored the idea of definition in this description. We here rigorously follow the authors in saying that definition is a convenience but not a necessity and so need not be considered part of the theoretical development. And so although we too shall at times use its shorthand, we do not encumber our theoretical survey with it.

[6] Truth values, truth functions, and our primitive truth tables are described in *Whitehead and Russell, 1910*, pp. 8 and 120, but the general notion of truth table is not introduced. This notion is quite precise with Jevons and Venn (see *Lewis 1918*, pp. 74 and 175ff., respectively) and has its foundation in the formula for the expansion of logical functions first given by Boole (*1854*, especially pp. 72–76). For the relation to Schröder see footnote 7.

we have a means of calculating the truth values of $\sim p$ and $p \vee q$ from those of their arguments. Now consider any function $f(p_1, p_2, \ldots, p_n)$ in our system of functions, which we will designate by F. Then since f is built up of combinations of $\sim$'s and $\vee$'s, if we assign any particular set of truth values to the p's, successive application of the above two primitive tables will enable us to calculate the corresponding truth value of f. So corresponding to each of the 2^n possible truth configurations of the p's a definite truth value of f is determined. The relation thus effected we shall call the truth table of f.

For example consider the function

$$\sim(\sim(\sim p \vee q) \vee \sim(\sim q \vee p)),$$

which is the ultimate definition of the function $p \equiv q$ of *Principia*. We have when p is $+$ and q is $+$ the following truth values of the successive components of the function and so finally of the function:

$$p:+,\quad \sim p:-,\quad \sim p \vee q:+,\quad \sim(\sim p \vee q):-$$
$$q:+,\quad \sim q:-,\quad \sim q \vee p:+,\quad \sim(\sim q \vee p):-$$
$$\sim(\sim p \vee q) \vee \sim(\sim q \vee p):-,\quad \sim(\sim(\sim p \vee q) \vee \sim(\sim q \vee p)):+,$$

the successive truth values being found by direct application of the primitive tables. In the same way the truth values for $p+$, $q-$, and so forth can be calculated and so we finally get the truth table of $p \equiv q$, that is,

p, q	$p \equiv q$
$+\ +$	$+$
$+\ -$	$-$
$-\ +$	$-$
$-\ -$	$+$.

It is needless to say that in actual work this amount of detail is quite unnecessary.

We shall call the number of variables which appear in a function the order of that function as well as that of its truth table. It is evident that there are 2^{2^n} tables of the nth order. We now prove the

THEOREM. *To every truth table of whatever order there corresponds at least one function of F which has it for its truth table.*

For, first, corresponding to the four tables of the first order $\begin{smallmatrix}+\\-\end{smallmatrix}\big|\begin{smallmatrix}+\\-\end{smallmatrix}$, $\begin{smallmatrix}+\\-\end{smallmatrix}\big|\begin{smallmatrix}+\\+\end{smallmatrix}$, $\begin{smallmatrix}+\\-\end{smallmatrix}\big|\begin{smallmatrix}-\\-\end{smallmatrix}$, $\begin{smallmatrix}+\\-\end{smallmatrix}\big|\begin{smallmatrix}-\\+\end{smallmatrix}$ we have the functions $p \vee p$, $p \vee \sim p$, $\sim(p \vee \sim p)$, $\sim p$. Now assume there is a function for each mth-order table. Then in any table of order $m + 1$ the configurations for which p_{m+1} is $+$ constitute an mth-order table for which there is some function $f_1(p_1, p_2, \ldots, p_m)$. Likewise corresponding to $p_{m+1}-$ we obtain $f_2(p_1, p_2, \ldots, p_m)$. Let $p.q$ stand for $\sim(\sim p \vee \sim q)$, a function which has the truth table

p, q	$p.q$
$+\ +$	$+$
$+\ -$	$-$
$-\ +$	$-$
$-\ -$	$-$.

Then it easily follows that the function

$$p_{m+1} \cdot f_1(p_1, p_2, \ldots, p_m) \,.\!\vee\!.\, \sim p_{m+1} \cdot f_2(p_1, p_2, \ldots, p_m)$$

has for its truth-table the given $(m + 1)$th-order table.

The functions of F can then be classified according to their tables as follows: those which have all their truth values $+$, all $-$, or some $+$ and some $-$. We shall call these functions respectively positive, negative, and mixed. This classification is of great importance in connection with the process of substitution, which is so fundamental in the postulational development. We shall say that any function obtained from another by the process of substitution is contained in that function. We then have the

THEOREM. *Every function contained in a positive function is positive; every function contained in a negative function is negative; every mixed function contains at least one function for every possible truth table.*

The first two results are immediate. In the third case note that any mixed function $f(p_1, p_2, \ldots, p_n)$ has at least one configuration which yields $+$ and one which yields $-$. Let the truth value of p_i in the positive configuration be denoted by t_i and in the negative by t_i', and construct a function $\varphi_i(p)$ with the truth table

p	$\varphi_i(p)$
$+$	t_i
$-$	t_i'

Then $\psi(p) = f(\varphi_1(p), \varphi_2(p), \ldots, \varphi_n(p))$ will be $+$ when p is $+$ and $-$ when p is $-$. But by our first theorem there is at least one function $g(q_1, q_2, \ldots, q_m)$ corresponding to any table of order m. Hence $\psi[g(q_1, q_2, \ldots, q_m)]$ is a function contained in $f(p_1, p_2, \ldots, p_n)$ corresponding to that table.

COROLLARY. *Every mixed function contains at least one positive function and one negative function.*

3. *The fundamental theorem*[7]

A necessary and sufficient condition that a function of F be asserted as a result of the postulates II, III, IV is that all its truth values be $+$.

[7] The method for testing propositions embodied in this theorem is essentially the same as that given by Schröder (*1891*, § 32) for the logical system he has developed. But we believe the range of significance of the proof we have given to be quite different from that of the work of Schröder. For first, as has been emphasized by Lewis (*1918*, chap. IV), formal and informal logic are inextricably bound together in Schröder's development to an extent that prevents the system as a whole from being completely determined. As a result the necessity of the condition of the theorem, which evidently requires such a complete determination if it is to be proved, remains unproved. As for the sufficiency, parts E and C of our proof appear in the proof for the expression of functions given by Schröder (*1890*). Part A, however, seems not to have been given explicitly, while corresponding to part D are all the theoretical difficulties met with in passing from the theory of classes to that of propositions when the development is not strictly formal. Hence the sufficiency of the condition is only incompletely proved. The theorem as given by Schröder is therefore of only partial significance even in his own system; and when transplanted to the system of *Principia* requires independent proof. Finally we may mention that the applications we have made of the theorem depend for their significance on those parts of the proof which do not appear, and could not appear, in Schröder.

Note first that each of the primitive assertions of IV is a positive function. Furthermore from the assertion of positive functions we can only get positive functions. For the only method we have of producing new assertions from old is through the use of II and III. Now II can only produce positive functions since every function contained in a positive function is positive. As for III, if P is $+$ and Q is $-$, $\sim P \vee Q$ is $-$, so that so long as P is a positive function and $\sim P \vee Q$ is a positive function Q must be positive, so that *III* can produce only positive functions. Hence every asserted function is positive and we have proved the condition necessary.

In order to prove it also sufficient we give a method for deriving the assertion of any positive function. It will simplify the exposition to introduce the other two defined functions of *Principia* besides $p \, . \, q$ (p and q) given above, namely,

$$p \supset q \, .{=}. \sim p \vee q \quad Df,^{8} \qquad p \equiv q \, .{=}. \, p \supset q \, . \, q \supset p \quad Df,$$

read "p implies q" and "p is equivalent to q", respectively, and having the tables

p, q	$p \supset q$
$+ \; +$	$+$
$+ \; -$	$-$
$- \; +$	$+$
$- \; -$	$+$

p, q	$p \equiv q$
$+ \; +$	$+$
$+ \; -$	$-$
$- \; +$	$-$
$- \; -$	$+$.

It will be noticed that if we have "$\vdash f_1(p_1, \ldots, p_n) \equiv f_2(p_1, \ldots, p_n)$" this asserted equivalence must have a positive table by the first part of our theorem, and so f_1 and f_2 must have the same truth values for the same configurations, that is, they must have the same truth table.

The proof is most conveniently given in four stages.

A. We prove the theorem $p \equiv q \, .{\supset}. \, f(p) \equiv f(q)$ where the function f may involve other arguments besides the one indicated and need not involve that. By means of this theorem we shall be able to replace a constituent of a given function by any equivalent function, and have the result equivalent to the given function.

It becomes necessary for the first time to introduce the notion of the rank of a function, which we define inductively as follows: the unmodified variable p will be said to be of rank zero, the negative of a function of rank m will be of rank $m + 1$; the logical sum of two functions the rank of one of which equals and the other does not exceed m will be of rank $m + 1$. Each function of F then is of finite rank as well as of finite order.[9] Returning now to the theorem, we notice that it is true for a function of rank zero, since it reduces either to $p \equiv q \, .{\supset}. \, p \equiv q$, which follows from $p \supset p$[10] by II, or to $p \equiv q \, .{\supset}. \, r \equiv r$, which follows from $p \supset \, . \, q \supset p$, $r \equiv r$, III and II. Assume now that the theorem holds for functions of rank m and lower. Then it also holds for functions of rank $m + 1$. For, if f is of rank $m + 1$, it can be written in the form $\sim f_1(p)$ or $f_2(p) \, . \vee \, f_3(p)$, where f_1, f_2, and f_3 are at most of rank m; and then the theorem follows by using $p \equiv q \, .{\supset}. \sim p \equiv \sim q$, $p \equiv q \, .{\supset}{:}. \, r \equiv s \, {:}{\supset}{:} \, p \vee r \, .{\equiv}. \, q \vee s$ along with $p \supset q \, {:}{\supset}{:} \, q \supset r \, .{\supset}. \, p \supset r$, III and II.

[8] III can now be written: "$\vdash P$" and "$\vdash P \supset Q$" produce "$\vdash Q$".

[9] But whereas the number of functions of given order is infinite, those of given rank are finite ⟦in number⟧.

[10] This as well as all other particular assertions that we use without an indication of proof appear in *Whitehead and Russell 1910*, part I, sec. A.

B. Consider now any function $f(p_1, p_2, \ldots, p_n)$. Using $\sim(p \vee q) \equiv . \sim p . \sim q$ and $\sim\sim p \equiv p$, with the aid of the equivalence theorem of A and $p \equiv q :\supset: q \equiv r .\supset. p \equiv r$ we finally obtain $f(p_1, p_2, \ldots, p_n)$ equivalent to a function $f'(p_1, p_2, \ldots, p_n)$ which is expressed merely through combinations of p's and $\sim p$'s by .'s and $\vee$'s.

C.[11] If we then apply the distributive law of logical multiplication to f', it will be reduced to an equivalent function consisting of successive logical sums of successive logical products of the p's and the $\sim p$'s. If any of these products has neither p_n nor $\sim p_n$ as a factor, we can introduce them through the propositions $p \vee \sim p$ and $p :\supset: q . \equiv . p . q$, whence $q :\equiv: (p \vee \sim p) . q :\equiv: p . q . \vee . \sim p . q$. Now apply the commutative and associative laws of logical multiplication along with $p . p . \equiv . p$, so that each product has at most one p_i and one $\sim p_i$. Again using the distributive law for purposes of factorization along with the commutative and associative laws of addition, we finally obtain f equivalent to

$$f_1(p_1, p_2, \ldots, p_{n-1}) . p_n . \sim p_n :\vee: f_2(p_1, \ldots, p_{n-1}) . p_n . \vee . f_3(p_1, \ldots, p_{n-1}) . \sim p_n,$$

where one or more of the terms and arguments may not appear.

D. Suppose now that the original function is positive; then this equivalent function will be positive. If in particular it be of first order, it can only be $p \vee \sim p$ or $p . \sim . p \vee . p \vee \sim p$. The first is an asserted function; likewise the second through $p .\supset. q \vee p$. Hence also $f(p)$ will be asserted through $p \equiv q .\supset. q \supset p$; and so every positive first-order function is asserted. Assume now that this is true for all mth and lower-ordered functions and let f be any positive $(m + 1)$th-order function. The reduced function being then positive, both f_2 and f_3 will be positive and hence will be asserted. From the use of $p :\supset: q .\supset. p \equiv q, p . r . \vee . p . \sim r :\equiv: p . (r \vee \sim r) . p :\supset: s .\supset. p . s$, and $p .\supset. q \vee p$, the reduced function will be asserted and so finally f. Hence every positive function can be asserted and so the proof is complete.

We thus see that given any function the theorem gives a direct method for testing whether that function can or cannot be asserted; and if the test shows that the function can be asserted the above proof will give us an actual *method for immediately writing down a formal derivation of its assertion by means of the postulates of* Principia.

Before we pass on to theorems about the system itself irrespective of truth tables, we give the following definitions, which apply directly to the system: a true function is one that can be asserted as a result of the postulates, any other is false; a completely false function is a false function such that every function therein contained is false—otherwise we call it incompletely false. We then have the

COROLLARY. *The sets of true, completely false, and incompletely false functions are identical with the sets of positive, negative, and mixed functions, respectively.*

4. Consequences of the fundamental theorem

In the above development the truth values $+$, $-$ were arbitrary symbols which were found related in certain suggestive ways through the fundamental theorem. We are now in a position to give direct definitions of these truth values in terms of the postulational development. In fact we shall define $+$ to be the set of true functions, $-$ the set of completely false functions. The truth value of a function will then exist

[11] This portion of the proof is essentially that given by Whitehead (*1898*, p. 46).

when and only when it is true or completely false, and it will be defined as that class $(+, -)$ of which it is a member. The content of the fundamental theorem consists now of these two theorems:

1. The truth values of $\sim p$ and $q \vee r$ exist whenever the truth values of p, q, and r exist and depend only on those truth values as given by the primitive tables. It therefore follows that the same is true of any function of F and that the truth table of such a function can be directly calculated from the primitive tables.

2. The fundamental theorem as stated, or else in the form: if f_1 and f_2 is any pair of positive and negative functions respectively, then a necessary and sufficient condition that a function $f(p_1, p_2, \ldots, p_n)$ be asserted is that each of the 2^n contained functions found by substituting f_1 and f_2 for the p's is asserted. It will be noticed that Theorem 1 tells us how to determine whether these latter are asserted.

We now pass on to several theorems about the system.

THEOREM. *It is possible to find 2^{2^n} functions of order n such that no two of them are equivalent and such that every other function of order n is equivalent to one of these.*

For we can find 2^{2^n} functions corresponding to the 2^{2^n} different tables of order n. The equivalence of any two of these will then not have a positive table and so will not be asserted. On the other hand any other nth-order function will have the same table as one of the 2^{2^n} possible tables, and so the corresponding equivalence will be positive and hence asserted.

THEOREM. *An incompletely false function contains at least one function, for each given function, which is equivalent to that given function.*

COROLLARY. *An incompletely false function contains at least one true function and one completely false function.*

THEOREM. *The negative of a completely false function is true.*

For a completely false function has a negative truth table, and so its negative will have a positive table and hence be asserted. It is worth noticing that although this theorem is immediate once we have the fundamental theorem it would be quite difficult without it.

COROLLARY. *Every function of F is either true, or its negative is true, or it contains both a true function and one whose negative is true.*

THEOREM. *The system of elementary propositions of* Principia *is consistent.*

For if it were inconsistent we would have both a function and its negative asserted. But then both the function and its negative would have to have positive tables whereas if a function has a positive table its negative has a negative table.[12]

THEOREM. *Every function of the system either can be asserted by means of the postulates or else is inconsistent with them.*

For, if a function be not asserted as a result of the postulates, it will contain a function whose negative can be so asserted. If then we assert the original function, the contained function will be asserted, so that we have asserted both a function and its negative, that is, we have a contradiction.

COROLLARY. *A function is either asserted as a result of the postulates or else its assertion will bring about the assertion of every possible elementary proposition.*

For by the theorem we would obtain the assertion of both a function and its

[12] This argument requires merely the first part of the fundamental theorem, which was proved quite simply.

negative and so by $\sim p\ .\supset.\ p \supset q$ the assertion of the unmodified variable q. But q then represents any elementary proposition.

In conclusion let us note that, while the fundamental theorem shows that the postulates bring about the assertion of those and only those theorems which should belong to the system, this last theorem enables us to say that they also automatically exclude the very possibility of any added assertions.

Generalization by truth tables

5. General survey of the systems generated

The system we have studied in the preceding sections is a particular system depending upon the two primitive functions $\sim p$ and $p \vee q$. Two modes of attack have presented themselves. On the one hand we have the original postulational method, on the other the truth table development. In passing to a general study of systems of the kind discussed these two methods present themselves as instruments of generalization. We reserve the postulational generalization for the next portion of our paper and now take up the truth table generalization.

To gain complete generality let us assume for our primitives μ arbitrary functions with an arbitrary number of arguments, which we will designate by

$$f_1(p_1, p_2, \ldots, p_{m_1}), f_2(p_1, p_2, \ldots, p_{m_2}), \ldots, f_\mu(p_1, p_2, \ldots, p_{m_\mu}),$$

and let us attach an arbitrary truth table to each. By successive combinations of these functions with different or repeated arguments we generate the set of derived functions, which as before we designate by F. Again each function of F will possess a truth table in virtue of the tables of the primitive functions of which each is composed. Denote the set of truth tables thus generated by T. Then, whereas in the system of *Principia* T consists of all possible truth tables, this will not necessarily be the case here.

In another paper[13] we completely determine all the possible systems T and show that there are 66 *systems that can be generated by tables of third and lower order, and* 8 *infinite families of systems that are generated by the introduction of fourth- and higher-order tables.*

If two systems have the same truth tables, the primitives of each can evidently be expressed in terms of those of the other so that truth tables are preserved. We can then say that each system has a representation in the other and the two are equivalent. In particular *every truth system has a representation in the system of* Principia while *every complete system,* that is, one having all possible truth tables, *is equivalent to it.* In the aforementioned paper we also determine the ways in which a complete system may be generated, and it turns out that one table alone is sufficient to generate it, and it can be either of these two

$$\begin{array}{cc|c} + & + & - \\ + & - & + \\ - & + & + \\ - & - & + \end{array} \qquad \begin{array}{cc|c} + & + & - \\ + & - & - \\ - & + & - \\ - & - & +, \end{array}$$

a result first given by Sheffer as stated in the introduction.

[13] [[*Post 1920a.* See also *1941,* whose Introduction gives the history of Post's work on the subject.]]

The truth table development for complete systems is essentially the same as that given in Section 2. It is easy to prove for all systems the

THEOREM. *Every function contained in a positive function is positive; every function contained in a negative function is negative; every mixed function contains a function for every table of the system.*

6. Postulates for a complete system

We now show how to construct a set of postulates for any complete system such that: *the set of asserted functions is identical with the set of positive functions, while the assertion of any other function brings about the assertion of every elementary proposition,* a property which also characterized the system of *Principia.*

Let $\sim' p$ and $p \vee' q$ be functions in the given complete system with the tables of $\sim$ and $\vee$. Out of $\sim'$ and $\vee'$ we then construct $p \supset' q$ and $p \equiv' q$ as $p \supset q$ and $p \equiv q$ are found from $\sim$ and $\vee$, and also $f_1'(p_1, \ldots, p_{m_1}), \ldots, f_\mu'(p_1, \ldots, p_{m_\mu})$ with the same tables as $f_1(p_1, \ldots, p_{m_1}), \ldots, f_\mu(p_1, \ldots, p_{m_\mu})$. This is possible since $\sim$ and $\vee$, and so $\sim'$ and $\vee'$, can generate a complete system. All the functions $\sim'$, $\vee'$, $\supset'$, $\equiv'$, $f_1', \ldots,$ f_μ' are ultimately expressed in terms of the f's and so belong to the system. Construct now the following set of postulates:

I. If $p_1, \ldots, p_m$ are elementary propositions, $f_1(p_1, \ldots, p_{m_1})$ is.

. . . .

If $p_1, \ldots, p_{m_\mu}$ are elementary propositions, $f_\mu(p_1, \ldots, p_{m_\mu})$ is.

II. The assertion of a function involving a variable p produces the assertion of any function found from the given one by substituting for p any other variable, q, or $f_1(q_1, \ldots, q_{m_1}), \ldots,$ or $f_\mu(q_1, \ldots, q_{m_\mu})$.

III. "$\vdash P$" and "$\vdash P \supset' Q$" produces "$\vdash Q$".

IV. (1) $\vdash : p \vee' p . \supset' p$

. . . .

(5) $\vdash \ldots$

(*a*) $\vdash . f_1(p_1, p_2, \ldots, p_{m_1}) \equiv' f_1'(p_1, p_2, \ldots, p_{m_1}),$

. . . .

(*u*) $\vdash . f_\mu(p_1, p_2, \ldots, p_{m_\mu}) \equiv' f_\mu'(p_1, p_2, \ldots, p_{m_\mu}).$

where (1)–(5) are the assertions of IV in Section 1 with $\sim'$ and $\vee'$ in place of $\sim$ and $\vee$.

That all asserted functions are positive can be verified as in the proof of Section 4. As for the converse, note that III and IV (1)–(5), being of the same form as III and IV of Section 4, will yield the assertion of all positive functions expressed in terms of $\sim'$ and $\vee'$. By the use of (*a*)–(*u*) every function can be shown to be equivalent ($\equiv'$) to some function expressed by $\sim'$ and $\vee'$, and so every positive function will be asserted. In the same way the assertion of any nonpositive function will bring about the assertion of a nonpositive function in $\sim'$ and $\vee'$ alone, and so of any proposition.

We thus see that complete systems are equivalent to the system of *Principia* not only in the truth table development but also postulationally. As other systems are in a sense degenerate forms of complete systems we can conclude that no new logical systems are introduced.

7. Application to Nicod's postulate set

Although, as in most existence theorems, the above set of postulates may not be the simplest in any one case, it can be used to advantage in showing that a given set has the same property as it possesses. For this purpose we show directly that all asserted functions are positive and then that (*a*) by means of the given postulates each of our formal postulates may be derived and (*b*) the results derivable by our informal postulates can also be derived by the given ones.[14]

As an example we consider the set of postulates given by Nicod for the theory of elementary propositions in terms of the single primitive function of Sheffer's, which Nicod denotes by $p \mid q$ and is termed incompatibility by Russell (*1919*, chap. XIV). It is the first of the two functions given in Section 5 as generating a complete system. Nicod gives the definitions

$$\sim p .=. p \mid p \quad Df, \qquad p \vee q .=. p/p \mid q/q \quad Df,$$

which we take to be our $\sim' p$ and $p \vee' q$ respectively. His $p \supset q .=. p \mid q/q$ *Df*, however, is not our $p \supset' q$, which is $\sim' p \vee' q$. The primary distinction of his system is that he uses but one formal primitive proposition.

In carrying out the proof suggested we merely note that by means of his informal proposition

$$\text{"} \vdash P\text{" and "} \vdash P \mid R/Q\text{" produce "} \vdash Q\text{"}$$

we get the effect of

$$\text{"} \vdash P\text{" and "} \vdash P \mid Q/Q\text{" (that is, "} \vdash P \supset Q\text{") produce "} \vdash Q\text{"}$$

when $R = Q$. Since he has $p \supset' q .\supset. p \supset q.$, we thus get the effect of

$$\text{"} \vdash P\text{" and "} \vdash P \supset' Q\text{" produce "} \vdash Q\text{"},$$

our III. Likewise each function IV is proved with, however, $\supset$ in place of $\supset'$. But by means of $p \supset q .\supset. p \supset' q$ this too is remedied. We then easily complete the proof of the

THEOREM. *If in Nicod's system we give to* $p \mid q$ *the table*

p, q	$p \mid q$
+ +	−
+ −	+
− +	+
− −	+,

then the set of asserted functions is identical with the resulting set of positive functions; and the assertion of any other function would bring about the assertion of every elementary proposition.

[14] That the informal postulates of a system must be proved effectively replaced by others in another system is a precaution rarely taken in discussions of equivalence or dependence of logical systems. Such a discussion is unnecessary in ordinary mathematical systems since their distinctive postulates are all formal, the informal ones being those of a common logic. But in comparing logical systems, which usually do contain different informal postulates, such a discussion is fundamental.

Generalization by postulation

8. The generalized set of postulates

As in the truth table development we assume arbitrary primitive functions of propositions

$$f_1(p_1, p_2, \ldots, p_{m_1}), \ldots, f_\mu(p_1, p_2, \ldots, p_{m_\mu});$$

but in place of the arbitrary associated truth tables we have a set of postulates of the following form. We have tried to preserve all the informal properties of the postulates of *Principia* (and of Section 5) but generalize the formal properties completely.

I. (As in Section 5.)
II. (As in Section 5.)
III. "$\vdash g_{11}(P_1, P_2, \ldots, P_{k_1})$" ... "$\vdash g_{\kappa 1}(P_1, P_2, \ldots, P_{k_\kappa})$"

. . . .

"$\vdash g_{1\kappa_1}(P_1, P_2, \ldots, P_{k_1})$" ... "$\vdash g_{\kappa\kappa_\kappa}(P_1, P_1, \ldots, P_{k_\kappa})$"
produce ... produce
"$\vdash g_1(P_1, P_2, \ldots, P_{k_1})$" ... "$\vdash g_\kappa(P_1, P_2, \ldots, P_{k_\kappa})$",

where the P's are any combinations of f's, including the special case of the unmodified variable, while the g's are particular combinations of this kind which need not have all the indicated arguments.

IV. $\vdash h_1(p_1, p_2, \ldots, p_{l_1})$
$\vdash h_2(p_1, p_2, \ldots, p_{l_2})$
. . . . . .
$\vdash h_\lambda(p_1, p_2, \ldots, p_{l_\lambda})$,

where the h's are particular combinations of the f's.

The retention of I and II, which are characteristic of the theory of elementary propositions, is our justification for giving that name to the systems that may be generated by the above set of postulates. In what follows we give what we consider to be merely an introduction to the general theory.

9. Definition of consistency and related concepts

The prime requisite of a set of postulates is that it be consistent. Since the ordinary notion of consistency involves that of contradiction, which again involves negation, and since this function does not appear in general as a primitive in the above system, a new definition must be given.

Now an inconsistent system in the ordinary sense will involve the assertion of a pair of contradictory propositions, which, as we have seen, will bring about the assertion of every elementary proposition through the assertion of the unmodified variable p. Conversely, since p stands for any elementary proposition, its assertion would yield the assertion of contradictory propositions and so render the system inconsistent. The two notions are thus equivalent in ordinary systems; and since one retains significance in the general case we are led to the

DEFINITION. *A system will be said to be inconsistent if it yields the assertion of the unmodified variable p.*

In a consistent system we may then define a true function as one that can be asserted as a result of the postulates. Instead of defining a false function as one not true, we give the following

DEFINITION. *A false function is one such that if its assertion be added to the postulates the system is rendered inconsistent.*

We can then state that in the system of *Principia* every function is true or false. This suggests the

DEFINITION. *If every function of a consistent system is true or false, the system will be said to be closed.*[15]
As a justification of this name we may note that the postulates of such a system automatically exclude the possibility of any added assertions—a state of affairs we believe to be highly desirable in the final form of a logical theory.

10. Properties of consistent systems

In all that follows we assume that the system discussed is consistent. If it be inconsistent one could hardly say anything more about it.

We turn to a theorem which will give us most of the results of this section. But first we must state two lemmas which we do not further prove.

LEMMA 1. If a given set of functions gives rise to some other function in accordance with II and III and if these functions involve certain letters $r_1, r_2, \ldots, r_i$ upon which no substitution is made in the process, then the same deductive process will be valid if we have given the original functions with an arbitrary substitution of the r's as described in II provided this substitution is also made throughout the process.

LEMMA 2. The most general process of obtaining an assertion from a given set of assertions in accordance with II and III can be reduced to first asserting a number of functions in accordance with II and then applying II and III in such a way that no substitutions are made on the arguments of those functions.[16]

THEOREM. *Every false function contains a finite set of untrue first-order functions* $\varphi_1(p), \varphi_2(p), \ldots, \varphi_\nu(p)$ *such that whenever p is replaced by an untrue function at least one of these functions remains untrue.*

By the definition of false functions there must be some deductive process whereby from the given false function and true functions we assert p. By Lemma 2 we can replace this process by another where from the given false function and true functions we obtain certain contained functions from which without substitution of the

[15] Had the name not been in use in a different connection we should have introduced the term "categorical".

[16] ⟦Concerning this passage Post wrote later (*1943*, footnote 2, p. 197): "We take this opportunity to make the following emendation: Lemma 1 thereof requires the added condition that the expressions replacing the r's do not involve any letter upon which a substitution is made in the given deductive process. This necessitates several minor changes in the proof of the theorem there following. Actually, both Lemma 1 and its companion Lemma 2 admit of further simplification, with the proof of the theorem then being valid as it stands". See also footnote 17 in *Post 1941a*, pp. 347–348.⟧

arguments we obtain p. Now first by Lemma 1 we can equate to p all the arguments thus appearing and still have a valid deductive process for obtaining p. Denote the resulting untrue functions which are contained in the original false function by $\varphi_1(p), \varphi_2(p), \ldots, \varphi_\nu(p)$. Then secondly by Lemma 1 we can replace p by any function ψ and still have a valid process, which now consists in obtaining ψ from certain true functions and $\varphi_1(\psi), \ldots, \varphi_\nu(\psi)$. If then each $\varphi_i(\psi)$ were true, ψ, being obtained from true functions in accordance with II and III, would be true. It follows that, if ψ be untrue, some $\varphi_i(\psi)$ must be untrue.

THEOREM. *Every false function contains an infinite number of untrue first-order functions; and if the system has at least one false function of order greater than one, then each false function contains an infinite number of untrue functions of every order.*

By the above theorem the false function contains at least one untrue function $\varphi_{i_1}(p)$. By the same theorem some $\varphi_{i_2}\varphi_{i_1}(p)$ must be untrue, and so on, through $\varphi_{i_j}\varphi_{i_{j-1}}\cdots\varphi_{i_1}(p)$. These are all different, being of different rank, and are all contained in the given function.

The last part of the theorem may then be proved by showing that by replacing equal by unequal variables in the infinity of functions thus gotten from the false function of order greater than one we get untrue functions of every order, and so by the above method an infinite number of every order in every false function.

We have immediately the

THEOREM. *A necessary and sufficient condition that a function of a closed system be true is that all contained first-order functions be true.*

COROLLARY. *It is also necessary and sufficient that all those of rank greater than some finite integer ρ be true.*

In analogy with corresponding ideas in the system of *Principia*, define a completely untrue function as one in which all contained functions are untrue, with a similar definition for completely false. We then have the interesting

THEOREM. *If a system has a completely untrue function, then every false function contains a completely untrue function.*

Every function contained in the completely untrue function makes at least one $\varphi_i(p)$ of a false function untrue. If ψ is such a contained function which makes say $\varphi_{i_1}(p)$ true, then ψ will be completely untrue, and all contained functions will make $\varphi_{i_1}(p)$ true yet some remaining $\varphi_i(p)$ untrue. By repeating this process we finally obtain a function ψ' such that all contained functions make each $\varphi_i(p)$ of a set that remains untrue. Each such $\varphi_i(\psi')$ will then be a completely untrue function in the given one.

COROLLARY. *If a closed system has a completely false function, every false function contains a completely false function.*

If we call such a system completely closed, we have the stronger

THEOREM. *In a completely closed system every false function $f(p_1, p_2, \ldots, p_n)$ contains a completely false function $f(\psi_1(p), \psi_2(p), \ldots, \psi_n(p))$, where each $\psi_i(p)$ is either true or completely false.*

By equating all variables to p in the function of the corollary we get such a completely false function where some ψ's may be incompletely false. These are then eliminated by successively substituting for p functions which make them true.

COROLLARY. *A necessary and sufficient condition that a function of a completely closed system be true is that all contained first-order functions found by substituting true or completely false functions for the arguments be true.*

This property begins to approximate to the truth table method. It leads us easily to the following criterion for a completely closed postulational system being a truth system, which we state without proof.

THEOREM. *A necessary and sufficient condition that a completely closed postulational system be a truth system is that a true first-order function remain true whenever we replace a true or completely false constituent function by any other true or completely false first-order function, respectively.*[17]

m-VALUED TRUTH SYSTEMS[18]

11. The generalized ($\sim$, $\vee$) *system*

We have seen that the truth-table generalization, at least with regard to complete systems, is included in the postulational development. We now show that the latter is more general by presenting a new class of systems, distinct from the two-valued systems of symbolic logic, which can be generated by a completely closed set of postulates.

In these systems instead of the two truth values $+$, $-$ we have m distinct "truth values" $t_1, t_2, \ldots, t_m$, where m is any positive integer. A function of order n will now have m^n configurations in its truth table, so that there will be m^{m^n} truth tables of order n. Calling a system having all possible tables complete, we now show that the following two tables generate a complete system.

p	$\sim_m p$
t_1	t_2
t_2	t_3
$\ldots$	$\ldots$
t_m	t_1

p, q	$p \vee_m q$	
$t_1 t_1$	t_1	
$\ldots$	$\ldots$	
$t_{i_1} t_{j_1}$	t_{i_1}	$i_1 \leq j_1$
$\ldots$	$\ldots$	$i_2 \geq j_2$
$t_{i_2} t_{j_2}$	t_{j_2}	
$\ldots$	$\ldots$	
$t_m t_m$	t_m.	

We see that $\sim_m p$, the generalization of $\sim p$, permutes the truth values cyclically, while $p \vee_m q$, the generalization of $p \vee q$, has the higher of the two truth values.[19]

To construct a function for any first-order table, of which there are m^m, note that

$$t_1(p) .=. p \vee \sim_m p . \vee_m \sim_m^2 p : \vee_m \ldots \sim_m^{m-1} p \quad Df,$$

where $\sim^2 p .=. \sim\sim p$ *Df*, and so forth, has all its truth values t_1. Then

$$\tau_{m_1}(p) .=. \sim_m^{m-1}(\sim_m^{m-1}(\sim_m t_1(p) . \vee_m . p) : \vee_m . \sim_m^{m_1} p) \quad Df$$

[17] In making a more complete study of the postulational generalization it would be desirable to classify all the systems that may result more or less in the way in which we have classified truth systems through the associated systems of truth tables. In this connection we might define the order of a set of postulates as the largest number of premisses used in deriving a conclusion in III, and the order of a system as the lowest order a set of postulates deriving it can have. It is then of interest to note that *whereas the set of postulates of the system of* Principia *is of the second order, the system itself is of the first order.*

[18] See *Lewis 1918*, p. 222, for the term "two-valued algebra".

[19] The higher truth value has here the smaller subscript.

has all values t_m except the first, which is t_{m_1}. Any first-order table

p	$f(p)$
t_1	t_{m_1}
t_2	t_{m_2}
...	...
t_m	t_{m_m}

can then be constructed by the function

$$\tau_{m_1}(p) \,.\vee_m.\, \tau_{m_2}(\sim_m^{m-1}p) :\vee_m.\, \tau_{m_3}(\sim_m^{m-2}p) :.\vee_m \ldots \tau_{m_m}(\sim_m p).$$

Construct now a function for the table

p	$\eqsim_m p$
t_1	t_m
t_2	t_{m-1}
...	...
t_m	t_1

and define $p \,._m q \,.{=}.\, \eqsim_m(\eqsim_m p \,.\vee_m.\, \eqsim_m q)$ *Df*, which is the generalization of $p.q$ and has the lower of the two truth values of its arguments. We can now construct a table all of whose values are t_m except for one configuration $t_{m_1}, t_{m_2}, \ldots, t_{m_n}$ when it is $t_{m_{m_1 m_2 \cdots m_n}} = t_\mu$ by the function

$$\tau_\mu(\sim_m^{m-m_1+1}p_1) \,._m \tau_\mu(\sim_m^{m-m_2+1}p_2) \,._m \ldots \tau_\mu(\sim_m^{m-m_n+1}p_n),$$

and so any table by constructing such a function for each configuration and then "summing up" by $\vee_m$.

12. Classification of functions—the m-dimensional space analogy

The generalization of the classification of functions into positive, negative, and mixed is afforded us by the following

THEOREM. *A function contains at least one function for every truth table whose values are contained among the values of the given table.*

Let $t_{m_1}, \ldots, t_{m_\mu}$ be the truth values that appear in the table of a given function $f(p_1, p_2, \ldots, p_n)$. Then we can pick out μ configurations having these values respectively. Construct functions $\varphi_i(p)$ such that, when p has the value t_{m_j} of one of these configurations, $\varphi_i(p)$ have the value of p_i in that configuration. It is then easily seen that $f(\varphi_1(p), \ldots, \varphi_n(p))$ has the value t_{m_j} whenever p has the value t_{m_j}. If then $\psi(q_1, q_2, \ldots, q_l)$ have a table whose values are among the t_{m_j}, $f(\varphi_1(\psi), \ldots, \varphi_n(\psi))$ will be a function contained in the given function with that table.

We are thus led to a classification of functions by means of their truth tables such that the set of tables of functions contained in a given function is the same for all functions in a given class. We then have m classes of functions where but one truth value appears, $[m(m-1)]/2!$ with two truth values, ..., $[m(m-1)\ldots(m-\mu+1)]/\mu!$ with μ truth values, ..., one class with all m truth values. We thus have $2^m - 1$ classes of functions, which when $m = 2$ reduces to the three classes of positive, negative, and mixed functions.

These formulas suggest an analogy which, if well founded, is of great interest. For this purpose replace the set of functions having all of a given set of μ truth values by all functions whose values are among these μ values. If then we compare the functions of our complete system to the points of a space of m dimensions,[20] the m classes of functions with but one truth value would correspond to the m coordinate axes, the $[m(m - 1)]/2!$ classes of functions with no more than two truth values to the $[m(m - 1)]/2!$ coordinate planes, and so on, so that except for the absence of an origin all properties of determination and intersection within the coordinate configurations go over. If then we attach the name "m-dimensional truth space" to our system, we observe the following difference, that whereas the highest-dimensioned intuitional point space is three, the highest-dimensioned intuitional proposition space is two. But just as we can interpret the higher-dimensioned spaces of geometry intuitionally by using some other element than point, so we shall later interpret the higher-dimensioned spaces of our logic by taking some other element than proposition.

13. Truth table characteristics of asserted functions

The following analysis presupposes that in constructing a set of postulates for the system we at least wish to impose the

CONDITION. *If a function is asserted, all functions with the same truth table will be asserted.*

It follows from the theorem of the preceding section that under the given condition, *if a function is asserted, every function of the truth space it determines is asserted.*

We can now prove that, *if the system is to be completely closed, its asserted functions must constitute a single truth space contained in the given truth space.* For, if there were at least two such spaces, then a function having all their truth values would be false, and so would contain a completely false function. This in turn would contain functions with but one truth value; and these being therefore in one of the two given spaces would be true, which contradicts their being in a completely false function.

No loss of generality ensues if we take the truth values of this contained truth space of asserted functions to be $t_1, t_2, \ldots, t_\mu$, where, to avoid degenerate cases, $0 < \mu < m$. We now show that a completely closed set of postulates can be constructed for all such systems.

14. A completely closed set of postulates for the systems

I and II are determined directly as in the general case. To obtain III, construct a function $p \supset^\mu_m q$ whose table is given by the following: when the truth value of p is that of q or lower, $p \supset^\mu_m q$ will have the value t_1, while if the truth value of p is above that of q, then if the value of p is t_μ or higher, $p \supset^\mu_m q$ will have the value of q, while if it is below t_μ, say t_ν, and that of q is $t_{\nu'}$, then the truth value of $p \supset^\mu_m q$ will be $t_{\nu' - \nu + 1}$. III will then be simply

$$\text{“} \vdash P\text{” and “} \vdash P \supset^\mu_m Q\text{” produce “} \vdash Q\text{”.}$$

[20] Or we might take the truth table as element, in which case the system is perhaps smoother than before.

Now by generalizing each part A, B, C, D of the proof of the fundamental theorem of Section 3 it can be shown that by the assertion of a finite number of functions with values from t_1 to t_μ all such can be obtained.[21] If then we assert these functions in IV, we shall have every function in the μ-space asserted. Furthermore no others can be asserted, for by the use of II and III we can only get functions with values from t_1 to t_μ by means of functions similarly restricted. This is obvious in II, while in III if the value of P is from t_1 to t_μ while that of Q is below t_μ, then from the above definition of the table of $P \supset_m^\mu Q$ it would have the value of Q and so be below t_μ. But that contradicts the assumption that the premisses had values from t_1 to t_μ.

This set of postulates will then give the proper set of true functions. Furthermore let us suppose that we assert a function with at least one value below t_μ. This will contain a function $\varphi(p)$ with but one value, and that below t_μ. By II, $\varphi(p)$ will be asserted. Furthermore, since $\varphi(p) \supset_m^\mu \varphi(p) \supset_m^\mu \sim_m\varphi(p)$ has its value t_1, it will be asserted, and so we obtain by III $\sim_m\varphi(p)$. Repetition of this process will finally give us a function $\psi(p)$ with but one value t_m. But $\psi(p) \supset_m^\mu p$ is asserted having but one value t_1. We thus obtain the assertion of p. The system is therefore closed. And since all functions with values from $t_{\mu+1}$ to t_m are completely false, the system is completely closed.

15. Comparison of systems

As in the truth-table development we can generalize the systems by using arbitrary functions as primitives, and as was done there we can show how to generate a complete m-dimensioned system by one second-order function and how to give a completely closed set of postulates for all complete systems. The problem of determining all possible systems of m-dimensional truth tables, however, is one we have not considered, though its solution would throw considerable light on the ordinary problem.

We turn now to the following

DEFINITIONS. *A closed system S with primitives $f_1, f_2, \ldots, f_n$ has a representation in a closed system S' with primitives $f'_1, f'_2, \ldots, f'_{n'}$ if we can so replace the f's by functions in S' that a function in S will be true when and only when the correspondent in S' is true.*

Two systems are equivalent if each has a representation in the other.

Denote a complete m-dimensional truth system with the asserted functions forming a truth space of μ dimensions by ${}_\mu T_m$. We then have the

THEOREM. *Two complete truth systems ${}_\mu T_m$ and ${}_{\mu'}T'_{m'}$ are equivalent when and only when $\mu = \mu'$ and $m = m'$.*

The conditions are clearly sufficient since we can make truth values correspond. To prove them necessary suppose $m > m'$. If we construct m^m functions of first order in T with different truth tables, then there will be two, $\varphi_1(p)$, $\varphi_2(p)$, whose correspondents $\varphi'_1(p)$, $\varphi'_2(p)$ have the same truth tables since there are in T' only $m'^{m'}$ of first order. Let $\chi(p, q)$ have value t_1 when p and q have the same value and t_m otherwise. Then $\chi(\varphi_1, \varphi_1)$ is true; hence $\chi'(\varphi'_1, \varphi'_1)$ is. φ'_2 having the same table as φ'_1, $\chi'(\varphi'_1, \varphi'_2)$ is true, and hence $\chi(\varphi_1, \varphi_2)$ the correspondent. But that would make φ_1 have the same table as φ_2. Now suppose $\mu > \mu'$. If φ have all the values from t_1 to t_μ and no others, there are μ^μ functions with values t_1 to t_μ of the form $\psi\varphi(p)$. These will then

[21] Lack of space prevents us from giving the details.

be asserted and so the correspondents will be asserted and have values t'_1 to $t'_{\mu'}$. Since we can only have $\mu'^{\mu'}$ functions $\psi'\varphi'(p)$ with different tables, we can find two of the μ^μ correspondents with the same table. The above contradiction then results as before.

For representation we have only found the

THEOREM. *To represent* ${}_\mu T_m$ *in* ${}_{\mu'}T'_{m'}$ *it is necessary to have* $\mu \leq \mu'$, $m \leq m'$; *it is sufficient to have* $\mu \leq \mu'$, $m - \mu \leq m' - \mu'$.

COROLLARY. *A necessary and sufficient condition that* ${}_\mu T_m$ *have a representation in* ${}_\mu T'_{m'}$ *is that* $m \leq m'$.

It is of interest to note as a result that the only complete truth systems equivalent to the system of *Principia* are ${}_1T_2$'s; and though it can be represented in every complete truth system, only ${}_1T_2$'s can be represented in it. We have thus verified our statement that we obtain essentially new logical systems.

16. *Interpretation of m-valued truth systems in terms of ordinary logic*

Let the elementary proposition of the $(\sim_m, \vee_m)$ system be interpreted as an ordered set of $(m - 1)$ elementary propositions of ordinary logic $P = (p_1, p_2, \ldots, p_{m-1})$ such that if one proposition is true all those that follow are true. P will be then be said to have the truth value t_1 if all the p's are true, t_2 if all but one are true, and so forth. Also P will be said to be true if at most $(\mu - 1)$ p's are false.

If $P = (p_1, p_2, \ldots, p_{m-1})$, $Q = (q_1, q_2, \ldots, q_{m-1})$, we define

$$P \vee_m Q \;.{=}.\; (p_1 \vee q_1, p_2 \vee q_2, \ldots, p_{m-1} \vee q_{m-1}) \quad Df,$$

$$\sim_m P \;.{=}.\; (\sim(p_1 \vee p_2 \vee \ldots \vee p_{m-1}), \sim(p_1 \vee \ldots \vee p_{m-1}) \;.\vee.\; p_1 \cdot p_2, \ldots,$$
$$\sim(p_1 \vee \ldots \vee p_{m-1}) \;.\vee.\; p_{m-2} \cdot p_{m-1}) \quad Df.$$

We easily justify these definitions by showing first that $P \vee_m Q$ and $\sim_m P$ are "elementary propositions" when P and Q are, and secondly that they have the proper truth tables. Thus in $P \vee_m Q$ the first $p_i \vee q_i$ to be true is the first for which either p or q is true; also all later terms will have p or q true and so will be true. $P \vee_m Q$ is therefore elementary and has the required table.

But in spite of this representation ${}_1T_2$ still appears to be the fundamental system since its truth values correspond entirely to the significance of true and completely false, whereas in ${}_\mu T_m$, $m > 2$ either $\mu > 1$ or $m - \mu > 1$, and this equivalence no longer holds. We must however take into account the fact that our development has been given in the language of ${}_1T_2$ and for that very reason every other kind of system appears distorted. This suggests that *if* we translate the entire development into the language of any one ${}_\mu T_m$ by means of its interpretation, then it would be the formal system most in harmony with regard to the two developments.

The notion "definite" and the independence of the axiom of choice

ABRAHAM A. FRAENKEL

(*1922b*)

Russell (*1905a*, pp. 47–48) gave an amusing illustration of the axiom of choice. If we have $\aleph_0$ pairs of boots, we can, by any one of a number of rules, get a "choice set" that includes a boot of every pair: take every left boot, or take the left boot for the even pairs and the right boot for the odd pairs, and so forth. If, now, we have $\aleph_0$ pairs of socks, we may be unable to specify a rule of that sort.

In the present paper Fraenkel gives a precise form to this idea and uses it to obtain an independence proof of the axiom of choice. The independence proof has to be carried out for a definite set theory, which is known, or assumed, to be consistent; Fraenkel adopts Zermelo's axiomatization (*1908a*), with a modification that we shall discuss presently. Since the principle of choice is provable for any finite set of sets, the weakest form for which independence is in doubt deals with a denumerably infinite set of pairs, and this is the version that he considers.

Fraenkel introduces a denumerably infinite number of distinct objects, a_1, $\bar{a}_1$, a_2, $\bar{a}_2, \ldots$, that are not sets. The sets $\{a_k, \bar{a}_k\}$, called *cells* by Fraenkel, represent Russell's pairs of socks in the axiomatization. The following sets, and those exclusively, are assumed to exist: (1) the null set, 0; (2) the set $Z_0 = \{0, \{0\}, \{\{0\}\}, \ldots\}$ (in essence, the set of natural numbers); (3) the set $A = \{\{a_1, \bar{a}_1\}, \{a_2, \bar{a}_2\}, \ldots\}$ (the set of cells); (4) all sets obtained from these three sets by any finite number of applications of Zermelo's Axioms II–V (with Axiom III suitably modified).

The key step in the proof (Fundamental Theorem) is to show that for every set M there exists a subset A_M of A that contains all but a finite number of elements of A and is such that A_M remains unchanged when the two elements, a_k and $\bar{a}_k$, of any cell $\{a_k, \bar{a}_k\}$ of A_M are interchanged.

Since, in contradiction with the Fundamental Theorem, every choice set for the set A would change as soon as, for any k, a_k and $\bar{a}_k$ are interchanged, there exists no such choice set. (Here the assumption that a_k and $\bar{a}_k$ are not sets is, of course, decisive; if they were, there might obtain set-theoretic relations such that the interchange of a_k and $\bar{a}_k$ would leave the choice set unchanged.) Hence a model has been constructed in which the negation of the axiom of choice as well as Zermelo's axioms (other than the axiom of choice, of course) hold.

The proof of the Fundamental Theorem proceeds by an inductive argument that parallels the construction of the set M from the sets 0, Z_0, A, and the objects a_1, $\bar{a}_1$, a_2, $\bar{a}_2, \ldots$. At a certain stage in the argument Zermelo's Axiom III (axiom of separation) has to be considered.

As stated by Zermelo (above, p. 202), this axiom asserts that, given a set and a "definite property", the elements of the set that possess the property form a set. The definition of "definite property" given by Zermelo is imprecise, and Fraenkel could not carry out his proof with such a vague notion; at the beginning of his paper he had to reformulate the axiom of separation with the help of a specific notion of function. By means of a function new objects are constructed from initial ones through a finite number of applications of the axioms. Other writings of Fraenkel's from that period (*1921, 1922a, 1922c, 1923, 1923a, 1925*) also deal with that point.

At about the same time Skolem was proposing another method for making Zermelo's "definiteness" more precise. A property is definite in Skolem's sense if it is expressed by a well-formed formula of a specified logical system, more precisely still, by a well-formed formula of the simple predicate calculus of first order in which the sole predicate constants are ε and, possibly, $=$ (see below, p. 292). Weyl's way of solving the difficulty (*1910*, p. 112; *1918*, chap. 1, § 2, and p. 36) anticipates Skolem's and is very close to it. Skolem shows (*1929*, pp. 7–9) that Fraenkel's method can be reduced to his own.

Some years later, Zermelo (*1929*) considered Fraenkel's emendation of his formulation of the axiom of separation (he apparently did not know Skolem's proposal at the time); he found that he could not accept it, because it makes use of the notion of natural number, and this notion itself, he thought, should have a set-theoretic definition. He made the elements of his original characterization of "definite" more explicit, but Skolem (*1930*) showed that its weaknesses could hardly be remedied.

Today an axiomatization of set theory is usually embedded in a logical calculus, and it is Weyl's and Skolem's approach to the formulation of the axiom of separation that is generally adopted. The introduction by von Neumann of classes that are not sets, with appropriate axioms, can be regarded as a third method of making the notion "definite property" more precise (see below, p. 393).

Fraenkel's independence proof is carried out for a system in which the existence of a denumerably infinite number of objects that are not classes can be assumed without contradiction. Such objects, now technically known as *urelements*, have become a standard tool used in independence proofs for set theory (*Fraenkel 1937*, *Lindenbaum and Mostowski 1938*, *Mostowski 1939*). But, from a broader perspective, these non-classes are unsatisfactory. In set theory, after all, all mathematical notions (natural numbers, real numbers, functions, and so on) should turn out to be sets or at worst classes. But only at the cost of some unwelcome tampering with the axioms of set theory could these non-classes be replaced by (rather special) sets (*Bernays 1937–1954*, *Mendelson 1956*, *Specker 1957*). Cohen's method (*1963*, *1963a*, *1964*) opened another road.

The paper was translated by Beverly Woodward. Professor Fraenkel read and approved the translation. It is printed here with his kind permission and that of Walter de Gruyter and Co.

Whether the principle of choice, or the "axiom of choice",[1] which forms its core, is independent of the other principles of mathematics, that is, in effect, of the axioms of set theory, is a question to which no answer has yet been found. This question acquired increased importance when the principle of choice showed itself, after

[1] See *Zermelo 1904*, p. 516 [[above, p. 141]], *1908*, p. 110 [[above, p. 186]], and *1908a*, p. 266 [[above, p. 204]].

suggestions made by B. Levi, E. Schmidt, and especially E. Zermelo, to be an indispensable device for proofs in nearly all branches of mathematics. On the occasion of an investigation into the foundations of set theory and into the independence in general of one of the axiom systems that can be taken as a basis for that theory,[2] I found in particular an independence proof for the axiom of choice. Its realization proved to be closely connected with a sharp clarification of the notion "definite property" in the acceptation requisite for set theory; such a clarification had not been attained previously and is even absent from the pioneer work of Zermelo (*1908a*). A preliminary remark about this fundamental notion is therefore required for the following proof, but it will be held within the bounds necessary for this purpose. Moreover, the independence proof will, for the sake of simplicity, be carried out for the Zermelo axiom system (*1908a*), whose Axioms I, II, IV–VII will therefore remain unchanged, as will the "domain" $\mathfrak{B}$ with which set theory is concerned and the expressions $a \varepsilon b$ (a is an element of the set b) and $a \Subset b$ (a is a subset of the set b).

To arrive at a precise formulation of Axiom III (axiom of separation), which as formulated in *Zermelo 1908a* contains the imprecise notion "definite", we shall understand by *function* a rule of the following kind: an object $\varphi(x)$ shall be formed from a ("variable") object x that can range over the elements of a set, and possibly from further given ("constant") objects, by means of a prescribed application (repeated only a finite number of times, of course, and denoted by φ) of Axioms II–VI. For example, $\varphi(x) = \{\{\{x\}, \{0\}\}, \mathfrak{U}x + \{\{0\}\}\}$. Clearly this does not include any general notion of function or correspondence, or any new fundamental notion whatsoever. The axiom of separation now reads:

Axiom III. If a set M is given, as well as, in a definite order, two functions φ and ψ, then M possesses a subset $M_{\mathfrak{E}}$ (or a subset $M_{\mathfrak{E}'}$) containing as elements all the elements x of M for which $\varphi(x)$ is an element of $\psi(x)$ (or for which $\varphi(x)$ is not an element of $\psi(x)$), and no others.

The role of "variables" in the primitive functions defined by Axioms II and IV–VI is clear without further comment. Axiom III can define a function in such a way that M is variable or that one of the constant objects occurring in the functions φ and ψ is left variable. The choice function of $x = T$, defined by means of Axiom VI, has to do with *any* object whatsoever of the required kind and not, as is the case for the other primitive functions, with a uniquely determined object.

In this connection some questions (to be discussed in *Fraenkel 1925*) arise, namely, whether and how in legitimate set theory the formation of subsets proceeds according to this well-determined schema, that is, in particular, whether this schema is sufficiently comprehensive; further, whether it is possible, without a vicious circle, to link Axiom III and the notion of function with each other; and finally whether, in introducing the notion of function, we can limit ourselves to Axioms II–V, thus omitting Axiom VI. The last question remains outside the scope of the considerations that follow, which take no account of Axiom VI at all, while, as it turns out, the first two are partially clarified below.

To show now that the axiom of choice (VI) is independent of the other axioms (I–V and VII) we shall consider as belonging to the domain $\mathfrak{B}$ (retaining the usual meaning of the primitive relation $a \varepsilon b$) the following objects: first, the null set 0 and

[2] See *Fraenkel 1921*, also *1922a* and its continuation now in preparation [[*1925*]]).

also a denumerably infinite number of distinct objects, $a_1, \bar{a}_1, a_2, \bar{a}_2, a_3, \bar{a}_3, \ldots$, none of which is considered a set; second, the set $Z_0 = \{0,\{0\},\{\{0\}\},\ldots\}$ (*Zermelo 1908a*, p. 267 [[above, p. 204]]); third, the set $A = \{\{a_1, \bar{a}_1\}, \{a_2, \bar{a}_2\}, \{a_3, \bar{a}_3\}, \ldots\}$;[3] fourth, the (uniquely determined) sets obtained from the objects of $\mathfrak{B}$ on the basis of Axioms II–V—and no other objects.[4] An object (set) "exists" if it occurs in $\mathfrak{B}$. Axioms I–V and VII are then satisfied. All objects, except those mentioned in the first clause, are sets; every set can be obtained (in at least one way) from the objects mentioned in the first, second, and third clauses, the "*primitive objects*", by a finite number of applications of Axioms II–V.

The axiom of choice (VI) will be proved to be independent when the following has been shown:

Fundamental theorem: If M is a set, then, corresponding to M, there exists at least one subset $A_M \in A$ that contains *almost all* elements of A, is of the form $A_M = \{A_{n_1}, A_{n_2}, \ldots\}$, where $A_{n_k} = \{a_{n_k}, \bar{a}_{n_k}\}$, and is such that, if $a_{n_k} \varepsilon A_{n_k} \varepsilon A_M$ (that is, if $a_{n_k} \varepsilon \mathfrak{S}A_M$), M is mapped onto itself whenever a_{n_k} and $\bar{a}_{n_k}$ are interchanged.

If this is proved, there cannot be any choice in the sense of Axiom VI for the set A itself, for instance, since clearly every choice set changes as soon as any a_k is interchanged with $\bar{a}_k$.

In what follows, the sets $A_k = \{a_k, \bar{a}_k\}$ will be called *cells*, and a_k and $\bar{a}_k$ conjugate cell elements; every subset of A containing almost all elements of A will be called a *principal set*, so that the intersection of a finite number of principal sets is a principal set; a set obtained from an arbitrary set M by the interchange of the elements of a cell A_k will be said to be *conjugate* to M with respect to A_k and will be denoted by $\bar{M}^k$; if A_k is here allowed to be any arbitrary cell of a certain principal set B, then we shall write $\bar{M}^{k|B}$. A set conjugate to itself will be said to be *symmetric* (with respect to A_k). The fundamental theorem then states that for every set M there exists a principal set such that M is symmetric with respect to all cells of that set. For the primitive objects this is patently the case.

The proof of the fundamental theorem can be carried out on the basis of Theorems 1–5 below. In order to reveal the essence of the proofs of these theorems without becoming prolix and repeating the same fundamental ideas, it will suffice to carry out the proofs of Theorems 2 and 4, which are simpler but contain all the essential ideas. Only Theorem 1, which at first perhaps looks obvious, occupies a somewhat special place.

Theorem 1. If M is a set, then the set $\bar{M}^k$ conjugate to M with respect to an arbitrary cell A_k also exists.

Theorem 2. If, by the use of Axioms II, IV, and V alone, a set is formed from given objects in such a way that for each of these objects there is a principal set with respect to whose cells the object is symmetric (hence, for instance, from the primitive objects), then that set is symmetric with respect to the cells of a principal set.

Proof. Assume that the object M is symmetric with respect to the cells of the

[3] I am particularly indebted to Zermelo for his suggestion that it suffices to let the elements of A consist of *only two* elements each, as well as, in general, for his emphasis on the notion of symmetry (see below).

[4] See the "axiom of limitation" [["Beschränktheitsaxiom"]] in *Fraenkel 1922a*, p. 234, and *1922*, end of § 1, p. 163).

principal set A_M and that N is symmetric with respect to those of the principal set A_N. Then, first, $\{M\}$ is symmetric with respect to the cells of A_M, and $\{M, N\}$ is symmetric with respect to those of the principal set $[A_M, A_N]$, the intersection of the two sets. Second, let A_k be an arbitrary cell of A_M, so that $\overline{M}^k = M$, and let M_0 be $\in M$; the set $\overline{M}_0^k \in \overline{M}^k = M$, conjugate to M_0 with respect to A_k, exists by Theorem 1; hence we have $\overline{M}_0^k \, \varepsilon \, \mathfrak{U}M$; therefore $\mathfrak{U}M$ is symmetric with respect to the cells of A_M. Finally, if $\mu \, \varepsilon \, m \, \varepsilon \, M$ and further $A_k \, \varepsilon \, A_M$, then $\overline{\mu}^k \, \varepsilon \, \overline{m}^k \, \varepsilon \, \overline{M}^k = M$, that is, $\overline{\mu}^k \, \varepsilon \, \mathfrak{S}M$ follows from $\mu \, \varepsilon \, \mathfrak{S}M$; therefore $\mathfrak{S}M$ too is symmetric with respect to the cells of A_M. But the set considered in Theorem 2 is obtained, by a finite number of repeated applications of the primitive functions just used, from objects such that for each there is a principal set with respect to whose cells it is symmetric; so Theorem 2 is proved.

THEOREM 3. Under the assumptions of Theorem 2, let one of the given objects be left variable and let it be denoted by x, so that instead of a set a function $\varphi(x)$ is obtained. Then for φ there is a principal set such that, for every object x, $\varphi(\overline{x}^k)$ is conjugate to $\varphi(x)$ with respect to all cells A_k of that set.

THEOREM 4. If φ and ψ are given functions of the kind specified in Theorem 3 and if M is a set symmetric with respect to the cells of some principal set, then the subset $M_{\mathfrak{E}}$ (or $M_{\mathfrak{E}'}$) $\in M$ determined in accordance with Axiom III by $\varphi \, \varepsilon \, \psi$ (or φ not $\varepsilon \, \psi$) is symmetric with respect to the cells of some principal set.

Proof. Let A_M be a principal set for which $M = \overline{M}^{k|A_M}$; let A_φ and A_ψ be, in conformity with Theorem 3, principal sets such that, for example, for every $m \, \varepsilon \, M$,

$$\varphi(\overline{m}^{k|A_\varphi}) = \overline{\varphi(m)}^{k|A_\varphi} \quad \text{and} \quad \psi(\overline{m}^{k|A_\psi}) = \overline{\psi(m)}^{k|A_\psi}.$$

Let B denote the intersection $[A_M, A_\varphi, A_\psi]$. By the definition of $M_{\mathfrak{E}}$, we have for every $n \, \varepsilon \, M_{\mathfrak{E}}$ (and for no other element of M) $\varphi(n) \, \varepsilon \, \psi(n)$ or even $\overline{\varphi(n)}^k \, \varepsilon \, \overline{\psi(n)}^k$, where k from now on stands for the subscript of some arbitrary cell $A_k \, \varepsilon \, B$. Because of $B \in A_\varphi$ and $B \in A_\psi$ we have, by Theorem 3, $\overline{\varphi(n)}^k = \varphi(\overline{n}^k)$ and $\overline{\psi(n)}^k = \psi(\overline{n}^k)$, hence also $\varphi(\overline{n}^k) \, \varepsilon \, \psi(\overline{n}^k)$. Therefore—since, B being $\in A_M$, $\overline{n}^k$ is an element of M whenever n is—$\overline{n}^k \, \varepsilon \, M_{\mathfrak{E}}$ holds; that is, $M_{\mathfrak{E}}$ is symmetric with respect to the cells of the principal set B. Consequently, the same clearly holds for $M_{\mathfrak{E}'}$ too.

THEOREM 5. Under the assumptions of Theorem 4, let M or one of the constant objects that are arguments of φ or ψ be left variable and let it be denoted by x, so that instead of the constant set $M_{\mathfrak{E}}$ (or $M_{\mathfrak{E}'}$) a function $\chi(x)$ is obtained; then for $\chi(x)$ there is a principal set such that, for every object (x), $\chi(\overline{x}^k)$ is conjugate to $\chi(x)$ with respect to all cells A_k of that set.

Having obtained these results, let us state the following recursive definitions.

(A) A primitive object or a set generated from primitive objects by the use of Axioms II, IV, and V alone is said to be of the 0th class. Likewise an associated *function* in whose formation one of the primitive objects is replaced by a variable is said to be of the 0th class.

(B) A subset $M_{\mathfrak{E}}$ (or $M_{\mathfrak{E}'}$) $\in M$ defined by $\varphi \, \varepsilon \, \psi$ (or φ not $\varepsilon \, \psi$) is said to be of at most the $(p + 1)$th class if M, as well as the functions φ and ψ, is of at most the pth class. Likewise an associated *function* in which M or one of the constant objects that are arguments of φ or ψ is left variable is said to be of at most the $(p + 1)$th class.

By Theorem 2 for every object of the 0th class, and by Theorem 4 for every object of at most the 1st class, there is a principal set with respect to whose cells the object is symmetric. Let us assume that the same holds for every object of at most the pth class; further, assume that for every function φ of at most the pth class there exists a principal set such that, with respect to all cells A_k of that set, $\varphi(\bar{x}^k) = \overline{\varphi(x)}^k$ always holds, as is the case for $p = 0$ by Theorem 3 and for $p = 1$ by Theorem 5. Then, by a proof exactly analogous to that of Theorem 4, we obtain the result that every object of at most the $(p + 1)$th class also possesses this very same property of symmetry. Finally, the assertion of Theorem 5 is thereby easily carried over to the functions of at most the $(p + 1)$th class. Accordingly, since for every object M there is a p such that M is of at most the pth class, every object M is symmetric with respect to the cells of a principal set A_M, as the fundamental theorem asserts.

Some remarks on axiomatized set theory

THORALF SKOLEM

(*1922*)

This is the text of an address delivered before the Fifth Congress of Scandinavian Mathematicians (Helsinki, 4–7 August 1922). It deals with several questions encountered in the axiomatization of set theory and contains eight points, which are listed in the fourth paragraph. Point 2, on Zermelo's notion "definite property", was discussed in the introductory note to Fraenkel's paper (above, p. 285).

Point 3 presents a new proof of Löwenheim's theorem and discusses some of the implications of the theorem for formalized set theory. Löwenheim's theorem states that, if a formula of the predicate calculus of first order is satisfiable, it is $\aleph_0$-satisfiable. Skolem had proved the theorem earlier (*1920*) and generalized it to a denumerably infinite set of formulas. The 1920 proof (see above, pp. 256–259) makes use of the axiom of choice and of a result of Dedekind's about chains. The new proof dispenses with the axiom of choice and, instead of Dedekind's result, uses an argument whose first part is already in Löwenheim's paper. From the satisfiability of the formula both Löwenheim and Skolem conclude that there is, for every n, a solution of the nth level (for a definition of this expression see below, p. 508). At that point Löwenheim states without justification that the formula is $\aleph_0$-satisfiable (see above, p. 240). Skolem supplies the missing step (which Quine (*1955b*, p. 254) called the "law of infinite conjunction"), and he does so without using the axiom of choice (the gap is now frequently bridged with the help of König's infinity lemma (*1926*, *1927*)).

Skolem then uses Löwenheim's theorem to show that any formalized set theory has a denumerable model. Since natural numbers can be defined in terms of sets (for example, by means of Dedekind's "chains" or in Zermelo's way), he sees the possibility that there might be two distinct models of set theory giving rise to two distinct systems of natural numbers. In a later paper (*1929*, § 7) Skolem comes back to this question in greater detail; then (*1933*, *1934*) he actually exhibits a nonstandard model of arithmetic.

For Skolem the discrepancy between an intuitive set-theoretic notion and its formal counterpart leads to the "relativity" of set-theoretic notions. Thus, two sets are equivalent if there exists a one-to-one mapping of the first onto the second; but this mapping is itself a collection of ordered pairs of elements. If, in a formalized set theory, this collection exists as a set, the two given sets are equivalent in the theory; if it does not, the sets are not equivalent in the theory and, when one set is that of the natural numbers as defined in the theory, the other becomes "nondenumerable". The existence of such a "relativity" is sometimes referred to as the Löwenheim–Skolem paradox. But, of course, it is not a paradox in the sense of an antinomy;

it is a novel and unexpected feature of formal systems.

In Point 4 Skolem indicates a limitation of Zermelo's set theory: it does not ensure the existence of some "large" sets, such as the set $\{Z_0, Z_1, Z_2, \ldots\}$, where Z_0 is the set of natural numbers and Z_i the power set of Z_{i-1}. If the continuum hypothesis is assumed, this means that all of Zermelo's sets are of cardinalities less than $\aleph_\omega$. To remedy this deficiency Skolem proposes a new axiom, the axiom of replacement, which will become a standard part of set theory. At about the same time Fraenkel (*1922a*, but see also *1921* and *1922*) and Lennes (*1922*) make similar suggestions. (See above, p. 114, line 22, for an early formulation of the axiom of replacement in Cantor.)

In Point 6 Skolem deals with the non-categoricity of Zermelo's axioms, not, however, by using the Löwenheim–Skolem theorem, but by actually exhibiting various models that are subdomains of some initial model B (assumed to exist). In footnote 9 he invokes this non-categoricity to suggest that set theory is perhaps unable to solve all questions concerning cardinality and, in particular, that the continuum hypothesis may be independent of the axioms of set theory (see also below, p. 368).

These indications do not exhaust the content of a rich and clearly written paper, which when it was published did not receive the attention it deserved, although it heralded important future developments.

The translation is by Stefan Bauer-Mengelberg, and it is printed here with the kind permission of Professor Skolem.

Set theory in its original version led, as we know, to certain contradictions (antinomies), and no one has yet succeeded in giving a clarification of them that has won general acceptance. In view of this threat to set theory, attempts have been made to develop that theory by means of certain fundamental assumptions, or axioms, in such a way that the part presumed to be correct and useful would remain provable while the contradictions would be avoided.

Until now, so far as I know, only *one* such system of axioms has found rather general acceptance, namely, that constructed by Zermelo (*1908a*). Russell and Whitehead, too, constructed a system of logic that provides a foundation for set theory; if I am not mistaken, however, mathematicians have taken but little interest in it. In what follows I therefore concern myself almost exclusively with Zermelo's axiomatization, and I touch upon Russell and Whitehead's at only a single point, albeit a rather important one.

Zermelo considers a domain B of objects, among which are the sets. Between these objects there obtain relations of the form $a \varepsilon b$ (a is an element of b) and of the form $a = b$.[1] In this domain, then, seven axioms, for which I refer the reader to Zermelo's paper, are to be satisfied. I shall use the same numbering and nomenclature for the axioms here as does Zermelo.

In this address I wish to discuss the following eight points:

1. The peculiar fact that, in order to treat of "sets", we must begin with "domains" that are constituted in a certain way;

[1] If all objects of the domain are sets, this relation can be reduced to the relation $a \varepsilon b$ by means of the axiom of extensionality. For, if a and b are sets, then (see the definitions in Section 2) $a = b$ means the same as

$$\Pi_i((\overline{i \varepsilon a}) + (i \varepsilon b))((i \varepsilon a) + \overline{(i \varepsilon b)}).$$

2. A definition, much to be desired, that makes Zermelo's notion "definite proposition" precise;

3. The fact that in every thoroughgoing axiomatization set-theoretic notions are unavoidably relative;

4. The fact that Zermelo's system of axioms is not sufficient to provide a foundation for ordinary set theory;

5. The difficulties caused by the nonpredicative stipulations when one wants to prove the consistency of the axioms;

6. The nonuniqueness [[Mehrdeutigkeit]] of the domain B;

7. The fact that mathematical induction is necessary for the logical investigation of abstractly given systems of axioms;

8. A remark on the principle of choice.

1. If we adopt Zermelo's axiomatization, we must, strictly speaking, have a general notion of domains in order to be able to provide a foundation for set theory. The entire content of this theory is, after all, as follows: for every domain in which the axioms hold, the further theorems of set theory also hold. But clearly it is somehow circular to reduce the notion of set to a general notion of domain. If, however, it were maintained that only the notion of the single domain B is necessary, hence not a general notion "domain", this consideration could also be applied to the sets themselves; thus, if a proposition should be advanced about some unspecified set, we could say: no general notion "set" is needed, but only the idea of a single set that we assume to be given.

It must be noted, incidentally, that the domain B is not uniquely determined by the axioms (see Section 6 below). Moreover, it will appear clearly from what follows that one cannot undertake logical investigations of such domains without to a certain extent applying set-theoretic considerations to them as well, if, that is, one wishes to follow the purely set-theoretic method and avoid including the notion of number among the fundamental ones.

Furthermore, it seems to be clear that, when founded in such an axiomatic way, set theory cannot remain a privileged logical theory; it is then placed on the same level as other axiomatic theories.

2. A very deficient point in Zermelo is the notion "definite proposition". Probably no one will find Zermelo's explanations of it satisfactory. So far as I know, no one has attempted to give a strict formulation of this notion; this is very strange, since it can be done quite easily and, moreover, in a very natural way that immediately suggests itself. In order to explain this, and also with a view to later considerations, I mention the five basic operations of mathematical logic here, using Schröder's notation (*1890*):

($1_\times$) Conjunction, denoted by a dot or by juxtaposition;

(1_+) Disjunction, denoted by the sign $+$;

(2) Negation, denoted by a bar over the expression to be negated;

($3_\times$) Universal quantification, denoted by the sign Π;

(3_+) Existential quantification, denoted by the sign Σ.

As is well known, only three of these five operations are really needed, since ($1_\times$) and (1_+), like ($3_\times$) and (3_+), are mutually definable by means of (2).

By a definite proposition we now mean a finite expression constructed from elementary propositions of the form a ε b or a = b by means of the five operations mentioned. This is a completely clear notion and one that is sufficiently comprehensive to permit us to carry out all ordinary set-theoretic proofs. I therefore adopt this conception as a basis here.

3. This third point is the most important: *If the axioms are consistent, there exists a domain B in which the axioms hold and whose elements can all be enumerated by means of the positive finite integers.*

To prove this I must first explain a theorem proved by Löwenheim (*1915*). By a *first-order proposition* ⟦*Zählaussage*⟧ (Löwenheim says "first-order expression" ⟦"Zählausdruck"⟧) is meant a finite expression constructed from class and relative coefficients in the sense of Schröder (*1895*) by means of the five logical operations mentioned above. Then Löwenheim's theorem reads as follows:

If a first-order proposition is satisfied in any domain at all, it is already satisfied in a denumerably infinite domain.

Löwenheim's proof is unnecessarily complicated, and moreover the complexity is rather essential to his argument, since he introduces subscripts when he expands infinite logical sums or products; in this way he obtains logical expressions with a nondenumerably infinite number of terms. Thus he must make a detour, so to speak, through the nondenumerable. In a previous paper (*1920*) I therefore gave a simplified proof of Löwenheim's theorem, along with some generalizations of it. One of these generalizations, which is of importance here, reads:

Let there be given an infinite sequence U_1, U_2, . . . of first-order propositions numbered with the integers; if, now, it is consistent to assume that all these propositions hold simultaneously, they can all be simultaneously satisfied in the infinite sequence of the positive integers, 1, 2, 3, . . . , by a suitable determination of the class and relation symbols occurring in the propositions.

I proved these theorems by forming an intersection and using the principle of choice. The formation of the intersection can be avoided immediately by use of a recursive definition; but here, where we are concerned with an investigation in the foundations of set theory, it will be desirable to avoid the principle of choice as well. Therefore I now indicate very briefly how this can be done. It will also appear from the proof that general set-theoretic notions are unnecessary for the understanding of the content of these theorems.

Since the proof of the generalization mentioned can very easily be carried out in the same way as the proof of Löwenheim's theorem, I will concern myself only with the latter here.

As I have shown (*1920*), every first-order proposition can be put into what I called a *normal form*, provided some auxiliary classes and relations are defined. A first-order proposition in normal form has the following structure:

$$\prod_{x_1}\prod_{x_2}\ldots\prod_{x_m}\sum_{y_1}\sum_{y_2}\ldots\sum_{y_n} U_{x_1,\ldots,x_m,y_1,\ldots,y_n},$$

where $U_{x_1,\ldots,x_m,y_1,\ldots,y_n}$ is a proposition constructed from class and relation symbols (class and relative coefficients in the sense of Schröder) by means of the first three of the logical operations mentioned above. Let, therefore, such a first-order proposition in normal form be given; we assume it to be consistent.

The proof proceeds by way of an infinite sequence of steps. For the first step we choose $x_1 = x_2 = \ldots = x_m = 1$. Then it must be possible to choose $y_1, \ldots, y_n$ among the numbers $1, 2, \ldots, n + 1$ in such a way that $U_{1,1,\ldots,1,y_1,\ldots,y_n}$ is satisfied. Thus we obtain one or more solutions of the first step, that is, assignments determining the classes and relations in such a way that $U_{1,1,\ldots,1,y_1,\ldots,y_n}$ is satisfied. The second step consists in choosing, for $x_1, \ldots, x_m$, every permutation with repetitions of the $n + 1$ numbers $1, 2, \ldots, n + 1$ taken m at a time, with the exception of the permutation $1, 1, \ldots, 1$, already considered in the first step. For at least one of the solutions obtained in the first step, it must then be possible, for each of these $(n + 1)^m - 1$ permutations, to choose $y_1, \ldots, y_n$ among the numbers $1, 2, \ldots, n + 1 + n((n + 1)^m - 1)$ in such a way that, for each permutation $x_1, \ldots, x_m$ taken within the segment $1, 2, \ldots, n + 1$ of the number sequence, the proposition $U_{x_1,\ldots,x_m, y_1,\ldots,y_n}$ holds for a corresponding choice of $y_1, \ldots, y_n$ taken within the segment $1, 2, \ldots, n + 1 + n((n + 1)^m - 1)$. Thus from certain solutions gained in the first step we now obtain certain continuations, which constitute solutions of the second step. It must be possible to continue the process in this way indefinitely if the given first-order proposition is consistent.

In order now to obtain a uniquely determined solution for the entire number sequence, we must be able to choose a single solution from among all those obtained in a given step. To achieve this, we can always take the first from all of the solutions obtained in an arbitrary step, once they have been ordered in a sequence in the following way.

The relative coefficients occurring in the given first-order proposition can be linearly ordered so that the relative coefficients formed within the segment $1, 2, \ldots, n$ of the number sequence precede all new relative coefficients that are formed within the segment $1, 2, \ldots, n + 1$. For any two different solutions L and L' of an arbitrary step, write $L < L'$ if and only if $R_{ij}\ldots$ is equal to 0 in L and 1 in L', where $R_{ij}\ldots$ is the first relative coefficient having different values in L and L'.[2] From $L < L'$ and $L' < L''$ it then follows that $L < L''$; we can also readily see that for two solutions L_n and L'_n of the nth step that are, respectively, continuations of the solutions L_ν and L'_ν of the νth step $L_n < L'_n$ implies $L_\nu \leqq L'_\nu$.

Let $L_{1,n}, L_{2,n}, \ldots, L_{e_n,n}$ be the solutions of the nth step. If we now form the sequence $L_{1,1}, L_{1,2}, \ldots$ of the first solutions, we can verify without difficulty that they converge in the logical sense. For let $L_{1,n}$ be a continuation of $L_{a_\nu^n,\nu}$ $(n > \nu)$. Then, if $n' > n$, $a_\nu^n \leqq a_\nu^{n'}$. But, since the number a_ν^n can only have the values 1 to e_ν, it must remain constant for all sufficiently large n. Thus we can obtain as "limit" the fact that the first-order proposition is satisfied in the domain of the entire number sequence. Q. e. d.

Now the previously mentioned generalization of Löwenheim's theorem can be applied in the present case, that of Zermelo's axiom system, in the following way.

The definite first-order propositions can be enumerated according to their form by means of the positive integers; for we can order them according to how many symbols for sets occur in them, and for a given number of such symbols there is only a finite number of propositions, so that these in turn can be thought of as ordered

[2] Here 0 and 1 are Schröder's propositional values; 0 means "false" and 1 means "true".

according to some rule. Consequently, Axiom III (axiom of separation) can be replaced by an infinite sequence of simpler axioms—which, like the rest of Zermelo's axioms, are first-order propositions in the sense of Löwenheim—containing the two binary relations ε and $=$. We may then conclude: *If Zermelo's axiom system, when made precise, is consistent, it must be possible to introduce an infinite sequence of symbols* 1, 2, 3, ... *in such a way that they form a domain B in which all of Zermelo's axioms hold provided these symbols are suitably grouped into pairs of the form $a \varepsilon b$.*

This is to be understood in the following way: one of the symbols 1, 2, 3, ... will be the null set (that is, among the remaining symbols there is none that has the relation ε to the symbol in question); if a is one of the symbols, then $\{a\}$ is another;[3] if M is one of the symbols, then UM, SM, and DM are others;[4] and so forth.

So far as I know, no one has called attention to this peculiar and apparently paradoxical state of affairs. By virtue of the axioms we can prove the existence of higher cardinalities, of higher number classes, and so forth. How can it be, then, that the entire domain B can already be enumerated by means of the finite positive integers? The explanation is not difficult to find. In the axiomatization, "set" does not mean an arbitrarily defined collection; the sets are nothing but objects that are connected with one another through certain relations expressed by the axioms. Hence there is no contradiction at all if a set M of the domain B is nondenumerable in the sense of the axiomatization; for this means merely that *within* B there occurs no one-to-one mapping Φ of M onto Z_0 (Zermelo's number sequence). Nevertheless there exists the possibility of numbering all objects in B, and therefore also the elements of M, by means of the positive integers; of course, such an enumeration too is a collection of certain pairs, but this collection is not a "set" (that is, it does not occur in the domain B). It is also clear that the set UZ_0 cannot contain as elements arbitrarily definable parts of the set Z_0. For, since the elements of UZ_0 are to be found among the objects of the domain B, they can be numbered with the positive integers just like the elements of Zermelo's number sequence Z_0, and in a well-known way a new part of Z_0 can then be *defined*; but this part will not be a set, that is, will not belong to B.

Even the notions "finite", "infinite", "simply infinite sequence", and so forth turn out to be merely relative within axiomatic set theory. A set M is finite, according to Dedekind's definition, if it is not equivalent to any of its proper subsets. But the fact that the axioms hold does not rule out the possibility that we can define, first, parts of M that are not subsets or, second, correspondences that are not mappings, that is, "sets" of pairs. It is therefore quite possible that, within a domain B in which Zermelo's axioms hold, there exist sets that are "finite" in the sense of Dedekind and for which there are one-to-one mappings onto some of their proper parts; but these "mappings" are not sets of the domain.

Likewise, the notion "simply infinite sequence" or that of the Dedekind "chain" has only relative significance. If Z is a set having the property required by Axiom VII, Zermelo's number sequence Z_0 is defined as the intersection of all subsets of Z

[3] $\{a\}$ is the set containing a and only a as an element.

[4] UM, SM, and DM are, respectively, the power set, the union set, and the intersection set of M.

that have the same (chain) property. But being a subset of Z is not merely to be in some way definable, and nothing can prevent a priori the possibility that there exist two different Zermelo domains B and B' for which different Z_0 would result.

These peculiar relativities could easily be illustrated in more detail for simpler axiom systems; but I cannot go into that, lest I make this address too prolix.

Thus, *axiomatizing set theory leads to a relativity of set-theoretic notions, and this relativity is inseparably bound up with every thoroughgoing axiomatization.*

The relativity is due to the fact that to be an object in B means something different and far more restricted than merely to be in some way definable. That this relativity must be inseparably bound up with every thoroughgoing axiomatization is clear; for it rests upon the general theorems of mathematical logic mentioned above. In order to obtain something absolutely nondenumerable, we would have to have either an absolutely nondenumerably infinite number of axioms or an axiom that could yield an absolutely nondenumerable number of first-order propositions. But this would in all cases lead to a circular introduction of the higher infinities; that is, *on an axiomatic basis higher infinities exist only in a relative sense.*

With a suitable axiomatic basis, therefore, the theorems of set theory can be made to hold in a merely *verbal* sense, on the assumption, of course, that the axiomatization is consistent; but this rests merely upon the fact that the use of the *word* "set" has been regulated in a suitable way. We shall always be able to define collections that are not called sets; if we were to call them sets, however, the theorems of set theory would cease to hold.

4. It is easy to show that Zermelo's axiom system is not sufficient to provide a complete foundation for the usual theory of sets. I intend to show, for instance, that if M is an arbitrary set, it cannot be proved that M, UM, $U^2M, \ldots$, and so forth ad infinitum form a "set". To prove this I introduce the notion "level" ⟦"Stufe"⟧ of a set. Such sets as 0, $\{0\}$, $\{\{0\}\}$, $\{0,\{0\}\}, \ldots$ I call sets of the first level; they are characterized by the fact that there exists a nonnegative integer n such that $S^nM = 0$.[5] The set Z_0 (Zermelo's number sequence) already constitutes an example of a set that is not of the first level, since, for every n, $S^nZ_0 = Z_0$. By a set M of the second level I mean one that is not itself of the first level but for which there exists a nonnegative integer n such that all elements of S^nM are sets of the first level. Thus Z_0 is a set of the second level. In a similar way sets of the third, fourth, and higher levels can be defined. We need not discuss whether with every set a level number is associated.

Now let a domain B in which the axioms hold be given. Then the sets in B that are of the first or second level form a partial domain B', and it is easy to see that the axioms must hold in B' too. The set Z_0 belongs to B'; if the infinite sequence Z_0, UZ_0, $U^2Z_0, \ldots$ formed a set M in B', however, it would clearly not be of the second level but of the third, and such a set just does not occur in B'. For it is evident that, for every n, S^nM will contain the set Z_0 as an element. Thus the sets Z_0, UZ_0, U^2Z_0, $\ldots$ do not form the elements of a set in B', even though Zermelo's axioms hold in B'; that is to say, the existence of such a set is not provable.

[5] S^nM is the nth union; thus $S^{n+1}M = S(S^nM)$, SM is the (first) union associated with M, and $S^0M = M$.

In order to remove this deficiency of the axiom system we could introduce the following axiom: *Let U be a definite proposition that holds for certain pairs (a, b) in the domain B; assume, further, that for every a there exists at most one b such that U is true. Then, as a ranges over the elements of a set M_a, b ranges over all elements of a set M_b.*

The addition of such an axiom does not, of course, change anything so far as the relativity explained above is concerned.[6]

5. It has presumably not yet been proved that Zermelo's axiom system is consistent, and it will no doubt be very difficult to do so. In particular, the *nonpredicative* stipulations governing the formation of sets will cause difficulties. It is not far-fetched to think of the axioms as generating principles of some sort; we generate new sets according to certain rules from sets already known. If, now, the formation of sets were a mere matter of constructing each new set solely from a finite number of prior sets, we could easily ascertain through an infinite process whether the axioms are consistent. But the difficulty is that we have to form some sets whose existence depends upon *all* sets. We then have what is called a *nonpredicative definition*. Poincaré criticized this kind of definition and regarded it as the real logical weakness of set theory.[7]

In Russell and Whitehead's system this point has been formally taken into account, namely, in the theory of what they call the logical types, but they, too, simply content themselves with circumventing the difficulty by introducing a stipulation, the *axiom of reducibility*. Actually, this axiom decrees that the nonpredicative stipulations will be satisfied. There is no proof of that; besides, so far as I can see, such a proof must be impossible from Russell and Whitehead's point of view as well as from Zermelo's. For it could probably be carried out only through the actual construction of a domain B with the desired properties by a procedure similar to that used in the proof of Löwenheim's theorem given above. There, however, the idea of the finite and the recursive mode of thought are employed. But neither in Russell and

[6] That this axiom indeed suffices for the proof of the existence of sets of the type mentioned can be seen as follows. (I must content myself here with the sketch of a proof.)

The segments (as they are called) A of Z_0 form, according to Axioms III and IV, the elements of a set. On the other hand, we can consider sets C that have the following, obviously definite properties:

(1) C is finite;
(2) $M \varepsilon C$;
(3) If $N \varepsilon C$, there exists an $N_1 \varepsilon C$ such that $N = UN_1$, except if $N = M$.

The sets C are obviously of the form $\{M, UM, U^2M, \ldots, U^nM\}$.

Furthermore it can be shown by means of inductive inferences (which here are in turn provable on the basis of the definition of Z_0 as an "intersection" or a "chain") that every set C is equivalent to one and only one segment A of Z_0, and conversely. If we form the proposition "$C \sim A$; C satisfies (1), (2), and (3); A is a segment of Z_0", it too is definite and of the kind required by the axiom mentioned. Since a set exists that contains all segments A of Z_0, the sets C according to this axiom also constitute all the elements of a set T. Then the associated union, ST, is the desired set, which contains M, UM, $U^2M, \ldots$, and so forth ad infinitum, as elements.

[7] A typical nonpredicative stipulation is, for example, that the intersection of all sets that have an arbitrary definite property E again be a set. This in fact follows from the axioms. For, first, the intersections $[M, M']$, where M is a fixed set having the property E while M' ranges over all sets having this property, constitute according to Axioms III and IV all the elements of a set T; this set, then, is already nonpredicatively introduced. Second, the intersection DT associated with T must obviously be the intersection of all sets having the property E.

Whitehead's system nor in Zermelo's are these notions supposed to be taken as basic; rather, they in turn are supposed to have their foundation in set theory. Thus we would come to a vicious circle.

It is clear that Zermelo's Axioms I–VI (VII excluded) are consistent; for with these axioms the domain B need contain only finite sets, formed by means of Axiom II alone.

6. I do not know whether anyone has proved rigorously that Zermelo's domain B is not uniquely determined by his axioms. That it is not so determined is a priori very plausible. But it does not suffice (as it would, for example, in the case of a commutative field [[Rationalitätsbereich]]) to say that we need only adjoin one new object in order to obtain a more comprehensive domain by means of the axioms; for we could, after all, imagine (though it is very improbable) that the axioms might then lead to contradictions, even if this were not the case before. I now present some of my reflections on this question.

If M is an arbitrary set, we can construct sequences of the form

$$\ldots M_2 \,\varepsilon\, M_1 \,\varepsilon\, M$$

I call them *descending ε-sequences.*

We now see almost immediately that in a domain B the sets M for which every ε-sequence necessarily terminates after a finite number of terms must form a part B' of B in which the axioms still hold. If now B' is a proper part of B, that is, if there are sets in B for which infinite descending ε-sequences exist, we already have two distinct Zermelo domains B and B'. But if every ε-sequence of an arbitrary set M in B terminates after a finite number of terms, the sets for which every ε-sequence terminates with the null set again form a part B' of B in which the axioms hold. If, therefore, B and B' are different, we again have two distinct Zermelo domains.

Let us now consider a domain B for whose sets every ε-sequence is finite and terminates with 0. Then with every set M we can associate a corresponding ε-tree [[ε-Verzweigungssystem]] that specifies how M is built up of elements, these in turn of elements, and so forth. Such a tree, then, is so constituted that infinitely many branches may indeed originate at one node; but, if we follow a connected sequence of branches (an ε-sequence), we reach 0 after a finite number of steps. If, now, a is an object not belonging to B, we can form an extended Zermelo domain B' containing a as follows.

Let M be an arbitrary set, N an arbitrary subset of $S^nM + S^{n-1}M + \cdots + M + \{M\}$, with n arbitrary, F the tree associated with M, and G the part of this system that consists of the points corresponding to N. Then from F we form a tree F' by inserting into all points of G a branch ending with a; that is, we adjoin a as a new element to each element of N. The trees thus obtained can then be regarded as sets of a more comprehensive domain B'.

We could now prove, by going through the axioms, that they hold for B' if they hold for B. The proof is very easy for all the axioms except Axiom III, for which the general formulation of the proof becomes rather involved. I therefore do not go into the matter in more detail here.

It would in any case be of much greater interest if we could prove that a new subset of Z_0 could be adjoined[8] without giving rise to contradictions; but this would probably be very difficult.[9]

7. In the investigation of an axiom system with respect to the logical dependence of the individual axioms upon one another, it has been regarded as sufficient simply to take other theories as a basis. But we cannot always be so fortunate as to be able to proceed on the basis of theories developed previously; we must insist on having a procedure that allows us to investigate the logical character of the axioms directly.

That B follows from A, if these are statements concerning certain objects, would, of course, from the point of view of set theory have to be regarded as follows: always, that is, for every set, the proposition $\bar{A} + B$ holds. It is, of course, often possible to prove this by means of set-theoretic axioms; but in the first place the question of the consistency of the axioms of set theory then arises, and in the second place we must take into consideration the fact that, whenever an axiomatic foundation has been provided, set-theoretic notions are relative, with the result that collections may yet be definable for which $\bar{A} + B$ does not hold. Now, if we were to investigate the axioms of set theory themselves in this way, we would have to prove that "domains" exist for which the axioms in question hold. If we do not again want to take axioms for domains as a basis (and so on ad infinitum), I see no other way out than to pass on to considerations such as those employed above in the proof of Löwenheim's theorem, considerations in which the idea of the finite integer is taken as basic.

Besides, the notion that really matters in these logical investigations, namely, "proposition following from certain assumptions", also is an inductive (recursive) one: the propositions we consider are those that are derivable by means of an *arbitrary finite number* of applications of the axioms. Thus the idea of the *arbitrary finite* is essential, and it would necessarily lead to a vicious circle if the notion "finite" were itself based, as in set theory, on certain axioms whose consistency would then in turn have to be investigated.

Set-theoreticians are usually of the opinion that the notion of integer should be defined and that the principle of mathematical induction should be proved. But it is clear that we cannot define or prove ad infinitum; sooner or later we come to something that is not further definable or provable. Our only concern, then, should be that the initial foundations be something immediately clear, natural, and not open to question. This condition is satisfied by the notion of integer and by inductive inferences, but it is decidedly not satisfied by set-theoretic axioms of the type of Zermelo's or anything else of that kind; if we were to accept the reduction of the former notions to the latter, the set-theoretic notions would have to be simpler than mathematical induction, and reasoning with them less open to question, but this runs entirely counter to the actual state of affairs.

[8] As was explained above (Section 3), B, after all, does not have to contain every "definable" subset of Z_0.

[9] Since Zermelo's axioms do not uniquely determine the domain B, it is very improbable that all cardinality problems are decidable by means of these axioms. For example, it is quite probable that what is called the continuum problem, namely, the question whether $2^{\aleph_0}$ is greater than or equal to $\aleph_1$, is not solvable at all on this basis; nothing need be decided about it. The situation may be exactly the same as in the following case: an unspecified commutative field is given, and we ask whether it contains an element x such that $x^2 = 2$. This is just not determined, since the domain is not unique.

In a paper (*1922*) Hilbert makes the following remark about Poincaré's assertion that the principle of mathematical induction is not provable: "His objection that this principle could not be proved in any way other than by mathematical induction itself is unjustified and is refuted by my theory." But then the big question is whether we can prove this principle by means of simpler principles and *without using any property of finite expressions or formulas that in turn rests upon mathematical induction or is equivalent to it.* It seems to me that this latter point was not sufficiently taken into consideration by Hilbert. For example, there is in his paper (bottom of page 170), for a lemma, a proof in which he makes use of the fact that in any arithmetic proof in which a certain sign occurs that sign must necessarily occur for a first time. Evident though this property may be on the basis of our perceptual intuition of finite expressions, a formal proof of it can surely be given only by means of mathematical induction. In set theory, at any rate, we go to the trouble of proving that every ordered finite set is well-ordered, that is, that every subset has a first element. Now why should we carefully prove this last proposition, but not the one above, which asserts that the corresponding property holds of finite arithmetic expressions occurring in proofs? Or is the use of this property not equivalent to an inductive inference?

I do not go into Hilbert's paper in more detail, especially since I have seen only his first communication. I just want to add the following remark: It is odd to see that, since the attempt to find a foundation for arithmetic in set theory has not been very successful because of the logical difficulties inherent in the latter, attempts, and indeed very contrived ones, are now being made to find a different foundation for it —as if arithmetic had not already an adequate foundation in inductive inferences and recursive definitions.

8. So long as we are on purely axiomatic ground there is, of course, nothing special to be remarked concerning the principle of choice (though, as a matter of fact, new sets are *not* generated *univocally* by applications of this axiom); but if many mathematicians—indeed, I believe, most of them—do not want to accept the principle of choice, it is because they do not have an axiomatic conception of set theory at all. They think of sets as given by specification of arbitrary collections; but then they also demand that every set be definable. We can, after all, ask: What does it mean for a set to exist if it can perhaps never be defined? It seems clear that this existence can be only a manner of speaking, which can lead only to purely formal propositions —perhaps made up of very beautiful *words*—about objects *called* sets. But most mathematicians want mathematics to deal, ultimately, with performable computing operations and not to consist of formal propositions about objects called this or that.

Concluding remark

The most important result above is that set-theoretic notions are relative. I had already communicated it orally to F. Bernstein in Göttingen in the winter of 1915–16. There are two reasons why I have not published anything about it until now: first, I have in the meantime been occupied with other problems; second, I believed that it was so clear that axiomatization in terms of sets was not a satisfactory ultimate foundation of mathematics that mathematicians would, for the most part, not be very much concerned with it. But in recent times I have seen to my surprise that so many mathematicians think that these axioms of set theory provide the ideal foundation for mathematics; therefore it seemed to me that the time had come to publish a critique.

The foundations of elementary arithmetic established by means of the recursive mode of thought, without the use of apparent variables ranging over infinite domains

THORALF SKOLEM

(*1923*)

Written in 1919, after Skolem had studied *Principia mathematica*, and published in 1923, the present paper offers a new way of developing arithmetic, a way that avoids the difficulties presented by the paradoxes and the complexities of the theory of types. Unbounded quantifiers are not used, and the theorems become free-variable formulas. Of course, such formulas constitute a limited means of expression as compared to those of an arithmetic embedded in the full predicate calculus of first order. To take a simple example, we can state and prove in free-variable arithmetic that there exist infinitely many primes, in the sense that, for any prime x, there exists a prime y such that $x < y \leqq x! + 1$. The existential quantifier used here is bounded and can easily be dispensed with. But that there exist infinitely many pairs of twin primes cannot (today) be stated by means of a free-variable formula.

In the arithmetic developed by Skolem the limitations of free-variable formulas are overcome, so far as possible, by the use of primitive recursive definitions for the introduction of new functions or predicates and the use of the rule of mathematical induction for proofs. This is what Skolem calls the "recursive mode of thought". Bounded quantifiers are simply a convenient shorthand notation. This arithmetic is now known as primitive recursive arithmetic.

Relinquishing existence as a primitive notion or as the negation of universalization brings Skolem's work somewhat close to intuitionism. Skolem himself wrote (*1929a*, p. 13) that in the present paper, "independently of Brouwer and without knowing his writings, I set forth similar ideas, confining myself, to be sure, to elementary arithmetic". Actually, intuitionistic arithmetic, while it refuses to apply the principle of excluded middle to negations of universal propositions, does contain such negations (even as antecedents of conditionals) and is wider than primitive recursive arithmetic.

Now consider a formula F of primitive recursive arithmetic and replace the variables by numerals. The defining equations of the primitive recursive functions that F may contain allow us to compute the values of these functions for the arguments selected. When the functions of F are replaced by their values, F becomes a variable-free numerical formula, which we shall call an *instance* of F; the truth of this instance can be determined by the truth-table method. If all the instances of F are true, we say that F is *verifiable*. The existence of instances gives to free-variable formulas a transparent meaning that other formulas lack. It is true that free-variable formulas are equivalent, from the point of view of derivation and that of interpretation, to their universal closures. But prenex occurrences of universal quantifiers do not raise the same problems as those whose scope is not the whole formula. With $A(x) \supset B$, or the equivalent $(x)(A(x) \supset B)$, contrast $(x)A(x) \supset B$. This last formula has no instances; in fact, it is equivalent to $\sim B \supset (Ex) \sim A(x)$, and the existential quantifier involves an infinite search in the sequence of natural numbers. If we have a proof of a free-variable formula F in a free-variable system of arithmetic and if we consider an instance of F, the proof of F can be transformed into a proof of this instance of F: variables are replaced by numerals, function values are computed, and an application of the rule of mathematical induction is replaced by a sequence of conditionals (see *Hilbert and Bernays 1934*, pp. 298–299). Hence a proof of the formula F in free-variable arithmetic can be regarded as a schema that, by effective (and easy) transformations, yields proofs of the instances of F.

Skolem does not raise any metamathematical questions about primitive recursive arithmetic. First, he does not work within a formal system but simply in "naïve" arithmetic. Then, what he does is to derive theorems to see how far one can go with the limited means that he has adopted. He reaches the greatest common divisor and the least common multiple, as well as elementary theorems about prime numbers.

Hilbert and Bernays (*1934*, pp. 307–346) gave an ample presentation of Skolem's work. Their interest in it was apparently prompted by the fact that they considered the arguments carried out in this arithmetic to be "finitary". Gödel's constructive arguments in his incompleteness proof for arithmetic (*1931*) are carried out in what is actually primitive recursive arithmetic (see, for instance, definitions 1–45 below, pp. 603–606; see also footnote 34, p. 603). Hilbert and Bernays (*1934*, p. 325) remark that the theorems of this arithmetic are verifiable formulas and that this fact directly yields the consistency of the arithmetic. A consistency proof carried out in a formalized metalanguage was given by H. E. Rose (*1961*), together with some incompleteness results. For further results on, and applications of, primitive recursive arithmetic see *Church 1955*, *1957*, *1957a*, and *Rose 1962*.

Gödel (*1931*, pp. 180–181; below, pp. 602–603) and, following him but in greater detail, Hilbert and Bernays (*1934*, pp. 310–312) showed that in primitive recursive arithmetic a truth function of equations is equivalent to an equation of the form $\mathfrak{a} = 0$. Using this idea, Curry (*1941*) and Goodstein (*1941*, *1957*) were able to develop primitive recursive arithmetic as a logic-free calculus (see some historical notes on that point in *Goodstein 1957*, p. 187).

The translation is by Stefan Bauer-Mengelberg, and it is printed here with the kind permission of Professor Skolem and the Norske Videnskaps-Akademi i Oslo.

The fundamental notions of logic that are customarily regarded as necessary for the foundations of mathematics (see, for example, *Whitehead and Russell 1910*, *1912*, and *1913*) are, in the first place, the following: the notions *proposition* and *propositional function* of one, two, or more variables; the three operations of (1) *conjunction* (expressed in ordinary language by means of the word "and" or the words "—— as well as ——"), (2) *disjunction* (customarily expressed by means of the words "either —— or ——"), and (3) *negation* (indicated by the word "not"); and finally Russell and Whitehead's notions "always" and "sometimes". These last two notions express the idea that a proposition holds in *all cases* or in *at least one case*, respectively. To say that a proposition holds in at least one case is to state an existential proposition, and that is customarily done by means of the words "there exists". Throughout the present work I shall use capital letters as symbols for propositional functions, so that $A(x)$, $B(x)$, ... denote propositional functions of one variable, $A(x, y)$, $B(x, y)$, ... denote propositional functions of two variables, and so on. Furthermore, I employ Schröder's signs (*1890*) for conjunction, disjunction, and negation, so that, if A and B are propositions, AB means the proposition "A as well as B", $A + B$ the proposition "either A or B", and finally $\bar{A}$ the negation of A. In addition, however, Russell and Whitehead introduce the notion "descriptive function". A descriptive function is an expression having an unambiguously determined meaning; it is a kind of functional proper name. Finally, according to Russell and Whitehead, it is necessary to introduce as a sort of *functional assertions* propositions that hold generally. We say that we have a functional assertion when it is asserted that a proposition holds for the indeterminate case.

Now what I wish to show in the present work is the following: *If we consider the general theorems of arithmetic to be functional assertions and take the recursive mode of thought as a basis, then that science can be founded in a rigorous way without use of Russell and Whitehead's notions "always" and "sometimes"*. This can also be expressed as follows: A logical foundation can be provided for arithmetic without the use of apparent logical variables. To be sure, it will often be advantageous to introduce apparent variables; but we shall require that these variables range over only finite domains, and by means of recursive definitions we shall then always be able to avoid the use of such variables. All this will become clear in what follows.

I wish to remark, incidentally, that I really regard all functions as descriptive; the propositional functions are just characterized by the fact that they can have only the two values "true" and "false", which, of course, are also among the fundamental logical notions.

I regard the descriptive functions as functional proper names, that is, proper names whose meaning depends upon what we choose for one or more variables. For instance, I regard "$n + 1$", "the number following n", as being the name of a number, but one that, depending on what we choose for n, denotes different numbers.

According to Russell and Whitehead, descriptive functions, such as "the author of *Waverley*", do not really mean anything but are merely incomplete symbols. This conception does not seem to me to be beyond doubt; but even if it were correct for the descriptive functions of ordinary language, we would not need to adopt it for the descriptive functions of arithmetic.

Russell and Whitehead reason as follows. "The author of *Waverley*" cannot mean Scott; for then the proposition "Scott is the author of *Waverley*" would mean "Scott is Scott", and that is an entirely empty proposition. On the other hand, "the author of *Waverley*" cannot mean any other person; for then the proposition "Scott is the author of *Waverley*" would be false, which, as is well known, is not the case. Consequently, "the author of *Waverley*" means nothing; it is an incomplete symbol.

This proof, which has a rather philosophical character, does not, however, seem entirely compelling to me. What is to prevent us from regarding "the author of X" —let us write $V(X)$ for short—as a kind of variable proper name? Then surely, "the author of *Waverley*" names the same person as does "Scott". But the two names are distinct, and that is why the proposition "Scott is the author of *Waverley*" cannot be replaced by the proposition "Scott is Scott". The latter proposition is empty, but not the former, and the reason for this is that one already knows something in advance about the person called $V(X)$, since, after all, it is generally accepted that a person is to be called $V(X)$ if and only if he and no one else wrote X. The information conveyed is, in my opinion, of the same kind as in the following case: A man has two proper names, A and B. At one time we have heard something about A, for example that he has five children. On another occasion we are introduced to Mr. B, and we are told that B is Mr. A. This proposition then contains information about B, namely that B has five children, because we had some prior knowledge about A. The proposition "B is A" is therefore entirely different from the propositions "A is A" and "B is B"; the latter are completely empty, but the former is not, if something is already known in advance about A but not about B, or about B but not about A.

The use of the equal sign in what follows is always to be understood in the sense that two names or expressions mean or designate the same thing. That is also why I regard it as obvious that I can everywhere replace equals by equals, and I do this throughout.

The notions "natural number" and "the number $n + 1$ following the number n" (thus, the descriptive function $n + 1$) as well as the recursive mode of thought are taken as basis.[1]

§ 1. Addition

I wish to introduce a descriptive function of two variables, a and b, which I shall call the *sum* of a and b; I denote it by $a + b$ since for $b = 1$ it is to mean precisely the number following a, namely, $a + 1$. This function is therefore to be regarded as already defined for $b = 1$ for arbitrary a. To define it generally, I then need only define it for $b + 1$ for arbitrary a if it is already assumed to be defined for b for arbitrary a. This is accomplished by the following definition:

Definition 1. $a + (b + 1) = (a + b) + 1$.

Thus the sum of a and $b + 1$ is set equal to the number following $a + b$. If, therefore, addition is already defined for arbitrary values of a for a certain number b, then by Definition 1 addition is defined for arbitrary a for $b + 1$ and is thereby defined generally. This is a typical example of a recursive definition.

Theorem 1. (The associative law.) $a + (b + c) = (a + b) + c$.

Proof. The proposition holds for $c = 1$ by virtue of Definition 1. I assume that it holds for a certain c for arbitrary values of a and b. Necessarily, then, for arbitrary values of a and b,

$$(\alpha) \qquad a + (b + (c + 1)) = a + ((b + c) + 1),$$

[1] The frequently cumbersome formalism of propositional functions in what follows is due to the fact that the present paper was written as a kind of sequel ⟦in Anschluß⟧ to Russell and Whitehead's work.

since, according to Definition 1, $b + (c + 1) = (b + c) + 1$. But according to Definition 1 necessarily also

$$(\beta) \qquad a + ((b + c) + 1) = (a + (b + c)) + 1.$$

Now, according to the assumption, $a + (b + c) = (a + b) + c$, whence

$$(\gamma) \qquad (a + (b + c)) + 1 = ((a + b) + c) + 1.$$

According to Definition 1 we finally also have

$$(\delta) \qquad ((a + b) + c) + 1 = (a + b) + (c + 1).$$

From (α), (β), (γ), and (δ), it follows that

$$a + (b + (c + 1)) = (a + b) + (c + 1),$$

which proves the proposition for $c + 1$ for unspecified a and b. Thus the proposition holds generally. This is a typical example of a recursive proof (proof by mathematical induction).

Lemma. $a + 1 = 1 + a$.

Proof. The proposition holds for $a = 1$. I prove that it holds for $a + 1$, assuming that it holds for a. In fact, we obtain

$$(a + 1) + 1 = (1 + a) + 1 = 1 + (a + 1)$$

by virtue of the assumption made and Definition 1. Thus the proposition holds generally.

THEOREM 2. (The commutative law.) $a + b = b + a$.

Proof. In consequence of the lemma the proposition holds for $b = 1$. I assume that it is true for arbitrary a for a certain b and then prove that it is true for $b + 1$ for arbitrary a. This is done as follows:

$$a + (b + 1) = (a + b) + 1 = (b + a) + 1 = b + (a + 1)$$
$$= b + (1 + a) = (b + 1) + a,$$

by use of Theorem 1 and the lemma. The functional assertion $a + b = b + a$ is therefore true.

§ 2. THE RELATIONS < (LESS THAN) AND > (GREATER THAN)

Closely connected with addition are the relations "less than" and "greater than", denoted by $<$ and $>$, respectively. Since the latter relation is just the converse of the former, only the relation $<$ need be defined. This is customarily done by means of an apparent logical variable, with the use of the logical notion of existence, or Russell and Whitehead's "sometimes". The usual definition, in fact, has the form

$$(a < b) = \sum_x (a + x = b),$$

where Schröder's sign is used to express that a proposition holds in at least one case (Schröder then speaks of propositional summation ⟦Aussagensummation⟧ and denotes it by Σ). In words the definition is as follows: "a is said to be less than b if and only if *there exists a number* x such that $a + x = b$". Hence this definition involves the use of the logical notion of existence, or, in other words, the use of an apparent variable. But we can easily avoid using this notion by defining the relation "less than" recursively. This can, in fact, be done thus:

Definition 2. $a < 1$ is false. $(a < b + 1) = (a < b) + (a = b)$.

It is easily seen that this is a perfectly legitimate recursive definition; for, first, it specifies under what conditions $a < b$ if $b = 1$, namely, none; and, second, it specifies under what conditions the relation $<$ is to obtain between an arbitrary a and a certain $b + 1$ if it is already defined for b for arbitrary a. As we see, *no logical Σ-sign occurs in this definition.*

Definition 2.1. $(a > b) = (b < a)$.

THEOREM 3. $\overline{(a < b)(b < c)} + (a < c)$, or, which is the same thing, $\overline{(a < b)} + \overline{(b < c)} + (a < c)$. In words this theorem reads: If both $a < b$ and $b < c$ hold, then $a < c$.

Proof. The proposition holds for arbitrary a and b when $c = 1$; for, according to Definition 2, $b < 1$ is false, that is, $\overline{(b < 1)}$ is true. I therefore want to prove that the proposition is true for $c + 1$ whenever it is true for c (with the values of a and b unspecified). From $(a < b)(b < c + 1)$ it follows, according to Definition 2, either that $(a < b)(b < c)$ or that $(a < b)(b = c)$. But, according to the assumption, from $(a < b)(b < c)$ it follows that $a < c$, whence, by Definition 2, $a < c + 1$. From $(a < b)(b = c)$, of course, $a < c$ also follows, whence again $a < c + 1$. In every case, therefore, $a < c + 1$ if both $a < b$ and $b < c + 1$ hold, so that the proposition must also be true for arbitrary a and b for $c + 1$. This proves Theorem 3.

Definition 3. $(a \leqq b) = (a < b) + (a = b)$.

Definition 3.1. $(a \geqq b) = (a > b) + (a = b)$.

These definitions, like Definition 2.1, merely introduce notational variants; they are therefore theoretically superfluous, which is not the case for recursive definitions.

Lemma 1 (for Theorem 4). $\overline{(a < b)} + (a + 1 \leqq b)$.

In words: Either a is not less than b, or $a + 1$ is less than or equal to b. But it is better to state it thus: From $a < b$ it follows that $a + 1 \leqq b$.

Proof. For an unspecified a the proposition holds for $b = 1$. I assume that it is true for arbitrary a for a certain b and then prove that it is true for arbitrary a for $b + 1$. From $a < b + 1$ it follows (Definition 2) either that $a < b$, whence, in consequence of the assumption, $a + 1 \leqq b$, which in turn yields (Definition 2) $a + 1 \leqq b + 1$, or that $a = b$, whence clearly $a + 1 = b + 1$, which yields (Definition 3) $a + 1 \leqq b + 1$. The proposition thus holds, for arbitrary values of a, for $b + 1$ also and therefore holds generally.

Lemma 2 (for Theorem 4). $1 \leqq a$.

Proof. The proposition holds for $a = 1$. From the assumption that it holds for a, that is, that $1 \leqq a$, it follows (Definitions 2 and 3) that $1 < a + 1$, which yields (Definition 3) $1 \leqq a + 1$; that is, the proposition also holds for $a + 1$.

THEOREM 4. $(a < b) + (a = b) + (a > b)$.

Proof. The proposition holds for $b = 1$, since according to Lemma 2 necessarily either $a = 1$ or $a > 1$. I assume that the proposition is true for b with unspecified a and prove that it is true for $b + 1$ for arbitrary a. For, if it is not the case that

$a < b + 1$, neither $a < b$ nor $a = b$ is possible (Definition 2); but then, in consequence of the assumption, $a > b$, whence (Lemma 1) $a \geqq b + 1$.

Lemma. $(a < b)(a + 1 < b + 1) + \overline{(a < b)}\ \overline{(a + 1 < b + 1)}$.

This proposition can be expressed in words thus: From $a < b$ it follows that $a + 1 < b + 1$, and conversely.

Proof. From $a < b$ it follows (Lemma 1 for Theorem 4) that $a + 1 \leqq b$, whence (Definition 2) $a + 1 < b + 1$. From $a + 1 < b + 1$ it follows (Definition 2) either that $a + 1 < b$ or that $a + 1 = b$. Hence in either case $a < b$.

Theorem 5. $\overline{a < a}$; that is, no number is less than itself.

Proof. This is true for $a = 1$ (Definition 2). I assume that it is true for a certain a. From $a + 1 < a + 1$ it would follow, according to the last lemma, that $a < a$, which conflicts with the assumption made.

Corollary to Theorems 3 and 5. ⟦For distinct a and b⟧ $(a < b)\overline{(a > b)} + \overline{(a < b)}$ $(a > b)$.

In words: If $a < b$, then not $a > b$, and conversely.

For, if we were to have $a < b$ and at the same time $a > b$, it would follow (Theorem 3) that $a < a$.

Corollary. $\overline{(a < b)} + (a \neq b)$;[2] that is, if $a < b$, then a is not equal to b. For from $(a < b)(a = b)$ it would follow that $a < a$.

The three relations $a < b$, $a = b$, and $a > b$ are thus mutually exclusive, while on account of Theorem 4 in every case one of them must be satisfied.

Theorem 6. $(a < b)(a + c < b + c) + \overline{(a < b)}\ \overline{(a + c < b + c)}$.

Proof. According to the lemma for Theorem 5 this is certainly true when $c = 1$. We assume that it has already been proved to be true for arbitrary a and b for a certain c. From $a < b$ it then follows that $a + c < b + c$, whence, by virtue of Theorem 1 and the same lemma, $a + (c + 1) < b + (c + 1)$. Conversely, it follows from $a + (c + 1) < b + (c + 1)$ by virtue of Theorem 1 and this lemma that $a + c < b + c$, whence, according to the assumption, $a < b$.

Theorem 7. $\overline{(a < b)(c < d)} + (a + c < b + d)$.

That is, if both $a < b$ and $c < d$ hold, $a + c < b + d$.

Proof. From $a < b$ it follows (Theorem 6) that $a + c < b + c$. From $c < d$ it follows (Theorems 2 and 6) that $b + c < b + d$. From the assumption that both $a + c < b + c$ and $b + c < b + d$ hold it follows (Theorem 3) that $a + c < b + d$.

This is a typical example of a nonrecursive proof, one that consists merely of a finite combination of earlier theorems, whereas a proof by mathematical induction represents an infinite process. We have, incidentally, already had several other examples of nonrecursive proofs (namely, those of the lemma for Theorem 5 and of the corollaries to Theorems 3 and 5).

Theorem 8. $(a + c \neq b + c) + (a = b)$.

That is, from $a + c = b + c$ it follows that $a = b$. Clearly, the converse is also true.

Proof. If $a \neq b$, then necessarily (Theorem 4) either $a < b$ or $a > b$. From $a < b$, however, it follows (Theorem 6) that $a + c < b + c$ and from $a > b$, in the same way,

[2] I write $a \neq b$, as one customarily does, to express the fact that a is not equal to b, that is, that $\overline{(a = b)}$ holds.

that $a + c > b + c$, and both contradict (corollary to Theorem 5) the equation $a + c = b + c$.

The special case $c = 1$ of this theorem asserts that there can exist at most one number having a certain successor, or, in other words, each number can have *one* predecessor at most.

THEOREM 9. $a < a + b$.

Proof. True for $b = 1$ for unspecified a (Definition 2). We assume that the proposition is true for arbitrary a for a certain b. But from $a < a + b$ we obtain further (Definition 2) $a < (a + b) + 1$, that is (Definition 1), $a < a + (b + 1)$.

§ 3. MULTIPLICATION

Definition 4. $a \cdot 1 = a$. $a(b + 1) = ab + a$.

This is a recursive definition of a descriptive function ab of two variables, a and b, which is called the *product* of a and b.

THEOREM 10. (First distributive law.) $a(b + c) = ab + ac$.

Proof. The proposition holds (Definition 4) for $c = 1$. We therefore assume that it holds for arbitrary a and b for a certain c. We then obtain

$$\begin{aligned} a(b + (c + 1)) &= a((b + c) + 1) = a(b + c) + a \\ &= (ab + ac) + a = ab + (ac + a) = ab + a(c + 1), \end{aligned}$$

making use of Theorem 1, Definition 4, and the assumption made.

THEOREM 11. (The associative law.) $a(bc) = (ab)c$.

Proof. The proposition holds (Definition 4) for $c = 1$. We therefore assume that it is true for arbitrary a and b for a certain c. Then we obtain

$$a(b(c + 1)) = a(bc + b) = a(bc) + ab = (ab)c + ab = (ab)(c + 1),$$

applying Definition 4, Theorem 10, and the assumption made.

THEOREM 12. (Second distributive law.) $(a + b)c = ac + bc$.

Proof. When $c = 1$, the proposition is true (Definition 4). We assume that it is true for arbitrary a and b for a certain c. Then by applying Definition 4 and Theorems 1 and 2 we obtain

$$\begin{aligned} (a + b)(c + 1) &= (a + b)c + (a + b) = (ac + bc) + (a + b) = ((ac + bc) + a) + b \\ &= (ac + (bc + a)) + b = (ac + (a + bc)) + b = ((ac + a) + bc) + b \\ &= (a(c + 1) + bc) + b = a(c + 1) + (bc + b) = a(c + 1) + b(c + 1). \end{aligned}$$

Lemma. $1 \cdot a = a$.

Proof. True for $a = 1$ (Definition 4). If the proposition is true for a, then $1 \cdot (a + 1) = (1 \cdot a) + 1 = a + 1$; that is, it is also true for $a + 1$.

THEOREM 13. (The commutative law.) $ab = ba$.

Proof. This proposition holds (by the lemma) for arbitrary a when $b = 1$. From the assumption that it is true for b for arbitrary a it follows (Theorem 12 and the lemma) that

$$a(b + 1) = ab + a = ba + a = (b + 1)a.$$

THEOREM 14. $(a < b)(ac < bc) + \overline{(a < b)}\,\overline{(ac < bc)}$.

In words: From $a < b$ it follows that $ac < bc$, and conversely.

Proof. Clearly, the proposition is true for arbitrary a and b when $c = 1$. Therefore we now assume that it holds for arbitrary a and b for a certain c. Then from $a < b$ it follows that $ac < bc$, whence (Theorem 7) $ac + a < bc + b$, that is (Definition 4), $a(c + 1) < b(c + 1)$. The converse must also be true; for from $a = b$ it would follow that $ac = bc$, and from $a > b$, according to what has been proved, that $ac > bc$.

Corollary. $(ac \neq bc) + (a = b)$; that is, from $ac = bc$ it follows that $a = b$.

THEOREM 15. $a \leqq ab$.

Proof. True for $b = 1$ (Definition 4). From the assumption that the proposition is true for b it follows by Theorem 3, since (Theorem 9) $ab < ab + a$, that $a < a(b + 1)$; that is, the proposition is also true for $b + 1$.

Corollary. From $ab \leqq c$ it follows that $a \leqq c$, or, alternatively, $(ab > c) + (a \leqq c)$.

§ 4. THE RELATION OF DIVISIBILITY

Closely connected with multiplication is the notion of divisibility. It is generally defined by means of an apparent variable. For we say that a is divisible by b if there exists a number x such that $a = bx$. If we use Schröder's symbols and let $D(a, b)$ mean the propositional function "a is divisible by b", this definition takes the following form:

$$D(a, b) = \underset{x}{\Sigma}(a = bx).$$

Such a definition involves an infinite task—that means one that cannot be completed—since the criterion of divisibility is whether, *by successive trials through the entire number sequence*, we can find a number x such that $a = bx$.

Here, however, it is easy to free ourselves from the infinitude that adheres to this definition. For it is clear that a number x having the required property, if such a one exists at all, must occur among the numbers $1, 2, \ldots, a$; for from $bx = a$ it follows by Theorem 15 that $x \leqq a$. Therefore the relation of divisibility can equally well be defined as follows:

$$D(a, b) = \overset{a}{\underset{1}{\Sigma}}{}_x(a = bx) = ((a = b) + (a = 2b) + (a = 3b) + \cdots + (a = ab)).$$

To be sure, an apparent variable x still occurs here; *but it ranges over a finite domain only*, namely, just the values from 1 to a. Therefore this definition furnishes us with a finite criterion of divisibility; we can in every case, by completing a finite task, that is, by performing a finite number of operations, ascertain whether the proposition $D(a, b)$ holds or not. Since now the Schröder propositional sum in this definition is a finite one, it can itself in turn be defined by recursion, and thus the use of an apparent variable can ultimately be avoided altogether. To achieve this we need only define, as a first step, a ternary relation $\varDelta(a, b, c)$ that is to mean that a is equal to b multiplied by a number between 1 and c (both inclusive). The precise definition of $\varDelta(a, b, c)$ reads as follows:

Definition 5. $\varDelta(a, b, 1) = (a = b)$. $\varDelta(a, b, c + 1) = \varDelta(a, b, c) + (a = b(c + 1))$.

By means of the propositional function $\varDelta$ the divisibility relation D is defined thus:

Definition 6. $D(a, b) = \varDelta(a, b, a)$.

It is very easy to see that the propositional equivalence

$$\Delta(a, b, c) = \sum_{x\,1}^{c}(a = bx)$$

obtains, so that Definition 6 coincides completely with the finite definition of divisibility mentioned above. In fact this equivalence holds for $c = 1$, since the propositional sum on the right then reduces to the single term $a = b$, and from the assumption that it is true for c it follows that it is true for $c + 1$; for, $\sum_{x\,1}^{c+1}(a = bx)$, after all, means the same as $\sum_{x\,1}^{c}(a = bx) + (a = b(c + 1))$.

THEOREM 16. $(a \neq bc) + \Delta(a, b, c)$; that is, from $a = bc$ it follows that $\Delta(a, b, c)$. This immediately follows from Definition 5.

THEOREM 17. $\overline{\Delta(a, b, c)} + (c > c') + \Delta(a, b, c')$.

This can also be expressed thus: From $\Delta(a, b, c)(c \leqq c')$ it follows that $\Delta(a, b, c')$.

Proof. Clearly, the proposition is true for $c' = 1$. We assume that it is true for a certain c' for arbitrary values of a, b, and c. From $c \leqq c' + 1$ it follows either that $c \leqq c'$, which, together with $\Delta(a, b, c)$ yields, according to the assumption, $\Delta(a, b, c')$ and consequently (Definition 5) also $\Delta(a, b, c' + 1)$, or that $c = c' + 1$, which, together with $\Delta(a, b, c)$, of course yields $\Delta(a, b, c' + 1)$. Hence the proposition is also true for $c' + 1$ for arbitrary values of a, b, and c and thus holds in full generality.

THEOREM 18. $\overline{\Delta(a, b, c)} + D(a, b)$; that is, from $\Delta(a, b, c)$ it follows that $D(a, b)$.

Proof. From $\Delta(a, b, 1)$, that is, $a = b$, it follows by Theorem 17, since (Lemma 2 for Theorem 4) $1 \leqq a$, that $\Delta(a, b, a)$, that is, $D(a, b)$. Let us now assume that the proposition holds for c. From $\Delta(a, b, c + 1)$ it follows (Definition 5) either that $\Delta(a, b, c)$, whence, in consequence of the assumption, $D(a, b)$, or that $a = b(c + 1)$, whence (Theorem 15) $c + 1 \leqq a$; but from $\Delta(a, b, c + 1)(c + 1 \leqq a)$ it follows (Theorem 17) that $\Delta(a, b, a)$, that is, $D(a, b)$.

Corollary to Theorems 16 and 18. $(a \neq bc) + D(a, b)$; that is, from $a = bc$ it follows that $D(a, b)$.

THEOREM 19. $(a \neq bd) + \overline{\Delta(b, c, e)} + \Delta(a, c, de)$; that is, from $(a = bd)\Delta(b, c, e)$ it follows that $\Delta(a, c, de)$.

Proof. If $e = 1$, then $\Delta(b, c, e) = (b = c)$; from $(a = bd)\Delta(b, c, 1)$ it thus follows that $a = cd$, whence (Theorem 16) $\Delta(a, c, d)$. We therefore assume that the proposition is true for e (with arbitrary a, b, c, and d). From $(a = bd)\Delta(b, c, e + 1)$ we obtain (Definition 5) either $(a = bd)\Delta(b, c, e)$, whence $\Delta(a, c, de)$, whence in turn by Theorem 17 $\Delta(a, c, d(e + 1))$, since ($de < de + d$ by Theorem 9) $de < d(e + 1)$, or $(a = bd)(b = c(e + 1))$, whence $a = cd(e + 1)$, which again yields (Theorem 16) $\Delta(a, c, d(e + 1))$.

Corollary to Theorems 18 and 19. $(a \neq bd) + \overline{D(b, c)} + D(a, c)$; that is, from $(a = bd)D(b, c)$ it follows that $D(a, c)$.

For from $(a = bd)\Delta(b, c, b)$ it follows (Theorem 19) that $\Delta(a, c, bd)$, whence (Theorem 18) $D(a, c)$.

THEOREM 20. $\overline{\Delta(a, b, d)} + \overline{D(b, c)} + D(a, c)$.

Proof. From $\Delta(a, b, 1)D(b, c)$ it follows, of course, that $D(a, c)$, since $\Delta(a, b, 1) = (a = b)$. We assume that the proposition is true for arbitrary a, b, and c for a certain d. From $\Delta(a, b, d + 1)D(b, c)$ we obtain (Definition 5) either $\Delta(a, b, d)D(b, c)$, whence, in consequence of the assumption, $D(a, c)$, or $(a = b(d + 1))D(b, c)$, whence, according to the corollary to Theorems 18 and 19, it follows that $D(a, c)$. This proves that the proposition is true for $c + 1$.

Corollary. $\overline{D(a, b)D(b, c)} + D(a, c)$; that is, from $D(a, b)D(b, c)$ it follows that $D(a, c)$.

This is in fact just the special case that we obtain from Theorem 20 if we put $d = a$.

This last proposition, that $D(a, c)$ follows from $D(a, b)D(b, c)$, is closely related to the proposition that states that $a = cde$ follows from $(a = bd)(b = ce)$, but it has a quite different meaning. Even according to the ordinary conception that makes use of apparent variables ranging over an infinite domain, the two propositions are quite different in content, as becomes clear immediately the two are formulated precisely. The one proposition is, in Schröder's symbols,

$$(\alpha) \qquad \Pi_a \Pi_b \Pi_c (\overline{\Sigma_x (a = bx)} + \overline{\Sigma_y (b = cy)} + \Sigma_z (a = cz)).$$

The other proposition is

$$(\beta) \qquad \Pi_a \Pi_b \Pi_c \Pi_d \Pi_e ((a \neq bd) + (b \neq ce) + (a = cde)).$$

When, nevertheless, one proves that the proposition $\overline{D(a, b)} + \overline{D(b, c)} + D(a, c)$, or (α), is true by proving proposition (β), as is usually done, he does so on the basis of certain logical schemata concerning the use of the signs Π and Σ; in ordinary mathematical thinking, however, these schemata are passed over in silence.

Theorem 21. $\overline{\Delta(a, c, d)} + (b \neq ce) + \Delta(a + b, c, d + e)$.

Proof. From $\Delta(a, c, 1)(b = ce)$, or, in other words, $(a = c)(b = ce)$, it follows that $a + b = c(1 + e)$ and thence (Theorem 16) that $\Delta(a + b, c, 1 + e)$. The proposition is therefore true for $d = 1$. We assume that it is true for a certain d and prove as follows that it is true for $d + 1$. From $\Delta(a, c, d + 1)(b = ce)$ it follows either that $\Delta(a, c, d)(b = ce)$, whence, in consequence of the assumption, $\Delta(a + b, c, d + e)$, which in turn implies (Theorems 1 and 2, and Definition 5) $\Delta(a + b, c, (d + 1) + e)$, or that $(a = c(d + 1))(b = ce)$, whence $a + b = c((d + 1) + e)$, which again implies (Theorem 16) $\Delta(a + b, c, (d + 1) + e)$.

Corollary. $\overline{\Delta(a, c, d)} + \overline{\Delta(b, c, 1)} + \Delta(a + b, c, d + 1)$.

Theorem 22. $\overline{\Delta(a, c, d)} + \overline{\Delta(b, c, e)} + \Delta(a + b, c, d + e)$.

Proof. According to the corollary to Theorem 21 this holds for $e = 1$. We therefore assume that the proposition is true for a certain e and prove as follows that it is true for $e + 1$. From $\Delta(a, c, d)\Delta(b, c, e + 1)$ we obtain either $\Delta(a, c, d)\Delta(b, c, e)$ and thence, in consequence of the assumption, $\Delta(a + b, c, d + e)$, which in turn implies $\Delta(a + b, c, d + (e + 1))$, or $\Delta(a, c, d)(b = c(e + 1))$, whence (Theorem 21) $\Delta(a + b, c, d + (e + 1))$.

Corollary. $\overline{D(a, c)D(b, c)} + D(a + b, c)$.

In words: If both a and b are divisible by c, then $a + b$ too is divisible by c.

For from $\Delta(a, c, a)\Delta(b, c, b)$ we obtain, by Theorem 22, $\Delta(a + b, c, a + b)$.

Lemma. $\overline{\Delta(a + b, b, c + 1)} + \Delta(a, b, c)$.

Proof. From $\Delta(a + b, b, 2)$ it follows (Definition 5) either that $a + b = b$, which is, however, impossible, since necessarily (Theorem 9) $a + b > b$ and since (Theorem 5) the relations $>$ and $=$ are mutually exclusive, or that $a + b = b \cdot 2$, whence $b = a$ (since from $a + b = b + b$ it must follow, according to Theorem 8, that $a = b$). Thus our lemma is true for $c = 1$. We assume that it holds for c and then prove it for $c + 1$. From $\Delta(a + b, b, c + 2)$ it follows in fact either that $\Delta(a + b, b, c + 1)$, whence, in consequence of the assumption, $\Delta(a, b, c)$ and therefore also $\Delta(a, b, c + 1)$, or that $a + b = b(c + 2)$, whence (Theorem 8) $a = b(c + 1)$, which again implies $\Delta(a, b, c + 1)$.

By making use of subtraction, even before it is introduced (see § 5), we can also state this lemma in the following form: From $\Delta(a + b, b, c)(c > 1)$ it follows that $\Delta(a, b, c - 1)$. For this is true when $c = 1$, because the hypothesis is then false. Whenever the proposition holds for c, it holds also for $c + 1$; for from $\Delta(a + b, b, c + 1)$ we obtain, according to what has just been proved, $\Delta(a, b, c)$, and $c = (c + 1) - 1$ (see Definition 7 ⟦below⟧).

Theorem 23. $\overline{\Delta(a + bd, b, c + d)} + \Delta(a, b, c)$.

Proof. In case $d = 1$ we are back at the lemma. We therefore assume that the proposition is true for d and prove that it is true for $d + 1$. From $\Delta(a + bd + b, b, c + d + 1)$ it follows (Definition 5) either that $a + bd + b = b(c + d + 1)$ or that $\Delta(a + bd + b, b, c + d)$. In the first case, since (Theorem 10) $b(c + d + 1) = bc + b(d + 1)$, we obtain (Theorem 8) $a = bc$, whence (Theorem 16) $\Delta(a, b, c)$. In the second case, since necessarily (Lemma 2 for Theorem 4, and Theorem 9) $c + d > 1$, we obtain, by the lemma, $\Delta(a + bd, b, c + d - 1)$, whence $\Delta(a + bd, b, c + d)$, which, according to the assumption, implies $\Delta(a, b, c)$.

Theorem 24. $\overline{\Delta(a + b, c, d + e)} + \overline{\Delta(b, c, e)} + \Delta(a, c, d + e)$.

Proof. This is true for $e = 1$; for from $\Delta(a + b, c, d + 1)(b = c)$ we obtain, according to the lemma for Theorem 23, $\Delta(a, c, d)$ and therefore also $\Delta(a, c, d + 1)$. We shall therefore assume that the proposition is true for e (with arbitrary a, b, c, and d) and then prove that it is true for $e + 1$ (for arbitrary values of a, b, c, and d). From $\Delta(a + b, c, d + (e + 1))\Delta(b, c, e + 1)$ we obtain either $\Delta(a + b, c, (d + 1) + e)\Delta(b, c, e)$, which, according to the assumption, implies $\Delta(a, c, (d + 1) + e)$, in other words $\Delta(a, c, d + (e + 1))$, or $\Delta(a + b, c, d + (e + 1))(b = c(e + 1))$, whence (Theorem 23) $\Delta(a, c, d)$, which again implies (Theorem 17) $\Delta(a, c, d + (e + 1))$.

Corollary. $\overline{D(a + b, c)D(a, c)} + D(b, c)$.

For, first, $\Delta(b, c, a + b)$ must follow (Theorem 24) from $\Delta(a + b, c, a + b)\Delta(a, c, a)$ and, second, $D(b, c)$ is a consequence (Theorem 18) of $\Delta(b, c, a + b)$.

After the introduction of subtraction we shall be able to write this theorem also as follows: $\overline{(a > b)} + \overline{D(a, c)} + \overline{D(b, c)} + D(a - b, c)$.

That is, if both a and b are divisible by c and if a is greater than b (so that the difference $a - b$ exists), then $a - b$ is divisible by c.

Theorem 25. $\overline{\Delta(a, b, c)} + (a \geqq b)$.

Proof. Clearly, the proposition holds for $c = 1$. If we assume that it holds for c, we obtain from $\Delta(a, b, c + 1)$ either $\Delta(a, b, c)$, which thus yields $a \geqq b$, or $a = b(c + 1)$, from which (Theorem 15) $a \geqq b$ also follows.

Corollary 1. $\overline{D(a, b)} + (a \geqq b)$.

Corollary 2. $\overline{D(a, b)} + \overline{D(b, a)} + (a = b)$.

For from $(a \geqq b)(b \geqq a)$ it must follow (Theorem 5) that $a = b$.

THEOREM 26. $\Delta(a, b, d)\Delta(ac, bc, d) + \overline{\Delta(a, b, d)}\ \overline{\Delta(ac, bc, d)}$.

In words: From $\Delta(a, b, d)$ it follows that $\Delta(ac, bc, d)$, and conversely.

Proof. The proposition is true for $d = 1$; for from $ac = bc$ it follows, according to the corollary to Theorem 14, that $a = b$, and conversely from $a = b$ it of course follows that $ac = bc$. We assume that the proposition is true for d and prove that it is true for $d + 1$. From $\Delta(a, b, d + 1)$ we obtain either $\Delta(a, b, d)$, which according to the assumption implies $\Delta(ac, bc, d)$ and therefore also $\Delta(ac, bc, d + 1)$, or $a = b(d + 1)$, whence (Theorems 11 and 13) $ac = (b(d + 1))c = b((d + 1)c) = b(c(d + 1)) = (bc)(d + 1)$, and from $ac = bc(d + 1)$ it again follows (Theorem 16) that $\Delta(ac, bc, d + 1)$. Likewise we obtain from $\Delta(ac, bc, d + 1)$ either $\Delta(ac, bc, d)$, whence $\Delta(a, b, d)$ and therefore also $\Delta(a, b, d + 1)$, or $ac = bc(d + 1) = b(d + 1)c$, whence, in consequence of the corollary to Theorem 14, $a = b(d + 1)$, whence (Theorem 16) $\Delta(a, b, d + 1)$.

Corollary. $D(a, b)D(ac, bc) + \overline{D(a, b)}\ \overline{D(ac, bc)}$.

THEOREM 27. $(a \neq bd) + \overline{\Delta(d, c, e)} + \Delta(a, bc, e)$.

Proof. True for $e = 1$. We therefore assume that the proposition is true for e and prove that it is true for $e + 1$. From $\Delta(d, c, e + 1)$ we obtain either $\Delta(d, c, e)$, which together with $a = bd$ yields, according to the assumption, $\Delta(a, bc, e)$ and therefore also $\Delta(a, bc, e + 1)$, or $d = c(e + 1)$, which together with $a = bd$ yields (Theorem 11) $a = (bc)(e + 1)$, whence (Theorem 16) $\Delta(a, bc, e + 1)$.

§ 5. SUBTRACTION AND DIVISION. DESCRIPTIVE FUNCTIONS WITH A RESTRICTED DOMAIN OF EXISTENCE

Subtraction, as is well known, can be defined in the following manner:

Definition 7. $(c - b = a) = (c = a + b)$.

It is clear that in this way a descriptive function $c - b$, called the *difference*, is defined; for, by means of the equation $a + b = c$, a is uniquely determined by b and c. This, however, is a descriptive function with a restricted domain of existence; for, if $c \leqq b$, no equation of the form $c = a + b$ can obtain, and then according to Definition 7 the inequality $c - b \neq a$ obtains for every number a; that is, $c - b$ is not equal to any number. On the other hand it can be proved that $c - b$ will certainly have a value when $c > b$. One would be inclined to formulate this proposition thus:

$$\overline{(c > b)} + \sum_x (x + b = c),$$

where the propositional summation over x would have to be extended over "all" numbers from 1 to ∞. But it is not necessary to adduce such an actual infinity here either; we can in fact prove the following proposition:

$$\overline{(c > b)} + \sum_{1}^{c}{}_x (x + b = c),$$

which, after all, will more readily serve to secure the existence of a value for $c - b$. But the propositional function

$$\sum_{1}^{z}{}_u (u + y = x) = L(x, y, z)$$

of the three variables x, y, and z again can be defined recursively, so that we can avoid the apparent variable u altogether. The theorem that is to be proved is then obviously the following:

THEOREM 28. $\overline{c > b} + L(c, b, c)$.

In words this theorem reads as follows: If $c > b$, there exists among the numbers from 1 to c a number x such that $x + b = c$, or, in other words, $x = c - b$.

We need the recursive definition of the function L and a few simple theorems about it.

Definition 8. $L(x, y, 1) = (x = 1 + y)$. $L(x, y, z + 1) = L(x, y, z) + (x = (z + 1) + y)$.

THEOREM 29. $\overline{L(x, y, z)(z \leqq z')} + L(x, y, z')$.

Proof. True for $z' = 1$. I assume that it is true for z' and from that prove that it is true for $z' + 1$. From $z \leqq z' + 1$ it in fact follows either that $z \leqq z'$, which together with $L(x, y, z)$, according to the assumption, yields $L(x, y, z')$ and therefore (Definition 8) also $L(x, y, z' + 1)$, or that $z = z' + 1$, which with $L(x, y, z)$ obviously yields $L(x, y, z' + 1)$.

THEOREM 30. $\overline{L(x, y, z)} + L(x + 1, y, z + 1)$.

Proof. True when $z = 1$; for from $x = 1 + y$ it follows (Theorems 1 and 2) that $x + 1 = (1 + y) + 1 = 1 + (y + 1) = 1 + (1 + y) = (1 + 1) + y$. I assume that the proposition is true for z and prove that it is true for $z + 1$. From $L(x, y, z + 1)$ it follows (Definition 8) either that $L(x, y, z)$, whence, according to the assumption made, $L(x + 1, y, z + 1)$ and therefore also $L(x + 1, y, z + 2)$ follow, or that $x = (z + 1) + y$, whence (Theorems 1 and 2) $x + 1 = ((z + 1) + y) + 1 = (z + 1) + (y + 1) = (z + 1) + (1 + y) = (z + 2) + y$, whence (Definition 8) $L(x + 1, y, z + 2)$.

Now the proof of Theorem 28 can be carried out.

The proposition stated as Theorem 28 is certainly true for $c = 1$; for, in that case, already $\overline{1 > b}$ is true. I therefore assume that it is true for c and prove that it is true for $c + 1$. From $c + 1 > b$ we obtain (see Lemma 1 for Theorem 4, the lemma for Theorem 5, and Theorem 8) either $c > b$, which, according to the assumption made, gives us $L(c, b, c)$, whence it further follows (Theorem 30) that $L(c + 1, b, c + 1)$, or $c = b$, whence (Theorem 2) $c + 1 = b + 1 = 1 + b$; hence (Definition 8) $L(c + 1, b, 1)$, and thence (Theorem 29) again $L(c + 1, b, c + 1)$.

Division is introduced in a manner analogous to subtraction.

Definition 9. $(c/b = a) = (c = ab)$.

The expression c/b, called the *quotient*, is clearly again a function with a restricted domain of definition; for from $c = ab$ it follows (corollary to Theorems 16 and 18) that $D(c, b)$, so that c/b has a value only if $D(c, b)$ is true. Conversely, this is also sufficient; for, the proposition $D(c, b)$, or, in other words (see Definition 6), $\varDelta(c, b, c)$, is equivalent to the propositional sum

$$\overset{c}{\underset{1}{\Sigma_x}}(c = bx)$$

(see p. 311), and $(c = bx) = (c = xb)$, so that this propositional sum can be formulated in words thus: Among the numbers from 1 to c [[inclusive]] there exists a number x such that $c = xb$, or, in other words, $c/b = x$.

The propositional function $D(c, b)$ is therefore completely equivalent to the assertion that between 1 and c there exists a value for c/b.

I now give some simple theorems about differences and quotients. Presumably I do not have to give the proofs, which are trivial; these theorems are, after all, mere transformations of the simplest theorems about sums and products.

THEOREM 31_+. $(a - b) + b = a$.

$31_\times$. $(a/b)b = a$.

32_+. $(a - b) + c = (a + c) - b$.

$32_\times$. $(a/b)c = ac/b$.

33_+. $(a-b) - c = a - (b + c)$.

$33_\times$. $(a/b)/c = a/bc$.

34_+. $a - (b - c) = (a - b) + c$.

$34_\times$. $a/(b/c) = (a/b)c$.

35. $(a - b)c = ac - bc$.

36a. $(a/c) + (b/c) = (a + b)/c$.

36b. $(a/c) - (b/c) = (a - b)/c$.

§ 6. GREATEST COMMON DIVISOR AND LEAST COMMON MULTIPLE

The customary definitions of the greatest common divisor and the least common multiple of two numbers make use of apparent variables that range over an infinite domain. In Schröder's symbols these definitions in fact have the following form:

(c is the greatest common divisor of a and b)

$$= D(a, c)D(b, c)\Pi_x(\overline{D(a, x)} + \overline{D(b, x)} + D(c, x)),$$

(c is the least common multiple of a and b)

$$= D(c, a)D(c, b)\Pi_x(\overline{D(x, a)} + \overline{D(x, b)} + D(x, c)).$$

Since, however, $x \leqq a$ follows (Corollary 1 to Theorem 25) from $D(a, x)$, the infinite domain over which the variable ranges in the definition of the greatest common divisor can at once be cut down to a finite one, and we can just as well write

(c is the greatest common divisor of a and b)

$$= D(a, c)D(b, c)\overset{a}{\underset{1}{\Pi_x}}(\overline{D(a, x)} + \overline{D(b, x)} + D(c, x)).$$

A disadvantage of this definition is, however, that it is asymmetric with respect to a and b. But this, too, can easily be remedied, since over the sign Π we can write $a + b$ instead of the upper bound a, or even better Min(a, b), where $Min(a, b)$ is the minimum of the numbers a and b. For the definition of the least common multiple such a reduction of the range of the variable to a finite domain cannot be carried out quite so easily.

In what follows I shall introduce these notions in a different way, one that avoids apparent variables altogether. In doing this, however, I must make use of the recursive method of definition in a different way from before (though, as I shall soon show, the difference is purely superficial). Until now we have always given recursive definitions strictly as follows: we defined a notion for the number 1 and then, on the assumption that the definition for an arbitrarily given number n is already complete, we defined the notion for $n + 1$. Furthermore, a formal logical principle will be used here, namely, that we can give separate definitions for each one of mutually exclusive cases. I introduce two descriptive functions of two variables a and b, namely $a \wedge b$ and $a \vee b$, which I shall later show to be identical with the greatest common divisor and the least common multiple, respectively.

Definition 10. $((a \neq b) + (a \wedge b = a))(\overline{(a > b)} + (a \wedge b = (a - b) \wedge b))(\overline{(a < b)} + (a \wedge b = a \wedge (b - a)))$.

This is a perfectly legitimate recursive definition of the descriptive function $a \wedge b$; if we assume that it is already defined for those values of a and b for which $a + b < n$, then by Definition 10 it is defined for $a + b = n$. For, if a and b are two numbers such that $a + b = n$, then $a = b$ or $a > b$ or $a < b$, and these three cases are mutually exclusive, while Definition 10 specifies for each case what $a \wedge b$ is to mean. Moreover, Definition 10 gives us the value of $a \wedge b$ when $a + b = 2$, in which case necessarily $a = b = 1$.

In the proofs of some of the theorems that follow I shall also make use of mathematical induction in a different way from before (though, as I shall soon show, the difference is purely superficial). Until now this inductive inference has always been made thus: we prove a proposition for the number 1 and then we prove it for $n + 1$ on the assumption that it holds for n. Now I shall also prove inductively as follows: I prove a proposition for 1, and then, assuming that it holds for an arbitrary number $< n$, I prove it for n.

Lemma. $a \wedge 1 = 1$.

Proof. True for $a = 1$ in consequence of Definition 10. If we assume that the proposition is true for a, we obtain, according to Definition 10, since (Theorem 9) $a + 1 > 1$, $(a + 1) \wedge 1 = a \wedge 1$, and therefore also $(a + 1) \wedge 1 = 1$.

Likewise it can be proved that $1 \wedge a = 1$. In general, of course, $a \wedge b = b \wedge a$.

Theorem 37. $D(a, a \wedge b)D(b, a \wedge b)$.

Proof. This proposition is true for arbitrary b for $a = 1$ and for arbitrary a for $b = 1$; for, according to the lemma, $a \wedge 1 = 1 \wedge b = 1$. I assume that the proposition is true for $a + b < n$ and then prove that it is true for $a + b = n$. For if, first, $a = b$, then, according to Definition 10, $a \wedge b = a$, and consequently the proposition holds. If, second, $a > b$, then $(a - b) + b < n$, and then, in consequence of the assumption, $D(a - b, (a - b) \wedge b)D(b, (a - b) \wedge b)$ holds. But, according to Definition 10, $(a - b) \wedge b = a \wedge b$, and we therefore obtain $D(a - b, a \wedge b)D(b, a \wedge b)$, whence by the corollary to Theorem 22 and by Theorem 31_+, $D(a, a \wedge b)$ also follows. The proposition, therefore, is true in this case also. If, third, $a < b$, we can proceed in an exactly analogous way.

Theorem 38. $\overline{D(a, c)D(b, c)} + D(a \wedge b, c)$.

Proof. The proposition holds for arbitrary b when $a = 1$ and for arbitrary a when $b = 1$; for then it follows from $D(a, c)$, or $D(b, c)$, respectively, that $c = 1$. I assume that the proposition holds for such a and b as will make $a + b < n$, and I prove from that assumption that it holds for such a and b as will make $a + b = n$. Therefore, let a and b be numbers such that $a + b = n$. First, then, it may be that $a = b$; in this case, according to Definition 10, $a \wedge b = a$, and the proposition is therefore true. Second, it may be that $a > b$. Since $(a - b) + b < n$, $D((a - b) \wedge b, c)$ must, by the assumption, follow from $D(a - b, c)D(b, c)$; but $D(a - b, c)$ in turn is a consequence of $D(a, c)D(b, c)$ (corollary to Theorem 24), and finally (Definition 10) in this case $(a - b) \wedge b = a \wedge b$. Thus $D(a \wedge b, c)$ follows from $D(a, c)D(b, c)$. In the third case, $a < b$, we proceed in exactly the same way.

Theorems 37 and 38 together express the characteristic properties of the greatest common divisor.

THEOREM 39. $\overline{D(a, b)} + (a \wedge b = b)$.

Proof. In consequence of Theorem 37 we have $D(b, a \wedge b)$. On the other hand, $D(a \wedge b, b)$ follows (Theorem 38) from $D(a, b)D(b, b)$. From $D(b, a \wedge b)D(a \wedge b, b)$, however, it follows (Corollary 2 to Theorem 25) that $a \wedge b = b$.

Corollary. $ab \wedge a = a$.

THEOREM 40. $ac \wedge bc = (a \wedge b)c$.

Proof. The proposition is true for arbitrary a for $b = 1$ and for arbitrary b for $a = 1$, as immediately follows from the lemma for Theorem 37 and the corollary to Theorem 39. We therefore assume that we have already proved that the proposition holds for $a + b < n$, and on this basis we prove that it holds for such a and b that $a + b = n$. It is possible, first, that $a = b$. Then also $ac = bc$, and, according to Definition 10, we have $ac \wedge bc = ac$ and $a \wedge b = a$, which yield $ac \wedge bc = (a \wedge b)c$. The second case is $a > b$. Since $(a - b) + b < n$, we have, according to the assumption made, $(a - b)c \wedge bc = ((a - b) \wedge b)c$. But furthermore $a \wedge b = (a - b) \wedge b$, and since, then, also (Theorem 14) $ac > bc$, we necessarily, according to Definition 10, have $(ac - bc) \wedge bc = ac \wedge bc$. Finally (Theorem 35), $(a - b)c = ac - bc$. Thus the equation $ac \wedge bc = (a \wedge b)c$ holds. In the third case, when $a < b$, we of course proceed in the same way.

Definition. We say that two numbers a and b are *relatively prime* if $a \wedge b = 1$. I do not introduce a special symbol for this, since we can always use the short equation $a \wedge b = 1$ as an expression for it.

THEOREM 41. $\overline{D(ac, b)(a \wedge b = 1)} + D(c, b)$.

This is the well-known theorem that a number b that divides a product ac whereas it is relatively prime to the factor a must divide the other factor, c.

Proof. From $D(ac, b)D(bc, b)$ it follows (Theorem 38) that $D(ac \wedge bc, b)$. But from $a \wedge b = 1$ it follows (Theorem 40) that $ac \wedge bc = c$, so that $D(c, b)$ must be true.

THEOREM 42. $\overline{(a \wedge b = 1)D(a, a')} + (a' \wedge b = 1)$.

Proof. In consequence of Theorem 37, $D(a', a' \wedge b)D(b, a' \wedge b)$ holds; but from $D(a, a')D(a', a' \wedge b)$ it follows (corollary to Theorem 20) that $D(a, a' \wedge b)$, which together with $D(b, a' \wedge b)$ yields (Theorem 38) $D(a \wedge b, a' \wedge b)$. Then from $(a \wedge b = 1)D(a, a')$ it must therefore surely follow that $a' \wedge b = 1$; for from $D(1, \alpha)$ it follows (Corollary 1 to Theorem 25) that $\alpha = 1$.

THEOREM 43. $(a \wedge b \neq 1) + (a \wedge c \neq 1) + (a \wedge bc = 1)$.

Proof. If a number d divides a as well as bc, then necessarily (Theorem 42) $d \wedge b = 1$ as well as $d \wedge c = 1$, provided $(a \wedge b = 1)(a \wedge c = 1)$ obtains. Further, it must, according to Theorem 41, follow from $D(bc, d)(c \wedge d = 1)$ that $D(b, d)$. But from $D(b, d)$ it follows (Theorem 39) that $b \wedge d = d$; on the other hand, we had $b \wedge d = 1$; therefore $d = 1$. But now (Theorem 37) $a \wedge bc$ divides a as well as bc; therefore a $\wedge$ $bc = 1$.

Theorem 44. (Generalization of Theorem 41.) $\overline{D(ac, b)} + D(c, b/(a \wedge b))$.

Proof. In the first place it is clear that necessarily $(a/(a \wedge b)) \wedge (b/(a \wedge b)) = 1$; for, by Theorem 40, we have $((a/(a \wedge b)) \wedge (b/(a \wedge b)))(a \wedge b) = a \wedge b$. From $D(ac, b)$ it follows (corollary to Theorem 26) that $D(ac/(a \wedge b), b/(a \wedge b))$, whence, according to Theorem 41, $D(c, b/(a \wedge b))$, since $a/(a \wedge b)$ and $b/(a \wedge b)$ are relatively prime.

Definition 11. $a \vee b = ab/(a \wedge b)$.

Theorem 45. $\overline{\Delta(a, b, d)D(a, c)} + \Delta(a, b \vee c, d)$.

Proof. True when $d = 1$; for from $(a = b)D(a, c)$ it follows that $D(b, c)$, whence (Theorem 39) $b \wedge c = c$ and consequently, by Definition 11, $b \vee c = b$. I assume that the proposition is true for d and prove that it is true for $d + 1$. From $\Delta(a, b, d + 1)$ it follows either that $\Delta(a, b, d)$, which together with $D(a, c)$ yields, according to the assumption, $\Delta(a, b \vee c, d)$ and therefore also $\Delta(a, b \vee c, d + 1)$, or that $a = b(d + 1)$, which together with $D(a, c)$ yields (Theorem 44) $D(d + 1, c/(b \wedge c))$. But from $(a = b(d + 1))D(d + 1, c/(b \wedge c))$ we again obtain $\Delta(a, b \vee c, d + 1)$, by Theorem 27.

Theorem 46. $D(a \vee b, a)D(a \vee b, b)$.

Proof. According to Theorem 37, $D(a, a \wedge b)$ holds, whence (corollary to Theorem 26) $D(ab/(a \wedge b), b)$, that is, $D(a \vee b, b)$. Likewise $D(b, a \wedge b)$ implies that $D(a \vee b, a)$ is true.

Theorem 47. $\overline{D(c, a)D(c, b)} + D(c, a \vee b)$.

This is merely a special case of Theorem 45, which results when d and a are equated.

Theorems 46 and 47 together express the characteristic properties of the least common multiple of two numbers a and b. Thus $a \vee b$ denotes the least common multiple of a and b.

An important special case of Theorem 47 is the one in which $a \wedge b = 1$. From $D(c, a)D(c, b)(a \wedge b = 1)$ it follows that $D(c, ab)$.

We have, corresponding to Theorem 39,

Theorem 48. $\overline{D(a, b)} + (a \vee b = a)$.

That this proposition is true follows immediately from Theorem 39 and Definition 11.

We have, corresponding to Theorem 40,

Theorem 49. $ac \vee bc = (a \vee b)c$.

Proof. $ac \vee bc = (ac \cdot bc)/(ac \wedge bc) = (ac \cdot bc)/(a \wedge b)c = (ab \cdot c)/(a \wedge b) = (a \vee b)c$; here use is made of Theorems 11, 13, 32, and 40.

I want to conclude this section by giving a proof that the more involved kinds of recursive definition and recursive proof used above differ only formally, not actually, from the ordinary simple recursive procedure that goes from n to $n + 1$. The standard form of a recursive definition, after all, is as follows: a propositional function $U(x)$ is defined first for $x = 1$ and then, if $U(n)$ has already been defined, for $x = n + 1$. But above we also considered recursive definitions of the following kind: first we defined $U(1)$ and then, with $U(m)$ assumed to have been defined for arbitrary $m < n$, $U(n)$. It is clear that the value of the propositional function $U(y)(y \leqq x)$ for arbitrarily selected x and y will be known if $U(y)$ is defined for the y in question; but conversely it will also be known whether $U(y)$ is true or false for a $y \leqq x$ if $U(y)$

$(y \leqq x)$ is defined for the x and y in question. To define $U(x)$, therefore, it will suffice to define $U(y)(y \leqq x)$ for arbitrary x and y; here we can observe that, for every pair of values (x, y) for which $y > x$, $U(y)(y \leqq x)$ must necessarily be false. To specify the value of $U(y)(y \leqq x)$ generally, we can now apply the standard recursive procedure. For $x = 1$, $U(y)(y \leqq x)$ is necessarily false, according to the definition of the relation "less than", whenever $y > 1$; we thus need merely specify $U(1)$ in order to have the value of $U(y)(y \leqq 1)$ for arbitrary y. Further, we specify the value of $U(y)(y \leqq x)$ for $x + 1$ for arbitrary y when it has already been specified for x for arbitrary y. Whenever $y > x + 1$, $U(y)(y \leqq x + 1)$ is false. When $y \leqq x$, we already know the value of the propositional function; that is, we must merely specify $U(x + 1)$. Therefore, *to specify the value of $U(x + 1)$ when this function is regarded as known for arbitrary $y < x + 1$ amounts precisely to specifying the value of the propositional function $U(x)(y \leqq x)$ for $x = n + 1$ for arbitrary y when this propositional function is already known for $x = n$ for arbitrary y.* Thus the apparently divergent kind of recursive definition has been reduced to the standard form.

Likewise it is easy to see that the more complicated form of inductive inference, which consists in proving a proposition for n when it is assumed to be true for arbitrary $m < n$, differs only formally but not actually from the standard form of inductive inference, the inference from n to $n + 1$. To say that $U(y)$ has been proved for arbitrary $y < x$ is, after all, equivalent to saying that $(y \geqq x) + U(y)$ has been proved for this x for arbitrary y, and it is then not difficult to see that *to prove $U(x)$, with $U(y)$ assumed to be true for arbitrary $y < x$, is equivalent to proving $(y \geqq x + 1) + U(y)$ for arbitrary y when $(y \geqq x) + U(y)$ has already been proved for arbitrary y.*

§ 7. The notion of prime number

Definition 12. $P(x, 1)$ is true. $P(x, y + 1) = P(x, y)(\overline{(x = y + 1) + D(x, y + 1)})$.

Definition 13. $P(x) = P(x, x)(x \neq 1)$.

The propositional function $P(x)$ means "x is a prime number".

Definition 14. $De(x, 1)$ is false. $De(x, y + 1) = De(x, y) + D(x, y + 1)(y + 1 < x)$.

Definition 15. $Dp(x, 1)$ is false. $Dp(x, y + 1) = Dp(x, y) + D(x, y + 1)P(y + 1)$.

Definition 16. $T(1)$ is true. $T(x + 1) = T(x)Dp(x + 1, x + 1)$.

In the customary definition of the notion of prime number an apparent variable is used. If the relation of divisibility is regarded as already defined, the definition of prime number can be given as follows:

$$P(x) = (x \neq 1)\Pi_{y}(\overline{D(x, y)} + (y = 1) + (y = x)).$$

An infinite propositional product (or, in other words, Russell and Whitehead's "always") occurs here. But in Definitions 12 and 13 above no apparent variable occurs. The infinite propositional product can be avoided since it can immediately be replaced by a finite one. For, according to the corollary to Theorem 25, we necessarily have $y \leqq x$ if $D(x, y)$ is true, and thence it follows that in the infinite product all factors for which $y > x$ holds become inoperative. Consequently we can just as well define $P(x)$ thus:

$$P(x) = (x \neq 1)\overset{x}{\underset{1}{\Pi}}{}_{y}(\overline{D(x, y)} + (y = 1) + (y = x)).$$

The finite propositional product occurring here is nothing but the factor $P(x, x)$ on the right side of Definition 13; $P(x, y)$ is recursively defined by Definition 12 and is equivalent to the product

$$\prod_{z=1}^{y}(\overline{D(x, z)} + (z = x)).$$

Precisely because these propositional products are finite can they be defined recursively and thus apparent variables avoided altogether.

The remaining propositional functions introduced above, De, Dp, and T, have, as is easily seen, the following meanings in words:

$De(x, y)$ means "x is divisible by a number >1, $\leqq y$, and $<x$".

$Dp(x, y)$ means "x is divisible by a prime $\leqq y$".

$T(x)$ means that all numbers from 2 to x are divisible by at least one prime $\leqq x$.

THEOREM 50. $\overline{P(x, y)D(x, z)(z \leqq y)} + (z = 1) + (z = x)$.

Proof. When $y = 1$, the proposition is clearly true. I prove the proposition for $y + 1$, assuming that it holds for y. From $z \leqq y + 1$ it follows that either $z \leqq y$ or $z = y + 1$. From $z \leqq y$, together with $P(x, y)$ (which follows from $P(x, y + 1)$) and $D(x, z)$, we obtain, according to the assumption, either $z = 1$ or $z = x$. From $z = y + 1$, together with $(x = y + 1) + \overline{D(x, y + 1)}$ (which also follows from $P(x, y + 1)$) and $D(x, z)$, we obtain $z = x$.

Corollary. $\overline{P(x)D(x, y)} + (y = 1) + (y = x)$.

For from $P(x, x)(x \neq 1)D(x, y)$ it follows (Theorem 25) that $P(x, x)D(x, y)(y \leqq x)$, whence, according to the theorem just proved, $(y = 1) + (y = x)$.

THEOREM 51. $\overline{P(y)} + D(x, y) + (x \wedge y = 1)$.

Proof. Since $D(y, x \wedge y)$ is true (Theorem 37), we necessarily have, whenever $P(y)$ holds, either $x \wedge y = 1$ or $x \wedge y = y$, according to the corollary to Theorem 50; but from $x \wedge y = y$ it follows (Theorem 37) that $D(x, y)$.

THEOREM 52. $\overline{D(xy, z)P(z)} + D(x, z) + D(y, z)$.

Proof. According to the preceding theorem, if $P(z)$ is true, either $D(x, z)$ or $x \wedge z = 1$ holds. From $D(xy, z)P(z)$, therefore, either $D(x, z)$ or $D(xy, z)(x \wedge z = 1)$ must follow; but from $D(xy, z)(x \wedge z = 1)$ it in turn follows (Theorem 41) that $D(y, z)$.

THEOREM 53. $P(x, y) + De(x, y)$.

Proof. True for $y = 1$, since (Definition 12) $P(x, 1)$ already is true. I assume that the proposition is true for y and on this basis prove that it is true for $y + 1$. From $\overline{P(x, y + 1)}$ it follows (Definition 12) either that $\overline{P(x, y)}$, whence we have $De(x, y)$ and consequently (Definition 14) also $De(x, y + 1)$, or that $(x \neq y + 1)D(x, y + 1)$, whence (Corollary 1 to Theorem 25) $D(x, y + 1)(y + 1 < x)$, which, on account of Definition 14, yields $De(x, y + 1)$.

THEOREM 54. $\overline{(y \leqq y')Dp(x, y)} + Dp(x, y')$.

Proof. Clearly true when $y' = 1$. I assume that the proposition is true for y' and then prove that it is true for $y' + 1$. From $y \leqq y' + 1$ it follows either that $y \leqq y'$, which together with $Dp(x, y)$, according to the assumption, implies $Dp(x, y')$ and therefore (Definition 15) also $Dp(x, y' + 1)$, or that $y = y' + 1$, which together with $Dp(x, y)$ of course implies $Dp(x, y' + 1)$.

THEOREM 55. $\overline{D(x, y)Dp(y, z)} + Dp(x, z)$.

Proof. For $z = 1$ this proposition is true, since (Definition 15) $Dp(y, 1)$ is false. I therefore assume that the proposition holds for z and then prove that it holds for $z + 1$. From $Dp(y, z + 1)$ we obtain (Definition 15) either $Dp(y, z)$, which together with $D(x, y)$ implies $Dp(x, z)$ and consequently also $Dp(x, z + 1)$, or $D(y, z + 1) P(z + 1)$, which together with $D(x, y)$, according to the corollary to Theorem 20, makes $D(x, z + 1)$ true, whence, according to Definition 15, $Dp(x, z + 1)$.

THEOREM 56. $\overline{Dp(x, y)} + Dp(x, x)$.

Proof. According to Definition 15 this must be true for $y = 1$. I assume that it is true for y and prove that it is true for $y + 1$. From $Dp(x, y + 1)$ it follows either that $Dp(x, y)$, whence we have $Dp(x, x)$, or that $D(x, y + 1)P(y + 1)$, whence, according to Corollary 1 to Theorem 25, $y + 1 \leqq x$; but from $Dp(x, y + 1)(y + 1 \leqq x)$ it follows (Theorem 54) that $Dp(x, x)$.

THEOREM 57. $\overline{De(x, y)T(y)} + Dp(x, x)$.

Proof. The proposition holds for $y = 1$; for $De(x, 1)$ is false. I assume that the proposition holds for y and prove it for $y + 1$. From $De(x, y + 1)T(y + 1)$ it follows either that $De(x, y)T(y + 1)$, whence (Definition 16) $De(x, y)T(y)$, whence, in consequence of the assumption, $Dp(x, x)$, or that $D(x, y + 1)(y + 1 < x)T(y + 1)$, whence (Definition 16) $D(x, y + 1)(y + 1 < x)Dp(y + 1, y + 1)$, whence in turn, according to Theorem 55, $Dp(x, y + 1)(y + 1 < x)$, which implies (Theorem 54) that $Dp(x, x)$ holds.

THEOREM 58. $T(x)$.

Proof. True for $x = 1$ (Definition 16). I prove that $T(x + 1)$ holds on the assumption that $T(x)$ does. According to Definition 16 I need only prove $Dp(x + 1, x + 1)$ in order to accomplish this. According to Definition 13 we have either $P(x + 1)$ or $\overline{P(x + 1, x + 1)}$. From $P(x + 1)D(x + 1, x + 1)$ it follows, according to Definition 15, that $Dp(x + 1, x + 1)$. From $\overline{P(x + 1, x + 1)}$ it follows, according to Theorem 53, that $De(x + 1, x + 1)$, which must (see Definition 14) have $De(x + 1, x)$ as a consequence; but from $De(x + 1, x)T(x)$ it follows, according to Theorem 57, that $Dp(x + 1, x + 1)$.

Corollary. $Dp(x + 1, x + 1)$.

For from $T(x + 1)$ it follows (Definition 16) that $Dp(x + 1, x + 1)$.

This is the theorem to the effect that every number >1 is divisible by at least one prime. $Dp(x + 1, x + 1)$ really means that $x + 1$ is divisible by at least one prime $\leqq x + 1$.

Now, to be able to formulate and prove the theorem that every number >1 is a product of primes, I introduce a ternary relation $P(x, y, z)$ that is to mean "x is the product of y primes that are all $\leqq z$".

Definition 17. $P(x, y, z) = P(x, y, z - 1) + P(x/z, y - 1, z)P(z)D(x, z)$. $P(x, 1, z) = (x \leqq z)P(x)$. $P(x, y, 1)$ is false.

This is a doubly recursive definition, for it is recursive with respect to y as well as z. $P(x, 1, z)$ is defined for arbitrary z (and x) and, if it is assumed that $P(x, y - 1, z)$ is already defined for arbitrary z (and x), then, by means of the first equation of Definition 17 and the stipulation that $P(x, y, 1)$ is to be false, we have a recursive definition of the propositional function $P(x, y, z)$, recursive, that is, with respect to z. Through the last stipulation of Definition 17, $P(x, y, 1)$ is determined and, through the first equation, $P(x, y, z)$ is determined on the basis of the assumption that $P(x, y - 1, z)$ is already known.

I want to introduce three additional propositional functions.

Definition 18. $P'(x, 1, z) = P(x, 1, z)$. $P'(x, y + 1, z) = P'(x, y, z) + P(x, y + 1, z)$.

Definition 19. $\Pi(x) = P'(x, x, x)$.

Definition 20. $\Pi'(1)$ is true. $\Pi'(x + 1) = \Pi'(x)\Pi(x + 1)$.

The proposition $P'(x, y, z)$ obviously means that x is a product of at most y primes, each $\leqq z$. The proposition $\Pi(x)$ means that x is a product of at most x primes, each $\leqq x$. $\Pi'(x)$ means that every number y from 1 to x either is equal to 1 or is a product of at most y primes $\leqq y$, or, in other words, that every number y from 2 to x is a product of at most y primes $\leqq y$. The real purpose of the considerations that follow is to prove that $\Pi(x + 1)$ holds.

THEOREM 59. $\overline{P(x, y, z)(z \leqq z')} + P(x, y, z')$.

Proof. Clearly, the proposition is true when $z' = 1$. I prove it for $z' + 1$ on the assumption that it holds for z'. From $z \leqq z' + 1$ it follows either that $z \leqq z'$, which together with $P(x, y, x)$ has $P(x, y, z')$ and therefore also (Definition 17) $P(x, y, z' + 1)$ as consequences, or that $z = z' + 1$, and from $(z = z' + 1)P(x, y, z)$ it clearly again follows that $P(x, y, z' + 1)$.

Lemma 1 (for Theorem 60). $\overline{P(x_1, 1, z)P(x_2, 1, z)} + P(x_1x_2, 2, z)$.

Proof. True for $z = 1$; for (Definitions 17 and 13) $P(x, 1, 1)$ already is false. I prove the proposition for $z + 1$ on the assumption that it holds for z. From $P(x_1, 1, z + 1)$ $P(x_2, 1, z + 1)$ it follows according to Definition 17 that $P(x_1, 1, z)P(x_2, 1, z)$, whence we have $P(x_1x_2, 2, z)$, which in turn has $P(x_1x_2, 2, z + 1)$ as a consequence, or that $(x_1 = z + 1)P(x_1)P(x_2, 1, z + 1)$, or that $(x_2 = z + 1)P(x_2)P(x_1, 1, z + 1)$. But from $P(z + 1)P(x_1x_2/(z + 1), 1, z + 1)$ it follows (Definition 17) that $P(x_1x_2, 2, z + 1)$. We proceed analogously in the third case.

Lemma 2 (for Theorem 60). $\overline{P(x_1, y, z)P(x_2, 1, z)} + P(x_1x_2, y + 1, z)$.

Proof. According to Lemma 1 this is true for $y = 1$. I prove that the proposition is true for $y + 1$ on the assumption that it holds for y. Now, in case $z = 1$ we indeed obtain $P(x_1x_2, y + 2, z)$ from $P(x_1, y + 1, z)P(x_2, 1, z)$, for $P(x_2, 1, 1)$ is false. Therefore, on the assumption that the inductive step from y to $y + 1$ holds for z, I prove that it holds for $z + 1$. From $P(x_1, y + 1, z + 1)P(x_2, 1, z + 1)$ we obtain (Definition 17) three possibilities: (1) $P(x_1, y + 1, z)P(x_2, 1, z)$ already holds; then according to the assumption we have $P(x_1x_2, y + 2, z)$, whence (Definition 17) $P(x_1x_2, y + 2, z + 1)$. (2) We have $P(x_1/(z + 1), y, z + 1)P(z + 1)P(x_2, 1, z + 1)$; but from $P(x_1/(z + 1), y, z + 1)P(x_2, 1, z + 1)$ it follows, since our proposition was assumed to be true for y for arbitrary z, that $P(x_1x_2/(z + 1), y + 1, z + 1)$, which together with $P(z + 1)$ by virtue of Definition 17 implies $P(x_1x_2, y + 2, z + 1)$. (3) We have $P(x_1, y + 1, z + 1)(x_2 = z + 1)P(x_2)$; but from this we obtain (Definition 17) $P(x_1x_2, y + 2, z + 1)$.

THEOREM 60. $\overline{P(x_1, y_1, z)P(x_2, y_2, u)(u \leqq z)} + P(x_1x_2, y_1 + y_2, z)$.

Proof. On account of Lemma 2, when $y_1 = 1$ or $y_2 = 1$, this proposition is certainly true for arbitrary values of the remaining variables. Thus it surely holds for the least possible value of the sum $y_1 + y_2 + z + u$, namely 4, for which necessarily $y_1 = y_2$

$= z = u = 1$. I now prove that the proposition is true for values of y_1, y_2, z, and u such that $y_1 + y_2 + z + u = n$, assuming it to be true for those cases in which $y_1 + y_2 + z + u = n - 1$. $P(x_1, y_1, z)$ is equivalent to either $P(x_1, y_1, z - 1)$ or $P(x_1/z, y_1 - 1, z)P(z)D(x, z)$. Furthermore we have either $u \leqq z - 1$ or $u = z$. The first possibility is therefore $P(x_1, y_1, z - 1)P(x_2, y_2, u)(u \leqq z - 1)$. But, since $y_1 + y_2 + (z - 1) + u = n - 1$, it follows according to the assumption made that $P(x_1x_2, y_1 + y_2, z - 1)$ and thence (Definition 17) in turn that $P(x_1x_2, y_1 + y_2, z)$. The second possibility is $P(x_1, y_1, z - 1)P(x_2, y_2, z)$. But, since here, too, necessarily $y_1 + y_2 + (z - 1) + u = n - 1$, it follows that $P(x_1x_2, y_1 + y_2, z)$. The third possibility is $P(x_1/z, y_1 - 1, z)P(z)P(x_2, y_2, u)(u \leqq z)$. But thence it follows, since $(y_1 - 1) + y_2 + z + u = n - 1$, that $P(x_1x_2/z, y_1 + y_2 - 1, z)P(z)$, and thence (Definition 17) again that $P(x_1x_2, y_1 + y_2, z)$.

Corollary. $\overline{P(x_1, y_1, z)P(x_2, y_2, z)} + P(x_1x_2, y_1 + y_2, z)$.

Lemma. $\overline{P'(x_1, y_1, z)P(x_2, y_2, z)} + P'(x_1x_2, y_1 + y_2, z)$.

Proof. True for $y_1 = 1$ (Definition 18 and Lemma 2 for Theorem 60). I prove that the proposition is true for $y_1 + 1$ on the assumption that it holds for y_1. From $P'(x_1, y_1 + 1, z)$ it follows either that $P'(x_1, y_1, z)$, which together with $P(x_2, y_2, z)$ has, according to the assumption, $P'(x_1x_2, y_1 + y_2, z)$ and therefore also $P'(x_1x_2, y_1 + y_2 + 1, z)$ as consequences, or that $P(x_1, y_1 + 1, z)$, whence together with $P(x_2, y_2, z)$ it follows according to the corollary to Theorem 60 that $P(x_1x_2, y_1 + 1 + y_2, z)$, whence (Definition 18) again $P'(x_1x_2, y_1 + 1 + y_2, z)$.

THEOREM 61. $\overline{P'(x_1, y_1, z)P'(x_2, y_2, z)} + P'(x_1x_2, y_1 + y_2, z)$.

Proof. According to the lemma this is true for $y_2 = 1$. I prove that the proposition is true for $y_2 + 1$ on the assumption that it is true for y_2. From $P'(x_2, y_2 + 1, z)$ it follows (Definition 18) either that $P'(x_2, y_2, z)$, whence together with $P'(x_1, y_1, z)$, according to the assumption made, $P'(x_1x_2, y_1 + y_2, z)$, and therefore also $P'(x_1x_2, y_1 + y_2 + 1, z)$, or that $P(x_2, y_2 + 1, z)$, which together with $P'(x_1, y_1, z)$, according to the lemma, implies $P'(x_1x_2, y_1 + y_2 + 1, z)$.

Lemma. $\overline{P'(x, y, z)} + P'(x, y, z + 1)$.

Proof. By virtue of Definitions 17 and 18 this is true for $y = 1$. I prove the proposition for $y + 1$ on the assumption that it holds for y. From $P'(x, y + 1, z)$ it follows either that $P'(x, y, z)$, whence we obtain $P'(x, y, z + 1)$, hence (Definition 18) in turn $P'(x, y + 1, z + 1)$, or that $P(x, y + 1, z)$, whence (Definition 17) $P(x, y + 1, z + 1)$, hence (Definition 18) $P'(x, y + 1, z + 1)$.

THEOREM 62. $\overline{P'(x, y, z)(z \leqq z')} + P'(x, y, z')$.

Proof. Clearly true when $z' = 1$. I assume that the proposition is true for z' and then prove it for $z' + 1$. From $z \leqq z' + 1$ either it follows that $z \leqq z'$, and then from $P'(x, y, z)(z \leqq z')$ it follows, according to the assumption made, that $P'(x, y, z')$, and thence, according to the lemma, that $P'(x, y, z' + 1)$, or it follows that $z = z' + 1$, and thence together with $P'(x, y, z)$ it clearly follows that $P'(x, y, z' + 1)$.

THEOREM 63. $\overline{P'(x, y, z)(y \leqq y')} + P'(x, y', z)$.

Proof. Clearly true when $y' = 1$. I assume that the proposition is true for y'. From $y \leqq y' + 1$ it follows either that $y \leqq y'$ or that $y = y' + 1$. Now, according to the assumption it follows from $P'(x, y, z)(y \leqq y')$ that $P'(x, y', z)$ and therefore

(Definition 18) also that $P'(x, y' + 1, z)$. From $P'(x, y, z)(y = y' + 1)$ it follows, of course, that $P'(x, y' + 1, z)$. Thus the proposition holds for $y' + 1$ too.

THEOREM 64. $(x = 1) + (y = 1) + \overline{\Pi(x)} + \overline{\Pi(y)} + \Pi(xy)$.

Proof. Since according to Theorem 15 necessarily $x \leqq xy$ and likewise $y \leqq xy$, it follows according to Theorem 62 from $P'(x, x, x)P'(y, y, y)$, that is, from $\Pi(x)\Pi(y)$, that $P'(x, x, xy)P'(y, y, xy)$. But according to Theorem 61 it follows from $P'(x, x, xy)$ $P'(y, y, xy)$ in turn that $P'(xy, x + y, xy)$. If now in addition $(x > 1)(y > 1)$, then $x + y \leqq xy$, so that by using Theorem 63 we obtain $P'(xy, xy, xy)$, that is, $\Pi(xy)$. Thus from $(x > 1)(y > 1)\Pi(x)\Pi(y)$ it follows that $\Pi(xy)$, which was to be proved.

Lemma 1 (for Theorem 65). $\overline{\Pi'(x)(y \leqq x)} + \Pi'(y)$.

Proof. Clearly true for $x = 1$. I assume that the proposition is true for x. From $y \leqq x + 1$ it follows either that $y \leqq x$ or that $y = x + 1$. From $\Pi'(x + 1)$ it follows (Definition 20) that $\Pi'(x)$, and therefore it follows from $\Pi'(x + 1)(y \leqq x)$ first that $\Pi'(x)(y \leqq x)$ and thence further that $\Pi'(y)$. From $\Pi'(x + 1)(y = x + 1)$ it follows, of course, that $\Pi'(y)$.

Lemma 2 (for Theorem 65). $\overline{\Delta(x, y, z)\Pi(y)\Pi'(z)(y > 1)} + \Pi(x)$.

Proof. This is true when $z = 1$. I assume that the proposition is true for z. From $\Delta(x, y, x + 1)$ it follows (Definition 5) either that $\Delta(x, y, z)$ or that $x = y(z + 1)$. From $\Delta(x, y, z)\Pi(y)\Pi'(z + 1)(y > 1)$ we obtain, according to Definition 20, $\Delta(x, y, z)$ $\Pi(y)\Pi'(z)(y > 1)$ and thence, according to the assumption made, $\Pi(x)$. From $(x = y(z + 1))\Pi(y)\Pi'(z + 1)(y > 1)$ it follows (Definition 20) that $(x = y(z + 1))\Pi(y)$ $\Pi(z + 1)(y > 1)$ and thence, according to Theorem 64, that $\Pi(x)$. Thus the proposition holds for $z + 1$ too.

Lemma 3 (for Theorem 65). $\overline{De(x, y)\Pi'(x - 1)} + \Pi(x)$.

Proof. This holds for $y = 1$, as may be seen from Definition 14. I assume that the proposition holds for y. From $De(x, y + 1)$ it follows (Definition 14) either that $De(x, y)$ or that $D(x, y + 1)(y + 1 < x)$. From $De(x, y)\Pi'(x - 1)$ it must follow, according to the assumption, that $\Pi(x)$. According to Definition 2, $D(x, y + 1)(y + 1$ $< x)$ must be equivalent to $D(x, y + 1)(y + 1 \leqq x - 1)$; but from $\Pi'(x - 1)(y +$ $1 \leqq x - 1)$ it follows (Lemma 1) that $\Pi'(y + 1)$, whence (Definition 20) $\Pi(y + 1)$, and since furthermore, as is easily seen, we necessarily (see Definitions 5 and 6) have $D(x, y + 1) = \Delta(x, y + 1, x - 1)$, it follows from $D(x, y + 1)\Pi(y + 1)\Pi'(x - 1)$, according to Lemma 2, that $\Pi(x)$.

THEOREM 65. $\Pi'(x)$.

Proof. True for $x = 1$ (Definition 20). I assume that $\Pi'(x - 1)$ is true and then prove $\Pi(x)$, which also proves (Definition 20) $\Pi'(x)$. Either $P(x)$ or $\overline{P(x)}$ holds. From $P(x)$ it follows (Definition 17) that $P(x, 1, x)$, that is (Definition 18), that $P'(x, 1, x)$, whence, according to Theorem 63, $P'(x, x, x)$, that is, $\Pi(x)$. From $\overline{P(x)}(x > 1)$ it follows (Definition 13) that $\overline{P(x, x)}$, whence (Theorem 53) $De(x, x)$. But, according to Lemma 3, $De(x, x)\Pi'(x - 1)$ implies $\Pi(x)$.

Corollary. $\Pi(x + 1)$.

For, according to Definition 20, $\Pi(x + 1)$ follows from $\Pi'(x + 1)$. This, however, is the important theorem that every number >1 is a product of prime numbers.

§ 8. Some explicit uses of finite logical sums and products

If we are concerned with avoiding only the use of logical variables ranging over infinite domains, we can, of course, still make free use of variables ranging over *finite* domains; we need not trouble ourselves either about ways in which these might be avoided by means of recursive definitions or about ways in which the conclusions they allow us to reach might be derived, by mathematical induction, from those that hold for propositions not containing apparent variables.

If we take this approach, then the theory of prime factorization, for example, can be presented more simply, as I show below. I use the definition of prime numbers given on page 320:

Definition. $P(x) = (x \neq 1) \overset{x}{\underset{y=1}{\Pi}} (\overline{D(x, y)} + (y = 1) + (y = x))$.

Theorem 66. $\overset{n}{\underset{p=1}{\Sigma}} D(n, p)P(p) + (n = 1)$.

Proof. Holds for $n = 1$. Assume that the proposition holds for all $\nu < n$, with $n > 1$. Then either the proposition $\overset{n}{\underset{\nu=1}{\Sigma}} D(n, \nu)(\nu < n)(\nu > 1)$ or its negation $\overset{n}{\underset{\nu=1}{\Pi}} (\overline{D(n, \nu)} + (\nu = n) + (\nu = 1))$ holds (since here $\nu \geqq n$ is equivalent to $\nu = n$, and $\nu \leqq 1$ to $\nu = 1$). In the latter case we have $D(n, n)P(n)$. In the former case, according to the assumption, $\overset{n}{\underset{p=1}{\Sigma}} D(\nu, p)P(p)$ holds; consequently $\overset{n}{\underset{\nu=1}{\Sigma}} \overset{\nu}{\underset{p=1}{\Sigma}} D(n, \nu)D(\nu, p)P(p)$, whence $\overset{n}{\underset{p=1}{\Sigma}} D(n, p)P(p)$.

Since $D(n, p)$ is equivalent to the propositional sum $\overset{n}{\underset{\nu=1}{\Sigma}} (n = \nu p)$, we have the

Corollary. $(n = 1) + \overset{n}{\underset{\nu=1}{\Sigma}} \overset{n}{\underset{p=1}{\Sigma}} (n = \nu p)P(p)$.

I now define recursively a propositional function that in words is "n is equal to a product of μ prime factors, all of which are $\leqq n$".

Definition 21. $P(n, 1) = P(n)$; $P(n, \mu + 1) = \overset{n}{\underset{\nu=1}{\Sigma}} \overset{n}{\underset{p=1}{\Sigma}} (n = \nu p)P(\nu, \mu)P(p)$.

The theorem on the factorization of any number >1 into a product of primes can then be formulated as follows:

Theorem 67. $(n = 1) + \overset{n}{\underset{\mu=1}{\Sigma}} P(n, \mu)$.

Proof. Holds for $n = 1$. Assume that the proposition holds for $\nu < n$ when $n > 1$. Then, according to the corollary to Theorem 66, $\overset{n}{\underset{\nu=1}{\Sigma}} \overset{n}{\underset{p=1}{\Sigma}} (n = \nu p)P(p)$ holds. However, from $(n = \nu p)P(p)$ it follows (Theorem 15) that $\nu < n$ and thus, according to the assumption, $\overset{\nu}{\underset{\mu=1}{\Sigma}} P(\nu, \mu)$. Hence $\overset{n-1}{\underset{\mu=1}{\Sigma}} P(\nu, \mu)$ and further $\overset{n}{\underset{\nu=1}{\Sigma}} \overset{n}{\underset{p=1}{\Sigma}} \overset{n-1}{\underset{\mu=1}{\Sigma}} (n = \nu p)P(\nu, \mu)P(p)$, which is $= \overset{n-1}{\underset{\mu=1}{\Sigma}} P(n, \mu + 1)$, whence $\overset{n}{\underset{\mu=1}{\Sigma}} P(n, \mu)$.

I want to prove, further, the theorem on the existence of infinitely many primes. First I define the function $n!$ by

Definition 22. $1! = 1$; $(n + 1)! = n!(n + 1)$.

Lemma. $(m > n) + D(n!, m)$.

Proof. Clearly true for $n = 1$. Assume that it is true for a certain n. If not $m > n + 1$, then either $m = n + 1$ or $m < n + 1$ (that is, $m \leqq n$). According to Definition 22 and Theorem 18 we have $D((n+1)!, n+1)$. If $m \leqq n$, we have, according to the assumption, $D(n!, m)$, and, since (Theorems 16 and 18) $D((n+1)!, n!)$ also holds, we obtain (corollary to Theorem 20) $D((n+1)!, m)$.

THEOREM 68. $\sum_{p=1}^{n!+1} P(p)(p > n)$. (In words: For arbitrary n there is a prime $p > n$ and $\leqq n! + 1$.)

Proof. According to Theorem 66 the proposition $\sum_{p=1}^{n!+1} D(n!+1, p)P(p)$ holds. But from $D(n!+1, p)P(p)$ it follows that $p > n$. For, if we had $p \leqq n$, it would follow, according to the lemma, that $D(n!, p)$, and this together with $D(n!+1, p)$ would imply (corollary to Theorem 24) $D(1, p)$ and therefore $p = 1$, which is impossible on account of $P(p)$.

To prove the uniqueness of the factorization of a number into a product of prime factors I must first advance some considerations on sums and products of arbitrarily many terms or factors, as well as define a function $I(a, b; m, n)$.

Let $f(r)$ be an arbitrary descriptive function. I define the expressions $\sum_{r=1}^{n} f(r)$ and $\prod_{r=1}^{n} f(r)$ recursively as follows:

Definition 23. $\sum_{r=1}^{1} f(r) = f(1)$; $\sum_{r=1}^{n+1} f(r) = \sum_{r=1}^{n} f(r) + f(n+1)$. $\prod_{r=1}^{1} f(r) = f(1)$; $\prod_{r=1}^{n+1} f(r) = \prod_{r=1}^{n} f(r) \cdot f(n+1)$.

Instead of $f(r)$ we often write a_r, b_r, and so forth. For an arbitrary descriptive function a_r I state

Definition 24. $a_r^{(\nu)} = a_r$ if $r < \nu$, and $a_r^{(\nu)} = a_{r+1}$ if $r \geqq \nu$.

THEOREM 69. $(n = 1) + (\nu > n) + (\sum_{r=1}^{n} a_r = a_\nu + \sum_{r=1}^{n-1} a_r^{(\nu)})(\prod_{r=1}^{n} a_r = a_\nu \cdot \prod_{r=1}^{n-1} a_r^{(\nu)})$.

Proof. I need consider only the sum. The proposition holds when $n = 1$. Assume that it holds for n. To show that it then holds for $n + 1$ also, first let $\nu \leqq n$. Then

$$\sum_{r=1}^{n+1} a_r = \sum_{r=1}^{n} a_r + a_{n+1} = a_\nu + \sum_{r=1}^{n-1} a_r^{(\nu)} + a_{n+1}$$

and furthermore $a_n^{(\nu)} = a_{n+1}$. Consequently (according to Definition 24)

$$\sum_{r=1}^{n-1} a_r^{(\nu)} + a_{n+1} = \sum_{r=1}^{n-1} a_r^{(\nu)} + a_n^{(\nu)} = \sum_{r=1}^{n} a_r^{(\nu)}$$

and thence

$$\sum_{r=1}^{n+1} a_r = a_\nu + \sum_{r=1}^{n} a_r^{(\nu)}.$$

On the other hand, let $\nu = n + 1$. Then $a_r^{(\nu)} = a_r$ for $r \leqq n$. Consequently

$$\sum_{r=1}^{n+1} a_r = \sum_{r=1}^{n} a_r + a_{n+1} = a_{n+1} + \sum_{r=1}^{n} a_r^{(\nu)}.$$

In many derivations we make use of the argument that such a sum (or product) remains unchanged whenever the terms (or factors) a are replaced by numbers b, provided these are in some sequential order identical with the a. To be able to formulate this conveniently I introduce a propositional function $I(a, b; m, n)$, where a and b are signs for two arbitrary descriptive functions.

Definition 25. $I(a, b; 1, 1) = (a_1 = b_1)$. $I(a, b; 1, n)(n > 1)$ is false. $I(a, b; m, 1)$ $(m > 1)$ is false. $I(a, b; m + 1, n) = \sum_{\nu=1}^{n} (a_{m+1} = b_\nu)I(a, b^{(\nu)}; m, n - 1)$ ($b^{(\nu)}$ was defined in Definition 24).

$I(a, b; m, n)$ then means in words: "The numbers $a_1, \ldots, a_m$ are in some ordering identical with the numbers $b_1, \ldots, b_n$".

THEOREM 70. $\overline{I(a, b; m, n)} + (m = n)$.

Proof. True when $m = 1$, by Definition 25. Assume that it is true for m. From $I(a, b; m + 1, n)$ it follows that $\sum_{\nu=1}^{n} (a_{m+1} = b_\nu)I(a, b^{(\nu)}; m, n - 1)$. But $I(a, b^{(\nu)}; m, n - 1)$ implies $m = n - 1$. Hence $m + 1 = n$.

THEOREM 71. $\overline{I(a, b; m, n)} + (\sum_{r=1}^{m} a_r = \sum_{r=1}^{n} b_r)(\prod_{r=1}^{m} a_r = \prod_{r=1}^{n} b_r)$.

Proof. I consider only the sum. First, when $n = 1$, it follows that $m = n$ and the proposition clearly holds. Assume that it holds for $n - 1$. From $I(a, b; m, n)$ it follows that $\sum_{\rho=1}^{n} (a_n = b_\rho)I(a, b^{(\rho)}; n - 1, n - 1)$. But, according to the assumption, $I(a, b^{(\rho)};$ $n - 1, n - 1)$ yields $\sum_{r=1}^{n-1} a_r = \sum_{r=1}^{n-1} b_r^{(\rho)}$, and furthermore we have (Theorem 69) $b_\rho + \sum_{r=1}^{n-1} b_r^{(\rho)}$ $= \sum_{r=1}^{n} b_r$. From $a_n = b_\rho$ and $\sum_{r=1}^{n-1} a_r = \sum_{r=1}^{n-1} b_r^{(\rho)}$ it therefore follows that $\sum_{r=1}^{n} a_r = \sum_{r=1}^{n} b_r$.

THEOREM 72. $\overline{D(\prod_{r=1}^{n} a_r, p)P(p)} + \sum_{r=1}^{n} D(a_r, p)$.

Proof. The proposition holds for $n = 1$; assume that it holds for a certain n. From $D(a_{n+1} \cdot \prod_{r=1}^{n} a_r, p)P(p)$ it follows (Theorem 52) that $D(a_{n+1}, p) + D(\prod_{r=1}^{n} a_r, p)P(p)$. From $D(\prod_{r=1}^{n} a_r, p)P(p)$ it follows, according to the assumption, that $\sum_{r=1}^{n} D(a_r, p)$. Thus in any case $D(a_{n+1}, p) + \sum_{r=1}^{n} D(a_r, p)$, which is $= \sum_{r=1}^{n+1} D(a_r, p)$, holds.

THEOREM 73. $\overline{D(\prod_{r=1}^{n} p_r, q) \prod_{r=1}^{n} P(p_r)P(q)} + \sum_{r=1}^{n} (p_r = q)$.

Proof. By Theorem 72, from $D(\prod_{r=1}^{n} p_r, q)$ we derive $\sum_{r=1}^{n} D(p_r, q)$. From $D(p_r, q)P(p_r)$ $P(q)$, however, it follows that $((p_r = q) + (q = 1))P(q)$; hence $p_r = q$. Therefore $\sum_{r=1}^{n} (p_r = q)$.

The theorem on the uniqueness of the prime factorization now reads as follows:

THEOREM 74. $(\prod_{r=1}^{\mu} p_r \neq \prod_{s=1}^{\nu} q_s) + \sum_{r=1}^{\mu} \overline{P(p_r)} + \sum_{s=1}^{\nu} \overline{P(q_s)} + I(p, q; \mu, \nu)$.

Proof. I first prove the proposition for $\mu = 1$. If $(p_1 = \prod_{s=1}^{\nu} q_s)P(p_1) \prod_{s=1}^{\nu} P(q_s)$, then it follows, according to Theorem 73, that $\sum_{\sigma=1}^{\nu} (p_1 = q_\sigma)$. Further, from $p_1 = q_\sigma$ and $p_1 =$ $p_\sigma \prod_{s=1}^{\nu-1} q_s^{(\sigma)}$ the equation $1 = \prod_{r=1}^{\nu-1} q_s^{(\sigma)}$ follows, which is impossible if $\nu > 1$.

Assume that the proposition is true for a certain μ. Then from the proposition $(\prod_{r=1}^{\mu+1} p_r = \prod_{s=1}^{\nu} q_s) \prod_{r=1}^{\mu+1} P(p_r) \prod_{s=1}^{\nu} P(q_s)$ it follows that $D(\prod_{s=1}^{\nu} q_s, p_{\mu+1}) \prod_{s=1}^{\nu} P(q_s)P(p_{\mu+1})$, whence in turn, according to Theorem 73, $\sum_{\rho=1}^{\nu} (p_{\mu+1} = q_\rho)$. Furthermore $\prod_{s=1}^{\nu} q_s = q_\rho \prod_{s=1}^{\nu-1} q_s^{(\rho)}$, and we therefore obtain $\sum_{\rho=1}^{\nu} (\prod_{r=1}^{\mu} p_r = \prod_{s=1}^{\nu-1} p_s^{(\rho)})$, whence, according to the assumption, $\sum_{\rho=1}^{\nu} I(p, q^{(\rho)}; \mu, \nu - 1)(p_{\mu+1} = q_\rho)$, which is $= I(p, q; \mu + 1, \nu)$.

Finally I wish to engage in some reflections of a more general kind. We have the following

THEOREM 75a. $\overline{U(n)} + \sum_{\nu=1}^{n} \prod_{\mu=1}^{\nu} U(\nu)(\overline{U(\mu)} + (\mu = \nu))$, where U denotes an arbitrary propositional function.[3]

In words: If we know a number n for which the proposition U is true, there exists a least number for which U is true. It must be noted here that the propositions with which we are now concerned are always regarded as given without apparent variables ranging over infinite domains, so that it is always finitely decidable whether $U(x)$ is true or not for an arbitrary x.

Proof. For $n = 1$ the proposition is clearly true. Assume that it is true for all $x \leqq$ a certain n. Then if $U(n + 1)$ is true, so is either $\prod_{x=1}^{n} \overline{U(x)}$ or $\sum_{x=1}^{n} U(x)$. In the former case, therefore, $U(n + 1) \prod_{\nu=1}^{n+1} (\overline{U(\nu)} + (\nu = n + 1))$ holds. In the latter case it follows from $U(x)(x \leqq n)$, according to the assumption, that[4] $\sum_{y=1}^{x} \prod_{z=1}^{y} U(x)(\overline{U(z)} + (z = y))$ and therefore a fortiori that $\sum_{\nu=1}^{n+1} \prod_{\mu=1}^{\nu} U(\nu)(\overline{U(\mu)} + (\mu = \nu))$.

THEOREM 75b. From $U(a) \prod_{x=1}^{a} (\overline{U(x)} + (x = a))U(b) \prod_{y=1}^{b} (\overline{U(y)} + (y = b))$ it follows that $a = b$.

This theorem expresses the uniqueness of the least number.

Proof. If we had $a \neq b$, we would have $(a < b) + (a > b)$. But from $a < b$ it follows at once that $\overline{U(a)}$ and likewise, from $b < a$, that $\overline{U(b)}$.

Because of this uniqueness we can introduce a very important descriptive function of the general propositional function U, which denotes the least number for which U is true. This descriptive function, to be sure, has a restricted domain of definition since it has no value if the proposition U is true of no number. But it must be emphasized here that we are concerned only with the natural numbers up to certain upper bound, which, however, may be chosen arbitrarily large. The descriptive function, therefore, may be denoted here by $\mathrm{Min}(U, n)$. This means the least number among the numbers 1 to n for which U is true, and it has no meaning if U is false for all these

[3] Of course, we can, for example, also write

$$\overline{U(n)} + \sum_{\nu=1}^{n} \prod_{\mu=1}^{n} U(\nu)(\overline{U(\mu)} + (\mu \geqq \nu)).$$

[4] [[In the German text the matrix of the formula that follows is, erroneously, "$U(x)(\overline{U(y)} + (y = x))$".]]

numbers. Thus we are not dealing with a function Min(U), or Min(U, ∞), that would mean the least number satisfying the proposition U and would have no meaning if U is false for every number; for all this would require the "actual infinite", hence the use of apparent logical variables ranging over infinite domains. But the restriction that we must here impose on the meaning of this minimum function does no harm in practice; for, whenever we use the theorem asserting that in a class of positive integers there exists a least, we must in any case first have come to know a number n of this class, and then we can make do with this very function Min(U, n).

Finally I wish to make some remarks on the *notion of cardinality* [[zum *Anzahlbegriffe*]]. If certain objects (numbers, number pairs, number triples, and so forth, or perhaps descriptive functions) are mapped one-to-one onto all numbers $\leqq$ a certain number n, then I say that *their cardinality is n*.

In order to show that this number n is invariant with respect to the different mappings, we must prove the following

THEOREM 76. From $\prod_{x=1}^{m}\prod_{y=1}^{m}((x = y) + (f(x) \neq f(y)))\prod_{x=1}^{m}(f(x) \leqq n)\prod_{y=1}^{n}\sum_{z=1}^{m}(y = f(z))$, where f is an arbitrary descriptive function, it follows that $m = n$.

In words: Assume that the numbers that are $\leqq m$ are mapped one-to-one onto all numbers $\leqq n$; then $m = n$.

Proof. I first take the case in which $m = 1$. From $\prod_{y=1}^{n}\sum_{z=1}^{1}(y = f(z)) = \prod_{y=1}^{n}(y = f(1))$ it follows on account of the uniqueness of $f(1)$ that necessarily $n = 1$.

Assume that the proposition holds for a certain m. Then, if I take the value $m + 1$, in any case necessarily $n > 1$. Otherwise from $\prod_{x=1}^{m+1}(f(x) \leqq 1)$ we would obtain $f(1) = 1$ as well as $f(2) = 1$, whereas we were supposed to have had $f(1) \neq f(2)$. Consequently we can write $n + 1$ instead of n, and we must now prove that $m = n$.

Consider first the assumption $f(m + 1) = n + 1$. Then $\prod_{x=1}^{m}(f(x) \neq f(m + 1))$, or $\prod_{x=1}^{m}(f(x) \neq n + 1)$. Since furthermore $\prod_{x=1}^{m}(f(x) \leqq n + 1)$, it follows that $\prod_{x=1}^{m}(f(x) \leqq n)$. From $\prod_{y=1}^{n+1}\sum_{z=1}^{m+1}(y = f(z))$ it follows that $\prod_{y=1}^{n}\sum_{z=1}^{m+1}(y = f(z))(y \leqq n)$, that is,

$$\prod_{y=1}^{n}\sum_{z=1}^{m}((y = f(z)) + (y = f(m + 1))(y < f(m + 1))),$$

which is $= \prod_{y=1}^{n}\sum_{z=1}^{m}(y = f(z))$. By virtue of the assumption made, therefore, necessarily $m = n$.

Consider next the assumption $f(m + 1) = \nu < n + 1$. According to the assumption that $\prod_{y=1}^{n+1}\sum_{z=1}^{m+1}(y = f(z))$, we have $\sum_{z=1}^{m+1}(n + 1 = f(z))$, that is,

$$\sum_{\mu=1}^{m}(n + 1 = f(\mu)) + (n + 1 = f(m + 1)),$$

hence $\sum_{\mu=1}^{m}(n + 1 = f(\mu))$. Here the number μ is uniquely determined. Now I introduce a new descriptive function f', which is defined thus: $f'(x) = f(x)$ if $x \neq \mu$ and $\leqq m$.

Further, $f'(\mu) = f(m + 1) = \nu$; $f'(m + 1) = f(\mu) = n + 1$. We must then show merely that f' satisfies the same conditions as f.

From $(f(x) \leqq n + 1)(x \neq \mu)(x \leqq m)$ it follows that $f'(x) \leqq n + 1$, since in this case $f'(x) = f(x)$. Besides, clearly both $f'(\mu)$ and $f'(m)$ are $\leqq n + 1$. Thus $\prod_{x=1}^{m+1} (f'(x) \leqq n + 1)$.

From $\sum_{y=1}^{m+1} (x = f(y))$ it follows that

$$\sum_{y=1}^{m+1} (x = f(y))(y \neq \mu)(y < m) + \sum_{y=1}^{m+1} (x = f(y))((y = \mu) + (y = m + 1)),$$

whence $\sum_{y=1}^{m+1} (x = f'(y)) + (x = f'(m + 1)) + (x = f'(\mu))$, which is $= \sum_{\nu=1}^{m+1} (x = f'(y))$. Thus from $\prod_{x=1}^{n+1} \sum_{y=1}^{m+1} (x = f(y))$ it follows that $\prod_{x=1}^{n+1} \sum_{y=1}^{m+1} (x = f'(y))$ also.

According to the assumption, $\prod_{x=1}^{m+1} \sum_{y=1}^{m+1} ((x = y) + (f(x) \neq f(y)))$ holds. If we have

$$(f(x) \neq f(y))(x \neq \mu)(x \leqq m)(y \neq \mu)(y \leqq m),$$

then it follows immediately that $f'(x) \neq f'(y)$. If we have

$$(f(x) \neq f(y))((x = \mu) + (x = m + 1))(y \neq \mu)(y \leqq m),$$

then

$$((f(x) = f(\mu)) + (f(x) = f(m + 1)))(f(y) = f'(y))(f(y) \neq f(\mu))(f(y) \neq f(m + 1))$$

holds and consequently $f'(x) \neq f'(y)$. Similarly if

$$(f(x) \neq f(y)(x = \mu))(x \leqq m)((y = \mu) + (y = m)).$$

Finally,

$$(f(x) \neq f(y))((x = \mu) + (x = m + 1))((y = \mu) + (y = m + 1))$$

yields

$$(x = \mu)(y = m + 1) + (x = m + 1)(y = \mu),$$

and therefore also

$$(f'(x) = f(m + 1))(f'(y) = f(\mu)) + (f'(x) = f(\mu))(f'(y) = f(m + 1)),$$

and consequently $f'(x) \neq f'(y)$. Thus the proposition

$$\prod_{x=1}^{n+1} \prod_{y=1}^{m+1} ((x = y) + (f'(x) \neq f'(y)))$$

is true.

Remark. When a certain class of objects is given, one will be tempted to say: that these objects are n in number, n being finite, means that a one-to-one mapping of these objects onto the first n numbers *exists*. But an apparent logical variable (the mapping) occurs in this definition, and there is no a priori reason to expect that the domain over which this variable ranges is finite, unless one has in advance a theorem asserting that *the number of possible mappings is finite*. From the strictly finitist point of view adopted here, therefore, such a theorem would first have to be proved if the notion of cardinality is to be definable for the objects in question. This seems to be a vicious circle, but it is not: we do not need the general definition, given above, of the notion of cardinality in order to establish a special theorem asserting that certain objects are n in number, for we can establish such a theorem by *actually giving* a mapping; in this way we avoid treating the mapping as a logical variable. Thus even in the cases mentioned we can first establish a special theorem to the effect that the number of mappings that are possible at all is finite and after that give the general definition of the notion of cardinality for classes of objects.

For any class, the number of positive integers that are $\leqq n$ and belong to that class can be given in a general way by a function definable in the following manner.

Let U be an arbitrary propositional function; I state[5]

Definition 26. $(NU(1) = 1)U(1) + (NU(1) = 0)\overline{U(1)}$ and $(NU(n + 1) = NU(n) + 1)U(n + 1) + (NU(n + 1) = NU(n))\overline{U(n + 1)}$.

In words: The descriptive function $NU(x)$ will have the value 1 or 0 for $x = 1$ according to whether $U(1)$ is true or not. Furthermore $NU(n + 1)$ will be equal to $NU(n) + 1$ or $NU(n)$ according to whether $U(n + 1)$ is true or not.

It is easy to show that $NU(n)$ then in fact gives the number of numbers $x \leqq n$ for which $U(x)$ is true. I do not go into this more closely here.

It is frequently the case that certain objects are numbered by means of integers, but in such a way that several numbers are assigned as subscripts to each object. The number of *distinct* objects among them can then be given by means of a function $T(x)$ that I now define.

Let the objects be $a_1, \ldots, a_n$; we state

Definition 27. $T(1) = 1$; $(T(r + 1) = T(r))(\sum_{s=1}^{r} (a_{r+1} = a_s)) + (T(r + 1) = T(r) + 1) \prod_{s=1}^{r} (a_{r+1} \neq a_s)$.

Furthermore, by means of the function $\text{Min}(U, n)$, defined on page 329, a uniquely determined choice of a complete system of representatives of distinct objects a can be given. Besides, it is possible to define the associated frequencies ⟦Häufigkeitszahlen⟧, which indicate how often each distinct object occurs in the sequence $a_1, \ldots, a_n$; and so on. I do not want to elaborate this any further here.

Concluding remark

This paper was written in the autumn of 1919, after I had studied the work of Russell and Whitehead. It occurred to me that already the use of the logical variables that they call "real" would surely suffice to provide a foundation for large parts of mathematics. (In this connection, then, it must be noted that apparent variables ranging over finite domains can be removed by means of recursive definitions.) The justification for introducing apparent variables ranging over infinite domains therefore seems very problematic; that is, one can doubt that there is any justification for the actual infinite or the transfinite.

On the other hand, I myself am no longer satisfied with the way in which the

[5] Strictly speaking, the number 0 has not been introduced; but we can make the convention that the cardinality 0 is to mean that there are no "objects".

logical elaboration proceeds here, since by following the pattern of Russell and Whitehead's work I made it too cumbersome from the purely formal point of view. Yet, even in providing a foundation for mathematics it is the substance that is important, not the notation. Soon I shall publish another work on the logical foundations of mathematics that is free from this formal cumbrousness.[6] But that work, too, is a consistently finitist one; it is built upon Kronecker's principle that a mathematical definition [[Bestimmung]] is a genuine definition if and only if it leads to the goal by means of a *finite* number of trials.

[6] [[Never published.]]

On the significance of the principle of excluded middle in mathematics, especially in function theory

LUITZEN EGBERTUS JAN BROUWER

(*1923b*)

The text below is the translation of an address delivered in German on 21 September 1923 at the annual convention of the Deutsche Mathematiker-Vereinigung in Marburg an der Lahn. It had been delivered in Dutch at the 22nd Vlaamsch Natuur- en Geneeskundig Congres, in Antwerp in August 1923, in an approximately similar form (*Brouwer 1923a*).

§ 1 shows how the principles of logic, which have their origin in finite mathematics, came to be applied to discourse about the physical world and then to nonfinite mathematics; but in that last field there is not necessarily a justification for each of these principles. In particular, such a justification seems to be lacking for the principle of excluded middle and that of double negation.

§ 2 shows how several important results of classical analysis become unjustified once the principle of excluded middle is abandoned. Here Brouwer's critique is essentially negative, being based on counterexamples to classical theorems; but elsewhere he investigates which fragments of the Bolzano-Weierstrass theorem can be preserved in intuitionistic analysis (*1919*, sec. 1, and *1952a*; see also *Heyting 1956*, arts. 3.4.4 and 8.1.3) and gives an intuitionistic form of the Heine-Borel theorem (*1926a* and *1926b*; see also *Heyting 1956*, art. 5.2.2). There are further counterexamples to theorems of classical analysis in *Brouwer 1928a*.

§ 3 is an example of the "splitting" of a classical notion, that of a convergent sequence, into several overlapping but distinct intuitionistic notions, here positively convergent sequence, negatively convergent sequence, and nonoscillating sequence. These notions were further investigated by one of Brouwer's disciples, M. J. Belinfante, and we refer the reader to Belinfante's papers listed below, p. 630. In order to avoid a number of complications that arise in the theory of infinite sequences as elaborated by Brouwer and Belinfante, J. G. Dijkman found it convenient (*1948*) to introduce the notions of strictly negatively convergent sequence and of strictly nonoscillating sequence.

Two notes, "Addenda and corrigenda" and "Further addenda and corrigenda", published by Brouwer in 1954, are appended to the 1923 paper. They reflect the development of Brouwer's ideas in the intervening years. In the main paper below (*1923b*) Brouwer had introduced an infinite sequence whose definition depends upon the occurrence of a certain finite sequence of digits in the decimal expansion of π. In *1948* he introduced an infinitely proceeding sequence whose definition depends upon whether a certain mathematical problem has, or has not, been solved at a certain time: let α be a mathematical assertion that so far has not been tested, that is, such that neither $\neg\alpha$ nor $\neg\neg\alpha$ has been proved; then, if between the choice for c_{n-1} and the choice for c_n "the creating subject has experienced either the truth or the absurdity of α" (*1948*, p. 1246), a certain value is chosen for c_n; otherwise, another value is chosen for c_n. This method of definition, by which the choices for the constituents of an infinitely proceeding sequence "may, at any stage, be made to depend on possible future mathematical experiences of the creating subject" (*1953*, p. 2), allowed Brouwer to offer new counterexamples to classical theorems, in particular in analysis (*1948a*, *1948b*, *1949*, *1949a*, *1950*, *1950a*, *1951*, and *1952a*). It is in these conditions that he came to write the two appendices, *1954* and *1954a*; *1954b* and *1954c* constitute a sequel to *1954a*.

The translation of the main paper (*1923b*) is by Stefan Bauer-Mengelberg and the editor, and it is printed here with the kind permission of Professor Brouwer and Walter de Gruyter and Co. The first appended paper (*1954*) was translated by Stefan Bauer-Mengelberg, Claske M. Berndes Franck, Dirk van Dalen, and the editor; the second appended paper (*1954a*) was translated by Stefan Bauer-Mengelberg, Dirk van Dalen, and the editor.

§ 1

Within a specific finite "main system" we can always *test* (that is, either prove or reduce to absurdity) properties of systems, that is, test whether systems can be mapped, with prescribed correspondences between elements, into other systems; for the mapping determined by the property in question can in any case be performed in only a finite number of ways, and each of these can be undertaken by itself and pursued either to its conclusion or to a point of inhibition. (Here the principle of mathematical induction often furnishes the means of carrying out such tests without individual consideration of every element involved in the mapping or of every possible way in which the mapping can be performed; consequently the test even for systems with a very large number of elements can at times be performed relatively rapidly.)

On the basis of the testability just mentioned, there hold, for properties conceived within a specific finite main system, the *principle of excluded middle*, that is, the principle that for every system every property is either correct [[richtig]] or impossible, and in particular the *principle of the reciprocity of the complementary species*, that is, the principle that for every system the correctness of a property follows from the impossibility of the impossibility of this property.

If, for example, the union $\mathfrak{S}(p, q)$ of two mathematical species[1] p and q contains at least eleven elements, it follows on the basis of the principle of excluded middle (which in this case appears as "principle of disjunction") that either p or q contains at least six elements.

Likewise, if we have proved in elementary arithmetic that, whenever none of the positive integers $a_1, a_2, \ldots, a_n$ is divisible by the prime number c, the product $a_1a_2a_3 \ldots a_n$ is not divisible by c either, it follows on the basis of the principle of the reciprocity of the complementary species that, if the product $a_1a_2a_3 \ldots a_n$ is divisible by the prime number c, at least one of the factors of the product is divisible by c.

[1] [[For the definition of "species" see below, p. 454.]]

For properties derived within a specific finite main system by means of the principle of excluded middle it is always certain that we can arrive at their empirical corroboration if we have a sufficient amount of time at our disposal.

It is a natural phenomenon, now, that numerous objects and mechanisms of the world of perception, considered in relation to extended complexes of facts and events, can be mastered if we think of them as (*possibly partly unknown*) finite discrete systems that for specific known parts are bound by specific laws of temporal concatenation. Hence the laws of theoretical logic, including the principle of excluded middle, are applicable to these objects and mechanisms in relation to the respective complexes of facts and events, even though here a complete empirical corroboration of the inferences drawn is usually materially excluded a priori and there cannot be any question of even a partial corroboration in the case of (juridical and other) inferences about the past. To this incomplete verifiability of inferences that are nevertheless considered irrefutably correct, as well as to our partial ignorance of the representing finite systems and to the fact that theoretical logic is applied more often and by more people to such material objects than to mathematical ones we must probably attribute the fact that an a priori character has been ascribed to the laws of theoretical logic, including the principle of excluded middle, and that one lost sight of the conditions of their applicability, which lie in the projection of a finite discrete system upon the objects in question, so that one even went so far as to look to the laws of logic for a deeper justification of the completely primary and autonomous mental activity ⟦Denkhandlung⟧ that the mathematics of finite systems represents. Accordingly, in the logical treatment of the world of perception the appearance of a contradiction never led us to doubt that the laws of logic were unshakable but only to modify and complete the mathematical fragments projected upon this world.

An a priori character was so consistently ascribed to the laws of theoretical logic that until recently these laws, including the principle of excluded middle, were applied without reservation even in the mathematics of infinite systems and we did not allow ourselves to be disturbed by the consideration that the results obtained in this way are in general not open, either practically or theoretically, to any empirical corroboration. On this basis extensive incorrect theories were constructed, especially in the last half-century. The contradictions that, as a result, one repeatedly encountered gave rise to the *formalistic critique*, a critique which in essence comes to this: the *language accompanying the mathematical mental activity* is subjected to a mathematical examination. To such an examination the laws of theoretical logic present themselves as operators acting on primitive formulas or axioms, and one sets himself the goal of transforming these axioms in such a way that the linguistic effect of the operators mentioned (which are themselves retained unchanged) can no longer be disturbed by the appearance of the linguistic figure of a contradiction. We need by no means despair of reaching this goal,[2] but nothing of mathematical value will thus be gained: an incorrect theory, even if it cannot be inhibited by any contradiction that would refute it, is none the less incorrect, just as a criminal policy is none the less criminal even if it cannot be inhibited by any court that would curb it.

[2] For the unjustified application of the principle of excluded middle to properties of well-constructed mathematical systems can never lead to a contradiction (see *Brouwer 1908*, ⟦p. 157, or *1919a*, p. 11⟧).

§ 2

The following two fundamental properties, which follow from the principle of excluded middle, have been of basic significance for this incorrect "logical" mathematics of infinity ("logical" because it makes use of the principle of excluded middle), especially for the *theory of real functions* (developed mainly by the Paris school):

1. *The points of the continuum form an ordered point species*;[3]
2. *Every mathematical species is either finite or infinite.*[4]

The following example shows that the first fundamental property is incorrect. Let d_ν be the νth digit to the right of the decimal point in the decimal expansion of π, and let $m = k_n$ if, as the decimal expansion of π is progressively written, it happens at d_m for the nth time that the segment $d_m d_{m+1} \ldots d_{m+9}$ of this decimal expansion forms the sequence 0123456789. Further, let $c_\nu = (-\frac{1}{2})^{k_1}$ if $\nu \geqq k_1$, otherwise let $c_\nu = (-\frac{1}{2})^\nu$; then the infinite sequence $c_1, c_2, c_3, \ldots$ defines a real number r for which none of the conditions $r = 0$, $r > 0$, or $r < 0$ holds.[5]

When the first fundamental property ceases to hold, the Paris school's notion of integral, the notion of L-integral, as it is called, ceases to be useful, because this notion of integral is bound to the notion "measurable function" and, according to the above, not even a constant function satisfies the conditions of "measurability". For in the case of the function $f(x) = r$, where r represents the real number defined above, the values of x for which $f(x) > 0$ do not form a measurable point species.[6]

That the second fundamental property is incorrect is seen from the example provided by the species of the positive integers k_n defined above.

When the second fundamental property ceases to hold, so does the "extended disjunction principle", according to which, if a fundamental sequence of elements is contained in the union $\mathfrak{S}(p, q)$ of two mathematical species p and q, either p or q contains a fundamental sequence of elements; and when the extended disjunction principle ceases to hold, so does the Bolzano-Weierstrass theorem, which rests upon it and according to which every bounded infinite point species has a limit point.

The following two theorems are less basic and simple than the fundamental properties mentioned, yet they are equally indispensable for the construction of the "logical" theory of functions.

1. *Every continuous function $f(x)$ defined everywhere in a closed interval i possesses a maximum, that is, an abscissa value x_1 having a neighborhood α such that $f(x_1) \geqq f(x)$ for every x that belongs to the intersection of α and i.*

[3] That is, if on the one hand $a < b$ either holds or is impossible, or on the other $a > b$ either holds or is impossible, then one of the conditions $a < b$ or $a > b$ or $a = b$ holds.

[4] For according to the principle of excluded middle a species s either is finite or cannot possibly be finite. In the latter case s possesses an element, e_1; for otherwise, on the basis of the principle of excluded middle, s could not possibly possess an element and would therefore be finite, which is excluded. Furthermore s possesses an element, e_2, distinct from e_1; for otherwise s would not possibly possess an element distinct from e_1 and would therefore be finite, which is excluded. Continuing in this manner, we show that s possesses a fundamental sequence of distinct elements, $e_1, e_2, \ldots$. [[For the definition of "fundamental sequence" see below, p. 455.]]

[5] Of course, we can also define r by means of any other property x whose existence or impossibility can be derived for every definite positive integer, while we can neither determine a positive integer that possesses x nor prove the impossibility of x for all positive integers.

[6] However, the notion of R-integral, that is, the notion of Riemann integral, can be applied to $f(x)$ without further ado.

The incorrectness of this theorem appears from the following example: If we enumerate the irreducible binary fractions between 0 and 1 (excluding 0 and 1) by means of a fundamental sequence $\delta_1, \delta_2, \ldots$ in the ordinary way, that is, so that any fraction follows all those with a smaller denominator and fractions with the same denominator are ordered according to the magnitude of the numerator, if we assign to k_1 the same meaning as above, if by $f_n(x)$ we understand the function that has the value 2^{-n} for $x = \delta_n$ and vanishes for $x = 0$ as well as for $x = 1$, while it remains linear between $x = 0$ and $x = \delta_n$ as well as between $x = \delta_n$ and $x = 1$, and if we put $g_n(x) = f_n(x)$ for $n = k_1$, otherwise $g_n(x) = 0$, then the continuous function

$$g(x) = \sum_{n=1}^{\infty} g_n(x),$$

which is defined everywhere in the closed unit interval, possesses no maximum.

2. (Heine-Borel covering theorem.) *If a neighborhood is assigned to every point core*[7] *of the point species A formed by the points and the limit points of a bounded entire*[8] *point species B, then the whole point species A can be covered by a finite number of these neighborhoods.*

The incorrectness of this theorem appears from the following example: If we choose for B the number sequence $c_1, c_2, c_3, \ldots$, defined above, while we assign to the number c_ν, for $\nu \geqq k_1$, the interval $(c_\nu - 2^{-k_1-2}, c_\nu + 2^{-k_1-2})$, otherwise the interval $(c_\nu - 2^{-\nu-2}, c_\nu + 2^{-\nu-2})$, and to a limit point e (if any) of the sequence the interval $(e - \frac{1}{2}, e + \frac{1}{2})$, then A cannot be covered by a finite number of these neighborhoods.[9]

In view of the fact that the foundations of the logical theory of functions are indefensible according to what was said above, we need not be surprised that a large part of its results becomes untenable in the light of a more precise critique. As an example, we shall refute one of the best-known classical theorems in this domain, namely, the theorem that a monotonic continuous function defined everywhere is "almost everywhere" differentiable, by constructing a monotonic continuous function that is defined everywhere in the closed unit interval but is nowhere differentiable.

Let $0 \leqq x_1 < x_2 \leqq 1$. By the *elementary function corresponding to the interval* (x_1, x_2) we shall understand the continuous function, defined everywhere in the closed unit interval, that, for $x_1 \leqq x \leqq x_2$, is equal to

$$\frac{x_2 - x_1}{2\pi} \sin 2\pi \frac{x - x_1}{x_2 - x_1}$$

and, for $0 \leqq x \leqq x_1$ and $x_2 \leqq x \leqq 1$, is equal to 0; by $\lambda', \lambda'', \lambda''', \ldots$ we shall understand the intervals $(a/2^n, (a + 2)/2^n)$ (where a and n denote positive integers) belonging to the closed unit interval and enumerated in the customary way; and by $f_n(x)$ we shall understand the elementary function corresponding to $\lambda^{(n)}$. Furthermore we

[7] ⟦For the definition of "point core" see below, p. 458.⟧

[8] ⟦"The species of the points that coincide with points of the point species Q is called the *completing* ⟦*ergänzende*⟧ *point species* or, for short, the *completion* ⟦*Ergänzung*⟧ of Q. A point species that is identical with its completion is called an *entire* ⟦*ganze*⟧ Punktspecies." (*Brouwer 1919*, p. 6.) For the definition of "coincide" see below, p. 458, and for that of "identical with", p. 454.⟧

[9] Nor does the theorem hold for a closed bounded entire point species A. Counterexample: take for A a species of abscissas $(-2)^{-\nu}$ such that an abscissa $(-2)^{-\nu}$ belongs to A if and only if a natural number k_1 satisfying the characterization above is known and ν is a natural number $\leqq k_1$; then with each abscissa that may belong to A associate the same interval as in the text.

assign to k_1 the same meaning as above; we put $g_1(x) = x$ and (for $n \geqq 2$) $g_n(x) = f_n(x)$ for $n = k_1$, otherwise $g_n(x) = 0$. Then the function

$$g = \sum_{n=1}^{\infty} g_n$$

is a monotonic continuous function that is defined everywhere in the closed unit interval but is nowhere differentiable.

§ 3

As an example illustrating the fact that even older and more firmly consolidated theories in the field of the mathematics of infinity are affected by the rejection of the principle of excluded middle and the consequent rejection of the Bolzano-Weierstrass theorem, even if in much smaller measure than the theory of real functions, we take the notion of convergence of infinite series.

Let us say that an infinite series $u_1 + u_2 + u_3 + \cdots$ with real terms, for which the sum of the first n terms is denoted by s_n, is *nonoscillating* if for every $\varepsilon > 0$ it has been established that it is impossible to have at the same time an infinite sequence of positive integers $n_1, n_2, n_3, \ldots$ increasing beyond all bounds and an infinite sequence of positive integers $m_1, m_2, m_3, \ldots$ such that

$$|s_{n_\nu + m_\nu} - s_{n_\nu}| > \varepsilon \quad \text{for every } \nu;$$

then according to the classical theory on the basis of the principle of excluded middle such a nonoscillating series is:

1. *Negatively convergent*, that is, there exists a real number s with the property that for every $\varepsilon > 0$ it has been established that it is impossible to have an infinite sequence of positive integers $n_1, n_2, n_3, \ldots$ increasing beyond all bounds such that

$$|s - s_{n_\nu}| > \varepsilon \quad \text{for every } \nu;$$

2. *Bounded*, that is, there exist two real numbers g_1 and g_2 such that

$$g_1 < s_n < g_2 \quad \text{for every } n;$$

3. *Positively convergent*, that is, there exists a real number s with the property that for every $\varepsilon > 0$ there exists a positive integer n_ε such that

$$|s - s_n| < \varepsilon \quad \text{for every } n > n_\varepsilon.$$

Let us now consider the following five nonoscillating series (where k_1 again has the same meaning as above):

(a) $u_n = 1/2^n$ for every n;
(b) $u_n = 2 + 1/2^n$ for $n = k_1$, $u_n = -2 + 1/2^n$ for $n = k_1 + 1$, otherwise $u_n = 1/2^n$;
(c) $u_n = n + 1/2^n$ for $n = k_1$, $u_n = -n + 1/2^n$ for $n = k_1 + 1$, otherwise $u_n = 1/2^n$;
(d) $u_n = 1$ for $n = k_1$, otherwise $u_n = 1/2^n$;
(e) $u_n = n$ for $n = k_1$, otherwise $u_n = 1/2^n$.

The series (a) turns out to be positively convergent and therefore also negatively convergent and bounded; the series (b) to be negatively convergent and bounded, but not positively convergent; the series (c) to be negatively convergent, but not bounded and therefore not positively convergent either; the series (d) to be bounded, but not negatively convergent and therefore not positively convergent either; the series (e), finally, to be not bounded, not negatively convergent, and not positively convergent.

To illustrate the consequences of the distinction made above we shall consider the Kummer convergence criterion, which reads as follows: "If $B_1, B_2, \ldots$ are positive numbers and if, for the infinite series of positive terms $r = u_1 + u_2 + u_3 + \cdots$, we have

$$\lim \left\{ B_n \frac{u_n}{u_n + 1} - B_{n+1} \right\} > 0,$$

then r is positively convergent".

The proof of this convergence criterion is customarily carried out as follows.

On the basis of what has been assumed we select M and k in such a way that, for $n \geqq M$,

$$B_n \frac{u_n}{u_n + 1} - B_{n+1} > k,$$

$$B_n u_n - B_{n+1} u_{n+1} > k u_{n+1},$$

$$B_n u_n - B_{n+p} u_{n+p} > k(u_{n+1} + \cdots + u_{n+p}),$$

$$u_{n+1} + \cdots + u_{n+p} < \frac{B_n u_n}{k},$$

whence *boundedness* follows for the series $r_n = u_{n+1} + u_{n+2} + \cdots$ $(n \geqq M)$ and therefore also for the series $r = u_1 + u_2 + \cdots$. On the basis of this boundedness the series r is then declared to be not only nonoscillating, which is permitted for a series of positive terms, but also positively convergent.

The last inference, however, rests upon the Bolzano-Weierstrass theorem and must be rejected along with it.

Pringsheim (*1916*, p. 378) offers an altogether different and more instructive proof. After he has proved the positive convergence of r for the case of the positive convergence as well as for the case of the positive divergence of $b = 1/B_1 + 1/B_2 + \cdots$, he assumes that the series b must be either positively convergent or positively divergent, and for this reason he declares that the general criterion has been proved.

But the assumption mentioned is inadmissible; for it, too, rests upon the Bolzano-Weierstrass theorem.

It is worth noting, now, that Kummer himself expressed (*1835*) his criterion only with the auxiliary condition $\lim B_n u_n = 0$ and that with this auxiliary condition the positive convergence of the series r is actually ensured by the criterion, as is immediately evident from the proof above.

That not only the derivations of the Kummer convergence criterion without any auxiliary condition are inadequate[10] but also the criterion itself is incorrect is shown

[10] The inadequacy of these derivations, in contradistinction to the correctness of the proof originally carried out by Kummer himself for the restricted criterion, was indicated to me by my student M. J. Belinfante as an example of the significance of the principle of excluded middle for the theory of infinite series.

by the series (d) above, which is neither positively convergent nor negatively convergent. For, if we determine the successive B_n for this series from the relations

$$B_1 = 4 \quad \text{and} \quad B_n \frac{u_n}{u_n + 1} - B_{n+1} = 1 \quad \text{for every } n,$$

all B_n turn out to be positive, so that the extended convergence criterion is satisfied here, although positive convergence does not exist. This *omission of the Kummer auxiliary condition*, which took place after Kummer and was prompted by Dini, has thus considerably curtailed the scope of the convergence criterion in question.

ADDENDA AND CORRIGENDA

(*1954*)

Regarding my paper "Over de rol van het principium tertii exclusi in de wiskunde, in het bijzonder in de functietheorie" (*1923a*), published thirty years ago in volume 2 of *Wis- en Natuurkundig Tijdschrift*, which has since been discontinued, I would now like to make the following remarks.

I. Page 1, line 4 ⟦above, page 335, line 1⟧, the term "to test" ⟦"toetsen" (*1923a*), or "prüfen" (*1923b*)⟧ is used for either proving or reducing to absurdity. In subsequent intuitionistic literature, however, a property of a mathematical entity is said to be "tested" if either its contradictoriness or its noncontradictoriness is ascertained, and "judged" ⟦"geoordeeld"⟧ if either its presence or its absurdity is ascertained.

II. Page 3, footnote (*) ⟦above, page 336, footnote 2⟧, the noncontradictoriness of applications of the principle of excluded middle to the attribution of a property E to a well-constructed mathematical system was pointed out. In subsequent intuitionistic literature, however, it became apparent that for the simultaneous application of the principle mentioned to the attribution of a property E to each element of a mathematical species S noncontradictoriness remains ensured only for finite S. For infinite S the simultaneous attribution mentioned can very well be contradictory.

III. Page 3, footnote (****) ⟦above, page 337, footnote 5⟧, for the construction, given in the text, of a real number r for which none of the relations $r = 0$, $r > 0$, and $r < 0$ holds, we allowed every property x for which neither a finite number possessing x nor the impossibility of x for every finite number is known. To this we must add the condition that x can be judged for every finite number.

IV. Page 4, line 18 up ⟦above, page 338, line 12⟧, the classical Heine-Borel covering theorem was formulated for an arbitrary "closed" bounded point species. The intuitionistic critique of this theorem that follows there should have been preceded by an exposition of the intuitionistic splitting of the classical notion "closed". For, if in a Cartesian or in a "located" ⟦"afgebakende"⟧ compact topological space R we understand by a *core* the species of the points that coincide with a given point, by an *accumulation core* of a core species Q a core of which every neighborhood contains an infinitely proceeding sequence of cores of Q that are mutually apart, and by a *limit core* of a core species Q a core of which every neighborhood contains a core of Q, if we then say that a core species Q containing all of its accumulation cores is *α-closed* and that a core species Q that contains all of its limit cores is *β-closed*, if, accordingly, we call the union of a core species Q and its accumulation cores the *α-closure of* Q and the species of limit cores of Q the *β-closure of Q*, if we take the formulation cited above of the classical Heine-Borel covering theorem as applying to "closed" bounded core species Q, then this formulation is intuitionistically correct only if by "closed" is meant "β-closed" and if, moreover, Q is a core species *located in R*, that is to say, it is from every core of R at a distance that is computable with unlimited accuracy. In particular, therefore, with regard to the number sequence $c_1, c_2, c_3, \ldots$ referred to on page 4, line 13 up ⟦above, page 338, line 17⟧, which is bounded and is located in the number continuum, the classical covering theorem is intuitionistically valid only for its *β-closure*, that is to say, for its union with its limit number, but not for its *α-closure*, referred to on page 4, line 13 up ⟦above, page 338, line 19⟧, that is to say, for its union with the number 0, *if this number should turn out to be identical with the limit number.* Nor is the classical covering theorem intuitionistically valid for number core species that are β-closed and bounded *but not located in the number continuum*, as, for example, the union of the number cores $p_1, p_2, p_3, \ldots$, in which $p_\nu = 1$ for $\nu < k_1$ and $p_\nu = -1$ for $\nu \geqq k_1$.

V. The example given on page 5, lines 1–13 ⟦above, page 338, line 8u, to page 339, line 5⟧, of a monotonic, continuous, nowhere differentiable function defined everywhere in the closed unit interval possesses these properties exclusively as a function of the (classical) continuum of approximations made according to a law, not as a function of the (intuitionistic) continuum of more or less freely proceeding approximations. A connection between monotonicity and differentiability of full functions of the intuitionistic continuum can be found in my *1923*, p. 24.

FURTHER ADDENDA AND CORRIGENDA

(1954a)

With reference to point V of my *1954*, pp. 104–105 ⟦above, pp. 341–342⟧, I give below an example of a *continuous, monotonic, nowhere differentiable, real, full function of the intuitionistic closed unit continuum K.*[1]

For a natural number n we understand by $\chi_n(x)$ the real function of K that for the "even n-cores"[2] $x = a/n$ (a being an integer and $0 \leqq a \leqq n$) is equal to 0, for the "odd n-cores" $x = (2a + 1)/2n$ (a being an integer and $0 \leqq a \leqq n$) is equal to $1/4n$, and for every a ($0 \leqq a \leqq n$) is linear between $x = a/n$ and $x = (2a + 1)/2n$ as well as between $x = (2a + 1)/2n$ and $x = (a + 1)/n$.[3] Further we put $\psi_1(x) \equiv x$ and, for $n \geqq 2$, f being an opaque fleeing property and $\kappa_1(f)$ being its critical number,[4] we

[1] ⟦For the definitions of "continuous", "full", and "unit continuum" see below, pp. 458–459; see also *Brouwer 1953*, p. 3, line 2u, to p. 4, line 6.⟧

[2] ⟦For the definition of "core" see below, p. 458; see also *Brouwer 1953*, p. 3, line 2u, to p. 4, line 6.⟧

[3] ⟦From the intuitionistic point of view the definition of $\chi_n(x)$ does not seem unobjectionable; see Remark in 2.2.8 of *Heyting 1956*, p. 27.⟧

[4] ⟦"We shall call a hypothetical property f of natural numbers a *fleeing property* if it satisfies the following conditions:

(1) For each natural number it can be decided either that it possesses the property f or that it cannot possibly possess the property f;

(2) No method is known for calculating a natural number possessing the property f;

(3) The assumption of existence of a natural number possessing the property f is not known to lead to an absurdity.

In particular, a fleeing property is called *opaque* if the assumption of existence of a natural

put $\psi_n(x) \equiv \chi_n(x)$ if $n = \kappa_1(f)$, otherwise $\psi_n(x) \equiv 0$. Then

$$\psi(x) \equiv \sum_{\nu=1}^{\infty} \psi_\nu(x)$$

is a continuous, monotonic, nowhere differentiable, real, full function of K.

For one must take into account the possibility (α) that at some time it turns out that $\kappa_1(f)$ is nonexistent, so that, for all values of x, $\psi(x)$ possesses an ordinary derivative equal to 1.

But one must also take into account the possibility (β) that at some time a natural number $m = \kappa_1(f)$ will be found. In that case $\psi(x)$ has, for all values of x that lie apart[5] from the m-cores, an ordinary derivative, either equal to 3/2 or equal to 1/2; for all even m-cores x it has a right derivative (nonexistent for $x = 1$) equal to 3/2, and a left derivative (nonexistent for $x = 0$) equal to 1/2; and for all odd m-cores x it has a right derivative equal to 1/2 and a left derivative equal to 3/2, while for every value of x the possibility must be taken into account that at some time it shall turn out either to be an m-core or to lie apart from the m-cores.

Therefore, with respect to the existence of an ordinary derivative, or of a right and a left derivative, of $\psi(x)$ one must, *for every value of* x, *take into account possibilities lying mutually apart, so that for no single value of* x *an ordinary derivative can be calculated.*

By the nature of the case this function $\psi(x)$ is not "completely differentiable" in the sense of *Brouwer 1923*, § 3, p. 20.[6]

So far as the function $g(x)$, mentioned in *Brouwer 1923a*, p. 5 ⟦above, p. 339⟧, is concerned, it must, according to the explanations that follow below, be abandoned as an example of a continuous, monotonic, nowhere differentiable function, *even for the classical* closed unit continuum K_r.[7]

§ 2

By a $k^{(\nu)}$ we understand a closed $\lambda^{(4\nu+1)}$-interval;[8] for $\nu \geqq 0$, by an $h^{(\nu)}$ we understand a $k^{(\nu)}$ entirely or partially covered by K; further, after ordering the $h^{(\nu)}$ for all values of ν in a single fundamental sequence[9] $\theta', \theta'', \theta''', \ldots$, to be called F, by a

number possessing f is not known to be noncontradictory either." (*Brouwer 1952*, p. 141; see also *1928a*, p. 161.)

The *critical number* of a fleeing property is apparently what Brouwer (*1929a*, p. 161) calls the *Lösungszahl* of the property, that is, the (hypothetical) least natural number that possesses the property.⟧

[5] ⟦"We say that ⟦a number core⟧ *a* lies *apart* from ⟦a number core⟧ *b* if there is some natural number n such that $|b - a| > 2^{-n}$." (*Brouwer 1953*, p. 4.) See also below, p. 462, footnote 10a.⟧

[6] ⟦The definition of "completely differentiable" requires too many preliminary definitions to be reproduced here; we refer the reader to the passage indicated in the text.⟧

[7] A similar disappearance of a counterexample, due to the disappearance of the absence of a requisite algorithm, belongs to the realm of possibilities when one considers $\psi(x)$ *simply for a fixed fleeing property f.*

⟦The *classical continuum* is the species of predeterminate intuitionistic real numbers; see *Brouwer 1952*, p. 142, bottom half of first column, and p. 143, top of first column, as well as above, p. 342, V.⟧

[8] ⟦For the definition of "$\lambda^{(\nu)}$-interval" see below, p. 457.⟧

[9] ⟦For the definition of "fundamental sequence" see below, p. 455.⟧

unitary standard number we understand an infinitely proceeding sequence[10] $\theta^{(c_1)}$, $\theta^{(c_2)}$, $\theta^{(c_3)}, \ldots$ in which, for every ν, $\theta^{(c_\nu)}$ is an $h^{(\nu)}$ and $\theta^{(c_{\nu+1})}$ consists entirely of inner points of $\theta^{(c_\nu)}$. Then, the species of unitary standard numbers is identical with the species of accretion sequences[11] of a *dressed fan* w,[12] of which we can say—because every unitary number core, that is, every number core of K, coincides with a unitary standard number—that it *represents* K.

As a function of a variable number core x, either of K_r or of K, $g(x)$ is now obtained as follows.[13] Let f be a fleeing property; let $\kappa_1(f)$ be its critical number; let p_ν and q_ν be respectively the least and the greatest endcores of $\theta^{(\nu)}$; and let $\varphi_\nu(x)$ be the continuous function of K_r, or of K, that for the part of $\theta^{(\nu)}$ that belongs to K_r, or to K, is equal to

$$\frac{q_\nu - p_\nu}{2\pi} \sin 2\pi \frac{x - p_\nu}{q_\nu - p_\nu}$$

and, for $x \leqq p_\nu$ as well as for $x \geqq q_\nu$, is equal to 0. Then we put $g_\nu(x) \equiv x$ for $\nu = 1$, $g_\nu(x) \equiv \varphi_\nu(x)$ for $\nu = \kappa_1(f)$, and $g_\nu(x) \equiv 0$ for all other values of ν. Finally, we put

$$g(x) \equiv \sum_{\nu=1}^{\infty} g_\nu(x).$$

If we call a $\theta^{(\nu)}$ for which $\nu = \kappa_1(f)$ the *critical interval* of f and if we represent this by $i(f)$, then (at least for the current examples of F and f) not a single indication is at hand concerning the position of a possible $i(f)$; therefore, it seems at the outset that for every x every possibility of obtaining a guarantee for the nonbelonging to $i(f)$ is lacking, and so is for every unitary finite binary fraction[14] x every possibility of computing a ratio 1/3 for the lengths of the segments into which it would have to divide a possible $i(f)$ to which it would belong; therefore finally it seems that for every x every possibility of computing an ordinary derivative is lacking.

§ 3

This situation, however, changes when one intends to make the infinitely proceeding process of *the creation, by free choices, of a unitary standard number* u run parallel to the infinitely proceeding process of the successive judgments of the assignment of f to the successive natural numbers and moreover to take care that the creation process of u continually lags sufficiently far behind the process of judging that was just mentioned to prevent contact with an $i(f)$ that might possibly appear, so that there must come into existence a number core x of K for which $g(x)$ possesses an ordinary derivative equal to 1.

Once this insight has been obtained, it is not far-fetched to observe that the way, indicated here, in which u comes to exist is at hand for all the accretion sequences

[10] ⟦See the definition of "unbounded choice sequence" below, p. 446; "infinitely proceeding sequence" was used in *Brouwer 1952*, p. 142, bottom of first column; see also 3.1.1. in *Heyting 1956*, pp. 32–34.⟧

[11] ⟦"Accretion sequence" ("accretiereeks") is here apparently used for "infinitely proceeding sequence in a dressed spread".⟧

[12] ⟦For the definition of "dressed fan" see *Brouwer 1953*, p. 16, first paragraph.⟧

[13] ⟦The remark made in footnote 3 applies to the function $g(x)$.⟧

[14] ⟦For the definition of "finite binary fraction" see below, p. 457, footnote 1.⟧

of the elements of a subfan w' of w that is obtained from w by the deletion, from the species of constituents[15] that are admitted for the nodes of w, of a possible $i(f)$, as well of the two λ-intervals that are of the same length as $i(f)$ and are partially covered by $i(f)$. Therefore, for every number core x of K that is represented by this dressed fan w', $g(x)$ possesses an ordinary derivative.

By means of the same fan w' it is even possible to exhibit, for every natural number n, a measurable core species S_n that is contained in K, has a content greater than $1 - 2^{-4n}$, and in which $g(x)$ everywhere possesses an ordinary derivative.[16] For that one establishes first of all for every n one of the following facts: either for $\nu \leqq n$ no critical interval of f occurs among the $h^{(\nu)}$ or for some $m \leqq n$ a critical interval of f occurs among the $h^{(m)}$. Further, there is chosen for S_n, in the first case, the core species of K represented by w' and, in the second case, the species of the cores of K that lie apart from the two endcores of $i(f)$. If we further observe that the union of the infinitely proceeding sequence of the S_ν forms a measurable core species that is contained in K and has content 1, then $g(x)$ turns out to be a *continuous, monotonic, real, full function of K that is differentiable almost everywhere.*

And since the predeterminate elements of w' represent number cores of K_r, K_r also possesses an (everywhere dense, ever unfinished, and ever enumerable) core species in which $g(x)$ is everywhere differentiable.

15 [For the definition of "constituent" see *Brouwer 1953*, p. 7.]

16 [For the definitions of "measurable core species" and "content" see *Brouwer 1919*, pp. 26–33; see also *Heyting 1956*, chap. V, secs. 1 and 3.]

On the introduction of transfinite numbers

JOHN VON NEUMANN

(*1923*)

Cantor defined ordinal numbers as order types of well-ordered sets (*1897*). The order type of an ordered set M he defined as the notion obtained from M when, while retaining the succession of its elements, we disregard, by an act of abstraction, their nature, so that we obtain mere units in a definite order. This method of "definition by abstraction" was used by Cantor when dealing with objects between which an equivalence relation obtains. Subsequently (for instance, in *Russell 1903*) it was found to be more convenient to speak of equivalence classes. Since similarity is an equivalence relation, order types can be defined as equivalence classes of ordered sets under the relation of similarity, and the ordinals become equivalence classes of well-ordered sets. These equivalence classes being determined in the set of all well-ordered sets, we come close to regions of set theory where paradoxes lurk, and this approach is not without danger if set theory is not axiomatized; if it is, the proof that an equivalence class exists as a set in the theory may present difficulties.

In the paper below von Neumann offers an alternative method. Instead of taking the equivalence class as the ordinal, he selects a particular well-ordered set as a representative of the class. Ordinals are specific well-ordered sets, and any well-ordered set is similar to one of them. A standard scale is thus obtained for well-ordered sets: any ordinal is the set of ordinals preceding it; O being the null set, the ordinals are $O, \{O\}, \{O, \{O\}\}$, ... This is simply a description of the generating procedure, not a definition, since the principle of definition by transfinite induction is not at hand. The theory is presented in the language of naïve set theory; it can be incorporated in any of the usual axiomatic set theories, for instance, in von Neumann's own (*1925*, p. 228 (below, p. 402) and *1928a*, pp. 679, 710–721).

The idea of regarding an ordinal as the set of the preceding ordinals has been adopted by the authors of most recent set theories (Gödel (*1940*), Bernays (*1941*, *1958*), Quine (*1963*)). It has two great advantages: first, it is applicable to infinite ordinals as well as to finite ones, and, second, the order relation for ordinals is given simply by the relation ε. However, von Neumann's method of carrying out this idea is rather roundabout; he assumes the notions of well-ordered set and similarity, and he proves, for every well-ordered set, the existence and the uniqueness of the corresponding ordinal. Ordinals can be defined outright, and their theory can be constructed independently of the theory of ordered sets. In presenting a simplification of von Neumann's set theory, Raphael M. Robinson (*1937*) gave such a direct definition of ordinals: an ordinal u is a set[a]

---

[a] In the specific sense of "set", as distinct from that of "class", see below, p. 393.

such that

(1) If $x \varepsilon u$ and $y \varepsilon u$, then $x \varepsilon y$, $x = y$, or $y \varepsilon x$;

(2) If $x \varepsilon y$ and $y \varepsilon u$, then $x \varepsilon u$.

We see that the definition does not employ anything beyond the relation ε. According to Bernays (*1941*, pp. 6, 10), Zermelo had in 1915 obtained (but left unpublished) a theory of ordinals that is also independent of the notion of order; von Neumann (*1928*, p. 374) conjectures that Zermelo had difficulties with his theory because he did not have at his disposal the axiom of replacement (first published in 1921—see above, p. 291), which is indispensable for the proof that every well-ordered set is similar to an ordinal. For a comparison of the methods of R. M. Robinson, Gödel, Bernays, and Quine, see *Quine 1963*, p. 155, footnote 2.

At the end of the paper below (footnote 4) von Neumann sketches a proof of the principle of definition by transfinite induction. The proof is given in detail in a subsequent paper (*1928*), in which von Neumann presents his theory of ordinals in a new and succinct form, with interesting variations and additions.

The translation is by the editor. It is reproduced here with the kind permission of Mrs. Klara von Neumann-Eckart and the editorial board of *Acta scientiarum mathematicarum*.

The aim of the present paper is to give unequivocal and concrete form to Cantor's notion of ordinal number.

Ordinarily, following Cantor's procedure, we obtain this notion by "abstracting" a common property from certain classes of sets.[1] We wish to replace this somewhat vague procedure by one that rests upon unequivocal set operations. The procedure will be presented below in the language of naive set theory, but, unlike Cantor's procedure, it remains valid even in a "formalistic" axiomatized set theory. Thus our conclusions retain their full validity even in the framework of Zermelo's axiomatization (if we add Fraenkel's axiom[2]).

What we really wish to do is to take as the basis of our considerations the proposition: "Every ordinal is the type of the set of all ordinals that precede it". But, in order to avoid the vague notion "type", we express it in this form: "Every ordinal is the set of the ordinals that precede it". This is not a proposition proved about ordinals; rather, it would be a definition of them if transfinite induction had already been established. According to it, we have

$$\begin{aligned}
&0 = \mathrm{O},\\
&1 = (\mathrm{O}),\\
&2 = (\mathrm{O}, (\mathrm{O})),\\
&3 = (\mathrm{O}, (\mathrm{O}), (\mathrm{O}, (\mathrm{O}))),\\
&\quad \ldots\ldots\ldots,\\
&\omega = (\mathrm{O}, (\mathrm{O}), (\mathrm{O}, (\mathrm{O})), (\mathrm{O}, (\mathrm{O}), (\mathrm{O}, (\mathrm{O}))), \ldots),\\
&\omega + 1 = (\mathrm{O}, (\mathrm{O}), (\mathrm{O}, (\mathrm{O})), \ldots, (\mathrm{O}, (\mathrm{O}), (\mathrm{O}, (\mathrm{O})), \ldots)),\\
&\quad \ldots\ldots\ldots\ldots\ldots\ldots\ldots,
\end{aligned}$$

where O is the null set and $(a, b, c, \ldots)$ is the set whose elements are $a, b, c, \ldots$.

[1] See *Cantor 1895* and *1897*.

[2] See *Zermelo 1908a* and *Fraenkel 1922a*.

Fraenkel's axiom reads: "If Ξ is a set and every element x of Ξ is replaced by a ξ, then there exists a set Ξ' whose elements are the ξ". It fills a substantial gap in Zermelo's axiomatization.

But, of course, we do not assume that transfinite induction has already been established; rather, we assume as given only the notions "well-ordered set" and "similarity".[1] For the rest we shall proceed in a strictly formalistic way; we shall everywhere avoid the symbol "..." and similar devices.

Our notation is as follows: If Ξ and H are sets, then $\Xi \leqq H$ or $H \geqq \Xi$ means that Ξ is a subset of H, and $\Xi < H$ or $H > \Xi$ means that Ξ is a proper subset of H. If Ξ is a set, then $x \varepsilon \Xi$ means that x is an element of Ξ. Let O be the empty set; let (a), (a, b), and (a, b, c) be the sets whose elements are, respectively, a, a and b, and a, b, and c. If x and y are elements of an ordered set, $x \prec y$ or $y \prec x$ means that x comes before y in the given ordering.

Let $E(x)$ be a property and let $f(x)$ be a function defined for all x that have the property $E(x)$. Then let

$$M(f(x); E(x))$$

be the set of all [[values of]] $f(x)$ when x runs through all x that have the property $E(x)$.[3] Let Ξ be an ordered set and x an element of Ξ. Then we call the set

$$M(y; y \varepsilon \Xi, y \prec x)$$

of all elements y of Ξ that precede x the segment [[Abschnitt]] of x in Ξ, and we also denote it more briefly by $A(x, \Xi)$.

Chapter I

1. Let Ξ be a well-ordered set. We call a function $f(x)$ defined in Ξ a "numeration" [["Zählung"]] of Ξ if for all elements x of Ξ we have

$$f(x) = M(f(y); y \varepsilon A(x, \Xi)).$$

If $f(x)$ is a numeration of Ξ, then we call

$$M(f(x); x \varepsilon \Xi)$$

an "ordinal number" of Ξ. And if there exists any numeration of Ξ, we say that Ξ is "numerable" [["zählbar"]].

If x_1, x_2, x_3, and x_4 are the 1st, 2nd, 3rd, and 4th elements of Ξ, respectively, then clearly for every numeration $f(x)$ of Ξ we have

$f(x_1) = \mathrm{O}$,
$f(x_2) = (\mathrm{O})$,
$f(x_3) = (\mathrm{O}, (\mathrm{O}))$,
$f(x_4) = (\mathrm{O}, (\mathrm{O}), (\mathrm{O}, (\mathrm{O})))$;

consequently, if Ξ has 0, 1, 2, or 3 elements, its ordinal is, respectively,

O,
(O),
$(\mathrm{O}, (\mathrm{O}))$, or
$(\mathrm{O}, (\mathrm{O}), (\mathrm{O}, (\mathrm{O})))$.

2. Let Ξ be a well-ordered set. Two numerations $f(x)$ and $g(x)$ of Ξ are always identical.

[3] From the axiomatic point of view it is not at all certain that such a set exists. Rather, we must require that all x having the property $E(x)$ form a set. Then Fraenkel's axiom guarantees the existence of $M(f(x); E(x))$. This condition will always be satisfied in what follows.

For otherwise there would exist a first x for which $f(x) \neq g(x)$. For all $y \prec x$ we would have $f(y) = g(y)$, hence

$$M(f(y); y \varepsilon \Xi, y \prec x) = M(g(y); y \varepsilon \Xi, y \prec x),$$

and that means precisely that $f(x) = g(x)$, contrary to the assumption.

A numerable Ξ, therefore, has one and only one numeration. Hence numeration and ordinal number are univocally determined notions for all numerable Ξ. In what follows we shall denote the ordinal of Ξ by $OZ(\Xi)$.

3. Let Ξ be numerable, let $f(x)$ be the numeration of Ξ, and let $\bar{x}$ be an element of Ξ. Then $A(\bar{x}, \Xi)$ is numerable and its ordinal is $f(\bar{x})$.

For the function $f'(x)$, which is defined in $A(\bar{x}, \Xi)$ and is there equal to $f(x)$, is a numeration of $A(\bar{x}, \Xi)$. We have in fact for all elements y of $A(\bar{x}, \Xi)$

$$\begin{aligned} M(f'(y); y \varepsilon A(x, A(\bar{x}, \Xi))) &= M(f(y); y \varepsilon A(x, A(\bar{x}, \Xi))) \\ &= M(f(y); y \varepsilon A(x, \Xi)) = f(x) = f'(x). \end{aligned}$$

Hence $A(\bar{x}, \Xi)$ is numerable and its ordinal is

$$OZ(A(\bar{x}, \Xi)) = M(f'(x); x \varepsilon A(\bar{x}, \Xi)) = M(f(x); x \varepsilon A(\bar{x}, \Xi)) = f(\bar{x}).$$

4. Let all segments in Ξ be numerable. Then Ξ too is numerable.

For we define, for every x of Ξ,

$$f(x) = OZ(A(x, \Xi)).$$

(We can do this, since all $A(x, \Xi)$ are numerable.) Then $f(x)$ is a numeration of Ξ. For, if x belongs to Ξ and if $\varphi(y)$ is the numeration of $A(x, \Xi)$, then for every element y of $A(x, \Xi)$ we have

$$\varphi(y) = OZ(A(y, A(x, \Xi))) = OZ(A(y, \Xi)) = f(y),$$

so that

$$f(x) = OZ(A(x, \Xi)) = M(\varphi(y); y \varepsilon A(x, \Xi)) = M(f(y); y \varepsilon A(x, \Xi)).$$

5. Let Ξ be well-ordered. Then Ξ is numerable.

For otherwise not every segment in Ξ would be numerable. Hence there would exist a first x for which $A(x, \Xi)$ would not be numerable. Now every segment in $A(x, \Xi)$ is equal to

$$A(y, A(x, \Xi)) = A(y, \Xi)$$

for some element y of $A(x, \Xi)$. Since $y \prec x$, $A(y, \Xi)$ is then numerable. Hence all segments in $A(x, \Xi)$ are numerable, that is, $A(x, \Xi)$ is numerable, contrary to the assumption.

Thus we have proved that numeration and ordinal number are univocally determined notions for all well-ordered sets.

Chapter II

6. Let Ξ be well-ordered and let $f(x)$ be the numeration of Ξ. Then for no element x of Ξ is $f(x)$ an element of $f(x)$.

For if, for any element x of Ξ, $f(x)$ belonged to $f(x)$, there would exist a first such x.

Since $f(x)$ is the set of all $f(y)$ for $y \prec x$, we would then have $f(x) = f(y)$ for some $y \prec x$. But then $f(y)$ would belong to $f(y)$ and y would be $\prec x$, contrary to the assumption.

7. Let Ξ be well-ordered, let $f(x)$ be the numeration of Ξ, and let x and y be two elements of Ξ such that $x \prec y$. Then $f(x) < f(y)$.

For from $x \prec y$ it follows that $A(x, \Xi) < A(y, \Xi)$, hence

$$M(f(u);\, u\,\varepsilon\, A(x, \Xi)) \leqq M(f(u);\, u\,\varepsilon\, A(y, \Xi)),$$

and

$$f(x) \leqq f(y);$$

since $f(x)$ belongs to $f(y)$ (because $x \prec y$) but not to $f(x)$, $f(x)$ is distinct from $f(y)$. Therefore $f(x) < f(y)$.

8. Using an expression of Hessenberg's ⟦*1910*⟧, we say that a set of sets Ξ is "capable of being ordered by inclusion" ⟦"durch Subsumption ordnungsfähig"⟧ if for two distinct elements x and y of Ξ we always have $x < y$ or $x > y$. And, if this is the case, we define an ordering by stipulating that $x \prec y$ whenever $x < y$. We call this ordering the "inclusion ordering".

Now let Ξ be well-ordered. Then $OZ(\Xi)$ is always capable of being ordered by inclusion and is similar to Ξ under the inclusion ordering.

For let $f(x)$ be the numeration of Ξ. Any two elements P and Q of $OZ(\Xi)$ are then equal to $f(x)$ and $f(y)$, respectively, because

$$OZ(\Xi) = M(f(x);\, x\,\varepsilon\,\Xi).$$

And, since $x \prec y$ or $x \succ y$, we have $f(x) < f(y)$ or $f(x) > f(y)$, hence $P < Q$ or $P > Q$. Therefore $OZ(\Xi)$ is capable of being ordered by inclusion.

The assignment of x to $f(x)$ is clearly a mapping of Ξ onto $OZ(\Xi)$. And since it always follows from $x \prec y$ that $f(x) < f(y)$, and hence, by the inclusion ordering, that $f(x) \prec f(y)$, the mapping is also one-to-one and similar. Hence Ξ is similar to $OZ(\Xi)$.

9. P is an ordinal number if and only if

(1) It is a set of sets that is capable of being ordered by inclusion;

(2) Its inclusion ordering is a well-ordering;

(3) For every element ξ of P, always $\xi = A(\xi, P)$.

First, we assume that P is an ordinal number. Then let P be the ordinal number of the well-ordered set Ξ whose numeration is $f(x)$. P is a set of sets that, by the theorem just proved, is capable of being ordered by inclusion (hence (1) is satisfied) and is similar to the well-ordered Ξ, thus is itself well-ordered (hence (2) is satisfied). For every ξ of P, since necessarily $\xi = f(x)$ (x being an element of Ξ), we have

$$\xi = f(x) = M(f(y);\, y\,\varepsilon\,\Xi,\, y \prec x) = M(f(y);\, y\,\varepsilon\,\Xi,\, f(y) < f(x))$$
$$= M(\eta;\, \eta\,\varepsilon\, P,\, \eta \prec \xi) = A(\xi, P)$$

(hence (3) is also satisfied).

Second, we assume that P satisfies conditions (1), (2), and (3). Then it is capable of being ordered by inclusion and is well-ordered in the inclusion ordering. If by definition we take $f(x) = x$ for all elements x of P, then $f(x)$ is a numeration of P. In fact, because of (3), we have for all ξ of P,

$$f(\xi) = \xi = A(\xi, P) = M(\eta\,;\, \eta \,\varepsilon\, A(\xi, P)) = M(f(\eta)\,;\, \eta \,\varepsilon\, A(\xi, P));$$

accordingly we have

$$OZ(P) = M(f(\xi)\,;\, \xi \,\varepsilon\, P) = M(\xi\,;\, \xi \,\varepsilon\, P) = P;$$

hence P is an ordinal and indeed its own ordinal.

Chapter III

10. Let P be an ordinal. Then P is the set of all ordinals that are $< P$.

For let P be the ordinal of the well-ordered set Ξ whose numeration is $f(x)$. First, we assume that Q is an element of P. Because of (3), we then have

$$Q = A(Q, P) < P,$$

and, since necessarily $Q = f(x)$ (x being an element of Ξ) and we have

$$f(x) = OZ(A(x, \Xi)),$$

$f(x)$, hence also Q, is an ordinal.

Second, we assume that Q is an ordinal such that $Q < P$. We order P by inclusion. Let $\eta \prec \xi$, and let ξ belong to Q. Since, then, ξ belongs to Q and to P, and P and Q are ordinals, we have

$$\xi = A(\xi, Q) = A(\xi, P).$$

Since η belongs to P and $\eta \prec \xi$, η belongs to $A(\xi, P)$. And since

$$A(\xi, P) = A(\xi, Q) < Q,$$

it also belongs to Q. That is, $Q < P$. If $\eta \prec \xi$ and ξ belongs to Q, then η also belongs to Q. According to a well-known theorem about well-ordered sets (and P is well-ordered), Q is therefore a segment in P. For some element ξ of P we therefore have

$$Q = A(\xi, P) = \xi;$$

hence Q belongs to P.

We can express the theorem just proved thus: If P and Q are ordinals, $P \prec Q$ is equivalent to $P \,\varepsilon\, Q$. Hence for an ordinal P it is never the case that $P \,\varepsilon\, P$.

11. Let P and Q be two distinct ordinals. Then either $P < Q$ or $P > Q$.

Let R be the intersection of P and Q. Since P is capable of being ordered by inclusion and its inclusion ordering is a well-ordering (by (1) and (2)), since furthermore $R \leqq P$, the same statements hold of R; that is, R satisfies conditions (1) and (2). Every element ξ of R belongs to P and to Q, and, since P and Q are ordinals,

$$\xi = A(\xi, P) = A(\xi, Q).$$

$A(\xi, R)$ is the intersection of $A(\xi, P)$ and $A(\xi, Q)$, hence

$$\xi = A(\xi, R);$$

thus R also satisfies (3). Therefore R is an ordinal.

We have $R \leqq P$ and $R \leqq Q$. If $R = P$ or $R = Q$, then $P \leqq Q$ or $Q \leqq P$, hence $P < Q$ or $P > Q$ (since $P \neq Q$). But the antecedent always holds; for if we had $R < P$ and $R < Q$, then, since P, Q, and R are ordinals, we would have $R \varepsilon P$ and $R \varepsilon Q$. Hence we would have $R \varepsilon R$ (because R is the intersection of P and Q), which is impossible since R is an ordinal.

12. Let U be a set of ordinals. Two distinct elements P and Q of U are always ordinals; hence we have $P < Q$ or $P > Q$. That is, U is capable of being ordered by inclusion. The inclusion ordering of U, however, is a well-ordering.

To prove this we must show that every $V \leqq U$, with $V \neq 0$, has a first element. Let P be an element of V. (P is an ordinal number.) If there does not exist an element of V that is $< P$, then V has a first element, namely P. If such elements do exist, let their set be W. Since all elements of W are ordinal numbers (because $W \prec V \leqq U$) and are $< P$, they all belong to P; hence we have $W \leqq P$. Since $W \leqq P$, $W \neq O$, and P is well-ordered, W has a first element Q. (Because $Q \varepsilon W$, $Q < P$.) Since every element of V is either $\geqq P$ (hence $> Q$) or $< P$ (hence an element of W and consequently $\geqq Q$), Q is also the first element of V. Therefore V always has a first element.

13. P is an ordinal if and only if every element of P is an ordinal and is $\leqq P$.

First, let P be an ordinal. Every element of P is an ordinal and is $< P$.

Second, let P satisfy our conditions. As a set of ordinals, P is then capable of being ordered by inclusion and its inclusion ordering is a well-ordering; hence (1) and (2) are satisfied. Moreover, for every element ξ of P,

$$A(\xi, P) = M(\eta; \eta \varepsilon P, \eta \prec \xi);$$

hence, since all elements η of P are ordinals,

$$A(\xi, P) = M(\eta; \eta \varepsilon P, \eta \, OZ, \eta \prec \xi) = M(\eta; \eta \varepsilon P, \eta \varepsilon \xi);$$

and hence, because $\xi \leqq P$, also

$$A(\xi, P) = M(\eta; \eta \varepsilon \xi) = \xi;$$

thus (3) is also satisfied. P is therefore an ordinal.

If P is a set of ordinals, then, for every element ξ of P, $\xi \leqq P$ means (since P is an ordinal) that all ordinals that are $< \xi$ (that is, all elements of ξ) belong to P. We can therefore also express the theorem just proved thus: P is an ordinal if and only if all elements of P are ordinals and if, for every element ξ of P, all ordinals η such that $\eta < \xi$ are also elements of P.

Chapter IV

14. Let Ξ and H be well-ordered sets. Ξ and H are similar if and only if $OZ(\Xi) = OZ(H)$.

First, let $OZ(\Xi) = OZ(H)$. Since Ξ is similar to $OZ(\Xi)$, which is ordered by inclusion, and H is similar to $OZ(H)$, which is ordered by inclusion, Ξ is indeed similar to H.

Second, let Ξ be similar to H. Then let $\varphi(x)$ be a similar mapping of Ξ onto H and let $g(x')$ be the numeration of H. If by definition we take $f(x) = g(\varphi(x))$ for all elements x of Ξ, then $f(x)$ is a numeration of Ξ. In fact, for all x in Ξ,

$$\begin{aligned} M(f(y); y \,\varepsilon\, \Xi, y \prec x) &= M(g(\varphi(y)); y \,\varepsilon\, \Xi, y \prec x) \\ &= M(g(\varphi(y)); y \,\varepsilon\, \Xi, \varphi(v) \prec \varphi(x)) \\ &= M(g(y'); y' \,\varepsilon\, H, y' \prec \varphi(x)) = g(\varphi(x)) = f(x); \end{aligned}$$

but from this it follows that

$$OZ(\Xi) = M(f(x); x \,\varepsilon\, \Xi) = M(g(\varphi(x)); x \,\varepsilon\, \Xi) = M(g(x'); x' \,\varepsilon\, H) = OZ(H).$$

15. Ξ is similar to a segment in H if and only if $OZ(\Xi) < OZ(H)$.

First, let $OZ(\Xi) < OZ(H)$. Then, by 10,[3a] $OZ(\Xi)$ is an element of $OZ(H)$, hence a segment in H (its own). And since Ξ is similar to $OZ(\Xi)$ and H to $OZ(H)$, Ξ is also similar to a segment of H.

Second, let Ξ be similar to a segment in H. Let the numeration of H be $g(x)$; let the segment in question be $A(\bar{x}, H)$. Then, because of the similarity, we have

$$OZ(\Xi) = OZ(A(\bar{x}, H)) = g(\bar{x});$$

hence $OZ(\Xi)$ is an element of $OZ(H)$, and therefore, by 10,[3a]

$$OZ(\Xi) < OZ(H).$$

16. Since (because of 11) always one and only one of the following three cases obtains:

$$OZ(\Xi) < OZ(H), \quad OZ(\Xi) = OZ(H), \quad \text{and} \quad OZ(\Xi) > OZ(H),$$

always one and only one of the following three cases obtains too:

Ξ is similar to a segment of H,
Ξ is similar to H, and
H is similar to a segment of Ξ.

And these three cases are equivalent, respectively, to

$OZ(\Xi) < OZ(H)$, or $OZ(\Xi) \,\varepsilon\, OZ(H)$,
$OZ(\Xi) = OZ(H)$,
$OZ(\Xi) > OZ(H)$, or $OZ(H) \,\varepsilon\, OZ(\Xi)$.

17. Let Ξ be well-ordered. Then there is one and only one ordinal (ordered by inclusion) similar to Ξ, namely $OZ(\Xi)$.

$OZ(\Xi)$ is indeed similar to Ξ. And if the ordinal P is similar to Ξ, then let P be the ordinal of H. Since H is similar to P and P to Ξ, H is similar to Ξ, hence

$$P = OZ(H) = OZ(\Xi);$$

from this it also follows that two ordinals ordered by inclusion are similar if and only if they are identical.

From this point on it is easy to develop the theory of ordinals further. Addition and multiplication of ordinals are not difficult to establish. "Definition by transfinite induction" is, to be sure, admissible only when the following theorem has been proved:

"Let $f(x)$ be a function that is defined for all sets of objects of a domain B and

[3a] [[The German text has "12", which is apparently a misprint.]]

whose values are always objects of the domain B. Then there is one and only one function $\Phi(P)$, defined for all ordinals P and having values that are always objects of the domain B, such that for all ordinals P

$$\Phi(P) = f(M(\Phi(Q); Q\overline{OZ}, Q < P)) = f(M(\Phi(Q); Q \varepsilon P)),$$

where $Q\overline{OZ}$ means that Q is an ordinal."

The proof of this theorem, which is not at all self-evident, is, however, not difficult to supply.[4] Once this theorem is proved, the theory of exponentiation of ordinals, as well as that of "continuous" or "normal"[5] functions of ordinals, can also be developed without further ado.

[4] The theorem is proved (in strong analogy with the proofs in Chapter I, 1–5 ⟦the German text erroneously has "1–6"⟧) somewhat as follows:

(*a*) Let P be an ordinal. Is there then a function $\Psi(Q)$, defined for all ordinals $Q < P$, such that

$$\Psi(Q) = f(M(\Psi(R); R\overline{OZ}, R < Q))?$$

And how many such functions are there?

(*b*) If there is any, there is exactly one. In that case let P be said to be "normal" and let Ψ be called Ψ_P. Furthermore, for a normal P, let

$$\Phi(P) = f(M(\Psi_P(Q); Q\overline{OZ}, Q < P)).$$

(*c*) If P is normal, then every $Q < P$ is normal, and for all $R < Q$ we have

$$\Psi_Q(R) = \Psi_P(R), \quad \text{and} \quad \Phi(Q) = \Psi_P(Q).$$

(*d*) If all $Q < P$ are normal, then P is also normal. (We have $\Psi_P(Q) = \Phi(Q)$.)

(*e*) All P are normal. It follows immediately from (*d*) that

$$\Phi(P) = f(M(\Psi_P(Q); Q\overline{OZ}, Q < P)) = f(M(\Phi(Q); Q\overline{OZ}, Q < P)),$$

hence $\Phi(P)$ is the desired function.

(*f*) There is only one such function.

[5] See *Hausdorff 1914*.

On the building blocks of mathematical logic

MOSES SCHÖNFINKEL

(*1924*)

These ideas were presented before the Göttingen Mathematical Society by Schönfinkel on 7 December 1920 but came to be written up for publication only in March 1924, by Heinrich Behmann. The last three paragraphs are given over to supplementary remarks of Behmann's own.

The initial aim of the paper is reduction of the number of primitive notions of logic. The economy that Sheffer's stroke function had wrought in the propositional calculus is here extended to the predicate calculus, in the form of a generalized copula "U" of mutual exclusiveness. Then presently the purpose deepens. It comes to be nothing less than the general elimination of variables.

Examples of how to eliminate variables had long been known in logic. The inclusion notation "$F \subseteq G$" gets rid of the universally quantified "x" of "$(x)(Fx \supset Gx)$". The notation "$F \mid G$" of relative product gets rid of an existentially quantified "x", since "$(\exists x)(Fyx \cdot Gxz)$" becomes "$(F \mid G)yz$". These devices and others figured already in Peirce's 1870 algebra of absolute and relative terms, thus even antedating any coherent logic of the variable itself, for such a logic, quantification theory, came only in Frege's work of 1879. The algebra of absolute and relative terms, however, or of classes and relations, is not rich enough to do the work of quantifiers and their variables altogether (see above, p. 229).

Schönfinkel's notions, which do suffice to do the work of quantifiers and their variables altogether, go far beyond the algebra of classes and relations, for that algebra makes no special provision for classes of classes, relations of classes, classes of relations, or relations of relations. Schönfinkel's notions provide for these things, and for the whole sweep of abstract set theory. The crux of the matter is that Schönfinkel lets functions stand as arguments.

For Schönfinkel, substantially as for Frege,[a] classes are special sorts of functions. They are propositional functions, functions whose values are truth values. All functions, propositional and otherwise, are for Schönfinkel one-place functions, thanks to the following ingenious device (which was anticipated by Frege (*1893*, § 36)). Where F is what we would ordinarily call a two-place function, Schönfinkel reconstrues it by treating Fxy not as $F(x, y)$ but as $(Fx)y$. Thus F becomes a one-place function whose value is a one-place function. The same trick applies to functions of three or more places; thus "$Fxyz$" is taken as "$((Fx)y)z$". In particular, a dyadic relation, as an ostensibly two-place propositional function, thus becomes a one-place function whose value is a class. An example is U, above, where "UFG" means that F and G are exclusive classes; UFG

[a] See *Frege 1879*, § 9. What prompts my qualification "substantially" is that in later writings Frege draws a philosophical distinction between a function and its *Wertverlauf*.

becomes $(UF)G$, so that U becomes a function whose value, for a class F as argument, is the class UF of all classes G that share no members with F. Or better, using general variables "x", "y", and so forth hereafter for classes and other functions and all other things as well, we may say that U is the function whose value Ux is the class of all classes that share no members with x.

Schönfinkel assumes one operation, that of application of a function to an argument. It is expressed by juxtaposition as in "Ux" above, or "$(Ux)y$", or in general "zy". Also he assumes three specific functions, as objects: U above, C, and S. C is the *constancy* function, such that $(Cx)y$ is always x, and S is the *fusion* function, such that $((Sx)y)z$ is always $(xz)(yz)$. Any sentence that can be built up of truth functions and quantification and the "ε" of class membership can be translated into a sentence built up purely by the application operation, purely from "C", "S", "U", and whatever free variables the given sentence may have had. This is made evident in the course of the paper. The elimination of quantification and bound variables is thus complete. Since sentences with free variables are wanted finally only as ingredients of closed sentences, the notion of variables may indeed be said at this point to have been analyzed away altogether. All we have is C, S, U, and application.

Variables seem to survive in rules of transformation, as when $(Cx)y$ is equated to x, and $((Sx)y)z$ to $(xy)(xz)$. But here the variables may be seen as schematic letters for metalogical exposition. If one cared to formalize one's metalanguage in turn, one could subject it too to Schönfinkel's elimination of variables.

C, S, and U, economical as they are, are further reducible. It is now known that we can get U from C, S, and the mere *identity relation*[b]—which, by the doctrine of relations noted above, is the same as the unit-class function ι.

Schönfinkel himself contrives a more drastic but very curious reduction of C, S, and U. He adopts a new function J, interpreted as having U as value for C as argument, C as value for S as argument, and S as value for other arguments. Then he defines "S" as "JJ", "C" as "JS", and "U" as "JC". This trick reduces every closed sentence (of logic and set theory) to a string of "J" and parentheses.

In the second of his three added paragraphs, Behmann proposes an alternative reduction, which, at the cost of resting with C, S, U, and yet a fourth basic function instead of just J, would get rid of parentheses by maneuvering them all into a left-converging position, where they could be tacit. However, as Behmann recognized later in a letter to H. B. Curry,[c] there is a fallacy here; the routine can generate new parentheses and not terminate.

But, if we want to get rid of parentheses, we can easily do so by adapting an idea that was used elsewhere in logic by Łukasiewicz (*1929*, footnote 1, pp. 610–612; *1929a*; *1958*; *1963*). Instead of using mere juxtaposition to express the application of functions, we can use a preponent binary operator "o". Thus "xy", "$x(yz)$", and "$(xy)z$" give way to "oxy", "$oxoyz$", and "$ooxyz$". All Schönfinkel's sentences built of "J" and parentheses go over unambiguously into strings of "J" and "o".

Schönfinkel's reduction to J, with parentheses or whatever, is interesting for its effortlessness and its broad applicability. One would like to rule it out as spurious reduction, but where can a line be drawn? The only contrast I think of between serious reduction and this reduction to J is that in serious reduction the axioms tend to diminish along with

[b] This reduction figured in my seminar, from about 1952 on. The essential reasoning is recoverable from the last two pages of *Quine 1956*.

[c] See *Curry and Feys 1958*, p. 184.

the primitive ideas: what had been axiomatic connections reduce in part to definitions. Schönfinkel does not get to axioms, but any axioms he might have adopted regarding C, S, and U would obviously have been diminished none by his reduction of these functions to J.

Axioms for C, S, and U pose, as it happens, a major problem. Since these three functions cover set theory, the question of axioms for them is as broad as axiomatic set theory itself. Moreover, this present angle of analysis is so radically different from the usual that we cannot easily adapt the hitherto known ways of getting around the set-theoretic paradoxes. The quest for an optimum axiomatization of the Schönfinkel apparatus has accounted for much work and many writings from 1929 onward, mainly by H. B. Curry, under the head of *combinatory logic*.[d]

It was by letting functions admit functions generally as arguments that Schönfinkel was able to transcend the bounds of the algebra of classes and relations and so to account completely for quantifiers and their variables, as could not be done within that algebra. The same expedient carried him, we see, far beyond the bounds of quantification theory in turn; all set theory was his province. His C, S, U, and application are a marvel of compact power. But a consequence is that the analysis of the variable, so important a result of Schönfinkel's construction, remains all bound up with the perplexities of set theory.

It proves possible, thanks to the perspective afforded by Schönfinkel's pioneer work, to separate these considerations. Ordinary quantification theory, the first-order predicate calculus, can actually be reworked in such a way as to get rid of the variables without thus increasing the power of the notation. The method turns on adopting a few operators attachable to predicates. They are reminiscent of Schönfinkel's functions, but with the difference that they do not apply to themselves or one another. An analysis of the variable is obtained that is akin to Schönfinkel's but untouched by the problems of set theory.[e]

W. V. Quine

The translation is by Stefan Bauer-Mengelberg, and it is printed here with the kind permission of Springer Verlag.

[d] See *Curry and Feys 1958*.

[e] See *Bernays 1957* and *Quine 1960*.

§ 1

It is in the spirit of the axiomatic method as it has now received recognition, chiefly through the work of Hilbert, that we not only strive to keep the axioms as few and their content as limited as possible but also attempt to make the number of fundamental undefined *notions* as small as we can; we do this by seeking out those notions from which we shall best be able to construct all other notions of the branch of science in question. Understandably, in approaching this task we shall have to be appropriately modest in our demands concerning the simplicity of the initial notions.

As is well known, the fundamental *propositional connectives* of mathematical logic, which I reproduce here in the notation used by Hilbert in his lectures,

$$\bar{a},\ a \vee b,\ a \,\&\, b,\ a \rightarrow b,\ a \sim b$$

(read: "not a", "a or b", "a and b", "if a, then b", "a is equivalent to b"), cannot be defined at all in terms of any single one of them; they can be defined in terms of two of them only if negation and any one of the three succeeding connectives are taken as undefined elements constituting the base. (Of these three kinds of reduction Whitehead and Russell employed the first and Frege the third.)

That the reduction to a single fundamental connective is nevertheless entirely possible provided we remove the restriction that the connective be taken only from the sequence above was discovered not long ago by Sheffer (*1913*). For, if we take as the base, say, the connective "not a or not b", that is, "of the propositions a and b at least one is false", which can be written with the signs above in the two equivalent forms

$$\bar{a} \vee \bar{b} \quad \text{and} \quad \overline{a \,\&\, b},$$

and if we adopt

$$a \mid b$$

as the new sign for it, then obviously

$$\bar{a} = a \mid a \quad \text{and} \quad a \vee b = (a \mid a) \mid (b \mid b);$$

thus, because

$$a \,\&\, b = \overline{\bar{a} \vee \bar{b}}, \quad (a \rightarrow b) = \bar{a} \vee b, \quad \text{and} \quad (a \sim b) = (a \rightarrow b) \,\&\, (b \rightarrow a),$$

the reduction has in principle been accomplished.

It is remarkable, now, that it is possible to go beyond this and, by suitably modifying the fundamental connective, to encompass even the two higher propositions

$$(x)f(x) \quad \text{and} \quad (Ex)f(x),$$

that is, "all individuals have the property f" and "there exists an individual having the property f", in other words, the two operations (x) and (Ex), which, as is well known, together with those above constitute a system of fundamental connectives for mathematical logic that is complete from the point of view of the axiomatization.

For, if from now on we use

$$(x)[\overline{f(x)} \vee \overline{g(x)}], \quad \text{or} \quad (x)\overline{f(x) \,\&\, g(x)},$$

as a fundamental connective and if we write

$$f(x) \mid^{x} g(x)$$

for it, then evidently (since we can treat constants formally as functions of an argument)

$$\bar{a} = a \mid^{x} a, \quad a \vee b = (x)(a \vee b) = \bar{a} \mid^{x} \bar{b} = (a \mid^{y} a) \mid^{x} (b \mid^{y} b),$$

and

$$(x)f(x) = (x)(\overline{\overline{f(x)}} \vee \overline{\overline{f(x)}}) = \overline{f(x)} \mid^{x} \overline{f(x)} = (f(x) \mid^{y} f(x)) \mid^{x} (f(x) \mid^{y} f(x));$$

thus, because

$$(Ex)f(x) = \overline{(x)\overline{f(x)}}$$

the last assertion is proved also.

The successes that we have encountered thus far on the road taken encourage us to attempt further progress. We are led to the idea, which at first glance certainly appears extremely bold, of attempting to eliminate by suitable reduction the remaining fundamental notions, those of proposition, propositional function, and variable, from those contexts in which we are dealing with completely arbitrary, logically general propositions (for others the attempt would obviously be pointless). To examine this possibility more closely and to pursue it would be valuable not only from the methodological point of view that enjoins us to strive for the greatest possible conceptual uniformity but also from a certain philosophic, or, if you wish, aesthetic point of view. For a variable in a proposition of logic is, after all, nothing but a token ⟦Abzeichen⟧ that characterizes certain argument places and operators as belonging together; thus it has the status of a mere auxiliary notion that is really inappropriate to the constant, "eternal" essence of the propositions of logic.

It seems to me remarkable in the extreme that the goal we have just set can be realized also; as it happens, it can be done by a reduction to three fundamental signs.

§ 2

To arrive at this final, deepest reduction, however, we must first present a number of expedients and explain a number of circumstances.

It will therefore be necessary to leave our problem at the point reached above and to develop first a kind of *function calculus* ⟦*Funktionenkalkül*⟧; we are using this term here in a sense more general than is otherwise customary.

As is well known, by function we mean in the simplest case a correspondence between the elements of some domain of quantities, the argument domain, and those of a domain of function values (which, to be sure, is in most cases regarded as coinciding with the former domain) such that to each argument value there corresponds at most one function value. We now extend this notion, permitting functions themselves to appear as argument values and also as function values. We denote the value of a function f for the argument value x by simple juxtaposition of the signs for the function and the argument, that is, by

$$fx.$$

Functions of several arguments can, on the basis of our extended definition of function, be reduced to those of a single argument in the following way.

We shall regard

$$F(x, y),$$

for example, as a function of the single argument y, say, but not as a fixed given function; instead, we now consider it to be a variable function that depends on x for its form. (Of course we are here concerned with a dependence of the *function*, that is, of the correspondence itself; we are not referring to the obvious dependence of the function *value* upon the argument.) In mathematics we would say in such a case that the function depends upon a parameter, too, and we would write, say,

$$G_x(y).$$

We can regard this function G itself—its form, so to speak—as the value (function value) of a new function f, so that $G = fx$.

We therefore write

$$(fx)y$$

in our symbolism, or, by agreeing, as is customary in the theory of infinite series for example, that parentheses around the left end of such a symbolic form may be omitted, more simply,

$$fxy,$$

where the new function, f, must be clearly distinguished from the former one, F.

I should like to make the transformation just described more intelligible by applying it to the specific number-theoretic function $x - y$. If we regard the expression as a function of y alone, this function has the "form" $x-$ and therefore means "the difference between x and any given quantity"; in that case the expression is to be taken as $(x-)y$. The essential point here is that we must not think of a simultaneous substitution of values for x and y, but, to begin with, of the substitution of a value—a, for instance—for x alone, so that we first obtain the function $a - y$ (in short: the function $a-$) as an intermediate step; only then does the replacement of y—by the fixed value b, say—become admissible.

In the foregoing case, therefore, fx is the value of a function that, upon substitution of a value for x, does not yet yield an object of the fundamental domain (if indeed such an object was intended as the value of $F(x, y)$) but yields another function, whose argument now is y; that is, f is a function whose argument need not be subject to any restriction but whose function value is again a function. On functions of more than one variable we shall henceforth always carry out the transformation described above (or think of it as having been carried out), so that these will appear throughout in the form

$$fxyz\ldots,$$

which, as we have already stated, is to be taken as an abbreviation of

$$(((fx)y)z)\ldots.$$

§ 3

Now a sequence of *particular functions* of a very general nature will be introduced. I call them the identity function [[Identitätsfunktion]] I, the constancy function [[Konstanzfunktion]] C, the interchange function [[Vertauschungsfunktion]] T, the composition function [[Zusammensetzungsfunktion]] Z, and the fusion function [[Verschmelzungsfunktion]] S.

1. By the *identity function* I we mean that completely determined function whose argument value is not subject to any restriction and whose function value always coincides with the argument value, that function, in other words, by which each object, as well as each function, is associated with itself. It is therefore defined by the equation

$$Ix = x,$$

where the equal sign is not to be taken to represent logical equivalence as it is ordinarily defined in the propositional calculus of logic but signifies that the expressions on the left and on the right mean the same thing, that is, that the function value Ix is always the same as the argument value x, whatever we may substitute for x. (Thus, for instance, II would be equal to I.)

2. Now let us assume that the argument value is again arbitrary without restriction, while, regardless of what this value is, the function value will always be the fixed value a. This function is itself dependent upon a; thus it is of the form Ca. That its function value is always a is written

$$(Ca)y = a.$$

And by now letting a, too, be variable we obtain

$$(Cx)y = x, \quad \text{or} \quad Cxy = x,$$

as the defining equation of the *constancy function* C. This function C is obviously of the kind considered on page 360; for only when we substitute a fixed value for x does it yield a function with the argument y. In practical applications it serves to permit the introduction of a quantity x as a "blind" variable.

3. Conversely, we can obviously always think of an expression like

$$fxy$$

as having been obtained from

$$F(x, y),$$

where F is uniquely determined by the given f. If, however, we now rewrite this expression as

$$gyx,$$

taking y as a parameter, then this new function, too, is uniquely given by F and therefore indirectly also by f.

Hence we may think of the function g as the value of a function T for the argument value f. This *interchange function* T has as its argument a function of the form φxy, and the function value

$$\psi = T\varphi$$

is that function ψxy whose value ψxy coincides with φyx for all argument values x and y for which φyx has meaning. We write this definition briefly as

$$(T\varphi)xy = \varphi yx,$$

where the parentheses may again be omitted.

The function T makes it possible to alter the order of the terms of an expression, and in this way it compensates to a certain extent for the lack of a commutative law.

4. If in the argument place of a function f there occurs the value (dependent upon x) of another function, g, then

$$f(gx)$$

obviously also depends upon x and can in consequence be regarded as the value of a third function, F, which is uniquely determined by f and g. In analysis, as is well known, we speak loosely in such cases of a "function of a function"—strictly, it should be a "function of a function value"—and we call F the function "compounded" from f and g. The function F is thus itself the value of a certain function Z' of f and g.

We could therefore define

$$[Z'(\varphi, \chi)]x = \varphi(\chi x).$$

But, following our earlier convention, we prefer to replace Z' by the corresponding function of one argument, and we consequently obtain

$$Z\varphi\chi x = \varphi(\chi x)$$

as the defining equation of the *composition function* Z.

By means of the function Z parentheses can be shifted (not really eliminated, since they must always be thought of as still being there) within a more comprehensive expression; its effect is therefore somewhat like that of the associative law, which is not satisfied here either.

5. If in

$$fxy$$

we substitute the value of a function g for y, and in particular the value taken for the same x as that which appears as argument of f, we come upon the expression

$$fx(gx),$$

or, as we shall write it for the moment to make it clearer,

$$(fx)(gx).$$

This, of course, is the value of a function of x alone; thus

$$(fx)(gx) = Fx,$$

where

$$F = S'(f, g)$$

again depends in a completely determined way upon the given functions f and g. Accordingly we have

$$[S'(\varphi, \chi)]x = (\varphi x)(\chi x),$$

or, if we carry out the same transformation as in the preceding case,

$$S\varphi\chi x = (\varphi x)(\chi x),$$

as the defining equation of the *fusion function* S.

It will be advisable to make this function more intelligible by means of a practical example. If we take for fxy, say, the value $\log_x y$ (that is, the logarithm of y to the base x) and for gz the function value $1 + z$, then $(fx)(gx)$ obviously becomes $\log_x (1 + x)$, that is, the value of a function of x that is univocally associated with the two given functions precisely by our general function S.

Clearly, the practical use of the function S will be to enable us to reduce the number of occurrences of a variable—and to some extent also of a particular function—from several to a single one.

§ 4

It will prove to be relevant to the solution of the problem that we have raised concerning the symbolism of logic that the five particular functions of the function calculus that were defined above, I, C, T, Z, and S, are not mutually independent, that, rather, two of them, namely C and S, suffice to define the others. In fact, the following relations obtain here.

1. According to the definitions of the functions I and C,

$$Ix = x = Cxy.$$

Since y is arbitrary, we can substitute any object or any function for it, hence, for example, Cx. This yields

$$Ix = (Cx)(Cx).$$

According to the definition of S, however, this means

$$SCCx,$$

so that we obtain

$$I = SCC.^{1}$$

The last C, incidentally, does not occur in the expression SCC in an essential way. For, if above we put for y not Cx but an arbitrary function φx, we obtain in a similar way

$$I = SC\varphi,$$

where any function can then be substituted for φ.[2]

2. According to the definition of Z,

$$Zfgx = f(gx).$$

Furthermore, by virtue of the transformations already employed,

$$f(gx) = (Cfx)(gx) = S(Cf)gx = (CSf)(Cf)gx.$$

Fusion over f yields

$$S(CS)Cfgx;$$

therefore

$$Z = S(CS)C.$$

3. In an entirely analogous way,

$$Tfyx = fxy$$

can be further transformed thus:

$$\begin{aligned} fx(Cyx) &= (fx)(Cyx) = Sf(Cy)x = (Sf)(Cy)x = Z(Sf)Cyx \\ &= ZZSfCyx = (ZZSf)Cyx = (ZZSf)(CCf)yx = S(ZZS)(CC)fyx. \end{aligned}$$

Therefore we have

$$T = S(ZZS)(CC).$$

If we here substitute for Z the expression found above, T too will have been reduced to C and S.

[1] This reduction was communicated to me by Boskowitz; some time before that, Bernays had called the somewhat less simple one $(SC)(CC)$ to my attention.

[2] Only such a function, of course, as has meaning for every x.

§ 5

Let us now apply our results to a special case, that of the calculus of logic in which the basic elements are individuals and the functions are propositional functions. First we require an additional particular function, which is peculiar to this calculus. The expression

$$fx \mid^x gx,$$

where f and g are propositional functions of one argument—in view of an earlier remark we may confine ourselves to these—is obviously a definite function of the two functions f and g; thus it is of the form $U(f, g)$, or, by our principle of transformation, Ufg. Thus we have

$$Ufg = fx \mid^x gx,$$

where f and g, of course, now are propositional functions, as the defining equation of the *incompatibility function* U.

It is a remarkable fact, now, that every formula of logic can be expressed by means of our particular functions I, C, T, Z, S, and U alone, hence, in particular, by means solely of C, S, and U.

First of all, every formula of logic can be expressed by means of the generalized stroke symbol, with the bound variables (apparent variables) at the upper ends of the strokes. This holds without restriction; hence it holds for arbitrary orders of propositions ⟦Aussagenordnungen⟧ and also if relations occur. Furthermore, we can introduce the function U step by step in place of the stroke symbol by suitable use of the remaining constant functions.

We will not give the complete demonstration here but only explain the role of the different particular functions in this reduction.

By means of the function C we can see to it that the two expressions standing to the left and the right of the stroke become functions of the same argument.

Thus, for example, the expression

$$fx \mid^x gy,$$

which depends upon f, g, and y and in which x does not occur on the right, would have to be rewritten as

$$fx \mid^x C(gy)x.$$

If, however, x occurs on the right at some place other than the end, we can move it there by means of the function T; in doing this we must use the function Z to extricate it from parentheses, if there are any, and the function S to fuse it, if it occurs several times. Thus, for example,

$$fx \mid^x gxy = fx \mid^x Tgyx = Uf(Tgy).$$

Or, to take a somewhat more involved example,

$$(fxy \mid^y gxy) \mid^x (hxz \mid^z kxz) = U(fx)(gx) \mid^x U(hx)(kx).$$

Here, for instance, the expression preceding the stroke must be dealt with further as follows:

$$U(fx)(gx) = ZUfx(gx) = S(ZUf)gx.$$

The entire expression thus becomes

$$S(ZUf)gx\,|^x S(ZUh)kx,$$

which is equal to

$$U[S(ZUf)g][S(ZUh)k].$$

If in the last example f and g were identical, we would obtain the expression

$$S(ZUf)f.$$

To be able to achieve the fusion over f here, we make use of the function I by calculating further:

$$S(ZUf)f = S(ZUf)(If) = [ZS(ZU)f](If) = S[ZS(ZU)]If.$$

To give a practical example of the claim of this section we shall deal with the following proposition: "For every predicate there exists a predicate incompatible with it", that is, "For every predicate f there exists a predicate g such that the propositional function fx & gx is not true of any object x".

In Hilbert's symbolism this sentence is written

$$(f)(Eg)(x)\overline{fx \,\&\, gx}.$$

This becomes, first,

$$(f)(Eg)(fx\,|^x gx)$$

and then, if we write the particular judgment as the negation of a universal one,

$$(f)\overline{(g)\overline{fx\,|^x gx}},$$

or

$$(f)\overline{(g)\overline{fx\,|^x gx \,\&\, fx\,|^x gx}}.$$

This is

$$(f)\overline{(fx\,|^x gx)\,|^g (fx\,|^x gx)}.$$

If we proceed similarly for f as well, we obtain, further,

$$(f)\overline{(fx\,|^x gx)\,|^g (fx\,|^x gx) \,\&\, (fx\,|^x gx)\,|^g (fx\,|^x gx)},$$

which is equal to

$$[(fx\,|^x gx)\,|^g (fx\,|^x gx)]\,|^f [(fx\,|^x gx)\,|^g (fx\,|^x gx)].$$

From this point on the stroke symbol occurs as the sole logical connective. If we now introduce the incompatibility function U, we obtain first

$$[(Ufg)\,|^g (Ufg)]\,|^f [(Ufg)\,|^g (Ufg)]$$

and then

$$[U(Uf)(Uf)]\,|^f [U(Uf)(Uf)].$$

But now

$$U(Uf)(Uf) = (ZUUf)(Uf) = S(ZUU)Uf;$$

hence the expression above becomes

$$[S(ZUU)Uf]\,|^f [S(ZUU)Uf];$$

but this is

$$U[S(ZUU)U][S(ZUU)U].$$

§ 6

So far as we can see, we cannot carry the reduction to anything beyond the symbols C, S, and U without doing violence to it.

Purely schematically, to be sure, we could replace even C, S, and U by a single function if we were to introduce the new function J through the definitions

$$JC = U, \quad JS = C, \quad \text{and} \quad Jx = S,$$

where x is any object distinct from C and S. We ascertain, first, that J is itself distinct from C and S, since J takes on only three function values whereas C and S take on infinitely many. Consequently we have

$$JJ = S, \quad J(JJ) = JS = C, \quad \text{and} \quad J[J(JJ)] = JC = U,$$

which in fact accomplishes the reduction. But on account of its obvious arbitrariness it is probably without any real significance.

However,[3] we can in a certain sense free ourselves at least from the sign U in another, more natural way. Every formula of logic [[when transformed as indicated]] certainly contains the sign U and—quite in accordance with our earlier conclusion about an arbitrary symbol—can be written in terms of the particular functions of the general function calculus (hence, in particular, by means of C and S) in such a way that U occurs as the argument of the entire expression; the expression thus assumes the form FU, where F itself no longer contains U. If in writing the expression down we omit the U, regarding it as understood, we can in fact manage with C and S.

On the other hand we could, while relinquishing the most extreme reduction of the basic function symbols, demand that parentheses be entirely avoided. If now we take the form FU as a point of departure, then, by means of Z alone, F can be transformed in such a way that all parentheses disappear. By means of C, Z, and S, therefore, every formula of logic can be written without parentheses as a simple sequence of these signs and can therefore be characterized completely by a number written to the base 3.

So far as the *uniqueness* of the reduction considered is concerned, it is, purely from the point of view of the symbolism, quite out of the question, since every formula both of the old and of the new calculus can be transformed in various ways. Yet in a certain more limited sense we can ascertain a uniqueness of correspondence here. For if we use the term "equivalent", on the one hand, for the formulas of the old calculus that can be reduced to one another purely on the basis of the definitions, that is, without use of the logical axioms (in which, of course, the generalized Sheffer connective would now have to figure as the fundamental connective), and, on the other hand, for those that differ from one another only in the typography of the variables occurring in them, then to a given formula of the new calculus and, likewise, to any one that can be obtained from it through symbolic calculation there in fact correspond precisely those formulas of the old calculus that are equivalent to one another in the sense just defined. The reduction here considered of the formulas of logic has the remarkable peculiarity, therefore, of being independent of the axioms of logic.

[3] The considerations that follow are the [[German]] editor's.

On the infinite

DAVID HILBERT

(*1925*)

Hilbert devoted to the foundations of mathematics a series of papers (*1904, 1917, 1922, 1922a, 1925, 1927, 1928a, 1930, 1931*, all but the last read as addresses), to which must be added his paper on the axiomatization of the real number system (*1900*) and some parts of his address to the Paris International Congress of Mathematicians (*1900a*), namely, those on the continuum hypothesis, the well-ordering hypothesis, and the consistency of the real-number system.[a]

Among all these contributions the 1925 paper stands out as the most comprehensive presentation of Hilbert's ideas. It is the text of an address delivered in Münster on 4 June 1925 at a meeting organized by the Westphalian Mathematical Society to honor the memory of Weierstrass. In addition to the original version (*1925*), two modified ones (*1925a*; *1930a*, Appendix VIII) were published. These, however, are abbreviated and, besides other passages, omit entirely the section that deals with the continuum problem; the third version contains a number of minor emendations and additions. The translator followed the original version, but he used the third to clarify the meaning of a few passages, and some of the additions contained in this third version are reproduced in the translation below between square brackets.

The paper consists of two quite distinct parts. The first is a clear and forceful presentation of Hilbert's ideas at the time on the foundations of mathematics. It starts by recalling how Weierstrass eliminated references to "infinity" in analysis (Hilbert could have mentioned Cauchy and even D'Alembert) and reviews the role played by the infinite in physics, set theory, and, when we deal with general propositions, logic. Leaning on the example of arithmetic, it introduces the distinction between finitary and ideal propositions, and it undertakes a simultaneous formalization of logic and arithmetic (the system presented by Hilbert is described in greater detail in his 1927 paper, pp. 465–469 below; on this system see also Supplement IV in *Hilbert and Bernays 1939* and, for a critical study, *Asser 1957*).

The second part of the paper is the sketch of an attempted proof of the continuum hypothesis. Stated for the first time by Cantor at the very beginning of the development of set theory (*1878*, p. 258), presented by Hilbert as Problem no. 1 in his famous list of unsolved mathematical problems (*1900a*), the continuum problem for years resisted the efforts of mathematicians. The only partial result known at the time of Hilbert's attempted proof was König's: the power of the continuum is not a limit of denumerably many smaller cardinal numbers (see *Hausdorff 1914*, pp. 67–68); for a general survey of the continuum problem see *Gödel 1947*).

---

[a] Perhaps *1930b* should also be added.

Instead of the set of real numbers Hilbert considers the equivalent set of number-theoretic functions and undertakes to prove that there is a mapping of the set of ordinals of the second number class onto the set of these functions. By dealing with the definitions of functions rather than with the functions themselves the argument takes on a metamathematical aspect that is absent from Cantor's work. The fact that the values of a number-theoretic function may be made to depend on the solvability of a certain mathematical problem establishes a connection between the second part of the paper and the first, and Hilbert invokes his previously expressed conviction that every well-posed mathematical problem is solvable. If the definition of a function involves the solvability of a mathematical problem, it will contain an unbounded existential quantifier, or, in Hilbert's system, the ε-function. Convinced of the possibility of justifying the transfinite part of mathematics by metamathematical means, Hilbert states as Lemma I that such definitions of functions can be replaced by recursive definitions, in which, except for a string of prenex universal quantifiers, no quantifier occurs. (In *1927* Hilbert claims that his proof does not require Lemma I; see below, p. 476.)

Once all the methods of defining number-theoretic functions have been reduced to (substitution and) recursion, Hilbert considers two basic schemas of recursion, ordinary and transfinite (in which the individual variables range, respectively, over the natural numbers and the ordinals of the second number class). He extends these schemas by allowing the use of functionals, that is, of functions from previously introduced functionals to natural numbers, the initial functionals being the number-theoretic functions. He presents Ackermann's function in order to show that the introduction of functionals leads to a genuine extension of the recursion schema.

Hilbert then proceeds to show that, since recursion equations defining functions contain "variable-types" of increasing height, these equations parallel the definitions of the ordinals of the second number class. The argument is far from being a proof. Among several unproved statements, it contains in particular, as Lemma II, the assertion that in the definitions of number-theoretic functions transfinite recursion can be replaced by ordinary recursion.

Hilbert's attempt aroused few comments. Paul Lévy recalls (*1964*, p. 89): "Zermelo told me in 1928 that even in Germany nobody understood what Hilbert meant". After 1927 Hilbert himself laid the problem aside. The scale of "variable-types" has not been further investigated. The continuum hypothesis was proved by Gödel (*1938, 1939, 1940*) to be consistent with the customary axioms of set theory and by Cohen (*1963, 1963a, 1964*) to be independent of these axioms.

In a review of *Ackerman 1928* Skolem (*1928a*; see also *1929a*, pp. 18–19) stated again his previously expressed conviction (see above, footnote 9, p. 299) that, in view of the Löwenheim–Skolem theorem, some problems that concern cardinal numbers might be undecidable in any formalized set theory and that the continuum problem might be one of them. A number of remarks on Hilbert's attempted proof were made by Luzin (*1928, 1933, 1935*). In his review of Gödel's second paper on the consistency of the continuum hypothesis (*1939*), Bernays wrote (*1940*, p. 118): "The whole Gödel reasoning may also be considered as a way of modifying the Hilbert project for a proof of the Cantor continuum hypothesis, as described in 108*13* ⟦*Hilbert 1925*⟧, so as to make it practicable and at the same time generalizable to higher powers". About the relation of Gödel's work on the continuum problem to Hilbert's 1925 paper, Professor Gödel wrote, in a letter to the editor dated 8 July

1965: "There is a remote analogy between Hilbert's Lemma II and my Theorem 12.2 (*1940*, p. 54) for $\alpha = 0$. There is, however, this great difference that Hilbert considers only strictly constructive definitions and, moreover, transfinite iterations of the defining operations only up to constructive ordinals, while I admit, not only quantifiers in the definitions, but also iterations of the defining operations up to *any* ordinal number, no matter whether or how it can be defined. The term 'constructible set', in my proof, is justified only in a very weak sense and, in particular, only in the sense of 'relative to ordinal numbers', where the latter are subject to no conditions of constructivity. It was exactly by viewing the situation from this highly transfinite, set-theoretic point of view that in my approach the difficulties were overcome and a *relative* finitary consistency proof was obtained. Of course there is no need in this approach for anything like Hilbert's Lemma I. Hilbert probably hoped to prove it as a special case of a general theorem to the effect that transfinite modes of inference applied to a constructively correct system of axioms lead to no inconsistency."

Hilbert's paper gave an impulse to the study of the hierarchy of number-theoretic functions and to that of the various schemas for the recursive definitions of functions. In particular, Hilbert's work provides an approach to the problem of associating ordinals with number-theoretic functions defined by recursions (see *Kleene 1958* and *Péter 1951a, 1953*).

The translation is by Stefan Bauer-Mengelberg, and it is reproduced here with the kind permission of Springer Verlag.

Weierstrass, through a critique elaborated with the sagacity of a master, created a firm foundation for mathematical analysis. By clarifying, among other notions, those of minimum, function, and derivative, he removed the remaining flaws from the calculus, cleansed it of all vague ideas concerning the infinitesimal, and conclusively overcame the difficulties that until then had their roots in the notion of infinitesimal. If today there is complete agreement and certitude in analysis whenever modes of inference are employed that rest upon the notions of irrational number and of limit in general, and if there is unanimity on all results concerning the most complicated questions in the theory of differential and integral equations, despite the boldest use of the most diverse combinations of superposition, juxtaposition, and nesting of limits, this is essentially due to the scientific activity of Weierstrass.

Nevertheless, the discussions about the foundations of analysis did not come to an end when Weierstrass provided a foundation for the infinitesimal calculus.

The reason for this is that the significance of the *infinite* for mathematics had not yet been completely clarified. To be sure, the infinitely small and the infinitely large were eliminated from analysis, as established by Weierstrass, through a reduction of the propositions about them to ⟦propositions about⟧ relations between finite magnitudes. But the infinite still appears in the infinite number sequences that define the real numbers, and, further, in the notion of the real number system, which we conceive to be an actually given totality, complete and closed.

The forms of logical inference in which this conception finds its expression—namely, those that we employ when, for example, we deal with *all* real numbers having a certain property or assert that *there exist* real numbers having a certain property—are called upon quite without restriction and are used again and again by Weierstrass precisely when he is establishing the foundations of analysis.

Thus the infinite, in a disguised form, was able to worm its way back into Weierstrass' theory and escape the sharp edge of his critique; therefore it is the *problem of the infinite* in the sense just indicated that still needs to be conclusively resolved. And just as the infinite, in the sense of the infinitely small and the infinitely large, could, in the case of the limiting processes of the infinitesimal calculus, be shown to be a mere way of speaking, so we must recognize that the infinite in the sense of the infinite totality (wherever we still come upon it in the modes of inference) is something merely apparent. And just as operations with the infinitely small were replaced by processes in the finite that have quite the same results and lead to quite the same elegant formal relations, so the modes of inference employing the infinite must be replaced generally by finite processes that have precisely the same results, that is, that permit us to carry out proofs along the same lines and to use the same methods of obtaining formulas and theorems.

That, then, is the purpose of my theory. Its aim is to endow mathematical method with the definitive reliability that the critical era of the infinitesimal calculus did not achieve; thus it shall bring to completion what Weierstrass, in providing a foundation for analysis, endeavored to do and toward which he took the first necessary and essential step.

But in clarifying the notion of the infinite we must still take into consideration a more general aspect of the question. If we pay close attention, we find that the literature of mathematics is replete with absurdities and inanities, which can usually be blamed on the infinite. So, for example, some stress the stipulation, as a kind of restrictive condition, that, if mathematics is to be rigorous, only a *finite* number of inferences is admissible in a proof—as if anyone had ever succeeded in carrying out an infinite number of them!

Even old objections that have long been regarded as settled reappear in a new guise. So in recent times we come upon statements like this: even if we could introduce a notion safely (that is, without generating contradictions) and if this were demonstrated, we would still not have established that we are justified in introducing the notion. Is this not precisely the same objection as the one formerly made against complex numbers, when it was said that one could not, to be sure, obtain a contradiction by means of them, but their introduction was nevertheless not justified, for, after all, imaginary magnitudes do not exist? No, if justifying a procedure means anything more than proving its consistency, it can only mean determining whether the procedure is successful in fulfilling its purpose. Indeed, success is necessary; here, too, it is the highest tribunal, to which everyone submits.

Another author seems to see contradictions, like ghosts, even when nothing has been asserted by anyone at all, namely, in the concrete world of perception ⟦Sinnenwelt⟧ itself, whose "consistent functioning" is regarded as a special assumption. I, for one, have always believed that only assertions and, insofar as they lead to assertions by means of inferences, assumptions could contradict each other, and the view that facts and events themselves could come to do so seems to me the perfect example of an inanity.

By these remarks I wanted to show only that the definitive clarification of the *nature of the infinite* has become necessary, not merely for the special interests of the individual sciences, but rather for the *honor of the human understanding* itself.

The infinite has always stirred the *emotions* of mankind more deeply than any other question; the infinite has stimulated and fertilized reason as few other *ideas* have; but also the infinite, more than any other *notion*, is in need of *clarification*.

If we now turn to this task, to the clarification of the nature of the infinite, we must ever so briefly call to mind the contentual significance that attaches to the infinite in reality; first we see what we can learn about this from physics.

The initial, naive impression that we have of natural events and of matter is one of uniformity, of continuity. If we have a piece of metal or a volume of liquid, the idea impresses itself upon us that it is divisible without limit, that any part of it, however small, would again have the same properties. But, wherever the methods of research in the physics of matter were refined sufficiently, limits to divisibility were reached that are not due to the inadequacy of our experiments but to the nature of the subject matter, so that we could in fact view the trend of modern science as an emancipation from the infinitely small and, instead of the old maxim "natura non facit saltus", now assert the opposite, "nature makes leaps".

As is well known, all matter is composed of small building blocks, *atoms*, which, when combined and connected, yield the entire multiplicity of macroscopic substances.

But physics did not stop at the atomic theory of matter. Toward the end of the last century the atomic theory of electricity, which at first seemed much stranger, took its place beside that theory. Whereas until that time electricity had been considered a fluid and had been the very model of an agent with a continuous effect, it too now proved to be made up of particles, namely, positive and negative *electrons*.

Besides matter and electricity there is in physics still something else that is real, for which the law of conservation also holds, namely, energy. Now not even energy, as we know today, permits of infinite division in an absolute and unrestricted way; Planck discovered that energy comes in *quanta*.

And the net result is, certainly, that we do not find anywhere in reality a homogeneous continuum that permits of continued division and hence would realize the infinite in the small. The infinite divisibility of a continuum is an operation that is present only in our thoughts; it is merely an idea, which is refuted by our observations of nature and by the experience gained in physics and chemistry.

We find the second place at which the question of infinity confronts us in nature when we consider the universe as a whole. Here we must investigate the vast expanse of the universe to see whether there is something infinitely large in it.

For a long time the opinion that the world is infinite was dominant; until the time of Kant and even afterward no one had entertained any doubt whatsoever about the infinitude of space.

Here again it is modern science, especially astronomy, that raises this question anew and seeks to decide it, not by the inadequate means of metaphysical speculation, but through reasons that are supported by experience and rest upon the application of the laws of nature. And weighty objections against infinity have appeared. *Euclidean* geometry necessarily leads to the assumption that space is infinite. Now, to be sure, Euclidean geometry, as a structure and a system of notions, is consistent in itself, but this does not imply that it applies to reality. Whether that is the case, only observation and experience can decide. In the attempt to prove the infinitude of space in a speculative way, moreover, obvious errors were committed. From the fact that outside of a region of space there always is still more space it follows only that space is unbounded but by no means that it is infinite. Unboundedness and finitude, however, do not exclude each other. In the geometry usually referred to as *elliptic*, mathematical research furnishes the natural model of a finite world. And the abandonment of Euclidean geometry is today no longer merely a purely mathematical or philosophical speculation; rather, we have come to abandon it also on account of other considerations, which originally had nothing at all to do with the question of the finitude of the world. Einstein showed that it was necessary to relinquish Euclidean geometry. On the basis of his theory of gravitation he attacks the cosmological questions and shows that a finite world is possible, and all the results discovered by astronomers are compatible also with the assumption of an elliptic world.

We have now ascertained in two directions, toward the infinitely small and toward the infinitely large, that reality is finite. Yet it could very well be the case that the infinite has a well-justified place *in our thinking* and plays the role of an indispensable notion. We shall examine what the situation in the science of mathematics is in this respect, and we shall first consult the purest and most naive child of the human intellect, the theory of numbers. Let us here select any formula from the rich multitude of elementary formulas, for example,

$$1^2 + 2^2 + 3^2 + \cdots + n^2 = \tfrac{1}{6}n(n + 1)(2n + 1).$$

Since in it we may substitute any integer for n, for example, $n = 2$ or $n = 5$, this formula contains *infinitely many* propositions, and this is obviously what is essential about it; that is why it constitutes the solution of an arithmetic problem and why its proof requires a genuine act of thought, whereas each of the specific numerical equations

$$1^2 + 2^2 = \tfrac{1}{6}\cdot 2\cdot 3\cdot 5$$

and

$$1^2 + 2^2 + 3^2 + 4^2 + 5^2 = \tfrac{1}{6}\cdot 5\cdot 6\cdot 11$$

can be verified by computation and hence is of no essential interest when considered by itself.

We come upon quite another, wholly different interpretation, or fundamental characterization, of the notion of infinity when we consider the method—so extremely important and fertile—of *ideal elements*. The method of ideal elements has an application already in the elementary geometry of the plane. There the points and straight lines of the plane are initially the only real, actually existing objects. The axiom of connection, among others, holds for them: for any two points there is always one and only one straight line that goes through them. From this it follows that two straight lines intersect each other in one point at most. The proposition that any two straight lines intersect each other in some point, however, does not hold; rather, the two lines can be parallel. But, as is well known, the introduction of ideal elements, namely, points at infinity and a line at infinity, renders the proposition according to which two straight lines always intersect each other in one and only one point universally valid.

The ideal elements "at infinity" have the advantage of making the system of the laws of connection as simple and perspicuous as is at all possible. As is well known, the symmetry between point and straight line then yields the duality principle of geometry, which is so fecund.

The ordinary *complex* magnitudes of algebra likewise are an instance of the use of ideal elements; they serve to simplify the theorems on the existence and number of the roots of an equation.

Just as in geometry infinitely many straight lines, namely, those that are parallel to one another, are used to define an ideal point, so in higher arithmetic certain systems of infinitely many numbers are combined into a *number ideal*, and indeed probably no use of the principle of ideal elements is a greater stroke of genius than this. When this procedure has been carried out generally within an algebraic field, we find in it again the simple and well-known laws of divisibility, just as they hold for the ordinary integers 1, 2, 3, 4, ... Here we have already entered the domain of higher arithmetic.

Now we come to analysis, the structure that in mathematical science is the most elaborate and has branched out more delicately than any other. You know what a dominant role the infinite plays there, how in a sense mathematical analysis is but a single symphony of the infinite.

The mighty advances made in the infinitesimal calculus rest for the most part upon operations with mathematical systems of infinitely many elements. Since, now, it was extremely tempting to identify the infinite with the "very large", there soon arose inconsistencies, the paradoxes of the infinitesimal calculus, as they are called, which in part were already known to the Sophists in antiquity. A fundamental advance was made when it was recognized that many propositions valid for the finite—for example, that the part is smaller than the whole, that a minimum or a maximum exists, that the order of terms or factors can be changed—may not be directly carried over to the infinite. At the beginning of my lecture I mentioned the fact that, chiefly through the sagacity of Weierstrass, these questions have been completely clarified, and today analysis has within its domain become an infallible guide and at the same time a practical instrument for the use of the infinite.

But analysis alone does not yet give us the deepest insight into the nature of the infinite. Rather, this is conveyed to us only by a discipline that is closer to the general philosophical way of thinking and was destined to place the entire complex of questions concerning the infinite in a new light. This discipline is set theory, whose creator was Georg Cantor. Here, however, we are concerned only with what was truly unique and original in Cantor's theory and constituted its real core, namely, his theory of *transfinite numbers*. This appears to me to be the most admirable flower of the mathematical intellect and in general one of the highest achievements of purely rational human activity. Now what is it all about?

If we wanted to characterize briefly the new conception of the infinite that Cantor introduced, we could no doubt say: in analysis we deal with the infinitely small or the infinitely large only as a limit notion—as something that is becoming, coming to be, being produced—that is, as we say, with the *potential infinite*. But this is not the real infinite itself. That we have when, for example, we consider the totality of the numbers 1, 2, 3, 4,... itself as a completed entity, or when we regard the points of a line segment as a totality of objects that is actually given and complete. This kind of infinite is called the *actual infinite*.

Frege and Dedekind, two mathematicians who did highly meritorious work in the foundations of mathematics, already used—independently of each other—the actual infinite. Their specific aim was to make pure logic provide for arithmetic a foundation that would be independent of all intuition and experience as well as to derive arithmetic by means of logic alone. Dedekind even went so far as to refuse to draw upon intuition for the notion of finite number; instead, he strove to derive it by purely logical means, making essential use of the notion of infinite sets. It was Cantor, however, who systematically developed the notion of the actual infinite. If we look at the two examples of the infinite that we have mentioned, (1) 1, 2, 3, 4,... and (2) the points of the line segment from 0 to 1, or, what is the same, the totality of real numbers between 0 and 1, then the idea that suggests itself most readily is to consider them purely from the point of view of cardinality, and when we do this we observe surprising facts that are familiar to every mathematician today. For, if we consider the set of all rational numbers, hence of all fractions $\frac{1}{2}$, $\frac{1}{3}$, $\frac{2}{3}$, $\frac{1}{4}$, ..., $\frac{3}{7}$, ..., it turns out that, purely from the point of view of cardinality, this set is not larger than the set of integers; we say that the rational numbers can be denumerated in the ordinary way, or that they are denumerable. And the same still holds of the set of all numbers that can be obtained [[from the rational numbers]] by root extraction, and indeed of the set of all algebraic numbers. In our second example we have a similar situation: unexpectedly, the set of all points in a square or a cube is, purely from the point of view of cardinality, not larger than the set of points on the line segment from 0 to 1; indeed, the same still holds even of the set of all continuous functions. Someone who first learns of this might come to think that purely from the point of view of cardinality there is but a single infinite. No, the sets in our two examples, (1) and (2), are not "equivalent", as we say. Rather, the set in (2) cannot be denumerated; it is larger than the set in (1). Here Cantor's ideas take their distinctive turn. The points of the line segment cannot be denumerated in the ordinary way by means of 1, 2, 3,...! But, if we admit the actual infinite, we are not at all limited to this ordinary kind of denumeration or in any way compelled to leave off there. Rather, when we have counted 1, 2, 3,..., we can view the objects thus denumerated as an infinite set that has been completed in this definite order; if we denote the type of this ordering, as Cantor does, by ω, then the denumeration continues in a natural way with $\omega + 1$, $\omega + 2$, ... to $\omega + \omega$, or $\omega \cdot 2$, and then again $\omega \cdot 2 + 1$, $\omega \cdot 2 + 2$, $\omega \cdot 2 + 3$, ..., $\omega \cdot 2 + \omega$ $(=\omega \cdot 3)$, and further $\omega \cdot 2$, $\omega \cdot 3$, $\omega \cdot 4$, ..., $\omega \cdot \omega$ $(=\omega^2)$, $\omega^2 + 1$, ..., so that we finally obtain the following table:

$$
\begin{array}{l}
1, 2, 3, \ldots, \\
\omega, \omega + 1, \omega + 2, \ldots, \\
\omega \cdot 2, \omega \cdot 2 + 1, \omega \cdot 2 + 2, \ldots, \\
\omega \cdot 3, \omega \cdot 3 + 1, \omega \cdot 3 + 2, \ldots, \\
\omega^2, \omega^2 + 1, \ldots, \\
\omega^2 + \omega, \omega^2 + \omega \cdot 2, \omega^2 + \omega \cdot 3, \ldots, \\
\omega^2 \cdot 2, \ldots, \\
\omega^2 \cdot 2 + \omega, \ldots, \\
\omega^3, \ldots, \\
\omega^4, \ldots, \\
\omega^\omega, \ldots.
\end{array}
$$

These are Cantor's first transfinite numbers, the numbers of the second number class, as he calls them.[1] Thus we come to them simply by a transnumeration ⟦Hinüberzählen⟧ beyond the ordinary denumerable infinite, that is, by an entirely natural and unambiguously determined, systematic continuation of ordinary counting as it takes place in the finite. Just as till now we merely counted the 1st, 2nd, 3rd, ... object of a set, so we now also count the ωth, $(\omega + 1)$th, ..., ω^ωth object.

Given this state of affairs, the question obviously immediately arises whether by means of this transfinite counting one could now actually enumerate the elements of sets that are not denumerable in the ordinary sense.

Now Cantor, in following these thoughts, developed the theory of transfinite numbers in a most successful way and created a complete calculus for them. So, finally, through the gigantic collaboration of Frege, Dedekind, and Cantor the infinite was enthroned and enjoyed the period of its greatest triumph. In the boldest flight the infinite had reached a dizzy pinnacle of success.

The reaction did not fail to set in; it took very dramatic forms. Events took quite the same turn as in the development of the infinitesimal calculus. In their joy over the new and rich results, mathematicians apparently had not examined critically enough whether the modes of inference employed were admissible; for, purely through the ways in which notions were formed and modes of inference used—ways that in time had become customary—contradictions appeared, sporadically at first, then ever more severely and ominously. They were the paradoxes of set theory, as they are called. In particular, a contradiction discovered by Zermelo and Russell had, when it became known, a downright catastrophic effect in the world of mathematics. Confronted with these paradoxes, Dedekind and Frege actually abandoned their standpoint and quit the field; for a long time Dedekind had reservations about permitting a new edition of his epoch-making booklet (*1888*), and Frege, too, was forced to recognize that the tendency of his book (*1893*, *1903*) was mistaken, as he confesses in an appendix. From the most diverse quarters extremely vehement attacks were directed against Cantor's theory itself. The reaction was so violent that the commonest and most fruitful notions and the very simplest and most important modes of inference in mathematics were threatened and their use was to be prohibited. There were, to be sure, defenders of the old; but the defensive measures were rather feeble, and moreover they were not put into effect at the right place in a unified front. Too many remedies were recommended for the paradoxes; the methods of clarification were too checkered.

Let us admit that the situation in which we presently find ourselves with respect to the paradoxes is in the long run intolerable. Just think: in mathematics, this paragon of reliability and truth, the very notions and inferences, as everyone learns, teaches, and uses them, lead to absurdities. And where else would reliability and truth be found if even mathematical thinking fails?

But there is a completely satisfactory way of escaping the paradoxes without committing treason against our science. The considerations that lead us to discover this way and the goals toward which we want to advance are these:

(1) We shall carefully investigate those ways of forming notions and those modes

[1] ⟦Unlike Cantor, Hilbert takes the number classes to be cumulative; see his definition of the second number class below, p. 386.⟧

of inference that are fruitful; we shall nurse them, support them, and make them usable, wherever there is the slightest promise of success. No one shall be able to drive us from the paradise that Cantor created for us.

(2) It is necessary to make inferences everywhere as reliable as they are in ordinary elementary number theory, which no one questions and in which contradictions and paradoxes arise only through our carelessness.

Obviously we shall be able to reach these goals only if we succeed in completely clarifying *the nature of the infinite*.

We saw earlier that the infinite is not to be found anywhere in reality, no matter what experiences and observations or what kind of science we may adduce. Could it be, then, that thinking about objects is so unlike the events involving objects and that it proceeds so differently, so apart from all reality? Is it not clear, rather, that when we believed we had discovered that the infinite was in some sense real we were only allowing ourselves to be led to that belief by the circumstance that we so often actually encounter in reality such immeasurable dimensions in the large and in the small? And has the contentual logical inference ever deceived and abandoned us anywhere when we applied it to real objects or events? No, contentual logical inference is indispensable. It has deceived us only when we accepted arbitrary abstract notions, in particular those under which infinitely many objects are subsumed. What we did, then, was merely to use contentual inference in an illegitimate way; that is, we obviously did not respect necessary conditions for the use of contentual logical inference. And in recognizing that such conditions exist and must be respected we find ourselves in agreement with the philosophers, especially with Kant. Kant already taught—and indeed it is part and parcel of his doctrine—that mathematics has at its disposal a content secured independently of all logic and hence can never be provided with a foundation by means of logic alone; that is why the efforts of Frege and Dedekind were bound to fail. Rather, as a condition for the use of logical inferences and the performance of logical operations, something must already be given to our faculty of representation [[in der Vorstellung]], certain extralogical concrete objects that are intuitively [[anschaulich]] present as immediate experience prior to all thought. If logical inference is to be reliable, it must be possible to survey these objects completely in all their parts, and the fact that they occur, that they differ from one another, and that they follow each other, or are concatenated, is immediately given intuitively, together with the objects, as something that neither can be reduced to anything else nor requires reduction. This is the basic philosophical position that I consider requisite for mathematics and, in general, for all scientific thinking, understanding, and communication. And in mathematics, in particular, what we consider is the concrete signs themselves, whose shape, according to the conception we have adopted, is immediately clear and recognizable.

Let us call to mind the nature and methods of ordinary finitary number theory. It can certainly be developed through the construction of numbers by means solely of intuitive contentual considerations. But the science of mathematics is by no means exhausted by numerical equations and it cannot be reduced to these alone. One can claim, however, that it is an apparatus that must always yield correct numerical equations when applied to integers. But then we are obliged to investigate the structure of the apparatus sufficiently to make this fact apparent. And the only tool at our disposal in this investigation is the same as that used for the derivation of numerical equations in the construction of number theory itself, namely, a concern for concrete content, the finitist frame of mind. This scientific requirement can in fact be satisfied; that is, it is possible to obtain in a purely intuitive and finitary way, just like the truths of number theory, those insights that guarantee the reliability of the mathematical apparatus. Let us now consider number theory in more detail.

In number theory we have the numerals

$$1, 11, 111, 11111,$$

each numeral being perceptually recognizable by the fact that in it 1 is always again followed by 1 ⟦if it is followed by anything⟧. These numerals, which are the object of our consideration, have no meaning at all in themselves. In elementary number theory, however, we already require, besides these signs, others that mean something and serve to convey information, for example, the sign[2] 2 as an abbreviation for the numeral 11, or the numeral 3 as an abbreviation for the numeral 111; further we use the signs $+$, $=$, $>$, and others, which serve to communicate assertions. So $2 + 3 = 3 + 2$ serves to communicate the fact that $2 + 3$ and $3 + 2$, when the abbreviations used are taken into account, are the same numeral, namely, the numeral 11111. Likewise, then, $3 > 2$ serves to communicate the fact that the sign 3 (that is, 111) extends beyond the sign 2 (that is, 11), or that the latter sign is a proper segment of the former.

When communicating, we also use letters, such as $\mathfrak{a}$, $\mathfrak{b}$, $\mathfrak{c}$, for numerals. Accordingly, $\mathfrak{b} > \mathfrak{a}$ is the communication that the numeral $\mathfrak{b}$ extends beyond the numeral $\mathfrak{a}$. And likewise, from the present point of view, we would regard $\mathfrak{a} + \mathfrak{b} = \mathfrak{b} + \mathfrak{a}$ merely as the communication of the fact that the numeral $\mathfrak{a} + \mathfrak{b}$ is the same as $\mathfrak{b} + \mathfrak{a}$. Here, too, the contentual correctness of this communication can be proved by contentual inference, and we can go very far with this intuitive, contentual kind of treatment.

I should now like to show you a first example in which we go beyond the bounds of this contentual way of thinking. The largest prime number known up to now (39 digits) is

$$\mathfrak{p} = 170\ 141\ 183\ 460\ 469\ 231\ 731\ 687\ 303\ 715\ 884\ 105\ 727.$$

By means of Euclid's well-known procedure we can, completely within the framework of the attitude we have adopted, prove the theorem that between $\mathfrak{p} + 1$ and $\mathfrak{p}! + 1$ there certainly exists a new prime number. This proposition itself, moreover, is completely in conformity with our finitist attitude. For "there exists" here serves merely to abbreviate the proposition:

Certainly $\mathfrak{p} + 1$ or $\mathfrak{p} + 2$ or $\mathfrak{p} + 3$ or ... or $\mathfrak{p}! + 1$ is a prime number.

But let us go on. Obviously, to say

There exists a prime number that (1) is $> \mathfrak{p}$ and (2) is at the same time $\leqq \mathfrak{p}! + 1$

[2] ⟦In the present paragraph and the next, "Zeichen" is uniformly translated as "sign" and "Zahlzeichen" as "numeral", although Hilbert on occasion uses the former as a short form for the latter; the *Zahlzeichen* are, of course, included among the *Zeichen*. Furthermore, when Hilbert uses "Abkürzung", here rendered as "abbreviation", he wavers between two meanings: pure abbreviation of a symbol and name of that symbol.⟧

would amount to the same thing, and this leads us to formulate a proposition that expresses only a part of Euclid's assertion, namely: there exists a prime number that is $> \mathfrak{p}$. So far as content is concerned, this is a much weaker assertion, stating only a part of Euclid's proposition; nevertheless, no matter how harmless the transition appears to be, there is a leap into the transfinite when this partial proposition, taken out of the context above, is stated as an independent assertion.

How can that be? We have here an existential proposition with "there exists". To be sure, we already had one in Euclid's theorem. But the latter, with its "there exists", was, as I have already said, merely another, shorter expression for

"$\mathfrak{p} + 1$ or $\mathfrak{p} + 2$ or $\mathfrak{p} + 3$ or ... or $\mathfrak{p}! + 1$ is a prime number",

just as, instead of saying: This piece of chalk is red or that piece of chalk is red or... or the piece of chalk over there is red, I say more briefly: Among these pieces of chalk there exists a red one. An assertion of this kind, that in a finite totality "there exists" an object having a certain property, is completely in conformity with our finitist attitude. On the other hand, the expression

"$\mathfrak{p} + 1$ or $\mathfrak{p} + 2$ or $\mathfrak{p} + 3$ or ... ad infinitum is a prime number"

is, as it were, an infinite logical product,[3] and such a passage to the infinite is no more permitted without special investigation and perhaps certain precautionary measures than the passage from a finite to an infinite product in analysis, and initially it has no meaning at all.

In general, from the finitist point of view an existential proposition of the form "There exists a number having this or that property" has meaning only as a *partial proposition*, that is, as part of a proposition that is more precisely determined but whose exact content is unessential for many applications.

Thus we here encounter the transfinite when from an existential proposition we extract a partial proposition that cannot be regarded as a disjunction.[4] In like manner we come upon a transfinite proposition when we negate a universal assertion, that is, one that extends to arbitrary numerals. So, for example, the proposition that, if $\mathfrak{a}$ is a numeral, we must always have

$$\mathfrak{a} + 1 = 1 + \mathfrak{a}$$

is from the finitist point of view *incapable of being negated*. This will become clear for us if we reflect upon the fact that [from this point of view] the proposition cannot be interpreted as a combination, formed by means of "and", of infinitely many numerical equations, but only as a hypothetical judgment that comes to assert something when a numeral is given.

From this it follows, in particular, that in the spirit of the finitist attitude we

[3] ⟦It is rather a logical sum, or disjunction. In the version published in *Grundlagen der Geometrie* (*1930a*) "logisches Produkt" is replaced by "Oder-Verknüpfung".⟧

[4] ⟦The German text says: "Wir stoßen also hier auf das Transfinite durch Zerlegung einer existentialen Aussage, die sich nicht als eine Oder-Verknüpfung deuten läßt". In the version published in *Grundlagen der Geometrie* (*1930a*) this sentence has been amended thus: "Wir stoßen also hier auf das Transfinite durch Zerlegung einer existentialen Aussage in Teile, deren keiner sich als eine Oder-Verknüpfung deuten läßt". The English translation that we propose seems to be closer to what Hilbert's train of thought requires than either of these two German sentences.⟧

cannot make use of the alternative according to which an equation like the one above, in which an unspecified numeral occurs, is either satisfied by every numeral or refuted by a counterexample. For, being an application of the principle of excluded middle, this alternative essentially rests upon the assumption that the assertion of the validity of that equation is capable of being negated.

At all events we observe the following. In the domain of finitary propositions, in which we should, after all, remain, the logical relations that prevail are very imperspicuous, and this lack of perspicuity mounts unbearably if "all" and "there exists" occur combined or appear in nested propositions. In any case, those logical laws that man has always used since he began to think, the very ones that Aristotle taught, do not hold. Now one could attempt to determine the logical laws that are valid for the domain of finitary propositions; but this would not help us, since we just do not want to renounce the use of the simple laws of Aristotelian logic, and no one, though he speak with the tongues of angels, will keep people from negating arbitrary assertions, forming partial judgments, or using the principle of excluded middle. What, then, shall we do?

Let us remember that *we are mathematicians* and as such have already often found ourselves in a similar predicament, and let us recall how the method of ideal elements, that creation of genius, then allowed us to find an escape. I presented some shining examples of the use of this method at the beginning of my lecture. Just as $i = \sqrt{-1}$ was introduced so that the laws of algebra, those, for example, concerning the existence and number of the roots of an equation, could be preserved in their simplest form, just as ideal factors were introduced so that the simple laws of divisibility could be maintained even for algebraic integers (for example, we introduce an ideal common divisor for the numbers 2 and $1 + \sqrt{-5}$, while an actual one does not exist), so we must here *adjoin the ideal propositions to the finitary ones* in order to maintain the formally simple rules of ordinary Aristotelian logic. And it is strange that the modes of inference that Kronecker attacked so passionately are the exact counterpart of what, when it came to number theory, the same Kronecker admired so enthusiastically in Kummer's work and praised as the highest mathematical achievement.

How, then, do we come to the *ideal propositions*? It is a remarkable circumstance, and certainly a propitious and favorable one, that to enter the path that leads to them we need only continue in a natural and consistent way the development that the theory of the foundations of mathematics has already taken. Indeed, let us acknowledge that elementary mathematics already goes beyond the point of view of intuitive number theory. For the method of algebraic calculation with letters is not within the resources of contentual, intuitive number theory as we have hitherto conceived of it. This theory always uses formulas for communication only; letters stand for numerals, and the fact that two signs are identical is communicated by an equation. In algebra, on the other hand, we consider the expressions formed with letters to be independent objects in themselves, and the contentual propositions of number theory are formalized by means of them. Where we had propositions concerning numerals, we now have formulas, which themselves are concrete objects that in their turn are considered by our perceptual intuition, and the derivation of one formula from another in accordance with certain rules takes the place of the number-theoretic proof based on content.

Hence, as soon as we consider algebra, there is an increase in the number of finitary objects. Up to now these were only the numerals, such as 1, 11, ..., 11111. They alone had been the objects of our contentual consideration. But in algebra mathematical practice already goes beyond that. Yes, even when a proposition, so long as it is combined with some indication as to its contentual interpretation, is still admissible from our finitist point of view, as, for example, the proposition that always

$$\mathfrak{a} + \mathfrak{b} = \mathfrak{b} + \mathfrak{a},$$

where $\mathfrak{a}$ and $\mathfrak{b}$ stand for specific numerals, we yet do not select this form of communication but rather take the formula

$$a + b = b + a.$$

This is no longer an immediate communication of something contentual at all, but a certain formal object, which is related to the original finitary propositions

$$2 + 3 = 3 + 2$$

and

$$5 + 7 = 7 + 5$$

by the fact that, if we substitute numerals, 2, 3, 5, and 7, for a and b in that formula (that is, if we employ a proof procedure, albeit a very simple one), we obtain these finitary particular propositions. Thus we arrive at the conception that a, b, $=$, and $+$, as well as the entire formula

$$a + b = b + \mathrm{a},$$

do not mean anything in themselves, any more than numerals do. But from that formula we can indeed derive others; to these we ascribe a meaning, by treating them as communications of finitary propositions. If we generalize this conception, mathematics becomes an inventory of formulas—first, formulas to which contentual communications of finitary propositions [hence, in the main, numerical equations and inequalities] correspond and, second, further formulas that mean nothing in themselves and are the *ideal objects of our theory*.

Now, what was our goal? In mathematics, we found, first, finitary propositions that contain only numerals, like

$$3 > 2, \quad 2 + 3 = 3 + 2, \quad 2 = 3, \quad \text{and} \quad 1 \neq 1,$$

which according to our finitist conception are immediately intuitive and directly intelligible. These are capable of being negated, and the result will be true or false; one can manipulate them at will, without any qualms, in all the ways that Aristotelian logic allows. The law of contradiction holds; that is, it is impossible for any one of these propositions and its negation to be simultaneously true. The principle of "excluded middle" holds; that is, of the two, a proposition and its negation, one is true. To say that a proposition is false is equivalent to saying that its negation is true. Besides these elementary propositions, which are of an entirely unproblematic character, we encountered finitary propositions of problematic character, for example, those that were not decomposable ⟦into partial propositions⟧. Now, finally, we have introduced the ideal propositions to ensure that the customary laws of logic again hold one and all. But since the ideal propositions, namely, the formulas, insofar as they do not express finitary assertions, do not mean anything in themselves, the logical operations cannot be applied to them in a contentual way, as they are to the finitary propositions. Hence it is necessary to formalize the logical operations and also the mathematical proofs themselves; this requires a transcription of the logical relations into formulas, so that to the mathematical signs we must still adjoin some logical signs, say

$\&$,	$\vee$,	$\rightarrow$,	$\overline{}$,
and	or	implies	not

and use, besides the mathematical variables, $a, b, c, \ldots$, also logical variables, namely, variable propositions $A, B, C, \ldots$.

How can this be done? It is our good fortune to find the same preestablished harmony here that we observe so often in the history of science, a harmony that benefited Einstein when he found the general calculus of invariants fully developed for his theory of gravitation; we discover that considerable spadework has already been done: the *logical calculus* has been developed. To be sure, it was originally created in an entirely different context, and, accordingly, its signs were initially introduced for purposes of communication only; but we will be consistent in our course if we now divest the logical signs, too, of all meaning, just as we did the mathematical ones, and declare that the formulas of the logical calculus do not mean anything in themselves either, but are ideal propositions. In the logical calculus we possess a sign language that is capable of representing mathematical propositions in formulas and of expressing logical inference through formal processes. In a way that exactly corresponds to the transition from contentual number theory to formal algebra we regard the signs and operation symbols of the logical calculus as detached from their contentual meaning. In this way we now finally obtain, in place of the contentual mathematical science that is communicated by means of ordinary language, an inventory of formulas that are formed from mathematical and logical signs and follow each other according to definite rules. Certain of these formulas correspond to the mathematical axioms, and to contentual inference there correspond the rules according to which the formulas follow each other; hence contentual inference is replaced by manipulation of signs [[äußeres Handeln]] according to rules, and in this way the full transition from a naive to a formal treatment is now accomplished, on the one hand, for the axioms themselves, which originally were naively taken to be fundamental truths but in modern axiomatics had already for a long time been regarded as merely establishing certain interrelations between notions, and, on the other, for the logical calculus, which originally was to be only another language.

Let me still explain briefly just how a *mathematical proof* is formalized. As I said, certain formulas, which serve as building blocks for the formal edifice of mathematics, are called axioms. A mathematical proof is an array that must be given as such to our perceptual intuition; it consists of inferences according to the schema

$$\begin{array}{c} \mathfrak{S} \\ \mathfrak{S} \rightarrow \mathfrak{T} \\ \hline \mathfrak{T}. \end{array}$$

where each of the premisses, that is, the formulas $\mathfrak{S}$ and $\mathfrak{S} \rightarrow \mathfrak{T}$ in the array, either is an axiom or results from an axiom by substitution, or else coincides with the end formula of a previous inference or results from it by substitution. A formula is said to be provable if it is the end formula of a proof.

Through our program the choice of axioms for our proof theory is already indicated. Although the choice of axioms is to a certain extent arbitrary, they nevertheless fall into a number of qualitatively distinct groups, just as in geometry; we cite a few examples from each.[5]

I. Axioms of implication:

$$A \rightarrow (B \rightarrow A) \text{ (introduction of an assumption)},$$

$$(B \rightarrow C) \rightarrow \{(A \rightarrow B) \rightarrow (A \rightarrow C)\} \text{ (elimination of a proposition)}.$$

II. Axioms of negation:

$$\{A \rightarrow (B \,\&\, \bar{B})\} \rightarrow \bar{A} \text{ (principle of contradiction)},$$

$$\bar{\bar{A}} \rightarrow A \text{ (principle of double negation)}.$$

[From the principle of contradiction the formula

$$(A \,\&\, \bar{A}) \rightarrow B$$

follows, and from the principle of double negation the principle of excluded middle,

$$\{(A \rightarrow B) \,\&\, (\bar{A} \rightarrow B)\} \rightarrow B,$$

follows.] These axioms of groups I and II are nothing but the axioms of the propositional calculus.

III. Transfinite axioms:

$(a)A(a) \rightarrow A(b)$ (inference from the universal to the particular, Aristotle's dictum),

$\overline{(a)}A(a) \rightarrow (Ea)\bar{A}(a)$ (if a predicate does not hold of all individuals, then there exists a counterexample),

$\overline{(Ea)}A(a) \rightarrow (a)\bar{A}(a)$ (if there is no individual for which a proposition holds, then the proposition is false for all a).

Here, moreover, we come upon a very remarkable circumstance, namely, that all of these transfinite axioms are derivable from a single axiom, one that also contains the core of one of the most attacked axioms in the literature of mathematics, namely, the axiom of choice:

$$A(a) \rightarrow A(\varepsilon(A)),$$

where ε is the transfinite logical choice function.

In addition there are the specifically mathematical axioms.

IV. Axioms of equality:

$$a = a,$$

$$a = b \rightarrow (A(a) \rightarrow A(b)),$$

and finally

[5] ⟦The full system of axioms is given below, pp. 465–469.⟧

V. Axioms of number:

$$a + 1 \neq 0$$

and

$$[\{A(0) \mathbin{\&} (x)(A(x) \to A(x'))\} \to A(a),]$$

the axiom of mathematical induction.

In this way we are able to develop our proof theory and to construct the system of provable formulas, that is, the science of mathematics.

But in our joy over the fact that we have, in general, been so successful and that, in particular, we found ready-made that indispensable tool, the logical calculus, we must nevertheless not forget the essential prerequisite of our procedure. For there is a condition, a single but absolutely necessary one, to which the use of the method of ideal elements is subject, and that is the *proof of consistency*; for, extension by the addition of ideals is legitimate only if no contradiction is thereby brought about in the old, narrower domain, that is, if the relations that result for the old objects whenever the ideal objects are eliminated are valid in the old domain.

In the present situation, however, this problem of consistency is perfectly amenable to treatment. As we can immediately recognize, it reduces to the question of seeing that "$1 \neq 1$" cannot be obtained as an end formula from our axioms by the rules in force, hence that "$1 \neq 1$" is not a provable formula. And this is a task that fundamentally lies within the province of intuition just as much as does in contentual number theory the task, say, of proving the irrationality of $\sqrt{2}$, that is, of proving that it is impossible to find two numerals $\mathfrak{a}$ and $\mathfrak{b}$ satisfying the relation $\mathfrak{a}^2 = 2\mathfrak{b}^2$, a problem in which it must be shown that it is impossible to exhibit two numerals having a certain property. Correspondingly, the point for us is to show that it is impossible to exhibit a proof of a certain kind. But a formalized proof, like a numeral, is a concrete and surveyable object. It can be communicated from beginning to end. That the end formula has the required structure, namely "$1 \neq 1$", is also a property of the proof that can be concretely ascertained. The demonstration ⟦that "$1 \neq 1$" is not a provable formula⟧ can in fact be given, and this provides us with a justification for the introduction of our ideal propositions.

At the same time we experience the pleasant surprise that this gives us the solution also of a problem that became urgent long ago, namely, that of proving *the consistency of the arithmetic axioms*.[6] For the problem of proving consistency arises wherever the axiomatic method is used. After all, in selecting, interpreting, and manipulating the axioms and rules we do not want to have to rely on good faith and pure confidence alone. In geometry and the physical theories the consistency proof is successfully carried out by means of a reduction to the consistency of the arithmetic axioms. This method obviously fails in the case of arithmetic itself. By making this important final step possible through the method of ideal elements, our proof theory forms the necessary keystone in the edifice of axiomatic theory. And what we have experienced twice, first with the paradoxes of the infinitesimal calculus and then with the paradoxes of set theory, cannot happen a third time and will never happen again.

But our proof theory as it is sketched here is not only able to secure the foundations

[6] ⟦In his various papers on the foundations of mathematics Hilbert means by "arithmetic axioms" at times axioms for the real number system, at times axioms for number theory.⟧

of the science of mathematics; I believe, rather, that it also opens up a path that, if we follow it, will enable us to deal for the first time with general problems of a fundamental character that fall within the domain of mathematics but formerly could not even be approached.

Mathematics in a certain sense develops into a tribunal of arbitration, a supreme court that will decide questions of principle—and on such a concrete basis that universal agreement must be attainable and all assertions can be verified.

Even the assertions of the recent doctrine called "intuitionism", modest though they may be, can in my opinion obtain their certificate of justification only from this tribunal.

As an example of the way in which fundamental questions can be treated I would like to choose the thesis that every mathematical problem can be solved. We are all convinced of that. After all, one of the things that attract us most when we apply ourselves to a mathematical problem is precisely that within us we always hear the call: here is the problem, search for the solution; you can find it by pure thought, for in mathematics there is no *ignorabimus*. Now, to be sure, my proof theory cannot specify a general method for solving every mathematical problem; that does not exist. But the demonstration that the assumption of the solvability of every mathematical problem is consistent falls entirely within the scope of our theory.

I would still like to play a last trump. The final test of every new theory is its success in answering preexistent questions that the theory was not specifically created to answer. By their fruits ye shall know them—that applies also to theories. As soon as Cantor had discovered his first transfinite numbers, the numbers of the second number class as they are called, the question arose, as I have already mentioned, whether by means of this transfinite counting one could actually enumerate the elements of sets known in other contexts but not denumerable in the ordinary sense. The line segment was the first and foremost set of this kind to come under consideration. This question, whether the points of the line segment, that is, the real numbers, can be enumerated by means of the numbers of the table constructed above, is the famous problem of the continuum, which was formulated but not solved by Cantor. Some mathematicians believed that they could dispose of this problem by denying its existence. The considerations that follow show how mistaken this attitude is. The problem of the continuum is distinguished by its originality and inner beauty; in addition it is characterized by two features that raise it above other famous problems: its solution requires new ways, since the old methods fail in its case, and, besides, this solution is in itself of the greatest interest on account of the result to be determined.

The solution of the continuum problem can be carried out by means of the theory I have developed, and indeed the first and most important step toward this solution is precisely the demonstration that every mathematical problem can be solved. The answer turns out to be affirmative: the points of a line segment can be enumerated by means of the numbers of the second number class, that is, by mere transnumeration beyond the denumerable infinite, to express it in a popular way. I should like to call this assertion itself the continuum theorem and offer a brief intuitive presentation here of the fundamental idea of its proof.

Instead of the set of real numbers we consider—which is evidently the same here —the set of number-theoretic functions, that is, of those functions of an integral argument whose values are also integers. If we want to order the set of these functions in the way required by the problem of the continuum, we must consider how an individual function is generated. Now a function of one argument can be defined in such a way that the values of the function for some or even all values of the argument are made to depend upon whether certain well-defined mathematical problems have a solution, for example, whether certain diophantine problems are solvable, whether prime numbers having certain properties exist, or whether a given number, say $2^{\sqrt{2}}$, is irrational. In order to avoid the difficulty inherent in this we make use of precisely the assertion mentioned above, namely, that every well-posed mathematical problem is solvable. This assertion is a general lemma belonging to *metamathematics*, as I would like to call the contentual theory of formalized proofs. To the part of the lemma that is of relevance here I now give the following precise formulation:

LEMMA I. If a proof of a proposition contradicting the continuum theorem is given in a formalized version with the aid of functions defined by means of the transfinite symbol ε (axiom group III), then in this proof these functions can always be replaced by functions defined, without the use of the symbol ε, by means merely of ordinary and transfinite recursion, so that the transfinite appears only in the guise of the universal quantifier.

Further, I need to make a few stipulations in order to carry out my theory.

For *variable* ⟦*atomic*⟧ *propositions* (indeterminate formulas) we always use capital ⟦italic⟧ Latin letters, but for *constant* ⟦*atomic*⟧ *propositions* (specific formulas) capital Greek letters; for example,

$\mathsf{Z}(a)$: "a is an ordinary integer";

$\mathsf{N}(a)$: "a is a number of the second number class".

For *mathematical variables* we always use lower-case ⟦italic⟧ Latin letters, but for *constant mathematical objects* (specific functions) lower-case Greek letters.[7]

Concerning the procedure of *substitution* the following general conventions hold.

For propositional variables only other indeterminate or constant propositions (formulas) may be substituted.

Any array may be substituted for a mathematical variable; however, when a mathematical variable occurs in a formula, the constant proposition that specifies of what kind it is, together with the implication sign, must always precede; for example,

$$\mathsf{Z}(a) \rightarrow (\ldots a \ldots)$$

and

$$\mathsf{N}(a) \rightarrow (\ldots a \ldots).$$

[7] ⟦Hilbert's terms require some comments. Capital italic Latin letters, like A or B, stand for *propositional variables* (*Aussagenvariablen*) or predicate variables; when provided with parentheses and individual variables, they represent what Hilbert calls *variable Aussagen* (*variable propositions* in the translation); these "propositions" are in fact propositional functions. Capital Greek letters stand for predicate constants and, when provided with parentheses and individual variables, represent what Hilbert calls *individuelle Aussagen* or *Individualaussagen* (*constant propositions* in the translation); they are in fact propositional functions, and an example is $\mathsf{Z}(a)$, "a is a natural number". Lower-case italic Latin letters stand for what Hilbert calls *mathematische Variablen* (*mathematical variables* in the translation); they are individual variables ranging over the natural numbers, the numbers of the second number class, or the number-theoretic functions, as specified in each case. Lower-case Greek letters stand for *individuelle mathematische Gebilde* (*constant mathematical objects* in the translation); they are individual constants that stand for numbers or functions (*spezielle Funktionen*, or *specific functions* in the translation).

The effect of this convention is that only ordinary numbers or numbers of the second number class come into consideration after all as substituends, for example for a in $\mathsf{Z}(a)$ or $\mathsf{N}(a)$, respectively.

German capital as well as lower-case letters have *reference* and are used only to convey information.

It must still be remarked that by "*array*" we are to understand a perceptually given object composed of primitive signs.

In order to understand the idea of the proof of the continuum theorem we must above all gain a precise understanding of the notion of mathematical variable in its most general sense. Mathematical variables are of two kinds: (1) *primitive variables* [[*Grundvariablen*]] and (2) *variable-types* [[*Variablentypen*]].

(1) Now, while in all of arithmetic and analysis [[the variable that ranges over]] the ordinary integers suffices as sole primitive variable, with each of Cantor's transfinite number classes there is associated a *primitive variable* that ranges over precisely the ordinals of that class. Accordingly, to each primitive variable there corresponds a proposition that characterizes it as such; this proposition is implicitly characterized by the axioms, for example,

$$\mathsf{Z}(0),$$
$$\mathsf{Z}(a) \rightarrow \mathsf{Z}(a+1),$$
$$\{A(0) \mathbin{\&} (a)(A(a) \rightarrow A(a+1))\} \rightarrow \{\mathsf{Z}(a) \rightarrow A(a)\}$$
(formula of ordinary induction),
$$\mathsf{N}(0),$$
$$\mathsf{N}(a) \rightarrow \mathsf{N}(a+1),$$
$$(n)\{\mathsf{Z}(n) \rightarrow \mathsf{N}(a(n))\} \rightarrow \mathsf{N}(\lim a(n)),$$

and, in addition, the formula of transfinite induction for the numbers of the second class.

With each kind of primitive variable there is associated one kind of recursion, by means of which we define functions whose argument is a primitive variable of that kind. The recursion associated with the number-theoretic variable is "ordinary recursion", by means of which a function of a number-theoretic variable n is defined when we indicate what value it has for $n = 0$ and how the value for $n + 1$ is obtained from that for n. The generalization of ordinary recursion is transfinite recursion; it rests upon the general principle that the value of the function for a value of the variable is determined by the preceding values of the function.

(2) From the primitive variables we derive further kinds of variables by applying logical connectives to the propositions associated with the primitive variables, for example, to Z and N. The variables thus defined are called *variable-types*, and the propositions defining them *type-propositions*; for these, new constant signs are introduced each time. Thus the formula

$$(a)\{\mathsf{Z}(a) \rightarrow \mathsf{Z}(f(a))\}$$

offers the simplest instance of a variable-type; for this formula defines the function variable f and, as type-proposition, will be denoted by $\Phi(f)$ ("being-a-function"). A further example is the formula

$$(f)\{\Phi(f) \rightarrow \mathsf{Z}(g(f))\};$$

it defines the "being-a-function-of-a-function" $\Psi(g)$, where the argument g is the new function-of-a-function variable.[8]

To characterize the higher variable-types we must provide the type-propositions themselves with subscripts; a type-proposition thus provided with a subscript is defined by recursion, the place of equality ($=$) being taken by logical equivalence ($\sim$).

In all of arithmetic and analysis only the function variable, the function-of-a-function variable, and so on in finite iteration, are used as higher variables. A variable-type that goes beyond these simplest examples is offered by the variable g that assigns a numerical value $g(f_n)$ to a sequence f_n consisting of

a function f_1 of an integer, $\Phi(f_1)$,
a function f_2 of a function, $\Psi(f_2)$,
a function f_3 of a function-of-a-function,
and so forth.

The corresponding type-proposition $\Phi_\omega(g)$ is represented by means of the following equivalences:

$$\Phi_0(a) \sim \mathsf{Z}(a),$$
$$\Phi_{n+1}(f) \sim (b)\{\Phi_n(b) \rightarrow \mathsf{Z}(f(b))\},$$
$$\Phi_\omega(g) \sim \{(n)\Phi_n(f_n) \rightarrow \mathsf{Z}(g(f))\};$$

these at the same time offer an example of the definition of a type-proposition by means of recursion.

The variable-types can be classified according to their "*heights*". Among those of height 0 we include all number-theoretic constants and among those of height 1 all those functions whose arguments and values have the property of a primitive variable, for example, the property Z or the property N. A function whose argument and value have certain heights possesses a height greater by 1 than the greater, or possibly than both, of those two heights. A sequence of functions of various heights has the limit of those heights as its height.

After these preparations we return to our task and recall that, in order to prove the continuum theorem, it is essential to correlate those definitions of number-theoretic functions that are free from the symbol ε one-to-one with Cantor's numbers of the second number class, or at least to establish a correspondence between them in such a way that every number-theoretic function is associated with at least one number of the second number class.

Clearly, the elementary means that we have at our disposal for forming functions are *substitution* (that is, replacement of an argument by a new variable or function) and *recursion* (according to the schema of the derivation of the function value for $n + 1$ from that for n).

One might think that these two processes, substitution and recursion, would have to be supplemented with other elementary methods of definition, for example, definition of a function by specification of its values up to a certain argument value, beyond which the function is to be constant, and also definition by means of elementary processes obtained from arithmetic operations, such as that of the remainder in

[8] [["Function-of-a-function" is used here in the sense of "functional", not "compound function".]]

division, say, or of the greatest common divisor of two numbers, or even the definition of a number as the least of finitely many given numbers.

It turns out, however, that any such definition can be represented as a special case of the use of substitutions and recursions. The method of search for the recursions required is in essence equivalent to that reflection by which one recognizes that the procedure used for the given definition is finitary.

Having ascertained this, we must now survey the results of the two operations of substitution and recursion. On account of the variety of ways in which we can pass from n to $n + 1$, however, we cannot, as it turns out, bring the recursions that are to be used into a standard form if we confine ourselves to operating with ordinary number-theoretic variables. This difficulty already becomes apparent in the following example.

Let us consider the functions $a + b$; from them we obtain by n-fold iteration and equating

$$a + a + \cdots + a = a \cdot n.$$

In the same way we pass from $a \cdot b$ to

$$a \cdot a \cdot \cdots \cdot a = a^n,$$

and, further, from a^b to

$$a^{(a^a)}, a^{(a^{(a^a)})}, \ldots.$$

Thus we successively obtain the functions

$$\begin{aligned} a + b &= \varphi_1(a, b), \\ a \cdot b &= \varphi_2(a, b), \\ a^b &= \varphi_3(a, b). \end{aligned}$$

$\varphi_4(a, b)$ is the bth term in the sequence

$$a, a^a, a^{(a^a)}, a^{(a^{(a^a)})}, \ldots.$$

In a corresponding way we arrive at $\varphi_5(a, b)$, $\varphi_6(a, b)$, and so on.

To be sure, we could now define $\varphi_n(a, b)$ for variable n by means of substitutions and recursions, but these recursions would not be ordinary, stepwise ones; rather, we would be led to a manifold simultaneous recursion, that is, a recursion on different variables at once, and a resolution of it into ordinary, stepwise recursions would be possible only if we make use of the notion of function variable; the function $\varphi_a(a, a)$ is an instance of a function, of the number-theoretic variable a, that cannot be defined by substitutions and ordinary, step-wise recursions alone if we admit only number-theoretic variables.[9] How we can define the function $\varphi_n(a, b)$ by using the function variable is shown by the following formulas:

$$\begin{aligned} \iota(f, a, 1) &= a, \\ \iota(f, a, n + 1) &= f(a, \iota(f, a, n)); \\ \varphi_1(a, b) &= a + b, \\ \varphi_{n+1}(a, b) &= \iota(\varphi_n, a, b). \end{aligned}$$

[9] This assertion was proved by W. Ackermann [[*1928*]].

Here ι stands for a specific function of three arguments, of which the first is itself a function of two ordinary number-theoretic variables.

Another example of a more complicated recursion is this:

$$\varphi_0(a) = \mathfrak{a}(a),$$
$$\varphi_{n+1}(a) = \mathfrak{f}(a, n, \varphi_n(\varphi_n(n + a))),$$

where $\mathfrak{a}$ stands for a known expression containing one argument, and $\mathfrak{f}$ for a known expression containing three arguments. What is characteristic of this recursion is that here we cannot obtain a numerical value for $n + 1$ from one for n but must make use of the range of the function φ_n in determining φ_{n+1}.

The difficulties that come to light in these examples are overcome when we make use of variable-types; then the general recursion schema reads as follows:

$$\rho(\mathfrak{g}, \mathfrak{a}, 0) = \mathfrak{a},$$
$$\rho(\mathfrak{g}, \mathfrak{a}, n + 1) = \mathfrak{g}(\rho(\mathfrak{g}, \mathfrak{a}, n), n);$$

here $\mathfrak{a}$ is a given expression of arbitrary variable-type; $\mathfrak{g}$ likewise is a given expression, having two arguments, of which the first is of the same variable-type as $\mathfrak{a}$ and the second is a number; the additional condition that $\mathfrak{g}$ must satisfy is that its value again be of the same variable-type as $\mathfrak{a}$. Finally, ρ is the expression to be defined by the recursion; it depends on three arguments and, after the substitutions for $\mathfrak{g}$, $\mathfrak{a}$, and n have been made, is of the same variable-type as $\mathfrak{a}$; in addition, other arbitrary parameters are permitted to occur in $\mathfrak{a}$ and $\mathfrak{g}$, and consequently also in ρ.

From this general schema we obtain specific recursions through substitution. Thus we obtain the recursions of our examples by considering, in the first example, f and a as parameters and by representing, in the second, the transition from $\varphi_n(a)$ to $\varphi_{n+1}(a)$ as a transition, mediated by the function-of-a-function $\mathfrak{g}$, from a function φ_n to the function φ_{n+1}, so that a is not regarded as a parameter at all in the recursion. Compared with elementary recursion, the recursion used in our two examples is of a wider kind, since in one case we introduce a higher parameter that is not an ordinary integer and in the other choose a function for $\mathfrak{a}$ and a function-of-a-function for $\mathfrak{g}$.

The variable-types form the link that makes possible the correspondence between the functions of a number-theoretic variable and the numbers of the second number class. Indeed, we arrive at such a correspondence between the numbers of the second number class and certain variable-types if we compare the two generating processes for the numbers of the second number class, namely, the process of adding 1 and the limit process for denumerable sequences, with the way in which variable-types increase in height. Let us establish a correspondence between the process of adding 1 and the taking of a function (that is, substitution of a given variable-type into a function as argument) and between the limit process and the aggregating of the denumerable sequence associated with a variable-type into a new variable-type, and let us designate the variable-types that in this way come to correspond to the numbers of the second number class specifically as Z-*types*. Thus, when the Z-types are formed, we use, besides the logical operations, only ordinary (not transfinite) recursions, that is, just those necessary for the denumeration of a type sequence as a preparatory step for the limit process. If we order these Z-types according to their heights, we have a one-to-one correspondence by means of which the variable-types of a given height are associated with a number of the second number class.

But therewith we also arrive at a one-to-one correspondence between the numbers of the second number class and the functions defined by means of the Z-types. To make that clear the following considerations suffice. Whenever we begin with variable-types up to a given height only and then form functions solely by means of substitution and recursion, we obtain only denumerably many functions. We can even formalize this denumeration rigorously; in particular, we can do this by first constructing a recursion function ρ that subsumes all the recursions in question and, consequently, contains a parameter that exceeds the variable-types admitted up to that point. In the definition of ρ we apply the general recursion schema in a way that makes essential use of this higher variable-type. Now we order according to their heights the relevant specializations of the variable-types occurring in ρ and therewith obtain the various initial substitutions. These we arrange in an enumerated sequence. Once this enumeration has been established, the introduction of an ordering according to the number of substitutions to be performed yields the functions that were to be defined.

In the argument just described I essentially presupposed the theory of the numbers of the second number class. I introduced the numbers of the second number class simply as resulting from transnumeration beyond the denumerable infinite, and the constant proposition N, "to be a number of the second number class", was characterized afterwards by a listing of axioms. But these axioms furnish only the general framework of a theory. To provide a more precise foundation for it, it is necessary to determine how the process of transnumeration beyond the denumerable infinite is to be formalized. This is done by applying the process of transnumeration to a sequence; this sequence can be given only by means of an ordinary recursion, and for these recursions certain types are in turn necessary.

This circumstance seems to present a difficulty, but in fact it turns out that reflecting upon precisely this point enables us to make the correspondence between the numbers of the second number class and the functions of a number-theoretic variable a much closer one. For the variable-types that we need for the production of the numbers of the second number class are obtained by formal replacement of the sign Z by the sign N at one or more places in the defining type-propositions that we have up to this point. We shall call the variable-types that then result N-*types*; as is apparent, corresponding Z-types and N-types have the same height. Now we do not need to associate with a given number of the second number class all functions of the same height, but we can let the numbers of the second number class and the functions correspond to each other according to the heights of the variable-types required for their definition. Let us now come to a more precise formulation of this correspondence.

If in the Z-types we go up to a certain height only, the heights of the corresponding N-types are also bounded. From the numbers of the second number class that can be produced with these types we can, by means of an increasing sequence, obtain a greater number of the second number class, which is defined by means of a higher variable-type. If, on the other hand, we have N-types up to a certain height, the functions definable by means of the corresponding Z-types can also be denumerated —in the way described above, according to the number of substitutions. As is well known, we obtain from such an enumeration $\varphi(a, n)$ by means of Cantor's diagonal procedure (for example, by forming $\varphi(a, a) + 1$) a function that differs from all of the enumerated functions and therefore could not be defined by means of the variable-types previously admitted.

Thus we have made it possible to establish a one-to-one correspondence between the numbers of the second number class that are definable at the height in question, but not at any lower one, and the denumerably many functions that are definable at the same height, and in this way each function is associated with at least one number of the second number class.

With this, however, the proof of the continuum theorem is not yet complete; rather, it still requires essential supplementation. For in our entire investigation up to now we have, in order to establish the correspondence, made restrictive assumptions in two respects: first, our general recursion schema for ρ embodies only the case of ordinary recursion (the number-theoretic variable being the variable on which the recursion proceeds), and, second, we also restricted the variable-types to those that result from transnumeration beyond denumerated sequences. It is certain that, in general, transfinite recursions and, accordingly, higher variable-types are necessarily used in mathematical investigations, for example, for the formation of certain kinds of functions of real variables. But here in our problem, where it is a matter of forming functions of a number-theoretic variable, we do not in fact require such higher recursions and variable-types; for the following lemma holds.

LEMMA II. In the formation of functions of a number-theoretic variable transfinite recursions are dispensable; in particular, not only does ordinary recursion (that is, the one that proceeds on a number-theoretic variable) suffice for the actual formation process of the functions, but also the substitutions call merely for those variable-types whose definition requires only ordinary recursion. Or, to express ourselves with greater precision and more in the spirit of our finitist attitude, if by adducing a higher recursion or a corresponding variable-type we have formed a function that has only an ordinary number-theoretic variable as argument, then this function can always be defined also by means of ordinary recursions and the exclusive use of Z-types.

The following typical example will make the meaning and scope of this lemma clear to us.

If we imagine that the correspondence between the functions of a number-theoretic argument and the numbers of the second number class has been formalized, then by this very fact we have a certain function $\zeta(a, n)$ that associates an ordinary number with an arbitrary number a of the second number class and the ordinary number n (for fixed a and variable n, $\zeta(a, n)$ represents precisely the function associated with a). If we now substitute for a a number α_n, of the second number class, that depends on n, the sequence [[of the α_n]] being defined by ordinary or even transfinite recursion, for example,

$$\alpha_{n+1} = \omega^{\alpha_n},$$

then $\zeta(\alpha_n, n)$ is a function of a number-theoretic variable n; and our Lemma II now asserts of this function that it can be defined through ordinary recursion by means of

Z-types, whereas it is certainly impossible to define $\zeta(a, n)$ by these means, since, after all, the contrary assumption clearly leads to a contradiction.

I should like to note expressly once more that the presentation given here of the proof of the continuum theorem contains only the fundamental ideas; its complete realization would require, besides the proofs of the two lemmas, a recasting strictly faithful to the finitist attitude.

Finally let us recall our real subject and, so far as the infinite is concerned, draw the balance of all our reflections. The final result then is: nowhere is the infinite realized; it is neither present in nature nor admissible as a foundation in our rational thinking—a remarkable harmony between being and thought. We gain a conviction that runs counter to the earlier endeavors of Frege and Dedekind, the conviction that, if scientific knowledge is to be possible, certain intuitive conceptions ⟦Vorstellungen⟧ and insights are indispensable; logic alone does not suffice. The right to operate with the infinite can be secured only by means of the finite.

The role that remains to the infinite is, rather, merely that of an idea—if, in accordance with Kant's words, we understand by an idea a concept of reason that transcends all experience and through which the concrete is completed so as to form a totality—an idea, moreover, in which we may have unhesitating confidence within the framework furnished by the theory that I have sketched and advocated here.

In closing I should like to express my thanks to P. Bernays for his sympathetic collaboration and the valuable aid that he extended to me in questions both of matter and of form, especially in the proof of the continuum theorem.

An axiomatization of set theory

JOHN VON NEUMANN

(*1925*)

In the twenties von Neumann wrote a cluster of papers on set theory (*1923, 1925, 1926, 1928, 1928a, 1929*). The first paper presents, in naive set theory, a new theory of ordinal numbers, which is easily adapted to various axiomatized set theories. The fourth paper, which offers a justification of definition by transfinite induction, is connected with the first. The second paper, given below in translation, offers a new axiomatization of set theory. The presentation is terse. A detailed development of the system, with proofs, is given in the fifth paper. The third paper, in Hungarian, is von Neumann's doctoral dissertation; it contains the main points of the fifth paper. In the sixth, von Neumann discusses questions of relative consistency.[a]

Recoiling from the pitfalls of the newly discovered paradoxes, Zermelo had produced what seemed a relatively safe system (although, of course, its consistency had not been proved). The Burali-Forti and Russell paradoxes cannot be reproduced in this system because the set of all ordinals and the set of all sets that do not contain themselves as elements do not exist in the system. But, being chary of sets, the system is in some respects remote from intuition. It has no universal set, and, what is more troublesome, a sequence of sets of the system is not necessarily a set of the system. Thus, if we assume the continuum hypothesis, it has no set of cardinality $\geqq \aleph_\omega$. With the introduction of the axiom of replacement in 1921–1922 (see above, p. 291), many more sets were at hand; also at that time the vagueness attached to a notion used by Zermelo, that of definite property, was eliminated (see above, p. 285). Thus amended, the system became known as the Zermelo–Fraenkel set theory.

The set theory proposed by von Neumann is still further removed from Zermelo's original system and represents a new departure. Its fundamental innovation is the introduction of two kinds of collections: sets and classes.[b] A set is a class, but there are classes—Quine will call them *ultimate* classes—that are not sets; an ultimate class is not an element of any other class. Sets and classes are characterized by axioms. As von Neumann points out (below, p. 397), classes can be regarded as Zermelo's "definite properties", now precisely delineated (see also *Gödel 1940*, p. 2). There is a universal class, the class of all sets, and there is a class of all ordinals; these two classes are both ultimate, so that Russell's and Burali-Forti's paradoxes cannot be reproduced in the system.

The present paper, however, is not couched in the language of sets and classes. The basic notion is that of

---

[a] Concerning the circumstances in which *von Neumann 1925* was written, see *Ulam 1958*, p. 10, footnote 3.

[b] Prior to von Neumann, Cantor (see above, p. 114) and König (see above, p. 148) had distinguished two kinds of collections in order to avoid the paradoxes.

function; more specifically, the paper deals with I-objects (arguments), II-objects (characteristic functions of classes), and I-II-objects, that is, objects that are at once I-objects and II-objects (characteristic functions of sets). This adventitious idiom is easily translated into a more customary language.[c]

An important feature of von Neumann's system is that, unlike Zermelo's, it is finitely axiomatizable. No axiom schema, such as that of separation, is required.

Having formulated his system, von Neumann devotes a great deal of attention to the question of categoricity and the import of the Löwenheim–Skolem theorem for set theory. In his search for a minimal model that could be uniquely characterized, he is led to consider models that are contained in other models. Such models, subsequently called inner models by Shepherdson (*1951–1953*), have been particularly useful in investigations of relative consistency (*von Neumann 1929*, *Mostowski 1939*, *Gödel 1938*, *1939*, *1940*, *Cohen 1963*, *1963a*, *1964*).

The system presented by von Neumann was simplified, revised, and expanded by R. M. Robinson (*1937*), Bernays (*1937–1954*, *1958*), and Gödel (*1940*), and it has come to be known as the von Neumann–Bernays–Gödel set theory. For the relation of this system of set theory to other systems, in particular to the Zermelo–Fraenkel set theory, we refer the reader to Chapter 14 in *Quine 1963*.

Errata pointed out by the author in a *Berichtigung* have been incorporated in the text. The translation is by Stefan Bauer-Mengelberg and Dagfinn Føllesdal, and it is printed here with the kind permission of Mrs. Klara von Neumann-Eckart and Walter de Gruyter and Co.

[c] A paper of Skolem's (*1938*, pp. 32–34) contains a translation of von Neumann's axioms into the first-order predicate calculus.

I. THE AXIOM SYSTEM

§ 1. Fundamentals concerning the axiomatization of set theory

The aim of the present work is to give a logically unobjectionable axiomatic presentation of set theory. I would like to make a few preliminary remarks about the difficulties that have made such a construction of set theory desirable.

It is well known that set theory in its first, "naive" version, due to Cantor, led to contradictions. These are the well-known antinomies of the set of all sets that do not contain themselves (Russell), of the set of all transfinite ordinal numbers (Burali-Forti), and of the set of all finitely definable real numbers (Richard).[1] Since naive set theory undeniably led to these contradictions, while on the other hand a certain part of its propositions seemed to be precise and reliable, and since, moreover, the modern formulation of mathematics absolutely required a set-theoretic foundation, there has been no lack of attempts to "rehabilitate" set theory.

In discussing these we must draw a fundamental distinction between two different tendencies. A number of writers were led by the antinomies of set theory to submit the entire logical foundation of the exact sciences to a critique. They made it their goal to put all of exact science on a new, universally evident basis from which they could again reach what is "correct" in mathematics and set theory, but from which they had to exclude a priori, by seeking a foundation in immediate intuition, what is self-contradictory. Here Russell, J. König, Weyl, and Brouwer must be mentioned.[2] They arrived at entirely different results, but the over-all effect of their activity seems to me outright devastating. In Russell, all of mathematics and set theory seems to rest upon the highly problematic "axiom of reducibility", while Weyl and Brouwer systematically reject the larger part of mathematics and set theory as completely meaningless. What we have here is not a rehabilitation of set theory at all, but rather a very sharp critique of the modes of inference hitherto used in elementary logic, in particular, of the principle of "excluded middle" and of the principles governing "all" and " there exists".

The other group, Zermelo, Fraenkel, and Schoenflies, has eschewed so radical a revision.[3] The methods of logic are not criticized to any extent, but are retained; only the (no doubt useless) naive notion of set is prohibited. To replace this notion the axiomatic method is employed; that is, one formulates a number of postulates in which, to be sure, the word "set" occurs but without any meaning. Here (in the spirit of the axiomatic method) one understands by "set" nothing but an object of which one knows no more and wants to know no more than what follows about it from the postulates. The postulates are to be formulated in such a way that all the desired theorems of Cantor's set theory follow from them, but not the antinomies. In these axiomatizations, however, we can never be perfectly sure of the latter point. We see only that the known modes of inference leading to the antinomies fail, but who knows whether there are not others? A standard consistency proof is obviously altogether unthinkable in this context. For the method by which Hilbert (*1899*) proved the consistency of the various geometries—reduction to the consistency of arithmetic and analysis—fails here. A direct consistency proof (without a shift of the problem) would, however, obviously belong to the domain of the first group mentioned above, since there still remains much to be clarified in logic itself. The announced works of Hilbert[4] have this proof as their goal.

The second group, then, while painstakingly avoiding the naive notion of set

[1] See, for example, *Poincaré 1908*.

[2] *Whitehead and Russell 1910, 1912, 1913, König 1914, Weyl 1920, Brouwer 1919b.*

[3] *Zermelo 1908a, Fraenkel 1922b, 1923, 1923a, 1925, Schoenflies 1921, 1922.*

[4] *1922, 1922a.* [[These "announced" works appeared between the writing and the publication of the present paper.]]

(Cantor's), want to specify an axiom system from which set theory (without its antinomies) follows. Their investigations can, to be sure, never settle the real problem so completely as those by the first group, but their goal is much clearer and closer at hand. Of course, it can never be shown in this way that the antinomies have really been eliminated, and much arbitrariness always attaches to the axioms.[5] But one thing can be attained here with certainty, namely, a precise determination of what the rehabilitated part of set theory is and what is at issue when henceforth we speak of the complete "formalistic set theory".

The present work is in the spirit of the second group. Below we shall specify a system of postulates from which all that is known in set theory follows in a logically unobjectionable way. "Logically unobjectionable", it is true, must be understood in the sense in which it has hitherto been understood in mathematics. There will be no attempt to make derivations unobjectionable also in the sense of the intuitionism of Brouwer and Weyl. I would like to remark, nevertheless, that this, too, could be attained rather easily (through a few insignificant modifications); but I forgo this as a matter of principle, since the axiomatic method is in itself contrary to the essence of intuitionism. (From the point of view of intuitionism an axiom system would have any meaning only if the consistency proof for it had itself been carried out in an intuitionistically correct way. To judge by a remark of Brouwer's, not even that would do.[6])

A number of essential questions concerning fundamentals will not be dealt with until Part II, since they presuppose knowledge of the axiom system.

§ 2. General remarks on the axiomatization

The task of our axiomatization is obviously to produce, by means of a finite number of purely formal operations (and that these can be performed is guaranteed precisely by the postulates), all the sets that we want to see formed. We must, however, avoid forming sets by collecting or separating elements ⟦durch Zusammenfassung oder Aussonderung von Elementen⟧, and so on, as well as eschew the unclear principle of "definiteness" that can still be found in Zermelo.

We prefer, however, to axiomatize not "set" but "function". The latter notion certainly includes the former. (More precisely, the two notions are completely equivalent, since a function can be regarded as a set of pairs, and a set as a function that can take two values.) The reason for this departure from the usual way of proceeding is that every axiomatization of set theory uses the notion of function (axiom of separation, axiom of replacement, see pages 400 and 403), and thus it is formally simpler to base the notion of set on that function than conversely. The intuitive picture of the axiomatized system is as follows.

We consider two domains of objects, that of "arguments" and that of "functions". (Both words are, of course, to be taken in a purely formal way, as if they had no

[5] There is, to be sure, a certain justification for the axioms in the fact that they go into evident propositions of naive set theory if in them we take the word "set", which has no meaning in the axiomatization, in the sense of Cantor. But what is omitted from naive set theory—and to circumvent the antinomies some omission is essential—is absolutely arbitrary.

[6] *Brouwer 1919b*. See *Fraenkel 1923b*, p. 98.

meaning.) The two domains are not identical, but they partly overlap. (There are "argument-functions", which belong to both domains.)

Now a two-variable operation $[x, y]$ (read "the value of the function x for the argument y"), whose first variable x must always be a "function" and whose second variable y must always be an "argument", is defined in these domains. What is formed by means of it is always an "argument", $[x, y]$.

The operation $[x, y]$ corresponds to a procedure that is encountered everywhere in mathematics, namely, the formation, from a function f (which must be carefully distinguished from its values $f(x)$!) and an argument x, of the value $f(x)$ of the function f for the argument x. Instead of $f(x)$ we write $[f, x]$ to indicate that f, just like x, is to be regarded as a variable in this procedure. Through the use of $[x, y]$ we replace, as it were, all one-variable functions by a single two-variable function. In this scheme the elements of the domain of "functions" answer to the functions (conceived naively) that are defined for the "arguments" and whose values are "arguments". (Later, certain of these "functions" x will be distinguished as "sets", namely, those for which $[x, y]$ can take only two given values as y runs through all "arguments"; see also § 3 and § 4.)

For $[x, y]$, moreover, the "axiom of extensionality" ⟦"Bestimmtheitsaxiom"⟧ holds in the following sense:

If a and b are "functions" and if for every argument x we have $[a, x] = [b, x]$, then $a = b$. (See Axiom I 4.)

A "function" a is thus determined without any ambiguity by its "values" $[a, x]$, but there is no reason to assume that these "values" can be assigned to it arbitrarily. (For then we would, after all, relapse into naive set theory.) Rather, the following question arises: what operations are available for the production of functions? It will be answered in axiom groups II and III.

But one more question presents itself: what "functions" are at the same time "arguments"? Obviously the most convenient answer would be: all of them. But with the ways of producing "functions" that are listed in axiom groups II and III this would again entangle us in the antinomies of naive set theory, in Russell's first of all. And, since we need all these possible ways of producing functions, we must forgo treating certain functions as arguments.

We impose this unavoidable restriction by going in the direction pointed out by Zermelo.

We arbitrarily select an "argument", A, and in effect declare that a function a is also an argument if and only if it does *not too often*, that is, *for too many arguments* x, take "values" $[a, x]$ differing from A. (Since a "set" will be defined as a function that can take two values only, one of these being A, this is a reasonable adaptation of Zermelo's point of view.)

We now make the expression "too often" more precise as follows: The "function" a will fail to be an "argument" if and only if the totality of arguments x for which $[a, x] \neq A$ can be mapped onto the totality of all arguments. (The mapping must, however, be accomplished by one of our "functions"; that is, it must have the form $y = [b, x]$. It must be single-valued, but it need not be one-to-one.) (See Axiom IV 2.)

This definition has the advantage that it guarantees that we can treat as "argument" any "function" a that is $\neq A$ less often than a "function" b that has already been recognized to be an "argument". "Less often" means that the totality of all x for which $[a, x] \neq A$ is an image of the totality of all x for which $[b, x] \neq A$ (or of a part of it).[7] The definition thus contains the axiom of separation, as it is called, which is due to Zermelo, and the axiom of replacement, due to Fraenkel.[8] But it accomplishes more: it also contains the well-ordering theorem and hence makes the principle of choice superfluous.

That the well-ordering theorem appears in this new context is due to the fact that we can construct an unobjectionable theory of ordinal numbers. "Naively", the totality of all ordinals leads to the Burali-Forti antinomy; in our system this has the consequence that a function having values $\neq A$ for all ordinals is not an argument. Accordingly, it must be possible to map the totality of all ordinals onto the totality of all "arguments", and this, of course, yields a well-ordering of the totality of all "arguments". (This reasoning, of course, must and can be carried through rigorously.)

I should like to make a further remark. It may seem odd that notions such as "totality" and "function" are used naively in an axiomatization of set theory (contrary to what was said in § 1). But that is done only here, in § 2, in order to give an intuitive picture of the system; of course, nothing of the kind will happen in the precise formulation, in § 3.

§ 3. The axioms and their meaning

Concerning what has been said above we must still make the following remarks.

Besides the universal two-variable operation $[x, y]$ already introduced, we must introduce another two-variable operation (x, y) (read "the ⟦ordered⟧ pair x, y"), whose variables x and y must both be "arguments" and which itself produces an "argument" (x, y). Its most important property is that $x_1 = x_2$ and $y_1 = y_2$ follow from $(x_1, y_1) = (x_2, y_2)$. (That property does not appear explicitly among the axioms, since it follows from Axioms II 3 and II 4.) This operation does not have the fundamental character of $[x, y]$ at all; it is necessary only because the notion "pair" has to be introduced.

Further, in addition to the "argument" A, already mentioned, another arbitrary "argument", B, will be introduced from the very beginning. (A and B will be the two values of those functions that represent the "sets".)

And, finally, we shall always speak of "I-objects", "II-objects", and "I-II-objects" instead of "arguments", "functions", and "argument-functions", respectively.

The axiom system reads as follows:

We are concerned with *I-objects*, *II-objects*, the two distinct objects A and B, and the two operations $[x, y]$ and (x, y). The following axioms hold.

[7] ⟦Hence "less often" should be understood as "not more often".⟧

[8] The "axiom of replacement" can be expressed as follows: Let $\mathfrak{M}$ be a set and $f(x)$ a function (defined in $\mathfrak{M}$); then there exists a set $\mathfrak{N}$ that contains, for every element x of $\mathfrak{M}$, the value of $f(x)$ ⟦and contains nothing else⟧. This axiom is due to Fraenkel (*1922a*). He was the first to point out that without this axiom the existence of sets of cardinality $\aleph_\omega$ is unprovable in Zermelo's axiomatization. (In his more recent work, however, he tries to manage with a weaker axiom.) In fact, I believe that no theory of ordinals is possible at all without this axiom.

I. Introductory axioms

1. A and B are I-objects.
2. $[x, y]$ has meaning if and only if x is a II-object and y a I-object; it is itself always a I-object.
3. (x, y) has meaning if and only if x and y are I-objects; it is itself always a I-object.
4. Let a and b be II-objects; if $[a, x] = [b, x]$ for all I-objects x, then $a = b$.

No further comment is needed concerning these axioms; all of them were already discussed in § 2.

II. Arithmetic construction axioms

1. There exists a II-object a such that always $[a, x] = x$.
2. Let u be a I-object; then there exists a II-object a such that always $[a, x] = u$.
3. There exists a II-object a such that always $[a, (x, y)] = x$.
4. There exists a II-object a such that always $[a, (x, y)] = y$.
5. There exists a II-object a such that always (if x is a I-II-object) $[a, (x, y)] = [x, y]$.
6. Let a and b be II-objects; then there exists a II-object c such that always $[c, x] = ([a, x], [b, x])$.
7. Let a and b be II-objects; then there exists a II-object c such that always $[c, x] = [a, [b, x]]$.

All these are ways of producing functions. They have been put together in such a manner that the functions in a certain sense form a group. For they have as a consequence (we do not go into the very simple proof) the following

Reduction theorem. Let $\mathfrak{A}(x_1, x_2, \ldots, x_n)$ be an expression constructed from the variables $x_1, x_2, \ldots, x_n$ and some constants $a_1, a_2, \ldots$ (I-objects or II-objects) by means of the operations $[x, y]$ and (x, y). Such an expression need not always have meaning in case $x_1, x_2, \ldots, x_n$ are I-objects (for example, $[x_1, x_2]$ has meaning only if x_1 is a I-II-object); but let us assume that, whenever it is meaningful, it is a I-object. (This excludes the case in which $\mathfrak{A}(x_1, x_2, \ldots, x_n)$ is simply a constant a that is a II-object but not a I-object.) Then there exists a II-object a such that, for all I-objects $x_1, x_2, \ldots, x_n$ for which $\mathfrak{A}(x_1, x_2, \ldots, x_n)$ has meaning, $\mathfrak{A}(x_1, x_2, \ldots, x_n) = [a, ((\ldots((x_1, x_2), x_3), \ldots), x_n)]$.

Thus, in $[a, ((\ldots((x_1, x_2), x_3), \ldots), x_n)]$ we now have a normal form for expressions in n variables; it is the general function of n variables, just as $[a, x]$ is the general function of one variable.

III. Logical construction axioms

1. There exists a II-object a with the following property: $x = y$ if and only if $[a, (x, y)] \neq A$.
2. Let a be a II-object; then there exists a II-object b with the following property: $[b, x] \neq A$ if and only if, for all y, $[a, (x, y)] = A$.
3. Let a be a II-object; then there exists a II-object b such that, whenever $[a, (x, y)] \neq A$ for a unique y, $[b, x]$ is equal to that y.

These ways of producing functions complement those of group II. They make it possible to bring every logical condition on the I-objects $x_1, x_2, \ldots, x_n$ into the normal form $[a, ((\ldots((x_1, x_2), x_3), \ldots), x_n)] \neq A$ and also to bring every I-object y that is logically uniquely determined by $x_1, x_2, \ldots, x_n$ into the normal form $[a, ((\ldots((x_1, x_2), x_3), \ldots), x_n)]$. For III 1 makes the "reduction to normal form" possible for the identity relation $x = y$, and III 2 for the notions "all" and "there exists"; III 3, finally, permits the y that is uniquely determined by an implicit condition $[a, (x, y)] \neq A$ to be explicitly represented as $y = [b, x]$.

Incidentally, while the axioms of group II are mutually independent, they are no longer so after group III has been added; thus, for example, Axiom II 1 follows from III 1 and III 3.

IV. I-II-objects

1. There exists a II-object a with the following property: a I-object x is a I-II-object if and only if $[a, x] \neq A$.

2. A II-object a is not a I-II-object if and only if there exists a II-object b such that for every I-object x there exists a y for which both $[a, y] \neq A$ and $[b, y] = x$.

We have here two formally analogous axioms; IV 1 states when a I-object is a I-II-object, and IV 2 does the same for a II-object. As for their content, however, IV 1 and IV 2 are fundamentally different.

In IV 1 we could more simply have required that every I-object be a I-II-object (that is, that all the objects considered be functions; Zermelo makes no such stipulation, but Fraenkel does; more will be said about this in Part II). But for the time being we keep the axioms as general as possible; we shall take up this restriction, as well as some others, in Part II. IV 1, however, requires nothing meritorious; it merely demands that for a I-object the property "being a I-II-object", just like any other property, be capable of being brought into normal form.

IV 2, on the other hand, is a very essential axiom, with many consequences. It was already discussed in § 2; Zermelo's axiom of separation, Fraenkel's axiom of replacement, and the well-ordering theorem follow from it. Zermelo and Fraenkel use no such general criterion to decide when a set is "too big". Also, departing from their procedure, we obtain the well-ordering theorem from this axiom and not from the principle of choice (the multiplicative axiom in Zermelo).

For the next group of axioms it is practical (but by no means necessary) to introduce a number of signs:

Let a be a II-object. For $[a, x] \neq A$ we also write $x \varepsilon a$. Let a and b be II-objects. If $x \varepsilon b$ follows from $x \varepsilon a$, then $a \lesssim b$. For $b \lesssim a$ we also write $a \gtrsim b$. If $a \lesssim b$ and $a \gtrsim b$, we write $a \sim b$. If $a \lesssim b$ but not $a \gtrsim b$, we write $a < b$; if $a \gtrsim b$ but not $a \lesssim b$, we write $a > b$. (See also § 4.)

V. Axioms of infinity

1. There exists a I-II-object a with the following properties: there exist I-II-objects x for which $x \varepsilon a$; if $x \varepsilon a$ for a I-II-object x, there exist I-II-objects $y \varepsilon a$ for which $x < y$,

2. Let a be a I-II-object; then there exists a I-II-object b for which, from $x \varepsilon y$ and $y \varepsilon a$ (y thus being a I-II-object), $x \varepsilon b$ follows.

3. Let a be a I-II-object; then there exists a I-II-object b with the following property: if $x \lesssim a$ for a I-II-object x, there exists a I-II-object y for which both $x \sim y$ and $y \varepsilon b$.

These three relatively complicated axioms occur in Zermelo and Fraenkel, too, and specifically as the "axiom of infinity" (V 1), the "axiom of the union" (V 2), and the "axiom of the power set" (V 3). But we call all three of them axioms of infinity since they are necessary only for the specific theory of infinite cardinalities. Not only the theory of finite sets and finite ordinals (that is, nonnegative integers) but even partly that of the continuum can be constructed without them, solely on the basis of the axioms of groups I–IV.

The axioms of infinity are ways of producing I-II-objects (analogous to groups II and III, which were ways of producing II-objects). Their meaning (naively formulated) is approximately as follows:

There is an infinite set a that is not too big.

If a is a set, not too big, of sets that are themselves not too big, then the set b of the elements of the elements of a is not too big either.

If a is a set that is not too big, then the set b of all subsets of a is not too big either.

To speak of functions rather than of sets, it is true, somewhat complicates the formulation of the axioms (in particular of V 3); nevertheless, the existence of infinite sets, as well as that of unions and power sets, readily follows by means of the axioms of group V (as soon as the notion "set" is strictly defined as a special case of "function"). Axiom V 1, incidentally, deviates from Zermelo's (and Fraenkel's) version of the axiom of infinity; in V 1 the existence of some infinite set is required, but not that of a specific one, as is the case in their version. This circumstance, however, is of no consequence.

The axioms of groups I–V constitute our axiom system. Outwardly, the grouping and formulation differ widely from those of Zermelo and Fraenkel; nevertheless, there are many analogies. Especially if we compare them with Fraenkel's axioms and definitions, we see that most of our axioms have analogues among his. There are, however, some quite essential differences. That we speak of "functions" rather than of "sets" is no doubt a superficial difference; it is essential, however, that the present set theory deals even with sets (or "functions") that are "too big", namely, those II-objects that are not I-II-objects. Rather than being completely prohibited, they are only declared incapable of being arguments (they are not I-objects!). This suffices to avoid the antinomies. At the same time, the existence of those sets is necessary for certain modes of inference. Axiom IV 2, finally, deviates quite essentially from what Zermelo and Fraenkel have, and indeed it is the distinctive feature of our axiomatization. It is, to be sure, related in a certain sense to the axioms of separation and replacement, but it goes much further. On the one hand it guarantees the existence of subsets and image sets, and in general it makes possible the theory of ordinals and alephs (which can hardly be developed successfully in an axiom system that lacks the axiom of replacement); yet all that could essentially be achieved by the axiom of replacement alone. But beyond this, IV 2 occupies an altogether central position in the axiom system; in several cases it enables us to prove that a set is "not too big", and finally it yields the well-ordering theorem.

Axiom IV 2, to be sure, requires something more than what was up to now regarded as evident and reasonable for the notion "not too big". One might say that it somewhat overshoots the mark. But, in view of the confusion surrounding the notion "not too big" as it is ordinarily used, on the one hand, and the extraordinary power of this axiom on the other, I believe that I was not too crassly arbitrary in introducing it, especially since it enlarges rather than restricts the domain of set theory and nevertheless can hardly become a source of antinomies. (We shall consider this latter point more closely in Part II.)

§ 4. On the derivation of set theory

In § 2 and § 3 an axiomatization of set theory was described, with emphasis on the fundamental points of view that governed its formulation. In what follows, something will be said about the derivation of set theory from these axioms.

Such a derivation of set theory would be arranged as follows:

1. *General set theory.* Here we have to take care of those general theorems and definitions that proceed more from the nature of the axiomatization than from that of set theory and are quite trivial in naive set theory. They are theorems like those on the existence of the union and the intersection of sets, the existence of the power set, and so forth. There is no mention here of order or of cardinality.

2. *Order and well-ordering.* We define what is to be understood by these notions and, further, by certain auxiliary notions like similarity, segment (initial section), and so on. Some quite trivial theorems on order and well-ordering follow, for example: if two ordered sets are similar to a third, they are similar to each other; if of two similar ordered sets one is well-ordered, so is the other; and so forth.

3. *Ordinal numbers.* Here the theory proper begins. A strict definition of ordinals is given, and then the most important properties of these ordinals are developed, among others the comparability of well-ordered sets and the admissibility of definition by transfinite induction.[9]

4. *The well-ordering theorem.* By means of the ordinals and Axiom IV 2 the well-ordering theorem is proved (without the principle of choice).

5. *Cardinalities.* When the theory of ordinals and the well-ordering theorem are at hand, the theory of alephs (cardinal numbers or cardinalities) can easily be developed. (This arrangement, which puts the cardinalities after well-ordering, deviates from the one generally adopted; but it leads to the goal more quickly.)

6. *Infinitude.* Finally the definition of finitude and infinitude for sets follows. The simplest properties of finitude and infinitude are developed and the existence of infinite sets is proved. The least infinite ordinal ω is defined.

In what follows we shall not carry through this derivation (which already exists in finished form). With all proofs it would take up a great deal of space and go beyond the scope of this presentation. It will be set forth in detail in another paper ⟦*1928a*⟧.

Let us introduce only the most important terms here.

[9] See *von Neumann 1923* ⟦and *1928*⟧.

The exact definition of "set" (as a special case of "function") reads as follows:

A II-object a is called a *class* [[*Bereich*]] if $[a, x]$ is always equal to A or B.

A I-II-object satisfying that condition is called a *set*.

That is, "sets" are the sets that (in the earlier terminology) are "not too big", and "classes" are all totalities [[*Gesamtheiten*]] irrespective of their "size". A class is "capable of being an argument" (that is, a I-II-object) if and only if it is a set.

The definitions of the relations $x \varepsilon a$ (x belongs to a, x is an element of a, a contains x), $a \lesssim b$ (a is a part of b), $a < b$ (a is a proper part of b), and $a \sim b$ (a and b are of the same size) have already been given in § 3. (If a and b are classes, $a = b$ of course follows from $a \sim b$ because of I 4; for II-objects in general, on the other hand, this is not a necessary consequence.)

If a and b are classes, $a + b$, $a \cdot b$, and $a - b$ are their union, intersection, and difference classes, respectively (in the same sense as these are understood in naive set theory; of course, the existence proofs must first be produced for all these classes; this applies also to what follows). If a is a class and its elements are sets, $S(a)$ and $D(a)$ are respectively the union and intersection classes of (the elements of) a. If a is a class and c a II-object, $|[c, a]|$ is the image of a mediated by c. If, finally, a is a class, $P(a)$ is the power class of a (and contains all sub*sets* of a; classes that are not sets are, after all, "incapable of being arguments"). Moreover, if a and b are sets, $a + b$ is also a set; if a or b is a set, $a \cdot b$ is also a set; and if a is a set, $S(a)$, $D(a)$, and $P(a)$ are also sets.

If a is a class, we call all classes b that have the following properties *orderings* of a:

Every x that is an element of b has the form $x = (u, v)$, where $u \varepsilon a$, $v \varepsilon a$, and $u \neq v$. If u and v are two distinct elements of a, then (u, v) or (v, u) belongs to b.[10] If (u, v) and (v, w) belong to b, then (u, w) also belongs to b.

If (u, v) belongs to b, we also write $u \overset{(b)}{<} v$ or $v \overset{(b)}{>} u$ (u precedes v in the ordering b, v follows u in the ordering b).

And, should a be a set, all its orderings are sets and they, too, form a set, $O(a)$.

The remaining definitions (those of similarity, well-ordering, ordinal number, cardinality, equivalence, finitude, and infinitude) will not be given here; some of them are obvious. In another paper (*1923*) I have already set forth the theory of ordinals that would be appropriate here (I used the language of naive set theory, to be sure; but a translation into the present axiomatization would not cause any difficulty).

II. INVESTIGATION OF THE AXIOMS

§ 1. Statement of the problem. Fundamentals

From the axioms, as they were presented in Part I, all the well-known theorems of set theory follow; on the other hand, these axioms (insofar as they are not of a restrictive kind) are nothing but trivial facts of naive set theory. In this sense, then, we could say that neither too much nor too little has been required by our axioms.

For several reasons, however, such an appraisal would not be sound. The first is that, while our axioms do enable us, starting with finite and denumerable sets, to

[10] [[The German text erroneously has "a" instead of "b".]]

construct the familiar sets $a + b$, $S(a)$, $P(a)$, and $|[c, a]|$, as well as to use "separation", they do not guarantee that besides the sets thus obtained there are no further sets that cannot be reached in this way. A priori there might very well exist sets of this kind too, for example, a set having a single element that is the set itself, $a = \{a\}$, or a "descending sequence of sets", $a_1 = \{a_2\}$, $a_2 = \{a_3\}, \ldots$ ($\{a\}$ means the set whose sole element is a).[11] It would now be desirable to eliminate all these superfluous sets, and this is certainly not accomplished by the axioms that we have adopted so far.

In order to fill this gap, Fraenkel regards as desirable the introduction of a further axiom (the axiom of restriction ⟦Beschränktheitsaxiom⟧), which has not yet been precisely formulated; it has no analogue among our axioms, and it would read approximately as follows:

Besides the sets (or, in our system, the I-objects and II-objects) whose existence is absolutely required by the axioms, there are no further sets.

We wish to express this axiom in our formalism in a precise way. To this end we introduce the following definitions:

Let Σ be the system of I-objects and II-objects. Let Σ' be a subsystem of Σ. Let $\mathrm{I}_{\Sigma'}$-objects and $\mathrm{II}_{\Sigma'}$-objects be the I-objects and II-objects, respectively, that are in Σ'. Let $[x, y]_{\Sigma'}$ (where x is a $\mathrm{II}_{\Sigma'}$-object and y a $\mathrm{I}_{\Sigma'}$-object) mean $[x, y]$; let $(x, y)_{\Sigma'}$ (where x and y are $\mathrm{I}_{\Sigma'}$-objects) mean (x, y); let $A_{\Sigma'}$ be A and let $B_{\Sigma'}$ be B.

Now, if these $\mathrm{I}_{\Sigma'}$-objects and $\mathrm{II}_{\Sigma'}$-objects, the operations $[x, y]_{\Sigma'}$ and $(x, y)_{\Sigma}$, and the objects $A_{\Sigma'}$ and $B_{\Sigma'}$ also satisfy our axioms, we say for short that Σ' satisfies our axioms.

Then the axiom of restriction just mentioned simply requires that besides Σ itself no other subsystem Σ' of Σ shall satisfy Axioms I–V.

This formulation makes it clear that two serious objections can immediately be raised against such an axiom (this, of course, is equally true in Fraenkel's system).

First, this axiom is of a type quite different from the previous ones, since, contrary to the principle we have followed hitherto, it does not avoid the notions of naive set theory. For what are we to understand by "subsystems" of Σ? Certainly not sets or classes in the sense of the previous axioms, for these can have only I-objects as elements, while Σ' and Σ contain I-objects and II-objects. But what else, then, since after all the naive notion of set was to be strictly prohibited? Such an axiom would make the whole process of axiomatization circular!

This difficulty could, however, be eliminated if we were to suppose, for example, that there was given a larger system P (of I_P-objects and II_P-objects, with the operations $[x, y]_P$ and $(x, y)_P$, and two objects A_P and B_P, the system also satisfying our Axioms I–V) such that all I-objects and II-objects are I_P-objects and that both $[x, y]$ and (x, y) can in P be brought into the normal form $[a, (x, y)_P]_P$ (where a is a II_P-object) of P. Then Σ is a class in P and its "subsystems" Σ' are simply to be taken as its subclasses (in P). We would have overlaid Σ with a "higher set theory" P, in which even objects incapable of being arguments in Σ are arguments. This in itself is not absurd. If we make the sets that are "too big" and incapable of being arguments capable of being arguments in a new system P, we can still circumvent the antinomies if in turn we admit the sets that are formed from all of these and are "still bigger"

[11] *Mirimanoff 1917*, p. 42. ⟦"(a)" has been changed to "$\{a\}$", the latter being the notation used by von Neumann below.⟧

(that is, too big in P) but declare them incapable of being arguments. The idea is partly the same as the one upon which Russell's "hierarchy of types" rests.

In such a "higher set theory" P it would thus make sense to ask whether the axiom of restriction[12] mentioned above is satisfied (for the "lower set theory" Σ). In what follows we shall, to be sure, use the terminology of naive set theory for the sake of simplicity, but in doing so we must always keep in mind that the existence of a "higher" system P has been assumed. Without such a hypothesis (which is somewhat more problematic still than that of the consistency of set theory) one just cannot undertake an investigation of the systems Σ' satisfying the axioms, unless one wishes to find oneself uncritically using the terminology of (inconsistent) naive set theory.

And now the second difficulty appears. It is easy to demonstrate that a system satisfying Axioms I–V need not satisfy the axiom of restriction (see above). We would, therefore, have to know something like this: If the system Σ satisfies Axioms I–V, there exists among those of its subsystems that also satisfy them at least one smallest one Σ', that is, one having the property that besides Σ' itself no subsystem of Σ' satisfies Axioms I–V.

Hence this subsystem would satisfy the axiom of restriction. For example, the common part (intersection) of all the subsystems Σ' of Σ that satisfy Axioms I–V might be such a smallest system (and it would be if it, too, should satisfy Axioms I–V). This, however, is not necessarily the case. A closer examination now shows that the only known way by which such a subsystem might be produced fails. Later we shall identify the circumstance to which this is due. For these reasons we believe that we must conclude, first, that the axiom of restriction absolutely has to be rejected and, second, that one cannot possibly succeed in formulating an axiom to the same effect. That, incidentally, is also connected with the fact that the axiom system I–V lacks "categoricity"; more will be said about this in § 5.

But, even if it is not possible to find a smallest subsystem of Σ satisfying Axioms I–V, we still wish to investigate what subsystems of Σ satisfying Axioms I–V can exist. In doing so we encounter a most peculiar phenomenon, which was first noticed by Löwenheim and Skolem.[13]

§ 2. On subsystems

Let a subsystem Σ' be given. We now wish to give a precise formulation to the condition that Σ' satisfies the axioms.

For Axioms I 1–3 this is very easy. They simply read as follows:

1. A and B belong to Σ'.
2. If x and y belong to Σ', then $[x, y]$ and (x, y) also belong to Σ'.

[12] ⟦The German text has "Bestimmtheitsaxiom" in place of what should clearly be "Beschränktheitsaxiom".⟧

[13] *Löwenheim 1915*, *Skolem 1920*, *1922*; in this connection see *Fraenkel 1923b*.

In the first two of these papers a general proof is given of the theorem whose special case for set theory is the subject of § 2 below: Every axiom system that is satisfiable at all is already satisfiable by denumerable systems. In the third paper Skolem draws (unfavorable) conclusions from this about set theory.

Although the content of § 2 and § 3 is therefore not new, we believe that it is not useless to call attention again in some detail to this interesting circumstance.

In connection with II 1–7, III 1, and IV 1, there is a certain difficulty. II 1, for instance, requires that there belong to Σ' a II-object a such that, for all II-objects x of Σ', $[a, x] = x$. Similarly for II 2–IV 1 ⟦of the list just given, that is, not including III 2 and III 3⟧.

To avoid this difficulty, however, we shall simply require something more in each case, namely:

3. There belongs to Σ' a II-object a such that always (that is, for all I-objects in Σ; similarly below) $[a, x] = x$.

4. Assume that the I-object u belongs to Σ'; then there belongs to Σ' a II-object a such that always $[a, x] = u$.

5. There belongs to Σ' a II-object a such that always $[a, (x, y)] = x$.

6. There belongs to Σ' a II-object a such that always $[a, (x, y)] = y$.

7. There belongs to Σ' a II-object a such that always $[a, (x, y)] = [x, y]$.

8. If the II-objects a and b belong to Σ', then there belongs to Σ' a II-object c such that always $[c, x] = ([a, x], [b, x])$.

9. If the II-objects a and b belong to Σ', then there belongs to Σ' a II-object c such that always $[c, x] = [a, [b, x]]$.

10. There belongs to Σ' a II-object a with the following property: $[a, (x, y)] \neq A$ if and only if $x = y$.

11. There belongs to Σ' a II-object a with the following property: $[a, x] \neq A$ if and only if x is a I-II-object.

(That there exist such II-objects at all is guaranteed by the corresponding axioms from II 1 to IV 1.) From here on, therefore, we have sufficient conditions but no longer necessary ones. And therewith the possibility of finding a smallest subsystem satisfying the axioms vanishes, since the conditions to be formulated all go too far.

A further difficulty appears in connection with Axioms III 2 and III 3. Let us, for example, consider III 2; it reads as follows:

Assume that the II-object a belongs to Σ'; then there exists a II-object b in Σ' with the following property: for all I-objects x in Σ', $[b, x] \neq A$ if and only if, for all I-objects y in Σ', $[a, (x, y)] = A$.

Again Σ' enters in the definition of the b that is to belong to Σ'; for in order to know whether something holds of "all I-objects y in Σ'" we must, of course, know all of Σ'. However, we can overcome this difficulty by forcing "for all I-objects y in Σ'" to mean the same as "for all I-objects y" in the cases that come under consideration. And this we achieve simply as follows.

12. (Preliminary condition.) Assume that both the II-object a and the I-object x belong to Σ'; if there exists any I-object y such that $[a, (x, y)] \neq A$, then at least one such y also is to belong to Σ'.

And now, for III 2, we can require the condition that follows (here again, as with II 1–IV 1, it is a trifle too much).

13. (Principal condition.) Assume that the II-object a belongs to Σ'; then there belongs to Σ' a II-object b with the following property: $[b, x] \neq A$ if and only if always $[a, (x, y)] = A$.

In the case of III 3 the situation is the same. To express the "uniqueness" of y here, we must add the following condition to the "preliminary condition" 12:

14. Assume that the II-object a and the I-objects x and y belong to Σ'; if

$[a, (x, y)] \neq A$ and if there exists a y' such that $[a, (x, y)] \neq A$ and $y' \neq y$, then one such y' also belongs to Σ'.

And now the principal condition reads:

15. Assume that the II-object a belongs to Σ'; then there belongs to Σ' a II-object b such that, whenever $[a, (x, y)] \neq A$ for a unique y, $[b, x] = y$.

In the case of Axiom I 4 we again require a "preliminary condition":

16. Assume that the II-objects a and b belong to Σ'; if there exists an x such that $[a, x] \neq [b, x]$, then one such x also belongs to Σ'.

A principal condition is not necessary here, since we are not stipulating that anything exist. (All these conditions are sufficient but not necessary.)

Finally there remain Axioms IV 2 and V 1–3. Here, too, we must formulate corresponding preliminary conditions (17–19 for IV 2, since "all" and "there exists" are nested three times; 20 and 21 for V 1; 22 for V 2; and 23 for V 3), but we shall not actually write them down one by one.

As for principal conditions, one is necessary for one half of IV 2 ("If a is not a I-II-object, then there exists a b such that..."). It reads as follows:

24. Let a be a II-object that is not a I-II-object and assume that it belongs to Σ'; then there belongs to Σ' a II-object b such that for every x there exists a y such that both $[a, y] \neq A$ and $[b, y] = x$.

But for the other half of IV 2 ("If there exists a b for which..., then a is not a I-II-object") no principal condition is necessary at all. The reason is the same as in the case of Axiom I 4; after all, we are not demanding that anything exist.

Finally, we do not need any principal conditions in the case of Axioms V 1–3 either. For, while we do indeed demand that some objects exist, these are I-II-objects. Now, that $\text{II}_{\Sigma'}$-objects with the required properties exist follows already from the axioms of groups I–IV, that is, for Σ', from conditions 1–19 and 24. But, since these $\text{II}_{\Sigma'}$-objects certainly are I-II-objects in Σ (Σ, after all, satisfies the axioms of group V), they must also be $(\text{I-II})_{\Sigma'}$-objects. (For $(\text{I-II})_{\Sigma'}$-objects are defined as I-II-objects that belong to Σ').

We are now in a position to summarize.

For Σ' to satisfy the axioms it is certainly sufficient that conditions 1–24 be satisfied. Each of these conditions has the following form:

Assume that the I-objects or II-objects $u_1, u_2, \ldots, u_n$ belong to Σ'; if they satisfy the condition $A(u_1, u_2, \ldots, u_n)$, then any I-object or II-object v satisfying the condition $B(u_1, u_2, \ldots, u_n, v)$ also belongs to Σ'.

Here (besides $u_1, u_2, \ldots, u_n$ and $u_1, u_2, \ldots, u_n, v$) only properties of Σ enter in conditions A and B; Σ' does not occur in them. For certain of these conditions (1, 3, 5–7, 10, and 11) we must set $n = 0$; that is, it is required that any v having the property $B(v)$ belong to Σ'.

But we can easily satisfy such conditions. There obviously exists a smallest Σ' satisfying these conditions (which, however, go farther than the original requirement that Σ' satisfy the axioms). We need only apply the following procedure:

Take all of the conditions 1–24 in which $n = 0$ (see above) and construct the I-objects and II-objects $v_1, v_2, \ldots, v_\mu$ that are postulated by them. Then take all of the conditions 1–24 in which $n \geqq 1$ (that is, that actually contain some of the variables $u_1, u_2, \ldots, u_n$). Substitute all possible combinations of $v_1, v_2, \ldots, v_\mu$ in these conditions and construct the I-objects and II-objects $v'_1, v'_2, \ldots, v'_{\mu'}$ that are then postulated. Then substitute all possible combinations of $v_1, v_2, \ldots, v_\mu, v'_1, v'_2, \ldots, v'_{\mu'}$ in the conditions and construct the I-objects and II-objects $v''_1, v''_2, \ldots, v''_{\mu''}$ that are then postulated, and so on, and so on.

If we now choose Σ' as the system of the objects $v_1, v_2, \ldots, v_\mu, v'_1, v'_2, \ldots, v'_{\mu'}, v''_1, v''_2, \ldots, v''_{\mu''}, \ldots$, we have a system that satisfies our axioms.

§ 3. Denumerability

The Σ' obtained in § 2 has a most surprising property: it is obviously denumerable. But attention must be paid here to the meaning of the word "denumerable". Σ' is not denumerable in the sense that it has cardinality $\aleph_0$ as a class in the system Σ (or Σ'), that is, that it can be mapped one-to-one onto the first infinite ordinal ω by means of a II-object in Σ.[14] That, of course, is out of the question, for Σ' is not a class at all; it also contains II-objects (see § 1). Moreover, parts of it (namely, all nondenumerable subclasses of Σ') are "nondenumerable" in the sense that this notion has in the system Σ'. But it is "denumerable" if we regard it (and together with it all the subclasses mentioned above) as a class in the "higher" system P, or, in the words of naive set theory, if its elements can *de facto* be written as a sequence.

It is essential to prevent any misunderstanding here. The system Σ' contains a number of sets and mappings. They satisfy the formal requirements of set theory. For every possible cardinality there are sets of that cardinality. But all these cardinalities are specious; they are cardinalities only with respect to the group of mappings that belong to the system. For (in spite of its formal completeness) the system Σ' does not by any means contain all conceivable mappings; any "higher" system P must already contain new mappings, for example, mappings that map all (infinite) sets in Σ' onto one another. For as parts of Σ', which is denumerable (in P), all of these sets are, of course, denumerable, hence of the same cardinality (in P). One might perhaps believe that this would contradict Axiom IV 2; the II-object ω (or any infinite I-II-object) can be mapped onto the class of all I-objects, and it is nevertheless a I-II-object! But it is clear what the answer must be: the mapping in question belongs to P (is a II_P-object) and not to Σ', and Axiom IV 2, of course, refers only to II-objects in Σ'.

This relativity of cardinalities is very striking evidence of how far abstract formalistic set theory is removed from all that is intuitive. One can indeed construct systems like Σ' that, by satisfying certain formal axioms, faithfully represent set theory down to the last detail and that, therefore, to all intents and purposes are formal set theory itself. In these systems, all known cardinalities occur, in their infinite multiplicity, which is larger than any cardinality. But as soon as one applies finer instruments of investigation ("higher" systems P) all this fades away to nothing. Of all the cardinalities only the finite ones and the denumerable one remain. Only these have real meaning; everything else is formalistic fiction.

This circumstance, incidentally, is by no means a special feature of our axiomatization. We could apply the procedure used in § 2 in almost exactly the same way (see

[14] This is indeed equivalent to "denumerable", since according to our definition of ordinal number (see *von Neumann 1923*) ω is a denumerable set.

the papers of Löwenheim and Skolem) if our axioms were replaced by any other logical conditions. The construction just carried out stamps every axiomatic set theory with the mark of unreality (or, to employ a frequently used word, "impredicativity" [["Imprädikativität"]]).

§ 4. Models of set theory

We now know that, if it is at all possible to find a system Σ satisfying the axioms, we can also find some such system in which there are only denumerably many I-objects and denumerably many II-objects. But this tends to suggest that we can find a model of set theory by purely arithmetic means. We could, perhaps, make an attempt along the following lines.

Let the I-objects be the integers 1, 2, ...

Let the II-objects all be functions f in a set Φ of functions (Φ, of course, is likewise denumerable) whose domain of definition and range of values are the integers.

Let (x, y) be a given function $p(x, y)$, defined for $x, y = 1, 2, \ldots$.

Let $[x, y]$ be $f(y)$ if x is the function f in Φ and $y = 1, 2, \ldots$.

Let A be 1 and let B be 2.

But an amendment must still be made concerning the I-II-objects; for, as these specifications read at this point, the I-objects all are distinct from the II-objects. This can, however, easily be remedied. We assume that all I-objects are also I-II-objects. This extension of Axiom IV 1 is nonessential; we can show that, if the axioms are consistent, they are still so after the extension. Then we must simply indicate for each I-object, that is, for each number 1, 2, ..., with which II-object, that is, with which function in Φ, it is to be identified. For this purpose we need a further two-variable function $\varphi(x, y)$; then we shall assign to each number x the function $\varphi(x, y)$ considered as a function of y (the function must therefore belong to Φ). To summarize:

Let the function $\varphi(x, y)$ be defined for $x, y = 1, 2, \ldots$. For a fixed x, $\varphi(x, y)$ is a function of y; as such it belongs to Φ for every x.

The following question remains: what conditions must be met by the two functions p and φ and the set of functions Φ (only these are still arbitrary in our specifications) for the axioms to be satisfied?

We can show, to begin with, that for $p(x, y)$ (which is $= (x, y)$) we may choose, say, $2^{x-1}(2y - 1)$ (from $p(x_1, y_1) = p(x_2, y_2)$ it follows that $x_1 = x_2$ and $y_1 = y_2$), since this function plays no particularly deep role. Essential, however, is what we choose for $\varphi(x, y)$ and Φ.

Now we can formulate the conditions for these two. We shall not carry this out in detail here but only indicate something of their form.

The conditions are rendered substantially more strict and complex (as against those in § 3) by the fact that now the system is no longer part of a larger one that satisfies the axioms (as Σ' was a part of Σ). Consequently, the conditions will no longer have the characteristic of being easily satisfiable. Axiom IV 2, in particular, raises problems. For it requires that no function that maps a set $\varphi(x, y) \neq 1$ (x fixed, y being the element) onto the set of all numbers can belong to Φ. Now, while most of the conditions prescribe a lower bound for Φ (that is, require that certain functions must belong to Φ), this axiom prescribes an upper bound for it. We have no guarantee, initially, that the two bounds do not collide; if they did, there would be new antinomies. This, however, is not as weighty an objection against the choice of Axiom IV 2 as one might at first believe. For, instead of IV 2, one would in any case have to adopt at least Zermelo's axiom of separation (and even that would not suffice for many purposes), say, in the following form:

IV 2. If b is a I-II-object, a is a II-object, and $a \lesssim b$, then a too is a I-II-object.

This axiom corresponds to the axiom of separation. For, that a is a II-object is precisely what Zermelo means when he says that a is "determined by a definite property". From this, together with $a \lesssim b$, where b is a I-II-object, the admissibility of a (that is, that a is a I-II-object) must follow.

And, as we can easily convince ourselves, this (absolutely unavoidable) axiom would create exactly the same difficulties as ⟦the original⟧ IV 2. The place of IV 2 absolutely must be occupied by some specifically "impredicative" axiom; and such an axiom is bound to cause difficulties in the construction of a model.

The conditions that are obtained in this way are so difficult to survey and so complex that we cannot specify any model and indeed cannot even determine whether they are compatible at all, although the construction must be feasible if set theory can in any way be erected on a nonintuitionistic basis.

I would like to make one final remark. If we omit the axioms of group V (infinity), the remaining axioms provide an adequate foundation for the theory of finite numbers; indeed, even the theory of real numbers remains possible to a limited extent: they are infinite classes, hence not sets (I-II-objects). We obtain a mathematics in which a theory of real numbers that rests on fundamental sequences is possible, the theorems on the convergence of sequences and series hold, and the theory of continuous functions, algebra, analysis, and the Riemann integral are possible. But, since sets of II-objects are impossible, the Weierstrass theorem on the upper bound (for sets of numbers, not sequences) becomes meaningless, as do the general notion of function, the well-ordering of the continuum, and the Lebesgue integral.

Since the axioms now considered deal with finite sets only (all I-II-objects are, after all, finite), we can specify a model for them. $\Phi(x, y)$ must be so chosen that we can represent all functions that are $\neq 1$ for only finitely many numbers.

For example, assume that the prime numbers have been arranged in order of increasing magnitude, $p_1, p_2, \ldots$. If

$$x = \prod_{n=0}^{\infty} p_n^{a_n}$$

(all $a_n \geqq 0$, only finitely many > 0), then

$$f(x, y) = a_y + 1.$$

The choice of Φ is then easy; we have only constructive conditions, as we can see rather easily (for in this case IV 2 is automatically satisfied; the collision of the two bounds on Φ that were mentioned above does not occur).

We thus obtain, as further evidence of what was said in § 3 about denumerability, a denumerable model of a pseudomathematics that agrees with the "real" one in many essential respects. To be sure, the model of the "big" set theory is unknown, but it, too, must exist if a formalistic set theory is at all possible.

§ 5. Categoricity

Now we must still investigate whether our axiom system is categorical ⟦kategorisch⟧, that is, whether it uniquely determines the system it describes.[15] We shall now explain this notion in more detail.

It is known that from Euclid's axioms without the fifth postulate nothing follows about this fifth postulate. That is, there can be two systems—both satisfying these axioms—of which the first satisfies the postulate while the second does not. Such a situation, however, can never prevail once geometry is properly axiomatized; a geometric proposition that is true in one system satisfying the axioms of geometry is true also in any other such system. (We shall for the moment disregard the fact that the axioms of geometry in the last analysis depend upon those of set theory—on account of "continuity".) That this is so is due to the following

Isomorphism theorem. Let A_1 and A_2 be two systems that satisfy the axioms of geometry. Then there is a one-to-one mapping of A_1 onto A_2 under which the relations underlying the axioms are preserved, that is, under which incident points and lines, segments of equal length, congruent triangles, and so on go over into their likes (that is what "isomorphism" means).

From this theorem, which is easy to prove, it obviously follows that, if a proposition formulated in terms of these basic relations is satisfied in A_1, it is also satisfied in A_2.

Therefore an axiom system of this latter kind, one for which an isomorphism theorem analogous to the theorem just stated holds, uniquely determines the logical properties of the systems described by it; it is said to be *categorical.* Now, is our axiom system of this kind?

This is most important. For we know only that the propositions of set theory that have already been settled follow from it. But those that have not yet been decided, for example, the continuum problem, might (if categoricity is lacking) be true in one system satisfying the axioms and false in another. That is, we would have no assurance at all that these axioms suffice to settle, say, the continuum problem.

It is clear that the axioms in their present form are far too broad to be categorical. After all, we do not know, for example, whether there exist I-objects that are not I-II-objects, whether A and B are sets, whether $(A, B) = A$ or $\neq A$, and so forth. This can, however, easily be remedied. We require the following (it can be proved that these axioms do not generate any contradiction if the previous ones are consistent[16]).

VI

1. All I-objects are I-II-objects. (This makes IV 1 superfluous.)
2. $A = O$ and $B = \{O\}$. (O is the set without elements, $\{O\}$ contains only O.)
3. $(u, v) = \{\{u, v\}, \{u\}\}$. ($\{\alpha, \beta\}$ is the set with the elements α and β. From $\{\{u_1, v_1\}, \{u_1\}\} = \{\{u_2, v_2\}, \{u_2\}\}$ it follows that $u_1 = u_2$ and $v_1 = v_2$, as we can easily prove.)

We can also remove a further obstacle, to which we already alluded in § 1, namely, the possibility that there might exist "inaccessible" sets, such as, for example, "descending sequences of sets" (see § 1). In § 1 we presented the reasons why it is

[15] This notion of "categoricity" is due to Veblen (*1904*, p. 346).
[16] ⟦This problem is treated in *von Neumann 1929.*⟧

impossible to achieve this directly through an "axiom of restriction". But, formally, the exclusion of "descending sequences of sets", too, is sufficient (we do not enter into the proof here). Hence

4. There exists no II-object α such that, for every finite ordinal (that is, integer) n, $[\alpha, n + 1] \varepsilon [\alpha, n]$.

(Here, too, it can be proved that Axiom VI 4 cannot generate any new contradiction. Strictly speaking, one further source of noncategoricity would still have to be excluded, namely, the possible existence of what are called regular initial numbers with a limit index.[17] This leads us too far into the field of special set theory, however, to be treated here; moreover, this difficulty, too, can be eliminated.)

But even now that this axiom group VI has been added our axiom system is in reality most likely still not categorical. Notwithstanding all our efforts, the required isomorphic mapping onto each other of two systems Σ_1 and Σ_2 that satisfy the axioms cannot be successfully constructed. It is wrecked, ultimately, on the fact that Axiom VI 4 does not exclude all "descending sequences of sets" but only those having the normal form of sequences in the system Σ_1 (or Σ_2), namely, $[a, 1]$, $[a, 2]$, $[a, 3]$, ... (where a is a II-object in Σ_1 (or Σ_2)). However, it is, of course, always possible that there remain some "outside the system". In the interest of the isomorphic mapping we would have to forbid these, too; that is, we would again have to reach for a system P "higher" than both of the systems Σ_1 and Σ_2. And this is impossible with Axiom VI 4, which has to do with Σ_1 (or Σ_2) *singly*.

The consequence of all this is that no categorical axiomatization of set theory seems to exist at all; for probably no axiomatization will be able to avoid the difficulty connected with the axiom of restriction and the "higher" systems. And since there is no axiom system for mathematics, geometry, and so forth that does not presuppose set theory, there probably cannot be any categorically axiomatized infinite systems at all. This circumstance seems to me to be an argument for intuitionism.

In connection with the question of categoricity, let us still mention the following. Let two systems Σ_1 and Σ_2 be given, let a_1 and a_2 be sets in them, and let the elements of a_1 in Σ_1 be exactly the same as the elements of a_2 in Σ_2. We know that it may then very well happen that a_1 is nondenumerable in Σ_1 while this is not the case for a_2 in Σ_2 (if Σ_2 is "higher" than Σ_1). But perhaps the situation is the same even for finitude. Since the notions "all" and "there exists" occur with reference to the entire system (Σ_1 or Σ_2), the definition of finitude is in any case such that we cannot say anything definite.[18] For well-ordering the situation is the same. Thus we must take the relativity of cardinalities into consideration not only upward (from the denumerable; see § 3) but also downward (in the finite). In any case, elementary notions such as finitude and well-ordering depend upon the system chosen (Σ_1 or Σ_2); and it is not precluded that this dependence is of an essential nature, that a set a appears to be well-ordered (or to be finite) in the system Σ_1 and turns out to be not well-ordered (or to be infinite) in the "finer" system Σ_2, merely because a certain part of a, namely

[17] ⟦This possibility was discussed for the first time by Hausdorff (*1908*, pp. 443–444, and *1914*, p. 131).⟧

[18] Finitude can, for example, be defined as follows:

A class a is said to be finite if there is no class b possessing the following properties: b has elements that are $\lesssim a$; if x is an element of b and is $\lesssim a$, then b also has elements that are $\lesssim a$ and at the same time $< x$.

b, that has no first element was not a set in the system Σ_1 and thus went unnoticed there, while it is a set in the system Σ_2 (analogously for finitude).

Indeed, in the last analysis it would be conceivable that *every* system Σ_1 could be further "refined" in such a way that finite (or well-ordered) sets turn out to be infinite (or not well-ordered). (For "nondenumerable" this is surely the case.) Of the notion of finitude, too, nothing but the shell of its formal characterization would then remain (just as was the case for nondenumerability). It is difficult to say whether this would militate more strongly, so far as finitude is concerned, against its intuitive character, championed by intuitionism, or its formalization as given by set theory.

It is really an objection against both; after all, a new difficulty appears here, one that is essentially different from those pointed out by Russell and Brouwer. The denumerable infinite as such is beyond dispute; indeed, it is nothing more than the general notion of the positive integer, on which mathematics rests and of which even Kronecker and Brouwer admit that it was "created by God". But its boundaries seem to be quite blurred and to lack intuitive, substantive meaning ⟦anschaulich-inhaltliche Bedeutung⟧. Upward, in the "nondenumerable", this is quite certain in view of Löwenheim's and Skolem's investigations. Downward, in the "finite", it is at least very plausible, for categoricity is lacking, as is any foothold that would enable us to make the definition of "finite" determinate. Moreover, *even Hilbert's approach is powerless here*, since this objection does not concern consistency but the univocality (categoricity) of set theory.

At present we can do no more than note that we have one more reason here to entertain reservations about set theory and that for the time being no way of rehabilitating this theory is known.

On the principle of excluded middle

ANDREI NIKOLAEVICH KOLMOGOROV

(*1925*)

To a large extent, this paper anticipated not only Heyting's formalization of intuitionistic logic, but also results on the translatability of classical mathematics into intuitionistic mathematics. It provides an important link between intuitionism and other works on the foundations of mathematics.

Two formal systems $\mathfrak{B}$ and $\mathfrak{H}$ (Brouwer's and Hilbert's) of the propositional calculus are proposed. Both contain a (somewhat uneconomical but) complete set of axioms of the positive implicational calculus (I, § 5, (1), Axioms 1, 2, 3, 4) and rules of modus ponens and substitution. In addition, $\mathfrak{B}$ contains also (II, § 6):

Axiom 5. $(A \rightarrow B) \rightarrow ((A \rightarrow \overline{B}) \rightarrow \overline{A})$.

The system $\mathfrak{H}$ is obtained from $\mathfrak{B}$ by the addition (II, § 7) of

Axiom 6. $\overline{\overline{A}} \rightarrow A$.

It is proved that $\mathfrak{H}$ is equivalent to Hilbert's formulation of the classical propositional calculus.

The system $\mathfrak{B}$ is nowadays known as the minimal calculus and differs from Heyting's system in that the latter contains, in addition to Axioms 1–5, also

Axiom *h*. $A \rightarrow (\overline{A} \rightarrow B)$.

The status of *h* in intuitionistic logic is not without question. Thus, according to Kolmogorov, it "is used only in a symbolic presentation of the logic of judgments; therefore it is not affected by Brouwer's critique" (I, § 6), and it "does not have and cannot have any intuitive foundation since it asserts something about the consequences of something impossible: we have to accept B if the true judgment A is regarded as false" (II, § 4). Heyting appears rather diffident about defending the inclusion of *h*. Thus, according to him (*1956*, p. 102), if we have deduced a contradiction from the supposition that the construction A was carried out, "then, in a sense, this can be considered as a construction.... I shall interpret implication in this wider sense".

Hence it is fair to say that, as a codification of Brouwer's ideas, $\mathfrak{B}$ is no less reasonable than Heyting's propositional calculus.

Kolmogorov makes no attempt in this paper to formalize intuitionistic quantification theory completely. Rather, he just lists as intuitively obvious the rule **P** of generalization, which allows us to prefix (a) to any given formula, and four axioms, I, II, III, IV (V, § 3). In the same context, however, he also argues to the effect that

Axiom *g*. $(a)A(a) \rightarrow A(t)$

is intuitively true. If, therefore, to the system $\mathfrak{B}$ we adjoin the explicitly stated rule **P** and Axioms I–IV plus Axiom *g*, we obtain an adequate axiomatization $\mathfrak{BQ}$ of intuitionistic logic that differs from Heyting's system only in the omission of the questionable Axiom *h*. This is the extent of the anticipation of Heyting's formalization.

The main purpose of the paper is to

prove that classical mathematics is translatable into intuitionistic mathematics. For this purpose, with each formula $\mathfrak{S}$ of mathematics there is associated a translation $\mathfrak{S}^*$ in a perfectly general manner (IV, § 2). If $\overline{}$, $\rightarrow$, (a), and (Ea) are the only symbols we use for forming new formulas from given formulas, the definition amounts to: for atomic $\mathfrak{S}$, $\mathfrak{S}^*$ is its double negation $\overline{\overline{\mathfrak{S}}}$, or $n\mathfrak{S}$; $(\overline{B})^*$ is $n(\overline{B^*})$; $(A \rightarrow B)^*$ is $n(A^* \rightarrow B^*)$; $((a)A(a))^*$ is $n(a)(A(a))^*$; $((Ea)A(a))^*$ is $n(Ea)(A(a))^*$.

The following results are proved exactly (III, § 3):

Lemma 1. $\vdash_{\mathfrak{B}} n\overline{A} \rightarrow \overline{A}$.

Lemma 2. If $\vdash_{\mathfrak{B}} nA \rightarrow A$ and $\vdash_{\mathfrak{B}} nB \rightarrow B$, then $\vdash_{\mathfrak{B}} n(A \rightarrow B) \rightarrow (A \rightarrow B)$.

Theorem I. If $\mathfrak{S}_1, \ldots, \mathfrak{S}_k$ are all the atomic formulas in A and A is constructed from $\mathfrak{S}_1, \ldots, \mathfrak{S}_k$ by negation and implication only, then, provided

$$n\mathfrak{S}_1 \rightarrow \mathfrak{S}_1, \ldots, n\mathfrak{S}_k \rightarrow \mathfrak{S}_k$$

are true (or taken as additional axioms), (1) $\vdash_{\mathfrak{B}} nA \rightarrow A$ and (2) $\vdash_{\mathfrak{B}} A$ if $\vdash_{\mathfrak{H}} A$.

If $\mathfrak{U} = \{\mathfrak{U}_1, \ldots, \mathfrak{U}_k\}$ is a set of axioms and $\mathfrak{U}^*$ is $\{\mathfrak{U}_1^*, \ldots, \mathfrak{U}_k^*\}$, then (IV, § 3):

Theorem II. If $\mathfrak{U} \vdash_{\mathfrak{H}} \mathfrak{S}$, then $\mathfrak{U}^* \vdash_{\mathfrak{B}} \mathfrak{S}^*$.

The proof of this theorem uses Lemma 2 for the rule of modus ponens and (in IV, § 4) the fact that the translations $\mathfrak{S}_1^*, \ldots, \mathfrak{S}_m^*$ of the atomic formulas have the property

$$\vdash_{\mathfrak{B}} n\mathfrak{S}_1^* \rightarrow \mathfrak{S}_1^*, \ldots, \vdash_{\mathfrak{B}} n\mathfrak{S}_m^* \rightarrow \mathfrak{S}_m^*.$$

This last fact and Theorem I yield the result that the translation of Axioms 1–6 all are theorems of $\mathfrak{B}$.

Strictly speaking, Theorem II is established only for the case in which $\mathfrak{S}$ is built up by implication and negation (in particular, $\mathfrak{S}$ does not contain quantifiers). However, Kolmogorov does envisage a much stronger result and illustrates by an example the treatment of axioms about quantifiers, axioms that he tends to take as being on the same footing as axioms about numbers and sets (IV, § 5).

If we take his system $\mathfrak{B}\mathfrak{Q}$ (Axioms 1–5, I–IV, g, with rules of inference) of quantification theory and extend his illustration to cover also the rule **P** and the remaining axioms, and if we let $\mathfrak{H}\mathfrak{Q}$ be $\mathfrak{B}\mathfrak{Q}$ plus Axiom 6, then we have also:

Theorem III. If $\mathfrak{U} \vdash_{\mathfrak{H}\mathfrak{Q}} \mathfrak{S}$, then $\mathfrak{U}^* \vdash_{\mathfrak{B}\mathfrak{Q}} \mathfrak{S}^*$.

In that case, all the derivations in V, § 4, can be dispensed with because they would follow from Theorem III.

A very suggestive remark (beginning of IV, § 5, and last two paragraphs of IV, § 6) is that every axiom A of mathematics is of type $\mathfrak{R}$, that is, A^* is (intuitionistically) true. From this it would seem to follow that all classical mathematics is intuitionistically consistent (V, § 1). As we know, however, this conclusion, even today, has not yet been firmly established so far as classical analysis and set theory are concerned. On the other hand, it seems not unreasonable to assert that Kolmogorov did foresee that the system of classical number theory is translatable into intuitionistic theory and therefore is intuitionistically consistent. In fact, it is not hard to work out his general indications and verify such a conclusion.

This completes the summary of the anticipations of results on the intuitionistic consistency of classical mathematics. It remains to mention a few of the incidental remarks.

Kolmogorov (V, § 1.1) states that, contrary to a remark by Brouwer, a finitary conclusion established by nonintuitionistic methods is (intuitionistically) true.

Two new examples are given of propositions not provable without the help of illegitimate uses of the principle of excluded middle. One of them, suggested by Novikov, is that every point in the complement C of a closed set is contained in an interval in C (V, § 5). The other example is the Cantor-Bendixson theorem (V, § 6).

With respect to system $\mathfrak{B}$, it is stated explicitly (II, § 6) that "the question whether this axiom system is a complete axiom system for the intuitionistic general logic of judgments remains open".

At the end of II, § 2, the question of the classical completeness of Axioms 1–4 for implication alone is raised. As we know, the answer to the question is negative. In fact, it is now a familiar result that it is necessary to add Peirce's law, $((A \rightarrow B) \rightarrow A) \rightarrow A$, to render the positive implicational calculus classically complete.

Hao Wang

The paper was translated by the editor. When informed of the projected publication of the translation of his paper, Professor Kolmogorov signified his acquiescence.

Introduction

Brouwer's writings have revealed that it is illegitimate to use the principle of excluded middle in the domain of transfinite arguments. Our task here will be to explain why this illegitimate use has not yet led to contradictions and also why the very illegitimacy has often gone unnoticed.

Only the finitary conclusions of mathematics can have significance in applications. But transfinite arguments are often used to provide a foundation for finitary conclusions. Brouwer considers, therefore, that even those who are interested only in the finitary results of mathematics cannot ignore the intuitionistic critique of the principle of excluded middle.

We shall prove that all the finitary conclusions obtained by means of a transfinite use of the principle of excluded middle are correct and can be proved even without its help.

A natural question is whether the transfinite premisses that are used to obtain correct finitary conclusions have any meaning.

We shall prove that every conclusion obtained with the help of the principle of excluded middle is correct provided every judgment that enters in its formulation is replaced by a judgment asserting its double negation. We call the double negation of a judgment its "pseudotruth". Thus, in the mathematics of pseudotruth it is legitimate to apply the principle of excluded middle.

The necessity of introducing such notions as "pseudoexistence" and "pseudotruth" has long been felt in mathematics, if only in connection with the question of Zermelo's axiom. It is only now, however, that one of the forms of pseudotruth has received a strict determination and has been given a firm basis through axioms used in the domain of pseudotruth but not used for truth proper.

I. Formalistic and intuitionistic points of view

§ 1. From the formalistic point of view mathematics is a collection of formulas.[1] Formulas are combinations of elementary symbols taken from a definite supply. At the basis of mathematics lie a certain group of formulas, called axioms, and certain rules that enable us to construct new formulas from given formulas; as rules of this kind we have at the present time the inference according to the schema $\mathfrak{S}, \mathfrak{S} \rightarrow \mathfrak{T} \mid \mathfrak{T}$ and the rule of substitution of particular values for the symbols of variables of various kinds.

[1] See *Hilbert 1922a*, p. 152.

In contradistinction to the axioms, wittingly taken as "true", a certain group of formulas are wittingly taken as "false". A system of axioms is said to be "consistent" if no formula considered "false" can be obtained from them by a derivation carried out according to the rules.

§ 2. The formalistic point of view in mathematics asserts that the selection of the axioms constituting its basis is arbitrary and subject only to considerations of practical convenience that lie outside of mathematics and are, of course, more or less conventional.[2] The sole absolute demand made upon every mathematical system is, from the point of view now considered, the demand that the axioms constituting its basis be consistent.

The formulas proved on the basis of the axioms are said to be true, those leading to a contradiction false. The question of the truth or falsity of a consistent but unprovable formula has no meaning from the formalistic point of view. The existence of such formulas shows that the system of axioms is incomplete. We can complete an incomplete system of axioms, if for some reason this is desirable, by taking as an axiom an unprovable and consistent formula, or, with the same right, its contradictory. The selection of the formula taken as the new axiom, from each pair of contradictory formulas, is thus subject only to considerations of convenience.

§ 3. The point of departure of the intuitionistic conception is the recognition of the real meaning of mathematical propositions. The axioms that constitute the basis of mathematics are adopted in order to express facts given to us. This conception tolerates the formalistic method in the study of mathematical constructions as one among other possible methods but goes against the formalistic conception of mathematics as a whole.

To the question of the nature of unprovable but consistent propositions the intuitionistic conception gives an answer completely different from that given by the purely formalistic conception. Suppose that a system of axioms is given for a certain branch of mathematics, geometry for example. These axioms express properties of the subject under investigation, in this particular case, of space. Suppose, further, that a certain proposition of the branch selected cannot be proved on the basis of the given axioms but does not lead to a contradiction either. From the intuitionistic point of view two cases can occur. First, it may happen that the truth or falsity of the proposition considered follows from direct examination; in that case the given proposition, if it is true—or, if it is false, its contradictory—can be taken as a new axiom. Second, it may happen that the proposition is indeterminate, that is, that its truth or falsity does not follow from direct examination; in that case the only thing that we can do is to try to derive the proposition in question from others that are immediately obvious; if this does not succeed, the proposition has to be regarded as indeterminate, since it is possible that we shall subsequently have to adopt as obviously true axioms from which its truth or falsity can be derived; but whether this will be the case is precisely what we do not know at the present time.

[2] See *Whitehead and Russell 1910*, Introduction. Hilbert, too, is close to this point of view; for him absolute truths (*absolute Wahrheiten*) are propositions of "metamathematics" only, that is, assertions of consistency, but, on the other hand, the formulas of ordinary mathematics (*eigentliche Mathematik*), too, are in his opinion expressions of certain thoughts (*Gedanken*). (See *Hilbert 1922a*, pp. 152–153.)

§ 4. The formalistic point of view is advanced in mathematical logic too. In the present paper we confront it precisely in the field of logic. To deny any real meaning to mathematical propositions is, however, what constitutes the basis of the formalistic point of view in mathematical logic. In fact, no one would propose applying to reality any logical formula that has no real meaning. Thus, so long as mathematical logic is regarded only as a formal system whose formulas have no real meaning, it diverges from general logic; the formalistic point of view can exist only in mathematics and mathematical logic but not in the ordinary logic that lays claim to significance in applications to reality.

As for us, we do not isolate a special "mathematical logic" from general logic, but we admit only that the originality of mathematics as a science creates for logic special problems that are investigated by a specialized "mathematical logic". Only in this logic does a doubt arise concerning the unconditional applicability of the principle of excluded middle.

§ 5. The difference between the two points of view presented manifests itself even in the domain of the logic of judgments. In what follows we understand by the general logic of judgments the science that investigates the properties of arbitrary judgments independently of their content, so far as their truth, their falsity, and the ways in which they are derived are concerned. (Each judgment is regarded as an unanalyzable element in the investigation.) The general logic of judgments is formally expressed with the help of symbols for arbitrary judgments, A, B, $C, \ldots$, of the symbol for implication, $A \rightarrow B$, and of the symbol for negation, $\overline{A}$.

Hilbert (*1922a*, p. 153) offered the following system of axioms for the logic of judgments:

Axioms of implication

$$(1) \qquad \begin{cases} 1. \ A \rightarrow (B \rightarrow A), \\ 2. \ \{A \rightarrow (A \rightarrow B)\} \rightarrow (A \rightarrow B), \\ 3. \ \{A \rightarrow (B \rightarrow C)\} \rightarrow \{B \rightarrow (A \rightarrow C)\}, \\ 4. \ (B \rightarrow C) \rightarrow \{(A \rightarrow B) \rightarrow (A \rightarrow C)\}. \end{cases}$$

Axioms of negation

$$(2) \qquad \begin{cases} 5. \ A \rightarrow (\overline{A} \rightarrow B), \\ 6. \ (A \rightarrow B) \rightarrow \{(\overline{A} \rightarrow B) \rightarrow B\}. \end{cases}$$

The inner consistency of these axioms can be proved in an extremely elementary way.[3] From the formalistic point of view this is sufficient for accepting them as a basis for the general logic of judgments.

Moreover, Hilbert's system is complete: no new independent axiom can be added without contradiction. More precisely, for every formula written with the symbols of the logic of judgments, even such a formula as

$$\overline{\overline{(A \rightarrow B)}} \rightarrow (\overline{\overline{A}} \rightarrow \overline{\overline{B}}),$$

[3] See *Ackermann 1924*.

either the formula can be proved on the basis of Hilbert's axioms or the consequence

$$A,$$

that is, the truth of an arbitrary judgment, can be derived from it by means of the same axioms.

§ 6. From the intuitionistic point of view the mutual consistency of Hilbert's axioms is by no means sufficient for their acceptance. In the next chapter we shall analyze the source of their significance for judgments in general and for particular forms of judgments.

One of Hilbert's two axioms of negation, Axiom 6, expresses the principle of excluded middle in a somewhat unusual form. Brouwer proved that the application of this principle to arbitrary judgments is without any foundation.[4] Axiom 5 is used only in a symbolic presentation of the logic of judgments; therefore it is not affected by Brouwer's critique, especially since it has no intuitive foundation either.

Thus, together with a critique of Hilbert's axioms, we shall have to present new axioms of negation, whose applicability to arbitrary judgments has been ascertained.

II. Axioms of the logic of judgment

§ 1. The axioms of the general logic of judgments lay claim to having significance for all judgments; therefore, they must follow from the general properties of judgments. To be sure, what comes immediately below is not at all a definition of fundamental notions or a proof of the axioms of the logic of judgments, but a search for their intuitive sources that already uses all the notions and devices of logic.

In the logic of judgments the judgment is considered the ultimate element in the investigation. When we consider the judgment independently of the synthesis of subject and predicate that it contains, there remains the sole characteristic property of a judgment, the one that distinguishes it from other forms of expression and was stated by Aristotle:[5] it can be appraised from the point of view of truth or falsity. It is natural to try to derive the axioms of the general logic of judgments without going beyond its own boundaries, that is, purely from the property of judgments that

[4] See *Brouwer 1923d*, p. 252. Hilbert, too, thinks that the principle of excluded middle is not intuitively obvious when applied to infinite collections of objects. In that case he expresses it symbolically by the two formulas

$$\overline{(a)}A(a) \ \ddot{a}q.\ (Ea)\bar{A}(a),$$

$$\overline{(Ea)}A(a) \ \ddot{a}q.\ (a)\bar{A}(a).$$

(See *Hilbert 1922a*, p. 155.) So far as the principle of excluded middle in the general logic of judgments (Axiom 6) is concerned, Hilbert does not say anything about the question of its intuitive obviousness; apparently he considers this obviousness to be indubitable. These views of Hilbert's are not inseparably linked with the fundamental and purely formal task that he set himself, the investigation of consistency; they seem incorrect to us.

First, Axiom 6 is not intuitively obvious. Its relation to finitary logic (*finite Logik*) is only illusory; while the truth of the axioms of implication (1–4) is perceived independently of the content of the judgments, the truth of Axiom 6, as will be explained in the next chapter, demands for its justification that the content of the judgments be considered, and this content may be transfinite.

Second, if Axioms 1–6 are adopted, the two formulas given above can be proved with the help of a few axioms whose intuitive obviousness cannot be questioned. We shall give the proof in Chapter V, but the first argument is sufficient to justify the investigations that now follow.

[5] *De interpretatione*, 4; *De anima*, III, 6.

was just mentioned. In the next sections of the present chapter we shall investigate to what extent this is possible.

§ 2. The meaning of the symbol $A \rightarrow B$ is exhausted by the fact that, once convinced of the truth of A, we have to accept the truth of B too. Or, in the formalistic interpretation: if formula A is written down, we can also write down formula B.[6] Thus, the relation of implication between two judgments does not establish any connection between their contents.

Hilbert's first axiom of implication, which means that "the true follows from anything", results from such a formalistic interpretation of implication: once B is true by itself, then, after having accepted A, we also have to regard B as true. The truth of the remaining three axioms of implication is seen just as easily on the basis of the interpretation given for the notion of implication. Moreover, the character of the judgments considered is not in the least affected; consequently, no doubt can arise about the possibility of applying these axioms to arbitrary judgments.

The question of the completeness of the system of the four axioms of implication is interesting. After what has been said concerning the completeness of Hilbert's full system of axioms for the logic of judgments, the question has to be put thus: A formula proved by means of the axioms of implication and the axioms of negation is said to be true; can every true formula written with the help of only the symbols for arbitrary judgments and implication, without the symbol for negation, be proved on the basis of the four axioms of implication alone?

§ 3. So far as a completed judgment, considered as a whole, is concerned, negation is merely the interdiction from regarding the judgment as true. We can obtain a fuller view of what negation is by considering the judgment as a statement attributing a predicate to a subject; negation then is the assertion that the predicate is incompatible with the subject.

The symbol $\bar{A}$ of the logic of judgments, of course, expresses the first interpretation of negation, that is, the interdiction from considering the judgment A true. However, the usual tradition in logic has been to pass from the first interpretation to the second, regarded as more primitive.[7] In the application to mathematical judgments this turns out to be impossible.

In so far as the negation of a judgment is the product of direct examination, the second interpretation, which takes its point of departure in the idea of the impossibility of the synthesis that creates the judgment, is actually closer to the substance of the matter than the first, which rests upon the purely formal idea of interdiction. But, when a negation is obtained as the result of a derivation, the reduction of the first interpretation to the second is no longer necessary and, in the case of mathematical judgments, is sometimes even impossible. In fact, many negative judgments in mathematics are proved by means of a reduction to a contradiction, according to the schema $\mathfrak{S} \rightarrow \mathfrak{T}, \overline{\mathfrak{T}} \mid \overline{\mathfrak{S}}$,[8] and cannot be proved in any other way.

[6] This is precisely what is expressed by the schema $\mathfrak{S}, \mathfrak{S} \rightarrow \mathfrak{T} \mid \mathfrak{T}$ in Hilbert's metamathematics. Sigwart, too, regards this schema as the most general schema in any inference (see *Sigwart 1908*, p. 372, [[or *1904*, p. 434]]).

[7] See, for instance, *Sigwart 1908*, pp. 135ff., [[or *1904*, pp. 155ff.]]

[8] See § 6 on the principle of contradiction.

Thus, the first interpretation of negation is independent. It was originally introduced by Brouwer (*1923d*), who defines negation as absurdity. It rests upon the second, since to derive a negative judgment by reduction to a contradiction we must already have some negative judgments, but at the same time it is broader than the second.

§ 4. Hilbert's first axiom of negation, "Anything follows from the false", made its appearance only with the rise of symbolic logic, as did also, incidentally, the first axiom of implication. But, while the first axiom of implication follows with intuitive obviousness from a correct interpretation of the idea of logical implication, the axiom now considered does not have and cannot have any intuitive foundation since it asserts something about the consequences of something impossible: we have to accept B if the true judgment A is regarded as false.

Thus, Hilbert's first axiom of negation cannot be an axiom of the intuitionistic logic of judgments, no matter which interpretation of negation we take as a point of departure. This, of course, does not exclude the possibility that the axiom can be a formula proved on the basis of other axioms.

§ 5. Hilbert's second axiom of negation expresses the principle of excluded middle. The principle is expressed here in the form in which it is used for derivations: if B follows from A as well as from $\overline{A}$, then B is true. Its usual form, "Every judgment is either true or false",[9] is equivalent to that given above.[10]

Clearly, from the first interpretation of negation, that is, the interdiction from regarding the judgment as true, it is impossible to obtain the certitude that the principle of excluded middle is true; incidentally, no such attempts have been made. Consequently, to justify the principle we must turn to the structure of the judgment, the relation of predicate to subject. Even in the very simple case of a judgment of the type "All A are B" the relations of all possible A, the supply of which can be infinite, to the predicate B inevitably enter into consideration. Brouwer showed[11] that in the case of such transfinite judgments the principle of excluded middle cannot, precisely for this reason, be considered obvious.

§ 6. Thus, from the intuitionistic point of view neither of Hilbert's two axioms of negation can be taken as an axiom of the general logic of judgments. We offer here the following axiom, which we shall call the principle of contradiction:

$$(3) \qquad 5. \ (A \rightarrow B) \rightarrow \{(A \rightarrow \overline{B}) \rightarrow \overline{A}\}.$$

[9] This is Leibniz's very simple formulation (see *Nouveaux essais*, IV, 2). The formulation "A is either B or not-B" has nothing to do with the logic of judgments.

[10] Symbolically the second form is expressed thus:

$$A \vee \overline{A},$$

where $\vee$ means "or". The equivalence of the two forms is easily proved on the basis of the axioms of implication and the following axioms, which determine the meaning of the symbol $\vee$ and are taken from *Ackermann 1924*:

1. $A \rightarrow A \vee B$,
2. $B \rightarrow A \vee B$,
3. $(A \rightarrow C) \rightarrow (B \rightarrow C) \rightarrow (A \vee B) \rightarrow C$.

⟦In these formulas $\vee$ binds more strongly than $\rightarrow$, and for $\rightarrow$ there is association on the right.⟧

[11] See *Brouwer 1923d* or the example of a proposition unprovable except by an illegitimate use of the principle of excluded middle, discussed in detail in *Brouwer 1920*.

Its meaning is: If both the truth and the falsity of a certain judgment B follow from A, the judgment A itself is false.

The usual principle of contradiction, "A judgment cannot be true and false', cannot be formulated in terms of an arbitrary judgment, implication, and negation. Our principle contains something else; namely, from it, together with the first axiom of implication, there follows the principle of *reductio ad absurdum*: If B is true and if the falsity of B follows from A, then A is false.

The truth of the proposed axiom follows from the simplest interpretation of negation, the interdiction from regarding a judgment as true, and does not depend upon whether the content of the judgments is considered.

The system of five axioms, the four axioms of implication (1) and the axiom of negation (3) just adopted, I shall call the system $\mathfrak{B}$. We do not know any formula of the general logic of judgments that possesses intuitive obviousness when applied to arbitrary judgments but is not provable on the basis of this system of axioms. Nevertheless, the question whether this axiom system is a complete axiom system for the intuitionistic general logic of judgments remains open.

§ 7. Although, as we have seen, the principle of excluded middle cannot be regarded as an axiom of the general logic of judgments, it has validity in the limited domain of the judgments that Brouwer calls finitary judgments. We shall not investigate here what the boundaries of the domain of finitary judgments are; this task is not as easy as it may seem. We therefore limit ourselves to the recognition of the fact that such a domain exists.

Besides the principle of excluded middle, the principle of double negation, which is expressed symbolically by

$$(4) \qquad 6.\ \overline{\overline{A}} \rightarrow A,$$

has validity in the domain of the finitary.[12]

It is self-evident that all five axioms of the general logic of judgments (the system $\mathfrak{B}$) are valid in the domain of the finitary too. The system of axioms that consists of the axioms of the system $\mathfrak{B}$—that is, (1) and (3)—and the axiom of double negation (4) we shall call the system $\mathfrak{H}$.

The system $\mathfrak{H}$ is equivalent to the system consisting of Hilbert's axioms (1) and (2). The axioms of implication are common to both. For the proof it is therefore sufficient to prove formulas (3) and (4) on the basis of the formulas (2), and conversely, the axioms of implication being used in both cases. We shall not carry out the proof of formulas (3) and (4) on the basis of Hilbert's axioms (1) and (2), but the converse proof, which rests upon axioms (3) and (4), introduced here for the first time, is carried out in the next section.

For us the system $\mathfrak{H}$ has the advantage of being obtained from the system $\mathfrak{B}$ of the general logic of judgments by the addition solely of the axiom of double negation; this considerably facilitates further investigation.

It is clear that the system $\mathfrak{H}$, just like Hilbert's system, is complete. In it we can derive all the formulas of the traditional logic of judgments. They are all true, if only we replace in them the symbols for arbitrary judgments, A, B, C, ..., by symbols for

[12] The formula $A \rightarrow \overline{\overline{A}}$ is provable on the basis of the system $\mathfrak{B}$. See formula (34) below.

arbitrary finitary judgments, A^f, B^f, C^f, ... The proof of this fact meets with some difficulties, which are explained and overcome in the next chapter.[13]

§ 8. We shall designate the axioms of the system $\mathfrak{H}$, (1), (3), and (4), by the numbers 1–6. The numbers of formulas that rest upon Axiom 6 are doubly underlined to indicate that these formulas have validity only in the domain of the finitary, while the others are valid for arbitrary judgments.

(5) $$\frac{A \rightarrow (B \rightarrow A)}{\bar{B} \rightarrow (A \rightarrow \bar{B})}$$ Axiom 1

(6) $$\frac{\{A \rightarrow (B \rightarrow C)\} \rightarrow \{B \rightarrow (A \rightarrow C)\}}{[(B \rightarrow C) \rightarrow \{(A \rightarrow B) \rightarrow (A \rightarrow C)\}] \rightarrow [(A \rightarrow B) \rightarrow \{(B \rightarrow C) \rightarrow (A \rightarrow C)\}]}$$ Axiom 3

$$[(B \rightarrow C) \rightarrow \{(A \rightarrow B) \rightarrow (A \rightarrow C)\}] \rightarrow [(A \rightarrow B) \rightarrow \{(B \rightarrow C) \rightarrow (A \rightarrow C)\}]$$ (6)

(7) $$\frac{(B \rightarrow C) \rightarrow \{(A \rightarrow B) \rightarrow (A \rightarrow C)\}}{(A \rightarrow B) \rightarrow \{(B \rightarrow C) \rightarrow (A \rightarrow C)\}}$$ Axiom 4

(8) $$\frac{(A \rightarrow B) \rightarrow \{(B \rightarrow C) \rightarrow (A \rightarrow C)\}}{\{\bar{B} \rightarrow (A \rightarrow \bar{B})\} \rightarrow [\{(A \rightarrow \bar{B}) \rightarrow \bar{A}\} \rightarrow (\bar{B} \rightarrow \bar{A})]}$$ (7)

$$\{\bar{B} \rightarrow (A \rightarrow \bar{B})\} \rightarrow [\{(A \rightarrow \bar{B}) \rightarrow \bar{A}\} \rightarrow (\bar{B} \rightarrow \bar{A})]$$ (8)

(9) $$\frac{B \rightarrow (A \rightarrow \bar{B})}{\{(A \rightarrow \bar{B}) \rightarrow \bar{A}\} \rightarrow (\bar{B} \rightarrow \bar{A})}$$ (5)

$$(A \rightarrow B) \rightarrow \{(A \rightarrow \bar{B}) \rightarrow \bar{A})$$ Axiom 5

(10) $$\frac{\{(A \rightarrow \bar{B}) \rightarrow \bar{A}\} \rightarrow (\bar{B} \rightarrow \bar{A})}{(A \rightarrow B) \rightarrow (\bar{B} \rightarrow \bar{A})}$$ (9)

(11) $$\frac{(A \rightarrow B) \rightarrow (\bar{B} \rightarrow \bar{A})}{(B \rightarrow A) \rightarrow (\bar{A} \rightarrow \bar{B})}$$ (10)

$$A \rightarrow (B \rightarrow A)$$ Axiom 1

(12) $$\frac{(B \rightarrow A) \rightarrow (\bar{A} \rightarrow \bar{B})}{A \rightarrow (\bar{A} \rightarrow \bar{B})}$$ (11)

(13) $$\frac{A \rightarrow (\bar{A} \rightarrow \bar{B})}{A \rightarrow (\bar{A} \rightarrow \bar{\bar{B}})}$$ (12)

(14) $$\frac{\bar{\bar{A}} \rightarrow A}{\bar{\bar{B}} \rightarrow B}$$ Axiom 6

(15) $$\frac{(B \rightarrow C) \rightarrow \{(A \rightarrow B) \rightarrow (A \rightarrow C)\}}{(\bar{\bar{\bar{B}}} \rightarrow B) \rightarrow \{(A \rightarrow \bar{\bar{\bar{B}}}) \rightarrow (A \rightarrow B)\}}$$ Axiom 4

$$(\bar{\bar{\bar{B}}} \rightarrow B) \rightarrow \{(A \rightarrow \bar{\bar{\bar{B}}}) \rightarrow (A \rightarrow B)\}$$ (15)

[13] See Chapter III, § 4.

$$\underline{\underline{(16)}}\qquad \frac{\bar{\bar{B}} \to B}{(A \to \bar{\bar{B}}) \to (A \to B)}\qquad \underline{\underline{(14)}}$$

$$\underline{\underline{(17)}}\qquad \frac{(A \to \bar{\bar{B}}) \to (A \to B)}{(\bar{A} \to \bar{\bar{B}}) \to (\bar{A} \to B)}\qquad \underline{\underline{(16)}}$$

$$A \to (\bar{A} \to \bar{\bar{B}})\qquad (13)$$

$$\underline{\underline{(18)}}\qquad \frac{(\bar{A} \to \bar{\bar{B}}) \to (\bar{A} \to B)}{A \to (\bar{A} \to B).}\qquad \underline{\underline{(16)}}$$

Thus Hilbert's first axiom of negation is proved.

$$(19)\qquad \frac{(A \to B) \to (\bar{B} \to \bar{A})}{(\bar{A} \to B) \to (\bar{B} \to \bar{\bar{A}})}\qquad (10)$$

$$(20)\qquad \frac{(A \to B) \to \{(A \to \bar{B}) \to \bar{A}\}}{(\bar{B} \to \bar{A}) \to \{(\bar{B} \to \bar{\bar{A}}) \to \bar{\bar{B}}\}}\qquad \text{Axiom 5}$$

$$(A \to B) \to (\bar{B} \to \bar{A})\qquad (10)$$

$$(21)\qquad \frac{(\bar{B} \to \bar{A}) \to \{(\bar{B} \to \bar{\bar{A}}) \to \bar{\bar{B}}\}}{(A \to B) \to \{(\bar{B} \to \bar{\bar{A}}) \to \bar{\bar{B}}\}}\qquad (20)$$

$$(22)\qquad \frac{(A \to B) \to \{(B \to C) \to (A \to C)\}}{\{(\bar{A} \to B) \to (\bar{B} \to \bar{\bar{A}})\} \to [\{(\bar{B} \to \bar{\bar{A}}) \to \bar{\bar{B}}\} \to \{(\bar{A} \to B) \to \bar{\bar{B}}\}]}\qquad (7)$$

$$\{(\bar{A} \to B) \to (\bar{B} \to \bar{\bar{A}})\} \to [\{(\bar{B} \to \bar{\bar{A}}) \to \bar{\bar{B}}\} \to \{(\bar{A} \to B) \to \bar{\bar{B}}\}]\qquad (22)$$

$$(23)\qquad \frac{(\bar{A} \to B) \to (\bar{B} \to \bar{\bar{A}})}{\{(\bar{B} \to \bar{\bar{A}}) \to \bar{\bar{B}}\} \to \{(\bar{A} \to B) \to \bar{\bar{B}}\}}\qquad (19)$$

$$(A \to B) \to \{(\bar{B} \to \bar{\bar{A}}) \to \bar{\bar{B}}\}\qquad (21)$$

$$(24)\qquad \frac{\{(\bar{B} \to \bar{\bar{A}}) \to \bar{\bar{B}}\} \to \{(\bar{A} \to B) \to \bar{\bar{B}}\}}{(A \to B) \to \{(\bar{A} \to B) \to \bar{\bar{B}}\}}\qquad (23)$$

$$\underline{\underline{(25)}}\qquad \frac{(A \to \bar{\bar{B}}) \to (A \to B)}{\{(\bar{A} \to B) \to \bar{\bar{B}}\} \to \{(\bar{A} \to B) \to B\}}\qquad \underline{\underline{(16)}}$$

$$(A \to B) \to \{(\bar{A} \to B) \to \bar{\bar{B}}\}\qquad (24)$$

$$\underline{\underline{(26)}}\qquad \frac{\{(\bar{A} \to B) \to \bar{\bar{B}}\} \to \{(\bar{A} \to B) \to B\}}{(A \to B) \to \{(\bar{A} \to B) \to B\}}.\qquad \underline{\underline{(25)}}$$

Thus Hilbert's second axiom, too, is proved.

Among the formulas proved with the help of axioms $\mathfrak{B}$, without the axiom of double negation, formulas (12) and (24) are those that are closest to Hilbert's axioms of negation. Formula (24), which comes close to the principle of excluded middle, means: if B follows from the truth of A and from its falsity as well, then it cannot be false. Indeed, if we assume that B is false, A cannot be true, since B would follow from A; but from the falsity of A the truth of B would follow.

III. The special logic of judgments and its domain of applicability

§ 1. The formulas provable on the basis of axioms $\mathfrak{B}$ constitute the general logic of judgments. We shall call the totality of formulas provable on the basis of the six axioms $\mathfrak{H}$ the special logic of judgments.[14] The content of the special logic of judgments is richer than that of the general logic, but its domain of applicability is narrower. Everything below is devoted to the explanation of what the domain of applicability of the special logic of judgments is. This domain is perhaps even somewhat narrower than the domain in which the principle of excluded middle in the Hilbert form is applicable.

§ 2. Let us introduce symbols, $A^{\cdot}$, $B^{\cdot}$, $C^{\cdot}, \ldots$, to denote arbitrary judgments for which the judgment itself follows from its double negation. The finitary judgments are of that kind. All true judgments are also of that kind; this, however, will not have any application in what follows. Brouwer proved (*1923d*) that all negative judgments are of that kind. The proof, given below, rests only upon the axioms of the system $\mathfrak{B}$.

On the basis of the axioms of implication it is easy to prove the formula

(27) $$A \to A.$$

(28) $$\frac{(A \to B) \to \{(A \to \bar{B}) \to \bar{A}\}}{(\bar{A} \to A) \to \{(\bar{A} \to \bar{A}) \to \bar{\bar{A}}\}} \qquad \text{Axiom 5}$$

(29) $$\frac{A \to (B \to A)}{A \to (\bar{A} \to A)} \qquad \text{Axiom 1}$$

$$A \to (\bar{A} \to A) \qquad (29)$$

(30) $$\frac{(\bar{A} \to A) \to \{(\bar{A} \to \bar{A}) \to \bar{\bar{A}}\}}{A \to \{(\bar{A} \to \bar{A}) \to \bar{\bar{A}}\}} \qquad (28)$$

(31) $$\frac{\{A \to (B \to C)\} \to \{B \to (A \to C)\}}{[A \to \{(\bar{A} \to \bar{A}) \to \bar{\bar{A}}\}] \to \{(\bar{A} \to \bar{A}) \to (A \to \bar{\bar{A}})\}} \qquad \text{Axiom 3}$$

$$[A \to \{(\bar{A} \to \bar{A}) \to \bar{\bar{A}}\}] \to \{(\bar{A} \to \bar{A}) \to (A \to \bar{\bar{A}})\} \qquad (31)$$

(32) $$\frac{A \to \{(\bar{A} \to \bar{A}) \to \bar{\bar{A}}\}}{(\bar{A} \to \bar{A}) \to (A \to \bar{\bar{A}})} \qquad (30)$$

(33) $$\frac{A \to A}{\bar{A} \to \bar{A}} \qquad (27)$$

$$(\bar{A} \to \bar{A}) \to (A \to \bar{\bar{A}}) \qquad (32)$$

[14] The general logic of judgments also has another, real interpretation (see Chapter I, § 5). The special logic of judgments can for the present be defined merely formally, since the real meaning of its formulas will be established only in what follows.

$$\frac{\bar{A} \to \bar{A}}{A \to \bar{\bar{\bar{A}}}} \tag*{(34) \quad (33)}$$

$$\frac{(A \to B) \to (\bar{B} \to \bar{A})}{(A \to \bar{\bar{A}}) \to (\bar{\bar{\bar{A}}} \to \bar{A})} \tag*{(35) \quad (10)}$$

$$(A \to \bar{\bar{A}}) \to (\bar{\bar{\bar{A}}} \to \bar{A}) \tag*{(35)}$$

$$\frac{A \to \bar{\bar{A}}}{\bar{\bar{\bar{A}}} \to \bar{A}}. \tag*{(36) \quad (34)}$$

The last formula shows that all negative judgments are judgments of type $A^{\cdot}$.

The axiom system $\mathfrak{H}$ differs from the system $\mathfrak{B}$, which is universally applicable, only by the axiom of double negation. For judgments of type $A^{\cdot}$ this axiom is expressed by the following formula:

$$\bar{\bar{A}}^{\cdot} \to A^{\cdot} \tag*{(37)}$$

We consider only this formula to be true; we consider formula (4) to be unfounded.

But it does not yet follow from what has been said that all formulas of the special logic of judgments are true for judgments of type $A^{\cdot}$; in fact, in the derivation of these formulas the axiom of double negation (4) is applied not only to elementary judgments—which for our case, that of judgments of type $A^{\cdot}$, is justified by formula (37)—but also to complex formulas; whether a formula of type $A^{\cdot} \to B^{\cdot}$, for example, is a formula of type $A^{\cdot}$ is, however, not yet apparent.

§ 3. We shall now prove that every formula expressed by means of the symbols $A^{\cdot}$, $B^{\cdot}$, $C^{\cdot}, \ldots$, and the symbols of implication and negation is a formula of type $A^{\cdot}$. For that it suffices to consider two very simple cases.

First, every negative judgment is a judgment of type $A^{\cdot}$ by virtue of Brouwer's formula (36).

Second, we now prove that a judgment of type $A^{\cdot} \to B^{\cdot}$ is also a judgment of type $A^{\cdot}$.

$$\frac{A \to A}{(A \to B) \to (A \to B)} \tag*{(38) \quad (27)}$$

$$\frac{\{A \to (B \to C)\} \to \{B \to (A \to C)\}}{\{(A \to B) \to (A \to B)\} \to [A \to \{(A \to B) \to B\}]} \tag*{(39) \quad Axiom 3}$$

$$\{(A \to B) \to (A \to B)\} \to [A \to \{(A \to B) \to B\}] \tag*{(39)}$$

$$\frac{(A \to B) \to (A \to B)}{A \to \{(A \to B) \to B\}} \tag*{(40) \quad (38)}$$

$$\frac{(A \to B) \to (\bar{B} \to \bar{A})}{(\bar{B} \to \bar{A}) \to (\bar{\bar{A}} \to \bar{\bar{B}})} \tag*{(41) \quad (10)}$$

$$(A \to B) \to (\bar{B} \to \bar{A}) \tag*{(10)}$$

$$\frac{(\bar{B} \to \bar{A}) \to (\bar{\bar{A}} \to \bar{\bar{B}})}{(A \to B) \to (\bar{\bar{A}} \to \bar{\bar{B}})} \tag*{(42) \quad (41)}$$

$$(43)\qquad \frac{(A \to B) \to (\bar{\bar{\bar{A}}} \to \bar{\bar{\bar{B}}})}{\{(A \to B) \to B\} \to \{\overline{\overline{\overline{(A \to B)}}} \to \bar{\bar{\bar{B}}}\}} \qquad (42)$$

$$A \to \{(A \to B) \to B\} \qquad (40)$$

$$(44)\qquad \frac{\{(A \to B) \to B\} \to \{\overline{\overline{\overline{(A \to B)}}} \to \bar{\bar{\bar{B}}}\}}{A \to \{\overline{\overline{\overline{(A \to B)}}} \to \bar{\bar{\bar{B}}}\}} \qquad (43)$$

$$(45)\qquad \frac{\{A \to (B \to C)\} \to \{B \to (A \to C)\}}{[A \to \{\overline{\overline{\overline{(A \to B)}}} \to \bar{\bar{\bar{B}}}\}] \to \{\overline{\overline{\overline{(A \to B)}}} \to (A \to \bar{\bar{\bar{B}}})\}} \qquad \text{Axiom 3}$$

$$[A \to \{\overline{\overline{\overline{(A \to B)}}} \to \bar{\bar{\bar{B}}}\}] \to \{\overline{\overline{\overline{(A \to B)}}} \to (A \to \bar{\bar{\bar{B}}})\} \qquad (45)$$

$$(46)\qquad \frac{A \to \{\overline{\overline{\overline{(A \to B)}}} \to \bar{\bar{\bar{B}}}\}}{\overline{\overline{\overline{(A \to B)}}} \to (A \to \bar{\bar{\bar{B}}})}. \qquad (44)$$

This formula is true for arbitrary judgments A and B. Replacing A by A and B by $B^{\cdot}$ and making use of formula (37), we easily obtain the formula

$$(47)\qquad \overline{\overline{(A^{\cdot} \to B^{\cdot})}} \to (A^{\cdot} \to B^{\cdot}),$$

which shows precisely that judgments of type $A^{\cdot} \to B^{\cdot}$ are of type $A^{\cdot}$.

By gradually passing to more complex formulas, we can prove the assertion made at the beginning of the present section.

§ 4. We can now assert that all formulas of the special logic of judgments are true for judgments of type $A^{\cdot}$, including all finitary and all negative judgments. In fact, the symbols $A^{\cdot}, B^{\cdot}, C^{\cdot}, \ldots, A^{\cdot} \to B^{\cdot}$, and $\bar{A}^{\cdot}$ allow all the operations that the symbols of the general logic of judgments do: substitution for the symbols $A^{\cdot}, B^{\cdot}, C^{\cdot}, \ldots$ of an arbitrary formula written by means of the symbols considered and inference according to the schema $\mathfrak{S} \to \mathfrak{T}, \mathfrak{T} \mid \mathfrak{T}$; moreover, all six axioms $\mathfrak{H}$ are true for them.

The precise boundary of the domain in which the special logic of judgments is applicable has thus been found; this domain coincides with the domain in which the formula of double negation (4) is applicable.

IV. The mathematics of pseudotruth

§ 1. In the preceding chapter we established that all the formulas of the traditional logic of judgments can actually be proved as formulas of the special logic of judgments. We must merely recognize that they deal only with judgments of type $A^{\cdot}$. Moreover, these formulas themselves turn out to be formulas of type $A^{\cdot}$.

The following question now arises: Can we in a similar way, after we have placed some restrictions on their real interpretation, again give a meaning to all those formulas of mathematics that are proved by an illegitimate use (that is, a use outside the domain in which they are applicable) of formulas of the special logic of judgments, in particular by use of the principle of excluded middle? It turns out that this task can be fulfilled.

§ 2. We shall construct, alongside of ordinary mathematics, a new mathematics, a "pseudomathematics" that will be such that to every formula of the first there corresponds a formula of the second and, moreover, that every formula of pseudomathematics is a formula of type $A^{\cdot}$. For the time being we are not concerned with the question of the truth of the formulas of pseudomathematics; we shall turn to it in § 5 of the present chapter.

A symbol, simple or complex, that expresses a judgment is called a formula. Formulas of which no part is a formula will be called elementary formulas, or formulas of the first order; $a = a$ is such a formula. A formula whose parts are formulas of the $(n\text{-}1)$th order at most will be called a formula of the nth order. For example, the formula

$$a = b \rightarrow \{A(a) \rightarrow B(a)\}$$

is a formula of the third order since its constituent part $A(a) \rightarrow B(a)$ is a formula of the second order.

To an elementary formula $\mathfrak{S}$ there corresponds in pseudomathematics the formula $\mathfrak{S}^*$, which expresses the double negation of $\mathfrak{S}$:

$$\mathfrak{S}^* \equiv \overline{\overline{\mathfrak{S}}}. \tag{48}$$

In what follows we shall, for convenience, denote the double negation of $\mathfrak{S}$ by $n\mathfrak{S}$.

To the formula of the nth order $F(\mathfrak{S}_1, \mathfrak{S}_2, \ldots, \mathfrak{S}_k)$, where $\mathfrak{S}_1, \mathfrak{S}_2, \ldots, \mathfrak{S}_k$ are formulas of the $(n-1)$th order at most, there corresponds in pseudomathematics the formula $F(\mathfrak{S}_1, \mathfrak{S}_2, \ldots, \mathfrak{S}_k)^*$ such that

$$F(\mathfrak{S}_1, \mathfrak{S}_2, \ldots, \mathfrak{S}_k)^* \equiv nF(\mathfrak{S}_1^*, \mathfrak{S}_2^*, \ldots, \mathfrak{S}_k^*), \tag{49}$$

$\mathfrak{S}_1^*, \mathfrak{S}_2^*, \ldots, \mathfrak{S}_k^*$ being regarded as already determined. For example, to the formula

$$a = b \rightarrow \{A(a) \rightarrow B(a)\}$$

there corresponds in pseudomathematics the formula

$$n[n(a = b) \rightarrow n\{nA(a) \rightarrow nB(a)\}].$$

To every symbol that is not a formula there also corresponds a definite symbol of pseudomathematics. To a symbol, simple or complex, of which no part is a formula there corresponds in pseudomathematics a symbol that is identical with it. To a complex symbol in which formulas enter there corresponds a symbol in which each formula $\mathfrak{S}$ is replaced by the formula $\mathfrak{S}^*$.

§ 3. All the formulas of mathematics are derived from axioms,[15] which we denote by $\mathfrak{A}_1, \mathfrak{A}_2, \ldots, \mathfrak{A}_k$, with the help of the operations of substitution of particular values for variables and of inference according to the schema $\mathfrak{S}, \mathfrak{S} \rightarrow \mathfrak{T} \mid \mathfrak{T}$. To the axioms there correspond in pseudomathematics the formulas $\mathfrak{A}_1^*, \mathfrak{A}_2^*, \ldots, \mathfrak{A}_k^*$. We shall prove that every formula of pseudomathematics that corresponds to a formula proved on the basis of the axioms $\mathfrak{A}$ is a consequence of the formulas $\mathfrak{A}^*$. For the proof it suffices to establish the following two facts.

First, if, when particular values are substituted for variables in a formula $\mathfrak{S}$, we obtain a formula $\mathfrak{T}$, then, when the corresponding formulas and symbols are substituted at the corresponding places in the formula $\mathfrak{S}^*$, we obtain the formula $\mathfrak{T}^*$.

[15] Here all the axioms of logic are included among the axioms of mathematics.

Second, in analogy with the schema $\mathfrak{S}$, $\mathfrak{S} \to \mathfrak{T} \mid \mathfrak{T}$, the schema

$$\mathfrak{S}^*,\ (\mathfrak{S} \to \mathfrak{T})^* \mid \mathfrak{T}^* \tag{50}$$

is correct.

In fact,

$$(\mathfrak{S} \to \mathfrak{T})^* \equiv \overline{\overline{(\mathfrak{S}^* \to \mathfrak{T}^*)}}; \tag{51}$$

since $\mathfrak{S}^*$ and $\mathfrak{T}^*$ are formulas of type $A^{\cdot}$, we have, according to formula (47),

$$\overline{\overline{(\mathfrak{S}^* \to \mathfrak{T}^*)}} \to (\mathfrak{S}^* \to \mathfrak{T}^*), \tag{52}$$

$$\left\{\begin{array}{c} \mathfrak{S}^* \\ (\mathfrak{S} \to \mathfrak{T})^* \to (\mathfrak{S}^* \to \mathfrak{T}^*) \\ \hline \mathfrak{T}^*. \end{array}\right. \tag{53}$$

Thus, we see that to every correct proof in the domain of ordinary mathematics there corresponds a correct proof in the domain of pseudomathematics. From that follows the truth of the proposition advanced at the beginning of the present section.

§ 4. To the five axioms of the general logic of judgments there correspond in pseudomathematics the following formulas:

$$\left\{\begin{array}{l} 1.\ n\{nA \to n(nB \to nA)\}, \\ 2.\ n[n\{nA \to n(nA \to nB)\} \to n(nA \to nB)], \\ 3.\ n[n\{nA \to n(nB \to nC)\} \to n\{nB \to n(nA \to nC)\}], \\ 4.\ n[n(nB \to nC) \to n\{n(nA \to nB) \to n(nA \to nC)\}], \\ 5.\ n[n(nA \to nB) \to n\{(nA \to n(\overline{nB})) \to n(\overline{nA})\}]. \end{array}\right. \tag{54}$$

These formulas can be obtained from

$$\left\{\begin{array}{l} 1.\ n\{A^{\cdot} \to n(B^{\cdot} \to A^{\cdot})\}, \\ \cdot\quad\cdot\quad\cdot\quad\cdot\quad\cdot\quad\cdot\quad\cdot\quad\cdot\quad\cdot\quad\cdot\quad\cdot\quad\cdot\quad\cdot\quad\cdot \\ \cdot\quad\cdot\quad\cdot\quad\cdot\quad\cdot\quad\cdot\quad\cdot\quad\cdot\quad\cdot\quad\cdot\quad\cdot\quad\cdot\quad\cdot\quad\cdot \\ 5.\ n[n(A^{\cdot} \to B^{\cdot}) \to n\{n(A^{\cdot} \to n\overline{B}^{\cdot}) \to n\overline{A}^{\cdot}\}] \end{array}\right. \tag{55}$$

by substitution of nA, nB, and nC for $A^{\cdot}$, $B^{\cdot}$, and $C^{\cdot}$, respectively. Since formulas (55) are formulas of the special logic of judgments, we have the right to prove them by using all the axioms of $\mathfrak{H}$, or all of Hilbert's axioms. Their proof does not present any difficulty. Thus, all the formulas (54) turn out to be true. It follows from this that all the formulas of pseudomathematics that correspond to the formulas of the general logic of judgments are true.

§ 5. All the axioms of mathematics that we know possess the same property as the axioms of the general logic of judgments, namely, that the formulas corresponding to them in the domain of pseudomathematics are true. For example, to the axiom

$$(a)A(a) \to A(a)$$

there corresponds the true formula

$$n\{n(a)nA(a) \to nA(a)\}.$$

We shall call the axioms that possess the property formulated above axioms of type $\mathfrak{K}$. Further, let us call the formulas provable on the basis of the axioms of type $\mathfrak{K}$ formulas of type $\mathfrak{K}$. All the axioms and formulas of mathematics that we know are of type $\mathfrak{K}$.[16]

By virtue of what was said above, the part of pseudomathematics whose formulas correspond to the formulas of type $\mathfrak{K}$ acquires a real meaning: all its formulas are true, since they are consequences of the true formulas that in pseudomathematics correspond to the axioms of type $\mathfrak{K}$. The name "pseudomathematics" becomes inappropriate for this part, the only one that for the time being exists; as a collection of true formulas, it is part of genuine mathematics.

We shall say that a judgment is pseudotrue if its double negation is true. A judgment of the form $n\mathfrak{S}$ thus asserts the pseudotruth of the judgment $\mathfrak{S}$. The formulas of pseudomathematics always express only judgments about pseudotruth. We have the right, therefore, to call the part of pseudomathematics that has real meaning the mathematics of pseudotruth.

§ 6. In the usual presentation of mathematics a number of conclusions are obtained by an illegitimate use of formulas of the special logic of judgments, for example, by use of the principle of excluded middle. All these cases, as has been shown, can be reduced to the use of the principle of double negation,

$$6.\ \overline{\overline{A}} \rightarrow A. \tag{4}$$

Among these conclusions let us consider those that, except for the illegitimate formula (4), rest only upon axioms of type $\mathfrak{K}$. The formulas that express them we shall call formulas of type $\mathfrak{K}'$.

Let us construct the formulas of pseudomathematics that correspond to the formulas $\mathfrak{K}'$. They will all follow from the formulas $\mathfrak{A}^*$ that correspond to axioms of type $\mathfrak{K}$ and from the formula

$$n\{\overline{\overline{n(nA)}} \rightarrow nA\}, \tag{56}$$

which corresponds to formula (4).

Formula (56) is true. In fact, by virtue of formula (34), it follows from

$$\overline{\overline{n(nA)}} \rightarrow nA. \tag{57}$$

Formula (57) can be obtained by means of substitution from a formula of the special logic of judgments,

$$\overline{\overline{n(A^{\cdot})}} \rightarrow A^{\cdot}; \tag{58}$$

in the special logic of judgments, as we know, an even number of negations reduces to affirmation.

Thus, while the formulas of ordinary mathematics that we have considered are based on an illegitimate use of formula (4), the corresponding formulas of pseudomathematics rest upon the true formula (56).

So we finally obtain the following result: None of the conclusions of ordinary

[16] In mathematics formulas whose truth is not obvious, for example, what is called Zermelo's axiom, are sometimes taken as axioms. But they, too, possess the property that $\mathfrak{A} \rightarrow \mathfrak{A}^*$.

mathematics that are based on the use, outside the domain of the finitary, of the formula of double negation and of other formulas that depend upon it (like the principle of excluded middle) can be regarded as firmly established. But, for those conclusions whose derivations require, besides this formula ⟦that of double negation⟧, only axioms of type $\mathfrak{K}$ (and no other axioms are known at present), the corresponding formulas of pseudomathematics are true and consequently become part of the mathematics of pseudotruth.

In other words, all the conclusions that rest upon axioms of type $\mathfrak{K}$ and on the formula of double negation are correct if we understand every judgment that enters into them in the sense of the affirmation of its pseudotruth, that is, of its double negation.

V. Addenda

§ 1. After having shown that it is illegitimate to apply the principle of excluded middle to transfinite judgments, Brouwer set himself the task of providing a foundation for mathematics without the help of that principle, and to a considerable extent he carried out this task.[17] But it then turned out that there exist a number of mathematical propositions that cannot be proved without the help of the principle of excluded middle, rejected by Brouwer. We consider below a few examples of such propositions.

We proved in the preceding chapters that, alongside of the development of mathematics without the help of the principle of excluded middle, we can also retain the usual development. To be sure, a limited interpretation then has to be given to all propositions; namely, every judgment of ordinary mathematics has to be replaced by the affirmation of its pseudotruth. But this development nevertheless retains two remarkable properties.

1. If a finitary conclusion is obtained with the help of arguments based on the use, even in the domain of the transfinite, of the principle of excluded middle, the conclusion is true in the usual sense. In view of what precedes, it can in fact be proved as a conclusion about pseudotruth; but, in the domain of the finitary, pseudotruth coincides with ordinary truth.

2. The use of the principle of excluded middle never leads to a contradiction. In fact, if a false formula were obtained with its help, then the corresponding formula of pseudomathematics would be proved without its help and would also lead to a contradiction.[18]

The first of these assertions contradicts a remark by Brouwer (*1923d*, p. 252, footnote), who thinks that finitary conclusions based on a transfinite use of the principle of excluded middle must also be considered unreliable.

§ 2. Generally, the propositions that we do not know how to prove without an illegitimate use of the principle of excluded middle ordinarily rest directly, not upon the principle of excluded middle of the logic of judgments, but upon another principle that bears the same name. In fact, from the principle of excluded middle in the form

[17] See, for example, *Brouwer 1918* and *1919*.

[18] We assume that all the axioms considered are axioms of type $\mathfrak{K}$ and, moreover, that any formula wittingly taken as false, $\mathfrak{S}$, is such that the corresponding $\mathfrak{S}^*$ is also false.

peculiar to the logic of judgments, namely, "Every judgment is either true or false", we can obtain further conclusions by following the schema that corresponds to Hilbert's formula: If B follows from A as well as from $\bar{A}$, then B is true. But in the case that interests us, that of transfinite judgments, it is difficult to obtain any positive conclusion B from the pure negation $\bar{A}$; for that we must first transform the judgment $\bar{A}$ into some other.

The following type of transfinite judgment is the most customary: for all a, $A(a)$ is true; symbolically, this judgment is written $(a)A(a)$. When we want to derive a positive conclusion from the negation $\overline{(a)}A(a)$ of this judgment, we put it in the form $(Ea)\bar{A}(a)$, that is, there exists an a for which $A(a)$ is not true. The equivalence of the last assertion to the simple negation of the judgment $(a)A(a)$ is expressed by the following two formulas:

$$\overline{(a)}A(a) \rightarrow (Ea)\bar{A}(a), \tag{59}$$

$$(Ea)\bar{A}(a) \rightarrow \overline{(a)}A(a). \tag{60}$$

In § 4 of the present chapter we shall demonstrate that formula (59) requires for its proof only the acceptance of the principle of double negation (4) in addition to formulas whose intuitive obviousness is unquestionable, among which the axioms of the general logic of judgments are included. But formula (60) can be proved without the help of that principle.

If we adopt formulas (59) and (60), we shall have the right to formulate the principle of excluded middle for judgments of the form $(a)A(a)$ as follows:

$$(a)A(a) \vee (Ea)\bar{A}(a),$$

that is, either $A(a)$ is true for all a or there exists an a for which $A(a)$ is not true.

To formulas (59) and (60) Hilbert adds (*1922a*, p. 157) the following:

$$\overline{(Ea)}A(a) \rightarrow (a)\bar{A}(a), \tag{61}$$

$$(a)\bar{A}(a) \rightarrow \overline{(Ea)}A(a). \tag{62}$$

In accordance with what was said above, he deems that formulas (59)–(62) justify the application of the principle of excluded middle to infinite collections of objects a. Unlike formulas (59) and (60), formulas (61) and (62) can both be proved on the basis only of axioms that are intuitively obvious. The proof is given in § 4.

Thus, formulas (60)–(62) simply are true formulas, while formula (59) is proved on the basis of the principle of double negation (4); everything presented in the preceding chapter, therefore, applies to the conclusions that rest upon that principle.

§ 3. Let us remark first of all that, having determined the meaning of the symbol $(a)A(a)$ as "for all a, $A(a)$ is true", we understand by "for all a" the same thing as by "for each a", that is, the meaning is that, whatever a may be given, we can assert that $A(a)$ is true.

Every formula of the general logic of judgments, when written by itself, means that it is true for all possible judgments, A, B, C, ... Thus, the formula $A \rightarrow \bar{\bar{A}}$ means that for any judgment the double negation of the judgment follows from its truth.

Accordingly, it is impossible to assert that the symbolic expression $(a)A(a)$, when introduced, is the first to lead us out of the domain of the finitary; the notion "for all a" is contained in hidden form in all formulas in which there are symbols for variables.

In general, the formula $A(a)$ written by itself means that, whatever the particular meaning of a may be, $A(a)$ is true. From this the following principle, which cannot be expressed symbolically, results: whenever a formula $\mathfrak{S}$ stands by itself ⟦in a derivation⟧, we can write the formula $(a)\mathfrak{S}$. When we have to refer to this principle, we denote it by **P**.[19]

Furthermore, we adopt the following axioms,

$$(63)\qquad \begin{cases} \text{I. } (a)\{A(a) \to B(a)\} \to \{(a)A(a) \to (a)B(a)\}, \\ \text{II. } (a)\{A \to B(a)\} \to \{A \to (a)B(a)\}, \\ \text{III. } (a)\{A(a) \to C\} \to \{\ Ea)A(a) \to C\}, \\ \text{IV. } A(a) \to (Ea)A(a). \end{cases}$$

We believe that all these axioms are intuitively obvious. The choice of these axioms and their number are exclusively determined by our goal, which is to prove formulas (59)–(62).

§ 4.

$$(A \to B) \to \\ \to (\bar{B} \to \bar{A}) \qquad (11)$$

$$(64)\qquad \frac{A(a) \to (Ea)A(a)}{\overline{(Ea)}A(a) \to \bar{A}(a)} \qquad \text{Axiom IV}$$

$$(65)\qquad (a)\{\overline{(Ea)}A(a) \to \bar{A}(a)\} \qquad (64)\ \mathbf{P}$$

$$(a)\{A \to B(a)\} \to \\ \to \{A \to (a)B(a)\} \qquad \text{Axiom II}$$

$$(66)\qquad \frac{(a)\{\overline{(Ea)}A(a) \to \bar{A}(a)\}}{\overline{(Ea)}A(a) \to (a)\bar{A}(a).} \qquad (65)$$

Thus formula (61) is proved.

$$\{A \to (B \to C)\} \to \\ \to \{B \to (A \to C)\} \qquad \text{Axiom 3}$$

$$(67)\qquad \frac{A \to (\bar{A} \to \bar{B})}{\bar{A} \to (A \to \bar{B})} \qquad (12)$$

$$(68)\qquad \frac{\bar{A} \to (A \to \bar{B})}{\bar{A}(a) \to \{A(a) \to \overline{(Ea)}A(a)\}} \qquad (67)$$

$$(69)\qquad (a)[\bar{A}(a) \to \{A(a) \to \overline{(Ea)}A(a)\}] \qquad (68)\ \mathbf{P}$$

[19] The symbol (a) can also occur in front of a formula that does not actually contain the variable a. Instead of the principle **P** we can introduce the axiom $(a)V$, where V denotes a true judgment, and the following rule of substitution, not formulated symbolically: any formula that stands by itself can be substituted for V.

$$(a)\{A(a) \to B(a)\} \to \to \{(a)A(a) \to (a)B(a)\} \qquad \text{Axiom I}$$

$$(70) \quad \frac{(a)[\overline{A}(a) \to \{A(a) \to \overline{(Ea)}A(a)\}]}{(a)\overline{A}(a) \to (a)\{A(a) \to \overline{(Ea)}A(a)\}} \qquad (69)$$

$$(71) \quad \frac{(a)\{A(a) \to C\} \to \{(Ea)A(a) \to C\}}{(a)\{A(a) \to \overline{(Ea)}A(a)\} \to \{(Ea)A(a) \to \overline{(Ea)}A(a)\}} \qquad \text{Axiom III}$$

$$(A \to B) \to \to \{(A \to \overline{B}) \to \overline{A}\} \qquad \text{Axiom 5}$$

$$(72) \quad \frac{A \to A}{(A \to \overline{A}) \to \overline{A}} \qquad (27)$$

$$(73) \quad \frac{(A \to \overline{A}) \to \overline{A}}{\{(Ea)A(a) \to \overline{(Ea)}A(a)\} \to \overline{(Ea)}A(a)} \qquad (72)$$

$$(a)\overline{A}(a) \to (a)\{A(a) \to \overline{(Ea)}A(a) \qquad (70)$$

$$(a)\{A(a) \to \overline{(Ea)}A(a)\} \to \{(Ea)A(a) \to \overline{(Ea)}A(a)\} \qquad (71)$$

$$(74) \quad \frac{\{(Ea)A(a) \to \overline{(Ea)}A(a)\} \to \overline{(Ea)}A(a)}{(a)\overline{A}(a) \to \overline{(Ea)}A(a).} \qquad (73)$$

Thus formula (62) is proved.

$$(75) \quad \frac{(a)\overline{A}(a) \to \overline{(Ea)}A(a)}{(a)\overline{\overline{A}}(a) \to \overline{(Ea)}\overline{A}(a)} \qquad (74)$$

$$(76) \quad \frac{A \to \overline{\overline{A}}}{A(a) \to \overline{\overline{A}}(a)} \qquad (34)$$

$$(77) \quad (a)\{A(a) \to \overline{\overline{A}}(a)\} \qquad (76)\ \mathbf{P}$$

$$(a)\{A(a) \to B(a)\} \to \to \{(a)A(a) \to (a)B(a)\} \qquad \text{Axiom I}$$

$$(78) \quad \frac{(a)\{A(a) \to \overline{\overline{A}}(a)\}}{(a)A(a) \to (a)\overline{\overline{A}}(a)} \qquad (77)$$

$$(a)A(a) \to (a)\overline{\overline{A}}(a) \qquad (78)$$

$$(79) \quad \frac{(a)\overline{\overline{A}}(a) \to \overline{(Ea)}\overline{A}(a)}{(a)A(a) \to \overline{(Ea)}\overline{A}(a)} \qquad (75)$$

$$(A \to B) \to \to (\overline{B} \to \overline{A}) \qquad (11)$$

$$(80)\qquad \frac{(a)A(a) \to \overline{(Ea)}\overline{A}(a)}{\overline{\overline{\overline{(Ea)}}}\overline{A}(a) \to \overline{(a)}A(a)} \qquad (79)$$

$$(81)\qquad \frac{A \to \overline{\overline{A}}}{(Ea)\overline{A}(a) \to \overline{\overline{\overline{(Ea)}}}\overline{A}(a)} \qquad (34)$$

$$(Ea)\overline{A}(a) \to \overline{\overline{\overline{(Ea)}}}\overline{A}(a) \qquad (81)$$

$$(82)\qquad \frac{\overline{\overline{\overline{(Ea)}}}\overline{A}(a) \to \overline{(a)}A(a)}{(Ea)\overline{A}(a) \to \overline{(a)}A(a).} \qquad (80)$$

Thus formula (60) is proved.

The proof of formula (59) cannot be carried out without the help of the axiom of double negation. The numbers of the formulas that rest upon this axiom are doubly underlined.

$$(83)\qquad \frac{\overline{(Ea)}A(a) \to (a)\overline{A}(a)}{\overline{(Ea)}\overline{A}(a) \to (a)\overline{\overline{A}}(a)} \qquad (66)$$

$$\underline{\underline{(84)}}\qquad \frac{\overline{\overline{A}} \to A}{\overline{\overline{A}}(a) \to A(a)} \qquad \text{Axiom 6}$$

$$\underline{\underline{(85)}}\qquad (a)\{\overline{\overline{A}}(a) \to (a)\} \qquad (84)\ \mathbf{P}$$

$$(a)\{A(a) \to B(a)\} \to$$
$$\to \{(a)A(a) \to (a)B(a)\} \qquad \text{Axiom I}$$

$$\underline{\underline{(86)}}\qquad \frac{(a)\{\overline{\overline{A}}(a) \to A(a)\}}{(a)\overline{\overline{A}}(a) \to (a)A(a)} \qquad \underline{\underline{(85)}}$$

$$\overline{(Ea)}\overline{A}(a) \to (a)\overline{\overline{A}}(a) \qquad (83)$$

$$\underline{\underline{(87)}}\qquad \frac{(a)\overline{\overline{A}}(a) \to (a)A(a)}{\overline{(Ea)}\overline{A}(a) \to (a)A(a)} \qquad \underline{\underline{(86)}}$$

$$(A \to B) \to \qquad (11)$$
$$\to (\overline{B} \to \overline{A})$$

$$\underline{\underline{(88)}}\qquad \frac{\overline{(Ea)}\overline{A}(a) \to (a)A(a)}{\overline{(a)}A(a) \to \overline{\overline{\overline{(Ea)}}}\overline{A}(a)} \qquad \underline{\underline{(87)}}$$

$$\underline{\underline{(89)}}\qquad \frac{\overline{\overline{A}} \to A}{\overline{\overline{\overline{(Ea)}}}\overline{A}(a) \to (Ea)\overline{A}(a)} \qquad \underline{\underline{\text{Axiom 6}}}$$

$$\overline{(a)}A(a) \to \overline{\overline{\overline{(Ea)}}}\overline{A}(a) \qquad \underline{\underline{(88)}}$$

$$(90)\qquad \frac{\overline{\overline{\overline{(Ea)}}}\overline{A}(a) \to (Ea)\overline{A}(a)}{\overline{(a)}A(a) \to (Ea)\overline{A}(a).} \qquad \underline{\underline{(89)}}$$

Thus formula (59) is proved by means of the axiom of double negation.

§ 5. An excellent example of a proposition unprovable without the help of an illegitimate use of the principle of excluded middle is given by Brouwer (*1920*); he shows that it cannot be considered proved that every real number has an infinite decimal expansion. He even exhibits a definite number for which it is not known whether it has a first digit in its decimal expansion.

Another example is the proposition stating that the complement of a closed set is a region, that is, that every point not belonging to a given closed set is contained in some interval that does not contain any point of the set.[20] The proof, as is well known, is carried out in the following way: by the principle of excluded middle in the form given in § 2 of the present chapter, either all intervals containing the point considered contain points of the set or there exists at least one interval that does not contain any such point; the first assumption leads to a contradiction, since it implies that the point belongs to the set, and therefore the second proposition is true. This example differs from Brouwer's in that here we do not know how to exhibit a definite closed set and a definite point exterior to it for which the existence of the required interval is doubtful.

§ 6. The following example is interesting: without the help of the principle of excluded middle it is impossible to prove any proposition whose proof usually comes down to an application of the principle of transfinite induction. For example, a proposition of that kind is: every closed set is the sum of a perfect set and a denumerable set.

The proof of such propositions is often carried out without the help of the principle of transfinite induction. But all these proofs rest upon the principle of excluded middle, applied to infinite collections, or upon the principle of double negation.

It is important to observe that the principle of transfinite induction itself can be derived without any assumption that, from the point of view of the theory of point sets, is new, but that the principle of excluded middle is necessarily used. It suffices merely to formulate the principle of transfinite induction without the use of the term "transfinite number", whose introduction would demand new axioms. Let us consider, instead, sets of rational numbers that are completely ordered from left to right. For such a set, a part that starts on the left from some point, which may or may not belong to the set, will be called a segment of the set. A segment, too, will always be a completely ordered set. The set of segments will itself be also completely ordered. Let us say that a segment is proper if there exists a point of the set that does not belong to it. The principle of transfinite induction can now be formulated in the following way.

Let a certain property J, which may or may not obtain of a completely ordered set of rational numbers, satisfy the following conditions:

(1) Sets consisting of one point alone have property J;

(2) If all the proper segments of a set have property J, then the set itself has it.

Under these conditions all completely ordered sets of rational numbers possess property J.

This formulation of the principle of transfinite induction can be used in the same cases as the ordinary one. Its proof is carried out thus: either all sets possess property

[20] This example was pointed out by P. S. Novikov.

J or there exists a set E that does not possess it; the second assumption leads to a contradiction, for among the segments of E there must be a first one that does not possess property J, and the existence of such a segment contradicts the conditions.

The examples adduced suffice to show that, alongside of the development of mathematics presented by Brouwer without the help of the principle of excluded middle, we must preserve the usual development, which uses this principle, if only as the development of mathematics of pseudotruth.

Formal proofs and undecidability

PAUL FINSLER

(*1926*)

This paper, which was received for publication on 28 November 1925, presents an example of a proposition that, although false, is formally undecidable. Finsler establishes the undecidability of the proposition by suitably modifying the argument used by Richard in stating his famous paradox (see above, pp. 143–144); first, the argument now bears on the notion of provability, rather than on that of finite definability, and, second, it keeps clear of the paradox and thus reaches a valid conclusion. Because of Finsler's result as well as his method, the paper has often been regarded as a forerunner of Gödel's 1931 paper. However, Finsler's conception of formal provability is so profoundly different from Gödel's that the affinity between the two papers should not be exaggerated.

Finsler first introduces a distinction between "formal" and "conceptual". The "formal" domain consists of all the expressions of a certain language, whose dictionary and grammar have been cleansed of vagueness; this language may be all-embracing, containing all the linguistic resources that have been or will ever be used. Outside this domain there is the "purely conceptual" domain, which contains nonlinguistic objects: unwritten and unwritable propositions, definitions, and proofs. Finsler elucidates the distinction by means of an example. For every real number α there is a proposition, true or false, of the form "α is a transcendental number". Since there are nondenumerably many real numbers and since only denumerably many such propositions can be expressed in language, there are propositions of that form that belong to the purely conceptual realm and not to the formal domain.

Finsler then considers the list of infinite binary sequences such that each is definable by a linguistic, hence finite, expression. The *antidiagonal* sequence of the list is defined as the sequence whose kth term is $1 - c_{kk}$, where c_{ij} is the jth term of the ith sequence in the list. From the conclusion that the antidiagonal sequence is finitely definable and should appear in the list he escapes by denying that this sequence has a finite definition: the sequence has been characterized by a grammatically correct sentence of the language, but, since this finite characterization demands that the sequence be not finitely definable, it contradicts itself, hence is no definition at all—at least so long as we remain in the formal domain. If the definition is regarded as belonging to the conceptual domain, it becomes unobjectionable; no longer formal, it is no longer a finite definition; hence it does not contradict itself. What we now have is a sequence, namely, the antidiagonal sequence, that is definable but not formally definable.

A parallel argument applies to proofs. Consider a formal proof establishing that, in a given binary sequence, 0 occurs infinitely many times or that, in the sequence, 0 does not occur infinitely many times. All such proofs can be ordered, say lexicographically. To each proof corresponds the binary sequence to which the proof refers. The binary sequences corresponding to proofs are therefore ordered once the proofs are. No formal proof corresponds to the antidiagonal sequence. Hence the proposition "In the antidiagonal sequence 0 does not occur infinitely many times" is not formally decidable. Nevertheless this proposition is false: there are infinitely many formal proofs establishing that 0 does not occur infinitely many times in the binary sequence 111..., and to each of these proofs there corresponds a 0 in the antidiagonal sequence.

Clearly, the argument bears a strong resemblance to Richard's. We also observe how it foreshadows Gödel's. Gödel's proof of his incompleteness result rests upon an argument that can be regarded as having been adapted from the Liar paradox, by a shift from "true" to "provable", but also, as Gödel himself notes (below, p. 598), from the Richard paradox. Consider the well-formed formulas of Gödel's system P that contain exactly one free variable, say x. These formulas can be ordered, say lexicographically; let $F_i(x)$ be the ith such formula. We define a_{ij} and b_{ij} by

$$a_{ij} = \begin{cases} 1 \text{ if } F_i(\mathrm{j}) \text{ is true,} \\ 0 \text{ if } F_i(\mathrm{j}) \text{ is not true,} \end{cases}$$

$$b_{ij} = \begin{cases} 1 \text{ if } F_i(\mathrm{j}) \text{ is provable,} \\ 0 \text{ if } F_i(\mathrm{j}) \text{ is not provable.} \end{cases}$$

(Here j is the representation of the number j in the system P). For $j = 0, 1, 2, \ldots$ we have the sequence

$$a_{i0}, a_{i1}, a_{i2}, \ldots;$$

then for $i = 0, 1, 2, \ldots$ we have a sequence of sequences, that is, a table of 0's and 1's that is infinite in two directions. Similarly for b_{ij}. The general terms of the antidiagonal sequences corresponding to these two tables are respectively $1 - a_{jj}$ and $1 - b_{jj}$. Now there is no k such that for every j

$$(1) \qquad 1 - a_{jj} = a_{kj};$$

the existence of such a k would lead to

$$(2) \qquad 1 - a_{kk} = a_{kk},$$

which expresses the Richard paradox. But there may very well be a k such that for every j

$$(3) \qquad 1 - a_{jj} = b_{kj}.$$

This is precisely where Gödel's shift from "true" to "provable" occurs. The antidiagonal sequence is not located on a line of its own table, which would lead to the paradox, but on a line of some other table. (That a k satisfying (3) for every j actually exists has to be proved, and a large part of Gödel's paper is devoted to the proof of just this statement, expressed in other terms.) By substitution, (3) yields

$$(4) \qquad 1 - a_{kk} = b_{kk}.$$

Since P is assumed to be consistent and its intended interpretation is assumed to be sound, no nontruth is provable; hence $a_{ij} \geqq b_{ij}$, and (4) yields $a_{kk} = 1$ and $b_{kk} = 0$, that is, the existence of an unprovable true proposition.

Finsler, too, considers a table of 0's and 1's, and from the fact that the corresponding antidiagonal sequence cannot occur on any line of the table he concludes that there is no formal proof corresponding to the antidiagonal sequence; hence the proposition that corresponds to this sequence, although false, is formally undecidable. Both Gödel's and Finsler's arguments, with their similarities and their differences, skirt the Richard paradox without falling into it; both exploit Richard's argument to obtain new and valid conclusions.

Finsler's paper, however, remains a sketch. The point that begs hardest for clarification is the nature of the purely conceptual realm. Undoubtedly, in introducing this domain Finsler remains faithful to the conviction, long unchallenged among mathematicians, that mathematics cannot be reduced to its linguistic expression. According to him, a definition or a proof that leads to a linguistic contradiction "becomes unobjectionable as soon as we transfer it from the formal to the purely conceptual realm and leave the formal out of consideration". The phrase is repeated but not discussed at length or elucidated. The distinction that Finsler introduces between a sentence written on a blackboard and the same sentence spoken in front of the blackboard would suggest that the relation of the formal domain to the purely conceptual realm may resemble that of a language to an associated metalanguage. But, since Finsler views the formal language as possibly all-embracing, this interpretation can hardly be sustained. It would also conflict with Finsler's own comments (*1944*) on Gödel's 1931 results. The nonformal argument by means of which Gödel's undecidable proposition is recognized as true Finsler considers to be "formal", since it can be expressed—and fairly simply at that—in a language. Hence, according to Finsler, Gödel has not exhibited a formally undecidable proposition at all: Gödel's true undecidable proposition is formally decidable and is simply a true and provable proposition that turns out to be unprovable in some artificially limited system, namely, Gödel's system P. This clearly shows the limits of Finsler's anticipation of Gödel's results. While Gödel puts the notion of formal system at the very center of his investigations, Finsler attempts to lay bare the fallacy hidden in any paradox and to remove this fallacy by means adapted to the specific case, without a general reconstruction of language. Other papers by Finsler (*1923*, *1927*, *1944* with its discussion of the Liar) complement the one reproduced here.

The paper was translated by Stefan Bauer-Mengelberg. Professor Finsler read the translation and suggested a number of emendations. The translation is reproduced here with the kind permission of Professor Finsler and Springer Verlag.

I. Statement of the problem

1. In order to demonstrate the consistency of certain systems of axioms, Hilbert[1] employs a theory of mathematical proofs in which these proofs are thought of as strictly formalized—as arrays, composed of certain signs, that are actually written down. "A proof is an array that must be given as such to our perceptual intuition."[2] And further, "A formula will be said to be provable either if it is an axiom or results from an axiom by substitution, or if it is the end formula of a proof".[3] The goal then is to show that on the basis of a given axiom system a formula representing a (likewise formalized) contradiction is certainly not provable. Axiom systems for which this can be demonstrated are said to be "consistent".[4] In what follows, once we have adopted a certain general notion of formal proof, we shall refer to these systems somewhat more precisely as *formally consistent* systems.

2. The question to what extent this formal consistency permits us to infer a genuine absence of internal contradictions in the axiom system under consideration is related to the problem of decidability.

For, if every mathematical assertion whose truth or falsity represents a logical

[1] See *Hilbert 1922*, *1922a*, *1925*, *Bernays 1922*, and *Ackermann 1924*.

[2] *Hilbert 1922a*, p. 152.

[3] *Hilbert 1922a*, pp. 152–153.

[4] *Hilbert 1922a*, p. 157, and *1925*, p. 179 [[above, p. 383]].

consequence of the axioms were decidable by means of purely formal proofs, we could at once infer absolute consistency from formal consistency.

But this ceases to be the case as soon as there exists a proposition that, though its truth is from the logical point of view unambiguously decided by the axiom system, can be neither proved nor refuted by purely formal means. For then we can add this proposition, or also its contradictory, to the system as a new axiom and thus obtain two formally consistent axiom systems, of which one certainly contains an internal contradiction.

3. Now this possibility in fact exists in the case of the axiom system for the real numbers if we have only a finite or denumerably infinite number of signs available for use in the formal proofs, as is certainly the case for proofs that are actually written down.

Then there can exist only denumerably many formal proofs altogether, since each individual one must consist of a finite arrangement of these signs.[5] But from the axiom system for the real numbers there follow, purely logically, more than denumerably many propositions. Consider, for example, the propositions of the form "α is a transcendental number". For each particular value of α this is a particular proposition, which is either true or false. It is not possible for every single one of these nondenumerably many propositions to be formally decidable, since to each specific proposition there would have to correspond a specific proof, which might have to be obtained from a general proof through substitution or specialization. It follows from this, however, that there exist propositions that are not formally decidable.

4. To be sure, it would still be conceivable that all those propositions that could themselves be formally represented were formally decidable.[6] In that case the axiom to be newly added would not be formally representable.

One might, moreover, readily be led to conjecture that such a nonformal contradiction could never be noticed and hence were harmless—that one could always pretend, therefore, that contradictions of this kind did not exist.

But we shall now show by means of an example that we can in fact exhibit propositions that are not formally decidable by general methods, that are therefore formally consistent, but in which we can nevertheless recognize a contradiction in another way. It follows, therefore, that the proof of the formal consistency of a system does not afford a guarantee against recognizable contradictions.

Since, however, these matters are closely connected with the "paradox of finite definability", we shall first give a precise statement and an explanation of this paradox.[7]

II. The paradox of finite definability

5. Let a fixed system $\mathfrak{S}$ of finitely many[8] signs be given. This system shall contain, in particular, all signs used for mathematical purposes; it may even contain all signs

[5] *Ackermann 1924*, p. 9.

[6] That we cannot formally represent arbitrarily chosen objects—simply by introducing a new sign, say—is to be understood here as follows: we must start from a fixed fundamental system and carry out all further steps, including the definition of new signs, *formally*, lest the representation not be *purely* formal. (See footnote 9.)

[7] See also *Finsler 1923*, p. 143.

[8] Without essential changes we could also assume denumerably infinitely many.

used until now (or even in the future) for writing or printing. A fixed ordering of these signs will be taken as the "alphabetic" one.

Further, assume that there is given a fixed "dictionary" $\mathfrak{B}$, including a "grammar", that unambiguously specifies the meaning associated with certain finite combinations of these signs, with the "words". Assume that words having a finite number of different significations in ordinary linguistic usage have been unambiguously distinguished, say, by means of subscripts. It is also possible, however, that certain signs or combinations of signs are defined as "variables", which will obtain an unambiguous meaning in context by virtue of the grammar. In particular, let $\mathfrak{B}$ contain all words occurring in the present paper, with the meanings employed here, and besides, say, all other words written or printed anywhere until now (or even in the future). The "definitions" of the dictionary need not consist of signs of the system $\mathfrak{S}$; they may be thought of as given in a purely conceptual way.

An arrangement of signs for which $\mathfrak{B}$ does not yield an unambiguous meaning is to be regarded as meaningless.

Any object will be said to be *finitely definable* if there is a finite arrangement of signs of the system $\mathfrak{S}$ such that this arrangement has a meaning, to be ascertained by means of $\mathfrak{B}$, that unambiguously determines the object.

6. Let us call the sequences formed from the numbers 0 and 1, including the sequences 0, 0, 0, . . . and 1, 1, 1, . . . , *binary sequences.*

Two binary sequences are to be regarded as equal if and only if every term of one sequence coincides with the corresponding term of the other.

According to a familiar argument of Cantor's, the totality of all binary sequences is not denumerable.

For, if an arbitrary sequence of binary sequences is given, it will not contain the unambiguously determined binary sequence—let us call it the *antidiagonal sequence* —in which for every n the nth term differs from the nth term of the nth binary sequence. Hence no sequence of binary sequences can contain all binary sequences.

The totality of all binary sequences that are *finitely definable* (by means of $\mathfrak{B}$) is, however, denumerable.

For all the finite combinations that can be formed from signs of the system $\mathfrak{S}$ can be arranged in a denumerable sequence: we put first, of any two of these combinations, the one containing fewer signs and observe "alphabetic" order for those having the same number of signs. This ordering establishes a denumerable sequence of all the combinations that unambiguously define any binary sequence and therewith also of the finitely definable binary sequences themselves. The same binary sequence can occur several times in that sequence, but that is immaterial here.

7. It follows from these considerations that there exist binary sequences that are not finitely definable (by means of $\mathfrak{B}$).

This is not in itself surprising, for, once the totality of the binary sequences has been completely defined, it would be a restriction to demand further that every single one of them have itself a finite definition as well.

But what is paradoxical is that we can unambiguously specify a certain one of the binary sequences that are not finitely definable, namely, *the antidiagonal sequence associated with the sequence of finitely definable sequences that was defined above.* This sequence seems hereby to have been finitely defined after all.

In reality, however, this is not the case. The definition just given does, to be sure, consist of signs of the system $\mathfrak{S}$ and of words that are listed in $\mathfrak{B}$. But that is precisely the reason why it is not logically unobjectionable.

For every object that is unambiguously characterized by a definition consisting of words of $\mathfrak{B}$ must be finitely definable. But a binary sequence cannot have this property and at the same time the other property required by the definition above. This definition, therefore, demands something that is impossible; hence there does not exist a binary sequence that satisfies it.

This result is really obvious since, of course, every attempt to define in a given way something that is not definable in that way must necessarily fail.

The given definition, however, becomes unobjectionable as soon as we transfer it from the formal to the purely conceptual realm and leave the formal out of consideration ⟦*und vom Formalen abstrahiert*⟧. Then it indeed unambiguously defines a certain binary sequence that is not finitely definable by means of $\mathfrak{B}$.[9]

8. The situation obtaining here can be made still clearer by means of a simple example.[10]

Let us write on a blackboard the numbers 1, 2, 3, and the phrase "the least natural number that is not listed on this blackboard". The phrase on the blackboard cannot define a natural number in an unobjectionable way. But just as we cannot infer from this the nonexistence of the number 4, so we cannot infer from the paradox above the nonexistence of binary sequences that are not finitely definable. The same holds also for real numbers that are not finitely definable.

III. A formally undecidable proposition

9. We shall now construct a formally undecidable proposition. To this end we state the following definitions, using the notions already introduced.

A *formal proof* is a finite combination of signs of the system $\mathfrak{S}$ such that this combination has a meaning, to be ascertained by means of $\mathfrak{B}$, that constitutes a logically unobjectionable proof.

A proposition is said to be *formally undecidable* if no formal proof is possible either for this proposition or for its contradictory.

Now consider all the combinations, of signs of the system $\mathfrak{S}$, that constitute a formal proof of the fact that in a certain binary sequence the number 0 occurs infinitely many times, or, alternatively, that it does not occur infinitely many times. Then with every such proof there is associated an unambiguously determined binary sequence, namely, precisely the one for which the proof holds. There may, however, be several such proofs for the same binary sequence.

A denumerable sequence of these proofs can be established by means of the ordering defined in Article 6; then the associated binary sequences, too, form a denumerable sequence. Now take the antidiagonal sequence associated with this sequence and construct the proposition:

[9] Nor can this binary sequence be represented purely formally through the introduction of a new sign (see footnote 6), particularly if the notions that will be formally represented in the future, too, are already contained in $\mathfrak{B}$ in their entirety.

[10] See footnote 7.

In the antidiagonal sequence just defined the number 0 *does not occur infinitely many times.*

This proposition is formally undecidable since the associated binary sequence cannot belong to the sequence established above. We can therefore say that the proposition is *formally consistent.*

10. But, for all that, we can see that this proposition is false, hence inconsistent.

For the number 0 must nevertheless occur infinitely many times in the antidiagonal sequence under consideration since we can obviously prove formally, by means of arbitrarily many words, that in the binary sequence whose first term is 1, whose second term also is 1, and whose succeeding terms likewise all are 1, the number 0 will not occur infinitely many times. But to every such proof there corresponds a zero in the antidiagonal sequence; hence this sequence certainly contains infinitely many zeros.

11. There seems to be a contradiction here, in that the formally undecidable proposition apparently was decided formally after all.

In reality, however, this is not the case. The formal proof just given, which consists of words of $\mathfrak{B}$, is not unobjectionable as a formal proof; for, in referring to a binary sequence that cannot occur in the sequence defined above, it implicitly demands that it should not itself be included among the valid formal proofs. Hence, if it is to represent such a proof after all, this brings about an internal contradiction in the proof itself, which is therefore invalid. Nor could we by any other means succeed in formally deciding the proposition under consideration in an unobjectionable way.

But the proof becomes unobjectionable as soon as we transfer it from the formal to the purely conceptual realm and leave the formal out of consideration.

The formal definition of the antidiagonal sequence under consideration is certainly unobjectionable in itself since it contains no such contradiction.

Thus there actually does exist a formally representable proposition that is formally consistent but logically false.

12. Here, too, we can offer a simpler example, corresponding to that considered in Article 8, for purposes of comparison.

Let us write the proofs of the following four theorems on a blackboard:

$$\sqrt{1} \text{ is rational; } \sqrt{2} \text{ is irrational; } \sqrt{3} \text{ is irrational; } \sqrt{4} \text{ is rational.}$$

Furthermore let us write the following on the same blackboard:

"*Definition.* Let m be the least natural number for which it is not decided by any of the proofs appearing on this blackboard whether its square root is rational or irrational.

"*Assertion.* $\sqrt{m}$ is irrational.

"*Proof.* Since on this blackboard it is decided for the numbers 1, 2, 3, and 4 whether their square roots are rational or irrational, $m > 4$. But, since the proofs appearing on the blackboard concern at most five numbers, certainly $m < 9$. Therefore $2 < \sqrt{m} < 3$. Now the square root of a natural number is either an integer or an irrational number. Since $\sqrt{m}$ cannot be an integer, it therefore follows that $\sqrt{m}$ is irrational."

In this example m is necessarily equal to 5, since for $m > 5$ the number 5 would still be the least natural number about which nothing was proved on the blackboard.

And indeed $m = 5$, since the last proof appearing on the blackboard is not unobjectionable as a proof. The fact that it appears on the blackboard constitutes a demand that it be invalid, for otherwise m would not be the correct number.

But the given proof is unobjectionable if we take it as purely conceptual, that is, if it does not itself appear on the blackboard.

The definition of m that appears on the blackboard is certainly unobjectionable in itself; it unambiguously defines the number 5.

If in addition we were to write the proposition "$\sqrt{m}$ is rational" on the blackboard, it would be a proposition that appears on the blackboard but is not decided, and therefore also not refuted, there; nevertheless the proposition would be false. It would be impossible, moreover, to write on the blackboard any proof that would refute this proposition.

On the domains of definition of functions

LUITZEN EGBERTUS JAN BROUWER

(*1927*)

In a series of papers published from 1918 onward, Brouwer set forth an intuitionistic "set theory" and on this basis an intuitionistic reconstruction of point-set topology and analysis. The text below is part of a paper published in 1927 (received for publication on 28 April 1926) and contains the proof that every function that is (in the intuitionistic sense) everywhere defined on the closed interval [0, 1] of the continuum is uniformly continuous (Theorem 3). In the course of the argument Brouwer proves the fundamental theorem on "sets" that he later (*1953*) called the bar theorem, as well as its corollary, the fan theorem (Theorem 2).[a]

The text brings together and reworks previous expositions of these results. The uniform-continuity theorem had been asserted earlier (*1923*, p. 4), with only an indication of the fan theorem. The bar theorem and the fan theorem were proved, again for the sake of uniform continuity, in a subsequent paper (*1924* (or its German translation, *1924a*), amended and added to in *1924b* (or *1924c*)).

The intuitionistic theory of the continuum is based on Brouwer's own notion of set (see below, p. 453). Brouwer was dissatisfied, it seems, with the treatment of the continuum by earlier constructivist mathematicians. They either abandoned their constructivism at this point and adopted an axiom of completeness or, as is done in theories of the continuum based on ramified type theory, rejected any means of quantifying over more than a denumerable subset of real numbers at a time (see *Brouwer 1952*, p. 140, and *1953*, p. 1). So what was required was an intuitionistic interpretation of quantification over all sequences of natural numbers or over all sequences satisfying some condition.

To explain the classical conception of such quantification, one sometimes pictures an arbitrary sequence as that which results from one choice for each term, these infinitely many choices being conceived *sub specie aeternitatis*, so that questions about the sequence as a whole (such as whether for some n the nth term is zero) are always objectively determined. Brouwer's idea was to substitute for this the picture of an *infinitely proceeding sequence* of choices that is such that at the nth choice one could restrict one's freedom as to future choices by laying down some (not necessarily deterministic) law. This is presented as a process in time: only so much about the sequence is determined at a given stage of its generation as follows from what the initial segment up to that stage is and from the laws that have been laid down. If there are *no* such laws, nothing will be true of the sequence but what is determined to be true on the basis of some

[a] The earliest printed use of "fan theorem" is, it seems, in *Brouwer 1952*, p. 143; Brouwer used "waaierstelling" earlier in lectures.

initial segment of it. This means that functions whose arguments are free-choice sequences will be continuous.

The conception that gives mathematical form to this picture is Brouwer's notion of *set*, or *spread*.[b] The definition can be expressed as follows (see *Heyting 1956*, pp. 34–35). We have a law Λ_M that characterizes certain finite sequences of natural numbers as *admissible* for the spread M and is such that

(1) Every finite sequence of natural numbers is either admissible or not (this means that the law enables us to *decide* of a given sequence whether it is admissible or not);[c]

(2) If $\langle a_1, \ldots, a_{n+1}\rangle$ is admissible, so is $\langle a_1, \ldots, a_n\rangle$;

(3) An admissible sequence of length 1 can be specified;

(4) If $\langle a_1, \ldots, a_n\rangle$ is admissible, either an m can be found so that $\langle a_1, \ldots, a_n, m\rangle$ is admissible or there is no such m (termination of the process).

Then we have a second law Σ_M that to each sequence admissible for M assigns a definite mathematical object.[d]

Thus, given a sequence $a_1, a_2, \ldots$ of natural numbers such that, for every n, $\langle a_1, \ldots, a_n\rangle$ is admissible, we obtain a corresponding sequence $\xi_1 = \Sigma(\langle a_1\rangle)$, $\xi_2 = \Sigma(\langle a_1, a_2\rangle), \ldots$. A sequence such as $\xi_1, \xi_2, \ldots$ is what Brouwer calls an *element* of the spread M.[e]

As an example, we consider the spread of *points of the continuum*, discussed in the paper below. There are many ways of defining a spread that corresponds to the intuitive notion of the continuum. Brouwer defines a point of the continuum as an infinitely proceeding sequence of intervals that are of the form $I_{m,n} = [m/2^n, (m+2)/2^n]$ and are such that each one lies in the interior of its predecessor. The spread of such points could be defined formally as follows. Let $p_1, p_2, \ldots$ be an enumeration of the pairs of natural numbers, and let $\rho_1(x)$ and $\rho_2(x)$ be such that, if $p_i = \langle r, s\rangle$, $r = \rho_1(i)$ and $s = \rho_2(i)$. Then any sequence $\langle n\rangle$ of length 1 is admissible and is assigned the interval $I_{\rho_1(n), \rho_2(n)}$. If $\langle a_1, \ldots, a_n\rangle$ is admissible and is assigned the interval $I_{r,s}$, then $\langle a_1, \ldots, a_n, k\rangle$ is admissible if and only if

$$\frac{r}{2^s} < \frac{\rho_1(k)}{2^{\rho_2(k)}} < \frac{\rho_1(k)+2}{2^{\rho_2(k)}} < \frac{r+2}{2^s},$$

and then the interval $I_{\rho_1(k), \rho_2(k)}$ is assigned to $\langle a_1, \ldots, a_n, k\rangle$.

Brouwer says that a spread is *finitary* if for every n there can be determined a k_n such that the nth term of an admissible sequence, if it exists, is always less than k_n (below, p. 454). This is equivalent to the following condition: there exists a k_0 such that $\langle n\rangle$ is admissible only if $n < k_0$, and for any admissible $\mathfrak{a}$ there exists a $k_{\mathfrak{a}}$ such that $\mathfrak{a} * \langle m\rangle$ is admissible only if $m < k_{\mathfrak{a}}$.[f] That is, at

[b] See below, p. 453.

[c] In *1953*, p. 8, Brouwer admits a species of admissible sequences that is "not necessarily predeterminate". Although what this means is not altogether clear, he apparently does not intend to relax the requirement that we have stated. It seems, however, that Brouwer intends to allow the definition of a not necessarily predeterminate species to contain a free-choice parameter. Then, as Kreisel points out (*1964*, pp. 0.35–0.36), an example due to Kleene implies that the statement of the bar theorem given in *Brouwer 1953*, p. 14, requires modification.

[d] Brouwer's definition specifies a "sign series" (below, p. 453); this would suggest that species and free-choice sequences are not allowed. It seems that this restriction is not held to in subsequent intuitionistic writings. In *1952*, p. 142, Brouwer speaks of "*infinitely proceeding sequences* $p_1, p_2, \ldots$, whose terms are *chosen more or less freely* from mathematical entities *previously acquired*", which would seem to allow anything compatible with a step-by-step generation of mathematical entities. But see *Brouwer 1942*.

[e] The sequence is finite if an n is reached for which $\langle a_1, \ldots, a_n\rangle$ is terminal (that is, the second side of the alternative under (4) holds). Clearly, we cannot say of a given sequence that it either comes to an end or does not.

[f] We shall use lower-case German letters as variables ranging over finite sequences of natural numbers. $\mathfrak{a}*\mathfrak{b}$ is the concatenation of $\mathfrak{a}$ and $\mathfrak{b}$; if $\mathfrak{a} = \langle a_1, \ldots, a_n\rangle$ and $\mathfrak{b} = \langle b_1, \ldots, b_m\rangle$, then $\mathfrak{a}*\mathfrak{b} = \langle a_1, \ldots, a_n, b_1, \ldots, b_m\rangle$.

each stage there are only finitely many choices.

The *unit* continuum can be represented by a finitary spread if one sets a limit on how much smaller an interval can be than its predecessor (and on how small the initial interval may be), for then there are only finitely many choices.

Before we go on, the reader's attention is called to Brouwer's conception of a species. The definition that he gives in 1925 will be found below, p. 454. Later, species are defined as "properties supposable for mathematical entities previously acquired and satisfying the condition that, if they hold for a certain mathematical entity, they also hold for all mathematical entities that have been defined to be *equal* to it" (*1953*, p. 2). Since two species are equal if they have the same members, the notion of species has about the same role as that of class in nonintuitionistic mathematics.

Before we can explain the results of the paper below, we need to ask the question what meaning we can give to the notion of a *function* that maps one spread into another or into the natural numbers. It is by the analysis of this notion that Brouwer obtains the information necessary to prove the bar theorem and the uniform-continuity theorem.

The essence of the analysis is that, when a function is defined on a spread and has definite objects such as natural numbers as its values, its value for a given sequence that is an element of the spread must be determined by a finite number of terms of the sequence. If the value itself is to be a free-choice sequence, a certain initial segment of the argument must suffice to determine the first term of the value, a certain further segment to determine the second term, and so on.[g]

We must state this point with precision in order to avoid ambiguities. Consider a functional F that is defined on a spread M and whose values are natural numbers, as in the bar theorem. Then for every element α of M there exists a number n such that, if β agrees with α on the first n terms, that is, if the sequence

---

[g] The first of these two statements is the continuity requirement that Brouwer makes in his proofs of the bar and fan theorems and of uniform continuity. But what in Theorem 1 below is stated to follow directly from the intuitionistic conception of a full function is a condition, weaker than the second statement, of *negative* continuity. It is puzzling why Brouwer states negative continuity with some fanfare and then goes on, at the beginning of § 2, to state quietly a stronger continuity requirement. The explanation seems to be as follows. Theorem 1 states the negative continuity of a full function in terms of the definition of § 1, that is, of the definition of negative continuity for a function of point *cores*. If f is a function that maps point cores onto point cores, it induces a function f_0 that maps *points* onto *points*. From the assumption of § 2 it follows immediately that, if f is a full function, f_0 is positively continuous. But it is not quite immediate that f is positively continuous.

Let ξ_0 be a point core; we must prove that f is continuous at ξ_0. Let p_0 be a point belonging to ξ_0; then the point $f_0(p_0)$ belongs to $f(\xi_0)$. Note that a point p is a sequence of intervals $p(n)$; we can denote by "$\overline{p}(m)$" the sequence of the first m intervals of p. From the continuity requirements of § 2 it follows that for any n we can find an m such that, if $\overline{p}(m) = \overline{p}_0(m)$, then $\overline{[f_0(p)]}(n) = \overline{[f_0(p_0)]}(n)$.

Given $\varepsilon > 0$, chose n_0 so that the diameter of $\overline{[f_0(p_0)]}(n_0) < \varepsilon$. For each m let a_m and b_m be the end points of $p_0(m)$. Let m_0 be the m obtained as above with $n = n_0$. Now let

$$\delta = \tfrac{1}{2}(\min(a_{m_0+1} - a_{m_0}, b_{m_0} - b_{m_0+1})).$$

δ is positive since $p_0(m_0 + 1)$ lies entirely within $p_0(m_0)$. Suppose that $|\xi - \xi_0| < \delta$ and that p is a point of ξ. Let k be such that the diameter of $p(k) < \delta$. Then $p(k)$ lies entirely within $p(m_0)$. Therefore p coincides with the point $q = p_0(0), \ldots, p_0(m_0), p(k), p(k+1), \ldots$, and $f_0(p)$ coincides with $f_0(q)$. Since $\overline{[f_0(q)]}(n_0) = \overline{[f_0(p_0)]}(n_0)$, and $f_0(q)$ is a point of $f(\xi)$, we have $|f(\xi) - f(\xi_0)| < \varepsilon$, q. e. d.

Since the number δ obtained for a given ξ_0 and a given ε depends on the particular point p_0 that represents ξ_0, the argument does not show that there is a *function* giving δ in terms of ξ_0 and ε. The existence of such a function is equivalent to uniform continuity in some neighborhood containing ξ_0 and presumably cannot be proved without the fan theorem.

of natural numbers of length n is the same for both, $F(\alpha) = F(\beta)$.

This implies that F can be represented by f, a function from finite sequences of natural numbers to natural numbers, in the following sense: if, in attempting to compute $F(\alpha)$ on the basis of the choices for α prescribed by the sequence $\mathfrak{a}$, we reach a point at which we lack the information about α needed to continue the computation, we set $f(\mathfrak{a}) = 0$. If we can complete the computation without meeting such an obstacle, we set $f(\mathfrak{a}) = F(\alpha) + 1$. This equation will hold for any α generated by a sequence of choices beginning with the choices of $\mathfrak{a}$.

The species μ_1 of § 2 in the text below can be identified with the species of admissible sequences $\mathfrak{a}$ for which $f(\mathfrak{a}) \neq 0$, but $f(\mathfrak{b}) = 0$ for any proper initial segment $\mathfrak{b}$ of $\mathfrak{a}$. Brouwer says that a sequence is *secured* if it belongs to μ_1, has an initial segment belonging to μ_1, or is inadmissible. The argument for the bar theorem, as well as for later results along the same lines, turns on an analysis of the species of *unsecured* sequences. If $\langle a_1, \ldots, a_n \rangle$ is secured and either $n = 1$ or $\langle a_1, \ldots, a_{n-1} \rangle$ is unsecured, $\langle a_1, \ldots, a_n \rangle$ is said to be *immediately secured.*

The important mathematical content of the paper is contained in the bar theorem. The fan theorem and the uniform-continuity theorem are corollaries. Below, the bar theorem is stated in the fourth paragraph of § 2. The claim is that, for a functional F from a spread M to the natural numbers, the species of unsecured sequences is capable of a certain kind of well-ordered construction. In order to explain this, we must make some remarks about Brouwer's theory of well-ordering.

The basic definitions are given below, pp. 456–457. I use the terminology presented in these definitions. It follows from the definition of a well-ordered species that with each well-ordered species S there is associated a species S' of finite sequences of natural numbers, the members of S' being the subscript sequences for the constructional underspecies of S. We can specify S, up to isomorphism, by giving S' and stating, for each sequence associated with a primitive species, whether the element of that species is a full or a null element.

Consider now the following ordering of finite sequences: $\langle a_1, \ldots, a_n \rangle \prec \langle b_1, \ldots, b_m \rangle$ if either

(1) $m < n$ and $a_i = b_i$ for $i = 1, \ldots, m$

or

(2) for some $i < \min(m, n)$, $a_j = b_j$ for all $j \leqq i$, while $a_{i+1} < b_{i+1}$;

that is, the sequences are so ordered by $\prec$ that an extension of a sequence *precedes* it and that otherwise two sequences are ordered lexicographically. Clearly, this is a primitive recursive linear ordering.[h] If $S_{\mathfrak{a}}$ and $S_{\mathfrak{b}}$ are the constructional underspecies of S with the subscript sequences $\mathfrak{a}$ and $\mathfrak{b}$ respectively, then $\mathfrak{a} \prec \mathfrak{b}$ if and only if $S_{\mathfrak{a}}$ precedes $S_{\mathfrak{b}}$ in the construction of S, that is, $S_{\mathfrak{a}}$ is a constructional underspecies of $S_{\mathfrak{b}}$, or every element of $S_{\mathfrak{a}}$ precedes every element of $S_{\mathfrak{b}}$ in the ordering of S. Since the latter side of the alternative must hold if $S_{\mathfrak{a}}$ and $S_{\mathfrak{b}}$ are primitive species, the ordering $\prec$ restricted to the subscript sequences of primitive species is isomorphic (as a linear ordering) to the ordering of S. The ordering $\prec$ restricted to S' satisfies the condition that every descending chain is finite, by virtue of the well-founded nature of the construction of S. Thus it is a well-ordering according to the classical conception.

The bar theorem can now be stated as follows. Let F be a functional that to each element α of a spread M assigns a natural number. With F is associated the species T of its unsecured and immediately secured sequences. Then T is the species S' of subscript sequences of

[h] We identify the sequence $\langle a_1, \ldots, a_n \rangle$ with the number $\prod_{i<n} p_i^{a_{i+1}+1}$.

some well-ordered species S. The elements of S can be taken to be the immediately secured sequences of F. The sequence $\mathfrak{a}$ is a full element if it is admissible (and a fortiori if $f(\mathfrak{a}) \neq 0$) and a null element otherwise.[1]

All this means that T can be inductively defined as follows:

(1) If $\mathfrak{a}$ is immediately secured, then $\mathfrak{a} \varepsilon T$;

(2) If $\mathfrak{a}*\langle n \rangle \varepsilon T$ for every n, then $\mathfrak{a} \varepsilon T$.

Hence we have the following *induction principle*: If a property holds of every immediately secured sequence and holds of $\mathfrak{a}$ if it holds of $\mathfrak{a}*\langle n \rangle$ for every n, then it holds of every sequence in T. In recent writings (*Spector 1961*, p. 9, *Kreisel 1963*) an essentially equivalent principle is formalized under the title "bar induction". Different versions of this principle are stated and compared in *Kleene and Vesley 1965*, § 6, where, however, the name "bar induction" is not used, and in *Howard and Kreisel 1966*.

An equivalent statement is that the representing function f of F belongs to a certain inductively generated species K of functions from finite sequences of natural numbers to natural numbers, defined as follows:

(1) Any constant function belongs to K;

(2) If H enumerates a sequence of functions in K, then

$$\lambda\langle a_0, \ldots, a_n\rangle\{H(a_0)\}(\langle a_1, \ldots, a_n\rangle) \varepsilon K.$$

Theorem 2, later called the fan theorem, states that, if the spread M on which F is defined is finitary, there can be found an n such that, for any α, the value of F depends only on the first n choices for α. This follows because one can show by an induction parallel to the generation of T that in this case there are only finitely many unsecured sequences. Clearly, the fan theorem in effect asserts the uniform continuity of F. Since, as we indicated above, the points of the unit continuum can be generated as a finitary spread, functions defined everywhere on the unit continuum, with points of the continuum as values, are uniformly continuous (Theorem 3).

The result of this analysis is that many definitions of functions of a real variable that, from the classical point of view, assign a value to the function for each value of the argument do not do so intuitionistically. The question arises whether one can single out certain subspecies of the continuum such that a function defined on such a subspecies will be analogous to a classically everywhere defined function. Such a function could be said to be *pseudo-full*. In the sections of the paper that are not reprinted here, Brouwer discusses a number of possible criteria for the domain D of a pseudo-full function of the unit continuum. Clearly, D should be a species that possesses a property classically equivalent to coincidence with the unit continuum. The one that Brouwer selects is *congruence*—that it is absurd that there should be a point of the unit continuum not coinciding with any point of the species (*1923d*, p. 255). The further requirement concerns measure: for *every* measure on the unit continuum D must be measurable and possess the measure 1.

A point that requires some discussion is the nature of the proof of the bar theorem. Even if we accept the continuity condition as expressing part of what we mean by a constructive functional on a spread, this is not sufficient for the proof. Brouwer goes on to exploit more fully than in any other intuitionistic argument the following peculiarity of intuitionistic mathematics: the supposition that a mathematical proposition is *true* is just the supposition that one has a (constructive) proof of it. In particular this will be the case, given a sequence $\mathfrak{a}$, for the statement that $\mathfrak{a}$ is *securable*, that

[1] This statement differs only in minor details from that of the text (fourth paragraph of § 2, p. 461 below).

is, that every sequence $\mathfrak{a}$, $\mathfrak{a}*\langle b_1\rangle, \ldots$, $\mathfrak{a}*\langle b_1, \ldots, b_m\rangle, \ldots$ contains a term that is secured. Brouwer goes on to make a controversial assumption about the possible form of a proof of such a statement. He claims that a proof of securability is based on the "givenness" of the secured sequences and on the relations between sequences that are formed by the composition of the relation of immediate succession, that is, the relation between $\mathfrak{a}$ and $\mathfrak{a}*\langle n\rangle$. Then the proof can be brought into a canonical form that uses only inferences resting upon the basic relations:

(1) If $\mathfrak{a}$ is (immediately) secured, it is securable;

(2) If $\mathfrak{a}$ is securable, so is $\mathfrak{a}*\langle n\rangle$ (ζ-inference);

(3) If $\mathfrak{a}*\langle n\rangle$ is securable for every n, $\mathfrak{a}$ is securable ($\digamma$-inference).

In the remainder of the argument it is shown that, for each unsecured sequence $\mathfrak{a}$ for which the canonical proof establishes that it is securable, there exists a well-ordered construction of the species $T_{\mathfrak{a}}$ of descendants of $\mathfrak{a}$ in T. It seems to follow, and indeed this is explicitly stated in *Brouwer 1953*, that the ζ-inferences are superfluous. We obtain a well-ordered construction in T itself by one more second generating operation (below, p. 456), taking as the nth term $T_{\langle n\rangle}$ if $\langle n\rangle$ is admissible and a species consisting of a single null element (below, p. 456) otherwise.

What can be said in justification of the claim that the proofs of securability have a canonical form? It does not seem to be at all evident. Indeed, it seems not to follow from Brouwer's remark that no other basis is available for the proofs than the relations of a sequence to those immediately issuing from it. For, perhaps, the proof might make some use of these relations that is not reducible to inferences directly involving the notion of securability.

Unfortunately, Brouwer's thesis that mathematics is independent of language and logic makes it difficult and perhaps impossible to consider possible counterexamples to his claim. However, it seems that his point of view implies the existence of *cut-free* canonical proofs, in which inductions are replaced by infinitary inferences (footnote 8 below, p. 460). This might be the answer to the objection that the assertion (represented as "q") of the securability of a sequence might be inferred from statements "p" and "$p \supset q$" while there might be no reason to expect the proofs of these premisses to contain the inferences that he claims. That modus ponens should be eliminable is suggested by Heyting's subsequent explanation of the conditional, together with the thesis that mathematics is independent of logic. According to Heyting "$p \supset q$" is the claim that there exists a method of reaching a proof of "q" from one of "p". Now, if the logical connectives are explained in terms of a notion of proof, one might expect that the "proofs" referred to in the explanation of a statement not containing the conditional should not themselves contain conditionals.

The general intuitionistic conceptions leave the meaning of quantification over free-choice sequences somewhat vague. It is made clearer by Brouwer's conception of a spread and by the continuity requirement, but there is room for further clarification. In the case of universal quantification over natural numbers, ordinary induction provides a very clear proof procedure, which arises directly from the inductive generation of the sequence of natural numbers itself. Nothing comparable is available for free-choice sequences. Indeed, it is not even certain whether they are to be regarded as individual mathematical objects at all or whether quantification over them is to be regarded as a *façon de parler*.

In the latter case we may be free to stipulate some criteria for the truth of statements containing such quantification. Even if free-choice sequences are individual mathematical objects, the notion seems vague enough, so that a stipulation as to truth or proof conditions of statements of certain forms may serve to clarify it. Indeed, Brouwer himself suggests in footnote 7, p. 460 below, that a stipulation underlies the bar theorem; he says that, "when carefully considered from the intuitionistic point of view", securability is just the property defined by an inductive definition like that given above for the species T.

In summary, Brouwer's justification for the bar theorem is certainly not evident or even satisfactory. His work makes clear that to obtain powerful results in intuitionistic analysis it is necessary to make some strong assumption that exploits the specifically intuitionistic force of quantification over free-choice sequences. If the assumptions that he makes are less evident than, say, the axiom of choice is in terms of the classical conception of set, this may be due to the newness of the whole subject.

Another point worth mentioning is that the well-ordering of unsecured sequences of continuous functionals was used in a classical context in descriptive set theory, prior to Brouwer's publication. Let S be a topological space, and suppose that we have a function that to each sequence $\mathfrak{a}$ assigns a subset $M_{\mathfrak{a}}$ of S. Then M is defined by the *operation* A applied to the sets $M_{\mathfrak{a}}$ if

$$M = \bigcup_{\alpha} \bigcap_{n=0}^{\infty} M_{\langle \alpha(0), \ldots, \alpha(n) \rangle},$$

that is, if

$$x \,\varepsilon\, M \equiv (\exists \alpha)(n)(x \,\varepsilon\, M_{\langle \alpha(0), \ldots, \alpha(n) \rangle})$$

(α ranging over one-place number-theoretic functions). Then we can say that a sequence $\mathfrak{a}$ is *secured* with respect to x if $x \notin M_{\mathfrak{a}}$; if α satisfies the condition[j]

$$\alpha(m) = \mathfrak{a}_m \text{ for } m < \tilde{\mathfrak{a}},$$

then in case $\mathfrak{a}$ is secured we have

$$(\exists n)(x \notin M_{\langle \alpha(0), \ldots, \alpha(n) \rangle}).$$

Hence, if every unsecured sequence is securable, we have

$$(\alpha)(\exists n)(x \notin M_{\langle \alpha(0), \ldots, \alpha(n) \rangle}),$$

that is, $x \notin M$. This will hold if and only if the ordering $\prec$ restricted to sequences unsecured with respect to x is a well-ordering. So we have

$$\overline{M} = \hat{x}(\prec \text{ well-orders } \hat{\mathfrak{a}}(x \,\varepsilon\, M_{\mathfrak{a}})).$$

Taking the real line as the set S, Suslin (*1917*) defines a class of sets called *A-sets*, which are those subsets of S that can be obtained by the operation A from systems of closed intervals. These sets are the same as the *analytic sets*, which can be defined in other ways, for example, as the images of S by functions continuous for all but countably many arguments or as the projections of Borel subsets of the plane. Suslin states that a set M is a Borel set if and only if both M and $\overline{M}$ are A-sets. A proof of this that uses the above definition of $\overline{M}$ in terms of unsecured sequences was given by Luzin and Sierpiński (*1918*, pp. 36–42). Extensive use of this method was made by Luzin in the twenties (*1927*; *1930*, chap. III), for example, to prove the first and second separation principles for analytic sets. Luzin (*1927*, pp. 2–3; *1930*, pp. 197–200) finds the germ of the idea in a construction by Lebesgue (*1905*).[k] From Brouwer's eminence as a topologist it would be plausible to conclude that he knew Luzin and Sierpiński's proof, but direct evidence is lacking.

The ideas of Brouwer and of the descriptive set theorists come together in Kleene's work on the hyperarithmetic and analytic hierarchies (*1955*). If we let the space S mentioned above be the set

---

[j] $\mathfrak{a}_m$ = the $(m + 1)$th term of $\mathfrak{a}$, $\tilde{\mathfrak{a}}$ = the least m for which $\mathfrak{a}_m$ does not exist.

[k] A special case of the operation A was introduced by Aleksandrov (*1916*). Although he does not give a definition of A-sets that is equivalent to Suslin's, he proves in effect that every Borel set is an A-set.

of natural numbers, the Σ_1^1 sets are just those that we obtain by applying the operation A, with the condition that "$x \varepsilon M_a$" be primitive recursive. In this case the ordering of unsecured sequences is also primitive recursive. Kleene seems to have derived the idea of the well-ordering from Brouwer by observing that in this case $(\alpha)(\exists n)(x \notin M_{\langle\alpha(0),\ldots,\alpha(n)\rangle})$ says that a certain recursive (and therefore continuous) functional is everywhere defined. He can thus show that any Π_1^1 set is recursive in the set O of notations for recursive ordinals. Spector (*1955*) shows that any Π_1^1 set is recursive also in the set W of Gödel numbers of recursive well-orderings. (Since W and O are Π_1^1, they are therefore recursive in each other.)

Further use by Kleene of the well-ordering (for example, to show that the hyperarithmetic sets are those that are Σ_1^1 and Π_1^1) is very close to that of descriptive set theory. The analogies are made very clear by Addison (*1958*), who, however, neglects to mention Kleene's debt to Brouwer.

As this example illustrates, the greatest influence of Brouwer's ideas has been on the development of theories of effectiveness and constructivity at higher types, in which the analysis of unsecured sequences is now a standard tool. Application of conceptions related to the bar theorem in proof theory has become quite extensive in recent work by Spector, Kreisel, and others. Although this work was motivated mainly by the consistency problem for classical analysis, it has also served to clarify the intuitionistic ideas themselves. The hope entertained by many that the idea of the bar theorem would yield a proof of the consistency of classical analysis led to a precise result by Spector (*1961*), who proved the consistency of classical analysis relative to that of a quantifier-free system containing functionals of arbitrary finite types and a schema for the definition of functionals by "bar recursion". But the form of "bar induction" that would serve to justify Spector's form of bar recursion is a generalization in which the "unsecured sequences" can be objects of any finite type. It is thus a substantial extension, for which no constructive foundation is known, of what has previously counted as intuitionistic mathematics.

Charles Parsons[1]

To help the reader in his study of the paper, we now print, in their logical order, the definitions, culled from Brouwer's writings, of a number of intuitionistic notions used in the paper.[m]

"A set ⟦*Menge*⟧ is a law on the basis of which, if repeated choices of arbitrary natural numbers ⟦Nummer⟧ are made, each of these choices either generates a definite sign series ⟦Zeichenreihe⟧, with or without termination of the process, or brings about the inhibition of the process together with the definitive annihilation of its result; for every $n > 1$, after every unterminated and uninhibited sequence of $n - 1$ choices, at least one natural number can be specified that, if selected as the nth number, does *not* bring about the inhibition of the process. Every sequence of sign series generated in this manner by an unlimited choice sequence ⟦unbegrenzten Wahlfolge⟧ (and hence generally not representable in a finished form) is called an *element of the set*. We shall also speak of the common mode of formation of the elements of a set M as, for short, *the set M*." (*Brouwer 1925*, pp. 244–245 (a footnote is omitted); see also *1918*, p. 3, and *1919b*, pp. 204–205, or in the reprint pp. 950–951.)

Subsequently, the terms that Brouwer uses for this notion are "spreiding" (*1947*, in Dutch) and "spread" (*1953*).

---

[1] I am indebted to Dirk van Dalen, Burton Dreben, J. J. de Iongh, Yiannis Moschovakis, and especially to Georg Kreisel and the editor, for their assistance and suggestions.

[m] I am grateful to Professor Richard E. Vesley for having helped me to understand a number of passages in Brouwer's writings.

"If for every n in ζ ⟦the sequence of natural numbers, $1, 2, 3, \ldots$⟧ a natural number k_n is determined such that the inhibition of the process takes place whenever a natural number lying in ζ above k_n is selected at the nth choice, then the set is said to be *finitary* ⟦*finit*⟧." (*Brouwer 1925*, p. 245.)

"Two set elements are said to be *equal* ⟦*gleich*⟧, or *identical* ⟦*identisch*⟧, if we are sure that for every n the nth choice generates the same sign series for the two elements.

"Two sets are said to be *equal*, or *identical*, if for every element of one set an equal element of the other can be specified.

"The set M is called a *subset* ⟦*Teilmenge*⟧ of the set N if for every element of M an equal element of N exists.

"Sets and elements of sets are called *mathematical entities*.

"By a *species* ⟦*Spezies*⟧ *of first order* we understand a property (defined in a conceptually complete form) that only a mathematical entity can possess, and, if it does, the entity is called an *element of the species of first order*. Sets constitute special cases of species of first order.

"Two species of first order are said to be *equal*, or *identical*, if for every element of one species an equal element of the other can be specified.

"By a *species of second order* we understand a property that only a mathematical entity or a species of first order can possess, and, if it does, the entity or the species is called an *element of the species of second order*.

"Two species of second order are said to be *equal*, or *identical*, if for every element of one species an equal element of the other can be specified.

"In an analogous manner we define *species of nth order*, as well as their equality, or identity, n representing an arbitrary element of A ⟦the set of natural numbers⟧.

"The species M is called a *subspecies* ⟦*Teilspezies*⟧ of the species N if for every element of M an equal element of N exists. If, moreover, it is possible to specify an element of N that cannot be equal to any element of M, then M is called a *proper subspecies* of N.

"Two set elements are said to be *distinct* ⟦*verschieden*⟧ if the impossibility of their equality has been established, that is, if we are sure that, in the course of their generation, their equality can never be proved.

"Two species are said to be *distinct* if the impossibility of their equality has been established." (*Brouwer 1925*, pp. 245–246.)

"A species of which any two elements can be recognized either to be equal or to be distinct is said to be *discrete*." (*Brouwer 1925*, p. 246.)

"The species that contains those elements that belong either to the species M or to the species N is called the *union* of M and N, and it is denoted by $\mathfrak{S}(M, N)$." (*Brouwer 1925*, p. 247.)

"Two species M and N are said to be *disjoint* ⟦*elementefremd*⟧ if they are distinct and it is impossible that there exist an element of M and an element of N that are identical with each other." (*Brouwer 1925*, p. 247.)

"If M' and M'' are disjoint subspecies of N and if $\mathfrak{S}(M', M'')$ and N are identical, then we say that N *splits* ⟦*zerlegt* ist⟧ into M' and M''; we call M' and M'' *conjugate splitting species of* N, and M', as well as M'', is called a *removable* ⟦*abtrennbare*⟧ *subspecies* of N." (*Brouwer 1925*, p. 247; see also *Brouwer 1918*, p. 4, *Heyting 1956*, p. 39, where, instead of "removable", "detachable" is used, and *Brouwer 1953*, p. 6.) If M' and M'' are removable subspecies of N, we can decide whether an arbitrary element of N belongs to M' or to M''.

"If between two species M and N there can be established a one-to-one relation, that is, a law that with every element of M associates an element of N in such a way that equal elements of N correspond to equal, and only to equal, elements of M and that every element of N is associated with an element of M, we write $M \sim N$ and say that M and N have the same *cardinality* ⟦*Mächtigkeit*⟧, or *cardinal number*, or are *equipollent* ⟦*gleichmächtig*⟧." (*Brouwer 1925*, p. 247.)

"The simplest example of an infinite set is the set A itself, and we denote its cardinal number by a. Species that possess that cardinal number are said to be *denumerably infinite* ⟦*abzählbar unendlich*⟧." (*Brouwer 1925*, p. 249; see also *Brouwer 1918*, pp. 6–7, and *Heyting 1956*, p. 39.)

"A species M ⟦of cardinal number m⟧ satisfying the formula $m \leqq a$ is said to be *denumerable* ⟦*abzählbar*⟧. In particular, it is said to be *numerable* ⟦*zählbar*⟧ if it can be mapped one-to-one onto a *removable* subspecies of A." (*Brouwer 1925*, p. 255; see also *Brouwer 1918*, p. 7, and *Heyting 1929* and *1956*, p. 40.)

"A species P is said to be *virtually ordered* ⟦*virtuell geordnet*⟧ if an asymmetric relation, which will be called the *ordering relation*, is defined for the elements of a subspecies of the species of pairs (a, b) of elements of P; we express this relation by '$a < b$', '*a before b*', '*a to the left of b*', '*a lower than b*', '$b > a$', '*b after a*', '*b to the right of a*', or '*b higher than a*', and, if we express the identity of two elements p and q of P by the formula '$p = q$', stipulate that it possess the following '*order properties*':

(1) The relations $r = s$, $r < s$, and $r > s$ are mutually exclusive;

(2) From $r = u$, $s = v$, and $r < s$ it follows that $u < v$;

(3) From the simultaneous absurdity ⟦Ungereimtheit⟧ of the relations $r > s$ and $r = s$ it follows that $r < s$;

(4) From the simultaneous absurdity of the relations $r > s$ and $r < s$ it follows that $r = s$;

(5) From $r < s$ and $s < t$ it follows that $r < t$." (*Brouwer 1925a*, p. 453; see also *Brouwer 1918*, p. 13, and *Heyting 1955*, p. 33, and *1956*, p. 107.)

"The virtually ordered species P is said to be *everywhere dense in the extended sense*, or, for short, *everywhere dense*, if between any two distinct elements of P there lie elements of P, and it is said to be *everywhere dense in the strict sense* if moreover an element of P can be specified and elements of P lie to the right as well as to the left of an arbitrary element of P." (*Brouwer 1925a*, p. 454; see also *Brouwer 1918*, p. 16.)

"If between two virtually ordered species P and Q there has been established a one-to-one relation that leaves the ordering relations invariant, we say that P and Q possess the same *ordinal number*, or are *similar*." (*Brouwer 1925a*, p. 455; see also *Brouwer 1918*, p. 14.)

"A virtually ordered species is said to be *ordered* if an ordering relation obtains for *every* pair (a, b) of *distinct* elements" (*Brouwer 1925a*, p. 455 (a footnote is omitted); see also *Heyting 1956*, p. 106.)

"A discrete ordered species is also said to be *completely ordered*." (*Brouwer 1925a*, p. 455.)

The ordinal number of the set A in its natural ordering is denoted by ω.

"Ordered species of ordinal number ω are also called *fundamental sequences* ⟦*Fundamentalreihen*⟧." (*Brouwer 1925a*, p. 455; see also *Brouwer 1918*, p. 14.)

"Let R be a virtually ordered species of virtually ordered species N such that equal elements e of $M = \mathfrak{S}(N)$ always belong only to equal species N and that equal species N are always virtually ordered in the same way. We denote the ordering relations of the given virtual orderings of R and the N by $>$ and $<$, and we define a virtual ordering of M as follows: We write $e' \gtrdot e''$, or $e'' \lessdot e'$, if either $N' > N''$ or both $N' = N''$ and $e' > e''$; we write $e' \underline{\gtrdot}\, e''$, or $e'' \underline{\lessdot}\, e'$, if $e' \lessdot e''$ is impossible; we write $e' > e''$ if $e' \underline{\gtrdot}\, e''$, and, moreover, $e' \neq e''$.... We call the species M, once it is virtually ordered as above, (or its ordinal number m) the *ordinal sum* ⟦*ordnungsgemäße*

Summe⟧ of the species N (or of their ordinal numbers n), and we call the formation of this sum the *addition* of the N (or of the n). In case R possesses a finite complete ordinal number or the ordinal number ω, the sign $+$ is used in the ordinary way to denote addition." (*Brouwer 1925a*, p. 456–457.)

"Let M be a denumerably infinite ordered species that is everywhere dense in the strict sense and whose elements are denumerated by the fundamental sequence $g_1, g_2, g_3, \ldots$. We put $\mathfrak{S}(g_1, g_2, \ldots, g_\nu) = s_\nu$. In M we establish an 'intercalation partition' ⟦'Einschaltungsteilung'⟧; that is, we generate in M, by an unlimited sequence of free choices, a left and a right subspecies (an arbitrary element of the left subspecies preceding an arbitrary element of the right one) in such a way that the left and the right subspecies are determined successively in $s_1, s_2, s_3, \ldots$, the procedure being such that only a single element g_{α_ν} of s_ν may remain excluded from these subspecies of s_ν and that for every ν the element $g_{\alpha_{\nu+1}}$, if it exists, is identical either with g_{α_ν} or with $g_{\nu+1}$. The species of the intercalation partitions t of M (*corresponding to arbitrary distinct denumerations of* M) that 'coincide' with a certain intercalation partition t_1 of M, in the sense that an element of the left subspecies of one partition never lies to the right of an element of the right subspecies of the other, is what we call an *intercalation element* of M, and t, as well as t_1, is a 'partition' of this intercalation element. We virtually order the species of the intercalation elements e of M by adopting the following stipulations: we write $e' \lessdot e''$ if we can specify a partition t' of e', a partition t'' of e'', and two elements g_σ and g_τ of M that belong to the right subspecies of t' and to the left subspecies of t'', respectively; we write $e' \underline{\lessdot} e''$ if $e' \gtrdot e''$ is impossible; we write $e' < e''$ if $e' \underline{\lessdot} e''$ and also $e' \neq e''$. The method already used several times above yields the result that these stipulations indeed entail the validity of the five ordering properties. We call the thus virtually ordered species of intercalation elements of M the *continuum over* M and denote it by $K(M)$." (*Brouwer 1925a*, p. 467.)

"The *well-ordered species* are ordered species that are defined on the basis of the following stipulations:

(1) An arbitrary element of a well-ordered species is either an *element of the first kind* and will be called a '*full element*' ⟦'*Vollelement*'⟧ or an *element of the second kind* and will be called a '*null element*' ⟦'*Nullelement*'⟧;

(2) A species with a unique element, once this element has been provided with either the predicate of being a full element or the predicate of being a null element, will be called a well-ordered species and, more particularly, a *primitive species* ⟦*Urspezies*⟧;

(3) From known well-ordered species further well-ordered species are derived through the *first generating operation*, which consists in the addition of a non-vanishing finite number of known well-ordered species, and through the *second generating operation*, which consists in the addition of a fundamental sequence of such species.

"Every well-ordered species that played a role in the construction of the well-ordered species F according to the preceding paragraph is called a *constructional underspecies* ⟦*konstruktive Unterspezies*⟧ of F. The constructional underspecies that played a role in the last generating operation of F are called the *constructional underspecies of first order* of F and are distinguished from one another by a subscript ν, hence are denoted by $F_1, F_2, \ldots, F_m$ or by $F_1, F_2, F_3, \ldots$. The constructional underspecies of the first order of an F_ν are called the *constructional underspecies of second order* of F and are denoted by $F_{\nu 1}$, $F_{\nu 2}$, $\ldots$, $F_{\nu m}$ or by $F_{\nu 1}$, $F_{\nu 2}$, $F_{\nu 3}, \ldots$. The constructional underspecies of the first order of a $F_{\nu_1 \cdots \nu_n}$ are called the *constructional underspecies of* $(n+1)$*th order* of F and are denoted by $F_{\nu_1 \cdots \nu_n 1}$, $F_{\nu_1 \cdots \nu_n 2}, \ldots, F_{\nu_1 \cdots \nu_n m}$ or by $F_{\nu_1 \cdots \nu_n 1}$, $F_{\nu_1 \cdots \nu_n 2}$, $F_{\nu_1 \cdots \nu_n 3}, \ldots$ (F itself is taken as the *constructional underspecies of 0th order* of F). Thus every primitive species used in the construction of F turns out to be a *constructional underspecies of some finite order* of F (although for suitably chosen primitive species of F this order can, of course, increase beyond any bound). To see that, we need only use the *inductive method*, that is, observe that the property in question is satisfied for primitive species, that, if $\xi = \xi_1 + \xi_2 + \cdots + \xi_m$ on the basis of the first generating operation and $\xi' = \xi_1 + \xi_2 + \cdots + \xi_{m-1}$ on the basis of the first generating operation ($m \geqq 2$), the property in question, in case it holds for $\xi_1, \xi_2, \ldots, \xi_m$ as well as for ξ', also obtains for ξ, and, finally, that, if $\xi = \xi_1 + \xi_2 + \xi_3 + \ldots$ on the basis of the second generating operation, the property in question, in case it holds for every ξ_ν, also obtains for ξ.

"By means of the inductive method we see that for an arbitrary well-ordered species F the species of the subscript sequences of the elements, as well as the species of the subscript sequences of the constructional underspecies, forms a removable subspecies of the species of finite sequences of natural numbers; further, that for an arbitrary constructional underspecies $F_{\nu_1 \cdots \nu_n}$ of F the cardinal number of the $F_{\nu_1 \cdots \nu_n \mu}$ is known." (*Brouwer 1926*, pp. 451–452; see also *Brouwer 1918*, pp. 22–23, and *Heyting 1955*, pp. 33–34.)

The translation is by Stefan Bauer-Mengelberg, and it is printed here with the kind permission of Professor Brouwer and Walter de Gruyter and Co.

§ 1

Following § 5 of my paper "Zur Begründung der intuitionistischen Mathematik I" we define the *κ-intervals* and the *λ-intervals* in the "naturally ordered" species of finite binary fractions;[1] in particular, we shall say that these intervals are *$\kappa^{(\nu)}$-intervals* or *$\lambda^{(\nu)}$-intervals*, respectively, if their length is equal to $2^{-\nu}$.

By a *point of the linear continuum* we understand an unlimited sequence of λ-intervals (the "generating intervals" of the point) such that each of them is contained, in the strict sense, in the preceding one; their size, therefore, converges positively[2] to zero.

If in a set every choice that does not lead to the inhibition of the process generates a λ-interval, while each of these λ-intervals is contained, in the strict sense, in the λ-interval generated by the preceding choice, the set is called a *point set of the linear continuum.*

Let the species of those points p of the linear continuum that "coincide" with a certain point p_1 of the linear continuum (by which we mean that every generating interval of p completely or partially covers every generating interval of p_1) be called

[1] *Brouwer 1925*, p. 253. ⟦"In the number continuum an interval with the end points $a \cdot 2^{-n}$ and $(a + 1)2^{-n}$ or with the end points $a \cdot 2^{-n}$ and $(a + 2)2^{-n}$ (where a is an arbitrary integer and n an arbitrary natural number) is called a κ-interval or a λ-interval, respectively."⟧

⟦"Naturally ordered" means in order of increasing magnitude. Binary fractions are fractions of the form $a \cdot 2^{-n}$, where a is an integer and n a natural number. Brouwer adds the word "finite" because he regards these fractions as finite sums of the form $b + \sum_{\nu=1}^{r} \alpha_\nu 2^{-\nu}$, where b is an integer, r is a natural number, α_ν, for $1 \leqq \nu \leqq r - 1$, is equal to 0 or 1, and α_r is equal to 1.⟧

[2] *Brouwer 1923b*, p. 6 ⟦above, p. 339⟧.

a *point core* [[*Punktkern*]] *of the linear continuum.* In what follows we shall denote the point cores of the linear continuum by y.

The points or point cores that are generated exclusively by λ-intervals partially covering the interval (0, 1) are called *points* or *point cores*, respectively, *of the unit continuum.* In what follows, the point cores of the unit continuum are denoted by x.

In a way similar to that followed in § 7 of my paper "Zur Begründung der intuitionistischen Mathematik II"[3] for the species of the intercalation elements of a denumerably infinite ordered species that is everywhere dense in the strict sense, the species of the point cores of the linear continuum or of the unit continuum can also be virtually ordered; once provided with this virtual ordering, this species is called the *linear continuum* or the *unit continuum*, respectively.

If the "naturally ordered" species of finite binary fractions between 0 and 1 is denoted by M, we shall say that the point core π of the unit continuum and the element e of $K(M)$[3] *coincide* if no element of the right, or left, subspecies of an intercalation partition of M that belongs to e can ever lie to the left, or right, respectively, of a generating interval of a point of π. These coincidence relations obviously determine a *similarity correspondence*[4] between the unit continuum and $K(M)$.

By a *real function*, or, for short, a *function*, $f(x)$ of x we understand a law that, with each of certain point cores of the unit continuum, which will be denoted by ξ and form the "domain of definition" of the function, associates one point core of the linear continuum, which will be denoted by $\eta = f(\xi)$.

A function $f(x)$ is said to be *negatively continuous for the value* ξ_0 if, for an arbitrary fundamental sequence $\xi_1, \xi_2, \ldots$ that converges positively to ξ_0, the fundamental sequence $f(\xi_1), f(\xi_2), \ldots$ converges negatively[2] to $f(\xi_0)$.

A function $f(x)$ is said to be *positively continuous for the value* ξ_0, or, for short, *continuous for the value* ξ_0, if for every positive rational ε a positive rational a_ε can be determined such that for $|\xi - \xi_0| < a_\varepsilon$ the inequality $|f(\xi) - f(\xi_0)| < \varepsilon$ holds.

A function that is negatively continuous or positively continuous for every ξ will be called, for short, a *negatively continuous* or a *continuous function*, respectively.

A function $f(x)$ is said to be *uniformly continuous* if for every positive rational ε a positive rational a_ε can be determined such that for $|\xi_2 - \xi_1| < a_\varepsilon$ the inequality $|f(\xi_2) - f(\xi_1)| < \varepsilon$ holds.[5]

A function $f(x)$ is said to be *discontinuous for the value* ξ_0 if a natural number n and a fundamental sequence $\xi_1, \xi_2, \ldots$ that converges positively[2] to ξ_0 can be specified such that $f(\xi_1), f(\xi_2), \ldots$ all differ from $f(\xi_0)$ by more than $1/n$.

[3] *Brouwer 1925a*, p. 467 [[above, p. 456]].

[4] See *Brouwer 1925a*, p. 455 [[above, p. 455]].

[5] It is only for the sake of simplicity that the definitions of continuity have been brought into the metric form above, of which they are independent so far as their content is concerned. To see that, we resort to the denumerably infinite, everywhere dense, ordered sets μ' and μ'', of ordinals $1 + \eta + 1$ and η, respectively, that generate the intercalation elements corresponding to the x and y, respectively; we denumerate μ' and μ'' by fundamental sequences $g_1', g_2', \ldots$ and $g_1'', g_2'', \ldots$, respectively, we denote $\mathfrak{S}(g_1', g_2', \ldots, g_\nu')$ and $\mathfrak{S}(g_1'', g_2'', \ldots, g_\nu'')$ by s_ν' and s_ν'', respectively, and we understand by an i_ν' or an i_ν'' a closed interval of μ' or μ'', respectively, whose end elements belong to s_ν' or s_ν'', respectively, but whose interior contains at most one element of s_ν' or s_ν'', respectively. On this basis, then, a *uniformly continuous function*, for example, is a function such that, given an arbitrary denumeration of μ' and an arbitrary denumeration of μ'', we can, for every natural number m, determine a natural number n such that, if ξ_1 and ξ_2 belong to the same i_n', $f(\xi_1)$ and $f(\xi_2)$ belong to the same i_m''.

A function that is discontinuous for any specific value belonging to its domain of definition is also said, for short, to be *discontinuous*.

A function $f(x)$ is said to be *full* [[*voll*]] if its domain of definition coincides with the unit continuum.

THEOREM 1. *Every full function is negatively continuous.*

Proof. Let $f(x)$ be a full function, let ξ_0 be an arbitrary point core x, and let ξ_1, $\xi_2, \ldots$ be a fundamental sequence of point cores x that converges positively to ξ_0. We now assume for the moment that there exist a natural number p and a fundamental sequence $p_1, p_2, \ldots$ of monotonically increasing natural numbers such that $|f(\xi_{p_\nu}) - f(\xi_0)| > 1/p$ for every ν, and we define a point core ξ_ω of the unit continuum by starting from an unlimited sequence F_1 of generating intervals of a point belonging to ξ_0 and then constructing, by means of an unlimited sequence of choices of λ-intervals, a point F_2 of the unit continuum in such a way that we temporarily choose, for every natural number n that we have already considered, the first n intervals identical with the first n intervals of F_1 but reserve the right to determine, at any time after the first, second, ..., $(m - 1)$th, and mth intervals have been chosen, the choice of all further intervals (that is, of the $(m + 1)$th, $(m + 2)$th, and so on) in such a way that either a point belonging to ξ_0 or one belonging to a certain ξ_{p_ν} is generated. Then the function $f(x)$ is not defined for the point core ξ_ω containing F_2; this brings us to a contradiction, and our assumption has proved to be illegitimate. But this means that the function $f(x)$ is negatively continuous.

Theorem 1, which is an immediate consequence of the intuitionistic point of view and has since 1918 frequently been mentioned by me in lectures and conversations, suggests the conjecture that Theorem 3 below, which asserts much more, is valid; I did not, however, succeed in proving it until much later.[6] The object of the following two sections is to present this proof in as lucid as way as possible.

§ 2

Let M be an arbitrary set, let μ be the denumerably infinite set of finite (inhibited or uninhibited) choice sequences $F_{sn_1\cdots n_r}$ upon which M is based (where s and the n_ν represent the natural numbers chosen one after the other for the choice sequence in question), and let a natural number β be associated with each element of M. Then there is distinguished in μ a removable numerable subset μ_1 of uninhibited finite choice sequences such that with an arbitrary element of μ_1 the same natural number β is associated for all elements of M issuing from μ_1, while furthermore a proof [[Beweisführung]][6a] h is given that makes it apparent, for an arbitrary uninhibited element of μ, that every uninhibited infinite choice sequence issuing from it possesses an [[initial]] segment belonging to μ_1. (For an uninhibited element of μ is to be taken as belonging to μ_1 if and only if for it—but for none of its proper segments —the decision with respect to β, according to the *algorithm* of the rule establishing the correspondence, is *not* postponed to further choices; it is of course by no means excluded here that we can afterward also specify elements of μ that neither belong

[6] See *Brouwer 1924* and *1924a* [[and also *1923*, pp. 3–5]].

[6a] [[In a paper published in English (*1953*) Brouwer uses the expression "mathematical argument".]]

to μ_1 nor possess a segment belonging to μ_1 but have the property that the same natural number is associated with all elements of M that issue from such an element of μ.)

If we say that an element of μ is *secured* when it either is inhibited or possesses a (proper or improper) segment belonging to μ_1, then μ splits into a numerable set τ of secured and a numerable set σ of unsecured finite choice sequences, and the proof h shows that an arbitrary F_s is *securable*, that is, that every infinite choice sequence that issues from it and is uninhibited for M possesses a certain segment belonging to μ_1.[7] Let $h_{sn_1\cdots n_r}$ be a proof in which the securability of the element $F_{sn_1\cdots n_r}$ of σ is derived; then what this securability and this proof rest upon is, if we leave aside the fact that μ_1 and the inhibited choice sequences of μ are given, exclusively the relations, obtaining between the elements of μ, that are formed by the composition of ⟦welche sich zusammensetzen aus⟧ *elementary relations* e of the kind obtaining between two elements $F_{mm_1\cdots m_g}$ and $F_{mm_1\cdots m_g m_{g+1}}$, of which one is an ⟦immediate⟧ extension ⟦Verlängerung⟧[7a] of the other. Now, if the relations employed in any given proof can be decomposed into basic relations, its "canonical" form (that is, the one decomposed into elementary inferences[8]) employs only basic relations; in the case of the canonical form $k_{sn_1\cdots n_r}$ of the proof $h_{sn_1\cdots n_r}$ we can therefore ultimately infer the securability of $F_{sn_1\cdots n_r}$ exclusively from a combination of the species $S_{sn_1\cdots n_r}$, formed from the elementary relations e connecting $F_{sn_1\cdots n_r}$ to $F_{sn_1\cdots n_{r-1}}$ and to the $F_{sn_1\cdots n_r\nu}$, with a property previously derived from arbitrary elementary relations e and also from the fact that μ_1 and the inhibited choice sequences are given. For the last step of $k_{sn_1\cdots n_r}$ we therefore must previously have established the securability either of $F_{sn_1\cdots n_{r-1}}$ or of *all* $F_{sn_1\cdots n_r\nu}$.

If we now call the derivation of the securability of an $F_{mm_1\cdots m_g}$ from that of

[7] When carefully considered from the intuitionistic point of view ⟦Intuitionistisch durchdacht⟧, this securability is seen to be nothing but the property defined by the stipulation that it shall hold for every element of μ_1 and for every inhibited element of μ, and that it shall hold for an arbitrary $F_{sn_1\cdots n_r}$ as soon as it is satisfied, for every ν, for $F_{sn_1\cdots n_r\nu}$. This remark immediately implies the well-ordering property for an arbitrary $F_{sn_1\cdots n_r}$. The proof carried out in the text for the latter property, however, seems to me to be of interest nevertheless on account of the propositions contained in its elaboration.

[7a] ⟦In a paper published in English (*1953*) Brouwer uses the expression "immediate descendant".⟧

[8] Just as, in general, well-ordered species are produced by means of the two generating operations from primitive species (*Brouwer 1926*, p. 451), so, in particular, mathematical proofs are produced by means of the two generating operations from null elements and elementary inferences that are immediately given in intuition (albeit subject to the restriction that there always occurs a last elementary inference). These *mental* mathematical proofs that in general contain infinitely many terms must not be confused with their linguistic accompaniments, which are finite and necessarily inadequate, hence do not belong to mathematics.

The preceding remark contains my main argument against the claims of Hilbert's metamathematics. A second argument is that the way in which Hilbert seeks to settle the question (which, incidentally, was taken over from intuitionism) of the reliability of the principle of excluded middle is a vicious circle; for, if we wish to provide a foundation for the correctness of this principle by means of the proof of its consistency, this implicitly presupposes the principle of the reciprocity of the complementary species and hence the principle of excluded middle itself (see *Brouwer 1923c*, p. 252) ⟦Concerning this passage Brouwer (*1953*, p. 14, footnote 1) writes: "The equivalence of the principles of the excluded third and of reciprocity of complementarity, mentioned there in a footnote by way of remark, subsequently has been recognized as nonexistent. In fact, as was also shown in the present paper, the fields of validity of these two principles have turned out to be essentially different".⟧.

$F_{mm_1\cdots m_{g-1}}$ a *ζ-inference* and the derivation of the securability of an $F_{mm_1\cdots m_g}$ from that of *all* $F_{mm_1\cdots m_g\nu}$ a *ϝ-inference*, then the proof $k_{sn_1\cdots n_r}$ forms a well-ordered species of which every full element is formed by an elementary inference that, in case it constitutes the derivation of the securability of an element of σ, represents either a ϝ-inference or a ζ-inference.

We now assert that every element $F_{sn_1\cdots n_r}$ of σ possesses the *well-ordering property* ⟦*Wohlordnungseigenschaft*⟧, that is, that the subset $M_{sn_1\cdots n_r}$ of M determined by $F_{sn_1\cdots n_r}$ splits into a species of subsets M_α that is similar to the species of the full elements of a well-ordered species $T_{sn_1\cdots n_r}$, each of these subsets being determined by a finite initial segment F_α of choices that contains $F_{sn_1\cdots n_r}$ and belongs to μ_1. The species $T_{sn_1\cdots n_r}$ is constructed by means of generating operations w of the second kind, of which each corresponds to the inversion of the continuation, by a new free choice, of a certain finite initial segment of choices that is uninhibited for M. Then to a new choice that is inhibited for M or terminates an element of μ_1 there corresponds, for the operation w in question, a primitive species consisting of a null element or a full element, respectively.

For the proof of this assertion we denote by $f_{sn_1\cdots n_r}$ the species of those elements of σ whose securability we ascertain in the course of $k_{sn_1\cdots n_r}$, and we say that a constructional underspecies u of $k_{sn_1\cdots n_r}$ possesses the well-ordering property if every element of σ whose securability we ascertain in the course of u possesses the property of being well-ordered. Further we shall say that the *preservation property* ⟦*Erhaltungseigenschaft*⟧ holds for a constructional underspecies u of $k_{sn_1\cdots n_r}$ if, *in case* every element of $f_{sn_1\cdots n_r}$ upon whose securability the proof u is based possesses the well-ordering property, every element of $f_{sn_1\cdots n_r}$ whose securability is derived in the course of u likewise possesses the well-ordering property. Then, as we observe the generation of $k_{sn_1\cdots n_r}$, we see by means of the inductive method that for every constructional underspecies of $k_{sn_1\cdots n_r}$, hence in particular for $k_{sn_1\cdots n_r}$ itself, the preservation property holds. But from the preservation property for $k_{sn_1\cdots n_r}$ the well-ordering property immediately follows for $k_{sn_1\cdots n_r}$, hence for $F_{sn_1\cdots n_r}$.[9]

In case M is a *finitary* set, the well-ordered species $T_{sn_1\cdots n_r}$ has the same content[9a] as a well-ordered species $Q_{sn_1\cdots n_r}$ that is constructed without the use of null elements and, moreover, in a way parallel to the construction, discussed above, of $T_{sn_1\cdots n_r}$ namely, in a way such that to each operation w_α used for the construction of $T_{sn_1\cdots n_r}$

[9] If the securability of $F_{sn_1\cdots n_r}$ is ascertained in several proofs $k_{sn_1\cdots n_r}$ or in several places of one and the same proof $k_{sn_1\cdots n_r}$, the corresponding $T_{sn_1\cdots n_r}$ all are generation-equivalent, as follows by virtue of the inductive method when we observe the generation of one of them. This remark, incidentally, is superfluous for the proof above. ⟦"Two well-ordered primitive species F' and F'' have *the same generating value* ⟦besitzen *denselben Erzeugungswert*⟧, or are said to be *generation-equivalent* ⟦*erzeugungsgleich*⟧, if the single element of which each of them consists is either a full element for both or a null element for both. Two well-ordered species F' and F'' are said to be *generation-equivalent* if for an arbitrary ν the two constructional underspecies of the first order F'_ν and F''_ν either both fail to exist or both exist and are generation-equivalent." (*Brouwer 1926*, p. 452.)⟧

[9a] ⟦"Two well-ordered species (or subspecies of well-ordered species) F' and F'' are said to *have the same content* ⟦heißen *inhaltsgleich*⟧ if the species of the full elements of F' and the species of the full elements of F'' are similar." (*Brouwer 1926*, p. 453.)⟧

there corresponds a generating operation v_α of the first kind used for the construction of $Q_{sn_1\cdots n_r}$, the terms of v_α being similar, in sequence, to the species of the full elements of those terms of w_α that contain full elements. The well-ordered species $Q_{sn_1\cdots n_r}$ is therefore constructed by the exclusive use of generating operations of the first kind. But from this it follows that the species of elements of $Q_{sn_1\cdots n_r}$, as well as the species of full elements of $T_{sn_1\cdots n_r}$, is *finite*, hence that in particular for every natural number s the species of the full elements of T_s is finite. Thus a natural number z can be specified such that an arbitrary element of μ_1 possesses at most z subscripts; therefore the natural number β_e associated with an arbitrary element e of M is completely determined by the first z generating choices of e, and we have established the property expressed in the following

THEOREM 2. *If with each element e of a finitary set M a natural number β_e is associated, a natural number z can be specified such that β_e is completely determined by the first z choices generating e.*

§ 3

In the unit continuum we now determine for every natural number ν the k_ν-intervals $k'_\nu, k''_\nu, \ldots, k_\nu^{(s_\nu)}$, that is, the $\lambda^{(4\nu+1)}$-intervals, ordered from left to right, that partially cover the interval $(0, 1)$. Then the finitary point set J formed by the nestings of intervals $k_1^{(\mu_1)}, k_2^{(\mu_2)}, k_3^{(\mu_3)}, \ldots$ (where each interval is contained, in the strict sense, in the preceding one) coincides with the species of the x; that is, every such nesting of intervals belongs to an x, and every x contains such a nesting of intervals.[10]

Now in the case of a full function $f(x)$ a nesting of λ-intervals $\lambda_1, \lambda_2, \ldots$ is associated with every nesting of intervals $k_1^{(\mu_1)}, k_2^{(\mu_2)}, \ldots$, and by Theorem 2 there exists, for every natural number ν, a natural number m_ν (of which we may assume that it does not decrease with increasing ν) such that λ_ν is determined by the choice of $k_1^{(\mu_1)}, k_2^{(\mu_2)}, \ldots, k_{m_\nu}^{(\mu_{m_\nu})}$. Hence for each ν only a finite number l_ν of λ-intervals can occur as λ_ν, and there exists for them a maximal width b_ν that converges to zero as ν increases beyond all bounds.

Let us denote by $t_\nu^{(\rho)}$ the interval that is concentric with $k_\nu^{(\rho)}$ and whose width is $\frac{3}{4}$ of the width of $k_\nu^{(\rho)}$, and let P_1 and P_2 be two arbitrary point cores of the unit continuum that are $\lessdot 2^{-4\nu-3}$, that is, $\lessdot \frac{1}{4}$ of the width of the k_ν-intervals, apart. Then a $t_\nu^{(\mu_\nu)}$ can be determined that contains both P_1 and P_2, and by means of this $t_\nu^{(\mu_\nu)}$ a nesting of intervals $k_1^{(\mu_1)}, \ldots, k_\nu^{(\mu_\nu)}, k_{\nu+1}^{(\sigma_1)}, k_{\nu+2}^{(\sigma_2)}, \ldots$ belonging to P_1 and a nesting of intervals $k_1^{(\mu_1)}, \ldots, k_\nu^{(\mu_\nu)}, k_{\nu+1}^{(\tau_1)}, k_{\nu+2}^{(\tau_2)}, \ldots$ belonging to P_2 can be determined.

Let ε be an arbitrary positive quantity that is positively different[10a] from zero. If we choose ν_ε so great that $b_{\nu_\varepsilon} < \varepsilon$ and if we put $2^{-4m_{\nu_\varepsilon}-3} = a_\varepsilon$, then, according to the

[10] If in an analogous way we consider a suitable *finitary set of pairs of points* that coincides with the species of the *pairs of point cores* of the unit continuum, then on the basis of Theorem 2 the *impossibility of splitting* ⟦*Unzerlegbarkeit*⟧ *the continuum* readily follows, that is, the property that, for an arbitrary splitting of the unit continuum into a discrete species of subspecies, one of these subspecies is identical with the unit continuum.

[10a] ⟦On this notion see *Brouwer 1919*, p. 3, lines 7u–5u, and *1923d*, p. 254, lines 3–5; compare Definition 1 in 2.2.3 of *Heyting 1956*, p. 19.⟧.

second paragraph of the present section, to any two elements of J for which μ_1, μ_2, $\ldots$, $\mu_{m_{\nu_\varepsilon}}$ are equal there correspond two "values" of $f(x)$ whose difference is less than b_{ν_ε}, hence less than ε, in absolute value. According to the third paragraph of the present section, therefore, it is also the case that to any two point cores P_1 and P_2 of the unit continuum that are $\lessdot a_\varepsilon$ apart there correspond two values of $f(x)$ whose difference is less than ε in absolute value, so that $f(x)$ turns out to be *uniformly continuous* and we have proved

THEOREM 3. *Every full function is uniformly continuous.*

The foundations of mathematics

DAVID HILBERT

(*1927*)

The present paper, the text of an address delivered in July 1927 at the Hamburg Mathematical Seminar, can be considered a sequel to Hilbert's 1925 paper. It restates the salient points of Hilbert's conception. The formal system used by Hilbert is described in greater detail here, and, by discussing a number of criticisms, the paper takes on a polemical tone absent from that of 1925. Hilbert comes back to his attempted proof of the continuum hypothesis and provides some explanations on Lemmas I and II. The paper ends with some remarks on Ackermann's proof of the consistency of arithmetic, and on this last point we refer the reader to the introductory note to Bernays's paper (below, pp. 485–486).

The translators followed the original version of the paper, but they used the version of *1930a* to clarify the meaning of a few passages, and some of the additions contained in this version are reproduced in the translation below between square brackets.

The translation is by Stefan Bauer-Mengelberg and Dagfinn Føllesdal, and it is printed here with the kind permission of Teubner Verlag.

It is a great honor and at the same time a necessity for me to round out and develop my thoughts on the foundations of mathematics, which I expounded here one day five years ago and which since then have constantly kept me most actively occupied.[1] With this new way of providing a foundation for mathematics, which we may appropriately call a proof theory, I pursue a significant goal, for I should like to eliminate once and for all the questions regarding the foundations of mathematics, in the form in which they are now posed, by turning every mathematical proposition into a formula that can be concretely exhibited and strictly derived, thus recasting mathematical definitions and inferences in such a way that they are unshakable and yet provide an adequate picture of the whole science. I believe that I can attain this goal completely with my proof theory, even if a great deal of work must still be done before it is fully developed.

No more than any other science can mathematics be founded by logic alone; rather, as a condition for the use of logical inferences and the performance of logical operations, something must already be given to us in our faculty of representation [[in der Vorstellung]], certain extralogical concrete objects that are intuitively [[anschaulich]] present as immediate experience prior to all thought. If logical inference is

[1] See my previous publications on this topic (*1922, 1922a, 1925*). [[The first is the text of the address delivered in Hamburg five years earlier, in the summer of 1922.]]

to be reliable, it must be possible to survey these objects completely in all their parts, and the fact that they occur, that they differ from one another, and that they follow each other, or are concatenated, is immediately given intuitively, together with the objects, as something that neither can be reduced to anything else nor requires reduction. This is the basic philosophical position that I regard as requisite for mathematics and, in general, for all scientific thinking, understanding, and communication. And in mathematics, in particular, what we consider is the concrete signs themselves, whose shape, according to the conception we have adopted, is immediately clear and recognizable. This is the very least that must be presupposed; no scientific thinker can dispense with it, and therefore everyone must maintain it, consciously or not.

I shall now present the fundamental idea of my proof theory.

All the propositions that constitute mathematics are converted into formulas, so that mathematics proper becomes an inventory of formulas. These differ from the ordinary formulas of mathematics only in that, besides the ordinary signs, the logical signs

$\rightarrow$,	&,	$\vee$,	$\overline{}$,	(x),	(Ex)
implies	and	or	not	all	there exists

also occur in them. Certain formulas, which serve as building blocks for the formal edifice of mathematics, are called axioms. A proof is an array that must be given as such to our perceptual intuition; it consists of inferences according to the schema

$$\frac{\begin{array}{c}\mathfrak{S}\\ \mathfrak{S}\rightarrow\mathfrak{T}\end{array}}{\mathfrak{T},}$$

where each of the premisses, that is, the formulas $\mathfrak{S}$ and $\mathfrak{S}\rightarrow\mathfrak{T}$ in the array, either is an axiom or results directly from an axiom by substitution, or else coincides with the end formula of an inference occurring earlier in the proof or results from it by substitution. A formula is said to be provable if it is either an axiom or the end formula of a proof.

The axioms and provable propositions, that is, the formulas resulting from this procedure, are copies ⟦Abbilder⟧ of the thoughts constituting customary mathematics as it has developed till now.

Through the program outlined here the choice of axioms for our proof theory is already indicated; we arrange them as follows.

I. Axioms of implication

1. $A \rightarrow (B \rightarrow A)$ (introduction of an assumption);
2. $(A \rightarrow (A \rightarrow B)) \rightarrow (A \rightarrow B)$ (omission of an assumption);
3. $(A \rightarrow (B \rightarrow C)) \rightarrow (B \rightarrow (A \rightarrow C))$ (interchange of assumptions);
4. $(B \rightarrow C) \rightarrow ((A \rightarrow B) \rightarrow (A \rightarrow C))$ (elimination of a proposition).

II. Axioms about & and $\vee$

5. $A \;\&\; B \rightarrow A$;
6. $A \;\&\; B \rightarrow B$;
7. $A \rightarrow (B \rightarrow A \;\&\; B)$;

8. $A \rightarrow A \vee B$;
9. $B \rightarrow A \vee B$;
10. $((A \rightarrow C) \,\&\, (B \rightarrow C)) \rightarrow ((A \vee B) \rightarrow C))$.

III. Axioms of negation

11. $(A \rightarrow B \,\&\, \bar{B}) \rightarrow \bar{A}$ (principle of contradiction);
12. $\bar{\bar{A}} \rightarrow A$ (principle of double negation).

The axioms of groups I, II, and III are nothing but the axioms of the propositional calculus. From 11 and 12 there follows, in particular, the formula

$$(A \,\&\, \bar{A}) \rightarrow B$$

and further the logical principle of excluded middle,

$$((A \rightarrow B) \,\&\, (\bar{A} \rightarrow B)) \rightarrow B.$$

IV. The logical ε-axiom

13. $A(a) \rightarrow A(\varepsilon(A))$.

Here $\varepsilon(A)$ stands for an object of which the proposition $A(a)$ certainly holds if it holds of any object at all; let us call ε the logical ε-function. [The substitution of any given formula for $A(a)$ will be made possible if, instead of $\varepsilon(A)$, we write more precisely $\varepsilon_x A(x)$, $\varepsilon_y A(y), \ldots$.] To elucidate the role of the logical ε-function let us make the following remarks.

In the formal system the ε-function is used in three ways.

1. By means of ε, "all" and "there exists" can be defined, namely, as follows:

$$(a)A(a) \rightleftarrows A(\varepsilon(\bar{A})),$$
$$(Ea)A(a) \rightleftarrows A(\varepsilon(A)).$$

Here the double arrow stands for a combination of two implication formulas; in its place we shall henceforth use the "equivalence" sign $\sim$.

On the basis of this definition the ε-axiom IV 13 yields the logical relations that hold for the universal and the existential quantifier, such as

$$(a)A(a) \rightarrow A(b) \text{ (Aristotle's dictum),}$$

and

$$\overline{(a)A(a)} \rightarrow (Ea)\overline{A(a)} \text{ (principle of excluded middle).}$$

2. If a proposition $\mathfrak{A}$ holds of one and only one object, then $\varepsilon(\mathfrak{A})$ is *the object* of which $\mathfrak{A}(a)$ holds.

The ε-function thus enables us to resolve a proposition such as $\mathfrak{A}(a)$, when it holds of only one object, so as to obtain

$$a = \varepsilon(\mathfrak{A}).$$

3. Beyond this, ε takes on the role of the choice function; that is, in case $A(a)$ holds of several objects, $\varepsilon(\mathfrak{A})$ is *some one* of the objects a of which $\mathfrak{A}(a)$ holds.

In addition to these purely logical axioms we have the following specifically mathematical axioms.

V. Axioms of equality

14. $a = a$;
15. $(a = b) \rightarrow (A(a) \rightarrow A(b))$.

VI. Axioms of number

16. $a' \neq 0$;
17. $(A(0) \;\&\; (a)(A(a) \rightarrow A(a'))) \rightarrow A(b)$ (principle of mathematical induction).

Here a' denotes the number following a, and the integers 1, 2, 3, ... can be written in the form $0'$, $0''$, $0'''$, ...

For the numbers of the second number class and of Cantor's higher number classes the corresponding induction axioms must be added; they would have to be combined, however, into a schema in agreement with Cantor's theory.

Finally, we also need *explicit definitions*, which introduce the notions of mathematics and have the character of axioms, as well as certain *recursion axioms*, which result from a general recursion schema. Before we discuss the formulation of these axioms, we must first lay down the rules that govern the use of axioms in general. For in my theory contentual inference is replaced by manipulation of signs ⟦äußeres Handeln⟧ according to rules; in this way the axiomatic method attains that reliability and perfection that it can and must reach if it is to become the basic instrument of all theoretical research.

First, the following stipulations hold.

For mathematical variables we always use lower-case ⟦italic⟧ Latin letters, but for constant mathematical objects (specific functions) lower-case Greek letters.

For ⟦variable atomic⟧ propositions (indeterminate formulas) we always use capital ⟦italic⟧ Latin letters, but for constant ⟦atomic⟧ propositions capital Greek letters, for example,

$Z(a)$ [a is a ⟦natural⟧ number]

and

$N(a)$ [a is a number of the second number class].[2]

Concerning the procedure of substitution, the following general conventions hold.

For propositional variables we may substitute only formulas, that is, arrays constructed from elementary formulas by means of the logical signs

$$\rightarrow,\ \&,\ \vee,\ \overline{},\ (x),\ (Ex).$$

The elementary formulas are the formula variables, possibly with arguments attached, and the signs for constant propositions, such as

$$Z,\ N,\ =,\ <,$$

with the associated argument places filled.

Any array may be substituted for a mathematical variable; however, when a mathematical variable occurs in a formula, the constant proposition that states of

[2] ⟦See footnote 7, p. 385, above.⟧

what kind the variable is, followed by the implication sign, must always precede, for example,

$$Z(a) \rightarrow a + 1 = 1 + a,$$

$$N(a) \rightarrow N(a').$$

This convention has the effect that only substituends that are ordinary numbers or numbers of the second number class come into consideration after all. In Axioms V and VI the propositions $Z(a)$ and $Z(b)$, which should precede, were omitted for the sake of brevity.

German capital and lower-case letters have reference and are used only to convey information.

The *mathematical variables* are of two kinds: (1) the *primitive variables* ⟦*Grundvariablen*⟧ and (2) the *variable-sorts* ⟦*Variablengattungen*⟧.

1. Now while in all of arithmetic and analysis the ordinary integer suffices as sole primitive variable, with each of Cantor's transfinite number classes there is associated a primitive variable that ranges over precisely the ordinals of that class. Hence to each primitive variable there corresponds a proposition that states of what kind it is; this proposition is implicitly characterized by axioms.

With each primitive variable there is associated one kind of recursion, by means of which we define functions whose argument is that primitive variable. The recursion associated with the number-theoretic variable is "ordinary recursion", by means of which a function of a number-theoretic variable n is defined when we indicate what value it has for $n = 0$ and how the value for n' is obtained from that for n. The generalization of ordinary recursion is transfinite recursion; it rests upon the general principle that the value of the function for a value of the variable is determined by means of the preceding values of the function.

2. From the primitive variables we derive further kinds of variables by applying logical connectives to the propositions associated with the primitive variables, for example, to Z. The variables thus defined are called variable-sorts, and the propositions defining them are called sort-propositions; for each of these a new particular sign is introduced. Thus the formula

$$\Phi(f) \sim (a)(Z(a) \rightarrow Z(f(a)))$$

offers the simplest instance of a variable-sort; this formula defines the sort of the function variable ("being-a-function"). A further example is the formula

$$\Psi(g) \sim (f)(\Phi(f) \rightarrow Z(g(f)));$$

it defines the "being-a-function-of-a-function";[3] the argument g is the new function-of-a-function variable.

To produce the higher variable-sorts we must provide the sort-propositions themselves with subscripts, thus making a recursion procedure possible.

We can now characterize what is to be understood by explicit definitions and by recursion axioms: an explicit definition is an equivalence or identity that on its left side has the sign to be defined (capital or lower-case Greek letter), along with certain

[3] ⟦See footnote 8, p. 387, above.⟧

variables as arguments, and on its right side has an array in which only these arguments occur as free variables and in which no signs for constants occur except those that have already been introduced.

In a corresponding way, the recursion axioms are formula systems that are modeled upon the recursive procedure.

These are the general foundations of my theory. To familiarize you with the way in which it is applied I would like to adduce some examples of particular functions as they are defined by recursion.

If we want to define the function $\iota(a)$, which is 0 for the argument 0 and everywhere else has the value 1, the equations

$$\iota(0) = 0,$$
$$\iota(a') = 1,$$

as we see, already form a recursion in themselves. How the sum, the product, and the function $a!$ can be defined by recursion is well known.

The function $\mu(a, b)$, too, whose value is the minimum of the two numbers a and b, is easily definable by recursion.

I mention, further, two more complicated examples, namely, the function

$$\tau(a) = 1 \text{ if } a \text{ is a prime number},$$
$$\tau(a) = 0 \text{ otherwise},$$

and the function $\pi(a)$, which gives the number of primes $\leqq a$.

In fact, these too can be defined by recursion; to this end we must first introduce the following two functions of three arguments each:

$$\begin{aligned}\varphi(a, b, c) &= 0 \text{ if } b \text{ is equal to one of the numbers } 1.a, 2.a, \ldots, c.a \ (b > 0),\\ &= 1 \text{ otherwise},\\ \psi(a, b, c) &= \text{the least of those of the numbers } 1, 2, \ldots, a \text{ that are divisors of } b \text{ and } > c,\\ &= b \text{ if none of these numbers has this property.}\end{aligned}$$

If we now begin to construct mathematics, we shall first set our sights upon elementary number theory; we recognize that we can obtain and prove its truths through contentual intuitive considerations. The formulas that we encounter when we take this approach are used only to impart information. Letters stand for numerals, and an equation informs us of the fact that two signs stand for the same thing.

The situation is different in algebra; in algebra we consider the expressions formed with letters to be independent objects in themselves, and the propositions of number theory, which are included in algebra, are formalized by means of them. Where we had numerals, we now have formulas, which themselves are concrete objects that in their turn are considered by our perceptual intuition, and the derivation of one formula from another in accordance with certain rules takes the place of the number-theoretic proof based on content.

Thus algebra already goes considerably beyond contentual number theory. Even the formula

$$1 + a = a + 1,$$

for example, in which a is a genuine number-theoretic variable, in algebra no longer merely imparts information about something contentual but is a certain formal object, a provable formula, which in itself means nothing and whose proof cannot be based on content but requires appeal to the induction axiom.

The formulas

$$1 + 3 = 3 + 1 \quad \text{and} \quad 1 + 7 = 7 + 1,$$

which can be verified by contentual considerations, can be obtained from the algebraic formula above only by a proof procedure, such as formal substitution of the numerals 3 and 7 for a, that is, by the use of a rule of substitution.

Hence even elementary mathematics contains, first, formulas to which correspond contentual communications of finitary propositions (mainly numerical equations or inequalities, or more complex communications composed of these) and which we may call the *real propositions* of the theory, and, second, formulas that—just like the numerals of contentual number theory—in themselves mean nothing but are merely things that are governed by our rules and must be regarded as the *ideal objects* of the theory.

These considerations show that, to arrive at the conception of *formulas as ideal propositions*, we need only pursue in a natural and consistent way the line of development that mathematical practice has already followed till now. And it is then natural and consistent for us to treat henceforth not only the mathematical variables but also the logical signs

$$\rightarrow, \ \&, \ \vee, \ \overline{}, \ (x), \ (Ex),$$

and the logical variables, namely, the propositional variables

$$A, B, C, \ldots,$$

just like the numerals and letters in algebra and to consider them, too, as signs that in themselves mean nothing but are merely building blocks for ideal propositions.

Indeed, we have an urgent reason for thus extending the formal point of view of algebra to all of mathematics. For it is the means of relieving us of a fundamental difficulty that already makes itself felt in elementary number theory. Again I take as an example the equation

$$a + 1 = 1 + a;$$

if we wanted to regard it as imparting the information that

$$\mathfrak{a} + 1 = 1 + \mathfrak{a},$$

where $\mathfrak{a}$ stands for any given number, then this communication could not be negated, since the proposition that there exists a number $\mathfrak{a}$ for which

$$\mathfrak{a} + 1 \neq 1 + \mathfrak{a}$$

holds has no finitary meaning; one cannot, after all, try out all numbers. Thus, if we adopted the finitist attitude, we could not make use of the alternative according to which an equation, like the one above, in which an unspecified numeral occurs either is satisfied for every numeral or can be refuted by a counterexample. For, as an application of the "principle of excluded middle", this alternative depends essentially on the assumption that it is possible to negate the assertion that the equation in question always holds.

But we cannot relinquish the use either of the principle of excluded middle or of any other law of Aristotelian logic expressed in our axioms, since the construction of analysis is impossible without them.

Now the fundamental difficulty that we face here can be avoided by the use of ideal propositions. For, if to the real propositions we adjoin the ideal ones, we obtain a system of propositions in which all the simple rules of Aristotelian logic hold and all the usual methods of mathematical inference are valid. Just as, for example, the negative numbers are indispensable in elementary number theory and just as modern number theory and algebra become possible only through the Kummer-Dedekind ideals, so scientific mathematics becomes possible only through the introduction of ideal propositions.

To be sure, one condition, a single but indispensable one, is always attached to the use of the method of ideal elements, and that is the proof of consistency; for, extension by the addition of ideal elements is legitimate only if no contradiction is thereby brought about in the old, narrower domain, that is, if the relations that result for the old objects whenever the ideal objects are eliminated are valid in the old domain.

In the present situation, however, this problem of consistency is perfectly amenable to treatment. For the point is to show that, when ideal objects are introduced, it is impossible for us to obtain two logically contradictory propositions, $\mathfrak{A}$ and $\overline{\mathfrak{A}}$. Now, as I remarked above, the logical formula

$$(A \ \& \ \overline{A}) \rightarrow B$$

follows from the axioms of negation. If in it we substitute the proposition $\mathfrak{A}$ for A and the inequality $0 \neq 0$ for B, we obtain

$$(\mathfrak{A} \ \& \ \overline{\mathfrak{A}}) \rightarrow (0 \neq 0).$$

And, once we have this formula, we can derive the formula $0 \neq 0$ from $\mathfrak{A}$ and $\overline{\mathfrak{A}}$. To prove consistency we therefore need only show that $0 \neq 0$ cannot be obtained from our axioms by the rules in force as the end formula of a proof, hence that $0 \neq 0$ is not a provable formula. And this is a task that fundamentally lies within the province of intuition just as much as does in contentual number theory the task, say, of proving the irrationality of $\sqrt{2}$, that is, of proving that it is impossible to find two numerals $\mathfrak{a}$ and $\mathfrak{b}$ satisfying the relation $\mathfrak{a}^2 = 2\mathfrak{b}^2$, a problem in which it must be shown that it is impossible to exhibit two numerals having a certain property. Correspondingly, the point for us is to show that it is impossible to exhibit a proof of a certain kind. But a formalized proof, like a numeral, is a concrete and surveyable object. It can be communicated from beginning to end. That the end formula has the required structure, namely "$0 \neq 0$", is also a property of the proof that can be concretely ascertained. The demonstration ⟦that "$0 \neq 0$" is not a provable formula⟧ can in fact be given, and this provides us with a justification for the introduction of our ideal propositions. At the same time we recognize that this also gives us the solution of a problem that became urgent long ago, namely, that of proving the consistency of the arithmetic axioms.

Wherever the axiomatic method is used it is incumbent upon us to prove the consistency of the axioms. In geometry and the physical theories this proof is successfully carried out by means of a reduction to the consistency of the arithmetic axioms.[4] This method obviously fails in the case of arithmetic itself. By making this important final step possible through the method of ideal elements, our proof theory forms the necessary keystone of the axiomatic system.

The final test of every new theory is its success in answering preexistent questions that the theory was not specifically created to answer. As soon as Cantor had discovered his first transfinite numbers, the numbers of the second number class as they are called, the question arose whether by means of this transfinite counting one could actually enumerate the elements of sets known in other contexts but not denumerable in the ordinary sense. The line segment was the first and foremost set of this kind to come under consideration. This question, whether the points of the line segment, that is, the real numbers, can be enumerated by means of the numbers of the second number class, is the famous problem of the continuum, which was formulated but not solved by Cantor. In my paper "On the infinite" (*1925*) I showed how through our proof theory this problem becomes amenable to successful treatment.

In order to show that this continuum hypothesis of Cantor's constitutes a perfectly concrete problem of ordinary analysis, I mention further that it can be expressed as a formula in the following way:

$$(Eh)\{(f)(\Phi(f) \rightarrow N(h(f))) \; \& \; (f)(g)[\Phi(f) \; \& \; \Phi(g) \rightarrow ((h(f) = h(g)) \rightarrow =(f, g))]\},$$

where, to abbreviate, we have put

$$\Phi(f) \quad \text{for} \quad (a)(Z(a) \rightarrow Z(f(a)))$$

and

$$=(f, g) \quad \text{for} \quad (a)(Z(a) \rightarrow (f(a) = g(a))).$$

In this formula there still occurs the proposition N, which is associated with the primitive variable of the second number class. But this can be avoided, since, as is well known, the numbers of the second number class can be represented by well-orderings of the number sequence—that is, by certain functions that have two number-theoretic variables and take the values 0 and 1—in such a way that the proposition in question takes the form of a proposition purely about functions.

I have already set forth the basic features of this proof theory of mine on different occasions, in Copenhagen [[*1922*]], here in Hamburg [[*1922*]], in Leipzig [[*1922a*]], and in Münster [[*1925*]]; in the meantime much fault has been found with it, and objections of all kinds have been raised against it, all of which I consider just as unfair as it can be. I would now like to elucidate some of these points.

Poincaré already made various statements that conflict with my views; above all, he denied from the outset the possibility of a consistency proof for the arithmetic axioms, maintaining that the consistency of the method of mathematical induction could never be proved except through the inductive method itself. But, as my theory shows, two distinct methods that proceed recursively come into play when the foundations of arithmetic are established, namely, on the one hand, the intuitive construction

[4] [[See footnote 6, p. 383, above.]]

of the integer as numeral (to which there also corresponds, in reverse, the decomposition of any given numeral, or the decomposition of any concretely given array constructed just as a numeral is), that is, *contentual* induction, and, on the other hand, *formal* induction proper, which is based on the induction axiom and through which alone the mathematical variable can begin to play its role in the formal system.

Poincaré arrives at his mistaken conviction by not distinguishing between these two methods of induction, which are of entirely different kinds. Regrettably Poincaré, the mathematician who in his generation was the richest in ideas and the most fertile, had a decided prejudice against Cantor's theory, which prevented him from forming a just opinion of Cantor's magnificent conceptions. Under these circumstances Poincaré had to reject my theory, which, incidentally, existed at that time only in its completely inadequate early stages.[5] Because of his authority, Poincaré often exerted a one-sided influence on the younger generation.

My theory is opposed on different grounds by the adherents of Russell and Whitehead's theory of foundations, who regard *Principia mathematica* as a definitively satisfying foundation for mathematics.

Russell and Whitehead's theory of foundations is a general logical investigation of wide scope. But the foundation that it provides for mathematics rests, first, upon the axiom of infinity and, then, upon what is called the axiom of reducibility, and both of these axioms are genuine contentual assumptions that are not supported by a consistency proof; they are assumptions whose validity in fact remains dubious and that, in any case, my theory does not require.

In my theory Russell's axiom of reducibility has its counterpart in the rule for dealing with function variables. But reducibility is not presupposed in my theory; rather, it is recognized as something that can be compensated for: the execution of the reduction would be required only in case a proof of a contradiction were given, and then, according to my proof theory, this reduction would always be bound to succeed.

Now with regard to the most recent investigations, the fact that research on foundations has again come to attract such lively appreciation and interest certainly gives me the greatest pleasure. When I reflect on the content and the results of these investigations, however, I cannot for the most part agree with their tendency; I feel, rather, that they are to a large extent behind the times, as if they came from a period when Cantor's majestic world of ideas had not yet been discovered.

In this I see the reason, too, why these most recent investigations in fact stop short of the great problems of the theory of foundations, for example, the question of the construction of functions, the proof or refutation of Cantor's continuum hypothesis, the question whether all mathematical problems are solvable, and the question whether consistency and existence are equivalent for mathematical objects.

Of today's literature on the foundations of mathematics, the doctrine that Brouwer advanced and called intuitionism forms the greater part. Not because of any inclination for polemics, but in order to express my views clearly and to prevent misleading conceptions of my own theory, I must look more closely into certain of Brouwer's assertions.

[5] [[Hilbert is apparently referring to his *1904* and its criticism by Poincaré (*1905*, pp. 17–34; *1908*, pp. 179–191).]]

Brouwer declares [just as Kronecker did in his day] that existence statements, one and all, are meaningless in themselves unless they also contain the construction of the object asserted to exist; for him they are worthless scrip, and their use causes mathematics to degenerate into a game.

The following may serve as an example showing that a mere existence proof carried out with the logical ε-function is by no means a piece of worthless scrip.

In order to justify a remark by Gauss to the effect that it is superfluous for analysis to go beyond the ordinary complex numbers formed with i, Weierstrass and Dedekind undertook investigations that also led to the formulation and proof of certain theorems. Now some time ago I stated a general theorem (*1896*) on algebraic forms that is a pure existence statement and by its very nature cannot be transformed into a statement involving constructibility. Purely by use of this existence theorem I avoided the lengthy and unclear argumentation of Weierstrass and the highly complicated calculations of Dedekind, and in addition, I believe, only my proof uncovers the inner reason for the validity of the assertions adumbrated by Gauss and formulated by Weierstrass and Dedekind.

But even if one were not satisfied with consistency and had further scruples, he would at least have to acknowledge the significance of the consistency proof as a general method of obtaining finitary proofs from proofs of general theorems—say of the character of Fermat's theorem—that are carried out by means of the ε-function.

Let us suppose, for example, that we had found, for Fermat's great theorem, a proof in which the logical function ε was used. We could then make a finitary proof out of it in the following way.

Let us assume that numerals

$$\mathfrak{p}, \mathfrak{a}, \mathfrak{b}, \mathfrak{c} \quad (\mathfrak{p} > 2)$$

satisfying Fermat's equation

$$\mathfrak{a}^{\mathfrak{p}} + \mathfrak{b}^{\mathfrak{p}} = \mathfrak{c}^{\mathfrak{p}}$$

are given; then we could also obtain this equation as a provable formula by giving the form of a proof to the procedure by which we ascertain that the numerals $\mathfrak{a}^{\mathfrak{p}} + \mathfrak{b}^{\mathfrak{p}}$ and $\mathfrak{c}^{\mathfrak{p}}$ coincide. On the other hand, according to our assumption we would have a proof of the formula

$$(Z(a) \mathbin{\&} Z(b) \mathbin{\&} Z(c) \mathbin{\&} Z(p) \mathbin{\&} (p > 2)) \rightarrow (a^p + b^p \neq c^p),$$

from which

$$\mathfrak{a}^{\mathfrak{p}} + \mathfrak{b}^{\mathfrak{p}} \neq \mathfrak{c}^{\mathfrak{p}}$$

is obtained by substitution and inference. Hence both

$$\mathfrak{a}^{\mathfrak{p}} + \mathfrak{b}^{\mathfrak{p}} = \mathfrak{c}^{\mathfrak{p}}$$

and

$$\mathfrak{a}^{\mathfrak{p}} + \mathfrak{b}^{\mathfrak{p}} \neq \mathfrak{c}^{\mathfrak{p}}$$

would be provable. But, as the consistency proof shows in a finitary way, this cannot be the case.

The examples cited are, however, only arbitrarily selected special cases. In fact, mathematics is replete with examples that refute Brouwer's assertions concerning existence statements.

What, now, is the real state of affairs with respect to the reproach that mathematics would degenerate into a game?

The source of pure existence theorems is the logical ε-axiom, upon which in turn the construction of all ideal propositions depends. And to what extent has the formula game thus made possible been successful? This formula game enables us to express the entire thought-content of the science of mathematics in a uniform manner and develop it in such a way that, at the same time, the interconnections between the individual propositions and facts become clear. To make it a universal requirement that each individual formula then be interpretable by itself is by no means reasonable; on the contrary, a theory by its very nature is such that we do not need to fall back upon intuition or meaning in the midst of some argument. What the physicist demands precisely of a theory is that particular propositions be derived from laws of nature or hypotheses solely by inferences, hence on the basis of a pure formula game, without extraneous considerations being adduced. Only certain combinations and consequences of the physical laws can be checked by experiment—just as in my proof theory only the real propositions are directly capable of verification. The value of pure existence proofs consists precisely in that the individual construction is eliminated by them and that many different constructions are subsumed under one fundamental idea, so that only what is essential to the proof stands out clearly; brevity and economy of thought are the *raison d'être* of existence proofs. In fact, pure existence theorems have been the most important landmarks in the historical development of our science. But such considerations do not trouble the devout intuitionist.

The formula game that Brouwer so deprecates has, besides its mathematical value, an important general philosophical significance. For this formula game is carried out according to certain definite rules, in which the *technique of our thinking* is expressed. These rules form a closed system that can be discovered and definitively stated. The fundamental idea of my proof theory is none other than to describe the activity of our understanding, to make a protocol of the rules according to which our thinking actually proceeds. Thinking, it so happens, parallels speaking and writing: we form statements and place them one behind another. If any totality of observations and phenomena deserves to be made the object of a serious and thorough investigation, it is this one—since, after all, it is part of the task of science to liberate us from arbitrariness, sentiment, and habit and to protect us from the subjectivism that already made itself felt in Kronecker's views and, it seems to me, finds its culmination in intuitionism.

Intuitionism's sharpest and most passionate challenge is the one it flings at the validity of the principle of excluded middle, for example, in the simplest case, at the validity of the mode of inference according to which, for any assertion containing a number-theoretic variable, either the assertion is correct for all values of the variable or there exists a number for which it is false. The principle of excluded middle is a consequence of the logical ε-axiom and has never yet caused the slightest error. It is, moreover, so clear and comprehensible that misuse is precluded. In particular, the principle of excluded middle is not to be blamed in the least for the occurrence of the well-known paradoxes of set theory; rather, these paradoxes are due merely to the introduction of inadmissible and meaningless notions, which are automatically excluded from my proof theory. Existence proofs carried out with the help of the principle of excluded middle usually are especially attractive because of their surprising brevity and elegance. Taking the principle of excluded middle from the mathematician would be the same, say, as proscribing the telescope to the astronomer or to the boxer the use of his fists. To prohibit existence statements and the principle of excluded middle is tantamount to relinquishing the science of mathematics altogether. For, compared with the immense expanse of modern mathematics, what would the wretched remnants mean, the few isolated results, incomplete and unrelated, that the intuitionists have obtained without the use of the logical ε-axiom? The theorems of the theory of functions, such as the theory of conformal mapping and the fundamental theorems in the theory of partial differential equations or of Fourier series—to single out only a few examples from our science—are merely ideal propositions in my sense and require the logical ε-axiom for their development.

In these circumstances I am astonished that a mathematician should doubt that the principle of excluded middle is strictly valid as a mode of inference. I am even more astonished that, as it seems, a whole community of mathematicians who do the same has now constituted itself. I am most astonished by the fact that even in mathematical circles the power of suggestion of a single man, however full of temperament and inventiveness, is capable of having the most improbable and eccentric effects.

Not even the sketch of my proof of Cantor's continuum hypothesis has remained uncriticized. I would therefore like to make some comments on this proof.

First, as for Lemma I, which I used there without proof, it is certainly very useful in fixing the train of thought, but it is dispensable for the proof itself. For the introduction of the ε-functions does not affect the denumerability of the objects that can be produced up to a certain height of variable-sorts.

We can, moreover, normalize these ε-functions in a certain way. In the domain of number-theoretic variables, for example, we need only take the particular function-of-a-function $\varepsilon(f)$, which is 0 if always $f(a) = 0$ and otherwise represents some number a for which $f(a) \neq 0$, in place of the general ε-function $\varepsilon(A)$, whose value is some individual of which the proposition A holds, if it holds of any at all.

Indeed, I also took this function-of-a-function as a point of departure when I wanted to formalize the principle of excluded middle in connection with the axiom of choice.[6]

In order to explain Lemma II, which I did not prove either, and at the same time to elucidate my conception of the numbers of the second number class, I will mention the following proposition, which is entailed by the assertion of the lemma. If a definition of a certain number of the second number class is given by means of transfinite recursion, then from this definition another one that uses only ordinary recursion on a number-theoretic variable can be obtained for the same number. The meaning of this proposition is that a recursively defined function of a number of the second number class can be computed for a given argument, just as a number-theoretic function defined by recursion can always be computed for a given numerical value.

In the proof of this the difficulty is, above all, in showing that, when a sequence $\alpha(n)$ of numbers of the second number class is given by a recursion

$$\alpha(n') = \varphi(\alpha(n)),$$

where φ is defined by transfinite recursion, this transfinite recursion can be eliminated.

[6] ⟦In *Hilbert 1922a*, pp. 158–159, but with τ instead of ε.⟧

In certain cases this elimination has been successfully carried out. Examples of this are the first ε-number and the first critical ε-number, as they have been introduced and named by Cantor.

The first ε-number is the limit number of the sequence $\alpha(n)$, where

$$\alpha(0) = \omega,$$
$$\alpha(n') = \omega^{\alpha(n)} \quad (n \text{ being an ordinary number}),$$

and ω^α is defined in the usual way by transfinite recursion.

Following Cantor, we understand by an ε-number a number α such that

$$\alpha = \omega^\alpha.$$

For the definition of the first ε-number by ordinary recursion we already need variable-sorts with numerical subscripts:

$$N_0(a) \sim N(a),$$
$$N_{n'}(a) \sim (b)(N_n(b) \rightarrow N_n(a(b))).$$

W. Ackermann recently succeeded in making considerable progress in the proof of consistency.[7] I would like to end my lecture with a very short report on that.

In proving consistency for the ε-function the point is to show that from a given proof of $0 \neq 0$ the ε-function can be eliminated, in the sense that the arrays formed by means of it can be replaced by numerals in such a way that the formulas resulting from the logical axiom of choice by substitution, the "critical formulas", go over into "true" formulas by virtue of those replacements.

The replacements are found by stepwise trials after the elimination of the free variables has been completed, and it must be shown that this process always terminates.

We here make the following particular assumptions:

1. Only Z and $=$ occur as signs for constant propositions.

2. Whenever the arrays occurring as arguments—we call them "functionals"—are free of the ε-function they either are themselves numerals or are constructed from numerals by means of signs for functions defined by recursion axioms.

For the case in which only one functional formed with ε and only a single critical formula occur, we can see in the following way that the process of stepwise replacements is finite. Let

$$\mathfrak{A}(\mathfrak{k}) \rightarrow \mathfrak{A}(\varepsilon_a \mathfrak{A}(a))$$

be the critical formula (where $\varepsilon_a \mathfrak{A}(a)$ may occur also in $\mathfrak{k}$). We first replace $\varepsilon_a \mathfrak{A}(a)$ everywhere by 0.

Then all functionals will be free of the ε-function; we can calculate everything and obtain numerical values for the functionals. Among the elementary propositions we can now distinguish between the "true" and the "false", since every Z-proposition is considered true and for equations the criterion is that the numerals on both sides be identical. In place of the critical formula we now have

$$A(\mathfrak{z}) \rightarrow \mathfrak{A}(0).$$

[7] [Hilbert is referring to the new version of Ackermann's proof, different from the published one (*1924*) and only communicated, by letter, to Bernays; see below, p. 486 and footnote 3, p. 489.]

Either this formula is true (then we have reached our goal) or $\mathfrak{A}(\mathfrak{z})$ is true. Hence we have then found a value $\mathfrak{z}$ for which $\mathfrak{A}$ holds.

We now make a new replacement, replacing $\varepsilon_a \mathfrak{A}(a)$ everywhere by the numeral $\mathfrak{z}$. If we carry out the computation of all functionals, the critical formula goes over into a formula,

$$\mathfrak{A}(\mathfrak{z}_1) \rightarrow \mathfrak{A}(\mathfrak{z}),$$

that is certainly true.

When several ε-functions occur, they may be compounded in complicated ways; in particular, this may take either the form of "embedding" [["Einlagerung"]], for example,

$$\varepsilon_a \mathfrak{A}(a, \varepsilon_b \mathfrak{K}(b)),$$

where $\varepsilon_b \mathfrak{K}(b)$ is free of the variable a, or that of "superposition" [["Überordnung"]],

$$\varepsilon_a \mathfrak{A}(a, \varepsilon_b \mathfrak{K}(a, b)).$$

In the case of mere embedding no difficulty of principle yet appears. We must take care to perform the replacements from the inside out and to give the axiom of equality its due, seeing to it, for example, that, if the replacements for $\varepsilon_b \mathfrak{C}(b)$ and $\varepsilon_b \mathfrak{K}(b)$ are the same in two ε-arrays,

$$\varepsilon_a \mathfrak{A}(a, \varepsilon_b \mathfrak{C}(\mathfrak{b}))$$

and

$$\varepsilon_a \mathfrak{A}(a, \varepsilon_b \mathfrak{K}(\mathfrak{b})),$$

we also replace the outer ε in the same way.

So long as the replacements for the inner ε's remain unchanged, the values found for the outer ε's are final. Hence they can become void only if a new value is found for an inner ε.

Consequently, in finding values for the ε's, we penetrate steadily further inward, so that finally values are found for the innermost ε's, provided the procedure does not already terminate earlier; these values are then final, and the maximal number of layers in the embedding is thus decreased by 1.

An upper bound for the total number of steps of replacement needed to transform any critical formula into a true one can be given beforehand in a simple way on the basis of the given proof array; this makes the finitary character of these considerations apparent.

The case of superposition is more difficult. Here, if we want to perform the replacement from the inside out, we cannot, for example, in

$$\varepsilon_a \mathfrak{A}(a, \varepsilon_b \mathfrak{K}(a, b))$$

replace the inner ε, that is, $\varepsilon_b \mathfrak{K}(a, b)$, by a number, but only by a function. As replacement functions we need take only those that have the value 0 everywhere except for a finite number of arguments. We always begin with the function that has the value 0 everywhere ("zero replacement").

It is now by no means immediately obvious, though it can be proved, that the replacement procedure terminates in this case also; again we obtain in an elementary way an upper bound for the number of steps required. In this it is essential that, whenever a new function replacement is made for the inner ε's, the replacement for the outer ε's begin anew with the zero replacement.

In this proof of finiteness it is assumed that the ε's occur only in the proof array itself but not in the recursion axioms introduced to define functions.

From my presentation you will recognize that it is the consistency proof that determines the effective scope of my proof theory and in general constitutes its core. The method of W. Ackermann permits a further extension still. For the foundations of ordinary analysis his approach has been developed so far that only the task of carrying out a purely mathematical proof of finiteness remains. Already at this time I should like to assert what the final outcome will be : mathematics is a presuppositionless science. To found it I do not need God, as does Kronecker, or the assumption of a special faculty of our understanding attuned to the principle of mathematical induction, as does Poincaré, or the primal intuition of Brouwer, or, finally, as do Russell and Whitehead, axioms of infinity, reducibility, or completeness, which in fact are actual, contentual assumptions that cannot be compensated for by consistency proofs.

I would like to note further that P. Bernays has again been my faithful collaborator. He has not only constantly aided me by giving advice but also contributed ideas of his own and new points of view, so that I would like to call this our common work. We intend to publish a detailed presentation of the theory soon.[8]

[8] [[Hilbert is apparently referring to *Hilbert and Bernays 1934* and *1939*.]]

Comments on Hilbert's second lecture on the foundations of mathematics

HERMANN WEYL

(*1927*)

In 1920 Weyl delivered several lectures on intuitionism before a mathematical colloquium in Zurich. These lectures constitute an illuminating exposition of some of Brouwer's ideas, and in the course of them Weyl declared: "I now give up my own attempt and join Brouwer" (*1920*, p. 56). Hilbert read the text of the lectures with attention, as references in his writings of the twenties indicate, and the attack upon intuitionism at the beginning of his 1922 lecture (delivered in Copenhagen and in Hamburg) can to some extent be regarded as a reply to Weyl. The present paper is the text of remarks made by Weyl immediately after Hilbert's 1927 lecture in Hamburg, and it is a new contribution to the controversy. Weyl's strict allegiance to intuitionism had been only momentary, and here, while defending Brouwer against some of Hilbert's criticisms, he attempts to bring out the significance of Hilbert's approach to the problems of the foundations of mathematics.

On one specific point Weyl opposes Hilbert. In his lecture Hilbert had taken up the accusation of circularity that, he claimed, Poincaré had directed against the metamathematical use of mathematical induction. Weyl, in his remarks, undertakes to defend Poincaré. The issue deserves some attention.

Commenting on Hilbert's first approach to the problem of the consistency of arithmetic (*1904*), Poincaré (*1905*, pp. 18–23) pointed out that, even if Hilbert could justify mathematical induction, this justification would come too late, since he had applied the principle in arguments leading up to this justification. Although the distinction between mathematics and metamathematics had not yet been clearly expressed by Hilbert, Poincaré recognized that mathematical induction in fact splits into two principles, one that applies to intuitive sequences (of signs or of sequences of signs) and one that is represented in an allegedly meaningless formal system; he drew the conclusion that, even if a consistency proof could justify the second principle, it would not justify the first. His objection to Hilbert's approach was that the first principle is not weaker than the second. Answering Poincaré years later (*1922*), Hilbert claims that the first of the two principles is not mathematical induction. He presents a proof of the commutative law of addition for numerals that are intuitively constructed as sequences of 1's, and he writes: "This proof, as I would still like to emphasize particularly, is a procedure that rests... exclusively upon the composition and decomposition of the numerals ⟦auf dem Auf- und Abbau der Zahlzeichen⟧ and is in essence different from the principle that, as the principle of mathematical induction, or inference from n to $n + 1$, plays such a prominent role in higher mathematics" (*1922*, p. 164). Later, in the lecture on which Weyl comments, Hilbert contrasts "contentual induction" with "formal induction proper" (above, p. 473); in preparing a reprint (*1928*) of the lecture, he crossed out the words "die inhaltliche Induktion" and "formale", leaving the first principle nameless and calling the second merely "induction proper". Ackermann (*1924*, p. 1) ranges mathematical induction among the questionable transfinite methods of inference and stresses that metamathematics deals only with concretely given objects, to which finitary methods of inference are applied. Herbrand gives a description of recursive metamathematical arguments (*1930*, pp. 3–4; see also *1930a*, p. 248, and *1931b*, footnote 3, p. 3) that is perhaps more precise than any given by Hilbert and his disciples. Metamathematical induction, according to Herbrand, "stops in the finite" ("cette récurrence qui s'arrête dans le fini", *1930*, p. 4) and differs from induction as it is applied in mathematics: in metamathematics an inductive argument "is never more than the indication, in one formula, of a procedure that will have to be applied a certain number of times in each particular case".

Thus to meet Poincaré's objection Hilbert and those inspired by him strove to show that metamathematical methods do not go beyond the typical case of a concretely given object. In fact, Hilbert wrote (*1922a*, p. 156) that arguments containing "all" belong to the transfinite—that is, nonfinitary—modes of inference. This position is hard to maintain. Metamathematics has to make universal statements. In particular, a consistency proof for a certain system, as Poincaré has already noted, involves all the propositions that are provable in the system. Answering Poincaré on precisely this point, Hilbert says (*1928a*, p. 140): "Hence, when I want to ascertain whether a formula, taken as an axiom, leads to a contradiction, the question is whether a proof that leads to a contradiction can be presented to me. If no such proof can be presented to me, all the better; I am spared the trouble of having to react. If such a proof is presented to me, I have the right to extract certain parts of it and to consider them in themselves, and, in particular, to decompose ⟦abbauen⟧ again the numerals that, composed and constructed, occur in them. This does not yet employ the inference from n to $n + 1$ in any way." But in a consistency proof the argument does not deal with one single specific formula; it has to be extended to all formulas. This is the point that Weyl has in mind when, in his remarks, he says: "One may here stress the 'concretely given'; on the other hand, it is just as essential that the contentual arguments in proof theory be carried out *in hypothetical generality*, on *any* proof, on *any* numeral". And the generality, he adds, is secured by induction.

Hilbert's idea seems to have been that in metamathematical arguments we can avoid the use of mathematical induction "proper" by appealing to the paradigmatic case.[a] To be sure, this is a method frequently used in mathematics. In algebra, to prove that a certain group is commutative, we may consider two elements of the group, x and y, and, reasoning about these two fixed but arbitrary individuals, we prove that $x \cdot y = y \cdot x$; since no assumption is made on x and y except that they are elements of the group, we come to the conclusion that the commutative law holds in the

[a] Herbrand writes (*1931b*, p. 3 (below, p. 622), footnote 3): "We never consider the totality of all the objects x of an infinite collection; and when we say that an argument (or a theorem) is true for all these x, we mean that, for each x taken by itself, it is possible to repeat the general argument in question, which should be considered to be merely the prototype of these particular arguments". Here "prototype" is just another word for "paradigm".

group. In geometry we take, say, a point P, and any conclusion reached about P is valid for any point so long as our argument does not rest upon any property that, while holding of P, does not hold of all points to which we extend the conclusion. In formalized logic the method of the paradigmatic case leads to the "rule of generalization" (Church), or the "rule of universal generalization" (Quine). In the case of intuitively constructed objects (intuitionistically defined numbers, Hilbert's numerals, formulas, sequences of formulas, and so on) this method blends with induction. Take the case of numerals and assume that a property P holds of 1 and that it holds of $\alpha 1$ whenever it holds of α. For a given numeral a, a logical proof that parallels the construction of a establishes that a has the property P. But, since nothing was assumed about a except that it is a numeral, the property P holds of every numeral.[b] The argument uses the method of the paradigmatic case but, in addition, rests squarely upon the fact that any numeral either is 1 or is constructed from 1 by writing 1 to the right of 1 one or more times, hence upon the notion of natural number.

The recursive inferences considered in Hilbert's metamathematics are not weaker than mathematical induction applied to intuitively constructed numbers. Since, according to Herbrand (*1930a*, p. 248), a property that is to be generalized by a metamathematical inductive argument must be such "that we can always test whether it holds or not for a given proposition and even that we must be told what actually are the operations required for that test", the method of the paradigmatic case can lead to the successive introduction of only initial universal quantifiers. Therefore, the induction used in Hilbert's metamathematics is weaker than the principle of mathematical induction as formalized in full number theory (say the system Z of *Hilbert and Bernays 1934*), where the formula to which induction is applied may contain, if written in prenex form, any sequence of universal and existential quantifiers. The arithmetic that allows only initial universal quantifiers is Skolem's primitive recursive arithmetic (*1923*), and Hilbert came to recognize this arithmetic as the proper framework for the formalization of metamathematics. To point out the distinction between Z and primitive recursive arithmetic would perhaps be Hilbert's final answer to Poincaré and Weyl. Gödel's incompleteness theorem (*1931*) and Gentzen's consistency proof for arithmetic (*1936* and *1938*; see also *1943*) showed that some metamathematical problems can be solved only by the use of forms of induction stronger than even the unrestricted principle of mathematical induction in full number theory.

The translation is by Stefan Bauer-Mengelberg and Dagfinn Føllesdal, and it is printed here with the kind permission of Teubner Verlag.

[b] This is, in essence, the intuitionistic justification of mathematical induction; see *Heyting 1956*, p. 14.

Permit me first to say a few words in defense of intuitionism.

Before Hilbert constructed his proof theory everyone thought of mathematics as a system of contentual [[inhaltliche]], meaningful [[sinnerfüllte]], and evident [[einsichtige]] truths; this point of view was the common platform of all discussions. When Poincaré claimed that *mathematical induction* is for mathematical thought an ultimate basis that cannot be reduced to anything more primal, he had in mind precisely the processes, of composition and decomposition of numerals, that Hilbert himself employs in his contentual considerations and that are completely transparent to our perceptual intuition. For after all Hilbert, too, is not merely concerned with, say, $0'$ or $0'''$, but with any $0''\cdots'$, with an *arbitrary concretely given* numeral. One may here stress the "concretely given"; on the other hand, it is just as essential that the contentual arguments in proof theory be carried out *in hypothetical generality*, on *any* proof, on *any* numeral. This, of course, is not to be taken as an objection, for the procedure of the "one after the other" can appeal to unshakable intuitive evidence; but, evident and primal though it be, we may nevertheless give it expression—not by formulating it as an "axiom", but simply by describing its concrete use—making its self-evidence and primal quality explicit, and we are no doubt justified in seeing in it the characteristic mark of contentual *mathematical* thought. It seems to me that Hilbert's proof theory shows Poincaré to have been completely right on this point. That in Hilbert's formalized mathematics the principle of mathematical induction in the Peano–Dedekind version occurs as an axiom, whose consistent compatibility with the other axioms is to be established by contentual considerations, is quite a different matter, of course; but Poincaré was not concerned with that at all.

Brouwer, like everyone else, required of mathematics that its theorems be (in Hilbert's terminology) "real propositions", meaningful truths. But he was the first to see exactly and in full measure how it had in fact everywhere far exceeded the limits of contentual thought. I believe that we are all indebted to him for this recognition of the limits of contentual thought. In the contentual considerations that are intended to establish the consistency of formalized mathematics Hilbert fully respects these limits, and he does so as a matter of course; we are really not dealing with artificial prohibitions here by any means. Accordingly, it does not seem strange to me that Brouwer's ideas have found a following; his position resulted of necessity from a thesis shared by all mathematicians before Hilbert proposed his formal approach and from a new, indubitable fundamental logical insight that even Hilbert acknowledged. That from this point of view only a part, perhaps only a wretched part, of classical mathematics is tenable is a bitter but inevitable fact. Hilbert could not bear this mutilation. And it is again a different matter that he succeeded in saving classical mathematics *by a radical reinterpretation of its meaning* without reducing its inventory, namely, by formalizing it, thus transforming it in principle from a system of intuitive results into a game with formulas that proceeds according to fixed rules.

Let me now by all means acknowledge the immense significance and scope of this step of Hilbert's, which evidently was made necessary by the pressure of circumstances. All of us who witnessed this development are full of admiration for the genius and steadfastness with which Hilbert, through his proof theory of formalized mathematics, crowned his axiomatic lifework. And, as I am very glad to confirm, there is nothing that separates me from Hilbert in the epistemological appraisal of the new situation thus created. He asserted, first of all, that the passage through ideal propositions is a legitimate formal device when real propositions are proved; this even the strictest intuitionist must acknowledge. We may perhaps still doubt whether this role makes the effort of the whole proof theory worth while (this is merely a question of economy). For in most cases the difficulty lies not so much in the discovery of a finitary proof as in finding real judgments at all that are acceptable to the intuitionist and yet can fill out the theorems of classical mathematics. For example, in order to accomplish this for the fundamental theorem of algebra on the existence of roots, we must specify the *finitary construction* that, as the coefficients are determined more and more precisely step by step, sets going an analogous process generating the values of the roots; once this construction is found, it is easy to see that the values it yields with ever closer approximation really are the roots of the given equation.

But Hilbert furthermore pointed with emphasis to the related science of *theoretical physics*. Its individual assumptions and laws have no meaning that can immediately be realized in intuition; in principle, it is not the propositions of physics taken in isolation, but only the theoretical system as a whole, that can be confronted with experience. What is achieved here is not a perceptual insight into particular or general states of affairs or a *description* that faithfully copies the given, but a theoretical, in the last analysis purely symbolic, *construction* of the world. It has been said that physics is concerned only with establishing pointer coincidences; Mach, especially, advocated a pure phenomenalism in the field of physics. But, if we are honest, we must admit that our theoretical interest does not attach exclusively or even primarily to the "real propositions", the reports that this pointer coincides with that part of the scale; it attaches, rather, to the ideal assumptions that *according to the theory* disclose themselves in such coincidences, but of which no perception directly gives the full meaning—as, for example, the assumption of the electron as a universal elementary quantum of electricity. According to Hilbert, already pure mathematics goes beyond the bounds of intuitively ascertainable states of affairs through such ideal assumptions.

What "truth" or objectivity can be ascribed to this theoretic construction of the world, which presses far beyond the given, is a profound philosophical question. It is closely connected with the further question: what impels us to take as a basis precisely the particular axiom system developed by Hilbert? Consistency is indeed a necessary but not a sufficient condition for this. For the time being we probably cannot answer this question except by asserting our belief in the reasonableness of history, which brought these structures forth in a living process of intellectual development—although, to be sure, the bearers of this development, dazzled as they were by what they took for self-evidence, did not realize how arbitrary and bold their construction was. Hilbert's appeal to the practical success of the method, too, seems to me to rest upon such a belief. Or is it his opinion that, the nearer we bring the construction of the axiomatic system to its completion, the more we shall eliminate arbitrariness and bring to the fore that which is unambiguously compelling? If Hilbert's view prevails over intuitionism, as appears to be the case, *then I see in this a decisive defeat of the philosophical attitude of pure phenomenology*, which thus proves to be insufficient for the understanding of creative science even in the area of cognition that is most primal and most readily open to evidence—mathematics.

Appendix to Hilbert's lecture "The foundations of mathematics"

PAUL BERNAYS

(*1927*)

In formalizing arithmetic Hilbert (*1922a*, but see also *1922*, p. 177) introduced a functional $\tau(f)$ whose argument is a number-theoretic function and whose value is a natural number. An interpretation of $\tau(f)$ is that, if $f(a) = 0$ for every a, the value of $\tau(f)$ is 0 and, otherwise, it is the least a for which $f(a) \neq 0$. Besides truth-functional and quantifier-free axioms, just one "transfinite" axiom, or rather axiom schema,

(1) $\qquad f(\tau(f)) = 0 \rightarrow f(a) = 0,$

is required for the formalization. The universal and existential quantifiers used in current formulations of first-order arithmetic can be defined in terms of $\tau(f)$, and the axioms governing them can be derived from these definitions and axiom schema (1).

Hilbert then takes up the problem of consistency for a system containing just one primitive function, φ, hence just one τ-term, $\tau(\varphi)$; φ has been introduced by recursion equations (not containing τ). Assume that we have a proof of $0 \neq 0$. In that proof we replace free variables by numerals; $\tau(\varphi)$ is tentatively replaced by 0 everywhere. Every formula of the proof becomes a numerical formula. The formula

(2) $\qquad \varphi(\tau(\varphi)) = 0 \rightarrow \varphi(a) = 0,$

which is an axiom obtained from schema (1), becomes

(3) $\qquad \varphi(0) = 0 \rightarrow \varphi(\mathfrak{z}) = 0,$

where $\mathfrak{z}$ is the numeral obtained from the term a by the substitution of 0 for $\tau(\varphi)$. From the equations defining φ we can decide whether $\varphi(\mathfrak{z}) = 0$. If so, formula (3) is true, and the substitution of 0 for $\tau(\varphi)$ is adopted. All the formulas in the proof have then become true numerical formulas, hence the end formula cannot be $0 \neq 0$. If $\varphi(\mathfrak{z}) \neq 0$, we adopt the substitution of $\mathfrak{z}$ for $\tau(\varphi)$, and formula (2) becomes, for some $\mathfrak{s}$,

$$\varphi(\mathfrak{z}) = 0 \rightarrow \varphi(\mathfrak{s}) = 0,$$

which is certainly true, and we can conclude as before.

Ackermann undertook (*1924*) to transform Hilbert's brief sketch into a full-fledged consistency proof of analysis. First, a new notation, used from then on by Hilbert and his disciples, is introduced. The functional $\tau(f)$ is replaced by $\varepsilon(A)$, or $\varepsilon_a A(a)$, and formula (1) by

(4) $\qquad A(a) \rightarrow A(\varepsilon_a A(a));$

$\varepsilon_a A(a)$ denotes a number of which the propositional function $A(a)$ holds if it holds of any number. Although Ackermann is still able to make use of Hilbert's argument, there are considerable differences between Hilbert's fragmentary example and the general case now treated. In particular, we now have more than one propositional function, hence more than one ε-term; therefore, in schema (4) some $\varepsilon_a B(a)$ may occur in $A(a)$, and whether $A(a)$ becomes a true or a false numerical formula will depend on what is substituted for the inner ε. This complication, as well as others connected with the equations defining functions and functionals, can be overcome and, after a finite number of steps, we can obtain a definite sequence of substitutions, called a *total replacement* (*Gesamtersetzung*), that transforms the proof into a sequence of numerical formulas. Let instances of schema (4) be called *critical* formulas. In any total replacement all noncritical formulas are transformed into formulas that are true. If a critical formula is transformed into a false formula, new total replacements are successively generated in a definite manner, and after a finite number of steps the process leads to a total replacement by which all critical formulas are transformed into true numerical formulas.

At the time of publication Ackermann came to realize that his proof was inadequate to insure the consistency of the full system of analysis. He added a footnote (*1924*, p. 9) restricting his rule of substitution and thus decreased the strength of his system (see also *von Neumann 1927*, pp. 41–46). Ackermann communicated a new version of his proof, still for the restricted system, to Bernays by letter, and this version was presented in *Hilbert and Bernays 1939*, pp. 93–130.

Hilbert's remarks above (pp. 477–479) and Bernays's paper below refer to this second version of Ackermann's proof. When several ε-terms occur, their composition, as Hilbert explains, can take the form of "embedding" or of "superposition". Hilbert shows that, in the case of embedding, $\varepsilon_a A(a)$ is replaced, at the first trial, by 0; if the substitution is not successful, that is, does not yield true numerical formulas, a second trial takes place; $\varepsilon_a A(a)$ is replaced by a numeral $\mathfrak{z} \neq 0$, and this substitution is certainly successful. In the case of superposition, Bernays introduces a function $\chi_0(a)$ that is equal to 0 for all values of the argument, and he is able to reduce, for the case of one ε-formula, superposition to embedding. Proceeding as Hilbert did in the case of embedding, he then obtains what he calls "one or two total replacements", $\mathfrak{E}_0$, or $\mathfrak{E}_0$ and $\mathfrak{E}'_0$, that correspond, respectively, to the cases of one or of two trials. Now, either the first or the second of these replacements is certainly successful for the ε-formula considered. If it is successful for all the other ε-formulas, it becomes final. If not, a new function, $\chi_1(a)$, which is 0 for all values of the argument but one, is tried. We thus obtain a sequence of functions, $\chi_0(a)$, $\chi_1(a)$, ... To any one of these functions, $\chi_p(a)$, there correspond either one replacement, $\mathfrak{E}_p$, or two, $\mathfrak{E}_p$ and $\mathfrak{E}'_p$. Bernays is able to show that the procedure terminates and to find an upper bound for the number of steps involved.

The translation is by Stefan Bauer-Mengelberg and Dagfinn Føllesdal, and it is printed here with the kind permission of Professor Bernays and Teubner Verlag.

1. To supplement the preceding paper [[by Hilbert]] let me add some more detailed explanations concerning the consistency proof by Ackermann that was sketched there.

First, as for an upper bound on the number of steps of replacement in the *case of embedding*, it is given by 2^n, where n is the number of ε-functionals distinct in form. The method of proof described furnishes yet another, substantially closer bound, which, for example, for the case in which there is no embedding at all yields the upper bound $n + 1$.[1]

[1] [[See in *Hilbert and Bernays 1939*, pp. 96–97, how this bound is obtained.]]

2. Let the argument by which we recognize that the procedure is finite in the *case of superposition* be carried out under simple specializing assumptions.

The assumptions are the following: Let the ε-functionals occurring in the proof be

$$\varepsilon_a\mathfrak{A}(a, \varepsilon_b\mathfrak{K}(a, b))$$

and

$$\varepsilon_b\mathfrak{K}(\mathfrak{a}_1, b), \varepsilon_b\mathfrak{K}(\mathfrak{a}_2, b), \ldots, \varepsilon_b\mathfrak{K}(\mathfrak{a}_n, b),$$

where $\mathfrak{a}_1, \ldots, \mathfrak{a}_n$ may contain $\varepsilon_a\mathfrak{A}(a, \varepsilon_b\mathfrak{K}(a, b))$ but no other ε-functional.

The procedure now consists in a succession of "total replacements"; each of these consists of a function replacement $\chi(a)$ for $\varepsilon_b\mathfrak{K}(a, b)$, by means of which $\varepsilon_a\mathfrak{A}(a, \varepsilon_b\mathfrak{K}(a, b))$ goes over into $\varepsilon_a\mathfrak{A}(a, \chi(a))$, and a replacement for $\varepsilon_a\mathfrak{A}(a, \chi(a))$, by means of which $\mathfrak{a}_1, \ldots, \mathfrak{a}_n$ go over into numerals $\mathfrak{z}_1, \ldots, \mathfrak{z}_n$ and the values

$$\chi(\mathfrak{z}_1), \ldots, \chi(\mathfrak{z}_n)$$

are obtained for

$$\varepsilon_b\mathfrak{K}(\mathfrak{a}_1, b, \ldots, \varepsilon_b\mathfrak{K}(\mathfrak{a}_n, b).$$

We begin with the function $\chi_0(a)$, which has the value 0 for all a ("zero replacement"), and accordingly also replace all the terms

$$\varepsilon_b\mathfrak{K}(\mathfrak{a}_1, b), \ldots, \varepsilon_b\mathfrak{K}(\mathfrak{a}_n, b)$$

by 0.

Holding this replacement fixed, we apply to

$$\varepsilon_a\mathfrak{A}(a, \chi_0(a))$$

the original testing procedure, which after two steps at most leads to the goal; that is, all the critical formulas corresponding to

$$\varepsilon_a\mathfrak{A}(a, \chi_0(a))$$

then become true.

Thus we obtain one or two total replacements, $\mathfrak{E}_0$, or $\mathfrak{E}_0$ and $\mathfrak{E}_0'$, respectively. Now either $\mathfrak{E}_0$ (or $\mathfrak{E}_0'$) is final or one of the critical formulas corresponding to $\varepsilon_b\mathfrak{K}(\mathfrak{a}_1, b)$, $\varepsilon_b\mathfrak{K}(\mathfrak{a}_2, b), \ldots$ becomes false. Assume that this formula corresponds to, say, $\varepsilon_b\mathfrak{K}(\mathfrak{a}_1, b)$ and that $\mathfrak{a}_1$ goes into $\mathfrak{z}_1$. Then we find a value $\mathfrak{z}$ such that

$$\mathfrak{K}(\mathfrak{z}_1, \mathfrak{z})$$

is true. Now that we have this value, we take as replacement function for

$$\varepsilon_b\mathfrak{K}(a, b)$$

not $\chi_0(a)$, but the function $\chi_1(a)$ defined by

$$\chi_1(\mathfrak{z}_1) = \mathfrak{z}$$
$$\chi_1(a) = 0 \quad \text{for } a \neq \mathfrak{z}_1$$

At this point we repeat the above procedure with $\chi_1(a)$, the values of the $\varepsilon_b\mathfrak{K}(\mathfrak{a}_\nu, b)$ $(\nu = 1, \ldots, n)$ now being determined only after a value has been chosen for

$$\varepsilon_a\mathfrak{A}(a, \chi_1(a)),$$

and thus we obtain one or two total replacements, $\mathfrak{E}_1$, or $\mathfrak{E}_1$ and $\mathfrak{E}_1'$.

Now either $\mathfrak{E}_1$ (or $\mathfrak{E}'_1$) is final or for one of the ε-functionals that result from

$$\varepsilon_b \mathfrak{K}(\mathfrak{a}_1, b), \ldots, \varepsilon_b \mathfrak{K}(\mathfrak{a}_n, b)$$

by the previous total replacement we again find a value $\mathfrak{z}'$, such that for a certain $\mathfrak{z}_2$

$$\mathfrak{K}(\mathfrak{z}_2, \mathfrak{z}')$$

is true, while

$$\mathfrak{K}(\mathfrak{z}_2, \chi_1(\mathfrak{z}_2))$$

is false. From this it directly follows that $\mathfrak{z}_2 \neq \mathfrak{z}_1$.

Now, instead of $\chi_1(a)$ we introduce $\chi_2(a)$ as replacement function by means of the following definition:

$$\begin{aligned} \chi_2(\mathfrak{z}_1) &= \mathfrak{z}, \\ \chi_2(\mathfrak{z}_2) &= \mathfrak{z}', \\ \chi_2(a) &= 0 \quad \text{for } a \neq \mathfrak{z}_1, \mathfrak{z}_2. \end{aligned}$$

The replacement procedure is now repeated with this function $\chi_2(a)$.

As we continue in this way, we obtain a sequence of replacement functions

$$\chi_0(a), \chi_1(a), \chi_2(a), \ldots,$$

each of which is formed from the preceding one by addition, for a new argument value, of a function value different from 0; and for every function $\chi_p(a)$ we have one or two replacements, $\mathfrak{E}_p$, or $\mathfrak{E}_p$ and $\mathfrak{E}'_p$. The point is to show that this sequence of replacements terminates. For this purpose we first consider the replacements

$$\mathfrak{E}_0, \mathfrak{E}_1, \mathfrak{E}_2, \ldots.$$

In these,

$$\varepsilon_a \mathfrak{A}(a, \varepsilon_b \mathfrak{K}(a, b))$$

is always replaced by 0; the $\varepsilon_b \mathfrak{K}(\mathfrak{a}_\nu, b)$ $(\nu = 1, \ldots, n)$ therefore always go over into the same ε-functionals; for each of these we put either 0 or a numeral different from 0, and this is then kept as a final replacement. Accordingly, at most $n + 1$ of the replacements $\mathfrak{E}_0, \mathfrak{E}_1, \mathfrak{E}_2, \ldots$ can be distinct.[2] If, however, $\mathfrak{E}_k$ is identical with $\mathfrak{E}_l$, then neither one has, or else each has, a successor replacement $\mathfrak{E}'_k$, or $\mathfrak{E}'_l$, and in these

$$\varepsilon_a \mathfrak{A}(a, \varepsilon_b \mathfrak{K}(a, b))$$

is then in both cases replaced by the same number found as a value, so that, for both replacements, the $\varepsilon_b \mathfrak{K}(a_\nu, b)$ $(\nu = 1, \ldots, n)$ also go over into the same ε-functionals.

Accordingly, of the replacements $\mathfrak{E}'_l$ for which $\mathfrak{E}_l$ coincides with a fixed replacement $\mathfrak{E}_k$ again at most $n + 1$ can be distinct. Hence there cannot be more than $(n + 1)^2$ distinct $\mathfrak{E}_p$, or $\mathfrak{E}_p$ and $\mathfrak{E}'_p$, altogether.

From this it follows, however, that our procedure comes to an end at the latest with the replacement function $\chi_{(n+1)^2}(a)$. For, the replacements associated with two distinct replacement functions $\chi_p(a)$ and $\chi_q(a)$, $q > p$, cannot coincide completely, since otherwise we would by means of $\chi_q(a)$ be led to the same value $\mathfrak{z}^*$ that has already been found by means of $\chi_p(a)$, whereas this value is already used in the

[2] [[See footnote 1.]]

definition of the replacement functions following $\chi_p(a)$, hence in particular also in that of $\chi_q(a)$.

3. Let us note, finally, that in order to take into consideration the axiom of mathematical induction, which for the purpose of the consistency proof may be given in the form

$$(\varepsilon_a A(a) = b') \rightarrow \overline{A}(b),$$

we need only, whenever we have found a value $\mathfrak{z}$ for which a proposition $\mathfrak{B}(a)$ holds, go to the least such value by seeking out the first true proposition in the sequence

$$\mathfrak{B}(0), \mathfrak{B}(0'), \ldots, \mathfrak{B}(\mathfrak{z})$$

of propositions that have been reduced to numerical formulas.[3]

[3] ⟦In *1935*, p. 213, end of footnote 1, Bernays writes that this last paragraph, on mathematical induction, should be deleted. In 1927 Hilbert and his collaborators had not yet gauged the difficulties facing consistency proofs of arithmetic and analysis. Ackermann had set out (in *1924*) to prove the consistency of analysis; but, while correcting the printer's proofs of his paper, he had to introduce a footnote, on page 9, that restricts his rule of substitution. After the introduction of such a restriction it was no longer clear for which system Ackermann's proof establishes consistency. Certainly not for analysis. The proof suffered, moreover, from imprecisions in its last part. Ackermann's paper was received for publication on 30 March 1924 and came out on 26 November 1924. In *1927*, received for publication on 29 July 1925 and published on 2 January 1927, von Neumann criticized Ackermann's proof and presented a consistency proof that followed lines somewhat different from those of Ackermann's. The proof came to be accepted as establishing the consistency of a first-order arithmetic in which induction is applied only to quantifier-free formulas. When he was already acquainted with von Neumann's proof, Ackermann communicated, in the form of a letter, a new consistency proof to Bernays. This proof developed and deepened the arguments used in Ackermann's 1924 proof, and, like von Neumann's, it applied to an arithmetic in which induction is restricted to quantifier-free formulas. It is with this proof of Ackermann's that Hilbert's remarks above (pp. 477–479) and Bernays's present comments are concerned. It was felt at that point, among the members of the Hilbert school, that the consistency of full first-order arithmetic could be established by relatively straightforward extensions of the arguments used by von Neumann or by Ackermann (see *Hilbert 1928a*, p. 137, lines 20–21; *1930*, p. 490, line 4u, to p. 491, line 2; *Bernays 1935*, p. 211, lines 4–7). These hopes were dashed by Gödel's *1931*. Ackermann's unpublished proof was presented in *Hilbert and Bernays 1939*, pp. 93–130. In *1940* Ackermann gave a consistency proof for full first-order arithmetic, using a principle of transfinite induction (up to ε_0) that is not formalizable in this arithmetic.⟧

Intuitionistic reflections on formalism

LUITZEN EGBERTUS JAN BROUWER

(*1927a*)

While logicism and intuitionism were too far apart to allow a dialogue between them, the emergence of Hilbert's metamathematics created between Hilbert and Brouwer a ground on which a discussion could proceed, however deep might be the disagreement between these two mathematicians on the role of consistency proofs. In 1912 Brouwer ended a presentation of intuitionism (*1912*, or *1912a*) on a pessimistic note, despairing of any communication between two groups of scholars who were not speaking the same tongue and could not learn each other's tongue. In the text below, which is § 1 of *1927a*, Brouwer lists four points concerning which he considers that intuitionism and formalism could enter into a dialogue. This, it seems, could be true of the first three points, which bring out the similarities between finitary metamathematics and a certain part of intuitionistic mathematics, but could hardly be true of the fourth point, which states that consistency proofs are unable to provide a foundation for mathematics. There is a commentary on § 1 of Brouwer's paper in *Heyting 1934*, pp. 54–57, or *Heyting 1955*, pp. 60–63.

The omitted § 2, which has no direct connection with § 1, investigates various intuitionistic versions of the principle of excluded middle as well as the conditions in which each of these versions is applicable; Brouwer returned to this question in *1948*, pp. 1243–1248, and *1953*, pp. 3–5.

The translation is by Stefan Bauer-Mengelberg, and it is printed here with the kind permission of Professor Brouwer and Walter de Gruyter and Co.

The disagreement over which is correct, the formalistic way of founding mathematics anew or the intuitionistic way of reconstructing it, will vanish, and the choice between the two activities be reduced to a matter of taste, as soon as the following insights, which pertain primarily to formalism but were first formulated in the intuitionistic literature, are generally accepted. The acceptance of these insights is only a question of time, since they are the results of pure reflection and hence contain no disputable element, so that anyone who has once understood them must accept them. Two of the four insights have so far been understood and accepted in the formalistic literature. When the same state of affairs has been reached with respect to the other two, it will mean the end of the controversy concerning the foundations of mathematics.

First insight. *The differentiation, among the formalistic endeavors, between a construction of the "inventory of mathematical formulas" (formalistic view of mathematics) and an intuitive (contentual) theory of the laws of this construction, as well as the recognition of the fact that for the latter theory the intuitionistic mathematics of the set of natural numbers is indispensable.*

SECOND INSIGHT. *The rejection of the thoughtless use of the logical principle of excluded middle, as well as the recognition, first, of the fact that the investigation of the question why the principle mentioned is justified and to what extent it is valid constitutes an essential object of research in the foundations of mathematics, and, second, of the fact that in intuitive (contentual) mathematics this principle is valid only for finite systems.*

THIRD INSIGHT. *The identification of the principle of excluded middle with the principle of the solvability of every mathematical problem.*

FOURTH INSIGHT. *The recognition of the fact that the (contentual) justification of formalistic mathematics by means of the proof of its consistency contains a vicious circle, since this justification rests upon the (contentual) correctness of the proposition that from the consistency of a proposition the correctness of the proposition follows, that is, upon the (contentual) correctness of the principle of excluded middle.*

1. The first insight is still lacking in *Hilbert 1904*, see in particular Section V, pp. 184–185 ⟦above, pp. 137–138⟧, which is in contradiction with it. After having been strongly prepared by Poincaré, it first appears in the literature in *Brouwer 1907*, where on pp. 173–174 the terms *mathematical language* and *mathematics of the second order* are used to distinguish between the parts of formalistic mathematics mentioned above and where the intuitive character of the latter part is emphasized.[1] This insight penetrated into the formalistic literature with *Hilbert 1922* (see in particular p. 165 and p. 174), where mathematics of the second order was given the name *metamathematics*. The claim of the formalistic school to have reduced intuitionism to absurdity by means of this insight, borrowed from intuitionism, is presumably not to be taken seriously.

2. The thoughtless use of the logical principle of excluded middle is still to be found in *Hilbert 1904* and *1917* (see, for example, *1917*, p. 413, ll. 11u–4u, and in particular *1904*: p. 182, ll. 16–19; p. 182, l. 2u, to p. 183, l. 2; p. 184, ll. 21u–13u ⟦above, p. 135, ll. 13u–11u; p. 136, ll. 5–7; p. 137, ll. 13–18⟧; in each of these places the principle of excluded middle is regarded as essentially equivalent to the principle of contradiction). The second insight is found in the literature for the first time in *Brouwer 1908* and then at greater or lesser length in *Brouwer 1912, 1914, 1917, 1919b, 1923b*, and *1923d*. Except for the recognition, most intimately connected with it, of the intuitionistic consistency of the principle of excluded middle, it penetrates the formalistic literature with *Hilbert 1922a*,[2] where, on the one hand, the limited contentual validity of the principle of excluded middle is acknowledged (see in particular pp. 155–156) and, on the other, the task is posed of consistently combining a logical formulation of the principle of excluded middle with other axioms in the framework of formalistic mathematics. The limited contentual validity of the principle of excluded middle is pointed out with particular eloquence in *Hilbert 1925* (pp. 173–174 ⟦above, pp. 378–379⟧), where, however, the goal is overshot when the area called into question is extended to include the remaining Aristotelian laws.

[1] An oral discussion of the first insight took place in several conversations I had with Hilbert in the autumn of 1909.

[2] After attention had already been paid to the principle of excluded middle in *Hilbert 1922*, p. 160.

3. During the period of the thoughtless use of the principle of excluded middle in the formalistic literature, the principle of the solvability of every mathematical problem is first advanced in *Hilbert 1900b*, p. 52, as an axiom or a conviction and then in *Hilbert 1917*, pp. 412–413, in two different forms (in which, instead of "solvability", "solvability in principle" and, after that, "decidability by means of a finite number of operations" are mentioned) as the object of problems still to be settled. But even after the discussion of the third insight in *Brouwer 1908*, p. 156, *1914*, p. 80, *1919b*, pp. 203–204, and the penetration of the second insight into the formalistic literature, we find that in *Hilbert 1925*, p. 180 [[above, p. 384]]—where the problem of the consistency of the axiom of the solvability of any mathematical problem is offered as an example of a "problem of a fundamental character that falls within the domain of mathematics but formerly could not even be approached" —this question is presented as still open, irrespective of whether the foundations of the science of mathematics (which also comprise the consistency of the principle of excluded middle) be secured or not.

4. The fourth insight is expressed in *Brouwer 1927*, p. 64 [[above, p. 460]]. No trace of it is to be found thus far in the formalistic literature but many an utterance contradicting it, for example in *Hilbert 1900b*, pp. 55–56, and above all in *Hilbert 1925*, where on pp. 162–163 [[above, p. 370]] we still find the exclamation: "No, if justifying a procedure means anything more than proving its consistency, it can only mean determining whether the procedure is successful in fulfilling its purpose."

According to what precedes, formalism has received nothing but benefactions from intuitionism and may expect further benefactions. The formalistic school should therefore accord some recognition to intuitionism, instead of polemicizing against it in sneering tones while not even observing proper mention of authorship. Moreover, the formalistic school should ponder the fact that in the framework of formalism *nothing* of mathematics proper has been secured up to now (since, after all, the metamathematical proof of the consistency of the axiom system is lacking, now as before), whereas intuitionism, on the basis of its constructive definition of set[3] and the fundamental property it has exhibited for finitary sets,[4] has already erected anew several of the theories of mathematics proper in unshakable certainty. If, therefore, the formalistic school, according to its utterance in *Hilbert 1925*, p. 180 [[above, p. 384]], has detected modesty on the part of intuitionism, it should seize the occasion not to lag behind intuitionism with respect to this virtue.

[3] [[Later Brouwer uses the word "spread" for this notion; here the word "Menge", translated as "set", suggests that Brouwer considers spreads to be constructive substitutes for classical sets. See above, p. 453.]]

[4] See *Brouwer 1927*, p. 66, Theorem 2 [[above, p. 462]].

On Hilbert's construction of the real numbers

WILHELM ACKERMANN

(*1928*)

The notion of primitive recursive function appeared, in *Dedekind 1888*, as a natural generalization of the recursive definitions of addition and multiplication. Primitive recursive functions are obtained from 0, the successor function, and the identity function by composition and the following schema:

$$A\begin{cases}(1)\quad f(x_1, x_2, \ldots, x_k, 0) \\ \qquad\qquad = g(x_1, x_2, \ldots, x_k),\\ (2)\quad f(x_1, x_2, \ldots, x_k, n+1) \\ \qquad = h(x_1, x_2, \ldots, x_k, n, f(x_1, x_2, \ldots, \\ \qquad\qquad x_k, n)),\end{cases}$$

where $g(x_1, x_2, \ldots, x_k)$ and $h(x_1, x_2, \ldots, x_{k+2})$ are previously introduced primitive recursive functions.[a]

If one looks, in turn, for a generalization of the notion of primitive recursive function, he can either introduce more involved combinations of functions and variables, thus abandoning the specific form of schema A, or preserve the form of schema A while extending the range of the variables occurring in it. The first method was explored by Skolem (*1923*, pp. 319–320 above), who showed that changing schema A to "course-of-values" recursion does not actually enlarge the class of functions obtained. The second method was followed by Hilbert in his attempt (*1925*, pp. 385–390 above) to classify number-theoretic functions. He introduced a hierarchy of functions by allowing the use of functionals. A function of type 1 is a number-theoretic function; a function of type i is a functional whose values are natural numbers and whose arguments include at least one function of type $i - 1$ and no function of a type greater than $i - 1$. An example of a function of type 2 is the iteration function, which in Ackermann's paper below is defined thus:

$$\rho_c(f(c), a, 0) = a,$$
$$\rho_c(f(c), a, n + 1) = f(\rho_c(f(c), a, n));$$

c is a dummy variable indicating that f is a function of one variable. (This function is the function $\iota(f, a, n)$ in *Hilbert 1925*, p. 388 above.)

This distinction of types leads to a

[a] Schema A is Gödel's schema of primitive recursion (*1931*; see below, p. 602), and it is the schema generally adopted today. Ackermann uses another schema, that of Hilbert (*1925*; see above, p. 389):

$$A'\begin{cases}(1')\quad f(x_1, x_2, \ldots, x_k, 0) = g(x_1, x_2, \ldots, x_k),\\ (2')\quad f(x_1, x_2, \ldots, x_k, n+1)\\ \qquad = \mathfrak{h}_{y_1 y_2 \cdots y_k}(x_1, x_2, \ldots, x_k, n, f(y_1, \\ \qquad\qquad y_2, \ldots, y_k, n),\end{cases}$$

where $\mathfrak{h}$ is a function obtained by composition from the initial functions (0, the successor function, the identity function), previously obtained primitive recursive functions, and the function $f(y_1, y_2, \ldots, y_k, n)$ considered as a function of $y_1, y_2, \ldots, y_k$, but *not* of n; the subscripts $y_1, y_2, \ldots, y_k$ indicate that the variables $y_1, y_2, \ldots, y_k$ merely mark argument places of f but do not actually occur in the function f; the argument places marked by y_1, $y_2, \ldots, y_k$ have been occupied by functions of the kind just prescribed. The equivalence of schemas A and A' was proved by Péter (*1934*; see also *1932*).

classification of number-theoretic functions. A number-theoretic function of level[b] n is defined by equations that are modeled upon schema A and in which essential use is made of a function of type n, but of no function of a greater type, for g or h. Primitive recursive functions are the functions of level 1. The first question raised by such an approach is whether there actually exist functions of a level greater than 1, that is, whether there exist, for instance, number-theoretic functions that cannot be defined, by equations modeled upon schema A, without the help of functionals.

Ackermann answered the question affirmatively by exhibiting a function of level 2. The example is already mentioned in *Hilbert 1925* (p. 388 above) but was not published till 1928 (received for publication on 20 January 1927). A function $\varphi(a, b, n)$ is defined by the equations:

$$B\begin{cases}(3) & \varphi(a, b, 0) = a + b,\\ (4) & \varphi(a, b, n+1)\\ & \qquad = \rho_c(\varphi(a, c, n), \alpha(a, n), b),\end{cases}$$

where ρ_c is the iteration function and $\alpha(a, n)$ is the primitive recursive function defined by

$$\alpha(a, 0) = 0,$$
$$\alpha(a, 1) = 1,$$
$$\alpha(a, n) = a \quad \text{for } n > 1.$$

Definition B follows schema A, except that a function of type 2, ρ_c, has been used in the construction of the function h. Ackermann then considers the function $\varphi(a, a, a)$. If $\varphi(a, b, n)$ were of level 1, that is, primitive recursive, so would be $\varphi(a, a, a)$. The proof that $\varphi(a, a, a)$ is not primitive recursive consists in showing that it increases faster than any function of level 1 (in the sense in which 2^a increases faster than any polynomial in a). The proof is rather long. It was simplified by Péter (*1935*); see also *Hilbert and Bernays 1934*, pp. 329–343 and *R. M. Robinson 1948*. Péter also presented an alternative proof, which makes use of the diagonal argument.[c]

In the course of Ackermann's proof it turns out that the function $\varphi(a, b, n)$ satisfies the equations:

$$C\begin{cases}(5) & \varphi(a, b, 0) = a + b,\\ (6) & \varphi(a, 0, n+1) = \alpha(a, n),\\ (7) & \varphi(a, b+1, n+1)\\ & \qquad = \varphi(a, \varphi(a, b, n+1), n).\end{cases}$$

These equations show that $\varphi(a, b, n)$ is effectively computable; if n is successively replaced by $0, 1, 2, \ldots$, equations (6) and (7) yield, for each n, equations for φ that obey the primitive recursion schema A' (see footnote a).

Equations C show also that functionals are not needed for the definition of the function φ. But, although they do not contain any functional, equations C do not follow the schema of primitive recursion, because the recursion proceeds on two variables simultaneously. Ackermann's proof establishes that such multiple recursions are not necessarily reducible to primitive recursion. (On multiple recursions see *Péter 1937*.)

The discovery of Ackermann's function gave a stimulus to the classification of recursion schemas; Péter's book (*1951*, *1957*) offers ample information on that point. The function also seems to have guided Herbrand in his search for a definition of the notion of effectively calculable function (see below, p. 624); in the definition of what became known as a recursive function, schema C is replaced by any finite set of equations satisfying certain rather broad conditions.

The translation is by Stefan Bauer-Mengelberg, and it is printed here with the kind permission of Professor Ackermann and Springer Verlag.

---

[b] Neither Hilbert nor Ackermann uses this notion; we introduce it here to facilitate the presentation.

[c] Ackermann's construction, too, makes use of the diagonal argument, but in a somewhat concealed form.

In order to provide a proof of Cantor's conjecture that the set of real numbers, that is, of the number-theoretic functions, can be enumerated by means of the numbers of the second number class, Hilbert makes use of a special construction of the number-theoretic functions. Essential in this construction is the notion "type of a function". A function of type 1 is one whose arguments and values are integers, hence an ordinary number-theoretic function. Functions of type 2 are functions-of-functions. A function of that kind assigns a number to each number-theoretic function. A function of type 3 in turn assigns numbers to the functions-of-functions, and so on. The definition of types can also be continued into the transfinite, but this is of no relevance to the subject of the present paper.[1]

All functions are defined by recursion on an ordinary number-theoretic variable. For instance,

$$\varphi(a, 0) = 0,$$

$$\varphi(a, n + 1) = \varphi(a, n) + a$$

is an example of a definition of a function of type 1. Obviously $\varphi(a, b)$ represents the function $a \cdot b$. The following recursion defines a function of type 2:

$$\rho_c(f(c), a, 0) = a,$$

$$\rho_c(f(c), a, n + 1) = f(\rho_c(f(c), a, n)).$$

$\rho_c(f(c), a, n)$ represents the n-fold iteration of the function f, taken on the argument a. (The subscript c on ρ is to mean that $\rho_c(f(c), a, n)$ depends on f and not on c.) Along with recursion, of course, the formation of functions by substitution, too, is admissible. If, for example, $\varphi(a, b)$ and $\psi(a, b)$ are number-theoretic functions of two variables, then $\varphi(\varphi(a, a), \psi(a, b))$, too, is a function of that kind.

The construction of the number-theoretic functions now proceeds in such a way that they can be sorted into certain groups. In the first group we put the functions for whose definition only type 1 is needed; in the second we put the functions for whose definition the use of a recursively defined function of type 2 is essential, while a higher type can be circumvented; and so on.

If the construction described is actually to yield all number-theoretic functions, it is essential that every group in fact bring about an extension of the inventory of functions. For it might, after all, be conceivable that we could obtain all functions for whose definition we use types 1 and 2 by means of type 1 alone.

Now in what follows we present a function for which we shall prove that it cannot be defined without type 2 or a higher type. By means of the method employed here it will be easy to prove for other groups, too, that they extend the inventory of functions.[2]

Incidentally, all the recursions are to be on one variable only. Simultaneous recursions are excluded. A function defined by simultaneous recursion can also be represented by ordinary recursion if one is prepared to use higher types.

[1] For details see *Hilbert 1925* [[pp. 386–388 above]].

[2] A paper having several points of contact with the present one will be published by G. Sudan [[*1927*]]. It deals with the definitions of numbers of the second number class, which can be classified similarly to the definitions of real numbers.

1

First we shall give the function just mentioned. When functions are constructed by means of recursion, the function $a + 1$ must be assumed to be known and cannot in turn be defined by recursion. Then the function $a + b$ is defined by the recursion

$$a + 0 = a,$$

$$a + (b + 1) = (a + b) + 1.$$

For the sake of simplicity we shall take two further functions as known initial functions, namely, $\lambda(a, b)$ and $\iota(a, b)$. $\lambda(a, b)$ equals 1 if $a = b$ and 0 if $a \neq b$. For $\iota(a, b)$ it is the other way round. But we can also define λ and ι by recursion. We shall write $\alpha(a, n)$ for $\iota(n, 1)\cdot\iota(n, 0)\cdot a + \lambda(n, 1)$. Here $a\cdot b$ is defined by

$$a\cdot 0 = a,$$

$$a\cdot(b + 1) = a\cdot b + a.$$

Thus $\alpha(a, n)$ is always equal to a, except for $n = 0$ and $n = 1$. $\alpha(a, 0) = 0$ and $\alpha(a, 1) = 1$. Let ρ stand for the iteration function mentioned above. Finally, a number-theoretic function φ of three variables is given by

$$\varphi(a, b, 0) = a + b,$$

$$\varphi(a, b, n + 1) = \rho_c(\varphi(a, c, n), \alpha(a, n), b).$$

As we can see from this, $\varphi(a, b, 1)$ is identical with $a\cdot b$. $\varphi(a, b, 2)$ coincides with a^b. $\varphi(a, b, 3)$ is the b-fold iteration of a^b, taken on a, and so on. We now obtain our desired function by setting all three arguments in $\varphi(a, b, c)$ equal. What we claim, then, is that $\varphi(a, a, a)$ cannot be defined without the use of type 2.[3]

The exclusion of simultaneous recursions is essential to our claim. For the following formulas hold:

$$\varphi(a, b, 0) = a + b,$$

$$\varphi(a, 0, n + 1) = \alpha(a, n),$$

$$\varphi(a, b + 1, n + 1) = \varphi(a, \varphi(a, b, n + 1), n).$$

(The last two formulas, incidentally, are proved immediately below.)

2

We shall prove our claim by showing that the function $\varphi(a, a, a)$ increases more rapidly than any function for whose definition only type 1 is used. In order to be able to show this, we must prove a number of properties and formulas for our function.

I. $\varphi(a, 0, n + 1) = \alpha(a, n)$.

Proof.

$$\varphi(a, 0, n + 1) = \rho_c(\varphi(a, c, n), \alpha(a, n), 0) = \alpha(a, n).$$

[3] The function that I have just presented was already mentioned in *Hilbert 1925*, p. 185 ⟦above, p. 388⟧.

II. $\varphi(a, b + 1, n + 1) = \varphi(a, \varphi(a, b, n + 1), n)$.

Proof.

$$\varphi(a, b + 1, n + 1) = \rho_c(\varphi(a, c, n), \alpha(a, n), b + 1),$$

$$\rho_c(\varphi(a, c, n), \alpha(a, n), b + 1) = \varphi(a, \rho_c(\varphi(a, c, n), \alpha(a, n), b), n),$$

$$\rho_c(\varphi(a, c, n), \alpha(a, n), b) = \varphi(a, b, n + 1),$$

whence the assertion is evident.

Furthermore, we need certain monotonicity properties of φ.

III. *If* $a \geqq 2$, *then* $\varphi(a, b + 1, n) > \varphi(a, b, n)$.

Proof. If $n = 0$ or $n = 1$, the assertion is true since

$$a + 2 + b + 1 > a + 2 + b$$

and

$$(a + 2)\cdot(b + 1) > (a + 2)\cdot b.$$

It remains to be proved that

$$\text{(A)} \qquad \varphi(a + 2, b + 1, n + 2) > \varphi(a + 2, b, n + 2).$$

We prove this formula, together with

$$\text{(B)} \qquad \varphi(a + 2, b, n + 2) > b,$$

by induction on n.

$$\varphi(a + 2, b + 1, 2) > \varphi(a + 2, b, 2),$$

since

$$(a + 2)^{b+1} > (a + 2)^b$$

and

$$\varphi(a + 2, b, 2) = (a + 2)^b > b.$$

Assume that both formulas have already been proved for n. We first prove (B) for $n + 1$ by induction on b. (B) is correct for $n + 1$ if $b = 0$ since for every n we have

$$\varphi(a + 2, 0, n + 3) = a + 2 > 0,$$

$$\varphi(a + 2, b + 1, n + 3) = \varphi(a + 2, \varphi(a + 2, b, n + 3), n + 2).$$

Hence, if $\varphi(a + 2, b, n + 3) \geqq b + 1$, then, since (A) holds for n,

$$\varphi(a + 2, \varphi(a + 2, b, n + 3), n + 2) \geqq \varphi(a + 2, b + 1, n + 2),$$

hence

$$\varphi(a + 2, b + 1, n + 3) \geqq \varphi(a + 2, b + 1, n + 2),$$

$$\varphi(a + 2, b + 1, n + 2) > b + 1,$$

$$\varphi(a + 2, b + 1, n + 3) > b + 1.$$

This proves (B) for $n + 1$.

$$\varphi(a + 2, b + 1, n + 3) = \varphi(a + 2, \varphi(a + 2, b, n + 3), n + 2)$$

$$> \varphi(a + 2, b, n + 3) \qquad \text{(by (B))}.$$

This proves formulas (A) and (B), hence also III.

The two formulas

IV. $\varphi(a + 1, b, n) \geqq \varphi(a, b, n)$,
V. $\varphi(a + 2, b + 3, n + 1) > \varphi(a + 2, b + 3, n)$

can be proved in a similar way.

Proof of IV. We readily convince ourselves that formula IV holds for $a = 0$ and $a = 1$. For $n = 0, 1, 2$ this can easily be checked. For $n \geqq 3$ we have the formulas

$$\varphi(0, b, n) \leqq 1,$$
$$\varphi(1, b, n) = 1,$$
$$\varphi(2, b, n) \geqq 1,$$

which we prove by using induction on n and distinguishing the cases $b = 0$ and $b \neq 0$. To prove IV we must still show that

$$\varphi(a + 3, b, n) \geqq \varphi(a + 2, b, n).$$

This holds for $n = 0$, and also for $b = 0$. Assume that it has already been shown for smaller n, as well as for the same n and smaller b.

$$\varphi(a + 3, b + 1, n + 1) = \varphi(a + 3, \varphi(a + 3, b, n + 1), n),$$
$$\varphi(a + 3, \varphi(a + 3, b, n + 1), n) \geqq \varphi(a + 2, \varphi(a + 3, b, n + 1), n),$$
$$\varphi(a + 3, b, n + 1) \geqq \varphi(a + 2, b, n + 1).$$

It then follows from III that

$$\varphi(a + 2, \varphi(a + 3, b, n + 1), n) \geqq \varphi(a + 2, \varphi(a + 2, b, n + 1), n).$$

But $\varphi(a + 2, \varphi(a + 2, b, n + 1), n)$ is the same as $\varphi(a + 2, b + 1, n + 1)$.

Proof of V. For $n = 0$ and $n = 1$ the formula holds. It remains to be shown that

$$\varphi(a + 2, b + 3, n + 3) > \varphi(a + 2, b + 3, n + 2).$$

Let us take instead the more general formula

$$\varphi(a + 2, b + 1, n + 3) > \varphi(a + 2, b + 1, n + 2).$$

This is correct for $b = 0$, since

$$\begin{aligned}\varphi(a + 2, 1, n + 3) &= \varphi(a + 2, \varphi(a + 2, 0, n + 3), n + 2)\\ &= \varphi(a + 2, a + 2, n + 2)\end{aligned}$$
$$\varphi(a + 2, a + 2, n + 2) > \varphi(a + 2, 1, n + 2) \qquad \text{(by III)}.$$

For $n = 0$ likewise. Now assume that it has already been proved for smaller n, as well as for the same n and smaller b.

$$\varphi(a + 2, b + 2, n + 4) = \varphi(a + 2, \varphi(a + 2, b + 1, n + 4), n + 3),$$
$$\varphi(a + 2, b + 1, n + 4) > \varphi(a + 2, b + 1, n + 3),$$
$$\varphi(a + 2, b + 1, n + 3) \geqq b + 2 \qquad \text{(by formula (B), p. 497)}.$$
$$\varphi(a + 2, \varphi(a + 2, b + 1, n + 4), n + 3) > \varphi(a + 2, b + 2, n + 3),$$
$$\varphi(a + 2, b + 2, n + 4) > \varphi(a + 2, b + 2, n + 3).$$

Thus V, too, is proved.

The most important of the properties used in our proof is the following.

VI. *Of the two formulas*

$$\rho_c(\varphi(c, c, n), a, a) \leqq \varphi(a, a, n + 3),$$
$$\rho_c(\varphi(c, c, n), a, a) \leqq 2$$

at least one is true.

(The first of the two formulas, incidentally, always holds except when $a = 1$ and $n = 0$.)

We now need a number of auxiliary formulas.

Formula 1: We first prove

$$\varphi(\varphi(a + 2, c, 3), \varphi(a + 2, c, 3), 1) \leqq \varphi(a + 2, c + 1, 3),$$

by induction on c. For $c = 0$ the formula is true, since the left side becomes equal to $\varphi(a + 2, a + 2, 1)$, that is, $(a + 2)\cdot(a + 2)$, and the right side to $(a + 2)^{(a+2)}$. If we make use of the fact that for $c = 1, 2, 3$ the function $\varphi(a, b, c)$ represents the specific functions that we know it does, we readily obtain the following formulas:

$$\begin{aligned} 2 &\leqq \varphi(a + 2, c, 3), \\ \varphi(\varphi(a + 2, c + 1, 3), \varphi(a + 2, c + 1, 3), 1) &= \varphi(a + 2, c + 1, 3)^2 \\ &= (a + 2)^{2\cdot\varphi(a+2,c,3)} \\ &\leqq (a + 2)^{\varphi(a+2,c,3)\cdot\varphi(a+2,c,3)} \\ &\leqq (a + 2)^{\varphi(a+2,c+1,3)}, \\ (a + 2)^{\varphi(a+2,c+1,3)} &= \varphi(a + 2, c + 2, 3). \end{aligned}$$

Formula 2:

$$\varphi(\varphi(a + 2, b, 3), \varphi(a + 2, c, 3), 2) \leqq \varphi(a + 2, c + 2, 3)$$

if $c \geqq b$.

Proof. (a) Let $b = 0$.

$$\varphi(a + 2, 0, 3) = a + 2,$$

$$\begin{aligned} \varphi(a + 2, \varphi(a + 2, c, 3), 2) &= (a + 2)^{\varphi(a+2,c,3)} \\ &= \varphi(a + 2, c + 1, 3) < \varphi(a + 2, c + 2, 3) \qquad \text{(by III)}. \end{aligned}$$

(b) $b \geqq 1$.

$$\begin{aligned} \varphi(\varphi(a + 2, b, 3), \varphi(a + 2, c, 3), 2) &= (a + 2)^{\varphi(a+2,b-1,3)\cdot\varphi(a+2,c,3)} \\ &\leqq (a + 2)^{\varphi(a+2,c,3)\cdot\varphi(a+2,c,3)} \\ &\leqq (a + 2)^{\varphi(a+2,c+1,3)} \end{aligned}$$

and by formula 1

$$(a + 2)^{\varphi(a+2,c+1,3)} = \varphi(a + 2, c + 2, 3).$$

Formula 3: Assume that the relation

$$\varphi(\varphi(a, b, n + 3), \varphi(a, c, n + 3), n + 2) \leqq \varphi(a, c + 2, n + 3)$$

holds for all $a \geqq 2$, $c \geqq b$, and $b \neq 0$ and for a fixed n. Then we assert that the relation

$$\varphi(\varphi(a, b, n + 4), m, n + 3) \leqq \varphi[a, \varphi(a, b - 1, n + 4) + 2m, n + 3]$$

holds for all such a, b, and c and for arbitrary m.

For $m = 0$ the right and left sides are equal.

$$\varphi(\varphi(a, b, n + 4), m + 1, n + 3)$$
$$= \varphi\{\varphi(a, b, n + 4), \varphi[\varphi(a, b, n + 4), m, n + 3], n + 2\}.$$

If the assertion already holds for smaller m, then it follows by virtue of formula III that the right side is not greater than

$$\varphi\{\varphi(a, b, n + 4), \varphi[a, \varphi(a, b - 1, n + 4) + 2m, n + 3], n + 2\},$$

with

$$\varphi(a, b, n + 4) = \varphi[a, \varphi(a, b - 1, n + 4), n + 3].$$

It then follows from the assumption that the last expression but one is less than or equal to

$$\varphi[a, \varphi(a, b - 1, n + 4) + 2m + 2, n + 3].$$

This proves the assertion.

Formula 4: Let $a \geqq 2$ and $c \geqq b$. Then

$$\varphi(\varphi(a, b, n + 3), \varphi(a, c, n + 3), n + 2) \leqq \varphi(a, c + 2, n + 3).$$

For $n = 0$ this formula coincides with formula 2. We now prove it for $n + 1$ on the assumption that it holds for n. First, assume that b is different from 0. According to formula 3 we have

$$\varphi(\varphi(a, b, n + 4), \varphi(a, c, n + 4), n + 3)$$
$$\leqq \varphi[a, \varphi(a, b - 1, n + 4) + 2 \cdot \varphi(a, c, n + 4), n + 3]$$
$$\leqq \varphi[a, 3 \cdot \varphi(a, c, n + 4), n + 3].$$

By induction on c it can easily be proved for $c \geqq 2$ that

$$3c \leqq \varphi(2, c, 3).$$

Since, furthermore, $\varphi(a, c, n + 4) \geqq 2$ because $a \geqq 2$, we have

$$3 \cdot \varphi(a, c, n + 4) \leqq \varphi(2, \varphi(a, c, n + 4), 3) \leqq \varphi(a, \varphi(a, c, n + 4), n + 3),$$
$$\varphi(a, \varphi(a, c, n + 4), n + 3) = \varphi(a, c + 1, n + 4),$$
$$\varphi(\varphi(a, b, n + 4), \varphi(a, c, n + 4), n + 3) \leqq \varphi(a, \varphi(a, c + 1, n + 4), n + 3),$$
$$\varphi(a, \varphi(a, c + 1, n + 4), n + 3) = \varphi(a, c + 2, n + 4).$$

This proves the formula for $b \neq 0$.

$$\varphi(\varphi(a, 0, n + 3), \varphi(a, c, n + 3), n + 2) = \varphi(a, \varphi(a, c, n + 3), n + 2)$$
$$= \varphi(a, c + 1, n + 3) \leqq \varphi(a, c + 2, n + 3).$$

Hence it is correct for $b = 0$ too.

Formula 5:

$$\rho_c(\varphi(c, c, n + 2), a + 2, b) \leqq \varphi(a + 2, 2b, n + 3).$$

Proof by induction on b. For $b = 0$, the formula reduces to

$$a + 2 \leqq a + 2.$$

$$\rho_c(\varphi(c, c, n + 2), a + 2, b + 1)$$
$$= \varphi\{\rho_c[\varphi(c, c, n + 2), a + 2, b], \rho_c[\varphi(c, c, n + 2), a + 2, b], n + 2\}.$$

By formulas IV and III, this last expression is

$$\leqq \varphi(\varphi(a + 2, 2b, n + 3), \varphi(a + 2, 2b, n + 3), n + 2).$$

By formula 4, in turn, the last term is

$$\leqq \varphi(a + 2, 2b + 2, n + 3).$$

This proves formula 5.

Proof of VI. (a) $a = 0$.

$$\rho_c(\varphi(c, c, n), 0, 0) = 0 \leqq \varphi(0, 0, 3).$$

(b) $a = 1$.

$$\rho_c(\varphi(c, c, n), 1, 1) = \varphi(1, 1, n),$$
$$\varphi(1, 1, 0) = 2,$$
$$\varphi(1, 1, n + 1) = 1.$$

The last equation is easy to prove by induction on n.

(c) $a \geqq 2$. VI is easy to verify for $n = 0$ and $n = 1$. It remains to be proved that for arbitrary a and n

$$\rho_c(\varphi(c, c, n + 2), a + 2, a + 2) \leqq \varphi(a + 2, a + 2, n + 5).$$

If in formula 5 we replace b by $a + 2$, we obtain

$$\rho_c(\varphi(c, c, n + 2), a + 2, a + 2) \leqq \varphi(a + 2, 2\cdot(a + 2), n + 3),$$
$$2(a + 2) \leqq (a + 2)^{a+2} = \varphi(a + 2, 1, 3),$$
$$\varphi(a + 2, 1, 3) \leqq \varphi(a + 2, a + 1, n + 4),$$
$$\rho_c(\varphi(c, c, n + 2), a + 2, a + 2) \leqq \varphi(a + 2, \varphi(a + 2, a + 1, n + 4), n + 3),$$
$$\varphi(a + 2, \varphi(a + 2, a + 1, n + 4), n + 3) = \varphi(a + 2, a + 2, n + 4),$$
$$\varphi(a + 2, a + 2, n + 4) \leqq \varphi(a + 2, a + 2, n + 5),$$
$$\rho_c(\varphi(c, c, n + 2), a + 2, a + 2) \leqq \varphi(a + 2, a + 2, n + 5), \quad \text{q. e. d.}$$

3

The formulas given above will make it possible for us to prove that $\varphi(a, a, a)$ cannot be defined without the use of type 2. If, to abbreviate, we denote by $\mathfrak{R}_a(g(a), f(a))$ the proposition that, from a certain argument on, the number-theoretic function $f(.)$ is not less than $g(.)$, the proof proceeds as follows.

Let $\psi(a)$ be any number-theoretic function defined without the use of type 2. We shall then show that there exists an n for which $\mathfrak{R}_a(\psi(a), \varphi(a, a, n))$, or, if we introduce an abbreviation, that $\mathfrak{G}_{an}(\psi(a), \varphi(a, a, n))$ is the case. This would complete the proof, since $\mathfrak{G}_{an}(\varphi(a, a, a), \varphi(a, a, n))$ does not hold, as follows from formula V.

Our initial functions λ, ι,[4] and $a + 1$ have property $\mathfrak{G}$. From these initial functions we successively obtain new functions by taking the functions already formed and using them in turn in the recursions. We shall prove by induction that all the newly formed functions will again have property $\mathfrak{G}$.

To speak more precisely, we prove the two theorems below. Assume that we have a

[4] ⟦The German text erroneously has "c" instead of "ι".⟧

finite domain of functions of one, two, or more number-theoretic variables, with the following properties:

1. If $\psi(a_1, a_2, \ldots, a_n)$ is a function of our domain, the domain also contains a function $\psi'(a_1, a_2, \ldots, a_n)$ that increases monotonically with respect to all variables (monotonicity here being understood to allow equality) and is such that for arbitrary $a_1, a_2, \ldots, a_n$

$$\psi'(a_1, a_2, \ldots, a_n) \geqq \psi(a_1, a_2, \ldots, a_n)$$

holds. Of course, ψ' may also be identical with ψ, if ψ itself is already monotonic.

2. For every function $\psi(a_1, a_2, \ldots, a_n)$ of the domain,

$$\mathfrak{G}_{an}(\psi(a, a, \ldots, a), \varphi(a, a, n))$$

holds.

Our two theorems then read as follows:

THEOREM 1. *If to our domain we add a new function obtained from the functions of our domain by substitution, then the new domain either again has both properties or can again be transformed, by the addition of further functions, into such a domain having both properties.*

THEOREM 2. *If, using functions of our domain, we form a new function ψ by recursion and add this function and possibly yet others to the domain, then both properties again hold of the newly obtained domain.*

Since our initial functions have properties 1 and 2, everything is settled once these two theorems are proved.

Proof of Theorem 1. Assume that a new function has been formed by substitution. If all the functions from which the new function was compounded are monotonic functions, then the new function, too, is monotonic with respect to all variables (monotonic here always being understood in the sense of monotonic nondecreasing). Hence the first property then obtains. If not all the functions used in compounding the new function are monotonic, then we replace every function in this combination by the associated monotonic one and add the function thus obtained to the domain. Then the first property obtains. So far as the second property is concerned, it suffices to show that it holds for all functions obtained from the monotonic functions of our domain by substitution. Furthermore, we can restrict ourselves to showing that, if $\psi(a_1, a_2, \ldots, a_m)$ is a function of our domain having m arguments and if $\chi_1(a), \ldots, \chi_m(a)$ are m functions, $\chi_i(a)$ with property $\mathfrak{G}_{an}(\chi_i(a), \varphi(a, a, n))$, then $\mathfrak{G}_{an}(\psi(\chi_1(a), \ldots, \chi_m(a)), \varphi(a, a, n))$ holds. For the general case is obtained from this by induction on the degree of nesting of the function that results from the functions of our domain by substitution. Since now, for $n \geqq m$, $\mathfrak{R}_a(\varphi(a, a, m), \varphi(a, a, n))$, since ψ is monotonic, and since, for an arbitrary i, $\mathfrak{G}_{an}(\chi_i(a), \varphi(a, a, n))$ is true, we have

$$\mathfrak{G}_{an}(\psi(\chi_1(a), \ldots, \chi_m(a)), \psi(\varphi(a, a, n), \ldots, \varphi(a, a, n))).$$

Since further

$$\mathfrak{G}_{an}(\psi(a, a, \ldots, a), \varphi(a, a, n)),$$

we also have

$$\mathfrak{G}_{an}(\psi(\chi_1(a), \ldots, \chi_m(a)), \varphi(\varphi(a, a, n), \varphi(a, a, n), n)),$$

$$\varphi(\varphi(a, a, n), \varphi(a, a, n), n) = \rho_c(\varphi(c, c, n), a, 2).$$

Because of

$$\mathfrak{R}_a(\rho_c(\varphi(c, c, n), a, 2), \rho_c(\varphi(c, c, n), a, a))$$

and formula VI we finally obtain

$$\mathfrak{G}_{an}(\psi(\chi_1(a), \ldots, \chi_m(a)), \varphi(a, a, n)), \text{ q. e. d.}$$

Proof of Theorem 2. (*a*) The proof of Theorem 2 is more complicated by far than that of Theorem 1, and it will constitute the remainder of this paper. First it will be necessary to discuss in more detail the recursion schema that does not use type 2. For a function of one variable the recursion looks as follows:

$$\psi(0) = \mathfrak{a},$$
$$\psi(a + 1) = \mathfrak{b}(a, \psi(a)),$$

where $\mathfrak{a}$ is a functional not containing any variable and $\mathfrak{b}$ is a function of two variables that might have been obtained by the nesting of several recursive functions. No recursive function of type 2 may occur in $\mathfrak{a}$ and $\mathfrak{b}$, but only functions already defined. For functions of two variables the schema reads analogously:

$$\psi(a, 0) = \mathfrak{a}(a),$$
$$\psi(a, b + 1) = \mathfrak{b}_c(a, b, \psi(c, b)).$$

Here the same remarks hold. It is to be noted especially that the function-of-a-function $\mathfrak{b}_c(a, b, f(c))$ must be formed without a recursive function of type 2. Of course, we can also form functions-of-functions by mere nesting, without recursion, for example, $f(f(a)), f(a) + f(b)$, and so on. But the recursion

$$\psi(a, 0) = \mathfrak{a}(a),$$
$$\psi(a, b + 1) = \rho_c(\psi(c, b), b, a).$$

is ruled out, while, on the other hand,

$$\psi(a, 0) = a + 1,$$
$$\psi(a, b + 1) = 1 + \psi[\psi(a + b, b), b].$$

is permitted. Correspondingly, for functions of n variables we have the schema:

$$\psi(a, b, \ldots, m, 0) = \mathfrak{a}(a, b, \ldots, m),$$
$$\psi(a, b, \ldots, m, n + 1) = \mathfrak{b}_{a_1 b_1 \cdots m_1}(a, b, \ldots, m, n, \psi(a_1, \ldots, m_1, n)).$$

Here we may assume that $\mathfrak{a}(\ldots)$ and $\mathfrak{b}(\ldots)$ are formed exclusively from monotonic functions. If the schema for ψ does not possess this property initially, let $\mathfrak{a}'(a, b, \ldots, m)$ be obtained from $\mathfrak{a}(a, b, \ldots, m)$ by replacement of all functions occurring in $\mathfrak{a}(a, b, \ldots, m)$ by the associated greater monotonic functions. We obtain $\mathfrak{b}'(\ldots)$ from $\mathfrak{b}(\ldots)$ in the same way. Then the function

$$\psi'(a, b, \ldots, m, 0) = \mathfrak{a}'(a, b, \ldots, m),$$
$$\psi'(a, b, \ldots, m, n + 1) = \mathfrak{b}'_{a_1 \cdots m_1}(a, \ldots, m, n, \psi(a_1, \ldots, m_1, n_1))$$

is of the desired kind. It may be considered in place of ψ, since it never has a value less than ψ. ψ' is monotonic with respect to $a, b, \ldots, m$, as we can readily convince ourselves by induction on n. In all the considerations that follow we shall tacitly assume that the transition from ψ to ψ' has already been made. We omit the prime for the sake of simplicity.

(*b*) We now show that instead of our general recursion schema we can take a more special one as a basis for our discussion. We shall, however, develop the considerations that follow only for functions of two variables. At the end we shall show that this restriction is unessential. Thus we consider the schema

$$\psi(a, 0) = \mathfrak{a}(a),$$
$$\psi(a, b + 1) = \mathfrak{b}_c(a, b, \psi(c, b)).$$

Here we may assume that $\mathfrak{b}_c(a, b, \psi(c, b))$ actually contains the sign ψ, since otherwise it is a case of mere substitution that would bring us back to Theorem 1. Let us for the time being assume that in $\mathfrak{b}_c(a, b, \psi(c, b))$ no ψ occurs within the parentheses associated with another ψ. In that case every ψ of $\mathfrak{b}$ has as its first argument a function $\alpha_i(a, b)$ that results from functions of our domain by substitution and therefore has the two properties of our domain. Let these functions α be $\alpha_1(a, b), \ldots, \alpha_m(a, b)$. Further, let $\alpha(a, b) = \text{Max}(\alpha_1(a, b), \ldots, \alpha_m(a, b))$. Obviously α, too, has the properties of the functions of our domain. If in the first arguments of all ψ in $\mathfrak{b}_c(a, b, \psi(c, b))$ we now substitute the function $\alpha(a, b)$ for the $\alpha_i(a, b)$ and if we form a recursion with the result, we obtain a function that is not less than $\psi(a, b)$ for any pair of values. But, if in $\mathfrak{b}_c(a, b, \psi(c, b))$ the function ψ always has $\alpha(a, b)$ as a first argument, we can also write $\mathfrak{b}_c(a, b, \psi(c, b))$ in the form $\omega(a, b, \psi(\alpha(a, b), b))$, where ω is a function belonging to our domain or resulting from functions of our domain by substitution. Thus we may confine ourselves to the consideration of the following schema:

$$\psi(a, 0) = \mathfrak{a}(a),$$
$$\psi(a, b + 1) = \omega(a, b, \psi(\alpha(a, b), b)).$$

Now assume that a twofold nesting of the ψ occurs in $\mathfrak{b}_c(a, b, \psi(c, b))$. Then the innermost ψ in $\mathfrak{b}$ again contain as the first argument functions $\alpha_i(a, b)$, which we can again replace by a single function $\alpha(a, b)$. The first arguments of the next ψ toward the outside are functions μ_i of a, b, and $\psi(\alpha(a, b), b)$. We can now use the maximum of the $\mu_i(a, b, c)$, just as before we took the maximum of the $\alpha_i(a, b)$. Thus, if ψ occurs in $\mathfrak{b}_c(a, b, \psi(c, b))$ in an at most twofold nesting, we may consider the following recursion schema:

$$\psi(a, 0) = \mathfrak{a}(a),$$
$$\psi(a, b + 1) = \omega\{a, b, \psi[\alpha(a, b), b], \psi(\mu(a, b, \psi[\alpha(a, b), b]), b)\}.$$

We can develop a similar line of reasoning for the case of an n-fold nesting. There it suffices to consider the schema

$$\psi(a, 0) = \mathfrak{a}(a),$$
$$\psi(a, b + 1) = \omega(a, b, \ldots).$$

Here the function ω has n arguments besides a and b, hence $n + 2$ altogether. These arguments look as follows: the third argument has the form $\psi(\alpha(a, b), b)$; if, furthermore, the first n arguments of ω are $\mathfrak{c}_1, \mathfrak{c}_2, \ldots, \mathfrak{c}_n$, then the $(n + 1)$th reads

$$\psi(\mu(\mathfrak{c}_1, \mathfrak{c}_2, \ldots, \mathfrak{c}_n), b),$$

where μ is a function belonging to our domain or compounded from functions of our domain.

(c) We must now show that the function $\psi(a, b)$, or a greater one, increases monotonically with respect to b even for fixed a, and that $\mathfrak{G}_{an}(\psi(a, a), \varphi(a, a, n))$ is the case. We first specify the greater monotonic function. We replace $\alpha(a, b)$ in the recursion by the function $\alpha'(a, b) = \text{Max}(a, \alpha(a, b))$. In place of each function μ we use $\mu'(a, b, \ldots, k) = \text{Max}(\mu(a, b, \ldots, k), k)$ and in place of $\omega(a_1, \ldots, a_{n+2})$ we take $\omega'(a_1, \ldots, a_{n+2}) = \text{Max}(\omega(a_1, \ldots, a_{n+2}), a_{n+2})$. We replace $\mathfrak{a}(a)$ by $\text{Max}(a, \mathfrak{a}(a)) = \sigma(a)$. The new function is, of course, again monotonic with respect to a, since all of the functions ω', σ, and α' and the functions μ' are monotonic. But this function is monotonic also with respect to b. We first prove: If, for all a, $\psi(a, b) \geqq a$, then, for all a, $\psi(a, b + 1) \geqq \psi(a, b)$ and $\psi(a, b + 1) \geqq a$.

For, with the assumption made, the following considerations hold. We have

$$\psi(a, b + 1) = \omega'(a, b, \psi[\alpha'(a, b), b], \ldots).$$

For the third argument, $\psi[\alpha'(a, b), b] \geqq a$ and $\psi[\alpha'(a, b), b] \geqq \psi(a, b)$.[5] For, since $\psi(a, b)$ is monotonic with respect to the first argument and since $\alpha'(a, b) \geqq a$, we have $\psi[\alpha'(a, b), b] \geqq \psi(a, b)$ and $\psi[\alpha'(a, b), b] \geqq a$. Assume that it has already been proved for an argument $\mathfrak{c}$ of ω' that $\mathfrak{c} \geqq \psi(a, b)$ and $\mathfrak{c} \geqq a$. These relations then hold for the next argument $\mathfrak{d}$, too. For

$$\mathfrak{d} = \psi[\mu(a, b, \ldots, \mathfrak{c}), b]$$
$$\psi[\mu(a, b, \ldots, \mathfrak{c}), b] \geqq \psi[\mu(a, b, \ldots, \psi(a, b)), b],$$
$$\mu(a, b, \ldots, \psi(a, b)) \geqq \psi(a, b),$$
$$\psi[\mu(a, b, \ldots, \psi(a, b)), b] \geqq \psi[\psi(a, b), b],$$

and, according to the assumption,

$$\psi(a, b) \geqq a,$$
$$\psi[\psi(a, b), b] \geqq \psi(a, b),$$

hence $\mathfrak{d} \geqq \psi(a, b)$ and $\mathfrak{d} \geqq a$.

Now since $\omega'(a, b, \ldots)$ is greater than or equal to the last argument of ω', we also have

$$\psi(a, b + 1) \geqq \psi(a, b) \quad \text{and} \quad \psi(a, b + 1) \geqq a, \text{ q. e. d.}$$

But $\psi(a, 0) = \sigma(a) \geqq a$. Consequently

$$\psi(a, b + 1) \geqq \psi(a, b)$$

holds generally. This shows that the function ψ has the first property of our domain.

Now it still remains to be shown that $\mathfrak{G}_{an}(\psi(a, a), \varphi(a, a, n))$. So long as in $\psi(a, b)$ we have $b \leqq a$, it suffices to consider a simplified recursion:

$$\psi(a, 0) = \sigma(a),$$
$$\psi(a, b + 1) = \omega'(a, a, \ldots).$$

Here a has been put for b everywhere except in the second argument of ψ. That this function is not less than the former one directly follows from the monotonicity of all

[5] ⟦In the last formula the German text erroneously has "a'" instead of "α'".⟧

functions that occur. We now denote $\omega'(a, a, \ldots, a)$ by $\omega''(a)$, $\alpha'(a, a)$ by $\alpha''(a)$, and $\mu_i'(a, \ldots, a)$ by $\mu_i''(a)$. In the formula $\psi(a, b + 1) = \omega'(a, a, \ldots)$ the arguments of ω' are all less [[that is, not greater]] than the last argument. For, first, we have $\psi(\alpha''(a), b) \geqq \psi(a, b) \geqq a$.[6]

Assume that it has already been shown that the arguments, to the kth one, never decrease. Let the kth argument be $\mathfrak{d}$. Then the $(k + 1)$th argument has the form $\psi(\mu(a, a, \ldots, \mathfrak{d}), b)$. Now we have

$$\mu(a, a, \ldots, \mathfrak{d}) \geqq \mathfrak{d}$$

and

$$\psi(\mu(a, \ldots, \mathfrak{d}), b) \geqq \psi(\mathfrak{d}, b) \geqq \mathfrak{d}.$$

We therefore do not obtain a smaller function if we set all arguments of ω' equal to the last one. The last argument is $\psi(\mu_{n-1}'(a, \ldots, \mathfrak{d}), b)$, if now $\mathfrak{d}$ is the last but one. Since all arguments of μ_{n-1}' are $\leqq \mathfrak{d}$, we have

$$\psi(\mu_{n-1}'(a, \ldots, \mathfrak{d}), b) \leqq \psi(\mu_{n-1}''(\mathfrak{d}), b).$$

$\mathfrak{d}$ in turn is less than or equal to $\psi(\mu_{n-2}'(\mathfrak{e}), b)$, where $\mathfrak{e}$ is the preceding argument. Finally, we do not obtain a smaller function if we take the recursion

$$\psi(a, 0) = \sigma(a),$$

$$\psi(a, b + 1) = \omega''(\psi\{\mu_{n-1}''(\ldots \psi[\mu_2''(\psi\{\mu_1''[\psi(\alpha''(a), b)], b\}), b], \ldots), b\}).$$

Now we must determine to which degree the functions ω'', μ_i'', α'', and σ are nested in $\psi(a, b + 1)$ when the recursion is decomposed. We see that, if $\psi(a, b)$ consists of an x-fold nesting, $\psi(a, b + 1)$ consists of an $(n(x + 1) + 1)$-fold nesting.[7] Thus the degree of nesting of $\psi(a, b)$ is equal to $2(n^b + n^{b-1} + \cdots + 1) - 1$. In place of this number we can take $2(b + 1)n^b$, which is not less. Furthermore there exists an m such that, from a certain a on, all of the functions μ'', ω'', α'', and σ are less than $\varphi(a, a, m)$.[8] Thus we have

$$\mathfrak{G}_{am}(\psi(a, a), \rho_c(\varphi(c, c, m), a, 2(a + 1)n^a).$$

$2(a + 1)n^a$ is a function of a obtained by substitution from functions having the properties of our domain. In consequence we have

$$\mathfrak{G}_{ak}(2(a + 1)n^a, \varphi(a, a, k)).$$

For every m and k, furthermore, we have

$$\mathfrak{R}_a(\rho_c(\varphi(c, c, m), a, \varphi(a, a, k)), \rho_c(\varphi(c, c, m), \varphi(a, a, k), \varphi(a, a, k))).$$

Further we have, by formula VI,

$$\rho_c(\varphi(c, c, m), \varphi(a, a, k), \varphi(a, a, k)) \leqq \varphi(\varphi(a, a, k), \varphi(a, a, k), m + 3),$$

or

$$\rho_c(\varphi(c, c, m), \varphi(a, a, k), \varphi(a, a, k)) \leqq 2.$$

[6] [[The German text erroneously has "α'" instead of "α''".]]

[7] [[For the "n" in this sentence, as well as for the next six occurrences of "n", the German text has "$\mathfrak{n}$".]]

[8] [[The German text erroneously has "σ'" instead of "σ".]]

But $\varphi(\varphi(a, a, k), \varphi(a, a, k), m + 3)$ is a function having the properties of our domain, since it is obtained by substitution from such functions. Thus we have

$$\mathfrak{G}_{an}(\psi(a, a), \varphi(a, a, n)), \quad \text{q. e. d.}$$

(*d*) This proves Theorem 2 for functions of two variables. We do not need to deal with functions of one variable separately. A function $\psi(a)$ of one variable can be considered as a special case, $\psi(a, 0)$, of a function of two variables. Now assume that the newly defined function has more than two arguments. This case can be reduced to that of two variables. Let

$$\psi(a, b, \ldots, m, 0) = \sigma(a, b, \ldots, m),$$
$$\psi(a, b, \ldots, m, n + 1) = \mathfrak{b}_{a_1 \cdots m_1}(a, b, \ldots, m, n, \psi(a_1, \ldots, m_1, n)).^9$$

In $\mathfrak{b}_{a_1 \cdots m_1}(\ldots)$ functions of $a, b, \ldots, m$ occur in the arguments of the innermost ψ. Hence the ψ have the following form:

$$\psi(\alpha_1(a, b, \ldots, m, n), \ldots, \alpha_r(a, \ldots, m, n), n).$$

Here we can substitute the maximum of all these functions for all arguments of ψ with the exception of the last one. In the same way we can arrange that in the next ψ toward the outside all arguments with the exception of the last one become equal, and so on. We now consider the recursion

$$\psi(a, \ldots, a, 0) = \sigma(a, a, \ldots, a),$$
$$\psi(a, a, \ldots, a, b + 1) = \mathfrak{b}_{a_1 b_1 \cdots m_1}(a, a, \ldots, a, b, \psi(a_1, \ldots, m_1, b)).$$

This is now a recursion schema for a function $\psi'(a, b)$ of two variables. Hence $\mathfrak{G}_{an}(\psi(a, a, \ldots, a), \varphi(a, a, n))$ holds. The monotonic function associated with ψ can even be specified. The monotonicity with respect to $a, b, \ldots, m$ is already given by the fact that, instead of the function σ and those occurring in $\mathfrak{b}$, we always take the associated monotonic ones. Now we had found a $\psi'(a, b)$ such that $\psi(a, a, \ldots, a, n) \leqq \psi'(a, n)$. We have $\psi'(a, n) \leqq \psi''(a, n)$, where now ψ'' is monotonic with respect to n also. The monotonic function associated with $\psi(a, b, \ldots, m, n)$ now is $\psi''(\operatorname{Max}(a, b, \ldots, m), n)$. For

$$\psi(a, b, \ldots, m, n) \leqq \psi(\operatorname{Max}(a, b, \ldots, m), \operatorname{Max}(\ldots), \ldots, \operatorname{Max}(\ldots), n)$$
$$\leqq \psi''(\operatorname{Max}(a, b, \ldots, m), n).$$

9 [[In the German text "ψ" is missing on the right side of the last equation.]]

On mathematical logic

THORALF SKOLEM

(*1928*)

This is the text of a lecture delivered before the Norwegian Mathematical Association on 22 October 1928. Skolem briefly presents Boolean algebra and the first-order predicate calculus, and makes some remarks about the second-order predicate calculus. He presents a few rules for the first-order predicate calculus but then says: "I do not go into this more deeply, especially since I believe that it is possible to deal with deduction problems in another, more expedient way". We shall see in a moment what this other way is.

In the first-order predicate calculus Skolem considers an arbitrary closed well-formed formula Z in prenex form. He puts the formula into what we shall call its functional form for satisfiability: all quantifiers are dropped, and the occurrences of each variable bound by an existential quantifier are replaced by a functional term[a] containing the variables that in Z are bound by universal quantifiers preceding the existential quantifier. A formula U is thus obtained. Skolem then introduces constants of various levels. "0" is the constant of the 0th level. The constants of the nth level are the expressions obtained when in each functional term of U the variables are replaced by constants that include at least one constant of, but no constant beyond, the $(n-1)$th level. Thus, the functional terms of U, indefinitely compounded, generate from "0" the *lexicon*[b] of U; the subset of the lexicon of U consisting of all constants of levels $\leqq n$ will be called the *lexicon of level n*.

If W is U or any truth-functional part of U, by a *lexical instance* of W will be meant the formula obtained when all the variables of W are replaced by elements from the lexicon of U. Skolem calls an assignment of truth-values to lexical instances of atomic parts of U a *solution of the nth level* if and only if the assignment verifies the conjunction of all the lexical instances of U formed from the lexicon of level $n-1$ of U. Clearly, any solution of the nth level is an extension of a solution of the $(n-1)$th level.[c]

Either for some n there is no solution of the nth level or for every n there is a solution of the nth level. In the first case, Skolem claims, Z "contains a contradiction" ("enthält einen Widerspruch"). If to the present paper we add the evidence furnished by other papers of Skolem's,[d]

---

[a] We find it convenient to use here the expression "functional term", although it is not Skolem's. A functional term is a formula consisting of a function letter followed by variables, with the proper parentheses and commas.

[b] This term, not in Skolem, is borrowed from *Quine 1955a*, p. 141.

[c] Compare the notions introduced here by Skolem with Herbrand's property C: there is a solution of the nth level for prenex Z if and only if $\sim Z$ does not have property C of order n (see below, p. 544).

[d] In *1920* "ist...ein Widerspruch" (p. 6, lines 7u–6u; "is a contradiction", above, p. 256, line 7u) is opposed to "innerhalb eines gegebenen Denkbereich für gewisse Werte der

we should take "contains a contradiction" to mean that Z is not satisfiable, and Skolem indeed establishes in those papers[e] that, if Z is satisfiable, there is, for every n, a solution of the nth level. (In the present paper Skolem's argument rests upon an implicit use of the axiom of choice; see below, p. 518.) In his other papers Skolem also establishes the (deeper) converse, namely, if for every n there is a solution of the nth level, Z is satisfiable.[f] (What he actually proves is that Z is $\aleph_0$-satisfiable.) We now see what Skolem's other way is. It is a proof procedure for the first-order predicate calculus: given a formula Z in prenex form and a number n, we can effectively decide whether or not there is a solution of the nth level for Z; if for some n there is no solution of the nth level, Z is not satisfiable, hence $\sim Z$ is valid. This approach furnishes an alternative to the axiomatic approach to the first-order predicate calculus (that is, the approach that follows the lines of *Whitehead and Russell 1910* or *Hilbert and Ackermann 1928*). Note that this alternative provides proofs that are cut-free and have the subformula property. What Skolem has in fact established in these various papers is that his proof procedure is sound and complete, in the sense of Quine (*1955a*, p. 145)—except for one minor point: it has to be shown that a formula is satisfiable if and only if one of its prenex forms is satisfiable. Skolem does not explicitly discuss this question, but the transformation rules that in the present paper (p. 130; below, p. 516) he gives as examples of rules for the first-order predicate calculus are precisely the rules that allow us to pass from a formula to any of its prenex forms and conversely.[g]

The present paper contains an obscure passage (below, p. 519, lines 6–17) that deals with the second case mentioned above, namely, the case in which for every n there is a solution of the nth level. Skolem seems to be arguing—and arguing finitistically—that, if such is the case, the "theory" obtained when Z is added as an axiom to quantification theory is syntactically consistent. (Two passages in *Skolem 1929*[h] very similar to the one under discussion and somewhat less obscure lend weight to this reading.) Thus Skolem seems to be arguing for a proposition that is the main consequence of Herbrand's fundamental theorem (see *Herbrand 1930*, p. 120 (below, p. 561), and also footnote 103 below, p. 561). Moreover, in § 5 and § 6 of *1929* Skolem uses this proposition to study consistency properties of first-order theories in a way quite analogous to Herbrand's. (See Kalmár's discussion of what he calls the Skolem-Herbrand lemma in *Kalmár 1951*.)

In *1962*, p. 47, Skolem writes: "The

Relative erfüllt ist" (p. 7, lines 9–10; "is satisfied in a given domain for certain values of the relatives", above, p. 257, lines 6–7). In *1922* "innerhalb irgend eines Denkbereichs erfüllt ist" (p. 220, lines 5–6; "is satisfied in any domain at all", above, p. 293, line 13) is equated to "ist...widerspruchsfrei" (p. 222, lines 4u–3u; "is consistent", above, p. 295, line 5). In *1929* the assumption that Z "is contradictory" ("ist...widerspruchsvoll", p. 23, line 3 of § 4) is taken to be the negation of "the assumption that Z is satisfied in a domain $\mathfrak{B}$" ("die Annahme des Erfülltseins von Z in einem Bereich $\mathfrak{B}$", p. 23, line 6u). In these papers of Skolem's there is a constant ambiguity whether the expressions that are equated are to be taken as synonymous or as logically equivalent by a tacit argument.

[e] *1922*, p. 221, lines 10–9u (above, p. 294, lines 1–17); *1929*, p. 25, lines 4–21, p. 26, lines 8–16, and especially p. 27, lines 12–1u. The first three passages deal with a formula in Skolem normal form, the last with an arbitrary prenex formula. Note that in the last two passages no use is made of the axiom of choice.

[f] *1922*, p. 221, line 8u, to p. 222, line 20 (above, p. 294, lines 18–7u); *1929*, p. 25, line 22, to p. 27, line 11, and p. 27, line 13u, to p. 28, line 9. The first two passages deal with a formula in Skolem normal form, the last with an arbitrary prenex formula.

[g] Compare these rules with Herbrand's rules of passage (below, pp. 528–529).

[h] P. 15, line 6u, to p. 16, line 4, and p. 29, line 10 to bottom of page.

theorem of Löwenheim says that, if F is a well-formed formula of the first-order predicate calculus with certain predicate variables $A, B, C, \ldots$, either $\sim F$ is provable or F can be satisfied in the natural number series by suitable determination of $A, B, C, \ldots$ in that domain of individuals". If "provable" is understood as derivable from axioms by rules of inference in an axiomatic version of quantification theory, that result was established by Gödel (*1930a*). If "provable" refers to some informal version of quantification theory, some of Skolem's arguments, when taken together, provide a sketch of a proof for the result. Finally, if by "$\sim F$ is provable" we understand, in the sense of the proof procedure, that for some n there is no solution of the nth level for F, the result was firmly established by Skolem in *1922* for any F in Skolem normal form and in *1929* (p. 27, line 12, to p. 28, line 9) for any prenex F.[i] Compare also 6.2 of *Herbrand 1930* (below, p. 558).

Let us observe that in these 1922 and 1929 proofs Skolem is able to dispense with the use of the axiom of choice by suitably ordering, for every n, the solutions of the nth level. In *1929*, p. 25, lines 9u–6u, he notes that use could be made of the following proposition of set theory: If a set M of finite sets is such that every element of M is a subset of another element of M, then M contains as a subset an infinite sequence $m_1, m_2, \ldots$ such that, for every i, m_i is a subset of m_{i+1}. This proposition,[j] which is related to König's infinity lemma (*1926, 1927*), can be regarded as a version of the axiom of dependent choices (*Tarski 1948*, p. 96; see also *Bernays 1937–1954, 7*, p. 86). Skolem adds that, if the use of the axiom of choice is permitted, there is a much simpler proof of the Löwenheim theorem, namely the proof that he gave in *1920* and, in a somewhat different version, in *1929* (p. 24, lines 20–10u).[k] In Gödel's completeness proof (*1930a*), when the argument requires that from the existence of an infinite sequence of satisfying assignments (Skolem's "solutions") the existence of a satisfying assignment in an infinite domain be inferred, an appeal is made to "familiar arguments" ("bekannten Schlußweisen"; see below, p. 589). In a conversation in April 1964 Professor Gödel told the editor that

---

[i] After having pointed out that their formulation of the first-order predicate calculus is not syntactically complete, that is, that it contains well-formed formulas, even closed well-formed formulas, such that neither the formula nor its negation is provable, Hilbert and Ackermann (*1928*, p. 68) raised the question of the completeness of the calculus in another sense: "Whether the system of axioms is complete at least in the sense that all the logical formulas that are true in every domain of individuals can actually be derived in it is a question that has not yet been solved. We can only say, in a purely empirical way, that this system of axioms has always been sufficient in all applications". Professor Gödel has noted that at the time when these lines were written a substantial part of this problem had implicitly been solved already by Skolem in his *1922*. On 14 August 1964 he wrote to the editor: "As for Skolem, what he could justly claim, but apparently does not claim, is that, in his 1922 paper, he implicitly proved: 'Either A is provable or $\sim A$ is satisfiable' ('provable' taken in an informal sense). However, since he did not clearly formulate this result (nor, apparently had made it clear to himself), it seems to have remained completely unknown, as follows from the fact that Hilbert and Ackermann in 1928 do not mention it in connection with their completeness problem".

[j] Compare *Sierpiński 1958*, p. 129, lines 5u–3u.

[k] This proof requires the general axiom of choice: we have a set, of arbitrary cardinality, of sets of arbitrary cardinality, and simultaneous choices have to be performed. It is known that the axiom of dependent choices does not imply the general axiom of choice (for systems with urelements see *Mostowski 1948*; for more general systems the result can be established with the help of Cohen's techniques (*1963, 1963a, 1964*). König (*1936*, p. 82, footnote 1) observes that his proof of the infinity lemma makes use of the principle of choice (actually, the axiom of dependent choices). But he adds: "In most applications of the infinity lemma, however, this principle can be avoided".

these words refer to König's infinity lemma.[1]

It should be observed that Herbrand's fundamental theorem (*1929a, 1930, 1931*; see below, p. 554) can be taken to establish the completeness of an axiomatic formulation of quantification theory once Skolem's proof procedure is seen to be complete. For Skolem's result of 1929, partially translated into Herbrand's terms (see above p. 508, footnote c), reads: If for no n a prenex formula Z has property C of order n, then $\sim Z$ is satisfiable. And Herbrand's theorem states, in part, that, if for some n the formula Z has property C of order n, Z is provable (in an axiomatic system), or, by contraposition, if Z is not provable, for no n does Z have property C of order n. Hence, by transitivity, if Z is not provable, $\sim Z$ is satisfiable, or, if Z is valid, Z is provable—which for prenex Z is Gödel's 1930 completeness result. The impact of Herbrand's errors upon this argument is discussed by Dreben in Note H below (pp. 578–580).

In the present paper Skolem approaches the decision problem from the point of view of his proof procedure. For certain classes of formulas we can effectively decide whether or not for every n there is a solution of the nth level. Skolem first considers a formula whose prefix consists of a single universal quantifier followed by any number of existential quantifiers, and he shows that in that case we are able to conclude to the satisfiability or nonsatisfiability of the given formula. The subcase of the prefix $(x)(Ey)$ had already been solved by Bernays and Schönfinkel (*1928*). A paper of Ackermann's (*1928a*) containing a solution of the case studied by Skolem was published soon after Skolem's lecture. (Ackermann's paper was received for publication on 9 February 1928 and the fascicle of the periodical was made ready for the press on 28 December 1928.) Ackermann's result goes beyond Skolem's; he obtains an upper bound for the number of individuals of the finite domain in which the formula is satisfiable, if it is satisfiable at all. He also solves the more general case of the prefix $(Ex_1)\ldots(Ex_m)(y)(Ez_1)\ldots(Ez_n)$ and finds an upper bound for that case too. Ackermann's method is somewhat similar to Skolem's. (Later, in *1954*, pp. 72–74, Ackermann follows an entirely different approach.) In a review of Ackermann's paper Skolem (*1928b*) gives an argument that simplifies Ackermann's proofs and yields a smaller upper bound. Skolem expands these remarks in a subsequent paper (*1935*); there he also presents his 1928 proof in a more detailed and precise form, and he generalizes his result in a new direction, namely, to formulas of the form

$$(x)(Ey_1)\ldots(Ey_n)K(x, y_1, \ldots, y_n) \;\&\; (x_1)(x_2)(x_3)L(x_1, x_2, x_3),$$

where $L(x_1, x_2, x_3)$ contains only binary predicates $R_1, R_2, \ldots, R_k$ and is a conjunction stating that each R_i, $1 \leqq i \leqq k$, is symmetric and transitive. (For the significance of this result see *Skolem 1935a*.) For the case of the prefix $(Ex_1)\ldots(Ex_m)(y)(Ez_1)\ldots(Ez_n)$ Herbrand was able (*1931*, pp. 45–47; see also *1930*, pp. 118–119 (below, pp. 558–560)), through the use of his fundamental theorem, to simplify Ackermann's proof and obtain a smaller upper bound; he also extended the result to a formula consisting of the prefix $(Ex_1)\ldots(Ex_m)$ followed by a conjunction of prenex formulas of the form $(y^i)(Ex_1^i))\ldots(Ez_{n_i}^i)\Phi_i$, $1 \leqq i \leqq l$.

The present paper shows also that it is possible to decide the satisfiability of a prenex formula if the prefix is of the form

[1] A number of remarks about Löwenheim's and Skolem's arguments, in particular about their use of the axiom of choice, can be found in *Skolem 1938*. Let us note that in that paper Skolem states the Löwenheim theorem thus: If a first-order expression is satisfiable ("réalisable"), it is satisfiable in a denumerable domain (p. 25). Skolem also says that the results obtained by Löwenheim and him concern "a predicate logic based upon set theory" ("une logique du prédicat fondée sur la théorie des ensembles", p. 25).

$(x_1)\ldots(x_m)(Ey_1)\ldots(Ey_n)$ and all of the variables $x_1, \ldots, x_m$ occur in each of the atomic parts of the matrix. A few years later (*1935*) Skolem again presents a proof of this result and points out that it can easily be extended to the case in which either all of the variables $x_1, \ldots, x_m$ or at least one of the variables $y_1, \ldots, y_n$ occurs in each atomic formula of the matrix. Below, Skolem observes that, for the prefix in question, the satisfiability of a formula can be decided if at least $m + 1$ distinct variables occur in each atomic part of the matrix. At the end of the paper Skolem presents Langford's result that the theory of open dense order is decidable.

Professor Skolem pointed out a number of corrections that should be introduced into the German text:

Page 132, line 9u, there should be a bar over the second occurrence of "V";

Page 136, line 1u, there should be a bar over the second occurrence of "B";

Page 137, line 7u, the third occurrence of "y" should have the subscript "1", not "n";

Page 137, line 6u, the first occurrence of "y" should have the subscript "1", and "..." should be deleted before the first and third occurrences of "y";

Page 139, lines 8u and 7u, the three occurrences of "$+$" should be deleted and the resulting spaces closed up;

Page 139, line 4u, "einfacheren" should be replaced by "gegebenen";

Page 139, the last three lines should be replaced by: "dadurch zu zeigen, dass man für jedes Paar x_1, x_2 ein $y > x_1$ und x_2 wählt und allgemein $A(x, y) = 1$ wählt so oft $x < y$, dagegen $A(x, y) = 0$ so oft $x \geqq y$".

These corrections have been incorporated into the translation below.

The translation is by Stefan Bauer-Mengelberg and Dagfinn Føllesdal, and it is printed here with the kind permission of Professor Skolem and the editing secretary of *Nordisk Matematisk Tidskrift*.

B. Dreben and the editor

Logic, as is well known, was established as a science by Aristotle. Everyone knows the Aristotelian syllogism. During the entire Middle Ages Aristotle's syllogistic figures constituted the principal content of logic. Kant is said to have remarked once that logic was the only science that had made no progress at all since antiquity. Perhaps this was true at the time, but today it is no longer so.

For recent times have seen the development of the calculus of logic, as it is called, or mathematical logic, a theory that has gone far beyond Aristotelian logic. It has been developed by mathematicians; professional philosophers have taken very little interest in it, presumably because they found it too mathematical. On the other hand, most mathematicians, too, have taken very little interest in it, because they found it too philosophical.

I do not intend to give a precise account of the history of this discipline here. I mention the names of Leibniz, Lambert, Boole, Peirce, Macfarlane, and Frege. The German mathematician Schröder wrote a large work, *Algebra der Logik* (three volumes: *1890*, *1891*, *1895*), which is very valuable. A different tendency was initiated by the Italian mathematician Peano, as well as by Frege, and carefully developed in all details by the Englishmen Russell and Whitehead. They wrote a large work, *Principia mathematica* (*1910*, *1912*, *1913*); in it they attempt to develop all of mathematics as a part of mathematical logic. Of the more recent writers I would like to mention Löwenheim, Behmann, Schönfinkel, Chwistek, Ramsey, and Langford. Finally, some of the investigations in the foundations of mathematics by Hilbert and his collaborators, Bernays, von Neumann, and Ackermann, belong here.

First I would like to discuss one of the simplest parts of the calculus of logic, the identical calculus [[identischen Kalkul]],[1] in some detail. It can be developed as a calculus of classes or regions (in a plane, say). We consider a totality of objects; this totality is called the universal class and is denoted by 1. On the other hand there is to be a null class, 0; it contains no element at all. If a and b are two arbitrary classes, it may happen that every object occurring in a also occurs in b; for that relation I write $a \leqq b$. If both $a \leqq b$ and $b \leqq a$, I write $a = b$. From the classes a and b we can separate the greatest common part; this class will be denoted by ab. Likewise we can combine the objects occurring in a or b into a class, which will be denoted by $a + b$. Finally, the objects not occurring in a form a class, which we shall denote by $\bar{a}$. Then we can develop an algebra in which $\leqq$ occurs as the fundamental relation and we have the three operations mentioned. Among others, the following laws hold:

$$0 \leqq a, \quad a \leqq 1, \quad a \leqq a, \quad a\bar{a} = 0, \quad a + \bar{a} = 1, \quad \overline{ab} = \bar{a} + \bar{b}, \quad \overline{a + b} = \bar{a}\bar{b},$$
$$a(b + c) = ab + ac, \quad a + bc = (a + b)(a + c),$$

and, besides, the rules of inference

$$(a \leqq b)(b \leqq c) \rightarrow (a \leqq c), \quad (c \leqq ab) \rightleftarrows (c \leqq a)(c \leqq b),$$
$$(a + b \leqq c) \rightleftarrows (a \leqq c)(b \leqq c).$$

We must remark, however, that the operations of identical multiplication and identical addition can be extended to infinitely many operands, too. By the identical product $\underset{\kappa}{\Pi} a_\kappa$ of arbitrarily many classes a_κ, where κ ranges over a certain supply of values, we understand the class of all elements that are common to all a_κ. By the identical sum $\underset{\kappa}{\Sigma} a_\kappa$ we understand the class of all objects that occur in at least one of the a_κ.

I shall not dwell on this calculus for long, but I would like to show, by way of an example, how equations containing unknowns can be solved in it. If an equation of the form

$$ax + b\bar{x} = 0$$

is given (every equation can be brought into the form $A = 0$, and every system of simultaneous equations can be written as a single equation), we find that $ab = 0$ is the necessary and sufficient condition for the existence of a solution and that the general solution is given by

$$x = \bar{a}u + b\bar{u},$$

where u stands for an arbitrary class.

Proof. $ab = ab(x + \bar{x}) = abx + ab\bar{x} \leqq ax + b\bar{x} = 0$. If $ax + b\bar{x} = 0$, then $x = (a + \bar{a})x = ax + \bar{a}x = \bar{a}x + b\bar{x}$, since both ax and $b\bar{x}$ are equal to 0. Hence every solution has the form $\bar{a}u + b\bar{u}$. If $x = \bar{a}u + b\bar{u}$, it follows that, when $ab = 0$, $ax = a\bar{a}u + ab\bar{u} = 0$.

[1] [[Schroder introduced (*1890*, p. 157) the expression "identical calculus" ("identischer Kalkul") for his version of Boolean algebra. He speaks of identical sum and identical product when we would say "logical sum" and "logical product". He avoided the word "logical" here because he regards the identical calculus as "an auxiliary discipline that precedes logic proper or runs parallel...and is of purely mathematical nature".]]

But we can also solve equations with an entirely arbitrary number of unknowns. For example, let the equation

$$\underset{\lambda}{\Sigma} a_\lambda x_\lambda + \underset{\mu,\nu}{\Sigma} a_{\mu,\nu}\bar{x}_\mu\bar{x}_\nu = 0 \quad (a_{\mu,\nu} = a_{\nu,\mu})$$

be given, where the subscripts range over a given supply of values; assume that the classes a are given but that the x are unknowns. Here we find that

$$\underset{\mu,\nu}{\Sigma} a_\mu a_\nu a_{\mu,\nu} = 0$$

is the necessary and sufficient condition for the existence of a solution and that

$$x_\kappa = \bar{a}_\kappa(u_\kappa + \underset{\lambda}{\Sigma} a_{\kappa,\lambda}(a_\lambda + \bar{u}_\lambda))$$

is the general solution.

Another example is

$$\underset{\mu,\nu}{\Sigma'} a_{\mu,\nu}(x_\mu x_\nu + \bar{x}_\mu\bar{x}_\nu) = 0 \quad (a_{\mu,\nu} = a_{\nu,\mu}),$$

where the prime on Σ means that μ and ν shall always take different values. The necessary and sufficient condition for the existence of a solution here is

$$\Sigma P = 0,$$

where P ranges over all possible finite products of the coefficients a in which the pairs of subscripts μ, ν form a cycle with an odd number of terms. Hence this shows that in such problems we are led to peculiar combinatorial questions.

But I shall not go into problems of this kind in more detail; rather, I shall proceed to discuss the general notions and operations of the more recent mathematical logic.

The fundamental notions of mathematical logic are the following:

(1) Propositions; true and false;

(2) Propositional functions of variables; properties (classes) and relations ⟦Beziehungen⟧ (relations ⟦Relationen⟧ and correspondences).[2]

The logical operations are

(1) The elementary ones: conjunction, disjunction, and negation;

(2) The higher ones: "all" and "there exists".

We assume that, if a sentence has meaning,[3] it is either true or false. If a proposition expresses something about an object x—I then write it $A(x)$—it may happen that it has meaning for all values of x within a certain domain. Then we say that $A(x)$ is a propositional function if x is a variable that can take arbitrary values in that domain. If a specific value is substituted for x, we obtain a proposition, which will be true or false, depending on the circumstances. In a similar way we can form propositional functions of two, three, or more variables: $A(x, y)$, $A(x, y, z)$, ... If a specific value is substituted for y in $A(x, y)$, we obtain a propositional function of x alone; if specific values are substituted for x as well as y, we obtain a proposition. The propositional

[2] ⟦Skolem's use of the two words "Beziehungen" and "Relationen" perhaps corresponds to his distinction between *properties* and *classes*, that is, between the intensional notion and the extensional notion. However, a few lines below he writes: "'Beziehungen' oder 'Relationen'".⟧

[3] ⟦The German text says: "Falls eine Aussage einen Sinn hat", and here the translators had to depart from the standard translation "proposition" for "Aussage".⟧

function "x is a prime number" is a function $P(x)$ of x; for example, $P(3)$ is true but $P(6)$ false. The propositional function "x and y are relatively prime" has two arguments; for $x = 8$ and $y = 15$, for example, it yields a true proposition, and for $x = 6$ and $y = 21$ a false one.

A propositional function like $A(x)$ represents what is called a "property"; we say that the objects x for which $A(x)$ becomes a true proposition form a "class". The propositional functions of several variables represent "relations" [["Beziehungen" oder "Relationen"]].

The three elementary operations are those that in ordinary language are rendered by the words "and", "or", and "not". In Schröder's notation the propositions "A as well as B", "either A or B", and "not A", where A and B are given propositions, are written AB, $A + B$, and $\overline{A}$, respectively.

Hence the expression $\overline{A} + B$ means "either not A or B". That is clearly the same as "if A holds, so does B", or "from A, B follows". This is what is called implication, and it is therefore reducible to the first three operations mentioned. Disjunction, moreover, can be reduced to conjunction and negation, since $A + B$ means the same as $\overline{\overline{A}\,\overline{B}}$.

These elementary operations are clearly applicable to propositions as well as to propositional functions. The two higher operations, which are expressed by the words "all" and "there exists", are, however, applicable only to propositional functions. They have the effect that for one or more variables substitutions become impossible, the variables being transformed from "real", or "free", variables into "apparent", or "bound", ones.

If $A(x)$ is a propositional function, we can form the propositions "$A(x)$ is true for all x" and "there exists an x such that $A(x)$ is true"; following Schröder, we write these as $\underset{x}{\Pi} A(x)$ and $\underset{x}{\Sigma} A(x)$, respectively. Clearly, no substitution can be made for x here; hence x has become an apparent variable, just like a variable of integration. From a propositional function $A(x, y)$ we can first form the propositional functions $\underset{y}{\Pi} A(x, y)$, $\underset{y}{\Sigma} A(x, y)$, $\underset{x}{\Pi} A(x, y)$, and $\underset{x}{\Sigma} A(x, y)$; the first two are functions of x (but not of y), the latter two are functions of y. From these, then, the propositions $\underset{x}{\Pi}\underset{y}{\Pi} A(x, y)$, $\underset{x}{\Sigma}\underset{y}{\Pi} A(x, y)$, and so forth can in turn be formed. We readily see that the proposition $\underset{x}{\Sigma}\underset{y}{\Pi} A(x, y)$ means that there exists an x such that, when this x is held constant, $A(x, y)$ is true for all y, while the proposition $\underset{y}{\Pi}\underset{x}{\Sigma} A(x, y)$ means that for every y there exists an x, which in general depends on y, such that $A(x, y)$ holds. Of course, we can form propositions such as $\underset{x}{\Pi}(A(x)\underset{y}{\Sigma} B(x, y))$, and so forth.

Here are some simple examples from arithmetic:

$$\underset{x}{\Pi}\underset{y}{\Sigma}(x < y), \quad \underset{x}{\Sigma}\underset{y}{\Pi}(x \leqq y), \quad \underset{x}{\Pi}\underset{y}{\Sigma}(x < y)\underset{z}{\Pi}((z \leqq x) + (y \leqq z)),$$

where the variables range over the sequence of natural numbers. The first proposition means that for every number there exists a greater one, the second that there exists a least natural number, and the third that among all the numbers that are greater than x there exists a least.

Now we can not only give a precise formulation to mathematical propositions but also represent mathematical proofs as transformations of such logical expressions according to certain rules. For example, we have the rules

$$\underset{x}{\Pi} A B(x) \rightleftarrows A \underset{x}{\Pi} B(x), \quad \underset{x}{\Sigma}(A + B(x)) \rightleftarrows A + \underset{x}{\Sigma} B(x),$$

$$\underset{x}{\Pi}(A + B(x)) \rightleftarrows A + \underset{x}{\Pi} B(x), \quad \underset{x}{\Sigma} A B(x) \rightleftarrows A \underset{x}{\Sigma} B(x),$$

$$\overline{\underset{x}{\Pi} A(x)} \rightleftarrows \underset{x}{\Sigma} \overline{A(x)}, \quad \overline{\underset{x}{\Sigma} A(x)} \rightleftarrows \underset{x}{\Pi} \overline{A(x)}.$$

But I do not go into this more deeply, especially since I believe that it is possible to deal with the deduction problems [[Deduktionsprobleme]] in another, more expedient way, to which I shall return in a moment.

If certain propositional functions are given, together with a range for the values of the variables, we are now in a position to form from them further propositional functions and propositions by means of the five logical operations. The values of the variables may be called "individuals". Propositional functions and propositions that have individuals as arguments and in which "all" and "there exists" are applied only to variable *individuals* are often called *first-order propositional functions* [[*Zählaussagenfunktionen*]] and *first-order propositions* [[*Zählaussagen*]], respectively.[4] It must be noted, however, that the logical construction of propositions does not end here. For we can have a variable propositional function in a propositional expression, and then "all" and "there exists" can be applied to it, too. So, for example,

$$\underset{x}{\Pi}(\overline{U(x)} + V(x)),$$

where U and V are variable propositional functions, is a proposition about U and V; that is, the expression is a propositional function of a higher logical kind, having first-order propositional functions as arguments. But we can also obtain functions of a higher kind whose arguments are still individuals. A simple example is

$$\underset{U}{\Pi}(U(x)U(y) + \overline{U(x)}\,\overline{U(y)});$$

this expression represents a propositional function whose arguments are x and y, since in it the universal quantifier governs the variable propositional function U. In this way the construction of propositional functions becomes quite complicated; and it cannot be denied that questions arise here that are very difficult to answer. If "all" and "there exists" are applied to variable propositional functions, the question arises: what is the totality of all propositional functions? It seems to me that only two conceptions are scientifically tenable here; these I now indicate.

1. We can start from certain actually given, that is, listed, primitive propositional functions. Other functions are formed from these, first, by means of the five operations in such a way that "all" and "there exists" are applied only to individuals. This yields the totality of propositional functions of the first level [[Stufe]] (the first-order functions [[Zählfunktionen]]). After this totality has been formed, it is meaningful, on

[4] It is, of course, understood that in all these expressions the logical operations are used only a finite number of times.

the one hand, to form functions whose arguments are functions of the first level; on the other hand, we can form functions that have individuals as arguments but that result from the application of "all" and "there exists" to functions of the first level. Thus we obtain functions of the second level, but of two essentially different kinds. Whenever the constructible functions of a certain "level", or "type", are collected into a totality, the formation of new functions is made possible.

2. We can attempt to introduce the notion of propositional function (which, after all, corresponds rather precisely to the notion "set") by means of an axiomatization, just as we axiomatize set theory. The axioms will then become first-order propositions, since the objects of the axiomatization (here the "propositional functions", in set theory the "sets") will assume the role of individuals. The relations between arguments and functions will then appear in the axiomatization as primitive functions, just as the relation ε (element of) appears as primitive function in axiomatic set theory.

I here permit myself a remark about the relation between the fundamental notions of logic and those of arithmetic. No matter whether we introduce the notion of propositional function in the first or the second way, we are confronted with the idea of the integer. For, even when the notion of propositional function is introduced axiomatically, we shall have to consider (for instance, in investigations concerning consistency) what we can derive by using the axioms an arbitrary finite number of times. On the other hand, it is not possible to characterize the number sequence logically without the notion of propositional function. For such a characterization must be equivalent to the principle of mathematical induction, and this reads as follows: If a propositional function $A(x)$ holds for $x = 1$ and if $A(x + 1)$ is true whenever $A(x)$ is true, then $A(x)$ is true for every x. In signs, it takes the form

$$\Pi_U(\overline{U(1)} + \Sigma_x U(x)\overline{U(x+1)} + \Pi_y U(y)).$$

This proposition clearly involves the totality of propositional functions. Therefore, the attempt to base the notions of logic upon those of arithmetic, or vice versa, seems to me to be mistaken. The foundations for both must be laid simultaneously and in an interrelated way.

I shall not go into these difficult questions more deeply; instead, I shall indicate how the deduction problems for first-order propositions can be reduced to a problem of combinatorial arithmetic ⟦eine arithmetisch-kombinatorische Frage⟧. If U and V are first-order propositions and if we pose the question whether V follows from U, this is equivalent to asking whether $U\overline{V}$ is a contradiction or not. It is therefore clear that everything depends upon our being able to decide whether a given first-order proposition is contradictory or not.

Let a first-order proposition $Z(A, B, C, \ldots)$ be given, where $A, B, C, \ldots$ are the given propositional functions occurring in it; I call these the primitive functions. I can assume that Z has the form

$$\Pi_{x_1}\Pi_{x_2}\ldots\Sigma_{y_1}\Sigma_{y_2}\ldots\Pi_{z_1}\Pi_{z_2}\ldots U(x_1, x_2, \ldots, y_1, y_2, \ldots, z_1, z_2, \ldots).$$

This axiom obviously means that, if $x_1, x_2, \ldots, z_1, z_2, \ldots$ are arbitrarily given, it is possible to introduce $y_1, y_2, \ldots, u_1, u_2, \ldots$, where the y depend only on the x, the u

only on the x and the z, and so on, and to determine the truth values of the functions $A, B, C, \ldots$ for these arguments in such a way that U turns out to be true. Since it does not matter what notation we use for the symbols, we can introduce the following symbolism, which is probably more advantageous for almost all investigations. Instead of $y_1, y_2, \ldots$, I write $f_1(x_1, x_2, \ldots), f_2(x_1, x_2, \ldots), \ldots$; instead of $u_1, u_2, \ldots$, I write $g_1(x_1, x_2, \ldots, z_1, z_2, \ldots), g_2(x_1, x_2, \ldots, z_1, z_2, \ldots), \ldots$; and so on. The given first-order proposition then means that

$$U(x_1, x_2, \ldots, f_1(x_1, x_2, \ldots), f_2(x_1, x_2, \ldots), \ldots, z_1, z_2, \ldots,$$
$$g_1(x_1, \ldots, z_1, \ldots), g_2(x_1, \ldots, z_1, \ldots), \ldots)$$

is true for arbitrary values of $x_1, x_2, \ldots, z_1, z_2, \ldots$ within a certain domain.

If we now wish to investigate whether this axiom is satisfiable or not, we can proceed as follows. Let us form, starting from the symbol 0, the symbols $f_1(0, 0, \ldots)$, $f_2(0, 0, \ldots), \ldots, g_1(0, 0, \ldots), g_2(0, 0, \ldots), \ldots$, and so on. 0 will be said to be the individual-symbol of the 0th level, while the symbols $f_1(0, 0, \ldots)$ and so on will be said to be of the 1st level. Once the individual-symbols are defined up to the nth level, the symbols of the $(n + 1)$th level shall be those that result from the insertion of symbols up to the nth level as arguments in the "functions" $f_1, f_2, \ldots, g_1, g_2, \ldots$ and that do not already occur among the symbols of the 0th to the nth level. The sequences $x_1, x_2, \ldots, z_1, z_2, \ldots$ formed by the symbols up to the nth level I call argument sequences of the nth level. Now we must try to assign values to the functions $A, B, C, \ldots$ in such a way that, for each argument sequence in turn, U comes out true. To translate everything into ordinary mathematical language I denote the propositional values "false" and "true" by 0 and 1, respectively. If I now denote the truth value of A by wA, the following relations hold:

$$w(AB) = \min(wA, wB), \quad w(A + B) = \max(wA, wB), \quad \text{and} \quad w\bar{A} = 1 - wA.$$

We thus come to the following purely arithmetic problem.

$A, B, C, \ldots$ are functions of a finite number of arguments; they can take only the values 0 and 1. Further U is a function constructed from $A, B, C, \ldots$ (these being written with certain arguments) by means of the operations min, max, and $1 -$; the arguments are $x_1, x_2, \ldots, z_1, z_2, \ldots, f_1(x_1, x_2, \ldots), \ldots, g_1(x_1, x_2, \ldots, z_1, z_2, \ldots), \ldots$. As all the argument sequences mentioned above come to be substituted successively for $x_1, x_2, \ldots, z_1, z_2, \ldots$, the point is to assign values to $A, B, C, \ldots$ in such a way that U takes the value 1 for each argument sequence.

First, we put 0 for $x_1, x_2, \ldots, z_1, z_2, \ldots$. Perhaps it is already impossible at this point to assign values to $A, B, C, \ldots$ in such a way that $U = 1$. In that case the first-order proposition is not satisfiable; there is a contradiction. Otherwise there exist some choices of values for $A, B, C, \ldots$ such that U will have the value 1.[5] These possibilities I call solutions of the 1st level. Next we have to insert all argument sequences of the 1st level. Perhaps it is impossible at this point to choose values for $A, B, C, \ldots$ in such a way—the values for the different argument sequences must, of course, be chosen in agreement with one another whenever these sequences have symbols in common—that U will always $= 1$; if so, we have a contradiction. Otherwise

[5] It is convenient to write U as an alternation of terms constructed by means of negation and conjunction (Boolean expansion).

there must again be some possibilities available; I call them solutions of the 2nd level. Every solution of the 2nd level is the continuation of a solution of the 1st level, in the sense that it contains such a solution within itself. In this way we continue indefinitely. The real question now is whether there are solutions of an arbitrarily high level or whether for a certain n there exists no solution of the nth level. In the latter case the given first-order proposition contains a contradiction. In the former case, on the other hand, it is consistent ⟦widerspruchslos⟧. The following will make this clear. Every consequence of the axiom results from repeated and combined uses of it. Every theorem derived can therefore be formulated as a proposition formed by means of the functions $A, B, C, \ldots$, and in these functions there will occur, on the one hand, indeterminate symbols $a, b, c, \ldots$ and, on the other, further symbols that have been obtained from these by possibly repeated substitutions in the functional expressions $f_1, f_2, \ldots, g_1, g_2, \ldots$. Every such proposition must, however, retain its validity when $a, b, c, \ldots$ all are replaced by 0. Thus, if a contradiction is derivable, a contradiction must be provable in which there occur both 0 and the symbols obtained from 0 by substitution in $f_1, f_2, \ldots, g_1, g_2, \ldots$ up to, say, the nth level. Hence there cannot then exist any solution of the nth level.

To be sure, this procedure is infinite; but there are some cases in which it is possible to make the procedure finite. For example, let a first-order proposition of the form

$$\Pi_x \Sigma_{y_1} \ldots \Sigma_{y_n} U(x, y_1, \ldots, y_n)$$

be given. Here, instead of $f_1(0), f_2(0), \ldots, f_n(0)$, I simply write $1, 2, \ldots, n$; instead of $f_1(1), \ldots, f_n(1)$, I write $n + 1, \ldots, 2n$; instead of $f_1(2), \ldots, f_n(2)$, I write $2n + 1, \ldots, 3n$; and so on, until, instead of $f_1(n), \ldots, f_n(n)$, I write $n^2 + 1, \ldots, n^2 + n$. First, let 0 be substituted for x and therefore $1, 2, \ldots, n$ for $y_1, y_2, \ldots, y_n$, respectively; the values that can then be chosen for $A, B, C, \ldots$ so as to make $U = 1$ will, when taken together, yield certain alternatives $\mathfrak{A}_1(0, 1, \ldots, n), \ldots, \mathfrak{A}_N(0, 1, \ldots, n)$. When at the next step the values $1, 2, \ldots, n$ are substituted for x, the only possibilities that we then have are those that result whenever an alternative $\mathfrak{A}_i(0, 1, \ldots, n)$ is compatible, for each r in the sequence $1, 2, \ldots, n$, with at least one of the alternatives $\mathfrak{A}_j(r, rn + 1, \ldots, (r + 1)n)$. Let $\mathfrak{A}'_i(0, 1, \ldots, n)$ be those of the alternatives $\mathfrak{A}_i(0, 1, \ldots, n)$ for which this is the case. If the continuation to the next level is possible, the only alternatives $\mathfrak{A}_j(r, rn + 1, \ldots, (r + 1)n)$ that have to be considered are the $\mathfrak{A}'_j(r, rn + 1, \ldots, (r + 1)n)$. Now again let $\mathfrak{A}''_i(0, 1, \ldots, n)$ be those, from among the alternatives $\mathfrak{A}'_i(0, 1, \ldots, n)$, that are compatible, for $r = 1, 2, \ldots, n$, with at least one of the alternatives $\mathfrak{A}'_j(r, rn + 1, \ldots, (r + 1)n)$. Then only the $\mathfrak{A}''_j(r, rn + 1, \ldots, (r + 1)n)$ will have to be considered, and let $\mathfrak{A}'''_i(0, 1, \ldots, n)$ be those, from among the alternatives $\mathfrak{A}''_i$, that are compatible, for $r = 1, 2, \ldots, n$, with at least one of the alternatives $\mathfrak{A}''_j(r, rn + 1, \ldots, (r + 1)n)$. And so on. Since only a finite number of alternatives ⟦$\mathfrak{A}_i(0, 1, \ldots, n)$ with $i = 1, 2, \ldots, N$⟧ is available, this process must actually terminate; that is, after a finite number of steps we arrive at a number s such that the alternatives $\mathfrak{A}_i^{(s+1)}$ are the same as the $\mathfrak{A}_i^{(s)}$. It may be that at this point no alternative is available any more; in that case the given first-order proposition is contradictory. But, if there is at least one alternative $\mathfrak{A}_i^{(s)}$, the first-order proposition is consistent. For we recognize that it is then possible, for each new argument sequence of the individual variables $x, y_1, \ldots, y_n$, to form a corresponding alternative $\mathfrak{A}_h^{(s)}(x, y_1, \ldots, y_n)$ that is consistently compatible with the alternatives $\mathfrak{A}^{(s)}$ set up for the earlier argument sequences. That it is so simple here is due to the fact that an argument sequence R_n, of the nth level, that does not already occur at the $(n-1)$th level has an element in common with only a single argument sequence R_{n-1}, of the $(n-1)$th level, if we disregard argument sequences of levels above the nth. Therefore the determination of the functions $A, B, C, \ldots$ for R_n is independent of the determination of these functions for all earlier argument sequences with the sole exception of R_{n-1}.

Example 1. Let us investigate whether or not the first-order proposition

$$\underset{y\ x}{\Pi\Sigma}(A(x, y) + A(x, x)\overline{A(y, y)})$$

is contradictory. Here we can put 1 for y if 0 is put for x, and in general $n + 1$ for y whenever n is put for x.

In the arithmetic formulation the problem is: Investigate whether it is possible to determine the function $A(x, y)$, whose values shall be restricted to 0 and 1, in such a way that for all n, $n = 0, 1, 2, \ldots$,

$$\max(A(n, n+1), \min(A(n, n), 1 - A(n+1, n+1))) = 1.$$

The alternatives $\mathfrak{A}_i(0, 1)$ here are these two:

$$\mathfrak{A}_1(0, 1)\colon A(0, 1) = 1; \quad \mathfrak{A}_2(0, 1)\colon A(0, 0) = 1, A(1, 1) = 0.$$

Hence

$$\mathfrak{A}_1(1, 2)\colon A(1, 2) = 1; \quad \mathfrak{A}_2(1, 2)\colon A(1, 1) = 1, A(2, 2) = 0.$$

Thus $\mathfrak{A}_1(0, 1)$ is here compatible with both $\mathfrak{A}_1(1, 2)$ and $\mathfrak{A}_2(1, 2)$, while $\mathfrak{A}_2(0, 1)$ is compatible only with $\mathfrak{A}_1(1, 2)$. But, since both $\mathfrak{A}_i(0, 1)$ can be continued and at the same time both $\mathfrak{A}_i(1, 2)$ can occur as continuations, the $\mathfrak{A}_i'$ coincide with the $\mathfrak{A}_i$ already at this point. Hence the given proposition is consistent. Indeed, we already obtain a solution of the arithmetic problem if we set $A(n, n + 1) = 1$ for all n, no matter what values the function has otherwise.

Example 2. Let us examine whether or not the first-order proposition

$$\underset{x\ y}{\Pi\Sigma}(A(x, y)B(x)\overline{B(y)} + A(x, x)\overline{A(y, y)}\,\overline{B(y)})$$

is consistent. Here we obtain

$$\mathfrak{A}_1(0, 1)\colon A(0, 1) = 1, B(0) = 1, B(1) = 0,$$

$$\mathfrak{A}_2(0, 1)\colon A(0, 0) = 1, A(1, 1) = 0, B(1) = 0$$

as alternatives of the first level. Hence

$$\mathfrak{A}_1(1, 2)\colon A(1, 2) = 1, B(1) = 1, B(2) = 0,$$

$$\mathfrak{A}_2(1, 2)\colon A(1, 1) = 1, A(2, 2) = 0, B(2) = 0.$$

Here $\mathfrak{A}_1(0, 1)$ is compatible with $\mathfrak{A}_2(1, 2)$ but not with $\mathfrak{A}_1(1, 2)$; $\mathfrak{A}_2(0, 1)$ cannot be continued at all. Thus there is only one $\mathfrak{A}_1'(0, 1)$, namely $\mathfrak{A}_1(0, 1)$; but there does not occur any $\mathfrak{A}_i''$ at all. Hence the proposition contains a contradiction.

There are some other propositions in $\Pi\Sigma$-form that can be treated in a precisely analogous way.[6] Let

$$\underset{x_1}{\Pi}\ldots\underset{x_m}{\Pi}\ \underset{y_1}{\Sigma}\ldots\underset{y_n}{\Sigma} U(x_1,\ldots,x_m,y_1,\ldots,y_n)$$

be such a first-order proposition Z. It can be written in such a way that we need take only *distinct* elements x into consideration. That is immediately the case if the truth of U for two identical x always follows from the truth of U for distinct x; otherwise we can transform the proposition. Let $x_1,\ldots,x_m$ be m distinct individuals; let $\xi_1^{(t)},\ldots,\xi_m^{(t)}$, $t = 1, 2,\ldots, m^m$, be the different permutations with repetitions of these individuals taken m at a time. According to the proposition Z there then exists, for each sequence $\xi_1^{(t)},\ldots,\xi_m^{(t)}$, a sequence of individuals $y_1^{(t)},\ldots,y_n^{(t)}$ such that $U(\xi_1^{(t)},\ldots,\xi_m^{(t)},y_1^{(t)},\ldots,y_n^{(t)})$ is true. Now let

$$V(x_1,\ldots,x_m,y_1^{(1)},\ldots,y_n^{(1)},\ldots,y_1^{(m^m)},\ldots,y_n^{(m^m)})$$
$$= U(\xi_1^{(1)},\ldots,\xi_m^{(1)},y_1^{(1)},\ldots,y_n^{(1)})U(\xi_1^{(2)},\ldots,\xi_m^{(2)},y_1^{(2)},\ldots,y_n^{(2)})\ldots.$$

Then Z can obviously be written thus:

$$\underset{x_1,\ldots,x_m}{\Pi}\ \underset{y_1^{(1)}}{\Sigma}\ldots\underset{y_n^{(m^m)}}{\Sigma} V(x_1,\ldots,x_m,y_1^{(1)},\ldots,y_n^{(m^m)}),$$

where now only the various *combinations* of individuals taken m at a time from the domain in question are to be substituted for $x_1,\ldots,x_m$. Hence we need only consider propositions of the form just mentioned, for example,

$$\underset{x_1,\ldots,x_m}{\Pi}\ \underset{y_1}{\Sigma}\ldots\underset{y_n}{\Sigma} U(x_1,\ldots,x_m,y_1,\ldots,y_n).$$

In order to investigate satisfiability we can first substitute the symbols $1, 2,\ldots, m$ for $x_1,\ldots,x_m$ and the numbers $m+1,\ldots,m+n$ for $y_1,\ldots,y_n$; this yields the argument sequence $1, 2,\ldots, m+n$ of the 1st level. Next, let all the combinations of the numbers $1, 2,\ldots, m+n$ taken m at a time that differ from the combination $1, 2,\ldots, m$ be formed and ordered (say lexicographically). To the first combination let us assign the numbers $m+n+1,\ldots,m+2n$ as y-values; to the second let us assign as y-values the numbers $m+2n+1,\ldots,m+3n$; and so on. Thus we obtain the argument sequences of the 2nd level. Then let us form all new combinations of all the numbers, taken m at a time, that we have considered so far (that is, all the combinations that have not yet occurred as values for the x); and so on.

I now make the assumption that all the primitive functions occurring in our first-order proposition possess at least m distinct argument symbols; then it is possible to decide the question of satisfiability by a finite procedure. This is due to the fact that every argument sequence of the νth level has, if we disregard argument sequences of higher levels, an argument system $a_1,\ldots,a_m$ in common with only a single argument sequence of the $(\nu-1)$th level.

Let $\mathfrak{A}_i(a_1,\ldots,a_m,b_1,\ldots,b_n)$ be the alternatives that are possible for the argument sequence $a_1,\ldots,a_m,b_1,\ldots,b_n$, where all a and b are distinct symbols. Further, let the

[6] I proved earlier (*1920*) that the problem whether or not a first-order proposition Z is satisfiable can be reduced to the same question for a first-order proposition Z' in $\Pi\Sigma$-form.

sequence $c_1, \ldots, c_m$ be any one of the combinations, of the symbols $a_1, \ldots, a_m, b_1, \ldots, b_n$ taken m at a time, that are different from the combination $a_1, \ldots, a_m$; with each such combination we associate a sequence $d_1, \ldots, d_n$ of n distinct symbols in such a way that the sequences $d_1, \ldots, d_n$ and $d'_1, \ldots, d'_n$ that correspond to two different combinations $c_1, \ldots, c_m$ are *completely* distinct and that all d are distinct from all a and b. Then we have to select from the set of alternatives $\mathfrak{A}_i(a_1, \ldots, a_m, b_1, \ldots, b_n)$ those $\mathfrak{A}'_i(a_1, \ldots, a_m, b_1, \ldots, b_n)$ that, for each combination $c_1, \ldots, c_m$, are compatible with at least one of the alternatives $\mathfrak{A}_j(c_1, \ldots, c_m, d_1, \ldots, d_n)$; to select from these alternatives $\mathfrak{A}'_i$ again the $\mathfrak{A}''_i(a_1, \ldots, a_m, b_1, \ldots, b_n)$ that, for each combination $c_1, \ldots, c_m$, are compatible with at least one of the alternatives $\mathfrak{A}'_j(c_1, \ldots, c_m, d_1, \ldots, d_n)$; and so on. Just as before, we arrive at a number s such that the alternatives $\mathfrak{A}_i^{(s+1)}$ coincide with the $\mathfrak{A}_i^{(s)}$. If at this point there exist alternatives $\mathfrak{A}_i^{(s)}$, the given proposition is consistent; otherwise it represents a contradiction.

I give a simple example. Assume that in the proposition

$$\prod_{x_1} \prod_{x_2} \sum_{y} U(x_1, x_2, y)$$

$U(x_1, x_2, y)$ has the following form:

$$A(x_1, x_2)A(x_1, y)\overline{A(x_2, y)} + A(x_2, x_1)A(x_2, y)\overline{A(x_1, y)} +$$
$$A(x_1, y)\overline{A(y, x_2)} + A(x_2, y)\overline{A(y, x_1)} + A(x_1, y)\overline{A(y, x_1)} + A(x_2, y)\overline{A(y, x_2)}.$$

Since $U(x_1, x_2, y)$ is symmetric with respect to x_1 and x_2, and $U(x_1, x_1, y)U(x_2, x_2, y)$ is ⟦logically⟧ contained in $U(x_1, x_2, y)$, the given proposition can at once be written in the form

$$\prod_{x_1, x_2} \sum_{y} U(x_1, x_2, y),$$

where the pair (x_1, x_2) must run through all unordered pairs of the domain considered. Here 1, 2, 3 is the argument sequence of the 1st level; the argument sequences of the 2nd level are 1, 3, 4 and 2, 3, 5. One of the alternatives $\mathfrak{A}_i(1, 2, 3)$ here is $\mathfrak{A}_1(1, 2, 3)$, which means $A(1, 3) = 1$ and $A(3, 2) = 0$. This alternative is compatible with $\mathfrak{A}_1(1, 3, 4)$, that is, $A(1, 4) = 1$ and $A(4, 3) = 0$, as well as with $\mathfrak{A}_1(2, 3, 5)$, that is $A(2, 5) = 1$ and $A(5, 3) = 0$. This already proves the consistency.

If on the other hand the proposition

$$\prod_{x_1} \prod_{x_2} \sum_{y} A(x_1, y)\overline{A(y, x_2)}$$

were given, then, by the rule of transformation explained above, it would turn into

$$\prod_{x_1, x_2} \sum_{y_1} \sum_{y_2} \sum_{y_3} \sum_{y_4} [A(x_1, y_1)\overline{A(y_1, x_2)}A(x_2, y_2)\overline{A(y_2, x_1)}$$
$$A(x_1, y_3)\overline{A(y_3, x_1)}A(x_2, y_4)\overline{A(y_4, x_2)}].$$

We can then show the consistency of this proposition by the systematic method; but it is simpler to show the consistency of the given proposition by choosing for each pair (x_1, x_2) a y greater than x_1 and x_2, and by taking in general $A(x, y) = 1$ whenever $x < y$ but $A(x, y) = 0$ whenever $x \geqq y$.

If all primitive functions of the $\Pi\Sigma$-proposition

$$\prod_{x_1}\prod_{x_2}\dots\prod_{x_m}\sum_{y_1}\dots\sum_{y_n} U(x_1, \dots, x_m, y_1, \dots, y_n)$$

have at least $m + 1$ distinct argument symbols, the proposition is consistent if and only if U is consistent. For, the distinct argument sequences that we obtain by the procedure applied on page 519 never have more than m symbols in common, so that values are assigned to the primitive functions for each of the argument sequences, as they successively appear, independently of all the others.

It is possible, of course, to pose further problems concerning first-order propositions. An interesting question is whether an axiom, or axiom system, given as a first-order proposition Z is *categorical* [[*kategorisch*]]; we say that Z is categorical if it is possible to show, for an arbitrary first-order proposition Z' formed by means of the same primitive functions as Z, that it is true or false.[7] C. H. Langford has recently (*1926*, *1926a*) given some examples of such categorical axiom systems. I would like to give a very simple but nevertheless not too trivial example of this here. I consider the following axioms:

1. $\sum_{x} R(x, x)$,
2. $\prod_{x}\prod_{y}(R(x, y) + R(y, x))$,
3. $\prod_{x}\prod_{y}\prod_{z}(\overline{R(x, y)} + \overline{R(y, z)} + R(x, z))$,
4. $\prod_{x}\sum_{y}\overline{R(x, y)}$,
5. $\prod_{x}\sum_{y}\overline{R(y, x)}$,
6. $\prod_{x}\prod_{y}\sum_{z}(R(x, y) + \overline{R(x, z)}\,\overline{R(z, y)})$.

If we write $I(x, y)$ for $R(x, y)R(y, x)$, we recognize on account of 3 that $\Phi(y)$ always follows from $I(x, y)\Phi(x)$, where Φ is an arbitrary first-order propositional function that can be formed from R, and likewise that $\Phi(x)$ follows from $I(x, y)\Phi(y)$. The propositional function I therefore deserves to be called "identity" in the domain of the first-order propositions derivable from R. Since x and y are no longer distinguished when $I(x, y)$ holds, it is easy to see that the axioms listed determine an ordering of all individuals in an open and dense sequence.

Let us now consider the expression

$$\sum_{y} U(x_1, x_2, \dots, x_n, y),$$

where U is an elementary propositional function[8] formed from R. I will show that this expression, by virtue of the axioms, is always equivalent to an elementary propositional function $V(x_1, \dots, x_n)$, that is, that both always have the same truth value. In order to show that, I of course need only consider the case in which U is constructed

[7] The word "categorical", incidentally, is used also in another sense.

[8] [[That is, a function constructed by means of the "elementary" logical operations, namely, conjunction, disjunction, and negation.]]

by means of negation and conjunction alone. Moreover, propositional conjuncts like $R(x_i, x_j)$ can be moved to the left of the $\underset{y}{\Sigma}$-sign. If the propositional conjunct $\overline{R(y, y)}$ occurs, the proposition is always false. If the conjunct $R(y, y)$ occurs, it can be omitted, since on account of 2 it is always true. If we let $\alpha_1, \ldots, \alpha_\mu, \beta_1, \ldots, \beta_\nu, \gamma_1, \ldots, \gamma_\rho, \delta_1, \ldots, \delta_\sigma$ denote the x that occur in U, we obtain

$$\underset{y}{\Sigma} R(\alpha_1, y) \ldots R(\alpha_\mu, y) R(y, \beta_1) \ldots R(y, \beta_\nu) \overline{R(\gamma_1, y)} \ldots \overline{R(\gamma_\rho, y)}\,\overline{R(y, \delta_1)} \ldots \overline{R(y, \delta_\sigma)}.$$

$R(\alpha, \beta)$, $\overline{R(\gamma, \alpha)}$, $\overline{R(\beta, \delta)}$, and $\overline{R(\gamma, \delta)}$ follow from this proposition for all α, β, γ, and δ. I assert that, conversely, the truth of the $\underset{y}{\Sigma}$-proposition follows from the conjunction of all $R(\alpha, \beta)$, $\overline{R(\gamma, \alpha)}$, $\overline{R(\beta, \delta)}$, and $\overline{R(\gamma, \delta)}$. For we can imagine that the subscripts of the letters are so chosen that $R(\alpha, \alpha_1)$ holds for all α, and likewise $R(\beta_1, \beta)$ for all β, $R(\gamma_1, \gamma)$ for all γ, and $R(\delta, \delta_1)$ for all δ. Now four cases are possible: (1) We have $R(\alpha_1, \delta_1)R(\gamma_1, \beta_1)$. Since $\overline{R(\gamma_1, \delta_1)}$ holds, there exists, by 6, a y such that $\overline{R(y, \delta_1)}\,\overline{R(\gamma_1, y)}$ obtains, from which we can easily derive the truth of the $\underset{y}{\Sigma}$-proposition. (2) $R(\alpha_1, \delta_1)\overline{R(\gamma_1, \beta_1)}$ holds. Then the $\underset{y}{\Sigma}$-proposition already holds if we substitute β_1 for y. (3) We have $\overline{R(\alpha_1, \delta_1)}R(\gamma_1, \beta_1)$. Then the $\underset{y}{\Sigma}$-proposition holds if we substitute α_1 for y. (4) $\overline{R(\alpha_1, \delta_1)}\,\overline{R(\gamma_1, \beta_1)}$ holds. If $R(\beta_1, \alpha_1)$ is true, the $\underset{y}{\Sigma}$-proposition holds if α or β is substituted for y. If $\overline{R(\beta_1, \alpha_1)}$ is true, there exists a y such that $\overline{R(\beta_1, y)}\,\overline{R(y, \alpha_1)}$ obtains, and from this the truth of the $\underset{y}{\Sigma}$-proposition follows.

It is easy to discover how the matter stands in the simpler cases in which not all four of the sequences α, β, γ, and δ occur.

Hence any propositional function $\underset{y}{\Sigma} U(x_1, \ldots, x_n, y)$ is false for all x, true for all x, or surely replaceable by an elementary function $V(x_1, \ldots, x_n)$. If we use negation, we see that the same holds for $\underset{y}{\Pi} U(x_1, \ldots, x_n, y)$. Consequently the apparent variables occurring in an arbitrary first-order proposition can gradually be eliminated. When all but one have been eliminated, the last elimination directly yields the result "true" or "false".

I hope hereby to have given a little insight into some of the most important problems of mathematical logic.

Investigations in proof theory: The properties of true propositions

JACQUES HERBRAND

(*1930*)

The present text is Chapter 5 of Herbrand's thesis. The date given at the end of the thesis is 14 April 1929, but the defense at the Sorbonne did not take place until 11 June 1930 (in October 1929 Herbrand was drafted for a year of service in the French army). Herbrand was granted his doctorate in mathematics with highest honors. There is no indication that he introduced revisions into the text of the thesis between April 1929 and June 1930; a paper of his (*1931*), dated September 1929, contains emendations to the thesis, and this shows, it seems, that the text of the thesis remained unchanged after April 1929.

Chapter 5 of the thesis contains the statement of the now famous Herbrand theorem (fundamental theorem, p. 554 below) and an attempted proof of it. The section devoted to the theorem and its proof is preceded by a study of certain properties (A, B, and C) of provable formulas of quantification theory and followed by examples of applications of the theorem (to several cases of the decision problem and to the study of the consistency of various systems). The chapter is relatively self-contained, and the definitions, culled from the previous chapters of the thesis, that we give a few paragraphs below should provide the reader with the necessary background information.

Herbrand's thesis bears the marks of hasty writing; this is especially true of Chapter 5. Some sentences are poorly constructed, and the punctuation is haphazard. Herbrand's thoughts are not nebulous, but they are so hurriedly expressed that many a passage is ambiguous or obscure. To bring out the proper meaning of the text the translators had to depart from a literal rendering, and more rewriting has been allowed in this translation than in any other translation included in the present volume.

In 1939 Bernays remarked that "Herbrand's proof is hard to follow" (*Hilbert and Bernays 1939*, footnote 1, p. 158), but he did not point out any specific difficulty. In the fall of 1963 Professor Gödel informed the editor that in the early forties he had discovered an essential gap in Herbrand's argument, but he never published anything on the subject. In the spring of 1963 counterexamples to the lemma of 3.3 and to Lemma 3 of 5.3, lemmas upon which Herbrand's proof of his fundamental theorem rests, had been published by Dreben, Andrews, and Aanderaa (*1963*). Dreben (*1963*) stated a substitute for the lemma of 3.3, and two notes (*Dreben, Andrews, and Aanderaa 1963a* and *Dreben and Aanderaa 1964*) outlined how Herbrand's argument could be repaired; Dreben and Denton (*1966*) gave a detailed proof of an emended version of the crucial lemma of 3.3. To deal with these questions and also to provide a commentary on the text, Dreben added a number of footnotes, given between square brackets, to the present translation, as well as ten Notes, A through J, given on pages 567–581 below.

The heading of Chapter 5 in Herbrand's thesis is "The properties of true propositions". Here "true" means provable in quantification theory, and Herbrand's intention is to investigate the three properties A, B, and C that hold of a formula if and only if the formula is provable. The statement and the proof of Herbrand's fundamental theorem are closely connected with these properties.

Herbrand's work can be viewed as a reinterpretation, from the point of view of Hilbert's program, of the results of Löwenheim and Skolem. Of his fundamental theorem Herbrand writes (*1931b*, p. 4) that it is "a more precise statement of the well-known Löwenheim–Skolem theorem". Having adopted Hilbert's finitistic viewpoint, Herbrand could not regard the notion "satisfiable" (or "valid"), which enters in the statement of the Löwenheim–Skolem theorem, as a proper metamathematical notion. He knew that from the assumption that for every p the conjunctive expansion of order p of a formula is satisfiable one can infer—"easily", as he says in *1931b*, p. 4—that the formula is $\aleph_0$-satisfiable, but for him this inference had no proper metamathematical content (see below: 6.2, p. 558; footnote 67, p. 552; and Note H, p. 578). On his attitude toward Hilbert's program Herbrand wrote amply (Introduction to his thesis, *1930a*, *1931a*).

Gentzen claims (*1934*, footnote 6, p. 409) that Herbrand's fundamental theorem is a special case of Gentzen's own *verschärfter Hauptsatz*. He interprets Herbrand's theorem as applying to a sequent whose antecedent is empty and whose succedent consists of a single prenex formula, while the *verschärfter Hauptsatz* applies to any sequent whose formulas are prenex. In this difference Gentzen sees the greater generality of his result. From the point of view of the standard version of quantification theory, however, Herbrand's theorem is more general than Gentzen's, since it applies to any (not necessarily prenex) formula, while Gentzen's theorem, once adapted to this version, applies only to formulas of a certain class. Also, with the study of the domains C_i (see below, p. 542), Herbrand supplies more information about the *Mittelsequenz*. Gentzen's *Hauptsatz* on the eliminability of the *Schnitt* schema (*1934*, p. 196) is, for Gentzen's system, the analogue of Herbrand's result that modus ponens is eliminable in quantification theory (see below, p. 558, and Note D). However, Gentzen's *Hauptsatz* (but not, of course, his *verschärfter Hauptsatz*) can be extended to intuitionistic logic and to various modal logics, while there is no similar extension of Herbrand's result.

In Chapter 1 of his thesis Herbrand presents the classical propositional calculus, which he calls the *theory of identities of the first kind*, and in Chapter 2 his version of quantification theory, which he calls the *theory of identities of the second kind*. The system presented in Chapter 2 is equivalent to any one of the standard formulations of quantification theory, and in Section 7 of Chapter 2 Herbrand proves that it is equivalent to the quantification theory used by Whitehead and Russell in *Principia mathematica*. In Chapter 3 Herbrand calls any system a *theory* if it is obtained from his system of Chapter 2 by the adjunction of some of the following:

(1) Function letters (called *descriptive functions* and intended to represent in the system functions from individuals to individuals) together with an extended rule of existential generalization,

(2) Predicate constants ($=$ or ε, for example), and

(3) New primitive formulas, called *hypotheses* or *axioms*.

The basic system used by Herbrand in Chapter 5 is, as he explains at the very beginning of the chapter (below, p. 529), a theory obtained from the quantification theory of Chapter 2 by the adjunction of descriptive functions. We shall call this system Q_H, and we now describe it, omitting details that are of no importance in Chapter 5.

Herbrand's statement of the rules of formation of Q_H lacks precision and is not always consistent with his usage; in a paper (*1931*) that is in many respects a sequel to his thesis he found it necessary to state his rules in an emended form. Before and after the emendations, the notation used by Herbrand frequently allows confusion between use and mention of letters, although, in practice, this confusion does not create any basic difficulty in the kind of problems that he is discussing. Here we shall not undertake a total reconstruction of Herbrand's symbolism but merely present it in a simple form that will allow the reader to grasp his formulas with ease.

For each $n \geqq 0$ there is a denumerable supply of n-adic predicate letters. These letters are left unspecified; they are mentioned but not used. The syntactic variables ranging over them are generally Φ and Ψ. There is a denumerable supply of quantifiable variables. The syntactic variables ranging over them are x, y, z, and occasionally t, with or without subscripts, sometimes with primes. For each $n \geqq 0$ there is a finite or denumerable supply of n-place function letters. These letters are left unspecified; the syntactic variable ranging over them is generally f or, if $n = 0$, a, with or without subscripts. An n-place *elementary descriptive function* (*fonction descriptive élémentaire*) is either a 0-place function letter or, for $n > 0$, an n-place function letter followed by parentheses enclosing n variables separated by commas and called the *arguments* of the function. The *descriptive functions* are obtained from the elementary descriptive functions by iterated substitution of descriptive functions for arguments. A function in which no variable occurs is a *constant*; a 0-place function letter is a *primitive constant* (*constante fondamentale*; in *1931* Herbrand uses the expression "*constante élémentaire*"). The word "argument" is used also for the primitive constants (and sometimes even more generally for the functions) that have been substituted for the arguments of a function.[a] Variables and descriptive functions are *individuals* (*individus*).

An *atomic propositional function* (*fonction propositionnelle élément*) is either a 0-adic predicate letter or, for $n > 0$, an n-adic predicate letter followed by parentheses enclosing n individuals separated by commas (Herbrand does not always use parentheses and commas, but we have found it convenient to reestablish them in the translation below); the individuals are the *arguments* of the function.

A *proposition* is a well-formed formula of Q_H; it is formed from atomic propositional functions by means of connectives and quantifiers. The primitive connectives are negation ($\sim$) and disjunction ($\vee$); conjunction (&), the conditional ($\supset$), and the biconditional ($\equiv$) are defined in terms of these two. Both the universal and the existential quantifiers, (x) and $(\exists x)$, are taken as primitive. The first is sometimes written $(+x)$ and the second $(-x)$. A quantifier left indeterminate is written $(\pm x)$. A string of quantifiers $(\pm x_1), (\pm x_2), \ldots, (\pm x_n)$ is sometimes written $(\pm x_1, \pm x_2, \ldots, \pm x_n)$. As in *Principia mathematica*, a bound variable is said to be *apparent* (*apparente*), and a free variable *real* (*réelle*). Herbrand's name for a quantifier is "symbole de variable apparente". In Chapter 5 he follows the rule that in any given proposition no variable occurs both free and bound and variables occuring in more than

[a] It is used also for the elements of the domains when these elements are substituted for the arguments of a function.

one quantifier are to be taken as distinct (even when the same syntactic letter is used). Hence he can call the scope of a quantifier $(\pm x)$ the scope of the variable x.

Being a well-formed formula, an atomic propositional function is a proposition, and more specifically an *atomic proposition* (*proposition-élément*). A quantifier-free proposition is an *elementary proposition* (*proposition élémentaire*). Herbrand uses the expression "fonction propositionnelle" in two different senses: (1) A proposition is said to be a *propositional function* of its real variables (and an atomic propositional function is a propositional function in that sense); (2) A proposition P is said to be a *propositional function* of any atomic proposition that occurs in P, the propositional function being *of the first kind* if it contains no quantifier, *of the second kind* otherwise (hence a propositional function of the first kind consists of atomic propositions connected by $\sim$ and $\vee$).[b]

Herbrand uses the letter p as a syntactic variable ranging over propositions, but, since p is also frequently used as a number ("order p"), in its first usage it has, in the translation below, been frequently replaced by Z; when variables have to be exhibited, the letters Φ and Ψ together with variables are generally used as syntactic variables ranging over propositions. "$\Phi(x)$" *denotes* a propositional function of x. However, the formula thus denoted may very well not contain x (in which case x occurs *fictitiously* in $\Phi(x)$) and may contain other variables. Herbrand may write $\Phi(x_1)$ or $\Phi(x_2)$ for one and the same formula according as x_1 or x_2 is the real variable to which he wants to draw attention.

Herbrand uses dots, single and multiple, to punctuate his formulas and also to conjoin propositions. But throughout the thesis, in particular in Chapter 5, he frequently flouts the rules stated in Chapters 1 and 2 concerning the use of these dots. This creates difficulties. In *1931* Herbrand found it convenient to use "$\times$" for conjunction. In the translation below "&" is used for conjunction, parentheses and brackets are used for the punctuation of formulas; when nesting of parentheses would make the reading of a formula difficult, we use, proceeding from the inside outward, ordinary parentheses (( )), square brackets ([ ]), curly brackets ({ }), and larger square brackets ($\Big[\ \Big]$).

An occurrence of a subproposition A in a proposition P (written in terms of the primitive connectives $\sim$ and $\vee$) is said to be *positive*, or A occurs *positively* in P, if this occurrence of A lies within the scopes of an even number of negation signs of P; an occurrence of A is *negative*, or A occurs *negatively*, in P if the number is odd.

If $(x)\Phi(x)$ occurs positively in P or $(\exists x)\Phi(x)$ occurs negatively in P, the variable x and its quantifier are said to be *general* in P. If $(x)\Phi(x)$ occurs negatively in P or $(\exists x)\Phi(x)$ occurs positively in P, the variable x and its quantifier are said to be *restricted* in P.

The axioms of Q_H are all the quantifier-free tautologies.

The rules of inference, or, as Herbrand calls them, the rules of reasoning (règles de raisonnement), are modus ponens (règle d'implication), simplification (from $Z \vee Z$ to infer Z, or rather—since Herbrand does not allow, in one formula, two quantifiers to contain the same variable—from $Z \vee Z'$ to infer Z, where Z' is an alphabetical variant of Z), universal generalization (from $\Phi(x)$ to infer $(x)\Phi(x)$), the extended rule of existential generalization (from $\Phi(\alpha)$, where α is an individual, to infer $(\exists y)\Phi(y)$), a rule allowing relettering, and the following twelve *rules of passage* (*règles de passage*):

[b] More generally, Herbrand occasionally speaks of a proposition P as being a propositional function of a not necessarily atomic proposition Q; what he means, then, is that Q is a well-formed part of P.

Let Z be a proposition that does not contain x; for $h = 1, \ldots, 6$, let J_h; K_h be the hth of the following six ordered pairs of propositions:

(1) $\sim (x)\Phi(x)$; $(\exists x) \sim \Phi(x)$,

(2) $\sim (\exists x)\Phi(x)$; $(x) \sim \Phi(x)$,

(3) $(x)\Phi(x) \vee Z$; $(x)[\Phi(x) \vee Z]$,

(4) $Z \vee (x)\Phi(x)$; $(x)[Z \vee \Phi(x)]$,

(5) $(\exists x)\Phi(x) \vee Z$; $(\exists x)[\Phi(x) \vee Z]$,

(6) $Z \vee (\exists x)\Phi(x)$; $(\exists x)[Z \vee \Phi(x)]$.

Also let J_{6+h}; K_{6+h} be K_h; J_h. Then a proposition T comes from a proposition S by one application of the ith rule of passage ($i = 1, \ldots, 12$) if T is the result of replacing one occurrence of J_i in S by one occurrence of K_i. (In Section 2 of Chapter 2 of Herbrand's thesis (4) and (6) are absent; the omission is noted and corrected in *1931*, p. 22, footnote 2.)

If a proposition is provable from the axioms by these rules, Herbrand says that it is *true* (*vraie*) and sometimes prefixes it with the sign $\vdash$. He also says that it is a *propositional identity*, or simply an *identity*, *of the first kind* if it is elementary, of the *second kind* otherwise; when the context prevents any misunderstanding, he speaks simply of an identity, without specifying the kind.

Herbrand uses the word "champ" for any domain and, in particular, for the domains C_i that play such an important role in his theory. For these C_i the word "field" has occasionally been used in English publications on logic. We have not followed this custom here, and "champ" has been uniformly translated by "domain". "Réduite" has been translated by "expansion", although the English word "reduction" has occasionally been used in this context.

Herbrand's text was translated by Burton Dreben and the editor. Wilbur Hart III, Marc Venne, Daniel Isaacson, and Dirk van Dalen read the translation with its editorial comments, and the translators are grateful to them for suggesting improvements. The translation is printed here with the kind permission of the Instytut matematyczny Polskiej Akademii nauk.

1. Thorough study of the rules of passage

1. Throughout the present chapter we shall work with the theory of Chapter 3, 1.4, which we obtained from the theory of Chapter 2 by adding descriptive functions. As we pointed out in Chapter 3, nearly all the results of Chapter 2 continue to hold for this extended theory. For the sake of simplicity we assume—and this will not change anything essential—that we have variables of one type only.[1]

In the present section we shall study the various forms that a proposition can take when the rules of passage are applied to it. Not all the results that we shall obtain will be used in the rest of the present chapter, but in deriving them we shall have an opportunity to introduce notions that we shall need in Section 2.

We recall the theorem of Chapter 2, 3.101, which states that by applying the rules of passage we can find, for any given proposition P, a certain proposition [[in prenex

[1] [[In Chapter 3 Herbrand generalizes his system of quantification theory and, indeed, his notion of an arbitrary theory to many-sorted systems. Each sort he calls a *type*. These types play no serious role in Chapter 5, except in 6.6. Below, in 1.1, Herbrand uses the word "type" in a totally different sense.]]

form⟧ equivalent to P and said to be *tied* to P. We refer the reader to the same place for the definition of *proposition in prenex form.*[2]

To avoid ambiguities we shall assume throughout the present chapter that no two quantifiers contain the same letter.

1.1. A sequence of quantifiers will be called a *type*. Two types will be said to be *similar* if they differ only in the variables that they contain; thus

$$+x_1,\ -x_2,\ +x_3$$

and

$$+y_1,\ -y_2,\ +y_3$$

are similar types.

1.11. The *type of a proposition* P will be the type formed by the quantifiers that occur at the beginning of the prenex proposition tied to P.

We shall now investigate in more detail ⟦than in Chapter 2⟧ what are the types of those prenex propositions obtained from a given proposition P by means of the rules of passage.[3] These types will be called the *types attached* ⟦*attachés*⟧ *to* P.

To say that $\pm x_1, \pm x_2, \ldots, \pm x_n$ is a type attached to a proposition P is, as we see, to say that we can, by means of the rules of passage, first put P into the form $(\pm x_1)P_1(x_1)$, then put $P_1(x_2)$ into the form $(\pm x_2)P_2(x_2)$, and so forth; this amounts to saying that we can pull the quantifiers

$$(\pm x_1), (\pm x_2), \ldots, (\pm x_n)$$

out of the proposition one after another.

1.12. Let us recall the following remark (proved in Chapter 2, 3.12[4]):

LEMMA 1. *There exist types attached to the proposition* $\Phi[(\pm x)A(x)]$ *that have* $(\pm x)$ *as their leftmost quantifier; here* $\Phi(Z)$ *is a propositional function of the first kind in* Z *and may, moreover, contain atomic propositions other than* Z.

1.2. Now, using ⟦just⟧ the rules of passage, we shall decrease the scope of x as much as we can. To decrease this scope step by step, we make use of the following two consequences of the rules of passage:

(1) If the scope of x is $\sim \Phi(x)$, we can decrease this scope so that it becomes $\Phi(x)$;

(2) If the scope of x is $\Phi(x) \vee Z$ or $Z \vee \Phi(x)$, we can decrease this scope so that it becomes $\Phi(x)$, provided Z does not contain x.

[2] ⟦"A proposition P is said to be *in prenex form* ⟦*de forme normale*⟧ if it has the form

$$(\pm x_1, \pm x_2, \ldots, \pm x_n)M(x_1, x_2, \ldots, x_n),$$

where $M(x_1, x_2, \ldots, x_n)$ is an elementary proposition, called the *matrix*. By applying the rules of passage we can put a proposition into prenex form (that is, find an equivalent proposition in prenex form). We can even find one, which ⟦is unique and⟧ will be said to be *tied to* P ⟦*liée à* P⟧, such that in it the quantifiers occur, from left to right, in the same order as in P."

Herbrand's "forme normale" has been translated by "prenex form" since below he uses "normal" in a different sense. (Occasionally, we use "prenex proposition" or "prenex form" for "proposition in prenex form".) Moreover, even if a proposition P is not in prenex form, Herbrand will call the elementary proposition that we obtain from P by deleting all quantifiers occurring in P the *matrix* of P.⟧

[3] [This question is answered in 1.5 below, p. 533.]

[4] ⟦"If $A(p_1, p_2, \ldots, p_n)$ is a propositional function of the first kind, we can, by applying the rules of passage, transform $A[(\pm x)\Phi(x), p_2, \ldots, p_n]$ into a prenex proposition in which the variable x occurs in the leftmost quantifier. This is readily proved by recursion on the construction of $A(p_1, p_2, \ldots, p_n)$."⟧

Thus we shall finally bring the proposition into a form in which the scope of x will be an atomic proposition, or a proposition of the form $\Phi(x) \vee \Psi(x)$, where $\Phi(x)$ and $\Psi(x)$ actually contain x, ⟦or a proposition of the form $(\pm y)\Phi(y)$⟧.

We now start from a given proposition P; we consider the rightmost quantifier and, using (1) and (2), reduce its scope as much as possible. We thus obtain its *minimal scope*. Once this is done, we perform the same operation on the second quantifier, counted from the right; we proceed thus until the scope of the leftmost quantifier has been minimized. The proposition has then taken what we shall call its *canonical form*, which is characterized by the fact that the scopes of all its variables are minimal ⟦with respect to the rules of passage⟧.

For example, the canonical form of

$$(x)(\mathfrak{A}y)(z)\{[\Phi(x) \;\&\; \Psi(x, y)] \vee \Phi(z)\},$$

is

$$(x)[\Phi(x) \;\&\; (\mathfrak{A}y)\Psi(x, y)] \vee (z)\Phi(z);$$

the minimal scope of z is $\Phi(z)$, that of y is $\Psi(x, y)$, and that of x is

$$\Phi(x) \;\&\; (\mathfrak{A}y)\Psi(x, y).^{5}$$

Merely by considering each rule of passage we see that, if a proposition results from another by the rules of passage, the two propositions have the same canonical form.

1.3. It is quite obvious that, if we consider the scopes of two different variables, either these scopes are disjoint or one scope is included in the other. A variable x will be *dominant* in a proposition P if ⟦in the canonical form of P⟧ the scope of x is not included in the scope of any other variable.

For example, x and z are dominant in the proposition given above.

LEMMA 2. *If $x_1, x_2, \ldots, x_n$ are the dominant variables of a given proposition and $\Phi_1(x_1), \Phi_2(x_2), \ldots, \Phi_n(x_n)$ are their respective minimal scopes, the proposition, once in canonical form, can be written $A[(\pm x_1)\Phi_1(x_1), (\pm x_2)\Phi_2(x_2), \ldots, (\pm x_n)\Phi_n(x_n)]$, where $A(Z_1, Z_2, \ldots, Z_n)$ is a propositional function of the first kind.*

We observe, incidentally, that if this lemma is true we can give a similar form to the $\Phi_i(x_i)$, and so on.

The lemma follows immediately from the fact that, except for the scopes of the dominant variables ⟦and their quantifiers⟧, a proposition in canonical form contains no signs other than $\sim$ and $\vee$. Since, as we just saw, the scopes of the x_i do not overlap, the proposition is a propositional function of these scopes ⟦together with their quantifiers⟧, hence cannot, relative to these scopes, be a propositional function of the second kind.

1.4. We can now define what we shall call the *scheme* ⟦*schème*⟧ *of a proposition*. It will be an array, formed of signed letters ⟦that is, each letter is preceded by the sign $+$ or $-$⟧ and braces, that will be defined by recursion on the number of apparent variables of the proposition.

First, the scheme of an elementary proposition is taken to be empty.

[5] ⟦In this example it must be assumed that $\Phi(x)$ and $\Psi(x)$ are such that no further application of (1) and (2) is possible; moreover, contrary to Herbrand's system, here & is taken as primitive.⟧

Then, the scheme of a [[nonelementary]] proposition P—which, as we just saw, can be put into the [[canonical]] form

$$A[(\pm x_1)\Phi_1(x_1), (\pm x_2)\Phi_2(x_2), \ldots, (\pm x_n)\Phi_n(x_n)],$$

where x_i is the dominant variable of $\Phi_i(x_i)$—is to be obtained from the schemes of the propositions $\Phi_1(x_1)$, $\Phi_2(x_2), \ldots, \Phi_n(x_n)$ thus: if the schemes of these propositions are respectively the arrays $S_1, S_2, \ldots, S_n$, the scheme of P will be

$$\begin{array}{l} \pm x_1 \,\{S_1 \\ \pm x_2 \,\{S_2 \\ \cdots\cdots\cdot \\ \pm x_n \,\{S_n. \end{array}$$

In this list the vertical ordering is of no consequence. The variables $x_1, x_2, \ldots, x_n$ will be called the *dominant variables of the scheme.*

For example, the scheme of the proposition already considered in 1.2 is

$$\begin{array}{l} +x\,\{-y \\ +z. \end{array}$$

That of

$$(\exists x)[(y)\Phi(x, y) \vee (z)\Phi(x, z)] \vee (x')(y')\Phi(x', y')$$

[[if this proposition is assumed to be in canonical form]] is

$$\begin{array}{l} -x\left\{\begin{array}{l}+y\\+z\end{array}\right. \\ +x'\,\{+y'. \end{array}$$

We shall agree to omit the brace when it separates two successive variables on the same line, as in the first scheme, which we now write

$$\begin{array}{ll} +x & -y \\ +z. & \end{array}$$

Hence we see that the scheme of a proposition is an array formed of signed letters; the letters are the apparent variables of the proposition, the sign that goes with each of them tells us whether it is general or restricted,[6] and each letter may be followed by a [[signed]] letter or by a brace joining several [[signed]] letters. [[Even if no proposition is given,]] such an array will, in what follows, still be called a *scheme.*[7]

[6] [[The convention that Herbrand makes here clearly conflicts with the one that he has previously made in Chapter 2, 1.5, namely, that "$(+x)$" stands for the universal quantifier "(x)" and "$(-x)$" for the existential quantifier "$(\exists x)$". Herbrand observes his new convention in Sections 1 and 2 of Chapter 5, then occasionally reverts to his previous convention in subsequent sections.]]

[7] It is convenient to agree that in a scheme no two letters shall ever be identical.

[When no proposition is given, Herbrand still calls the letters in the scheme that are not behind any brace the *dominant* variables of the scheme; moreover, he refers to the letters in the scheme that are preceded by " + " as the *general* variables of the scheme and to those that are preceded by "$-$" as the *restricted* variables of the scheme. This generalized notion of scheme becomes important in Section 2.

Herbrand (*1931*, p. 45) points out that his notion of scheme is related to Ackermann's "Stammbaum" (*1928a*, p. 646) and also to the notion of free group.]

A *line* of a scheme will be a sequence of signed letters of the scheme such that ⟦the first letter of the sequence is a dominant variable of the scheme and⟧ each letter ⟦in the sequence after the first⟧ occurs in the scheme immediately on the right of the letter that precedes it in the sequence or immediately on the right of the brace affixed to that letter. Hence every line of a scheme can be regarded as a type, and this type will be said to be the *type of the line* (see 1.1).

Thus the lines of the scheme of the proposition considered above are

$$-x \quad +y,$$

$$-x \quad +z,$$

and

$$+x' \quad +y'.$$

In particular, a type is a scheme that has a single line.

1.41. We see immediately that, according to Lemma 2 and the definition of the scheme ⟦of a proposition⟧, a variable y comes after a variable x in a line of the scheme ⟦of a proposition P⟧ if and only if ⟦in P⟧ the minimal scope of y is included in the minimal scope of x.

We shall say that in a proposition P a variable x *dominates* a variable y if and only if the minimal scope of y is included in the minimal scope of x. We shall also say that in a proposition P a variable x is *superior to* a variable y if the scope of y is included in the scope of x (hence x dominates y if and only if, when the proposition P is in canonical form (1.2), x is superior to y).

We readily see that, if x dominates y, y cannot be superior to x.

1.42. A type is said to be *deduced* from a scheme if the variables of the type are all the letters of the scheme and, whenever in some line of the scheme one letter precedes another, the first letter also precedes the second in the type. Thus, in the example considered above, we obtain a type deduced from the scheme by writing on one line the five signs $-x$, $+y$, $+z$, $+x'$, and $+y'$ in such a way that x' precedes y' and x precedes y and z (there are twenty such permutations).

Given any type deduced from the scheme ⟦of a proposition P⟧, we see that, if x dominates y ⟦in P⟧, then x precedes y in the type.

1.43. Before we proceed to the proof of the theorem that concludes the present section, let us observe that, given a proposition P and the proposition P' that we obtain from P by deleting a quantifier governing a dominant variable, we obtain the scheme of P' by deleting the variable ⟦and the brace that immediately follows it, if any⟧ from the scheme of P; this can be readily seen from the recursive definition of scheme and from Lemma 2.

1.5. Theorem. *The types attached to a proposition are the types deduced from the scheme of that proposition.*

This theorem, which answers the question raised at the beginning of the present section, is proved as follows.

(1) *Any type deduced from the scheme is a type attached to the proposition.*

For, to obtain a type deduced from the scheme, we can proceed thus: we delete a dominant variable ⟦and the brace affixed to it, if any⟧ from the scheme ⟦and write the variable somewhere⟧; from the new scheme so obtained we delete a variable that is dominant in this new scheme and write it to the right of the variable first considered;

we continue in this manner until we have exhausted all the variables of the original scheme. It follows from Remark 1.43 and Lemma 1 (1.12) that we can exhaust in this fashion all the [[apparent]] variables of the proposition in question, hence that the variables that we have thus successively selected form a type attached to the proposition.

(2) *Any type attached to the proposition is a type deduced from the scheme.*

To see this it suffices to observe that, if x is superior to y in a proposition, y cannot dominate x and to remember the characteristic property (1.42) of a type deduced from the scheme.

2. Property A

2. In the present section we shall study the most general properties that are sufficient for a proposition to be true;[8] in subsequent sections we shall see that these properties are also necessary.

For the sake of simplicity we assume, first, that *there are no descriptive functions in the propositions under investigation.*[9]

2.1. We begin by defining the notion of *normal identity*. Let P be a proposition with which, by means of the rules of passage, we have associated some prenex form,

$$(\pm x_1, \pm x_2, \ldots, \pm x_n)M(x_1, x_2, \ldots, x_n).$$

In the matrix M of this prenex form we replace each restricted variable either by a real variable of P (we can even assume that this real variable does not actually occur in P—in which case it will be said to be fictitious) or by a general variable that is superior (1.41) to the restricted variable in the prenex form (that is, a general variable whose quantifier, in the type of this prenex form, precedes that of the restricted variable considered). We shall say that the proposition P is a *normal identity* if [[for some prenex form P' of P]] at least one elementary proposition Π obtained [[from P']] in the manner just described is an identity of the first kind. We say also that such an identity is *associated with* P and that the type $\pm x_1, \pm x_2, \ldots, \pm x_n$ [[of P']] is a *normal* type.

For example,

$$(x)\Phi(x) \vee (\exists y) \sim \Phi(y)$$

is a normal identity; for we can turn it into

$$(x)(\exists y)[\Phi(x) \vee \sim \Phi(y)],$$

and, if y is replaced by x, the matrix becomes

$$\Phi(x) \vee \sim \Phi(x),$$

which is an identity.[10]

[8] [[Here "true" should be understood as "provable in Q_H".]]

[9] [[The case in which the proposition contains descriptive functions is taken up in 2.4, p. 539 below.]]

[10] [[In the last formula the French text has "y" for "x", which is a misprint.]]

2.11. We see that to pass from a proposition P to an identity Π associated with P we must repeatedly perform the following operations ⟦on some prenex form of P⟧:

(1) Delete a quantifier of a general variable, so that the variable becomes real;

(2) Delete a quantifier of a restricted variable and replace the restricted variable by a real variable, that is, pass from a proposition of the form $(\exists y)\Phi(x, y)$ to the proposition $\Phi(x, x)$.

In the example above we would first replace the ⟦prenex form of the⟧ proposition by

$$(\exists y)[\Phi(x) \vee \sim\Phi(y)]$$

and this in turn by

$$\Phi(x) \vee \sim\Phi(x).$$

We readily see that every normal identity is true:[8] by the rules of generalization, in each of the two operations just described the first proposition is true provided the second proposition is true.[11]

We can always actually decide whether a proposition P is a normal identity: we take all the types attached to P and ⟦for each type⟧ consider all the propositions obtained when in the matrix of P each restricted variable y is replaced ⟦by some real variable or⟧ by some general variable that precedes y in the type; the criterion of Chapter 1, 5.21,[12] then enables us to decide whether one of the propositions thus obtained is an identity. Thus we have to run through only a finite number of tests.

2.12. From a normal identity we obtain another normal identity (equivalent to the first, according to Chapter 2, 4.6[13]) if, in a normal type, we permute two successive restricted quantifiers or two successive general quantifiers. We obtain other normal identities (implied by the first) if, in a normal type, we shift a restricted quantifier to the right or a general quantifier to the left.

2.13. We obtain an example of a class of normal identities when we put nonelementary propositions for the atomic propositions in those identities of the first kind in which each atomic proposition has at most one positive occurrence and at most one negative occurrence.[14] To see this it suffices to look again at our proof in Chapter 2, Section 4,[15] that the identities of the second kind thus obtained are true.[8]

2.2. Henceforth we shall use the following notation: if $M(x_1, x_2, \ldots, x_n)$ is an elementary proposition (in general, the matrix of a proposition), a substitution on the x_i, that is, the replacement of $x_1, x_2, \ldots, x_n$ by other letters $y_1, y_2, \ldots, y_n$, will be denoted by a letter such as S, with or without subscripts, and the result of this substitution, namely $M(y_1, y_2, \ldots, y_n)$, by $M(S)$.

Let us now consider a scheme such that each of its lines is similar (1.1) to a type attached to some proposition P.[16] For each line of the scheme, let us take a similar

[11] [Thus a proposition is a normal identity if and only if it is provable from a quantifier-free identity by the use of at most the two rules of generalization and the rules of passage. Moreover, if the proposition is in prenex form, it can be proved from the quantifier-free identity by the two rules of generalization alone.]

[12] ⟦The truth-table method.⟧

[13] ⟦"A proposition in prenex form is equivalent to those that we obtain from it by permuting two ⟦successive⟧ general quantifiers or two ⟦successive⟧ restricted quantifiers; it implies those that we obtain from it by shifting a restricted quantifier to the right or a general quantifier to the left."⟧

[14] ⟦For the definition of positive occurrence and negative occurrence see above, p. 528.⟧

[15] [See Note A, p. 567 below.]

[16] [The scheme now being considered is not in general the scheme of P, nor need all the lines of this scheme be similar to one and the same type attached to P. See Note C, p. 569 below.]

type attached to P and correlate the kth letter of the line with the kth letter of the type. The replacement of the apparent variables of P by the corresponding letters of a line of the scheme defines a substitution; there are as many substitutions as there are lines. Let us denote these substitutions by $S_1, S_2, \ldots, S_m$; if M is the matrix of P and T is any type deduced from the scheme, the proposition

$$(T)[M(S_1) \vee M(S_2) \vee \cdots \vee M(S_m)]$$

is said to be a *proposition derived from* P. We readily see that this proposition does not ⟦essentially⟧ depend upon the particular type T selected (this, incidentally, is of no importance for what follows). The scheme is said to *generate* the derived proposition, and the proposition is said to be *derived from* P by the scheme.

For example, given the proposition

$$(\exists x)(y)M(x, y),$$

in which $M(x, y)$ is elementary, and the scheme

$$-x\begin{cases}+y\\+z,\end{cases}$$

we obtain the derived proposition

$$(-x, +y, +z)[M(x, y) \vee M(x, z)].$$

We see quite easily that, to pass from a proposition P to a proposition derived from P, we must repeatedly perform the following operation: if $(\pm x_1, \ldots, \pm x_n)M(x_1, \ldots, x_n)$ is any prenex form of P, we go from a proposition Q of the form

$$(\pm x_1, \ldots, \pm x_r)[A(x_1, \ldots, x_r) \vee (\pm x_{r+1}, \ldots, \pm x_n)M(x_1, \ldots, x_r, x_{r+1}, \ldots, x_n)]$$

to a proposition Q' of the form

$$(\pm x_1, \ldots, \pm x_r)[A(x_1, \ldots, x_r) \vee (\pm x_{r+1}, \ldots, \pm x_n)M(x_1, \ldots, x_r, x_{r+1}, \ldots, x_n) \\ \vee (\pm y_{r+1}, \ldots, \pm y_n)M(x_1, \ldots, x_r, y_{r+1}, \ldots, y_n)].$$

This operation enables us to pass from a proposition derived from P by a given scheme to a proposition derived from P by a new scheme that differs from the given scheme in having just one more line. Therefore, a proposition is equivalent to any proposition that is derived from it (because of the identity[17] $(P \vee P) \equiv P$ and Chapter 2, 6.1[18]).[19]

2.3. Let us now assume that we have found a proposition P' that is derived from P, is a normal identity, and—this further hypothesis is not essential—is such that one of the normal types (2.1) of P' is deduced (1.42) from the scheme generating P';[20]

[17] [See Note B, p. 568 below.]

[18] ⟦"If $F(p)$ is a propositional function of the proposition p, then $\vdash (p \equiv q) \supset [F(p) \equiv F(q)]$."⟧

[19] [See Note C, p. 569 below.]

[20] [The scheme Σ generating P' from P need not be the scheme of P'. Hence it might not be possible to deduce from Σ all the types attached to P'; in particular, it might not be possible to deduce from Σ any normal type of P'. For example, let P be $(\exists x_1)(y_1)Fx_1y_1 \supset (y_2)(\exists x_2)Fx_2y_2$, let P' be $(x_1)(\exists y_1)(y_2)(\exists x_2)(Fx_1y_1 \supset Fx_2y_2)$, and let Σ be

$$+x_1 \quad -y_1 \quad +y_2 \quad -x_2.$$

Then

$$+y_2 \quad +x_1 \quad -y_1 \quad -x_2$$

is a normal type of P', but neither this type nor any other normal type of P' can be deduced from Σ. See 1.42 above, p. 533.]

moreover, ⟦for that normal type⟧ let Π be some identity associated with P'. We shall then say that P *has property* A, that Π is an identity *associated with* P, that the scheme generating P' is *associated with* P,[21] and that the normal type of P' in question is *associated with* P.[22]

We see that every proposition having property A is an identity.[23] Furthermore, a normal identity has property A (the associated scheme is simply the normal type).

In the scheme ⟦associated with P⟧ let us put for each restricted variable the general or real variable by which we replaced it ⟦in P'⟧ in order to obtain Π. Then, as in 2.2, the lines of the ⟦resulting⟧ scheme determine substitutions, namely $T_1, T_2, \ldots, T_m$. If M is the matrix of P, we see that Π is nothing but $M(T_1) \vee M(T_2) \vee \cdots \vee M(T_m)$.

We observe that property A is not a property in the sense of the Introduction,[24] for, given an arbitrary proposition P, we do not know how to decide whether P has property A. Hence, only if we can actually show that a proposition satisfies our definition shall we say that the proposition has property A. The same remark will apply to properties B and C, to be defined in the next section.

2.31. Thus, to show that a proposition has property A we must first construct an associated scheme, then deduce from it an associated type ⟦if possible, otherwise find a normal type for the derived proposition⟧, and finally specify those general ⟦and real⟧ variables by which we must replace the restricted ones in order to obtain an associated identity (this will yield what we shall call *associated equations* between restricted and general variables).

For example, let P be a proposition that can, by means of the rules of passage, take the ⟦prenex⟧ forms

$$(y)(\exists x)(z)(\exists t)M(x, y, z, t)$$

and

$$(z)(\exists t)(y)(\exists x)M(x, y, z, t).$$

Let us assume that

$$M(z'', y, z, y') \vee M(y', y', z', z'') \vee M(y', y', z'', z'')$$

[21] [In general, the lines of the scheme generating P' cannot all be disjoint; for consider the following proposition P:

$$(\exists x)(y)(\exists z)[\{F(y) \& [G(y) \vee \sim G(z)]\} \vee \{\sim F(y) \& [H(x) \vee \sim H(z)]\}].$$

P has property A, an associated scheme being

$$-x \quad +y\begin{cases}-z\\ -z'.\end{cases}$$

But from P no normal identity can be generated by any scheme all of whose lines are disjoint.]

[22] [According to these definitions, a proposition P need not be in prenex form in order to have property A, but any normal identity P' generated from P by any associated scheme must be in prenex form. However, it is possible to generalize the notion of a generating scheme so that a normal identity associated with P need not be in prenex form. Such a generalization is important because of the falsity of the lemma of 3.3 below. See below: Notes E, p. 571, and F, p. 577, and also footnote 77, p. 555.]

[23] [See Note D, p. 571 below.]

[24] ⟦In the Introduction to Herbrand's thesis, on page 4, we read: "The properties to which we shall apply recursion arguments will be such that for a particular proof or proposition we should always be able to decide whether the properties hold or not."⟧

is an identity. Then P has property A, an associated scheme, as we can easily verify, being

$$\begin{array}{llll} +z & -t & +y & -x \\ +y' & -x' \left\{\begin{array}{l} +z' \\ +z'' \end{array}\right. & \begin{array}{l} -t' \\ -t''. \end{array} & \end{array}$$

An associated type is

$$+y',\ +z,\ -t,\ +y,\ -x',\ +z',\ +z'',\ -x,\ -t',\ -t'',$$

and the associated equations for that associated type are

$$t = y', \quad x' = y', \quad x = z'', \quad t' = z'', \quad \text{and} \quad t'' = z''.$$

The proposition derived from P is

$$(+y',\ +z,\ -t,\ +y,\ -x',\ +z',\ +z'',\ -x,\ -t',\ -t'')[M(x, y, z, t) \vee M(x', y', z', t') \vee M(x', y', z'', t'')],$$

which is a normal identity.

2.32. Remark 1. We can always replace an associated type by another differing only in that a given general variable has been shifted to the left (for this will not prevent us from retaining the same associated equations); this remark is the analogue of 2.12.

2.33. Remark 2. If a proposition has property A, we can assume that immediately behind each brace in the associated scheme only restricted variables occur. For, if an isolated fragment of the scheme is

$$\pm x \left\{\begin{array}{l} +y \\ +y', \end{array}\right.$$

we can replace this fragment by

$$\pm x \quad +y.$$

This can always be done, because, according to Remark 1 (2.32), from any type associated with P we can obtain one in which y' follows y; hence we can, by deleting y' in the first type, obtain a type that is associated with P but is deduced from the second scheme, since this deletion amounts to replacing y' by y in the associated identity.[25]

2.34. Remark 3. We have not obtained a uniform procedure that would enable us to decide whether any given proposition has property A, for in general we do not know how to find the associated scheme.

However, there is a class of propositions for which we have such a procedure, namely, the class of propositions such that the matrix of each is a disjunction of atomic propositions and of negations of atomic propositions. Using the same notation

[25] [Let us say that a scheme in which no general variable occurs immediately behind a brace is *reduced*. Then Remark 2 tells us that a proposition P has property A *only if* there is at least one reduced scheme that generates a normal identity from P. But, by the definition of property A, a proposition P has property A *if* there is at least one scheme generating a normal identity from P. Hence Herbrand has proved that in order to show that a proposition has property A we need consider only reduced schemes. See below: 4.3, p. 553; 5.1, p. 554; and footnote 74, p. 554.]

as above, in 2.2, let us assume that an identity associated with a given member P of this class is[26]

$$M(S_1) \vee \cdots \vee M(S_m).$$

It follows from Chapter 1, 5.22, that for two suitable subscripts, i and j, $M(S_i) \vee M(S_j)$ would also be an identity.[27] The two substitutions ⟦S_i and S_j⟧ would correspond to two lines of the scheme, and these lines together would form a new scheme. If the lines were $\pm x_1, \ldots, \pm x_\alpha, \pm y_1, \ldots, \pm y_\beta$ and $\pm x_1, \ldots, \pm x_\alpha, \pm z_1, \ldots, \pm z_\beta$, the new scheme would be

$$\pm x_1 \ldots \pm x_\alpha \begin{cases} \pm y_1 \ldots \pm y_\beta \\ \pm z_1 \ldots \pm z_\beta, \end{cases}$$

and clearly this scheme, too, would be associated with P.[28] But there are only finitely many such ⟦two-line⟧ schemes; hence by means of only a finite number of tests (this number can be computed quite easily as soon as the proposition P has been stated) we can decide whether P has property A.[29]

To carry out this test efficiently one should remember that not all the lines of a scheme need be similar to one and the same type attached to the proposition P (see 5.1).

2.4. We shall now assume that the propositions being studied contain descriptive functions. We shall give the same definitions as above for the notions "normal identity" and "property A", the only modification being that in a normal identity we can now replace a restricted variable either by a superior general variable, or by an (actually occurring or fictitious) real variable, as before, *or* by a descriptive function whose arguments are such variables. An equation between a restricted variable and a descriptive function of superior general or real variables will still be called an associated equation (2.31).[30]

Let us try to test whether a proposition P ⟦which may now contain descriptive functions⟧ is a normal identity. To do this we have to decide whether the type (1.11) of at least one prenex form of P is normal. Hence let P' be any prenex form of P. In P' we shall replace each restricted variable by a superior general variable, by an (actually occurring or fictitious) real variable, or by a function of such variables; ⟦for at least one replacement⟧ the proposition Π thus obtained must ⟦if the type of P' is

[26] ⟦The French text has "Σ" for "S"; but this is an oversight.⟧

[27] ⟦In Chapter 1, 5.22, Herbrand shows that in the propositional calculus a disjunction of atomic propositions and negations of atomic propositions is provable if and only if it is of the form

$$p \vee \sim p \vee A,$$

where p is a propositional letter.⟧

[28] ⟦In the French text the letters of the scheme are unsigned.⟧

[29] [In 6.3 below, p. 559, Herbrand remarks that the class of propositions just considered constitutes a solvable case of the decision problem because of the proof in Section 5 below that a proposition is provable if and only if it has property A.]

[30] ⟦In a given occurrence of a descriptive function $f(z_1, z_2, \ldots, z_k)$ in a proposition P, each of the variables $z_1, z_2, \ldots, z_k$ may be real or apparent, and, if apparent, general or restricted. When, in trying to obtain a normal identity, we replace a restricted variable y by the descriptive function $f(x_1, x_2, \ldots, x_k)$, the variables $x_1, x_2, \ldots, x_k$ must be general variables that are superior to y in P or (possibly fictitious) real variables of P, no matter what the variables $z_1, z_2, \ldots, z_k$ are in the occurrence of $f(z_1, z_2, \ldots, z_k)$ in P.⟧

normal⟧ turn out to be an identity of the first kind. We consider all individuals (variables and descriptive functions) that in P' are arguments of propositional functions. In Π, because of the associated equations, some of these arguments become identical (hence any proposition that is obtained from P' by the consideration of other equations, but is such that the same arguments turn out to be identical, is also an identity of the first kind). Thus we must find those ⟦associated⟧ equations between arguments that turn P' into an identity of the first kind. There are only a finite and determinate number of equations between individuals. We obtain the associated equations by first equating restricted variables to general variables, to (actually occurring or fictitious) real variables, or to functions of such variables; then we have only to check whether these equations are associated equations.

Now, to find an appropriate set of associated equations is easy, if such a set exists; it suffices, for each system of equations between arguments, to proceed by recursion, using one of the following procedures, which simplify the system of equations to be satisfied.[31]

(1) If one of the equations to be satisfied equates a restricted variable x to an individual, either this individual contains x ⟦or some other restricted variable⟧, and then the equation cannot be satisfied, or else the individual does not contain x ⟦or any other restricted variable, or any general variable that is not superior to x⟧, and then the equation will be one of the associated equations that we are looking for; in the other equations to be satisfied we replace x by the individual;

(2) If one of the equations to be satisfied equates a general variable to an individual that is not a restricted variable, the equation cannot be satisfied;

(3) If one of the equations to be satisfied equates $f_1(\varphi_1, \varphi_2, \ldots, \varphi_n)$ to $f_2(\psi_1, \psi_2, \ldots, \psi_m)$, either the elementary functions f_1 and f_2 are different, and then the equation cannot be satisfied, or they are the same, and then we turn to those equations that equate the φ_i to the ψ_i.

Therefore, if we successively consider each prenex form of P, we shall be able, after a finite and determinate number of steps, to decide whether the proposition P is a normal identity.

Similarly, given a proposition P and any scheme, we can test whether a proposition P' derived from P by the scheme is a normal identity, hence whether the scheme permits us to show that P has property A.[32]

3. Properties B and C

3.1. In what follows we shall consider certain collections of letters; these collections will be called *domains* ⟦*champs*⟧, and the letters the *elements* of the domains.

We shall also consider functions that will be either descriptive functions or new functions; the new functions will be called *index functions* (and will play a role anal-

[31] ⟦That an equation "is to be satisfied" means here that it is to be shown to be associated.⟧

[32] ⟦The last paragraph of Section 2 in the French text reads: "Au bout d'un nombre fini et déterminé d'opérations de ce genre, on arrivera à vérifier si une proposition est normale (ou donc si elle possède la propriété A pour un schème associé donné)". Herbrand's parenthesis was expanded into the last paragraph of the English text.⟧

ogous to that of the descriptive functions).[33] In general, we shall have a number of symbols,

$$f_i(x_1, x_2, \ldots, x_{n_i}), \quad i = 1, 2, \ldots, m,$$

which will be called the *elementary functions* (the x_i will be their arguments); from these we shall form further functions by the following procedure: if $\varphi_1, \varphi_2, \ldots, \varphi_n$ are functions and $f(x_1, x_2, \ldots, x_n)$ is an elementary function, $f(\varphi_1, \varphi_2, \ldots, \varphi_n)$ will also be a function.[34]

We now define the *height of a function*. The height of an elementary function is 1; the height of a nonelementary function $f(\varphi_1, \varphi_2, \ldots, \varphi_n)$ will be greater by 1 than the maximum of the heights of the functions φ_i. We shall also say that the height of an element of a domain is 0.

By definition the *height of a proposition* will be the maximum of the heights of the descriptive functions occurring in it.

If with a system consisting of an elementary function $f(x_1, x_2, \ldots, x_n)$ and n letters $a_{i_1}, a_{i_2}, \ldots, a_{i_n}$ taken from a given domain (these letters being in a definite order) we correlate an element of the domain, this element will be called the *value* of the function for these values of the arguments, or the *value* of $f(a_{i_1}, a_{i_2}, \ldots, a_{i_n})$. We define the *value* of a nonelementary function, when this value exists, in terms of the values of the elementary functions as follows: if the functions $\varphi_1, \varphi_2, \ldots, \varphi_n$ take respectively the values $a_{i_1}, a_{i_2}, \ldots, a_{i_n}$ when values are assigned to their arguments, then the value assigned to the function $f(\varphi_1, \varphi_2, \ldots, \varphi_n)$ is the value assigned to $f(a_{i_1}, a_{i_2}, \ldots, a_{i_n})$. We consider also functions of 0 argument; with each of them we correlate a fixed element of the domain, which will be its *value*. These functions are similar to constants (Chapter 3, 1.12[35]).

A function that ⟦whenever the values of its arguments are taken in a given domain⟧ has its value in the domain is said to be *attached* to the domain.

3.11. In what follows we shall often have to *equate*[36] two elements of a domain. We now define this operation.

Let C be any domain consisting of the elements $a_1, a_2, \ldots, a_n$. To equate a_α to a_β

[33] Observe the close relation between the index functions and Hilbert's logical functions (see *Hilbert 1927*). [See Note I, p. 580 below.]

[34] ⟦See above, p. 527.⟧

[35] ⟦This article states that in a theory there may be constants. See above, p. 527.⟧

[36] [The notion of equating elements of a domain will be used repeatedly by Herbrand. Although the definition he now gives is rather unclear, the underlying idea is simple.

A domain C with its attached functions can be viewed as a slight generalization of an abstract algebra (generalization because an attached function need not be defined for all elements in C). Let R be any congruence relation on C. Then any two elements a_α and a_β of C will be said to be *equated* (under R) if and only if they are congruent under R. (Of course, a_α and a_β can be congruent only if they are accepted as arguments by the same functions.) But then the domain Γ introduced by Herbrand can be taken to be the quotient algebra C/R. Hence the homomorphism in stipulation (4) is the natural homomorphism of C onto Γ. Finally, as Herbrand says, by picking one representative from each of the congruence classes, we can embed Γ in C, and the natural homomorphism becomes an endomorphism on C. (When, however, Herbrand comes to apply his notion of equating, he usually does *not* begin with a specification of the congruence R on C; rather, he first specifies an endomorphism on C and then uses this endomorphism to determine R. See especially the argument for (c) in 3.3 below, p. 548, as well as footnotes 51, 54, and 59.)]

in the domain C is to replace C by any domain Γ satisfying the following four conditions:

(1) Γ has the same attached functions as C;

(2) To each element of C there corresponds exactly one element of Γ and to each element of Γ there corresponds at least one element of C;

(3) Any two elements of C correspond to the same element of Γ if and only if they are the values of individuals obtained thus: we consider the individuals whose expressions, formed from the elementary functions attached to the domain, contain a_α or a_β, and in these expressions we replace a number of occurrences of a_α by a_β, or vice versa (thus, the elements a_α and a_β of C correspond to the same element b_i of Γ; if $f(x)$ is a function attached to the domain C, $f(a_\alpha)$ and $f(a_\beta)$ correspond to the same element b_j of Γ);

(4) If $f(x_1, x_2, \ldots, x_n)$ is a function attached to C ⟦hence, by (1), to Γ⟧, the *value* of $f(b_{i_1}, b_{i_2}, \ldots, b_{i_n})$ is taken to be the element of Γ that corresponds to $f(a_{j_1}, a_{j_2}, \ldots, a_{j_n})$, where the elements a_{j_s} of C correspond to the elements b_{i_s} of Γ (clearly, this assignment specifies a unique value).

This definition allows us to take for the letter b_j any letter that is one of the corresponding elements of C, for example the letter with the least subscript. Then Γ becomes a part of C, and the values of the attached functions ⟦for arguments in Γ⟧ are the same in the two domains. Henceforth this is what we shall do.

3.2 From now on we shall consider a proposition P in which there may occur descriptive functions (and even variables of various types,[1] if we so desire; the occurrence of such types would change nothing essential in what follows, but to simplify notation we shall always assume that there is only one type). We shall assume that P contains no real variable: if P does contain the real variables $x_1, x_2, \ldots, x_n$, we replace it by $(x_1)(x_2) \ldots (x_n)P$; see Chapter 2, 4.7;[37] in 3.221 we shall see that the order in which we take $x_1, x_2, \ldots, x_n$ is immaterial for the considerations that follow.

With each *general* variable y we correlate an index function whose arguments are the restricted variables that *dominate* y (1.41). In particular, to a real variable there corresponds a function of 0 argument (a "constant").

3.21. We now consider a domain C_1 consisting of the letters $a_1, a_2, \ldots, a_{n_1}$. In what follows we shall assume that this domain is fixed; later (3.51) we shall see that we can take any domain for C_1. A domain C_2 is then obtained thus: to each elementary function (index function or descriptive function) taken together with a given system of arguments chosen in C_1 we assign a letter that will be in C_2; this letter is to be the value of the function for the system of arguments, and the assignment is such that each letter of C_2 will be the value of just one function for just one system of arguments. Let $a_{n_1+1}, \ldots, a_{n_2}$ be the elements of C_2. In general, we construct C_k from $C_1, C_2, \ldots, C_{k-1}$ by assigning a letter of C_k to each elementary function taken together with a given system of arguments; these arguments are chosen in ⟦the union of⟧ $C_1, C_2, \ldots,$ and C_{k-1}, and at least one argument is taken in C_{k-1}; the assignment is so made that each element of C_k corresponds to just one function for just one system of arguments. The letter assigned is to be the value of the function for the system of arguments.

We thus construct $C_1, C_2, \ldots, C_{p+1}$; finally, if h is the height of the proposition P,

[37] ⟦This article states that $\Phi(x)$ is provable in Herbrand's system of quantification theory if and only if $(y)\Phi(y)$ is.⟧.

we construct $C_{p+2}, \ldots, C_{p+h+1}$ (omitting from these latter domains, if we so desire, elements corresponding to the index functions; it is easily seen that this will not affect our argument).[38]

For each k let $a_{n_{k-1}+1}, \ldots, a_{n_k}$ be the elements of C_k. We shall call the number k the *order* of C_k. Moreover, we shall write N for n_p.[39] We note that C_k contains the values of all the functions of height $k - 1$ when their arguments are taken in C_1.

We could construct similar domains by starting from any descriptive or index functions whatever, and not from ⟦just⟧ those supplied by the proposition P. The functions thus used will be said to *generate* the domains.

3.211. When we say that we *equate* (3.11) two elements in the domains $C_1, C_2, \ldots, C_p$, we mean that we equate them in the domain that is the union of these domains; using the convention made at the end of 3.11, we see that each domain C_i is replaced by a domain Γ_i that consists of elements of C_i.[40]

3.212. We shall say that we have a *system of logical values* in these domains whenever a logical value has been assigned to each atomic proposition that we obtain by replacing the arguments of the atomic propositions of P with elements of the domains. From such a system we can derive the logical value of any proposition that is free of both apparent variables and descriptive functions and contains as real variables only elements of the domains.

3.22. Any proposition included in a given proposition P, that is, any proposition of which P is a propositional function (the first proposition is always the scope of some logical sign), will be called a *part*, or a *subproposition*, of P. With each part of the proposition P we correlate a proposition, called an *expansion* ⟦*réduite*⟧ (Chapter 2, Section 8),[41] according to the following rules:[42]

(1) The expansion of an atomic proposition is the proposition itself;

(2) If the expansion of A is a, that of $\sim A$ is $\sim a$;

(3) If ⟦the expansion of A is a and⟧ that of B is b, that of $A \vee B$ is $a \vee b$;

(4) In case x is a general variable, if the expansion of $\Phi(x)$ is $\varphi(x)$, the expansion of $(\pm x)\Phi(x)$ is $\varphi[f_x(y_1, y_2, \ldots, y_n)]$, where $f_x(y_1, y_2, \ldots, y_n)$ is the index function corresponding to x (hence $y_1, y_2, \ldots, y_n$ are real variables of $\Phi(x)$);

[38] [The subscripts "p", "$p + 1$", and "$p + h + 1$" that Herbrand introduces here find their justification on page 544 below.]

[39] [Several points to be made in footnotes below can be stated more easily if we now say that each element of C_k is also of *order* k and write G_p for the union $C_1 \cup C_2 \cup \cdots \cup C_p$. Thus $N = n_p$ is the cardinality of G_p. (This use of G_p should not be confused with Herbrand's use of D_p, introduced on page 552 below; see footnote 71 below, p. 553.) Moreover it will henceforth be assumed that all the elements of G_p are arranged in some sequence in which, for each $k \leqq p$, no element of order k precedes any element of order less than k.]

[40] [Apparently, Herbrand is here stipulating that only elements of the same order (that is, elements belonging to the same domain C_i) are to be equated in $G_p = C_1 \cup C_2 \cdots \cup C_p$. (Let us say that such an equating is *order-preserving*.) But in all subsequent important applications of equating he flouts this stipulation (see footnote 51, footnote 59, and Note E).]

[41] [The *expansion* correlated in Section 8 of Chapter 2 with a proposition P must be sharply distinguished from the *expansion* now to be specified. The former is, for each finite domain, merely the truth-functional expansion of P that we obtain by turning existential quantifications into disjunctions and universal quantifications into conjunctions. Such an expansion of P over a finite domain will henceforth be called the *common expansion* of P over the domain.]

[42] [For each $p \geqq 1$ these rules specify the expansion of P with respect to the elements of G_p (see Rule 5, where N is the cardinality of G_p). To make this dependence on G_p explicit, we shall sometimes speak of *the expansion of P over G_p*.]

(5) In case x is a restricted variable, if the expansion of $\Phi(x)$ is $\varphi(x)$, the expansion of $(-x)\Phi(x)$ is

$$\varphi(a_1) \vee \varphi(a_2) \vee \cdots \vee \varphi(a_N),$$

and the expansion of $(+x)\Phi(x)$ is

$$\varphi(a_1) \mathbin{\&} \varphi(a_2) \mathbin{\&} \cdots \mathbin{\&} \varphi(a_N).$$

These rules enable us to associate an expansion with any proposition assumed to contain no real variables. Let us now replace by their values all the functions that occur in the expansion, so that no function remains in the expansion. Obviously we can do this, by introducing elements of the domains $C_{p+1}, C_{p+2}, \ldots, C_{p+h+1}$. We thus obtain a proposition Π. If Π is an identity of the first kind, we shall say that P has *property B of order p.*[43]

Let us perform the same operations again, but now *let the index functions that replace the general variables have for their arguments not the ⟦restricted⟧ variables dominating these general variables but the ⟦restricted⟧ variables superior (1.41) to these general variables.* If the proposition Π thus obtained is an identity, we shall say that P *has property C of order p.*[44]

3.221. We see immediately that—as was noted in 3.2—if we start from a proposition containing real variables, properties B and C are independent of the order in which we generalize these variables (hence do not depend upon which of these variables dominate a given variable). For the index functions of 0 argument that correspond to the real variables have their values in C_2; they play no role in the construction of the domains that come after C_2.

We see that property B and property C are identical if the proposition is in canonical form (1.2 and 1.41).

3.3. Lemma. *For every p, if a proposition has property B, or property C, of order p, any proposition derived from it by the rules of passage has property B, or property C, of order p, provided the domain C_1 is the same in both cases.*[45]

We shall show that the properties are preserved when any rule of passage is applied.

[43] [Herbrand will also call the expression Π, which is free of both quantifiers and functions, an *expansion of P*. Note that Π contains elements of $C_{p+1} \cup \cdots \cup C_{p+h+1}$, $h \geqq 0$, but the expansion of P determined just by Rules 1–5 contains elements only of $G_p = C_1 \cup \cdots \cup C_p$.

Note further that, if A is any subproposition of P, the expansion of A need not occur in the expansion of P. Indeed, if A lies within the scopes of $n > 0$ restricted quantifiers $\pm x_1, \ldots, \pm x_n$ and if either at least one variable x_i, $i = 1, \ldots, n$, occurs in A or at least one general quantifier occurs in A, the expansion of A does not occur in the expansion of P. Rather, with respect to each given system of values for $x_1, \ldots, x_n$, a substitution instance of the expansion of A occurs in the expansion of P. However, for convenience we shall continue to speak in subsequent footnotes of each such substitution instance as the *expansion of A in the expansion of P* (for a given set of values of $x_1, \ldots, x_n$). If $n = 0$, the expansion of A in the expansion of P *is* the expansion of A.]

[44] [In the present paragraph Herbrand has again, just as we noted in the first paragraph of footnote 43, specified two expressions, both of which he will again call expansions of P. We shall need to distinguish the two expansions defined in the preceding paragraph from the two expansions just defined. So, in analogy with Herbrand's usage on page 549 below, let us call each of the former a *first expansion of P over G_p*, and each of the latter a *second expansion of P over G_p*.]

[45] [This is the basic lemma of Herbrand's thesis (see, for example, his remarks below, p. 559). But for certain cases of property C it is false (see Note E, p. 571 below). To be able to carry out Herbrand's argument in the remainder of the present chapter—in particular, the argument for his Fundamental Theorem—we must substantially emend this lemma (see Note E).]

(1) *Case of property B.*

For the rules of passage governing the sign $\sim$, the lemma immediately follows from Chapter 1, 3.42.[46]

If we consider the rules of passage governing the sign $\vee$, we see immediately (by using Chapter 1, 5.34[47]) that $(x)[\Phi(x) \vee Z]$ and $(x)\Phi(x) \vee Z$ always have equivalent expansions; similarly for $(\exists x)[\Phi(x) \vee Z]$ and $(\exists x)\Phi(x) \vee Z$.

(2) *Case of property C.*

For the rules of passage governing the sign $\sim$, the result follows from Chapter 1, 3.42.[46]

Let us now consider the rules of passage governing the sign $\vee$.

(*a*) *Passage from* $(\pm x)\Phi(x) \vee Z$ *to* $(\pm x)[\Phi(x) \vee Z]$, *and inverse passage, in case x is a general variable.*

It is sufficient to observe that these two subpropositions have identical expansions.

(*b*) *Passage from* $(\exists x)\Phi(x) \vee Z$ *to* $(\exists x)[\Phi(x) \vee Z]$, *and inverse passage, in case x is a restricted variable* (hence the occurrence is positive).

In its first form P will be written P_1 and, in its second, P_2. In P_2 the index functions of the general variables of the subproposition Z are functions of x. If y is a general variable of Z and $x_1, x_2, \ldots, x_n$ are the restricted variables superior to x, let $f_y(x_1, x_2, \ldots, x_n, x)$ be the index function of y in P_2 and let $f_y(x_1, x_2, \ldots, x_n)$ be the index function of y in P_1.[48]

The descriptive and index functions of P_1 will, from C_1, generate the domains $C_2^{(1)}, C_3^{(1)}, \ldots$. The functions of P_2 differ from those of P_1 only in that, for each general variable y of Z, $f_y(x_1, x_2, \ldots, x_n, x)$ has taken the place of $f_y(x_1, x_2, \ldots, x_n)$, and they will, from C_1, generate the domains $C_2^{(2)}, C_3^{(2)}, \ldots$.

(1)[49] Let us observe that we can pass from the $C_i^{(2)}$ to the $C_i^{(1)}$ thus: for each general variable y of Z and for[50] each given system of elements $x_1, x_2, \ldots, x_n$ ⟦in the union⟧ of the $C_i^{(2)}$, the values of $f_y(x_1, x_2, \ldots, x_n, x)$ for all x in the $C_i^{(2)}$ are all equated (3.211) to one of these values, namely $f_y(x_1, x_2, \ldots, x_n, a_i)$, where a_i is an arbitrary element of the $C_i^{(2)}$. The domains $C_i^{(1)}$ can then be regarded as part of the domains $C_i^{(2)}$.[51]

[46] ⟦This article gives

$$\vdash (p_1 \,\&\, p_2 \,\&\, \cdots \,\&\, p_n) \equiv \sim(\sim p_1 \vee \sim p_2 \vee \cdots \vee \sim p_n),$$

$$\vdash \sim(p_1 \,\&\, p_2 \,\&\, \cdots \,\&\, p_n) \equiv (\sim p_1 \vee \sim p_2 \vee \cdots \vee \sim p_n),$$

and

$$\vdash \sim(p_1 \vee p_2 \vee \cdots \vee p_n) \equiv (\sim p_1 \,\&\, \sim p_2 \,\&\, \cdots \,\&\, \sim p_n).⟧$$

[47] ⟦Distributivity of logical products and sums.⟧

[48] [The notations $f_y(x_1, x_2, \ldots, x_n, x)$ and $f_y(x_1, x_2, \ldots, x_n)$ exhibit just the restricted variables that are now at the center of Herbrand's attention. For, in general, the index functions of y will have as arguments also those $s \geqq 0$ restricted variables $z_{y1}, z_{y2}, \ldots, z_{ys}$ whose quantifiers occur in Z and are superior to y.]

[49] [In this paragraph Herbrand considers the case of the passage from $(\exists x)[\Phi(x) \vee Z]$ to $(\exists x)\Phi(x) \vee Z$. Hence he proves that, if the expansion of P_2 over $G_p^{(2)}$ is an identity, so is the expansion of P_1 over $G_p^{(1)}$.]

[50] ⟦Herbrand writes "for each given system of elements $x_1, x_2, \ldots, x_n, y$", but "y" is extraneous.⟧

[51] [For $n > 0$ and $p > 2$ the element a_i cannot be completely arbitrary if $G_{p+1}^{(1)}$ is to be embedded in $G_{p+1}^{(2)}$. Let $x_1, x_2, \ldots, x_n, x$ be elements of $G_{p+1}^{(2)}$ that are all of order 1, but let a_i in $G_{p+1}^{(2)}$ be of order p. Then the element b in $G_{p+1}^{(2)}$ that is the value of $f_y(x_1, x_2, \ldots, x_n, x)$ is of order 2 and the element d in $G_{p+1}^{(2)}$ that is the value of $f_y(x_1, x_2, \ldots, x_n, a_i)$ is of order $p + 1$. Hence the element

So we see that we can pass from the expansion of P_2 to a proposition equivalent to the expansion of P_1 by doing the following: for [50] each given system of values of $x_1, x_2, \ldots, x_n$, we replace, in the expansion of P_2, the values of $f_y(x_1, x_2, \ldots, x_n, x)$ obtained for all x by the one value among them that is in the domains $C_i^{(1)}$. Hence, if the expansion of P_2 is an identity, so is the expansion of P_1.

(2) If the expansion Π_2 of P_2 is not an identity, there exists, according to Chapter 1, 5.21,[52] a system of logical values (3.212) falsifying Π_2; hence, as we shall now show, there exists a system of logical values falsifying the expansion Π_1 of P_1.[53]

To do this we replace the expansion Π_1 of P_1 by another expansion, obtained as follows. Let $a_1, a_2, \ldots, a_N$ be the elements of the domains $C_1^{(2)}, \ldots, C_p^{(2)}$; in these domains we define the value of $f_y(x_1, x_2, \ldots, x_n)$ to be the value of $f_y(x_1, x_2, \ldots, x_n, a_i)$ for some a_i, which we shall select shortly and which will depend upon $x_1, x_2, \ldots, x_n$. This definition will permit us to take subdomains of these domains as the domains $C_i^{(1)}$. The modification that we intend to introduce consists in using all N elements of ⟦the union of⟧ the $C_i^{(2)}$ to form the expansion of P_1, just as they are used to form that of P_2. In other words, of the five rules stated in 3.22 the fifth alone will undergo a modification, namely the following: the letters $a_1, a_2, \ldots, a_N$ will denote the elements of the domains $C_1^{(2)}, C_2^{(2)}, \ldots, C_p^{(2)}$, not those of the domains $C_1^{(1)}, C_2^{(1)}, \ldots, C_p^{(1)}$. The transformation described at the end of 3.22 ⟦that is, the replacement of functions by their values⟧ will, of course, be carried out in the $C_i^{(2)}$ and not in the $C_i^{(1)}$. We thus construct a new expansion Π_1' of P_1.[54]

If we observe that

$$[\varphi(a_{i_1}) \vee \varphi(a_{i_2}) \vee \cdots \vee \varphi(a_{i_m})] \supset [\varphi(a_1) \vee \varphi(a_2) \vee \cdots \vee \varphi(a_N)]$$

and

$$[\varphi(a_1) \,\&\, \varphi(a_2) \,\&\, \cdots \,\&\, \varphi(a_N)] \supset [\varphi(a_{i_1}) \,\&\, \varphi(a_{i_2}) \,\&\, \cdots \,\&\, \varphi(a_{i_m})]$$

in $G_{p+1}^{(2)}$ that is the value of $f_y(b, x_2, \ldots, x_n, x_1)$ is of order 3. But the value of $f_y(d, x_2, \ldots, x_n, x_1)$ is of order $p + 2$, hence is not an element of $G_{p+1}^{(2)}$. Therefore $C_3^{(1)}$ cannot be embedded in $G_{p+1}^{(2)}$. Hence a_i must always be of an order less than or equal to the maximum of the orders of $x_1, x_2, \ldots, x_n$. However, to satisfy this requirement we need only take a_i to be always a_1. (Interestingly enough, in the original French text "$f_y(x_1, \ldots, x_n, a_1)$" occurs; but in his list of errata Herbrand changed this formula to "$f_y(x_1, \ldots, x_n, a_i)$".) Note that even this emended equating is not order-preserving (see above: footnotes 36, p. 541, and 40, p. 543).]

[52] ⟦This article states that a proposition is an identity of the first kind if and only if it has the logical value "true" for any assignment of logical values to the atomic parts of the proposition.⟧

[53] [It is indeed true that, if the expansion of P_2 over $G_p^{(2)}$ is not an identity, neither is the expansion of P_1 over $G_p^{(1)}$, but the argument that Herbrand is about to give is faulty (see footnote 59).]

[54] [Herbrand is doing two different things in this paragraph. First, to each ordered n-tuple $\langle a_{i_1}, \ldots, a_{i_n} \rangle$ of elements of $G_p^{(2)}$ he assigns a unique element a_i of $G_p^{(2)}$ so that for all a in $G_p^{(2)}$ the value in $G_{p+1}^{(2)}$ of $f_y(a_{i_1}, \ldots, a_{i_n}, a)$ is equated to the value of $f_y(a_{i_1}, \ldots, a_{i_n}, a_i)$. (This a_i is specified on page 547, lines 19–27. Herbrand makes it depend upon both $a_{i_1}, \ldots, a_{i_n}$ and the particular system $\mathscr{S}$ of logical values that is assumed to falsify Π_2.) Herbrand concludes—falsely, as we shall see in footnote 59—that, as a result of this equating, $G_p^{(1)}$ is embedded in $G_p^{(2)}$, and so he takes the expansion Π_1 of P_1 as being formed over this subset of $G_p^{(2)}$. That is, the value of $f_y(a_{i_1}, \ldots, a_{i_n}, a)$ in $G_p^{(2)}$ is defined by Herbrand to be $f_y(a_{i_1}, \ldots, a_{i_n}, a_i)$. (See footnote 36 above, p. 541.)

However, this is not the modification that Herband intends to introduce. Rather, as he says in the latter part of the paragraph, he constructs a new expansion Π_1' of P_1 over *all* of $G_p^{(2)}$, still assuming that the same a_i has been chosen. (Remember, an expansion can be formed over any union of domains.) He will then in the next few paragraphs show that $\Pi_1 \supset \Pi_1'$ is an identity and argue that Π_1' is false under $\mathscr{S}$.]

are identities and apply the theorem of Chapter 2, 3.2,[55] we can conclude that $\Pi_1 \supset \Pi_1'$ is an identity. Hence, if Π_1' is not an identity, neither is Π_1.[56]

We now show that, if we have a system of logical values that falsifies Π_2, we can, by suitably choosing the value of each $f_y(x_1, x_2, \ldots, x_n)$, obtain a system of logical values that falsifies Π_1'.[57]

The expansions of the subpropositions $(\mathfrak{E}x)\Phi(x) \vee Z$ and $(\mathfrak{E}x)[\Phi(x) \vee Z]$ depend upon the ⟦restricted⟧ variables $x_1, x_2, \ldots, x_n$, which are superior to x. But we shall consider these expansions only for given values of these variables, namely, $a_{i_1}, a_{i_2}, \ldots, a_{i_n}$; then these expansions have only one occurrence in Π_2 and one in Π_1 ⟦rather, Π_1'⟧, as we can readily see.[58]

For these values, let $\pi(x)$ be the expansion ⟦in Π_2⟧ of Z when Z occurs in P_2 and let a_i be the ⟦still to be selected⟧ element such that, for each general variable y of Z, the value of $f_y(a_{i_1}, a_{i_2}, \ldots, a_{i_n}, a_i)$ is the value of $f_y(a_{i_1}, a_{i_2}, \ldots, a_{i_n})$. We readily see that, when Z occurs in P_1, its expansion ⟦in Π_1'⟧ is $\pi(a_i)$. Hence, when $(\mathfrak{E}x)\Phi(x) \vee Z$ occurs in P_1, its expansion ⟦in Π_1'⟧ is

$$[\varphi(a_1) \vee \varphi(a_2) \vee \cdots \vee \varphi(a_N)] \vee \pi(a_i), \tag{1}$$

but, when $(\mathfrak{E}x)[\Phi(x) \vee Z]$ occurs in P_2, its expansion ⟦in Π_2⟧ is

$$[\varphi(a_1) \vee \pi(a_1)] \vee [\varphi(a_2) \vee \pi(a_2)] \vee \cdots \vee [\varphi(a_N) \vee \pi(a_N)]. \tag{2}$$

To proposition (2) the system $\mathscr{S}$ of logical values now under consideration ⟦for Π_2⟧ assigns either the logical value "true" or the logical value "false".

1. If the logical value is "true", then either at least one of the $\varphi(a_j)$ has the logical value "true", and we set

$$f_y(a_{i_1}, a_{i_2}, \ldots, a_{i_n}) = f_y(a_{i_1}, a_{i_2}, \ldots, a_{i_n}, a_1),$$

or at least one of the $\pi(a_j)$, say $\pi(a_k)$, has the logical value "true", and we set

$$f_y(a_{i_1}, a_{i_2}, \ldots, a_{i_n}) = f_y(a_{i_1}, a_{i_2}, \ldots, a_{i_n}, a_k).$$

2. If the logical value is "false", we set

$$f_y(a_{i_1}, a_{i_2}, \ldots, a_{i_n}) = f_y(a_{i_1}, a_{i_2}, \ldots, a_{i_n}, a_1).$$

[55] ⟦"Let $F(p)$ be a propositional function of the proposition p such that the occurrences of p are all positive or all negative. If these occurrences are positive, then

$$\vdash (p \supset q) \supset [(F(p) \supset F(q)].$$

If they are negative, then

$$\vdash (p \supset q) \supset [F(q) \supset F(p)]."⟧$$

[56] [The inductive argument that Herbrand has just sketched in order to show that $\Pi_1 \supset \Pi_1'$ is an identity justifies the following more general result, which is used below in Note E, p. 571, and Note F, p. 577.

Let P be any proposition, Π an expansion of P over G, and Π' an expansion of P over G'. If G is a subset of G', then $\Pi \supset \Pi'$ is an identity. Thus, if P has property C of order p, it has property C of order q for each q greater than p. See 3.54, p. 550 below and footnote 66.]

[57] [Thus the specification of each a_i must satisfy the following two requirements: (1) $G_p^{(1)}$ is embeddable in $G_p^{(2)}$; (2) the extended expansion Π_1' of P_1 is false under a system of logical values if Π_2 is false under some system of logical values.]

[58] [See the second paragraph of footnote 43.]

Thus we see that propositions (1) and (2) will have the same logical value.[59] So the method just described ⟦of selecting a_i⟧ does indeed give the same logical value to Π_2 and Π_1.

(c) *Passage from* $(x)\Phi(x) \vee Z$ *to* $(x)[\Phi(x) \vee Z]$, *and inverse passage, in a negative occurrence.*[60]

The argument is the same as in (*b*). Only the end is slightly modified. We have to consider

$$[\varphi(a_1) \mathbin{\&} \varphi(a_2) \mathbin{\&} \cdots \mathbin{\&} \varphi(a_N)] \vee \pi(a_i) \tag{1}$$

and

$$[\varphi(a_1) \vee \pi(a_1)] \mathbin{\&} [\varphi(a_2) \vee \pi(a_2)] \mathbin{\&} \cdots \mathbin{\&} [\varphi(a_N) \vee \pi(a_N)]. \tag{2}$$

To proposition (2) the system of logical values under consideration ⟦for Π_2⟧ assigns either the logical value "true" or the logical value "false".

1. If the logical value is "true", either all the $\varphi(a_j)$ have the same logical value, and we put

$$f_y(a_{i_1}, a_{i_2}, \ldots, a_{i_n}) = f_y(a_{i_1}, a_{i_2}, \ldots, a_{i_n}, a_1),$$

or some $\pi(a_j)$, say $\pi(a_k)$, has the logical value "true", and we put

$$f_y(a_{i_1}, a_{i_2}, \ldots, a_{i_n}) = f_y(a_{i_1}, a_{i_2}, \ldots, a_{i_n}, a_k).$$

2. If the logical value is "false", some $\pi(a_j)$, say $\pi(a_k)$, has the logical value "false", and we put

$$f_y(a_{i_1}, a_{i_2}, \ldots, a_{i_n}) = f_y(a_{i_1}, a_{i_2}, \ldots, a_{i_n}, a_k).$$

The end of the argument is then unchanged.

3.31. Corollary. Since properties B and C of order p are identical for a proposition in canonical form (3.221), *these two properties are always equivalent.*[61]

3.4. By considering the relation between the expansion of an arbitrary proposition P and the expansion of a prenex form of P, we can give property B a simple formulation.[62] We consider the domains C_i (of 3.21) that are generated by the elementary descriptive functions and by the index functions corresponding to the general variables of the proposition P, the arguments of each of these index functions being the ⟦restricted⟧

[59] [In his specification of a_i Herbrand goes wrong. The difficulty is the same as the one described in footnote 51. If a_k is of too high an order, $G_p^{(1)}$ cannot be embedded in $G_p^{(2)}$. So let us once again modify Herbrand's argument and take a_i to be always a_1 (for all systems of logical values assigned to Π_2). But now we can no longer say, as Herbrand does, that, for each given set of elements $a_{i_1}, \ldots, a_{i_n}$, the expansion of $(\exists x)\Phi(x) \vee Z$ in Π_1' has the same logical value under $\mathscr{S}$ as the corresponding expansion of $(\exists x)[\Phi(x) \vee Z]$ in Π_2. However, we can say that, if any given expansion of $(\exists x)[\Phi(x) \vee Z]$ in Π_2 is false under $\mathscr{S}$, the corresponding expansion of $(\exists x)\Phi(x) \vee Z$ in Π_1' is also false under $\mathscr{S}$. But nothing more is required in order to prove that, if $\mathscr{S}$ falsifies Π_2, as we are assuming, then $\mathscr{S}$ falsifies Π_1'. For, by hypothesis, each expansion of $(\exists x)\Phi(x) \vee Z$ occurs positively in Π_1 and each expansion of $(\exists x)[\Phi(x) \vee Z]$ occurs positively in Π_2. Hence, if $\mathscr{S}$ verifies an expansion of $(\exists x)[\Phi(x) \vee Z]$ in Π_2 and proposition (1) is the corresponding expansion of $(\exists x)\Phi(x) \vee Z$ in Π_1', it does not matter for the falsity of Π_1' under $\mathscr{S}$ whether $\mathscr{S}$ verifies or falsifies proposition (1). So we need consider only those expansions of $(\exists x)[\Phi(x) \vee Z]$ in Π_2 that are falsified by $\mathscr{S}$. (Compare the proof of Lemma I in Note E, p. 574 below.)]

[60] [See Note E, p. 571 below.]

[61] [See Note F, p. 577 below.]

[62] ⟦In the French text the paragraph begins: "En mettant les propositions sous une forme normale, on peut mettre la propriété B sous une forme simple". The translators found it advisable to somewhat expand the text.⟧

variables dominating the general variable with which the function is correlated. In the matrix of P we replace each restricted variable by an element arbitrarily chosen in ⟦the union of⟧ $C_1, C_2, \ldots,$ and C_p; then we replace each general variable y by the value of the corresponding index function $f_y(a_{i_1}, a_{i_2}, \ldots, a_{i_n})$, the arguments $a_{i_1}, a_{i_2}, \ldots, a_{i_n}$ being the elements by which the restricted variables dominating y have been replaced; from the resulting proposition we eliminate each descriptive function by replacing it with its value. We now form the disjunction of all the elementary propositions that are thus obtained when we replace the restricted variables in all possible ways. To say that P has property B of order p amounts to saying that this disjunction is an identity; we can see this by simply going back to the definition of property B and applying it to a prenex form of P. This disjunction will henceforth be called *the first disjunction of order p associated with P.*

To show that P has property C we would now consider index functions that have superior ⟦restricted⟧ variables as arguments and we would construct the domains generated by the descriptive functions and these index functions. On P we would perform the same operations as above, but now we would replace each general variable y by the value of the corresponding index function $f_y(a_{i_1}, a_{i_2}, \ldots, a_{i_n})$, where the arguments $a_{i_1}, a_{i_2}, \ldots, a_{i_n}$ are the elements by which we replaced the restricted variables preceding y in the prefix (for these are the ⟦restricted⟧ variables superior to y in the prenex form). We thus obtain *the second disjunction of order p associated with P*, and this disjunction must be an identity if P is to have property C.[63]

3.5. We shall now make some very important remarks about properties B and C (the arguments are the same for both).

3.51. (1) *The number of elements in the domain C_1 is immaterial* (hence we can in general take it equal to 1).

(*a*) We can increase the number. For let $C_1, C_2, \ldots, C_p, \ldots$ be any domains, and assume that C_1 has n elements, $a_1, a_2, \ldots, a_n$. Let $\Gamma_1, \Gamma_2, \ldots, \Gamma_p, \ldots$ be some other domains generated by the same functions, and assume that Γ_1 has m elements, $a_1, a_2, \ldots, a_m$. Assume that $m < n$. The elements of Γ_i are denoted by the same letters as certain elements of C_i, so that a function $f(a_{i_1}, a_{i_2}, \ldots, a_{i_n})$ will have the same value on Γ_i as on C_i. Then the terms of the first associated disjunction of order p that corresponds to the domains Γ_i are some of the terms of the first associated disjunction of order p that corresponds to the domains C_i; hence, if the former is an identity, so is the latter, since $\vdash A \supset (A \vee B)$.

[63] [Since we know (see Note F, p. 577 below) that for an arbitrary proposition P, even if P has property C of order p, no prenex form of P need have property C of order p, to obtain a correct definition of the second disjunction of order p associated with P we must replace the second sentence of this paragraph by the following sentence:

On P we would perform the same operations as above, but now we would replace each general variable y by the value of the corresponding index function $f_y(a_{i_1}, a_{i_2}, \ldots, a_{i_n})$, where the arguments $a_{i_1}, a_{i_2}, \ldots, a_{i_n}$ are the elements by which we replaced the restricted variables *whose quantifiers have within their scopes the quantifier of y* (for these variables are the restricted variables superior to y).

But, when thus emended, the second disjunction of order p associated with P is what in Note E, p. 572 below, is called the standard expansion of P of order p. So, to avoid confusion in footnotes below, we shall preserve the term "second disjunction of order p associated with P" for the nonemended notion. (Note, however, that, if P is in prenex form, the (nonemended) second disjunction of order p associated with P is still the standard expansion of P of order p.)]

(*b*) We can decrease the number. Let us use the same notation. We can pass from C_i to Γ_i by equating (3.11; 3.211) $a_{m+1}, a_{m+2}, \ldots, a_n$ to, say, a_1. In the first associated disjunction of order p that corresponds to the domains C_i we replace $a_{m+1}, a_{m+2}, \ldots, a_n$ by a_1. We see immediately that the disjuncts thus obtained are disjuncts of the first associated disjunction of order p that corresponds to the domains Γ_i and are in fact all of these disjuncts. Hence, if the former disjunction is an identity, so is the latter, as we see by Remark 3.31 of Chapter 1.[64]

3.52. (2) *Without modifying properties B and C, we can assume that the domains that we construct in order to show that a proposition has property B, or property C, are generated not only by the functions introduced above but also by other functions (called "fictitious functions").* We can, in particular, introduce *fictitious* real variables, and this will lead to the introduction of fictitious *constants* (3.2).

Let $\varphi_i(x_1, x_2, \ldots, x_{n_i})$, with $i = 1, 2, \ldots, \alpha$, be the new functions that are introduced; let $\psi_j(x_1, x_2, \ldots, x_{m_j})$, with $j = 1, 2, \ldots, \beta$, be the original functions; let $C_1, C_2, \ldots, C_p, \ldots$ be the domains generated by the φ_i and the ψ_j. We equate all the elements that are values of the functions $\varphi_i(a_1^k, a_2^k, \ldots, a_{n_i}^k)$ to some element of C_1. We thus obtain the domains $C_1, \Gamma_2, \Gamma_3, \ldots, \Gamma_p$, generated by the functions ψ_j. By the argument used above in (*b* ⟦rather, *a*⟧) we see that, if the first associated disjunction of order p obtained by means of the former ⟦rather, latter⟧ domains is an identity, so is the disjunction obtained by means of the latter ⟦rather, former⟧ domains.

3.53. (3) ⟦*For every p,*⟧ *if both P and Q have property B* ⟦*of order p*⟧, *then P & Q has property B* ⟦*of order p*⟧.[65]

Let us construct the domains that correspond to the proposition P & Q, namely $C_1, C_2, \ldots, C_p, \ldots$. In these domains the first disjunction of order p associated with P is of the form

$$P_1 \vee P_2 \vee \cdots \vee P_N,$$

where each P_i results from the matrix of P by the operations described in 3.4. According to 3.52, this first disjunction is an identity. Similarly, the disjunction associated with Q is of the form

$$Q_1 \vee Q_2 \vee \cdots \vee Q_M.$$

Now, clearly, the disjunction associated with P & Q contains all terms of the form P_i & Q_j; according to Chapter 1, 5.34,[47] it is an identity.

3.54. (4) *For every p, if a proposition P has property B, or property C, of order p, then P has property B, or property C, of order q for every q greater than p.*[66]

The domains considered are in one case $C_1, C_2, \ldots, C_p$ and in the other $C_1', C_2', \ldots, C_p', \ldots, C_q'$. Let us use the same letters for the elements of C_i', $1 \leqq i \leqq p$, as for those of C_i. Then the disjuncts of the first disjunction of order p associated with P are

[64] ⟦"Since $\vdash p \equiv (p \vee p)$, we can always in a true[8] ⟦quantifier-free⟧ proposition replace $p \vee p$ by p, and conversely."⟧

[65] [Using the standard expansion of order p of P and that of Q, we can similarly prove:
If both P and Q have property C of order p, then P & Q has property C of order p. (This is needed for the proof of Lemma 3 in 5.3 below.)]

[66] [To appreciate the simplicity gained by the use of the notion of standard expansion compare the argument that Herbrand now gives for this result with the argument that is implicit in Herbrand's text on pp. 546–547 above (see footnote 56).]

among those of the first disjunction of order q associated with P; hence, if the former disjunction is an identity, so is the latter (by virtue of the identity $\vdash A \supset (A \vee B)$).

3.6. Consider the following example of a proposition assumed to have property C of order 2:

$$(x)(\exists y)(z)P(x, y, z, f(y, z)),$$

where $f(y, z)$ is an elementary descriptive function; this proposition is of height 1.

Two index functions have to be introduced: φ_x, with no argument, for x and $\varphi_z(y)$ for z.

The domain C_1 will consist of a_1.

The domain C_2 will consist of a_2, which will be φ_x,
a_3, which will be $\varphi_z(a_1)$,
and a_4, which will be $f(a_1, a_1)$.

The domain C_3 will consist of

a_5, a_6, a_7, which will be $\varphi_z(a_2)$, $\varphi_z(a_3)$, $\varphi_z(a_4)$,
a_8, a_9, a_{10}, which will be $f(a_2, a_2)$, $f(a_3, a_3)$, $f(a_4, a_4)$,
a_{11}, a_{12}, a_{13}, which will be $f(a_1, a_2)$, $f(a_1, a_3)$, $f(a_1, a_4)$,
a_{14}, a_{15}, a_{16}, which will be $f(a_2, a_3)$, $f(a_2, a_4)$, $f(a_3, a_4)$,
a'_{11}, a'_{12}, a'_{13}, which will be $f(a_2, a_1)$, $f(a_3, a_1)$, $f(a_4, a_1)$,
a'_{14}, a'_{15}, a'_{16}, which will be $f(a_3, a_2)$, $f(a_4, a_2)$, $f(a_4, a_3)$.

In C_4 we consider only a_{17}, which will be $f(a_3, a_6)$, a_{18}, which will be $f(a_4, a_7)$, and a_{19}, which will be $f(a_2, a_5)$.

For y we have to take successively a_1, a_2, a_3, and a_4.

The disjunction of order 2 is

$$P(a_2, a_1, a_3, a_{12}) \vee P(a_2, a_2, a_5, a_{19}) \vee P(a_2, a_3, a_6, a_{17}) \vee P(a_2, a_4, a_7, a_{18}).$$

We see that a_8, a_9, a_{10}, a_{11}, a_{13}, a_{14}, a_{15}, a_{16} [[and a'_{11}–a'_{16}]] play no role.

If the rules of passage permit us to put the proposition into the form

$$(x)(z)(\exists y)P(x, y, z, f(y, z)),$$

then we have to equate a_3, a_5, a_6, and a_7.

4. The infinite domains

We now introduce the fundamental notion of *infinite domain.*

4.1. Let C_1 be a domain consisting of the elements $a_1, a_2, \ldots, a_{n_1}$, and let us assume that we have a certain number of descriptive functions, index functions, and atomic propositional functions. We shall say that we have an *infinite domain* if we have a definite procedure for correlating with every number p: first, a domain C' that contains C_1; then, for the functions, a system of values (3.1) in C' that permits us to obtain in C' the value of any function of height not greater than p whenever the arguments are taken in C_1; and, finally, a system of logical values that are assigned to the atomic propositional functions whenever the variables of these functions are replaced by elements of C'. These functions, propositional functions, and logical values will be said to be *attached* to the infinite domain.

We shall sometimes consider also domains to which no propositional functions or logical values are attached; we shall always indicate explicitly when this is the case.

It is quite obvious that generally the number of elements of C' increases with p.

We shall say that C' is the *domain of order p* in the infinite domain; C_1 is the *domain of order 0*.

We observe that this definition differs from the definition that would seem the most natural only in that, as the number p increases, the new domain C' and the new values need not be regarded as forming an "extension" ⟦"prolongement"⟧ of the previous ones. Clearly, if we know C' and the values for a given number p, then for each smaller number we know a domain and values that answer to the number; but only a "principle of choice" could lead us to take a fixed system of values in an infinite domain.[67]

We now start from a proposition P of height h, with no real variables. With each restricted variable y of P we correlate an index function whose arguments are the general variables that *dominate* y. Let us consider an infinite domain for which the attached functions are these index functions and the elementary ⟦rather, descriptive⟧ functions of P, and the attached propositional functions are the atomic propositional functions occurring in P; let C_1 be the domain of order 0, D_p the domain of order p. ⟦We delete all quantifiers and⟧ in the matrix of P we replace each restricted variable of P by the corresponding index function. We thus obtain a proposition P'. Replacing its variables by some elements of D_p, we obtain another proposition, P''. We try to eliminate all the functions that occur in P'' by replacing them with their values. We thus obtain a proposition Π. (This is not always possible. But we carry out this replacement whenever we can, and we always can if p is greater than the height of the proposition P, for then it suffices to take elements of C_1 as arguments of the functions.)

We shall say that P is *true in the infinite domain* if for every number p we have a procedure enabling us to verify that each of the propositions ⟦obtained over D_p in the way Π was⟧ has "true" as its logical value (derived from the logical values that its atomic propositions take in the domain).

Let us start again from P; but now with each general variable y we correlate an index function whose arguments are the restricted variables that dominate y. We construct an infinite domain as above, using the new index functions instead of the old ones; in the matrix of P we now replace each general variable by the corresponding index function and perform the same operations as above; if all the propositions thus obtained have the logical value "false", we shall say that P is *false in this infinite domain*.

It is absolutely necessary to adopt such definitions if we want to give a precise sense to the words "true in an infinite domain", words that have frequently been used without sufficient explanation, and also if we want to justify a proposition proved by Löwenheim,[68] a proposition to which many refer without clearly seeing that

[67] [See 6.2, Remark 3 in 6.4, page 4 in *Herbrand 1931b* (below, p. 623, lines 8–11), and especially page 53 with its footnote in *Herbrand 1931*. From this last reference it seems clear that by a "principle of choice" Herbrand meant the axiom of choice and not an argument such as Skolem's (*1922*; above, p. 294), nor an application of König's infinity lemma (*1926, 1927*) such as Gödel's (*1930a*; below, p. 589). See the introductory note to *Skolem 1928* (above, p. 510) and also Note H, p. 578 below.]

[68] *1915*; see also *Skolem 1920*.

Löwenheim's proof is totally inadequate for our purposes (see 6.2[69]) and that, indeed, the proposition has no precise sense until such a definition has been given.

4.11. We observe at once that, if P is true in an infinite domain, $\sim P$ is false in it and, if P is false in it, $\sim P$ is true in it.

4.2. To clarify the notion of infinite domain let us for the moment ignore the associated logical values; we can then classify the elements of D_p in the following manner: first we take the elements of C_1; then we take all the elements that are values of the functions of height 1 when the arguments of these functions are chosen in C_1; these elements form a domain Γ_2 (which may have elements in common with C_1); then we take all the elements that are values of the functions of height 2 when the arguments of these functions are chosen in C_1; they form Γ_3; we thus form $\Gamma_1, \Gamma_2, \Gamma_3, \ldots, \Gamma_p, \ldots$. We shall have at least p terms; but, clearly, this sequence will terminate when some domain is entirely contained in the previous ones.[70]

Let us now look again at the domains considered in 3.21 for the proposition P; if we take C_1 as the first domain, we see that Γ_2 can be regarded as resulting from C_2 when some elements of C_2 are equated to some elements of C_1 or C_2 (see the conventions of 3.211); we see that in general Γ_k results from C_k when some elements of C_k are equated to some elements of $C_1, C_2, \ldots, C_{k-1}$, or C_k.

If we recall the theorem of Chapter 1, 5.21,[52] we obtain the following two basic theorems:

4.21. (1) *A proposition P that is false in some* ⟦*infinite*⟧ *domain cannot* ⟦*for any p*⟧ *have property B* ⟦*of order p*⟧.

(2) *If* ⟦*for every p*⟧ *a proposition P does not have property B* ⟦*of order p*⟧, *we can construct an infinite domain in which P is false.*

The first theorem follows from the fact that for ⟦the union of⟧ $C_1, C_2, \ldots,$ and C_{p+h+1} we can immediately obtain from the infinite domain a system of logical values that gives the logical value "false" to the first disjunction of order p associated with P (see 3.4). The second theorem follows from the fact that C_p determines precisely the domain of order p in the infinite domain that we are trying to find and that the logical values assigning the value "false" to this disjunction (according to Chapter 1, 5.21[52]) yield the attached logical values.[71]

To simplify we shall say (the meaning of the assertion being determined by Theorems (1) and (2) above):

The necessary and sufficient condition for a proposition not to have property B is that it be false in some infinite domain.[72]

4.3. The preceding considerations show that we need concern ourselves only with infinite domains of the following kind. Let C_1 be an arbitrary domain (consisting, for example, of one element). Let C_k be, for each $k > 1$, the domain obtained when, as in 3.21, with each function of height $k - 1$ that can be constructed from the elementary functions at our disposal and whose arguments are taken in C_1 we correlate one-to-one some element, which will be the value of the function. It then suffices, in defining the truth (or falsity) of a proposition of height h, to assume that the domain of order p in

[69] ⟦The French text has "6.4", but this seems to be an oversight.⟧

[70] There may be elements of D_p that will not be contained in any of these domains; these elements will play no role in what follows.

[71] [Thus, for each p, D_p can be taken to be $G_{p+h+1} = C_1 \cup C_2 \cup \cdots \cup C_{p+h+1}$.]

[72] [Compare 6.2.]

the infinite domain is given by the union of $C_1, C_2, \ldots, C_{p+h+1}$ (the attached functions being the descriptive functions and the index functions mentioned in 4.1) and by a system of logical values for the propositional functions having arguments in these domains. Such an infinite domain will be called a *reduced* domain. All the previous results still hold when the words "infinite domain" are replaced by "reduced domain". For we can always assume that two different functions have different values: if two different functions have an element b as their common value, we replace b by two other elements, b_1 and b_2; the value of the first function will be b_1, that of the second b_2; and we stipulate that, for every atomic propositional function φ and for every system of arguments, the logical values of $\varphi(b_1, x_2, x_3, \ldots, x_n)$ and $\varphi(b_2, x_2, x_3, \ldots, x_n)$ will be the same as that of $\varphi(b, x_2, x_3, \ldots, x_n)$. We can similarly assume that any function takes different values for different arguments.

4.31. We can even start from several propositions; to define their truth we construct the domains C_k that correspond to their logical product (to define their falsity, those that correspond to their logical sum); we see immediately that, if each proposition is true in a reduced domain associated with their product (or false in a reduced domain associated with their sum), this product (or sum) is true (or false) in this domain, and conversely.

5. Fundamental theorem

5. Theorem. (1) *If for some number p a proposition P has property B of order p, the proposition P is true.*[8] *Once we know any such p, we can construct a proof of P.*

(2) *If we have a true*[8] *proposition P, together with a proof of P, we can, from this proof, derive a number p such that P has property B of order p.*

We shall say: *The necessary and sufficient condition for a proposition P to be true*[8] *is that for some p the proposition P have property B of order p.*

5.1. Lemma 1. *A proposition P that* ⟦*for some p*⟧ *has property C* ⟦*of order p*⟧ *has property A.*

This will be proved when we have stated a scheme, a type, and a set of equations associated with the proposition P (see Section 2).

Assume that P has property C of order p ⟦and is of height h⟧; let N be the number of elements in ⟦the union of all the domains up to and including⟧ C_p (same notation as in 3.21). Consider some prenex form of P without real variables, and take property C in the form of 3.4.[73] We shall construct a scheme associated with P (2.2) such that each restricted variable ⟦of the scheme⟧ is immediately behind a brace joining N rows;[74] we shall say that the restricted variable occurring in the ith row immediately

[73] [Since, if P is not in prenex form, no prenex form of P need have property C of order p (see Note F, p. 577 below), to carry through the argument that Herbrand is about to give we must be able to calculate from the number p and the proposition P some number q such that some prenex form of P has property C of order q. Hence in the light of 3.31′ of Note F the first two sentences in the present paragraph should be replaced by the following paragraph:

Let P contain j quantifiers and d descriptive functions, the maximum height of these functions being h, and assume that P has property C of order p. Let P' be a prenex form of P without real variables. Then P' has property C of order $q = \delta(j, d, p, n_1)$. Let N be the number of elements in G_q, and let us consider the second disjunction of order q associated with P'.

Moreover, throughout the remainder of Herbrand's proof of Lemma 1 in 5.1 read "q" wherever the text has "p".]

[74] ⟦To Herbrand's definition of a scheme (above, p. 531) we must add here the stipulation that the dominant variables of the scheme are joined by an initial brace. A row should not be

on the right of the brace is *attached* to the element a_i. Hence, in an arbitrary line of the scheme, each restricted variable is attached to an element a_i, with $i \leqq N$.

We shall now define a correspondence between the elements of all the domains ⟦up to and including C_{q+h+1}⟧ and the general variables of the scheme (and descriptive functions of such variables, if any).[75] We stipulate that each general variable y shall correspond to the element that is the value of $f_y(a_{i_1}, a_{i_2}, \ldots, a_{i_n})$, where f_y is the index function of y, and $a_{i_1}, a_{i_2}, \ldots, a_{i_n}$ are the elements attached to the restricted variables that occur on the same line as y but to its left. With the elements of C_1 we correlate fictitious (in the sense of 2.1) general variables. Finally, with each element of the domains ⟦up to and including C_{q+h+1}⟧ we correlate either a general variable, in accordance with the preceding rule, or a descriptive function of general variables, stipulating that for a *descriptive* function f, if $a_{i_1}, a_{i_2}, \ldots, a_{i_n}$ correspond to the general variables or to the functions $\varphi_1, \varphi_2, \ldots, \varphi_n$ of general variables,[75] the value of $f(a_{i_1}, a_{i_2}, \ldots, a_{i_n})$ in the domains corresponds to the function $f(\varphi_1, \varphi_2, \ldots, \varphi_n)$. We readily see that to each element of the domains there thus corresponds one and *only one* general variable or function of general variables.

Let the order of a line ⟦in the scheme⟧ be the greatest of the orders of the domains of any of the elements that correspond to the restricted variables of that line.

We obtain the associated type by selecting, one after another, all the variables occurring in the lines of order 1, then all those occurring in the lines of order 2, then all those occurring in the lines of order 3, and so on up to and including order p.

We obtain the associated equations by equating each restricted variable to the general variable, or the function of general variables, that corresponds to the element attached to the restricted variable. We see that a restricted variable is always equated to a general variable, or to a function of general variables, occurring in lines whose order is less than its own; hence these equations yield associated equations that are compatible with the type considered.

Now we only have to see that the disjunction thus obtained is an identity. But, clearly, this disjunction comes from the second disjunction of order p associated with P (3.4)[76] when each element of the domains is replaced by its corresponding general variable or function of general variables (each line of the scheme corresponds to a disjunct of this identity).[77]

Corollary. *A proposition that ⟦for some number p⟧ has property B ⟦of order p⟧ is true.*[8]

confused with a line of the scheme. A row consists of just one signed letter immediately to the right of a brace.

Let us note that, if P contains r restricted quantifiers, the scheme now being constructed has N^r lines and is what in footnote 25 was called a reduced scheme.⟧

[75] ⟦See footnote 30, p. 539 above.⟧

[76] [Rather, of order q associated with P' (see footnote 73).]

[77] [Thus, under the correspondence between elements of G_{q+h+1} and members of the associated scheme, the standard expansion of the prenex proposition P' (see footnotes 63 and 73) yields the identity associated with P, hence shows that P has property A. But, then, a far more informative way of correcting Herbrand's argument for Lemma 1 than that given in footnote 73 would be to use the generalized notion of generating scheme mentioned in footnote 22 above, p. 537. For we would not need to consider a second disjunction of order q associated with P (that is, the standard expansion of P' of order q); rather, we would be able to set up a correspondence between the elements of G_{p+h+1} and members of the new scheme such that the standard expansion of P of order p would yield an identity associated with P.]

For we have seen that

(1) Property B is equivalent to property C (3.31)[61] and

(2) A proposition that has property A is true (2.3).[23]

This proves the first part of the theorem.

5.2. To prove the second part of the theorem we proceed by recursion on the proof of the proposition.

(1) ⟦For every p,⟧ if $P \vee P$ has property B ⟦of order p⟧, then P has property B ⟦of order p⟧. This is seen at once from the definition ⟦of property B⟧ given in 3.22.

(2) ⟦For every p,⟧ if $\Phi(x)$ has property B ⟦of order p⟧, then $(x)\Phi(x)$ has property B ⟦of order p⟧. For, as we saw in 3.2, in order to construct the expansions of $\Phi(x)$ and $(x)\Phi(x)$, we must first replace each of these propositions by the same proposition $(y_1, y_2, \ldots, y_n, x)\Phi(x)$, where $y_1, y_2, \ldots, y_n$ are the real variables of $(x)\Phi(x)$.

Lemma 2. *Property C is preserved when the second rule of generalization (in the form given in Chapter 3, 1.4*[78]*) is used.*[79]

⟦This second rule of generalization is:⟧ If t is an individual that depends upon ⟦possibly fictitious⟧ real variables, we can go from $\vdash \Phi(t)$ to $\vdash (\exists x)\Phi(x)$.

Let us take property C in the form given in 3.4.[80]

We know (3.221) that to the real variables there will correspond some fixed elements of C_2; so the element that will correspond to t will be in C_k for some $k \geqq 2$; let us call this element a_i. Suppose that the first proposition has property C of order p. For the second proposition we consider the second associated disjunction (see 3.4) of order $p + k - 1$, and in this disjunction we consider only the disjuncts in which x is replaced by a_i. We see immediately that their disjunction constitutes a second disjunction of order p associated with the proposition $\Phi(t)$, the first domain being C_1; hence it is a propositional identity; but then so is the original disjunction.

5.3. Lemma 3. *For every p, if the propositions P and P ⊃ Q have property C of order p, then Q has property C of order p.*[81]

According to 3.53, P & $(P \supset Q)$ has property C of order p.[82] We put P and Q into prenex form; they are then written

$$(\pm y_1, \pm y_2, \ldots, \pm y_r)M(y_1, y_2, \ldots, y_r, x_1, x_2, \ldots, x_n)$$

and

$$(\pm z_1, \pm z_2, \ldots, \pm z_s)N(z_1, z_2, \ldots, z_s, x_1, x_2, \ldots, x_n),$$

respectively, if we exhibit all the real variables (some may be fictitious (3.52) in P or in Q). The conjunction P & $(P \supset Q)$ can be written, when transformed according to the rules of passage,

$$(-y_1^{\alpha_1}, +y_1^{1-\alpha_1}, -y_2^{\alpha_2}, +y_2^{1-\alpha_2}, \ldots, -y_r^{\alpha_r}, +y_r^{1-\alpha_r}, \pm z_1, \pm z_2, \ldots, \pm z_s)$$
$$\{M(y_1^0, y_2^0, \ldots, y_r^0, x_1, x_2, \ldots, x_n) \;\&\; [M(y_1^1, y_2^1, \ldots, y_r^1, x_1, x_2, \ldots, x_n)$$
$$\supset N(z_1, z_2, \ldots, z_s, x_1, x_2, \ldots, x_n)]\}.$$

[78] ⟦"If $\Phi(f, x_1, x_2, \ldots, x_n)$ is true[8] (f being a real variable, a descriptive function whose arguments are real variables, or a constant), then $(\exists y)\Phi(y, x_1, x_2, \ldots, x_n)$ is true."⟧

[79] [In general the order does not remain invariant. If t is of height $k \geqq 2$ and $\Phi(t)$ has property C of order p, then $(\exists x)\Phi(x)$ has property C of some order less than or equal to $p + k - 1$.]

[80] [Throughout this proof of Lemma 2 we should use the standard expansion of a proposition instead of the second disjunction of order p associated with the proposition; see footnote 63.]

[81] [See Note G, p. 578 below.]

[82] [See footnote 65.]

In this proposition α_i is 0 or 1 according as y_i is restricted or general in P; moreover, for each pair of variables y_i^0 and y_i^1 we have a pair of successive quantifiers, of which the first is restricted and the second general (since, whenever y_1^0 is restricted, y_1^1 is general, and conversely, according to Chapter 2, 3.102[83]). For the proposition in question we take property C in the form given in 3.4. We therefore have domains $C_1, C_2, \ldots, C_k, \ldots, C_{p+h+1}$ generated by the descriptive and index functions of the proposition. We transform them into domains $C_1', C_2', \ldots, C_k', \ldots, C_{p+h+1}'$ as follows. First, C_1' is C_1. Then, for each $k > 1$, C_k' is obtained from C_k thus: let f_t be the index function of the ⟦general⟧ variable t; for each value of $y_i^{\alpha_i}$, we consider the elements of C_k that are values of $f_{y_i^{1-\alpha_i}}(y_1^{\alpha_1}, y_2^{\alpha_2}, \ldots, y_i^{\alpha_i})$ for all possible values of $y_1^{\alpha_1}, y_2^{\alpha_2}, \ldots, y_{i-1}^{\alpha_{i-1}}$, and ⟦in $C_1 \cup C_2 \cup \cdots \cup C_k$⟧ we equate all these elements to the value of $y_i^{\alpha_i}$; moreover, for each system of values of $z_{u_1}, z_{u_2}, \ldots, z_{u_j}$, we consider the elements of C_k that are values of $f_{z_j}(y_1^{\alpha_1}, y_2^{\alpha_2}, \ldots, y_r^{\alpha_r}, z_{u_1}, z_{u_2}, \ldots, z_{u_j})$ for all possible values of $y_1^{\alpha_1}, y_2^{\alpha_2}, \ldots, y_r^{\alpha_r}$ and ⟦in $C_1 \cup C_2 \cup \cdots \cup C_k$⟧ we equate all these elements to one of them.

Clearly, these new domains are the very domains that are generated from C_1 by the descriptive and index functions of Q. And we see, once the equating just presented is taken into account, that the second disjunction of order p associated with P & $(P \supset Q)$ ⟦over the original domains⟧ becomes the identity

$$\text{(I)} \qquad [P_1 \,\&\, (P_1 \supset Q_1)] \vee [P_2 \,\&\, (P_2 \supset Q_2)] \vee \cdots \vee [P_\gamma \,\&\, (P_\gamma \supset Q_\gamma)],$$

where P_i and Q_i, $1 \leqq i \leqq \gamma$, are the propositions obtained when the operations described in 3.4 are carried out over these new domains on the matrices of P and Q. Moreover,

$$\text{(II)} \qquad Q_1 \vee Q_2 \vee \cdots \vee Q_\gamma$$

differs from the ⟦second⟧ disjunction of order p associated with Q only in that certain disjuncts are repeated several times.[84]

Now, Theorem 5.21 of Chapter 1[52] shows that the truth of the disjunction (I) implies that of the disjunction (II), since, for each given system of logical values, there is an i such that the logical value "true" is assigned to P_i & $(P_i \supset Q_i)$, hence to $P_i \supset Q_i$, hence to Q_i, and hence to the disjunction (II).

If P should contain descriptive functions not occurring in Q, we would use Remark 3.52.

5.31. We now recall that

(1) Identities of the first kind, that is, with no apparent variables, have property B,

(2) Property B and property C are equivalent (3.31),[61] and

(3) For every p, if a proposition P has property B of order p, then P has property B of order q for every q greater than p (3.54).

Therefore, Lemma 2, Lemma 3,[81] the lemma of 3.3,[60] and the remarks of 5.2 prove the second part of the Fundamental Theorem.

Remarks. (1) Properties A, B, and C are therefore equivalent[85] and can be said

[83] ⟦If the quantifier (x) occurs positively, x is general; if it occurs negatively, x is restricted; the situation is the opposite for the quantifier (Ex).⟧

[84] [Throughout this last paragraph the prenex forms of P, Q, and P & $(P \supset Q)$ should be understood when Herbrand speaks simply of P, Q, and P & $(P \supset Q)$; see Note G, p. 578 below. However, Herbrand's argument could be slightly modified so that Q need not be in prenex form.]

[85] [See Note F and footnote 73.]

to be necessary and sufficient for the truth[8] of a proposition. The preceding theorem then shows that the proof of any true proposition can be reduced to a canonical form, namely, one in which the proposition is obtained from its associated identity (2.3) (which is the ⟦second⟧ disjunction of order p associated with the proposition).[86]

(2) We could easily show that, if in the proof of a proposition P the second rule of generalization is used n times, for individuals of heights $h_1, h_2, \ldots, h_n$, then P has property B of an order at most equal to $h_1 + h_2 + \cdots + h_n + n$.[87]

(3) If there should be more than one type in the proposition under study, we would have to distinguish, in each domain, elements of different types (the value of a function, in particular, is of the type of that function) and replace a variable only by elements of the same type. But, aside from these points, nothing would be changed in the previous considerations. (An index function is taken to be of the same type as the variable to which it is attached.)[88]

6. Consequences

A. Consequences concerning the decision problem

6.1. Corollary 1. We have seen[89] that the rule of implication is not necessary for the proof of a proposition having property A, provided we replace the rule of simplification by the following

Generalized rule of simplification. *If in a true*[8] *proposition the subproposition* $P \vee P$ *is replaced by* P, *we obtain another true proposition.*

Now, from our theorem and Lemma 1 in 5.1 it follows that *every true*[8] *proposition has property* A. Hence the only rules of reasoning that are needed are the rules of passage, the two rules of generalization and the generalized rule of simplification. The rule of implication drops out.

Because of the difficulties that the rule of implication might create in certain ⟦metamathematical⟧ demonstrations that proceed by recursion on ⟦formal⟧ proofs, we consider this result most important. It shows, moreover, that the rule of implication, whose origin, after all, is in the classical syllogism, is not necessary in building logic. (The rule remains necessary, however, in mathematical theories, which contain hypotheses.)

6.2. The theorems stated in Section 5 and in 4.21 permit us to set forth the following results:

Theorem 1. *If* P *is an identity,* $\sim P$ *is not true in any infinite domain*;

Theorem 2. *If* P *is not an identity, we can construct an infinite domain in which* $\sim P$ *is true.*

Similar results have already been stated by Löwenheim (*1915*), but his proofs, it seems to us, are totally insufficient for our purposes. First, he gives an intuitive

[86] [In this canonical form for a proof of P, some prenex form P' of P is first proved, and then P is obtained by use of the rules of passage. Hence the associated identity (2.3) of P is indeed a second disjunction associated with P, for it is the standard expansion of P' (see footnotes 63 and 73). Compare Remark 4 in 6.4 below, p. 560.]

[87] [This is false because of the falsity of the lemma of 3.3 and Lemma 3 in 5.3.]

[88] [Here Herbrand extends his Fundamental Theorem to many-sorted logics; see footnote 1 above and article 6.6 below.]

[89] [See 2.2, 2.3, and Note D, p. 571 below.]

meaning to the notion "true in an infinite domain", hence his proof of Theorem 2 does not attain the rigor that we deem desirable (indeed the [[required]] proof is contained in the considerations of Section 4 and of 5.1). Then—and this is the gravest reproach—because of the intuitive meaning that he gives to this notion, he seems to regard Theorem 1 as obvious. This is absolutely impermissible; such an attitude would lead us, for example, to regard the consistency of arithmetic as obvious. On the contrary, it is precisely the proof of this theorem (of which the lemma of 3.3 is a part) that presented us with the greatest difficulties.

We could say that Löwenheim's proof was sufficient in mathematics; but, in the present work, we had to make it "metamathematical" (see Introduction[90]) so that it would be of some use to us.

6.3. The two theorems that we just stated provide us with a method for investigating the decision problem. By means of this method we have succeeded in obtaining new solutions for all the particular cases already solved and even in somewhat extending these cases.[91] Moreover, we can now deal with a case that is essentially different from those already known, namely, the case of *a proposition whose matrix is a disjunction of atomic propositions and negations of atomic propositions* (the atomic propositions may contain descriptive functions). A proposition is an identity if and only if it has property A; in the present case, whether a proposition has property A can readily be decided by the use of 2.34 and 2.4.

6.4. Remark 1. We see without difficulty that a proposition true in some finite domain (Chapter 2, 8.3[92]) is also true in some infinite domain (here there is a bound, namely, the number of elements in the finite domain, on the number of elements in each field D_n in the infinite domain); hence the theorem of Chapter 2, Section 8,[93] immediately follows from Theorem 1.

Remark 2. The notion "satisfiability" ("Erfüllbarkeit") that Ackermann used in *1928a* does not seem to us to be defined in a sufficiently complete manner, for we have not been given a precise stipulation that would permit us to state that a proposition is "satisfiable".[94] Theorem 1 immediately allows us to remove this objection. Now, let P be a proposition of the form

$$(x_1, x_2, \ldots, x_n)(\exists z)(y_1, y_2, \ldots, y_p)A(x_1, x_2, \ldots, x_n, z, y_1, y_2, \ldots, y_p).$$

If P is not an identity, we can construct an infinite domain in which $\sim P$ is true;

[90] [[In the Introduction to his thesis, on page 6, Herbrand presents the distinction between mathematics and metamathematics. Here he suggests that the set-theoretic notion of satisfiability is not a proper object of metamathematical investigations.]]

[91] For example, to the case of a disjunction of propositions of the form

$$(x_1, x_2, \ldots, x_n)(\exists y)(z_1, z_2, \ldots, z_p)A(x_1, x_2, \ldots, x_n, y, z_1, z_2, \ldots, z_p).$$

[92] [In Section 8 of Chapter 2 Herbrand stipulates that a proposition is *true in a finite domain* if and only if the common expansion (see footnote 41) of the proposition over the domain is truth-functionally valid. Here, however, he apparently uses "true in a finite domain" to mean that the common expansion of the proposition is truth-functionally satisfiable. The same shift in meaning occurs on page 60 of the thesis.]

[93] [["If $\Phi(y_1, y_2, \ldots, y_p)$ is a true[8] proposition containing only $y_1, y_2, \ldots, y_p$ as real variables, the [[common]] expansion of $(y_1, y_2, \ldots, y_p)\Phi(y_1, y_2, \ldots, y_p)$ [[over any finite domain]], which no longer contains any apparent variable, is, relative to its atomic propositions, a propositional identity of the first kind."]]

[94] [Herbrand is here contrasting Ackermann's intuitive (that is, set-theoretic) notion "satisfiable" with his own metamathematical (that is, finitistic) notion "true in an infinite domain", which by Theorems 1 and 2 of 6.2 is equivalent to the notion "irrefutable in Q_H". Herbrand is *not*

clearly, merely by repeating Ackermann's argument, we see that there is also a finite domain, consisting of a known number of elements, in which $\sim P$ is true.[92] On the other hand, if P is an identity, $\sim P$ cannot be true in such a finite domain (according to Chapter 2, 8.1[93]); we therefore have a criterion that enables us to decide whether P is an identity or not.

Moreover, it is possible, and simpler, to show by a similar argument that, if a proposition P of the form considered has property C, it has this property with an order that we can determine in advance; and this provides a criterion equivalent to the previous one.[95]

Remark 3. We see the following: if to our rules of reasoning, listed at the beginning of Chapter 2,[96] we were to adjoin other rules that could not be derived from them, then we would be led to regard as true ⟦that is, provable⟧ some propositions that are in fact false in some infinite domain. We must acknowledge that such a consequence would be difficult to accept. This fact corresponds to what the Germans call the *Vollständigkeit* of our system of rules. (If we could *prove* that these additional rules lead us to regard as true a proposition P that, without them, would not be so, then, as we can readily see, the inconsistency of classical mathematics would follow, because we could construct a denumerable set over which P would be false.)[97]

Remark 4. We can view the theorem of Section 5 as giving us a canonical form for every mathematical proof; for, if we know the order p with which a proposition has property B, we can, according to 5.1, specify the proof of this proposition.[86] More generally, we can say that, if a theorem P is true ⟦that is, provable⟧ in a theory on the basis of hypotheses (containing no apparent variables[98]) whose logical product is H, the proof of P can be obtained in the following manner (according to 6.2 and Chapter 3, 2.4[99]): *we try to construct an infinite domain in which $H \supset P$ is false, and after a certain number of operations we see that the construction is impossible.*[100] Thus we can say

demanding that a decision procedure for satisfiability be given. Indeed, he is objecting to Ackermann's use of "satisfiability" with respect to a class K of propositions for which he knows that Ackermann (*1928a*) has shown that

> For each proposition P in K, there is effectively obtainable a number n such that $\sim P$ is satisfiable if and only if the common expansion of $\sim P$ over a domain of cardinality n is truth-functionally satisfiable.

Rather, Herbrand is demanding that the notion "satisfiability" not be used in metamathematics. (Of course, Herbrand permits "truth-functional satisfiability".) Hence, in this first paragraph of Remark 2, Herbrand exploits Ackermann's argument to prove that

> For each proposition P in K, there is effectively obtainable a number n such that P is provable in Q_H if and only if the common expansion of $\sim P$ over a domain of cardinality n is not truth-functionally satisfiable.]

[95] [See Herbrand's treatment of this solvable case (*1931*, pp. 45–46); see also *Dreben 1962*.]

[96] ⟦These are the rules of Herbrand's system of quantification theory, see above, p. 528.⟧

[97] [See Note H, p. 578 below.]

[98] [Rather, containing no real variables. Herbrand's argument here rests upon his version of the deduction theorem (see footnote 99 and Note I, p. 580 below).]

[99] ⟦In this article Herbrand states and proves the deduction theorem in the following form:
Let P be a proposition in a theory having only a finite number of hypotheses; let H be the product of these hypotheses. If H does not contain any real variable, P is provable in the theory if and only if $H \supset P$ is provable in Herbrand's Q_H.⟧

[100] [See the "no-counterexample" proof techniques of Beth (*1955*) and Hintikka (*1955*); see also *Skolem 1928* and *1929*, § 5. Kreisel (*1958*, p. 36) pointed out that Beth's constructive argument for modus ponens elimination (*1956a*, pp. 38–40; see also *1959*, pp. 278–280) contains an error. But it should be noted that no argument that proceeds, as does Beth's, just by recursion on the construction of the "eliminable formula" can succeed.]

that, if the proposition is indeed a theorem, we can always give a proof of it that does not use any artifice, the role of artifices being only to make proofs shorter.[101]

B. Consequences concerning the investigation of mathematical theories

6.5. The theorems obtained in the previous sections have many applications; they provide a general method for the investigation of mathematical theories. We shall briefly point out some of these applications, laying stress on the most important of them, the consistency of ordinary arithmetic.

(1) *If all the hypotheses of a theory are true in some infinite domain (4.31), the theory is consistent* (Chapter 3, 2.1[102]).[103]

For, if a theory is inconsistent, we see immediately that we can find hypotheses $H_1, H_2, \ldots, H_n$, which can always be assumed to contain no apparent variables,[98] such that $\sim(H_1 \,\&\, H_2 \,\&\, \cdots \,\&\, H_n)$ is an identity, hence has property B. But, according to 4.11, this is impossible, since $H_1 \,\&\, H_2 \,\&\, \cdots \,\&\, H_n$, by 4.31, is true in some infinite domain, and so $\sim(H_1 \,\&\, H_2 \,\&\, \cdots \,\&\, H_n)$ is false ⟦in this infinite domain⟧.

(2) *If a theory is consistent and has only a finite number of hypotheses,*[104] *we can construct an infinite domain in which these hypotheses are true.*[105]

It suffices, as we see immediately by virtue of 4.3, to make the logical product of these hypotheses true in a reduced domain (4.3).

We shall now briefly present some immediate consequences ⟦of (1) and (2)⟧. But first we note that, if a theory contains constants, that is, descriptive functions of 0 argument, we shall have to represent these constants in the infinite domains by certain elements that we include in the domain of order 1.

6.6. The theorems stated in 6.5 can be used to show that no proposition ⟦written in the notation of but⟧ unprovable in a given theory becomes provable when certain changes are made in the theory.[106]

Let us assume, for example, that we have a theory with a certain number of types; for each type we can introduce the propositional function $x = y$, of arguments x and y, to be read "x is equal to y", in which x and y will be of the type considered (hence we shall have different propositional functions, represented by arrays of signs that differ only in the type of the variables).

[101] [See *Gentzen 1934*, p. 177; in particular, consider Gentzen's "Er macht keine Umwege".]

[102] ⟦"A theory is said to be *consistent* if it does not contain any proposition Z such that both Z and $\sim Z$ are true ⟦that is, provable in the theory⟧."⟧

[103] [This is one of the most important consequences of Herbrand's Fundamental Theorem and supplies much of the motivation for it; indeed, the present statement (1) greatly simplifies the Hilbert–Ackermann–von Neumann approach to proving the consistency of theories (see 6.8 below, *Ackermann 1924*, *Hilbert 1925* and *1927*, *Bernays 1927*, *von Neumann 1927*, also Bernays's discussion of this point (*1934a*; *1936*, remark 1, pp. 115–116; *Hilbert and Bernays 1939*, pp. 38 and 48) as well as *Kleene 1952*, pp. 475–476).]

[104] It is easy but unnecessary to remove the latter restriction.

[105] This proposition is the "metamathematical" form of the proposition known as the "Skolem paradox"; for one should observe that ⟦metamathematical⟧ arguments about infinite domains, when translated into mathematical arguments, become arguments about denumerable sets. [See footnote 67, p. 552 above.]

[106] [Compare 6.6 with §§ 5–6 in *Skolem 1929* and with *Herbrand 1931*, pp. 30–33. Also, compare the last paragraph of 6.6 with *Craig 1957a* and *1960*.]

We add the following hypotheses, hereafter denoted by (1):

$$x = x, \quad (x = y) \supset (y = x), \quad \text{and} \quad [(x = y) \,\&\, (y = z)] \supset (x = z)$$

for each type,

$$[(x_1 = y_1) \,\&\, (x_2 = y_2) \,\&\, \cdots \,\&\, (x_n = y_n)] \supset [\Phi(x_1, x_2, \ldots, x_n) \supset \Phi(y_1, y_2, \ldots, y_n)]$$

for each atomic propositional function $\Phi(x_1, x_2, \ldots, x_n)$, and

$$[(x_1 = y_1) \,\&\, (x_2 = y_2) \,\&\, \cdots \,\&\, (x_n = y_n)] \supset [f(x_1, x_2, \ldots, x_n) = f(y_1, y_2, \ldots, y_n)]$$

for each elementary descriptive function $f(x_1, x_2, \ldots, x_n)$.

Clearly, the last two formulas of (1) can be proved in the extended theory for every propositional function and every descriptive function.

Let us further assume that the hypotheses (1), together with the original ones, bring about the truth ⟦that is, the provability in the theory⟧ of a certain proposition P, which does not contain the new propositional function and was not true ⟦that is, provable in the theory⟧ before the introduction of this new propositional function. Let $H_1, H_2, \ldots, H_n$ be the hypotheses of the original theory that are necessary for the proof of P; then the hypotheses $H_1, H_2, \ldots, H_n$, and $\sim P$, which were conjointly consistent, become inconsistent when taken together with the hypotheses (1). Now, we can construct an infinite domain in which $H_1, H_2, \ldots, H_n$, and $\sim P$ are true; moreover, to make the hypotheses (1) true in this infinite domain we merely have to agree that $a_i = a_j$ (where a_i and a_j are elements of the domain) has the logical value "true" just if the letters a_i and a_j are the same. But, since the hypotheses $H_1, H_2, \ldots, H_n$, $\sim P$, and (1) are true in some infinite domain, they cannot be conjointly inconsistent.

Therefore, *if a proposition is not true ⟦that is, provable⟧ before equality has been introduced, it cannot be so afterward.*

We could show by the same method that it is also possible to introduce into a theory the notion "couple", $\langle x, y\rangle$, along with the axiom

$$(\langle x, y\rangle = \langle x', y'\rangle) \equiv [(x = x') \,\&\, (y = y')]$$

(provided that a couple is not of the type of its arguments). This would show the consistency of the theory of rational numbers.[107]

Let us now go back to the case in which equality was introduced and see how we can obtain a proof of a proposition P, not containing equality, in a theory without equality ⟦once we have a proof of P in the extension of the theory obtained by adjunction of equality⟧. Let us assume that the proposition

$$(H_1 \,\&\, H_2 \,\&\, \cdots \,\&\, H_n \,\&\, H) \supset P,$$

where H is the product of those hypotheses (1) that we are using, has property B of order p. If we look at the proof of the results obtained in 4.21 and 4.3, we can conclude that

$$(H_1 \,\&\, H_2 \,\&\, \cdots \,\&\, H_n) \supset P$$

has property B of order p; from this last fact we can obtain a proof of P.

[107] And of Euclidean geometry, in Hilbert's axiomatization, without the completeness axiom. [See *Hilbert and Bernays 1939*, pp. 33–48, especially 38–48.]

6.7. According to Löwenheim (*1915*), the decision problem would be solved for the general case if it were solved for the case in which each ⟦atomic⟧ propositional function has only two arguments; the proof, which rested on his insufficiently established results, is now completely justified.[108] We can even go further and:

(*a*) Prove that we can restrict ourselves to the case in which there is no longer any constant or descriptive function. For this we use, primarily, Russell and Whitehead's theory of descriptions. It would suffice first to introduce equality as in 6.6, and then to replace each descriptive function $f(x_1, x_2, \ldots, x_n)$ by a propositional function $\Phi(y, x_1, x_2, \ldots, x_n)$, which is to be taken as $y = f(x_1, x_2, \ldots, x_n)$, and add the hypotheses $(\exists z)(y)[\Phi(y, x_1, x_2, \ldots, x_n) \supset (y = z)]$.

(*b*) Prove that we can restrict ourselves to the case in which either there is only one propositional function of three arguments ⟦rather, one ternary predicate letter⟧ or there are no more than three propositional functions of two arguments ⟦rather, three binary predicate letters⟧.

(*c*) Prove easily that we can dispense with apparent variables if, following Hilbert, we introduce the "logical function" (each proposition will give rise to such a function, with the real variables as arguments, and to a corresponding axiom; see *Hilbert 1927*). In particular, we can always assume that the hypotheses of a theory contain only real variables.[109]

6.8. *Consistency of arithmetic.*[110] The results of 6.5 enable us to solve the problem of the consistency of arithmetic much more completely ⟦that is, for more extensive subsystems⟧ than we did in Chapter 4.

Let us recall that in arithmetic there is one atomic propositional function, $x = y$, one primitive descriptive function, $x + 1$, and one primitive constant, 0; we can, moreover, introduce new descriptive functions, which are defined by recursion (Chapter 4, 8.5[111]).[112]

Let the domain A be the unbounded sequence of letters $a_0, a_1, a_2, \ldots, a_n, \ldots$; we stipulate that a_0 shall be the constant 0. With these letters we shall form the domains that we need; we stipulate that $a_i = a_j$ shall have the logical value "true" just if the subscripts i and j are identical.

Let $f(x_1, x_2, \ldots, x_n)$ be a (descriptive or index) function; we shall say that we know its value in the domain A if we have a procedure that enables us to assign a value to $f(a_{i_1}, a_{i_2}, \ldots, a_{i_n})$ for each system of arguments taken in the domain.

A proposition P will be said to be *true in the domain A* if, after correlating with each restricted variable x of P an index function whose arguments are the general variables superior to x and the real variables, we can so assign values in the domain A to these functions that the proposition obtained when the restricted variables of P are

[108] [See *Herbrand 1931*, pp. 33–41, for Herbrand's new proof of this result of Löwenheim and also for his proofs of the results announced here in (*a*) and (*b*).]

[109] [See footnote 33, p. 541 above, and Note I, p. 580 below.]

[110] [See Note J, p. 580 below.]

[111] [This article extends Theory 3 (see Note J) by introducing functions defined by primitive recursion. But the consistency proof that Herbrand is about to give works just for the unextended Theory 3. See footnotes 115 and 118.]

[112] [In the next five paragraphs of the text Herbrand gives a new and very simple proof, based on Theorem 1 of 6.5, of the consistency of Theory 3 (see Note J); this proof, however, does not show the decidability of Theory 3, as did the more complex argument of Chapter 4.]

replaced by their index functions has the logical value (calculated as in 4.1) "true", no matter which elements of the domain A are put for the general ⟦and real⟧ variables.

Clearly, if several propositions are true in the domain A, then A yields an infinite domain in which these propositions are true.

Now, if we stipulate that the value of $a_i + 1$ is a_{i+1}, we can immediately see that all the axioms of Section 5 of Chapter 4[110] are true in the domain A; moreover, the domain A yields an infinite domain once we agree that for each p the domain ⟦D_p⟧ of order p in this infinite domain consists of $a_0, a_1, \ldots, a_p$. Hence these axioms are consistent.

But now we can go further.[113]

I. *Introduction of recursive definitions.* Proceeding in a more general way than in Chapter 4, 8.5,[114] we shall assume that to the theory considered[115] we can adjoin new descriptive functions and new hypotheses in the following manner: we assume that the modifications introduced above ⟦in 6.6⟧ have already led to a theory T whose hypotheses are true in the domain A; we introduce a new descriptive function $f(y, x_1, \ldots, x_n)$, of arguments $y, x_1, \ldots, x_n$, and add hypotheses of the form

$$\Phi[f(0, x_1, \ldots, x_n)], \tag{1}$$

$$\Psi[f(y, x_1, \ldots, x_n), f(y + 1, x_1, \ldots, x_n)], \tag{2}$$

$$[(y = y') \ \& \ (x_1 = x'_1) \ \& \ \cdots \ \& \ (x_n = x'_n)] \supset [f(y, x_1, \ldots, x_n) = f(y', x'_1, \ldots, x'_n)], \tag{3}$$

provided that the propositions

$$(\exists x)\Phi(x) \tag{4}$$

and

$$(y)(\exists x)\Psi(y, x) \tag{5}$$

are true ⟦that is, provable⟧ in the theory T (Φ and Ψ may contain other variables $x_1, x_2, \ldots, x_n$[116]).

We shall prove the consistency ⟦of this extended theory⟧ on the very weak assumption that propositions (4) and (5) are true in the domain A (in other words, that they can be "interpreted" in that domain). It suffices to observe that propositions (1) and (2) are true in the domain A when, first, as index functions of the ⟦restricted⟧ variables of these propositions we take the index functions of the corresponding variables of (4) and (5) and, second, we choose the values of the function $f(y, x_1, x_2, \ldots, x_n)$ in the following way: $f(0, x_1, x_2, \ldots, x_n)$ shall be the index function of x in (4); if the index function of x in (5) is $\varphi(y, x_1, x_2, \ldots, x_n)$, we shall take the value of $\varphi(f(y, x_1, x_2, \ldots, x_n), x_1, x_2, \ldots, x_n)$ for that of $f(y + 1, x_1, x_2, \ldots, x_n)$; this completely defines the value of $f(y, x_1, x_2, \ldots, x_n)$ in the domain A. Proposition (3) is, of course, true in the domain.

[113] [Later (*1931b*; below, pp. 623–626) Herbrand gave much clearer and simpler proofs for results (I) and (II) obtained immediately below.]

[114] ⟦This article deals with the introduction of primitive recursive definitions into arithmetic.⟧

[115] [This theory has as its axioms just the first six axioms of Theory 3 (see Note J). It is called Theory 2 by Herbrand in Chapter 4.]

[116] ⟦This parenthetical remark is unclear. Formulas (4) and (5) contain no other free variables than $x_1, x_2, \ldots, x_n$, while their bound variables are x, y, and possibly others.⟧

II. *Introduction of the axiom of mathematical induction.* We recall (Chapter 4, 8.1[117]) that propositions of the form

$$\{\Phi(0) \mathbin{\&} (x)[\Phi(x) \supset \Phi(x+1)]\} \supset (y)\Phi(y) \tag{6}$$

must be added to the hypotheses.[118]

When we attempt to prove the consistency of the theory thus obtained,[119] we are confronted with difficulties that we shall not try to resolve here, our goal being only to show the fruitfulness of the notion "infinite domain". If hypotheses of the form (6) are introduced among the hypotheses ⟦referred to at the end of the preceding paragraph⟧, we shall prove the consistency ⟦of the theory thus obtained⟧ only in either of the following two cases (which, however, as can easily be seen, are the principal cases in which the axiom of mathematical induction is used).

(*a*) *Case in which* $\Phi(\mathrm{a})$ *is true in the domain* A *for every numeral* a. It is then quite obvious that $(x)\Phi(x)$ is true in the domain A and can be added to the hypotheses without contradiction.

(*b*) *Case in which* $\Phi(x)$ *contains no apparent variable.*[120] Let us now assume that to the hypotheses we add all the propositions $P_1, P_2, \ldots, P_m$ obtained when in (6) we successively substitute $\Phi_1(x), \Phi_2(x), \ldots, \Phi_m(x)$ for $\Phi(x)$. Corresponding to the ⟦restricted⟧ quantifier (x) in P_i, $1 \leqq i \leqq m$, let us introduce the new index function $f_i(y_1, y_2, \ldots, y_{n_i})$, where $y_1, y_2, \ldots, y_{n_i}$ are the real variables of $\Phi_i(x)$ other than x. We seek to construct an infinite domain in which all the hypotheses of the theory under consideration are true. Hence we must show that, in addition to the original hypotheses (those of Chapter 4 and those due to the recursive definitions), each proposition

$$\{\Phi_i(0) \mathbin{\&} [\Phi_i(f_i) \supset \Phi_i(f_i+1)]\} \supset \Phi_i(y), \tag{7}$$

which results from the matrix of (6) when x is replaced by its index function, also can be made true. (Whenever we wish to exhibit all the real variables, $\Phi_i(x)$ will be written $\Phi_i(x, y_1, y_2, \ldots, y_{n_i})$.)

⟦For an arbitrary $p \geqq 1$⟧ let E be the ⟦set of⟧ individuals of height not greater than p that are constructed from the constant 0, the original descriptive functions, and the new index functions.

We now have to construct an infinite domain in which all the hypotheses will come out true. The domain of order 0 will consist of a_0, and for each k the domain of order k will consist of letters from the domain A. Let Γ be the domain of order p ⟦in the infinite domain to be constructed⟧; every element of Γ is to be the value of an individual of E, and conversely every individual of E is to have a value in Γ. The hypotheses

[117] ⟦This article introduces the axiom schema of mathematical induction.⟧

[118] [These hypotheses are the first six axioms of Theory 3 (see Note J), together with propositions (1), (2), and (3) just given for the introduction of primitive recursive functions.]

[119] [This theory is Theory 4 (see Note J) with primitive recursive functions added. It is full elementary number theory. To understand the difficulties with which "we are confronted", of whose true nature Herbrand had as yet no inkling, see *Gödel 1931* and *Herbrand 1931b*, § 4 (below, pp. 626–628); see also *Gentzen 1936* and *1938*, *Ackermann 1940*, and *Schütte 1951* and *1960*.]

[120] [Although in *1931b* (below, pp. 623–626) Herbrand gives a much simpler proof of this result, the argument that he sketches here, once clarified, can be extended so as to establish the consistency of full elementary number theory (see *Dreben and Denton 1967*). Moreover, the present argument is closely related to the argument for (c') in Note E below, pp. 572–576.]

under consideration will be true in the infinite domain provided the following two conditions hold:

(1) The original descriptive functions have the same values as in A, and the propositional function $x = y$ has the same logical value as in A;

(2) For each given sequence of n_i elements, $u_1, u_2, \ldots, u_{n_i}$, in Γ, the function $f_i(u_1, u_2, \ldots, u_{n_i})$ has the value 0 [[that is, a_0]] unless there is a $j > 0$ such that

(α) a_j and a_{j+1} are in Γ,

(β) $\Phi_i(a_j, u_1, u_2, \ldots, u_{n_i})$ has the logical value "true", and

(γ) $\Phi_i(a_{j+1}, u_1, u_2, \ldots, u_{n_i})$ has the logical value "false",

in which case the function $f_i(u_1, u_2, \ldots, u_{n_i})$ has the value a_j.

Thus to actually construct the domain Γ we need only find the values that the new index functions have to take in Γ.

Every function (of whatever height) obtained from the original elementary descriptive functions will be called a function of the first kind; the new index functions (which are all of height 1) will be called functions of the second kind.

On any domain C [[contained in A]] in which the values of certain individuals of E are known, we shall perform, whenever possible, the following [[two-step]] operation:

(i) For each function of the first kind that is a function of individuals of E whose values in C are known, we add to C the value that this function has in the domain A (this value may or may not already occur in C; if it does, we simply have one more individual of E whose value in C is known);

(ii) Once this is done, for each function $f_i(y_1, y_2, \ldots, y_{n_i})$ of the second kind that is a function of individuals of E whose values $u_1, u_2, \ldots, u_{n_i}$ in C are known, we proceed thus:

(a) If $\Phi_i(0, u_1, u_2, \ldots, u_{n_i})$ is false in A, we assign the value 0 to $f_i(u_1, u_2, \ldots, u_{n_i})$;

(b) If there is a $j > 0$ such that a_j and a_{j+1} are in C and $\Phi_i(a_j, u_1, u_2, \ldots, u_{n_i})$ has the logical value "true" but $\Phi_i(a_{j+1}, u_1, u_2, \ldots, u_{n_i})$ has the logical value "false", then we assign the value a_j to $f_i(u_1, u_2, \ldots, u_{n_i})$.

So let us start from the domain consisting of a_0 and perform the preceding operation a sufficient number of times; we obtain larger and larger domains, in which an ever-increasing number of individuals of E have values. It is quite obvious that we shall reach a point (after a number of operations at most equal to the number of individuals of E) where we have a domain Γ in which it is no longer possible to perform the preceding operation. The individuals of E will then fall into three sets:

(1) e_1, [[the set of]] those that have a value in the domain Γ;

(2) e_2, [[the set of]] those that are functions of the second kind of elements of e_1 but do not satisfy either (a) or (b) above;

(3) e_3, [[the set of]] those that are functions of elements of e_2.

To the elements of e_2 we assign the value 0. The elements of e_3 are functions of elements of e_2; in these functions we replace the elements of e_2 by 0. The elements of e_3 remain individuals of E, and their heights are not increased. To each of these elements of e_3 that thus become elements of e_1 or e_2 we assign the value of the individual of E into which it is transformed. For those that become other elements of e_3 we repeat the same transformation.

Thus we finally assign values in Γ to all the individuals of E. Hence we obtain the domain of order p that we were looking for. We observe immediately that this domain satisfies conditions (1) and (2) stated above ⟦on page 566⟧; and so we come to the end of our proof.

This proof is a characteristic application of our method. We hope that our theorems will also enable us to prove the consistency of the theory that is obtained when the type of classes (Chapter 3, 2.3[121]) and the corresponding axioms are added to arithmetic; the difficulties that we have encountered in this direction are similar to those presented by the general case of the axiom of mathematical induction and are, it seems to us, closely related to the questions raised by Hilbert's Lemma I (*1925*; ⟦above, p. 385⟧). We believe, moreover, that it may be possible by this approach to arrive at a general theorem of which a particular case would be: *transcendental methods will not enable us to prove in arithmetic any theorem that cannot be proved without their help.*[122] Our theorem of Section 5, which provides a canonical form for every proof, would then enable us to give the purely arithmetic proof of these ⟦arithmetic⟧ theorems.

In the present Section 6 we have given some simple applications of our theorems. We hope that in a paper soon to be written we shall be able to show their usefulness in the study of "completely determined" theories (Chapter 3, 2.1[123]) and in that of the constructibility of mathematical objects.[124]

But it is quite certain that the most interesting line to pursue, and no doubt the most difficult, would be the one in which we search for a solution of the decision problem.[125] The solution of this problem would yield a general method in mathematics and would enable mathematical logic to play with respect to classical mathematics the role that analytic geometry plays with respect to ordinary geometry.

⟦The following ten Notes, A through J, were written by Burton Dreben. Each Note is connected with a particular passage of the text and, were it not for its length, would be appended there as a footnote. The passage to which a Note is attached is indicated by a footnote, and this is the footnote mentioned below the title of each Note. There are, moreover, cross references between footnotes and Notes.⟧

Note A

(*Footnote 15, p. 535 above*)

In Chapter 2, 4.1, Herbrand proves the following statement:

Let $A(p_1, \ldots, p_n)$ be an identity of the first kind in the distinct atomic propositions

[121] ⟦There, as an example of a theory with an infinite number of hypotheses, Herbrand presents "the simplest case of 'set theory': there are two types, the type of elements and the type of classes, and one atomic propositional function, $x \varepsilon \alpha$, in which the variable x is of the type of elements and the variable α of the type of classes; the hypotheses are all the propositions obtained when in

$$(\exists \alpha)(x)[x \varepsilon \alpha \equiv \Phi(x)]$$

we replace $\Phi(x)$ by any propositional function of x".⟧

[122] We ascertained that this theorem could be proved if the decision problem were solved and we had a proof of the consistency of Russell and Whitehead's theory (with the assumption that the multiplicative axiom and the axiom of infinity are true in it). [See *Herbrand 1931*, p. 54.]

[123] ⟦"A theory will be said to be *completely determined* if, for every proposition Z containing no real variable, Z or $\sim Z$ is true ⟦that is, provable in the theory⟧."⟧

[124] ⟦Parts of Herbrand's subsequent papers (*1931* and *1931b*) touch upon these questions, but the paper that he contemplates here was never written.⟧

[125] [See the last paragraph of the main text (p. 55) as well as the final Appendix (pp. 55–56) of *Herbrand 1931*.]

$p_1, \ldots, p_n$, $n \geqq 1$, where each p_i occurs at most once positively and at most once negatively. If $A(\Phi_1, \ldots, \Phi_n)$ results from $A(p_1, \ldots, p_n)$ when each p_i is replaced by a nonelementary proposition Φ_i, then $A(\Phi_1, \ldots, \Phi_n)$ is an identity of the second kind.

Herbrand's argument for this statement is inductive and rests upon repeated use of the following:

For some $i = 1, \ldots, n$, put Φ_i into prenex form, say $(\exists x)\Phi_i'(x)$, and, if p_i occurs twice in $A(p_1, \ldots, p_n)$, let $(\exists x)\Phi_i'(x)$ be the positive occurrence and $(\exists y)\Phi_i'(y)$ the negative occurrence of $(\exists x)\Phi_i'(x)$ in $A(\Phi_1, \ldots, (\exists x)\Phi_i'(x), (\exists y)\Phi_i'(y), \ldots, \Phi_n)$ (this formula displays the two different substitutions for the two occurrences of p_i). Now assume that $A(\Phi_1, \ldots, \Phi_i'(y), \Phi_i'(y), \ldots, \Phi_n)$ is provable. Then, by existential generalization, $(\exists x)A(\Phi_1, \ldots, \Phi_i'(x), \Phi_i'(y), \ldots, \Phi_n)$ is provable; then, by universal generalization, $(y)(\exists x)A(\Phi_1, \ldots, \Phi_i'(x), \Phi_i'(y), \ldots, \Phi_n)$ is provable; hence, by the rules of passage, $A(\Phi_1, \ldots, (\exists x)\Phi_i'(x), (\exists y)\Phi_i'(y), \ldots, \Phi_n)$ is provable.

The requirement that in $A(p_1, \ldots, p_n)$ each p_i occurs at most once positively and once negatively is essential to the proof just given. For take $A(p)$ to be $(p \vee p) \supset p$. It is obviously an identity of the first kind; but a quick check shows that the substitution instance

$$[(x)(\exists y)F(x, y) \vee (x)(\exists y)F(x, y)] \supset (x)(\exists y)F(x, y),$$

although an identity of the second kind (see Note B below) is *not* a normal identity.

Note B

(*Footnote 17, p. 536 above*)

It is important to note that, for an arbitrary proposition P, not only is the proposition $(P \vee P) \equiv P$ provable in Herbrand's system Q_H, but also each of the propositions $P \supset (P \vee P)$ and $(P \vee P) \supset P$ is provable in Q_H without the use of modus ponens (see Note D below; also compare the following argument with 5.3 in the text).

Assume that P has the prenex form

$$(\pm x_1) \ldots (\pm x_n) M(x_1, \ldots, x_n).$$

(1) We shall first see that $P \supset (P \vee P)$ is a normal identity, hence is provable in Q_H without the use of modus ponens (see footnote 11 above). Using the rules of passage, we turn $P \supset (P \vee P)$ into the provably equivalent prenex proposition Q:

$$(+x_1^{\alpha_1}, -x_1^{1-\alpha_1}, +x_2^{\alpha_2}, -x_2^{1-\alpha_2}, \ldots, +x_n^{\alpha_n}, -x_n^{1-\alpha_n}, \pm x_1^2, \ldots, \pm x_n^2)$$

$$\{M(x_1^0, x_2^0, \ldots, x_n^0) \supset [M(x_1^1, x_2^1, \ldots, x_n^1) \vee M(x_1^2, \ldots, x_n^2)]\},$$

where α_i $(i = 1, \ldots, n)$ is 0 or 1 according as x_i is restricted or general in P.

But clearly Q is a normal identity. For we get an elementary identity Π associated with Q simply by deleting all quantifiers, replacing each restricted variable $x_i^{1-\alpha_i}$ by the general variable $x_i^{\alpha_i}$, and replacing each restricted variable among the x_i^2 by the general variable $x_1^{\alpha_1}$.

(2) As we saw in Note A, $(P \vee P) \supset P$ is not in general a normal identity. However, the following prenex proposition R is a normal identity:

$$(+x_1^{\alpha_1}, -x_1^{1-\alpha_1}, +x_2^{\alpha_2}, -x_2^{1-\alpha_2}, \ldots, +x_n^{\alpha_n}, -x_n^{1-\alpha_n}, +y_1^{\beta_1}, -y_1^{1-\beta_1}, \ldots,$$
$$+y_n^{\beta_n}, -y_n^{1-\beta_n}, \pm x_1^2, \ldots, \pm x_n^2, \pm y_1^2, \ldots, \pm y_n^2)$$

$$\big[\{[M(x_1^0, \ldots, x_n^0) \lor M(y_1^0, \ldots, y_n^0)] \supset M(x_1^1, \ldots, x_n^1)\} \lor \{[M(x_1^2, \ldots, x_n^2) \lor M(y_1^2, \ldots, y_n^2)] \supset M(y_1^1, \ldots, y_n^1)\}\big],$$

where α_i is 0 or 1, and x_i^2 is general or restricted, according as x_i is restricted or general in P, and β_i is 0 or 1, and y_i^2 is general or restricted, according as x_i is restricted or general in P.

Here we get an elementary identity Π associated with R by deleting all quantifiers, replacing each restricted variable $x_i^{1-\alpha_i}$ by the general variable $x_i^{\alpha_i}$ and each restricted variable $y_i^{1-\beta_i}$ by the general variable $y_i^{\beta_i}$, and replacing each restricted variable among the x_i^2 and the y_i^2 by the general variable $x_1^{\alpha_1}$. (Thus Π has the form

$$[(p \lor q) \supset p] \lor [(r \lor s) \supset q].)$$

Now, by the rules of passage R is equivalent to

$$[(P \lor P) \supset P] \lor [(P \lor P) \supset P].$$

But then by the rule of simplification we get $(P \lor P) \supset P$.

Note C

(Footnote 19, p. 536 above)

Herbrand's argument in this paragraph is ingenious but rather briefly expressed. Since it plays an essential role in his proof that modus ponens is eliminable from quantification theory (see 6.1 on p. 558 above as well as Note D below), it merits expansion.

Let P be any proposition that contains $n \geqq 1$ bound variables $\pm x_1, \ldots, \pm x_n$, let $M(x_1, \ldots, x_n)$ be the matrix of P, and let P' be any proposition derived from P by a scheme Σ. Then $P \equiv P'$ is provable in Q_H.

Proof. Assume that the scheme Σ has $k \geqq 1$ lines, each containing n signed letters. For each $i \leqq k$ let the ith line of Σ be $\pm x_1^i, \ldots, \pm x_{r_i}^i, \pm x_{r_i+1}^i, \ldots, \pm x_n^i$, where the initial segment $\pm x_1^i, \ldots, \pm x_{r_i}^i$, $0 \leqq r_i < n$, is the longest initial segment common to both the ith and the $(i-1)$th lines of Σ. (Take $r_1 = 0$. Footnote 21 shows why nondisjoint lines are being considered.)

Now let Σ_i be the subscheme of Σ that consists of the first i lines of Σ, and let P_i be a proposition derived from P by Σ_i. Then the matrix M_i of the prenex proposition P_i has the form

$$M(x_1^1, x_2^1, \ldots, x_n^1) \lor M(x_1^2, x_2^2, \ldots, x_{r_2}^2, x_{r_2+1}^2, \ldots, x_n^2) \lor \cdots \lor M(x_1^i, x_2^i, \ldots, x_{r_i}^i, x_{r_i+1}^i, \ldots, x_n^i),$$

and the type T_i (that is, the prefix) of P_i can be taken to be

$$(\pm x_1^1, \pm x_2^1, \ldots, \pm x_n^1, \pm x_{r_2+1}^2, \ldots, \pm x_n^2, \ldots, \pm x_{r_i+1}^i, \ldots, \pm x_n^i).$$

But then, if we write r for r_{i+1}, a proposition P_{i+1} derived from P by the subscheme Σ_{i+1} consisting of the first $i+1$ lines of Σ can be given the form

$$(T_i, \pm x_{r+1}^{i+1}, \pm x_{r+2}^{i+1}, \ldots, \pm x_n^{i+1})[M_i \lor M(x_1^{i+1}, \ldots, x_r^{i+1}, x_{r+1}^{i+1}, \ldots, x_n^{i+1})].$$

Moreover, for $r>0$, the initial segment $\pm x_1^i, \pm x_2^i, \ldots, \pm x_r^i$ of the ith line is identical with the initial segment $\pm x_1^{i+1}, \pm x_2^{i+1}, \ldots, \pm x_r^{i+1}$ of the $(i+1)$th line and lies in the type T_i. So, if $r \leqq r_i = s$, by using the rules of passage first to turn P_i into canonical form and then to pull the quantifiers $\pm x_1^i, \ldots, \pm x_{r_i}^i$ out again, we can turn the proposition P_i into the provably equivalent proposition P_i^1 of the form

$$(\pm x_1^i, \pm x_2^i, \ldots, \pm x_s^i)[A(x_1^i, x_2^i, \ldots, x_s^i) \vee (\pm x_{s+1}^i, \ldots, \pm x_n^i)M(x_1^i, \ldots, x_n^i)],$$

and we can turn the proposition P_{i+1} into the provably equivalent proposition P_{i+1}^1 of the form

$$(\pm x_1^i, \pm x_2^i, \ldots, \pm x_s^i)[A(x_1^i, x_2^i, \ldots, x_s^i) \vee (\pm x_{s+1}^i, \ldots, \pm x_n^i)M(x_1^i, \ldots, x_n^i) \vee (\pm x_{s+1}^{i+1}, \ldots, \pm x_n^{i+1})M(x_1^{i+1}, \ldots, x_n^{i+1})].$$

And if $r > r_i = s$, then, similarly, P_i can be turned into P_i^1 of the form

$$(\pm x_1^{i+1}, \pm x_2^{i+1}, \ldots, \pm x_r^{i+1})[A(x_1^{i+1}, x_2^{i+1}, \ldots, x_r^{i+1}) \vee (\pm x_{r+1}^i, \pm x_{r+2}^i, \ldots, \pm x_n^i)M(x_1^i, \ldots, x_n^i)],$$

and P_{i+1} turned into P_{i+1}^1 of the form

$$(\pm x_1^{i+1}, \pm x_2^{i+1}, \ldots, \pm x_r^{i+1})[A(x_1^{i+1}, x_2^{i+1}, \ldots, x_r^{i+1}) \vee (\pm x_{r+1}^i, \pm x_{r+2}^i, \ldots, \pm x_n^i) M(x_1^i, x_2^i, \ldots, x_n^i) \vee (\pm x_{r+1}^{i+1}, \pm x_{r+2}^{i+1}, \ldots, \pm x_n^{i+1})M(x_1^{i+1}, x_2^{i+1}, \ldots, x_n^{i+1})].$$

But, by once again using the rules of passage, we can make the disjunct $(\pm x_{s+1}^{i+1}, \ldots, \pm x_n^{i+1})M(x_1^{i+1}, \ldots, x_n^{i+1})$ identical with the disjunct

$$(\pm x_{s+1}^i, \ldots, x_n^i)M(x_1^i, \ldots, x_n^i).$$

But then P_{i+1}^1 becomes provably equivalent to the proposition P_{i+1}^2 of the form

$$(\pm x_1^i, \ldots, \pm x_s^i)[A(x_1^i, \ldots, x_s^i) \vee (\pm x_{s+1}^i, \ldots, \pm x_n^i)M(x_1^i, \ldots, x_n^i) \vee (\pm x_{s+1}^i, \ldots, \pm x_n^i)M(x_1^i, \ldots, x_n^i)].$$

(A strictly analogous transformation of P_{i+1}^1 takes place if $r > r_i = s$.) But, for any Q, $\vdash (Q \vee Q) \equiv Q$ (see Note B, p. 568 above). Hence, by the theorem of Chapter 2, 6.1 (see footnote 18),

$$\vdash P_i^1 \equiv P_{i+1}^2,$$

hence, by the rules of passage,

$$\vdash P_i \equiv P_{i+1},$$

hence, by induction on the number of lines in Σ,

$$\vdash P \equiv P'.$$

Note that, by an argument similar to the one given in part 1 of Note B, $P \supset P'$ can easily be shown to be a normal identity, and hence provable in Q_H. Thus Herbrand's primary concern in the present argument is to show that $P' \supset P$ is provable in Q_H.

Note D

(Footnote 23, p. 537 above)

If a proposition P has property A, then there is, in Herbrand's system Q_H, a formal proof of P that can be divided into two parts. Part 1 is a proof of some prenex normal identity P'. Part 2 is a proof of P from P'. (In general, P' is assumed to be derivable from P by some scheme Σ. But any such derivation is not part of the formal proof of P in Q_H.)

Now we know—see footnote 11—that part 1 can always be put into a simple standard form consisting of a quantifier-free identity followed by the results of successive applications of the rules of generalization.

However, the discussion in Note C shows us that we can also give part 2 a standard form if to the system Q_H we add, as a new primitive rule of inference (see p. 558 above), the following *generalized rule of simplification*:

If a proposition R is like a proposition Q except for containing an occurrence of the subproposition P where Q contains an occurrence of the subproposition $P \vee P$, then from Q we can infer R.

For assume that P^2_{i+1} is proved in Q_H (see Note C). It has the form

$$(\pm x_1, \ldots, \pm x_r)[A(x_1, \ldots, x_r) \vee (\pm x_{r+1}, \ldots, \pm x_n)M(x_1, \ldots, x_n) \vee (\pm x_{r+1}, \ldots, \pm x_n)M(x_1, \ldots, x_n)].$$

Hence, by the generalized rule of simplification, we prove P^1_i of the form

$$(\pm x_1, \ldots, \pm x_r)[A(x_1, \ldots, x_r) \vee (\pm x_{r+1}, \ldots, \pm x_n)M(x_1, \ldots, x_n)].$$

But then, as Herbrand says in 6.1, a proposition P that has property A can be proved from a quantifier-free identity just by the use of the two rules of generalization, the rules of passage, and the generalized rule of simplification. The rule of implication (that is, modus ponens) is not needed. (In Section 5, pp. 554–558 above, Herbrand shows constructively that a proposition is provable in Q_H if and *only if* it has property A. Hence Q_H is equivalent to a system Q'_H that is just like Q_H except for containing the generalized rule of simplification in place of both the rule of simplification and the rule of implication. This constructively proved equivalence between Q_H and Q'_H is the basic result of Herbrand's thesis.)

Note E

(Footnote 60, p. 548 above)

The following example, taken from *Dreben, Andrews, and Aanderaa 1963*, shows that the first half of (*c*) is false. Indeed, it shows that to calculate, in general, a number q from a number p such that P_2 has property C of order q when P_1 has property C of order p we must know more about P_1 than just the number p. (P_2 contains $(x)[\Phi(x) \vee Z]$ in one place where P_1 has $(x)\Phi(x) \vee Z$.)

For each $p \geqq 3$ let A_p be

$$(y_2)(y_3)\ldots(y_{p+1})[\sim W(y_2, y_3) \vee \sim W(y_3, y_4) \vee \cdots \vee \sim W(y_{p-1}, y_p) \vee \sim H(y_2, y_{p+1}) \vee (\exists x_3)H(y_p, x_3)],$$

let $P_{1,p}$ be

$$A_p \vee (\exists x_1)(\exists x_2) \sim [\sim W(x_1, x_2) \vee (x) \sim H(x_1, x) \vee (\exists y_1)H(x_2, y_1)],$$

and let $P_{2,p}$ be

$$A_p \vee (\exists x_1)(\exists x_2) \sim \{\sim W(x_1, x_2) \vee (x)[\sim H(x_1, x) \vee (\exists y_1)H(x_2, y_1)]\}.$$

Now, for each $p \geqq 3$, $P_{1,p}$ has property C of order 3 and $P_{2,p}$ has property C of order p but of *no* smaller order.

We shall soon see, however, that the only additional information needed about P_1 to obtain q is the number of quantifiers and descriptive functions occurring in P_1. For, by combining Herbrand's argument with certain ideas of Hilbert, Ackermann, and Bernays (see *Hilbert 1927*, *Bernays 1927*, *Hilbert and Bernays 1939*, pp. 93–130, and *Ackermann 1940*, as well as pp. 485–486 and 565–567 above), we can prove:

(c') Let P_1 be any proposition that contains no real variable but does contain j quantifiers and d descriptive functions, $j, d \geqq 0$. Let P_2 be like P_1 except for containing $(x)[\Phi(x) \vee Z]$ in one negative occurrence where P_1 contains $(x)\Phi(x) \vee Z$. Assume that P_1 has property C of order p. Then P_2 has property C of some order less than or equal to $p(1 + N^j)^p$, where $N = n_p$ is the cardinality of $G_p^{(1)} = C_1^{(1)} \cup \cdots \cup C_p^{(1)}$. But then, since N can obviously be bounded by a primitive recursive function $\zeta(j, d, n_1)$, where n_1 is the cardinality of $C_1^{(1)}$, there is a primitive recursive function $\gamma(j, d, p, n_1)$ such that, if P_1 has property C of order p, then P_2 has property C of some order less than or equal to $\gamma(j, d, p, n_1)$.

Proof (adapted from *Dreben and Denton 1966*). We begin by defining the notion of a *standard expansion of a proposition.*

DEFINITION I. Let P be any proposition of height $h \geqq 0$. If y is any real or general variable of P, and $x_1^y, \ldots, x_{r_y}^y$, with $r_y \geqq 0$, are the restricted variables of P superior to y, call $f_y(x_1^y, \ldots, x_{r_y}^y)$ the *superiority index function correlated with* y.

Let $E(P)$ be the elementary proposition that we obtain from P by deleting all the quantifiers of P and replacing each real and each general variable of P with its correlated superiority index function. Call $E(P)$ the *elementary proposition associated with* P and note that the real variables of $E(P)$ are the restricted variables of P.

For each $c \geqq 1$, let $G(P, c)$ be the union of the domains $C_1, C_2, \ldots, C_c$ generated by a set of functions among which are all the descriptive functions of P and all the superiority index functions correlated with the general variables of P. (In the present Note we write $G(P, c)$ rather than G_c.)

Let K be either $E(P)$ or a subproposition of $E(P)$. A *substitution instance of* K *over* $G(P, p)$ is the elementary proposition that we obtain from K by first replacing each real variable of K with some member of $G(P, p)$ and then replacing each function in the resulting expression with its value in $G(P, p + h + 1)$. If $x_1 \ldots, x_k$ are all the real variables of K and $a_1, \ldots, a_k$ are any (not necessarily distinct) members of $G(P, p)$, then write $K[a_1, \ldots, a_k]$ for the substitution instance of K that we obtain by replacing x_i with a_i for $i = 1, \ldots, k$.

Now let $R(P, p)$ be the disjunction (in some order henceforth assumed fixed) of all the distinct substitution instances of $E(P)$ over $G(P, p)$, and call $R(P, p)$ the *standard expansion of* P *over* $G(P, p)$, or the *standard expansion of* P *of order* p. (End of Definition I.)

On pages 25–27 and in footnote 1 on page 42 of a paper (*1931*) that he wrote shortly after his thesis, Herbrand introduced both the notion of an elementary proposition associated with P and that of a standard expansion of P. (The expression "elementary proposition associated with P" comes from Herbrand. He had no name for a standard expansion; but see footnote 63 above and Note H below.) Herbrand did not show that the (far more perspicuous) notion of standard expansion is interchangeable with the notion of expansion given on pages 543–544 above. But, if we take into account the associative and distributive laws, it is not hard to see that the standard expansion of P over $G(P, p)$ is truth-functionally equivalent to the second expansion of P over $G(P, p)$. (For the term "second expansion of P" see footnote 44 above.) Hence a proposition P has property C of order p if and only if $R(P, p)$ is an identity. But then to prove (c') above it suffices to prove that, if $R(P_1, p)$ is an identity, $R(P_2, u)$, where $u = p(1 + N^j)^p$, is also an identity.

Let $r \geqq 1$ be the number of restricted quantifiers in P_1 and let $n \geqq 0$ be the number of restricted quantifiers $\pm x_1, \ldots, \pm x_n$ in P_1 within whose scopes $(x)\Phi(x) \vee Z$ lies. (Note that $n \leqq r \leqq j$.) Let $t \geqq 0$ be the number of restricted quantifiers in Φ and let $m \geqq 0$ be the number of restricted quantifiers in Z. We shall assume that Φ and Z each contains at least one general quantifier. (This innocuous assumption simplifies our notational conventions. Also, if Z contains no general quantifier, $E(P_1)$ and $E(P_2)$ are identical, and so P_2 obviously has property C of order p.) Let us write $\Phi_1 \vee Z_1$ for what $(x)\Phi(x) \vee Z$ becomes in $E(P_1)$, and $\Phi_2 \vee Z_2$ for what $(x)[\Phi(x) \vee Z]$ becomes in $E(P_2)$. When we wish to exhibit the real variables of Φ_1 and Φ_2, Φ_1 will be written $\Phi_1(x_1, \ldots, x_n, x, x_{n+1}, \ldots, x_{n+t})$ and Φ_2 will be written $\Phi_2(x_1, \ldots, x_n, x, x_{n+1}, \ldots, x_{n+t})$. Finally, if S is any set, let us write $[S]^k$ for the k-fold Cartesian product of S, $k \geqq 0$ ($[S]^0$ is the unit set of the null set, and an ordered 0-tuple is the null set).

By hypothesis, $R(P_1, p)$ is an identity. Let $q = p(1 + N^n)^p$. Since $n \leqq r \leqq j$, to prove (c') is suffices to prove that $R(P_2, q)$ is an identity. This we shall now do. The argument turns on three main definitions and two lemmas. (In what follows, the term "truth value" is used for Herbrand's "logical value", the term "truth-value assignment" for Herbrand's "system of logical values", and the sign "**1**" for the first element of $C_1^{(1)} = C_1^{(2)}$.)

Definition II. Let A be any truth-value assignment to $R(P_2, q + p)$. A function α from $[G(P_2, q + p)]^n$ into $G(P_2, q)$ is said to be *A-admissible* if and only if, for each ordered n-tuple $\langle a_1, \ldots, a_n\rangle$ in $[G(P_2, q + p)]^n$, either $\alpha(a_1, \ldots, a_n\rangle = \mathbf{1}$ or there is some ordered t-tuple $\langle a_{n+1}, \ldots, a_{n+t}\rangle$ in $[G(P_2, q)]^t$ such that A falsifies the substitution instance

$$\Phi_2[a_1, \ldots, a_n, \beta, a_{n+1}, \ldots, a_{n+t}]$$

of $\Phi_2(x_1, \ldots, x_n, x, x_{n+1}, \ldots, x_{n+t})$, where $\beta = \alpha(a_1, \ldots, a_n)$.

By an inductive construction on the orders of the elements we can easily associate with each function α from $[G(P_2, q + p)]^n$ into $G(P_2, q)$ a unique one-to-one mapping Δ_α of $G(P_1, p + h + 1)$ into $G(P_2, q + p + h)$ that satisfies the following three conditions:

(1) $\Delta_\alpha(a_e) = a_e$ for each element a_e in $C_1^{(1)} = C_1^{(2)}$.

(2) Let $\pm y^1$ be any general quantifier in P_1 and let $\pm y^2$ be the corresponding general quantifier in P_2.

(*i*) Assume that $\pm y^1$ does not occur in Z. If $b_1, \ldots, b_{r_{y^1}}$ are any r_{y^1} members of $G(P_1, p)$ and if $b = f_{y^1}(b_1, \ldots, b_{r_{y^1}})$, then $\Delta_\alpha(b) = f_{y^2}[\Delta_\alpha(b_1), \ldots, \Delta_\alpha(b_{r_{y^1}})]$.

(*ii*) Assume that $\pm y^1$ does occur in Z and lies within the scopes of s_{y^1} restricted quantifiers also occurring in Z. If $b_1, \ldots, b_n, b_{n+1}, \ldots, b_{n+s_{y^1}}$ are any $n + s_{y^1}$ members of $G(P_1, p)$ and if

$$b = f_{y^1}(b_1, \ldots, b_n, b_{n+1}, \ldots, b_{n+s_{y^1}}),$$

then

$$\Delta_\alpha(b) = f_{y^2}[\Delta_\alpha(b_1), \ldots, \Delta_\alpha(b_n), \beta, \Delta_\alpha(b_{n+1}), \ldots, \Delta_\alpha(b_{n+s_{y^1}})],$$

where $\beta = \alpha[\Delta_\alpha(b_1), \ldots, \Delta_\alpha(b_n)]$.

(3) Let $f(x_{i_1}, \ldots, x_{i_\theta})$ be any descriptive function of height $\nu \leqq h$ occurring in P_1. If $b_1, \ldots, b_\theta$ are any members of $G(P_1, p)$ and if $b = f(b_1, \ldots, b_\theta)$, then $\Delta_\alpha(b) = f[\Delta_\alpha(b_1), \ldots, \Delta_\alpha(b_\theta)]$.

Definition III. For each A-admissible function α, call the one-to-one mapping Δ_α of $G(P_1, p + h + 1)$ into $G(P_2, q + p + h)$ that satisfies the above three conditions an *A-admissible α-mapping.*

Note that, since Δ_α maps $G(P_1, p + h + 1)$ into $G(P_2, q + p + h)$, Δ_α maps $G(P_1, p + 1)$ into $G(P_2, q + p)$ and not merely into $G(P_2, q + p + 1)$.

We now come to our basic definition.

Definition IV. An A-admissible α-mapping Δ_α is said to be an *A-resolvent* if and only if the following two conditions are satisfied:

(1) Δ_α maps $G(P_1, p + h + 1)$ into $G(P_2, q + h)$;

(2) For each ordered n-tuple $\langle b_1, \ldots, b_n\rangle$ in $[G(P_1, p)]^n$, if A falsifies $\Phi_2[\Delta_\alpha(b_1), \ldots, \Delta_\alpha(b_n), \Delta_\alpha(b^*), a_1^*, \ldots, a_t^*]$ for some element b^* in $G(P_1, p)$ and for some ordered t-tuple $\langle a_1^*, \ldots, a_t^*\rangle$ in $[G(P_2, q)]^t$, then A falsifies $\Phi_2(\Delta_\alpha(b_1), \ldots, \Delta_\alpha(b_n), \beta, a_1, \ldots, a_t]$ for $\beta = \alpha[\Delta_\alpha(b_1), \ldots, \Delta_\alpha(b_n)]$ and for some ordered t-tuple $\langle a_1, \ldots, a_t\rangle$ in $[G(P_2, q)]$.

Lemma I. Let $R(P_1, p)$ be an identity. Let A be any truth-value assignment to $R(P_2, q + p)$. If A possesses an A-resolvent Δ_α, then A verifies $R(P_2, q)$.

Proof. Let $A(\Delta_\alpha)$ be the truth-value assignment induced on $R(P_1, p)$ by A and Δ_α thus: for each k-adic predicate letter W of P_1 and any k members $b_1, \ldots, b_k$ of $G(P_1, p + h + 1)$, $A(\Delta_\alpha)$ assigns to the atomic proposition $W[b_1, \ldots, b_k]$ the same truth value that A assigns to $W[\Delta_\alpha(b_1), \ldots, \Delta_\alpha(b_k)]$. By hypothesis, $R(P_1, p)$ is an identity. But it is also a disjunction. Hence $A(\Delta_\alpha)$ must verify at least one disjunct of $R(P_1, p)$, say the substitution instance $E(P_1)[b_1, \ldots, b_r]$. Using $E(P_1)[b_1, \ldots, b_r]$ and Δ_α, we shall specify a disjunct of $R(P_2, q)$ that A verifies.

Let

$$\Phi_1[b_{c_1}, \ldots, b_{c_n}, b^*, b_{c_{n+1}}, \ldots, b_{c_{n+t}}] \vee Z_1[b_{c_1}, \ldots, b_{c_n}, b_{u_1}, \ldots, b_{u_m}],$$

abbreviated $\Phi_1^1 \vee Z_1^1$, be the substitution instance of $\Phi_1 \vee Z_1$ that occurs in $E(P_1)[b_1, \ldots, b_r]$. Consider now the substitution instance $E(P_2)[\Delta_\alpha(b_1), \ldots, \Delta_\alpha(b_r)]$ of $E(P_2)$ over $G(P_2, q)$. Then

$$\Phi_2[\Delta_\alpha(b_{c_1}), \ldots, \Delta_\alpha(b_{c_n}), \Delta_\alpha(b^*), \Delta_\alpha(b_{c_{n+1}}), \ldots, \Delta_\alpha(b_{c_{n+t}})] \vee$$
$$Z_2[\Delta_\alpha(b_{c_1}), \ldots, \Delta_\alpha(b_{c_n}), \Delta_\alpha(b^*), \Delta_\alpha(b_{u_1}), \ldots, \Delta_\alpha(b_{u_m})],$$

abbreviated $\Phi_2^1 \vee Z_2^1$, is the substitution instance of $\Phi_2 \vee Z_2$ that occurs in $E(P_2)[\Delta_\alpha(b_1), \ldots, \Delta_\alpha(b_r)]$. We see at once that Φ_2^1 takes the same truth value under A as Φ_1^1 takes under $A(\Delta_\alpha)$. More generally, we see at once that

(†) If φ is any atomic proposition in $E(P_1)[b_1, \ldots, b_r]$ not occurring just in Z_1^1 and if χ is any homologously occurring atomic proposition in $E(P_2)[\Delta_\alpha(b_1), \ldots, \Delta_\alpha(b_r)]$, then A assigns to χ the same truth value as $A(\Delta_\alpha)$ assigns to φ.

However, we cannot see at once that A assigns to Z_2^1 the same truth value as $A(\Delta_\alpha)$ assigns to Z_1^1. For, by Definition III, $A(\Delta_\alpha)$ assigns to Z_1^1 the same truth value as A assigns to

$$Z_2^2 = Z_2[\Delta_\alpha(b_{c_1}), \ldots, \Delta_\alpha(b_{c_n}), \beta, \Delta_\alpha(b_{u_1}), \ldots, \Delta_\alpha(b_{u_m})],$$

where $\beta = \alpha[\Delta_\alpha(b_{c_1}), \ldots, \Delta_\alpha(b_{c_n})]$, and we have no right to assume that $\Delta_\alpha(b^*) = \beta$.

Nevertheless, it follows from (†) that A verifies $E(P_2)[\Delta_\alpha(b_1), \ldots, \Delta_\alpha(b_r)]$ provided that A assigns to $\Phi_2^1 \vee Z_2^1$ the same truth value as $A(\Delta_\alpha)$ assigns to $\Phi_1^1 \vee Z_1^1$. Moreover, it follows from (†) that A verifies $E(P_2)[\Delta_\alpha(b_1), \ldots, \Delta_\alpha(b_r)]$ provided that A falsifies $\Phi_2^1 \vee Z_2^1$ and $A(\Delta_\alpha)$ verifies $\Phi_1^1 \vee Z_1^1$. For, since $\Phi_1 \vee Z_1$ occurs negatively in $E(P_1)$, the substitution instance $\Phi_1^1 \vee Z_1^1$ occurs negatively in $E(P_1)[b_1, \ldots, b_r]$. Hence, if we replace $\Phi_1^1 \vee Z_1^1$ in $E(P_1)[b_1, \ldots, b_r]$ by any proposition that is false under $A(\Delta_\alpha)$, the resulting proposition Θ is still true under $A(\Delta_\alpha)$. But then, since $\Phi_2^1 \vee Z_2^1$ occurs negatively in $E(P_2)[\Delta_\alpha(b_1), \ldots, \Delta_\alpha(b_r)]$ and is assumed to be false under A, the propositions Θ and $E(P_2)[\Delta_\alpha(b_1), \ldots, \Delta_\alpha(b_r)]$ will agree in truth value. Hence A verifies $E(P_2)[\Delta_\alpha(b_1), \ldots, \Delta_\alpha(b_r)]$.

Thus we need consider only the case in which $A(\Delta_\alpha)$ falsifies $\Phi_1^1 \vee Z_1^1$ but A verifies $\Phi_2^1 \vee Z_2^1$. But here A falsifies Φ_2^1 and also Z_2^2. Hence, since Δ_α is an A-resolvent, A falsifies some substitution instance, namely

$$\Phi_2^2 = \Phi_2[\Delta_\alpha(b_{c_1}), \ldots, \Delta_\alpha(b_{c_n}), \beta, a_1, \ldots, a_t],$$

where $\langle a_1, \ldots, a_t\rangle$ is some ordered t-tuple in $[G(P_2, q)]^t$ and $\beta = \alpha[\Delta_\alpha(b_{c_1}), \ldots, \Delta_\alpha(b_{c_n})]$. Obviously, A falsifies $\Phi_2^2 \vee Z_2^2$. So, let $E^*(P_2)$ be the substitution instance of $E(P_2)$ over $G(P_2, q)$ that results from $E(P_2)[\Delta_\alpha(b_1), \ldots, \Delta_\alpha(b_r)]$ when we put $\Phi_2^2 \vee Z_2^2$ for $\Phi_2^1 \vee Z_2^1$. But then we see from (†) that A verifies $E^*(P_2)$. Q. e. d.

LEMMA II. Let A be any truth-value assignment to $R(P_2, q + p)$. Then A possesses an A-resolvent Δ_α.

Proof. We shall construct by induction a sequence $\sigma = \langle \alpha_1, \alpha_2, \ldots, \alpha_{q/p}\rangle$ of q/p (not necessarily distinct) A-admissible functions and then show that for some $z < q/p$ there is a function α_z in σ that determines an A-resolvent.

If α is any A-admissible function and Δ_α is the mapping determined by α, let η_α be the maximum of the orders of the elements in the range of α, and let δ_α be the maximum of the orders of those elements in the range of Δ_α that are images of elements in $G(P_1, p + 1)$. Call η_α the *order* of α, and δ_α the *order* of Δ_α. Now $\eta_\alpha \leqq q$. So $\delta_\alpha \leqq \eta_\alpha + p \leqq q + p$. (For the remainder of the proof of Lemma II we shall write Δ_d, η_d, and δ_d in place of Δ_{α_d}, η_{α_d}, and δ_{α_d}, respectively, when $d = 1, \ldots, q/p$.)

Begin the construction of σ by letting α_1 be the A-admissible function that assigns **1** to all n-tuples in $[G(P_2, q + p)]^n$. Since **1** is of order 1, the mapping Δ_1 has the order $p + 1$. Continue inductively as follows.

Consider successively each $g < q/p$ and assume that the A-admissible function α_g has an order $\eta_g \leqq p(g - 1) + 1$. Then the mapping Δ_g has an order $\delta_g \leqq p(g - 1) + 1 + p = pg + 1$. But $g < q/p$. So $pg + 1 \leqq q$ and then $\delta_g \leqq q$. If Δ_g is an A-resolvent, let $\alpha_{g+1} = \alpha_g$. Then Δ_{g+1} coincides with Δ_g and is also an A-resolvent. If, however, Δ_g is not an A-resolvent, since $\delta_g \leqq q$, there must be an ordered n-tuple $\langle b_1, \ldots, b_n \rangle$ in $[G(P_1, p)]^n$ such that,

(1) For some element b^* in $G(P_1, p)$ and some ordered t-tuple $\langle a_1^*, \ldots, a_t^* \rangle$ in $[G(P_2, q)]^t$, A falsifies $\Phi_2[\Delta_g(b_1), \ldots, \Delta_g(b_n), \Delta_g(b^*), a_1^*, \ldots, a_t^*]$, but

(2) For $\beta = \alpha_g[\Delta_g(b_1), \ldots, \Delta_g(b_n)]$ and *each* ordered n-tuple $\langle a_1, \ldots, a_t \rangle$ in $[G(P_2, q)]^t$, A verifies

$$\Phi_2[\Delta_g(b_1), \ldots, \Delta_g(b_n), \beta, a_1, \ldots, a_t].$$

Hence $\Delta_g(b^*) \neq \beta$. Moreover, since the function α_g is assumed to be A-admissible, it follows from Definition II that $\beta = \mathbf{1}$. Let α_{g+1} be the function that assigns $\Delta_g(b^*)$ to the ordered n-tuple $\langle \Delta_g(b_1), \ldots, \Delta_g(b_n) \rangle$ and agrees with α_g on all other elements of $G(P_2, q + p)$. But the order of $\Delta_g(b^*)$ is less than or equal to $\delta_g \leqq q$. So $\eta_{g+1} \leqq q$, and α_{g+1} is A-admissible.

Now, in the sequence $\sigma = \langle \alpha_1, \alpha_2, \ldots, \alpha_{q/p} \rangle$ just constructed, a function a_g is identical with the function α_{g+1} if and only if the mapping Δ_g is an A-resolvent. Hence, to complete the proof of Lemma II, we have only to show that there is at least one α_z in σ such that $\alpha_z = \alpha_{z+1}$.

For each $g \leqq q/p$ and each $k \leqq p$, let $\xi(g, k)$ be the number of ordered n-tuples $\langle b_1, \ldots, b_n \rangle$ in $[G(P_1, k)]^n$ such that $\alpha_g[\Delta_g(b_1), \ldots, \Delta_g(b_n)] = \mathbf{1}$, and let the *index* of α_g, abbreviated ind(α_g), be the ordered p-tuple $\langle \xi(g, 1), \xi(g, 2), \ldots, \xi(g, p) \rangle$. Then, for each g and each $m \leqq q/p$, write ind$(\alpha_g) <$ ind(α_m) if and only if, for some $k \leqq p$ and for all $j < k$, $\xi(g, k) < \xi(m, k)$ but $\xi(g, j) = \xi(m, j)$. We shall now establish:

(★) For each $g < q/p$ either $\alpha_g = \alpha_{g+1}$ or ind$(\alpha_{g+1}) <$ ind(α_g).

By the construction of σ, if $\alpha_g \neq \alpha_{g+1}$, there is a unique ordered n-tuple $\langle b_1, \ldots, b_n \rangle$ in $[G(P_1, p)]^n$ such that

$$\alpha_g[\Delta_g(b_1), \ldots, \Delta_g(b_n)] = \mathbf{1}$$

but

$$\alpha_{g+1}[\Delta_g(b_1), \ldots, \Delta_g(b_n)] \neq \mathbf{1}.$$

Moreover, the functions α_g and α_{g+1} agree on all other ordered n-tuples in $[G(P_2, q + p)]^n$. Hence, by the construction of an A-admissible α-mapping, if $k \leqq p$ is the maximum of the orders of $b_1, \ldots, b_n$, then, for each element b in $G(P_1, k)$, $\Delta_g(b) = \Delta_{g+1}(b)$. Therefore, for all $j < k$, $\xi(g + 1, j) = \xi(g, j)$. Also,

$$\alpha_{g+1}[\Delta_g(b_1), \ldots, \Delta_g(b_n)] = \alpha_{g+1}[\Delta_{g+1}(b_1), \ldots, \Delta_{g+1}(b_n)] \neq \mathbf{1}.$$

So $\xi(g + 1, k) < \xi(g, k)$, and ind$(\alpha_{g+1}) <$ ind(α_g). Thus (★) holds.

But from (★) Lemma II is proved. For there are fewer than $(1 + N^n)^p = q/p$ different possible indices. Hence, for some $z < q/p$, $\alpha_z = \alpha_{z+1}$. But then Δ_z is an A-resolvent. Q. e. d.

Thus, with Lemmas I and II, we have proved that, if $R(P_1, p)$ is an identity, every truth-value assignment to $R(P_2, q)$ verifies $R(P_2, q)$. But then $R(P_2, q)$ is an identity. Hence (c') is proved, since $q = p(1 + N^n)^p \leqq p(1 + N^j)^p$.

Note F

(Footnote 61, p. 548 above)

Corollary 3.31 is false, since the lemma in 3.3. is false with respect to property C. Indeed, the following argument, based on an example due to Stål Aanderaa, shows that for each $p \geqq 3$ and each $q \geqq p$ we can construct a proposition $P_{p,q}$ that has property C of order p but no prenex form of which has property C of order less than q. For $q \geqq p$ let P_q be the proposition

$$\begin{aligned}&(y_2)(y_3)\ldots(y_q)\big[[G(y_2, y_3) \;\&\; G(y_3, y_4) \;\&\; \cdots \;\&\; G(y_{q-1}, y_q)]\\ &\quad\supset \{[(\exists x_8)(\exists x_9)M(y_q, x_8, x_9) \;\&\; W(z)] \vee [(\exists x_6)(\exists x_7) {\sim} M(y_2, x_6, x_7) \;\&\; {\sim} W(z)]\\ &\quad \vee [(\exists x_{10}) {\sim} M(y_2, y_2, x_{10}) \;\&\; W(z)] \vee [(\exists x_{11})M(y_q, x_{11}, y_q)\\ &\quad \& {\sim} W(z)]\}\big] \vee (\exists x_1)(\exists x_2)\big[G(x_1, x_2) \;\&\; (\exists x)\{[(\exists x_3)M(x_1, x, x_3) \vee {\sim} W(z)]\\ &\quad \& [(y)M(x_1, x, y) \vee W(z)]\} \;\&\; (\exists x_4)\{[(y_1){\sim} M(x_2, y_1, x_4) \vee {\sim} W(z)]\\ &\quad \& [(\exists x_5){\sim} M(x_2, x_5, x_4) \vee W(z)]\}\big].\end{aligned}$$

The proposition P_q has property C of order 3, but no prenex form of P_q has property C of an order less than q. Now, for each $p \geqq 3$ and each $q \geqq p$ let $P_{p,q}$ be the conjunction P_q & $P_{2,p}$, where $P_{2,p}$ is the proposition defined at the beginning of Note E, p. 572 above. Then $P_{p,q}$ has property C of order p, but no prenex form of $P_{p,q}$ has property C of an order less than q.

Thus, whenever Corollary 3.31 is cited in Herbrand's text, it should be replaced by the relevant part of the following:

3.31′. Let P contain j quantifiers and d descriptive functions, and let the cardinality of C_1 be $n_1 \geqq 1$. Let $\gamma(j, d, p, n_1)$ be the primitive recursive function introduced in Note E, p. 572 above, and let $\delta(j, d, p, n_1)$ be the primitive recursive function defined thus:

For every $i \geqq 1$, let

$$\begin{aligned}\gamma_1(j, d, p, n_1) &= \gamma(j, d, p, n_1),\\ \gamma_{i+1}(j, d, p, n_1) &= \gamma(j, d, \gamma_i(j, d, p, n_1), n_1);\end{aligned}$$

then set

$$\delta(j, d, p, n_1) = \gamma_{j^2}(j, d, p, n_1).$$

(*i*) If P has property C of order p, it has property B of order p.

(*ii*) If P has property C of order p, any prenex form of P has property C of some order less than or equal to $\delta(j, d, p, n_1)$.

(*iii*) If P has property B of order p, it has property C of some order less than or equal to $\delta(j, d, p, n_1)$.

Part (*i*) of 3.31′ follows at once from the correct part of the lemma in 3.3 (in its original statement by Herbrand). Part (*ii*) follows from the lemma in 3.3 as emended (see Note E, p. 572 above) and the fact that to get any prenex form of P we need no more than j^2 crucial passages from $(x)\Phi(x) \vee Z$ to $(x)[\Phi(x) \vee Z]$, that is, passages in which $(x)\Phi(x) \vee Z$ occurs negatively and Z contains a general quantifier. (We are also using the fact that, if a proposition has property C of order p, it has property C of order q for each q greater than p. But this was justified in footnote 56.) Part (*iii*) follows from part (*ii*), the correct part of the lemma in 3.3 (as stated by Herbrand), and the fact that properties B and C are identical for a proposition in canonical form.

Note G

(*Footnote 81, p. 556 above*)

Lemma 3 in 5.3 is false, and unlike the lemma in 3.3 it remains false even when property B replaces property C. For consider the following example (adapted from Example 2 in *Dreben and Aanderaa 1964*), in which properties B and C coincide:

For each $s \geqq 4$ let P_s be

$$[(y_1)\sim H_1(y_1) \vee (\exists x_1)H_1(x_1)] \,\&\, [(y_2)\sim H_2(y_2) \vee (\exists x_2)H_2(x_2)] \,\&\, \cdots \,\&\, [(y_s)\sim H_s(y_s) \vee (\exists x_s)H_s(x_s)],$$

and let Q_s be

$$(\exists x)(y)\{H_s(y) \vee \sim H_1(y) \vee [H_1(x) \,\&\, \sim H_2(y)] \vee [H_2(x) \,\&\, \sim H_3(y)] \vee \cdots \vee [H_{s-1}(x) \,\&\, \sim H_s(y)]\}.$$

Now, for each $s \geqq 4$, P_s has property C of order 2, $P_s \supset Q_s$ has property C of order 3, and Q_s has property C of order s but of no smaller order.

Herbrand's argument for the lemma, however, goes wrong in only one point. Relying on the false Corollary 3.31 (see Note F, p. 577 above), Herbrand assumes that, since the conjunction $P \,\&\, (P \supset Q)$ has property C of order p, a certain prenex form of this conjunction has property C of order p (see the second and third sentences of his argument). So Herbrand does succeed in proving the very interesting sublemma:

5.3′. If a certain prenex form (specified in the third sentence of his argument) of $P \,\&\, (P \supset Q)$ has property C of order q, then Q has property C of order q.

Hence, if we now use part (*ii*) of 3.31′ (see Note F), then from 5.3′ we obtain as a replacement for Lemma 3 in 5.3:

5.3″. Let P contain j quantifiers and d descriptive functions. Let Q contain k quantifiers and e descriptive functions. (Both P and Q are assumed to contain no real variables.) If both P and $P \supset Q$ have property C of order p, then Q has property C of some order less than or equal to $\delta(2j + k, 2d + e, p, n_1)$.

Note H

(*Footnote 97, p. 560 above*)

Herbrand (*1931*, p. 54) states the following theorem: "If we assume that the decision problem has been solved and if we do not wish R + Infin Ax + Mult Ax to be inconsistent, we must not add any rule of reasoning to those already considered." ("R + Infin Ax + Mult Ax" is Herbrand's name for his version of the simple theory of types together with the infinity and multiplicative axioms.)

Immediately after the theorem Herbrand writes: "This theorem corresponds to what the Germans call the *Vollständigkeit* of our system of rules". (The French text of this sentence differs from that of the third sentence in Remark 3, on page 560 above, only in having "théorème" where the latter has "fait".)

Page 54 belongs to the part of Herbrand's 1931 paper that was written no later than September 1929. After Herbrand became acquainted with Gödel's incompleteness results (see *Gödel 1931*), Herbrand added an appendix dated April 1931, and in this appendix, on page 56, he restates the theorem above as follows:

"If the decision problem is solved for a proposition P, if the solution is formalizable in R + Infin Ax + Mult Ax (as it is for all the particular cases of the decision problem solved up to now) and if P is not an identity, then

(1) No new rule of reasoning that makes P an identity can be added without entailing a contradiction in R + Infin Ax + Mult Ax;

(2) P cannot be true [that is, provable] in R + Infin Ax + Mult Ax."

Thus it seems clear that by "Vollständigkeit" Herbrand did not primarily mean what has come to be called "semantic completeness", that is, the notion of completeness specified in *Hilbert and Ackermann 1928*, p. 68, which for our purposes is best formulated thus:

I. Every proposition of quantification theory either is provable in Herbrand's system Q_H or can be given a falsifying interpretation over a denumerable universe. (For a discussion of this notion of completeness see above, pp. 508–511, and below, pp. 582–583.)

Rather, the theorem quoted in the first paragraph of the present Note and the argument that Herbrand sketches for it (*1931*, pp. 52–54) should be compared with Problem V in *Hilbert 1928a*, pp. 140–141, and also with the discussion in *Hilbert and Bernays 1939*, pp. 234–253, of what is there called a finitistic sharpening of Gödel's completeness theorem (see also *Kleene 1952*, p. 395, Theorem 36).

Nevertheless, Herbrand's text contains an argument for the semantic completeness of his system Q_H. For, by 4.1 (see footnote 67), if a proposition is false in some infinite domain, then by *nonconstructive* means it can be given a falsifying interpretation over a denumerable universe. Moreover, by 4.21, either a proposition has property B or it is false in some infinite domain. But, by the corollary in 5.1 (above, p. 555), a proposition that has property B is provable. Hence every proposition either is provable or can be given a falsifying denumerable interpretation. (Naturally, this argument is not "metamathematical"; see 6.2 and footnote 94 above.)

Note, however, that the corollary in 5.1. rests upon the lemma in 3.3. So we cannot say that with the preceding argument Herbrand *proved* the semantic completeness of his system, unless we include in Herbrand's argument footnotes 51, 59, and especially Note E (or the alternative proof of Lemma 1 in 5.1 suggested in footnote 77). But, if we make one change in the definition, given above on page 552, of "false in an infinite domain"—a change that Herbrand himself makes later (*1931*, pp. 25–27)—we can say that Herbrand has proved the following:

II. Every *prenex* quantificational proposition either is provable in Q_H or can be given a falsifying denumerable interpretation.

For, if we replace the sentence of lines 30–31, on page 552 above, by "Let us start again from P; but now with each general variable y we correlate an index function whose arguments are the restricted variables that are *superior* to y", then we can avoid any appeal to 3.3 thus:

Let P be any prenex proposition. By considerations strictly analogous to those in 4.21, P either has property C or is false in some infinite domain. But by Lemma 1 in 5.1, if P has property C, it has property A. (Note that for *prenex* propositions

Herbrand's proof of Lemma 1 does *not* depend upon 3.3, hence is correct.) By 2.3 (see Note D, p. 571 above), if P has property A, it is provable. So P either is provable or is false in some infinite domain. But, if P is false in some infinite domain, it can be given a falsifying denumerable interpretation (again see footnote 67). Hence every prenex proposition either is provable or has a falsifying denumerable interpretation.

Now, by the rules of passage every proposition can be turned into a prenex one. But this, by itself, permits us to say only that Herbrand has shown that every proposition either is provable or is provably equivalent to a prenex proposition that has a falsifying denumerable interpretation. To establish proposition I, Herbrand has to show that, if a prenex proposition P has a falsifying interpretation, any proposition obtained from P solely by the rules of passage also has a falsifying interpretation. Of course, it is not hard to give semantic arguments showing this. And one such argument is easily extracted from *Hilbert and Ackermann 1928*, a book to which Herbrand refers in his thesis.

Note I

(Footnote 109, p. 563 above)

Let P be any proposition. The *Herbrand functional form* of P is the proposition $F(P)$ that we obtain from P by deleting all restricted quantifiers of P and replacing each occurrence of each restricted variable y of P by an occurrence of the index function $f_y(x_{y_1}, \ldots, x_{y_m})$, where $x_{y_1}, \ldots, x_{y_m}$ are all the general variables of P superior to y; the *strict Herbrand functional form of* P is that quantifier-free proposition $\mathscr{F}(P)$ that we obtain by deleting all (general) quantifiers of $F(P)$. (Thus the strict Herbrand functional form is a generalization of the Skolem functional form of P, since P need not be prenex; see p. 508 above.)

Call any proposition *proper* if it contains no index functions. Clearly, if P is any proposition and R is any proper proposition, the same elementary proposition can be associated with both $P \supset R$ and $F(P) \supset R$. (For the notion "elementary proposition associated with a proposition" see Definition I in Note E, p. 572 above. Note that, if a proposition P contains no real variables, the elementary proposition associated with P is the strict Herbrand functional form of $\sim P$.) Hence, if R is proper, $P \supset R$ is provable in Herbrand's Q_H if and only if $F(P) \supset R$ is provable in Q_H. But $F(P) \supset \mathscr{F}(P)$ is obviously provable in Q_H, and $F(P)$ is provable in $Q_H + \mathscr{F}(P)$. Thus, for any *proper* theory T, that is, the system Q_H with proper propositions added as hypotheses, we can constructively find an inessential extension all of whose hypotheses are quantifier-free, namely, the theory T' that we obtain from T by replacing each hypothesis with its strict Herbrand functional form. (See the discussion of *symbolische Auflösung* in *Hilbert and Bernays 1939*, pp. 1–18 and 130–137.)

Note J

(Footnote 110, p. 563 above)

In Chapter 4 of his thesis Herbrand used what has come to be called the method of "elimination of quantifiers" (see *Presburger 1929*, *Tarski 1948* and *1951*, and *Hilbert and Bernays 1934*, pp. 234–238) to show that the following subsystem of elementary number theory is consistent, formally complete (or, as Herbrand says, "completely determined"), and decidable (or, as Herbrand says, "resoluble"):

I. The primitive signs of the system, which Herbrand calls Theory 3, are (besides punctuation signs) truth functions, variables, quantifiers, the constant "0", the descriptive function "$x + 1$", and the atomic propositional function "$x = y$".

II. A numeral is defined as "$0 + 1 + 1 + \cdots + 1$", where "1" occurs zero or more times.

III. The axioms of Theory 3 are

1. $x = x$,
2. $(x = y) \supset (y = x)$,
3. $[(x = y) \,\&\, (y = z)] \supset (x = z)$,
4. $(x = y) \equiv (x + 1 = y + 1)$,
5. $x + 1 \neq 0$,
6. $x \neq x + \mathrm{a}$ for any numeral $\mathrm{a} \neq 0$,
7. $(x = 0) \vee (\exists y)(x = y + 1)$.

On page 73 of his thesis Herbrand states that von Neumann (*1927*) showed the consistency of Theory 3, but claims rightly that his argument in Chapter 4 is much simpler than von Neumann's. Moreover, on pages 81–82 Herbrand shows that Theory 3 is equivalent to another theory—call it Theory 4—consisting of the first five axioms of Theory 3 and the following axiom schema of mathematical induction

8. $\{\Phi(0) \,\&\, (x)[\Phi(x) \supset \Phi(x + 1)]\} \supset \Phi(y)$.

(In $\Phi(x)$ bound variables may occur, but no descriptive or propositional functions except those explicitly definable in terms of the primitive ones; Herbrand points out that his consistency proof does not work if functions definable by recursion but not explicitly definable are introduced into Theory 4.) Herbrand establishes this equivalence by showing that

(*a*) All instances of Axiom Schema 8 follow from the axioms of Theory 3, and

(*b*) Axioms 1–5 together with Axiom Schema 8 yield Axioms 6 and 7.

(See *Shepherdson 1963*, especially theorems 2.2 and 3.3.)

At the end of Chapter 4 Herbrand remarks that the method he has used to study Theory 3 has many applications and that, whenever it shows the consistency of a theory, it also establishes its decidability. He then concludes by saying: "It appears probable that it [that is, his method of elimination of quantifiers] would also permit us to obtain the consistency of the theory of real closed fields (*Artin and Schreier 1926*) [see *Tarski 1948* and *1951*, top of p. 54]. But the methods of the following chapter [that is, the present Chapter 5] would lead us to this result more easily."

One final point. Although Hilbert, Bernays, Ackermann, and von Neumann (see *Bernays 1927*, p. 489 above, and the editor's footnote 3 on the same page) thought —until they became acquainted with Gödel's incompleteness results (*1930b*, *1931*)— that their consistency arguments could readily be extended to full elementary number theory, Herbrand was clearly aware that attempts to so extend the arguments that he now gives in 6.8 would meet with difficulties (see pp. 565 and 567 above; but see also footnote 120, p. 565).

The completeness of the axioms of the functional calculus of logic[1]

KURT GÖDEL

(*1930a*)

In his doctoral dissertation at the University of Vienna (*1930*)[a] Gödel proved that the predicate calculus of first order is complete, in the sense that every valid formula is provable. The present paper is a rewritten version of this dissertation.

At the beginning of the paper Gödel mentions Whitehead and Russell's method of deriving logic and mathematics from axioms by means of purely formal rules, and he writes: "Of course, when such a procedure is followed the question at once arises whether the initially postulated system of axioms and principles of inference is complete". One must acknowledge that Whitehead and Russell had not shown much concern for that problem, any more than they did in general for questions that, being semantic in character, went beyond provability in their system.

The statement that the pure predicate calculus of first order is complete, that is, that every valid formula is provable, is equivalent to the statement that every formula is either refutable or satisfiable. Gödel actually proves a stronger statement, namely, that every formula is either refutable or $\aleph_0$-satisfiable. Hence his proof yields, besides completeness, the Löwenheim–Skolem theorem, which states that a satisfiable formula is $\aleph_0$-satisfiable. The proof makes use of a number of devices introduced by Löwenheim (*1915*) and Skolem (*1920*; see also *1922*, *1928*, and *1929*) but contains a step (Theorem VI) through which semantic arguments are connected with provability in a definite system. At about the same time, Herbrand, too, was expanding Löwenheim's and Skolem's work; the relation between Gödel's methods and results and those of Herbrand (*1930*) was brought out by Dreben (*1952*; see also above, pp. 510 and 579).

Gödel generalizes his result—that every formula is either refutable or satisfiable—in two directions, to the predicate calculus of first order with identity (in which some irrefutable formulas are finitely satisfiable without being $\aleph_0$-satisfiable) and to infinite sets of formulas. The second generalization (Theorem IX) is derived from the result now known as the finiteness, or compactness, theorem (Theorem X), of which Gödel gives a semantic proof.

By using the procedure of arithmetization introduced by Gödel in another paper (*1931*), Hilbert and Bernays (*1939*, pp. 205–253) were able to give what they call a "proof-theoretic" version of Gödel's completeness theorem: if a formula is irrefutable in the pure predicate calculus of first order, it is irrefutable

[1] I am indebted to Professor H. Hahn for several valuable suggestions that were of help to me in writing this paper.

[a] The degree was granted on 6 February 1930.

also in every consistent system S that remains consistent when the axioms of number theory, as well as any verifiable formulas of the theory, are added to the axioms of S (p. 252; see also Theorem 36 in *Kleene 1952*, p. 395).

This proof-theoretic form of Gödel's completeness theorem was extended to the case of infinitely many formulas by Wang (*1951*; see also *1950*, p. 449, Theorem 5, and *1962*, pp. 345–352) and modified so as to give a sharp form of the Löwenheim–Skolem theorem. If Con(Σ) is the usual formula expressing the consistency of a first-order system Σ, the result, called Bernays's lemma by Wang, says that, if we add Con(Σ) to number theory as a new axiom, we can prove in the resulting system arithmetic translations of all theorems of Σ. This lemma has been applied in several directions (see, for instance, *Wang 1952* and *1955*).

The Löwenheim–Skolem theorem shows that, if the predicates assigned to predicate letters in accordance with the definition of validity are just number-theoretic predicates, the class of formulas that turn out to be always true coincides with the class of valid formulas. Kleene (*1952*, pp. 394–395), making explicit some results obtained by Hilbert and Bernays in their arithmetization of Gödel's completeness proof, showed that the predicates can be further restricted to the class $\Sigma_2 \cap \Pi_2$ (in the hierarchy of arithmetic predicates). Putnam (*1961* and *1965*) refined Kleene's result; he restricted the predicates to the class Σ_1^*, the smallest class that contains the recursively enumerable predicates and is closed under truth functions; earlier Kreisel (*1953*) and Mostowski (*1955*) had shown that the predicates could not be restricted to the class of recursive predicates, and Putnam (*1956*) that they could not be restricted to the class $\Sigma_1 \cup \Pi_1$.

Another proof of the completeness of first-order logic, along lines somewhat different from those of Gödel's proof, was given by Henkin (*1949*).

The translation is by Stefan Bauer-Mengelberg, and it is printed here with the kind permission of Professor Gödel and Springer Verlag.

Whitehead and Russell, as is well known, constructed logic and mathematics by initially taking certain evident propositions as axioms and deriving the theorems of logic and mathematics from these by means of some precisely formulated principles of inference in a purely formal way (that is, without making further use of the meaning of the symbols). Of course, when such a procedure is followed the question at once arises whether the initially postulated system of axioms and principles of inference is complete, that is, whether it actually suffices for the derivation of every true logico-mathematical proposition, or whether, perhaps, it is conceivable that there are true propositions (which may even be provable by means of other principles) that cannot be derived in the system under consideration. For the formulas of the propositional calculus the question has been settled affirmatively; that is, it has been shown[2] that every true formula of the propositional calculus does indeed follow from the axioms given in *Principia mathematica*. The same will be done here for a wider realm of formulas, namely, those of the "restricted functional calculus";[3] that is, we shall prove

[2] See *Bernays 1926*.

[3] In terminology and symbolism this paper follows *Hilbert and Ackermann 1928*. According to that work, the restricted functional calculus contains the logical expressions that are constructed from propositional variables, X, Y, Z, ..., and functional variables (that is, variables for properties and relations) of type 1, $F(x)$, $G(x, y)$, $H(x, y, z)$, ..., by means of the operations $\vee$ (or), $\overline{}$ (not),

THEOREM I. *Every valid*[4] *formula of the restricted functional calculus is provable.*

We lay down the following system of axioms[5] as a basis:

Undefined primitive notions: $\vee$, $\overline{}$, and (x). (By means of these, &, $\rightarrow$, $\sim$, and (Ex) can be defined in a well-known way.)

Formal axioms:

1. $X \vee X \rightarrow X$,
2. $X \rightarrow X \vee Y$,
3. $X \vee Y \rightarrow Y \vee X$,
4. $(X \rightarrow Y) \rightarrow [Z \vee X \rightarrow Z \vee Y]$,
5. $(x)F(x) \rightarrow F(y)$,
6. $(x)[X \vee F(x)] \rightarrow X \vee (x)F(x)$.

Rules of inference:[6]

1. The inferential schema: From A and $A \rightarrow B$, B may be inferred;
2. The rule of substitution for propositional and functional variables;
3. From $A(x)$, $(x)A(x)$ may be inferred;
4. Individual variables (free or bound) may be replaced by any others, so long as this does not cause overlapping of the scopes of variables denoted by the same sign.

For what follows, it will be expedient to introduce some abbreviated notations.

(P), (Q), (R), and so on mean prefixes constructed in any way whatever, that is, finite sequences of signs of the form $(x)(Ey)$, $(y)(x)(Ez)(u)$, and the like.

Lower-case German letters, $\mathfrak{x}$, $\mathfrak{y}$, $\mathfrak{u}$, $\mathfrak{v}$, and so on, mean n-tuples of individual variables, that is, sequences of signs of the form x, y, z, or x_2, x_1, x_2, x_3, and the like, where the same variable may occur several times. The signs $(\mathfrak{x})$, $(E\mathfrak{x})$, and so on are to be understood accordingly. Should a variable occur several times in $\mathfrak{x}$, we must, of course, think of it as written only once in $(\mathfrak{x})$ or $(E\mathfrak{x})$.

Furthermore we require a number of lemmas, which are collected here. The proofs are not given, since they are in part well known, in part easy to supply.

1. For every n-tuple $\mathfrak{x}$

$$(a)\ (\mathfrak{x})F(\mathfrak{x}) \rightarrow (E\mathfrak{x})F(\mathfrak{x}),$$

$$(b)\ (\mathfrak{x})F(\mathfrak{x})\ \&\ (E\mathfrak{x})G(\mathfrak{x}) \rightarrow (E\mathfrak{x})[F(\mathfrak{x})\ \&\ G(\mathfrak{x})],$$

$$(c)\ (\mathfrak{x})\overline{F(\mathfrak{x})} \sim \overline{(E(\mathfrak{x})F(\mathfrak{x}))}$$

are provable.

2. If $\mathfrak{x}$ and $\mathfrak{x}'$ differ only in the order of the variables, then

$$(E\mathfrak{x})F(\mathfrak{x}) \rightarrow (E\mathfrak{x}')F(\mathfrak{x})$$

is provable.

(x) (for all), (Ex) (there exists), with the variable in the prefixes (x) or (Ex) ranging over individuals *only, not* over functions. A formula of this kind is said to be valid (tautological) if a true proposition results from every substitution of specific propositions and functions for X, Y, Z, ... and $F(x)$, $G(x, y)$, ..., respectively (for example, $(x)[F(x) \vee \overline{F(x)}]$).

[4] To be more precise, we should say "valid in every domain of individuals", which, according to well-known theorems, means the same as "valid in the denumerable domain of individuals". For a formula with free individual variables, $A(x, y, \ldots, w)$, "valid" means that $(x)(y)\ldots(w)A(x, y, \ldots, w)$ is valid and "satisfiable" that $(Ex)(Ey)\ldots(Ew)A(x, y, \ldots, w)$ is satisfiable, so that the following holds without exception: "A is valid" is equivalent to "$\overline{A}$ is not satisfiable".

[5] It coincides (except for the associative principle, which P. Bernays proved to be redundant) with that given in *Whitehead and Russell 1910*, *1 and *10.

[6] Although Whitehead and Russell use these rules throughout their derivations, they do not formulate all of them explicitly.

3. If $\mathfrak{x}$ consists entirely of distinct variables and if $\mathfrak{x}'$ has the same number of terms as $\mathfrak{x}$, then

$$(\mathfrak{x})F(\mathfrak{x}) \rightarrow (\mathfrak{x}')F(\mathfrak{x}')$$

is provable, even when a number of identical variables occur in $\mathfrak{x}'$.

4. If (p_i) means one of the prefixes (x_i) or (Ex_i) and if (q_i) means one of the prefixes (y_i) or (Ey_i), then

$$(p_1)(p_2)\ldots(p_n)F(x_1, x_2, \ldots, x_n) \ \& \ (q_1)(q_2)\ldots(q_m)G(y_1, y_2, \ldots, y_m) \sim (P)[F(x_1, x_2, \ldots, x_n) \ \& \ G(y_1, y_2, \ldots, y_m)]$$

is provable[7] for every prefix (P) that is formed from the (p_i) and the (q_i) and satisfies the condition that, for $i < k \leqq n$, (p_i) precedes (p_k) and, for $i < k \leqq m$, (q_i) precedes (q_k).

5. Every expression can be brought into normal form; that is, for every expression A there is a normal formula N such that $A \sim N$ is provable.[8]

6. If $A \sim B$ is provable, so is $\mathfrak{F}(A) \sim \mathfrak{F}(B)$, where $\mathfrak{F}(A)$ represents an arbitrary expression containing A as a part (see *Hilbert and Ackermann 1928*, chap. 3, § 7).

7. Every valid formula of the propositional calculus is provable; that is, Axioms 1–4 form a complete axiom system for the propositional calculus.[9]

We now proceed to the proof of Theorem I and first note that the theorem can also be stated in the following form:

Theorem II. *Every formula of the restricted functional calculus is either refutable*[10] *or satisfiable* (and, moreover, satisfiable in the denumerable domain of individuals).

That I follows from II can be seen as follows: Let A be a valid expression; then $\overline{A}$ is not satisfiable, hence according to II it is refutable; that is, $\overline{\overline{A}}$ is provable and, consequently, so is A. The converse is as apparent.

We now define a class $\mathfrak{K}$ of expressions K by means of the following stipulations:

1. K is a normal formula;
2. K contains no free individual variable;
3. The prefix of K begins with a universal quantifier and ends with an existential quantifier.

Then we have

Theorem III. *If every $\mathfrak{K}$-expression is either refutable or satisfiable,*[11] *so is every expression.*

Proof. Let A be an expression not belonging to $\mathfrak{K}$. Let it contain the free variables $\mathfrak{x}$. As is immediately obvious, the refutability of $(E\mathfrak{x})A$ follows from that of A, and conversely (by Lemma 1(c), and either Rule of inference 3 or, for the converse, Axiom 5); the same holds, according to the stipulation in footnote 4, for satisfiability. Let $(P)N$ be the normal form of $(E\mathfrak{x})A$, so that

$$(E\mathfrak{x})A \sim (P)N \qquad (1)$$

[7] An analogous theorem holds with $\vee$ instead of &.

[8] See *Hilbert and Ackermann 1928*, chap. 3, § 8.

[9] See *Bernays 1926*.

[10] "A is refutable" is to mean "$\overline{A}$ is provable".

[11] "Satisfiable" without additional specification here and in what follows always means "satisfiable in the denumerable domain of individuals". The same holds for "valid".

is provable. Further let

$$B = (x)(P)(Ey)[N \ \& \ \{F(x) \lor \overline{F(y)}\}].^{12}$$

Then

$$(P)N \sim B \tag{2}$$

is provable (on the basis of Lemma 4 and the provability of $(x)(Ey)[F(x) \lor \overline{F(y)}]$). B belongs to $\mathfrak{K}$ and thus according to the assumption is either satisfiable or refutable. But, by (1) and (2), the satisfiability of B entails that of $(E\mathfrak{x})A$, hence also that of A; the same holds for refutability. Thus A, too, is either satisfiable or refutable.

Because of Theorem III, therefore, it suffices to show that

Every $\mathfrak{K}$-expression is either satisfiable or refutable.

For this purpose we define the degree of a $\mathfrak{K}$-expression[13] to be the number of blocks in its prefix that consist of universal quantifiers and are separated from each other by existential quantifiers, and we first prove

Theorem IV. *If every expression of degree k is either satisfiable or refutable, so is every expression of degree $k + 1$.*

Proof. Let $(P)A$ be a $\mathfrak{K}$-expression of degree $k + 1$. Let $(P) = (\mathfrak{x})(E\mathfrak{y})(Q)$ and let $(Q) = (\mathfrak{u})(E\mathfrak{v})(R)$, where (Q) is of degree k and (R) of degree $k - 1$. Further let F be a functional variable not occurring in A. If we now put[14]

$$B = (\mathfrak{x}')(E\mathfrak{y}')F(\mathfrak{x}', \mathfrak{y}') \ \& \ (\mathfrak{x})(\mathfrak{y})[F(\mathfrak{x}, \mathfrak{y}) \rightarrow (Q)A]$$

and

$$C = (\mathfrak{x}')(\mathfrak{x})(\mathfrak{y})(\mathfrak{u})(E\mathfrak{y}')(E\mathfrak{v})(R)\{F(\mathfrak{x}', \mathfrak{y}') \ \& \ [F(\mathfrak{x}, \mathfrak{y}) \rightarrow A]\},^{15}$$

then a double application of Lemma 4 in combination with Lemma 6 yields the provability of

$$B \sim C; \tag{3}$$

furthermore,

$$B \rightarrow (P)A \tag{4}$$

is obviously valid. Now C is of degree k and by assumption is therefore either satisfiable or refutable. If it is satisfiable, so is $(P)A$ (by (3) and (4)). If it is refutable, so is B (by (3)); that is, $\bar{B}$ is then provable. In that case, if we substitute $(Q)A$ for F in $\bar{B}$, it follows that

$$\overline{(\mathfrak{x}')(E\mathfrak{y}')(Q)A \ \& \ (\mathfrak{x})(\mathfrak{y})[(Q)A \rightarrow (Q)A]}$$

is provable.

But since, of course

$$(\mathfrak{x})(\mathfrak{y})[(Q)A \rightarrow (Q)A]$$

is provable, so is $\overline{(\mathfrak{x}')(E\mathfrak{y}')(Q)A}$; that is, in that case $(P)A$ is refutable. $(P)A$ is therefore indeed either refutable or satisfiable.

[12] The variables x and y must not occur in (P).

[13] The term "degree of a prefix" is used in the same sense.

[14] An analogous procedure was used by Skolem (*1920*) in proving Löwenheim's theorem.

[15] The variable-sequences $\mathfrak{x}$, $\mathfrak{x}'$, $\mathfrak{y}$, $\mathfrak{y}'$, $\mathfrak{u}$, $\mathfrak{v}$ are, of course, assumed to be pairwise disjoint.

It now remains only to prove

THEOREM V. *Every formula of degree 1 is either satisfiable or refutable.*

A few definitions are required for the proof. Let $(\mathfrak{x})(E\mathfrak{y})A(\mathfrak{x}\,;\,\mathfrak{y})$ (abbreviated as $(P)A$) be any formula of degree 1. Let $\mathfrak{x}$ stand for an r-tuple and $\mathfrak{y}$ for an s-tuple of variables. We think of the r-tuples taken from the sequence $x_0, x_1, x_2, \ldots, x_i, \ldots$ as forming a sequence ordered according to increasing sum of the subscripts ⟦and for equal sums according to some convention⟧:

$$\mathfrak{x}_1 = (x_0, x_0, \ldots, x_0), \quad \mathfrak{x}_2 = (x_1, x_0, \ldots, x_0), \quad \mathfrak{x}_3 = (x_0, x_1, x_0, \ldots, x_0),$$

and so forth; we now define a sequence $\{A_n\}$ of formulas derived from $(P)A$ as follows:

$$A_1 = A(\mathfrak{x}_1\,;\,x_1, x_2, \ldots, x_s),$$
$$A_2 = A(\mathfrak{x}_2\,;\,x_{s+1}, x_{s+2}, \ldots, x_{2s})\ \&\ A_1,$$
$$\cdots\cdots\cdots\cdots$$
$$A_n = A(\mathfrak{x}_n\,;\,x_{(n-1)s+1}, x_{(n-1)s+2}, \ldots, x_{ns})\ \&\ A_{n-1}.$$

Let the s-tuple $x_{(n-1)s+1}, \ldots, x_{ns}$ be denoted by $\mathfrak{y}_n$, so that we have

$$A_n = A(\mathfrak{x}_n\,;\,\mathfrak{y}_n)\ \&\ A_{n-1}. \tag{5}$$

Further we define $(P_n)A_n$ by the stipulation

$$(P_n)A_n = (Ex_0)(Ex_1)\ldots(Ex_{ns})A_n.$$

As we can easily convince ourselves, it is precisely the variables x_0 to x_{ns} that occur in A_n; hence they all are bound by (P_n). Further it is apparent that the variables of the r-tuple $\mathfrak{x}_{n+1}$ already occur in (P_n) (and therefore certainly differ from those occurring in $\mathfrak{y}_{n+1}$). Denote by (P'_n) what remains of (P_n) when the variables of the r-tuple $\mathfrak{x}_{n+1}$ are omitted, so that, except for the order of the variables, $(E\mathfrak{x}_{n+1})(P'_n) = (P_n)$.

This notation once assumed, we have

THEOREM VI. *For every n*

$$(P)A \rightarrow (P_n)A_n$$

is provable.

For the proof we use mathematical induction.

I. $(P)A \rightarrow (P_1)A_1$ is provable, for we have

$$(\mathfrak{x})(E\mathfrak{y})A(\mathfrak{x}\,;\,\mathfrak{y}) \rightarrow (\mathfrak{x}_1)(E\mathfrak{y}_1)A(\mathfrak{x}_1\,;\,\mathfrak{y}_1)$$

(by Lemma 3 and Rule of inference 4) and

$$(\mathfrak{x}_1)(E\mathfrak{y}_1)A(\mathfrak{x}_1\,;\,\mathfrak{y}_1) \rightarrow (E\mathfrak{x}_1)(E\mathfrak{y}_1)A(\mathfrak{x}_1\,;\,\mathfrak{y}_1)$$

(by Lemma 1(a)).

II. For every n, $(P)A\ \&\ (P_n)A_n \rightarrow (P_{n+1})A_{n+1}$ is provable, for we have

$$(\mathfrak{x})(E\mathfrak{y})A(\mathfrak{x}\,;\,\mathfrak{y}) \rightarrow (\mathfrak{x}_{n+1})(E\mathfrak{y}_{n+1})A(\mathfrak{x}_{n+1}\,;\,\mathfrak{y}_{n+1}) \tag{6}$$

(by Lemma 3 and Rule of inference 4) and

$$(P_n)A_n \rightarrow (E\mathfrak{x}_{n+1})(P'_n)A_n \tag{7}$$

(by Lemma 2). Furthermore,

$$(\mathfrak{x}_{n+1})(E\mathfrak{y}_{n+1})A(\mathfrak{x}_{n+1}\,;\,\mathfrak{y}_{n+1})\ \&\ (E\mathfrak{x}_{n+1})(P'_n)A_n \rightarrow (E\mathfrak{x}_{n+1})[(E\mathfrak{y}_{n+1})A(\mathfrak{x}_{n+1}\,;\,\mathfrak{y}_{n+1})\ \&\ (P'_n)A_n] \quad (8)$$

(by Lemma 1(b) with the substitutions $(E\mathfrak{y}_{n+1})A(\mathfrak{x}_{n+1}\,;\,\mathfrak{y}_{n+1})$ for F and $(P'_n)A_n$ for G).

If we observe that the antecedent of the implication (8) is the conjunction of the consequents of (6) and (7), it is clear that

$$(P)A\ \&\ (P_n)A_n \rightarrow (E\mathfrak{x}_{n+1})[(E\mathfrak{y}_{n+1})A(\mathfrak{x}_{n+1}\,;\,\mathfrak{y}_{n+1})\ \&\ (P'_n)A_n] \quad (9)$$

is provable. Furthermore, from (5) and Lemmas 4, 6, and 2 the provability of

$$(E\mathfrak{x}_{n+1})[(E\mathfrak{y}_{n+1})A(\mathfrak{x}_{n+1}\,;\,\mathfrak{y}_{n+1})\ \&\ (P'_n)A_n] \sim (P_{n+1})A_{n+1} \quad (10)$$

follows. II follows from (9) and (10), and from II, together with I, Theorem VI follows.

Assume that the functional variables $F_1, F_2, \ldots, F_k$ and the propositional variables $X_1, X_2, \ldots, X_l$ occur in A. Then A_n consists of elementary components of the form

$$F_1(x_{p_1}, \ldots, x_{q_1}),\ F_2(x_{p_2}, \ldots, x_{q_2}), \ldots, X_1, X_2, \ldots, X_l$$

compounded solely by means of the operations $\vee$ and $\overline{}$. With each A_n we associate a formula B_n of the propositional calculus by replacing the elementary components of A_n by propositional variables, making certain that different components (even if they differ only in the notation of the individual variables) are replaced by different propositional variables. Furthermore, we understand by "satisfying system of level n [[Erfüllungssystem n-ter Stufe]] of $(P)A$" a system of functions $f_1^{(n)}, f_2^{(n)}, \ldots, f_k^{(n)}$ defined in the domain of integers z $(0 \leqq z \leqq ns)$ as well as of truth values $w_1^{(n)}, w_2^{(n)}, \ldots, w_l^{(n)}$ for the propositional variables $X_1, X_2, \ldots, X_l$ such that a true proposition results if in A_n the F_i are replaced by the $f_i^{(n)}$, the x_i by the numbers i, and the X_i by the corresponding truth values $w_i^{(n)}$. Satisfying systems of level n obviously exist if and only if B_n is satisfiable.

Each B_n, being a formula of the propositional calculus, is either satisfiable or refutable (Lemma 7). Thus only two cases are conceivable:

1. At least one B_n is refutable. Then, as we can easily convince ourselves (Rules of inference 2 and 3, Lemma 1(c)), the corresponding $(P_n)A_n$ is refutable also, and consequently, because of the provability of $(P)A \rightarrow (P_n)A_n$, so is $(P)A$.

2. No B_n is refutable; hence all are satisfiable. Then there exist satisfying systems of every level. But, since for each level there is only a finite number of satisfying systems (because the associated domains of individuals are finite) and since furthermore every satisfying system of level $n + 1$ contains one of level n as a part[16] (as is clear from the fact that the A_n are formed by successive conjunctions), it follows by

[16] That a system $\{f_1, f_2, \ldots, f_k\,;\, w_1, w_2, \ldots, w_l\}$ is part of another, $\{g_1, g_2, \ldots, g_k\,;\, v_1, v_2, \ldots, v_l\}$, is to mean that

1. The domain of individuals of the f_i is part of the domain of individuals of the g_i;
2. The f_i and the g_i coincide within the narrower domain;
3. For every i, $v_i = w_i$.

familiar arguments[16a] that in this case there exists a sequence of satisfying systems $S_1, S_2, \ldots, S_k, \ldots$ (S_k being of level k) such that each contains the preceding one as a part. We now define in the domain of *all* integers $\geqq 0$ a system $S = \{\varphi_1, \varphi_2, \ldots, \varphi_k; \alpha_1, \alpha_2, \ldots, \alpha_l\}$ by means of the following stipulations:

1. $\varphi_p(a_1, \ldots, a_i)$ $(1 \leqq p \leqq k)$ holds if and only if for at least one S_m of the sequence above (and then for all subsequent ones also) $f_p^{(m)}(a_1, \ldots, a_i)$ holds;

2. $\alpha_i = w_i^{(m)}$ $(1 \leqq i \leqq l)$ for at least one S_m (and then also for all those that follow).

Then it is evident at once that S makes the formula $(P)A$ true. In this case, therefore, $(P)A$ is satisfiable, which concludes the proof of the completeness of the system of axioms given above. Let us note that the equivalence now proved, "valid = provable", contains, for the decision problem, a reduction of the nondenumerable to the denumerable, since "valid" refers to the nondenumerable totality of functions, while "provable" presupposes only the denumerable totality of formal proofs.

Theorem I, as well as Theorem II, can be generalized in various directions. First, it is easy to bring the notion of identity (between individuals) into consideration by adding to Axioms 1–6 above two more:

$$7.\ x = x, \qquad 8.\ x = y \rightarrow [F(x) \rightarrow F(y)].$$

An analogue of what we had above now holds for the extended realm of formulas too:

THEOREM VII. *Every formula of the extended realm is provable if it is valid* (more precisely, if it is valid in every domain of individuals),

and, equivalent to VII,

THEOREM VIII. *Every formula of the extended realm is either refutable or satisfiable* (and, moreover, satisfiable in a finite or denumerable domain of individuals).

For the proof, let A denote an arbitrary formula of the extended realm. We construct a formula B as the product (conjunction) of A, $(x)(x = x)$, and all the formulas that we obtain from Axiom 8 by substituting for F the functional variables occurring in A, that is, more precisely,

$$(x)(y)\{x = y \rightarrow [F(x) \rightarrow F(y)]\}$$

for all singulary functional variables of A,

$$(x)(y)(z)\{x = y \rightarrow [F(x, z) \rightarrow F(y, z)]\} \ \&\ (x)(y)(z)\{x = y \rightarrow [F(z, x) \rightarrow F(z, y)]\}$$

for all binary functional variables of A (including "=" itself), and corresponding formulas for the n-ary functional variables of A for which $n \geqq 3$. Let B' be the formula resulting from B when the sign "=" is replaced by a functional variable G not otherwise occurring in B. Then the sign "=" no longer occurs in the expression B', which, therefore, according to what was proved above, is either refutable or satisfiable. If it is refutable, so is B, since it results from B' through the substitution of "=" for G. But B is the logical product of A and a subformula that is obviously provable from Axioms 7 and 8. In this case, therefore, A is also refutable. Let us now assume that B' is satisfiable in the denumerable domain Σ of individuals for a certain system S of functions.[17] From the way in which B' is formed it is clear that g (that is,

[16a] ⟦Apparently by König's infinity lemma (*1926*, p. 120; see also *1927*), which was becoming known among mathematicians at the time Gödel was writing.⟧

[17] If propositional variables also occur in A, S will, of course, have to contain, besides functions, truth values for these propositional variables.

the function of the system S that is to be substituted for G) is a reflexive, symmetric, and transitive relation; hence it generates a partition of the elements of Σ, in such a way, moreover, that a function occurring in the system S continues to hold, or not to hold, as the case may be, when elements of the same class are substituted for one another. If, therefore, we identify with one another all elements belonging to the same class (perhaps by taking the classes themselves as elements of a new domain of individuals), then g goes over into the identity relation and we have a satisfying system of B, hence also of A. Consequently, A is indeed either satisfiable[18] or refutable.

We obtain a different generalization of Theorem I by considering denumerably infinite sets of logical expressions. For these, too, an analogue of Theorems I and II holds, namely

THEOREM IX. *Every denumerably infinite set of formulas of the restricted functional calculus either is satisfiable* (that is, all formulas of the system are simultaneously satisfiable) *or possesses a finite subsystem whose logical product is refutable.*

IX follows immediately from

THEOREM X. *For a denumerably infinite system of formulas to be satisfiable it is necessary and sufficient that every finite subsystem be satisfiable.*

Concerning Theorem X we first note that in proving it we can confine ourselves to systems of normal formulas of degree 1, for, by repeated application of the procedure used in the proofs of Theorems III and IV to the individual formulas, we can specify for every system Σ of formulas a system Σ' of normal formulas of degree 1 such that the satisfiability of any subsystem of Σ is equivalent to that of the corresponding subsystem of Σ'.

Thus let

$$(\mathfrak{x}_1)(E\mathfrak{y}_1)A_1(\mathfrak{x}_1;\mathfrak{y}_1),\quad (\mathfrak{x}_2)(E\mathfrak{y}_2)A_2(\mathfrak{x}_2;\mathfrak{y}_2),\ldots,\quad (\mathfrak{x}_n)(E\mathfrak{y}_n)A_n(\mathfrak{x}_n;\mathfrak{y}_n),\ldots$$

be a denumerable system Σ of normal expression of degree 1, and let $\mathfrak{x}_i$ be an r_i-tuple, $\mathfrak{y}_i$ an s_i-tuple of variables. Let $\mathfrak{x}^i_1, \mathfrak{x}^i_2, \ldots, \mathfrak{x}^i_n, \ldots$ be the sequence of all r_i-tuples taken from the sequence $x_0, x_1, x_2, \ldots, x_n, \ldots$ and ordered according to increasing sum of the subscripts ⟦and for equal sums according to some convention⟧; furthermore, let $\mathfrak{y}^i_k$ be an s_i-tuple of variables, of the sequence above, such that the sequence of variables

$$\mathfrak{y}^1_1, \mathfrak{y}^1_2, \mathfrak{y}^2_1, \mathfrak{y}^1_3, \mathfrak{y}^2_2, \mathfrak{y}^3_1, \mathfrak{y}^1_4, \ldots$$

becomes identical with the sequence $x_1, x_2, \ldots, x_n, \ldots$ if every $\mathfrak{y}^i_k$ is replaced by the corresponding s_i-tuple of the x. Further we define, in a way analogous to what was done above, a sequence $\{B_n\}$ of formulas by means of the stipulations

$$B_1 = A_1(\mathfrak{x}^1_1;\mathfrak{y}^1_1),$$

$$B_n = B_{n-1}\ \&\ A_1(\mathfrak{x}^1_n;\mathfrak{y}^1_n)\ \&\ A_2(\mathfrak{x}^2_{n-1};\mathfrak{y}^2_{n-1})\ \&\ \ldots\ \&\ A_{n-1}(\mathfrak{x}^{n-1}_2;\mathfrak{y}^{n-1}_2)\ \&\ A_n(\mathfrak{x}^n_1;\mathfrak{y}^n_1).$$

We can easily see that $(P_n)B_n$ (that is, the formula that results from B_n when all

[18] And, moreover, in an at most denumerable domain (for it consists of disjoint classes of the denumerable domain Σ of individuals).

individual variables occurring in it are bound by existential quantifiers) is a consequence of the first n expressions of the system Σ given above. If, therefore, every finite subsystem of Σ is satisfiable, so is every B_n. But, if every B_n is satisfiable, so is the entire system Σ (as follows by the argument used in the proof of Theorem V (see p. 588)), and Theorem X is thus proved. Theorems IX and X can be extended without difficulty, by the procedure used in the proof of Theorem VIII, to systems of formulas containing the sign "=".

We can also give a somewhat different turn to Theorem IX if we confine ourselves to systems of formulas without propositional variables and regard them as systems of axioms whose primitive notions are the functional variables occurring in them. Then Theorem IX clearly asserts that every finite or denumerable axiom system in whose axioms "all" and "there exists" never refer to classes or relations but only to individuals[19] either is inconsistent (that is, a contradiction can be constructed in a finite number of formal steps) or possesses a model ⟦Realisierung⟧.

Finally, let us also discuss the question of the independence of Axioms 1–8. So far as Axioms 1–4 (those of the propositional calculus) are concerned, it has already been shown by P. Bernays[20] that none of them follows from the other three. That their independence is not affected even by the addition of Axioms 5–8 can be shown by means of the very same interpretations that Bernays uses, provided that, in order to extend them to formulas containing functional variables and the sign "=", we make the following stipulations:

1. The prefixes and individual variables are omitted;
2. In what remains of each formula the functional variables are to be treated just like propositional variables;
3. Only "distinguished" values may ever be substituted for the sign "=".

To demonstrate the independence of Axiom 5, we associate with each formula another one, which we obtain by replacing components of the form

$$(x)F(x),\ (y)F(y), \ldots;\ (x)G(x),\ (y)G(y), \ldots;\ \ldots,$$[21]

should such occur, by $X \vee \overline{X}$. Then Axioms 1–4 and 6–8 go over into valid formulas, and the same holds, as we can convince ourselves by mathematical induction, of all formulas derived from these axioms by Rules of inference 1–4; Axiom 5, however, does not possess this property. The independence of Axiom 6 can be shown in exactly the same way, except that here $(x)F(x)$, $(y)F(y), \ldots$, and so on must be replaced by $X \mathbin{\&} \overline{X}$. To prove the independence of Axiom 7 we note that Axioms 1–6 and 8 (and therefore also all formulas derived from them) remain valid if the identity relation is replaced by the empty relation, whereas this is not the case for Axiom 7. Similarly, the formulas derived from Axioms 1–7 remain valid when the identity relation is replaced by the universal relation, whereas this is not the case for Axiom 8 (in a domain of at least two individuals). We can also readily see that none of the Rules of inference 1–4 is redundant, but we shall not look into this more closely here.

[19] Hilbert's axiom system for geometry, without the axiom of continuity, can perhaps serve as an example.

[20] See *Bernays 1926*.

[21] That is, the singulary functional variables F, $G, \ldots$ preceded by a universal quantifier whose scope is just the F, $G, \ldots$ in question, along with the associated individual variable.

Some metamathematical results on completeness and consistency,
On formally undecidable propositions of Principia mathematica *and related systems I,* and *On completeness and consistency*

KURT GÖDEL

(*1930b*, *1931*, and *1931a*)

The main paper below (*1931*), which was to have such an impact on modern logic, was received for publication on 17 November 1930 and published early in 1931. An abstract (*1930b*) had been presented on 23 October 1930 to the Vienna Academy of Sciences by Hans Hahn.

Gödel's results are now accessible in many publications, but his original paper has not lost any of its value as a guide. It is clearly written and does not assume any previous result for its main line of argument. It is, moreover, rich in interesting details. We now give some indications of its contents and structure.

Section 1 is an informal presentation of the main argument and can be read by the nonmathematician; it shows how the argument, by dealing with the proposition that states of itself "I am not provable", instead of the proposition that states of itself "I am not true", skirts the Liar paradox, without falling into it. Gödel also brings to light the relation that his argument bears to Cantor's diagonal procedure and Richard's paradox (Herbrand, on pages 626–628 below, and Weyl (*1949*, pp. 219–235) particularly stress this aspect of Gödel's argument; see also above, p. 439).

Section 2, the longest, is the proof of Theorem VI. The theorem states that in a formal system satisfying certain precise conditions there is an undecidable proposition, that is, a proposition such that neither the proposition itself nor its negation is provable in the system. Before coming to the core of the argument, Gödel takes a number of preparatory steps:

(1) A precise description of the system P with which he is going to work. The variables are distinguished as to their types and they range over the natural numbers (type 1), classes of natural numbers (type 2), classes of classes of natural numbers (type 3), and so forth. The logical axioms are equivalent to the logic of *Principia mathematica* without the ramified theory of types. The arithmetic axioms are Peano's, properly transcribed. The identification of the individuals with the natural numbers and

the adjunction of Peano's axioms (instead of their derivation, as in *Principia*) have the effect that every formula has an interpretation in classical mathematics and, if closed, is either true or false in that interpretation; moreover, proofs are considerably shortened.

(2) An assignment of natural numbers to sequences of signs of P and a similar assignment to sequences of sequences of signs of P. The first assignment is such that, given a sequence, the number assigned to it can be effectively calculated, and, given a number, we can effectively decide whether the number is assigned to a sequence and, if it is, actually write down the sequence; similarly for the second assignment. By means of these assignments we can correlate number-theoretic predicates with metamathematical notions used in the description of the system; for example, to the notion "axiom" corresponds the predicate $Ax(x)$, which holds precisely of the numbers x that are assigned to axioms (the "Gödel numbers" of axioms, we would say today).

(3) A definition of primitive recursive functions (Gödel calls them recursive functions) and the derivation of a few theorems about them. These functions had already been used in foundational research (for example, by Dedekind (*1888*), Skolem (*1923*), Hilbert (*1925*, *1927*), and Ackermann (*1928*)); Gödel gives a precise definition of them, which has become standard.

(4) The proof that forty-five number-theoretic predicates, forty of them associated with metamathematical notions, are primitive recursive.

(5) The proof that every primitive recursive number-theoretic predicate is numeralwise representable in P. That is, the predicate holds of some given numbers if and only if a definite formula of P is provable whenever its free variables are replaced by the symbols that represent these numbers in P.

(6) The definition of ω-consistency.

Gödel can then undertake to prove Theorem VI. The scope of the theorem is enlarged by the addition of any ω-consistent primitive recursive class κ of formulas to the axioms of P. For each such κ a different system is thus obtained (in the present note, "P_κ", a notation not used by Gödel, will denote the system corresponding to a given κ). After the proof Gödel makes a number of important remarks:

(*a*) He points out the constructive content of Theorem VI.

(*b*) He introduces predicates that are *entscheidungsdefinit* (in the translation below these are called *decidable* predicates, at the author's suggestion). If we take into account the few lines added in proof at the end of a later note of Gödel's (*1934a*), these predicates are in fact those that today we call recursive (that is, general recursive) predicates. Gödel somewhat extends the result of Theorem VI by assuming only that κ is decidable, and not that it is primitive recursive.

(*c*) If κ is assumed to be merely consistent, instead of ω-consistent, the proof yields the existence of a predicate whose universalization is not provable but for which no counterexample can be given; P_κ is ω-incomplete, as we would say today.

(*d*) The adjunction of the undecidable formula Neg(17 Gen r) to κ yields a consistent but ω-inconsistent system.

(*e*) Even with the adjunction of the axiom of choice or the continuum hypothesis the system contains undecidable propositions.

The section ends with a review of the properties of P that are actually used in the proof and the remark that all known axiom systems of mathematics, or of any substantial part of it, have these properties.

Section 3 presents two supplementary undecidability results. Gödel establishes (Theorem VII) that a primitive recursive number-theoretic predicate is *arithmetic*, that is, can be expressed as a formula of first-order number theory (this yields a stronger result than the numeralwise representability of such predicates, as it was introduced and used in Section 2). Hence every formula of the form $(x)F(x)$, with $F(x)$ primitive recursive, is equivalent to an arithmetic formula; moreover, this equivalence is provable in P_κ: one can review the informal proof presented by Gödel and check that P_κ is strong enough to express and justify each of its steps. Since the proposition that was proved to be undecidable in Theorem VI is of the form $(x)F(x)$, with $F(x)$ primitive recursive, P_κ contains undecidable *arithmetic* propositions (Theorem VIII). For all its strength, the system P_κ cannot decide every first-order number-theoretic proposition. Theorem X states that, given a formula $(x)F(x)$, with $F(x)$ primitive recursive, one can exhibit a formula of the *pure* first-order predicate calculus, say A, that is satisfiable if and only if $(x)F(x)$ holds. Moreover, since P_κ contains a set theory, the equivalence

$$(x)F(x) \equiv (A \text{ is satisfiable})$$

is expressible in P_κ and, as one can verify by reviewing Gödel's informal argument, provable in P_κ. Therefore (Theorem IX) there are formulas of the pure first-order predicate calculus whose validity is undecidable in P_κ.

In Section 4 an important consequence of Theorem VI is derived. The statement "there exists in P_κ an unprovable formula", which expresses the consistency of P_κ, can be written as a formula of P_κ; but this formula is not provable in P_κ (Theorem XI). The main step in the demonstration of this result consists in reviewing the proof of the first half of Theorem VI and checking that all the statements made in that proof can be expressed and proved in P_κ. It is clear that this is the case, and Gödel does not go through the details of the demonstration. The section ends with various remarks on Theorem XI (its constructive character, its applicability to set theory and ordinary analysis, its effect upon Hilbert's conception of mathematics).

Gödel's paper immediately attracted the interest of logicians and, although it caused some momentary surprise, its results were soon widely accepted. A number of studies were directly inspired by it. By using a somewhat more complicated predicate than "is provable in P", Rosser (*1936*) was able to weaken the assumption of ω-consistency in Theorem VI to that of ordinary consistency. Hilbert and Bernays (*1939*, pp. 283–340) carried out in all details the proof of the analogue of Theorem XI for two standard systems of number theory, Z_μ and Z, and this proof can be transferred almost literally to any system containing Z. As Gödel indicates in a note appended to the present translation of his paper, Turing's work (*1937*) gave to the notion of formal system its full generality. The notes of Gödel's Princeton lectures (*1934*) contain the most important results of the present paper, in a more succinct form; they also make precise the notion of (general) recursive function, already suggested by Herbrand (see below, p. 618). In developing the theory of these functions, Kleene (*1936*) obtained undecidability results of a somewhat different character from those presented here. Gödel's work led to Church's negative solution (*1936*) of the decision problem for the predicate calculus of first order. Tarski (*1953*) developed a general theory of undecidability. The device of the "arithmetization" of metamathematics became an everyday tool of the research worker in foundations. Gödel's results, finally, led to a profound revision of Hilbert's program (on that point see, among other texts, *Bernays 1938*, *1954* and *Gödel 1958*).

These indications are far from giving a full account of the deep influence exerted in the field of foundations of mathematics by the results presented in the paper below and the methods used to obtain them. There is not one branch of research, except perhaps intuitionism, that has not been pervaded by this influence.

The translation of the paper is by the editor, and it is printed here with the kind permission of Professor Gödel and Springer Verlag. Professor Gödel approved the translation, which in many places was accommodated to his wishes. He suggested, in particular, the various phrases used to render the word "inhaltlich". He also proposed a number of short interpolations to help the reader, and these have been introduced in the text below between square brackets.

Below, on page 601, the author shows how a number-theoretic predicate can be associated with a given metamathematical notion and then used to represent the notion. In the German text such a predicate is denoted by the same word as the original notion, except that the word is printed in italics. Since in English italics are used for emphasis (while the German text uses letter spacing for that purpose), the translation below uses SMALL CAPITALS for the names of these predicates. This scheme of italicization (or small-capitalization), however, is used for only some of the number-theoretic predicates in question. According to Professor Gödel, "the idea was to use the notation only for those metamathematical notions that had been defined in their usual sense before, namely, those defined on pp. 599–601. From p. 607 up to the general considerations at the end of Section 2, and again in Section 4, every metamathematical term referring to the system P is supposed to denote the corresponding arithmetic one. But, of course, because of the complete isomorphism the distinction in many cases is entirely irrelevant".

Before the main text the reader will find a translation, by Stefan Bauer-Mengelberg, of its abstract (*1930b*); in that translation, at the author's suggestion, "entscheidungsdefinit", when referring to an axiom system, has been translated by "complete", and "Entscheidungsdefinitheit" by "completeness". A translation, by the editor, of *1931a*, a note dated 22 January 1931 and closely connected with *1931*, follows the main text. Both translations are printed here with the kind permission of Professor Gödel.

SOME METAMATHEMATICAL RESULTS ON COMPLETENESS AND CONSISTENCY

(*1930b*)

If to the Peano axioms we add the logic of *Principia mathematica*[1] (with the natural numbers as the individuals) together with the axiom of choice (for all types), we obtain a formal system S, for which the following theorems hold:

I. The system S is *not* complete ⟦entscheidungsdefinit⟧; that is, it contains propositions A (and we can in fact exhibit such propositions) for which neither A nor $\overline{A}$ is provable and, in particular, it contains (even for decidable properties F of natural numbers) undecidable problems of the simple structure $(Ex)F(x)$, where x ranges over the natural numbers.[2]

II. Even if we admit all the logical devices of *Principia mathematica* (hence in particular the extended functional calculus[1] and the axiom of choice) in metamathematics, there does *not* exist a *consistency proof* for the system S (still less so if we

[1] With the axiom of reducibility or without ramified theory of types.

[2] Furthermore, S contains formulas of the restricted functional calculus such that neither universal validity nor existence of a counterexample is provable for any of them.

restrict the means of proof in any way). Hence a consistency proof for the system S can be carried out only by means of modes of inference that are not formalized in the system S itself, and analogous results hold for other formal systems as well, such as the Zermelo-Fraenkel axiom system of set theory.[3]

III. Theorem I can be sharpened to the effect that, even if we add finitely many axioms to the system S (or infinitely many that result from a finite number of them by "type elevation"), we do *not* obtain a complete system, provided the extended system is ω-consistent. Here a system is said to be ω-consistent if, for no property $F(x)$ of natural numbers,

$$F(1),\ F(2), \ldots,\ F(n), \ldots \text{ ad infinitum}$$

as well as

$$(Ex)\overline{F(x)}$$

are provable. (There are extensions of the system S that, while consistent, are not ω-consistent.)

IV. Theorem I still holds for all ω-consistent extensions of the system S that are obtained by the addition of *infinitely many* axioms, provided the added class of axioms is decidable ⟦entscheidungsdefinit⟧, that is, provided it is metamathematically decidable ⟦entscheidbar⟧ for every formula whether it is an axiom or not (here again we suppose that the logic used in metamathematics is that of *Principia mathematica*).

Theorems I, III, and IV can be extended also to other formal systems, for example, to the Zermelo-Fraenkel axiom system of set theory, provided the systems in question are ω-consistent.

The proofs of these theorems will appear in *Monatshefte für Mathematik und Physik*.

[3] This result, in particular, holds also for the axiom system of classical mathematics, as it has been constructed, for example, by von Neumann (*1927*).

ON FORMALLY UNDECIDABLE PROPOSITIONS OF *PRINCIPIA MATHEMATICA* AND RELATED SYSTEMS I[1]

(*1931*)

1

The development of mathematics toward greater precision has led, as is well known, to the formalization of large tracts of it, so that one can prove any theorem using nothing but a few mechanical rules. The most comprehensive formal systems that have been set up hitherto are the system of *Principia mathematica* (*PM*)[2] on the one hand and the Zermelo-Fraenkel axiom system of set theory (further developed by J. von Neumann)[3] on the other. These two systems are so comprehensive that in

[1] See a summary of the results of the present paper in *Gödel 1930b*.

[2] *Whitehead and Russell 1925*. Among the axioms of the system *PM* we include also the axiom of infinity (in this version: there are exactly denumerably many individuals), the axiom of reducibility, and the axiom of choice (for all types).

[3] See *Fraenkel 1927* and *von Neumann 1925, 1928*, and *1929*. We note that in order to complete the formalization we must add the axioms and rules of inference of the calculus of logic to the set-theoretic axioms given in the literature cited. The considerations that follow apply also to the formal systems (so far as they are available at present) constructed in recent years by Hilbert and his collaborators. See *Hilbert 1922, 1922a, 1927, Bernays 1923, von Neumann 1927*, and *Ackermann 1924*.

them all methods of proof today used in mathematics are formalized, that is, reduced to a few axioms and rules of inference. One might therefore conjecture that these axioms and rules of inference are sufficient to decide *any* mathematical question that can at all be formally expressed in these systems. It will be shown below that this is not the case, that on the contrary there are in the two systems mentioned relatively simple problems in the theory of integers[4] that cannot be decided on the basis of the axioms. This situation is not in any way due to the special nature of the systems that have been set up but holds for a wide class of formal systems; among these, in particular, are all systems that result from the two just mentioned through the addition of a finite number of axioms,[5] provided no false propositions of the kind specified in footnote 4 become provable owing to the added axioms.

Before going into details, we shall first sketch the main idea of the proof, of course without any claim to complete precision. The formulas of a formal system (we restrict ourselves here to the system *PM*) in outward appearance are finite sequences of primitive signs (variables, logical constants, and parentheses or punctuation dots), and it is easy to state with complete precision *which* sequences of primitive signs are meaningful formulas and which are not.[6] Similarly, proofs, from a formal point of view, are nothing but finite sequences of formulas (with certain specifiable properties.) Of course, for metamathematical considerations it does not matter what objects are chosen as primitive signs, and we shall assign natural numbers to this use.[7] Consequently, a formula will be a finite sequence of natural numbers,[8] and a proof array a finite sequence of finite sequences of natural numbers. The metamathematical notions (propositions) thus become notions (propositions) about natural numbers or sequences of them;[9] therefore they can (at least in part) be expressed by the symbols of the system *PM* itself. In particular, it can be shown that the notions "formula", "proof array", and "provable formula" can be defined in the system *PM*; that is, we can, for example, find a formula $F(v)$ of *PM* with one free variable v (of the type of a number sequence)[10] such that $F(v)$, interpreted according to the meaning of the terms of *PM*, says: v is a provable formula. We now construct an undecidable proposition of the system *PM*, that is, a proposition A for which neither A nor *not-A* is provable, in the following manner.

[4] That is, more precisely, there are undecidable propositions in which, besides the logical constants $\overline{}$ (not), $\vee$ (or), (x) (for all), and $=$ (identical with), no other notions occur but $+$ (addition) and . (multiplication), both for natural numbers, and in which the prefixes (x), too, apply to natural numbers only.

[5] In *PM* only axioms that do not result from one another by mere change of type are counted as distinct.

[6] Here and in what follows we always understand by "formula of *PM*" a formula written without abbreviations (that is, without the use of definitions). It is well known that [in *PM*] definitions serve only to abbreviate notations and therefore are dispensable in principle.

[7] That is, we map the primitive signs one-to-one onto some natural numbers. (See how this is done on page 601.)

[8] That is, a number-theoretic function defined on an initial segment of the natural numbers. (Numbers, of course, cannot be arranged in a spatial order.)

[9] In other words, the procedure described above yields an isomorphic image of the system *PM* in the domain of arithmetic, and all metamathematical arguments can just as well be carried out in this isomorphic image. This is what we do below when we sketch the proof; that is, by "formula", "proposition", "variable", and so on, *we must always understand the corresponding objects of the isomorphic image*.

[10] It would be very easy (although somewhat cumbersome) to actually write down this formula.

A formula of *PM* with exactly one free variable, that variable being of the type of the natural numbers (class of classes), will be called a *class sign*. We assume that the class signs have been arranged in a sequence in some way,[11] we denote the nth one by $R(n)$, and we observe that the notion "class sign", as well as the ordering relation R, can be defined in the system *PM*. Let α be any class sign; by $[\alpha; n]$ we denote the formula that results from the class sign α when the free variable is replaced by the sign denoting the natural number n. The ternary relation $x = [y; z]$, too, is seen to be definable in *PM*. We now define a class K of natural numbers in the following way:

$$n \, \varepsilon \, K \equiv \overline{Bew}\,[R(n); n] \tag{1}$$

(where $Bew\, x$ means: x is a provable formula).[11a] Since the notions that occur in the definiens can all be defined in *PM*, so can the notion K formed from them; that is, there is a class sign S such that the formula $[S; n]$, interpreted according to the meaning of the terms of *PM*, states that the natural number n belongs to K.[12] Since S is a class sign, it is identical with some $R(q)$; that is, we have

$$S = R(q)$$

for a certain natural number q. We now show that the proposition $[R(q); q]$ is undecidable in *PM*.[13] For let us suppose that the proposition $[R(q); q]$ were provable; then it would also be true. But in that case, according to the definitions given above, q would belong to K, that is, by (1), $\overline{Bew}\,[R(q); q]$ would hold, which contradicts the assumption. If, on the other hand, the negation of $[R(q); q]$ were provable, then $\overline{q \, \varepsilon \, K}$,[13a] that is, $Bew\,[R(q); q]$, would hold. But then $[R(q); q]$, as well as its negation, would be provable, which again is impossible.

The analogy of this argument with the Richard antinomy leaps to the eye. It is closely related to the "Liar" too;[14] for the undecidable proposition $[R(q); q]$ states that q belongs to K, that is, by (1), that $[R(q); q]$ is not provable. We therefore have before us a proposition that says about itself that it is not provable [in *PM*].[15] The method of proof just explained can clearly be applied to any formal system that, first, when interpreted as representing a system of notions and propositions, has at

[11] For example, by increasing sum of the finite sequence of integers that is the "class sign", and lexicographically for equal sums.

[11a]. The bar denotes negation.

[12] Again, there is not the slightest difficulty in actually writing down the formula S.

[13] Note that "$[R(q); q]$" (or, which means the same, "$[S; q]$") is merely a *metamathematical description* of the undecidable proposition. But, as soon as the formula S has been obtained, we can, of course, also determine the number q and, therewith, actually write down the undecidable proposition itself. [This makes no difficulty in principle. However, in order not to run into formulas of entirely unmanageable lengths and to avoid practical difficulties in the computation of the number q, the construction of the undecidable proposition would have to be slightly modified, unless the technique of abbreviation by definition used throughout in *PM* is adopted.]

[13a] ⟦The German text reads $\overline{n \, \varepsilon \, K}$, which is a misprint.⟧

[14] Any epistemological antinomy could be used for a similar proof of the existence of undecidable propositions.

[15] Contrary to appearances, such a proposition involves no faulty circularity, for initially it [only] asserts that a certain well-defined formula (namely, the one obtained from the qth formula in the lexicographic order by a certain substitution) is unprovable. Only subsequently (and so to speak by chance) does it turn out that this formula is precisely the one by which the proposition itself was expressed.

its disposal sufficient means of expression to define the notions occurring in the argument above (in particular, the notion "provable formula") and in which, second, every provable formula is true in the interpretation considered. The purpose of carrying out the above proof with full precision in what follows is, among other things, to replace the second of the assumptions just mentioned by a purely formal and much weaker one.

From the remark that $[R(q); q]$ says about itself that it is not provable it follows at once that $[R(q); q]$ is true, for $[R(q); q]$ *is* indeed unprovable (being undecidable). Thus, the proposition that is undecidable *in the system PM* still was decided by metamathematical considerations. The precise analysis of this curious situation leads to surprising results concerning consistency proofs for formal systems, results that will be discussed in more detail in Section 4 (Theorem XI).

2

We now proceed to carry out with full precision the proof sketched above. First we give a precise description of the formal system P for which we intend to prove the existence of undecidable propositions. P is essentially the system obtained when the logic of PM is superposed upon the Peano axioms[16] (with the numbers as individuals and the successor relation as primitive notion).

The primitive signs of the system P are the following:

I. Constants: "$\sim$" (not), "$\vee$" (or), "Π" (for all), "0" (zero), "f" (the successor of), "(", ")" (parentheses);

II. Variables of type 1 (for individuals, that is, natural numbers including 0): "x_1", "y_1", "z_1", ...;

Variables of type 2 (for classes of individuals): "x_2", "y_2", "z_2", ...;

Variables of type 3 (for classes of classes of individuals): "x_3", "y_3", "z_3", ...;

And so on, for every natural number as a type.[17]

Remark: Variables for functions of two or more argument places (relations) need not be included among the primitive signs since we can define relations to be classes of ordered pairs, and ordered pairs to be classes of classes; for example, the ordered pair a, b can be defined to be $((a), (a, b))$, where (x, y) denotes the class whose sole elements are x and y, and (x) the class whose sole element is x.[18]

By a *sign of type* 1 we understand a combination of signs that has [any one of] the forms

$$a, fa, ffa, fffa, \ldots, \text{ and so on,}$$

where a is either 0 or a variable of type 1. In the first case, we call such a sign a *numeral*. For $n > 1$ we understand by a *sign of type n* the same thing as by a *variable of type n*. A combination of signs that has the form $a(b)$, where b is a sign of type n

[16] The addition of the Peano axioms, as well as all other modifications introduced in the system PM, merely serves to simplify the proof and is dispensable in principle.

[17] It is assumed that we have denumerably many signs at our disposal for each type of variables.

[18] Nonhomogeneous relations, too, can be defined in this manner; for example, a relation between individuals and classes can be defined to be a class of elements of the form $((x_2), ((x_1), x_2))$. Every proposition about relations that is provable in PM is provable also when treated in this manner, as is readily seen.

and a a sign of type $n + 1$, will be called an *elementary formula*. We define the class of *formulas* to be the smallest class[19] containing all elementary formulas and containing $\sim(a)$, $(a) \vee (b)$, $x\Pi(a)$ (where x may be any variable)[18a] whenever it contains a and b. We call $(a) \vee (b)$ the *disjunction* of a and b, $\sim(a)$ the *negation* and $x\Pi(a)$ a *generalization* of a. A formula in which no free variable occurs (*free variable* being defined in the well-known manner) is called a *sentential formula* ⟦Satzformel⟧. A formula with exactly n free individual variables (and no other free variables) will be called an *n-place relation sign*; for $n = 1$ it will also be called a *class sign*.

By Subst $a\binom{v}{b}$ (where a stands for a formula, v for a variable, and b for a sign of the same type as v) we understand the formula that results from a if in a we replace v, wherever it is free, by b.[20] We say that a formula a is a *type elevation* of another formula b if a results from b when the type of each variable occurring in b is increased by the same number.

The following formulas (I–V) are called *axioms* (we write them using these abbreviations, defined in the well-known manner: $.$, $\supset$, $\equiv$, (Ex), $=$,[21] and observing the usual conventions about omitting parentheses):[22]

I. 1. $\sim(fx_1 = 0)$,

2. $fx_1 = fy_1 \supset x_1 = y_1$,

3. $x_2(0) . x_1\Pi(x_2(x_1) \supset x_2(fx_1)) \supset x_1\Pi(x_2(x_1))$.

II. All formulas that result from the following schemata by substitution of any formulas whatsoever for p, q, r:

1. $p \vee p \supset p$,
2. $p \supset p \vee q$,
3. $p \vee q \supset q \vee p$,
4. $(p \supset q) \supset (r \vee p \supset r \vee q)$.

III. Any formula that results from either one of the two schemata

1. $v\Pi(a) \supset \text{Subst } a\binom{v}{c}$,
2. $v\Pi(b \vee a) \supset b \vee v\Pi(a)$

when the following substitutions are made for a, v, b, and c (and the operation indicated by "Subst" is performed in 1):

For a any formula, for v any variable, for b any formula in which v does not occur free, and for c any sign of the same type as v, provided c does not contain any variable that is bound in a at a place where v is free.[23]

[19] Concerning this definition (and similar definitions occurring below) see *Łukasiewicz and Tarski 1930*.

[18a] Hence $x\Pi(a)$ is a formula even if x does not occur in a or is not free in a. In this case, of course, $x\Pi(a)$ means the same thing as a.

[20] In case v does not occur in a as a free variable we put Subst $a\binom{v}{b} = a$. Note that "Subst" is a metamathematical sign.

[21] $x_1 = y_1$ is to be regarded as defined by $x_2\Pi(x_2(x_1) \supset x_2(y_1))$, as in *PM* (I, $*13$) similarly for higher types).

[22] In order to obtain the axioms from the schemata listed we must therefore
(1) Eliminate the abbreviations and
(2) Add the omitted parentheses
(in II, III, and IV after carrying out the substitutions allowed).
Note that all expressions thus obtained are "formulas" in the sense specified above. (See also the exact definitions of the metamathematical notions on pp. 603–606.)

[23] Therefore c is a variable or 0 or a sign of the form $f\ldots fu$, where u is either 0 or a variable of type 1. Concerning the notion "free (bound) at a place in a", see I A 5 in *von Neumann 1927*.

IV. Every formula that results from the schema

1. $(Eu)(v\Pi(u(v) \equiv a))$

when for v we substitute any variable of type n, for u one of type $n + 1$, and for a any formula that does not contain u free. This axiom plays the role of the axiom of reducibility (the comprehension axiom of set theory).

V. Every formula that results from

1. $x_1\Pi(x_2(x_1) \equiv y_2(x_1)) \supset x_2 = y_2$

by type elevation (as well as this formula itself). This axiom states that a class is completely determined by its elements.

A formula c is called an *immediate consequence* of a and b if it is the formula $(\sim(b)) \vee (c)$, and it is called an *immediate consequence* of a if it is the formula $v\Pi(a)$, where v denotes any variable. The class of *provable formulas* is defined to be the smallest class of formulas that contains the axioms and is closed under the relation "immediate consequence".[24]

We now assign natural numbers to the primitive signs of the system P by the following one-to-one correspondence:

"0" ... 1 "$\sim$" ... 5 "Π" ... 9
"f" ... 3 "$\vee$" ... 7 "(" ... 11
")" ... 13;

to the variables of type n we assign the numbers of the form p^n (where p is a prime number >13). Thus we have a one-to-one correspondence by which a finite sequence of natural numbers is associated with every finite sequence of primitive signs (hence also with every formula). We now map the finite sequences of natural numbers on natural numbers (again by a one-to-one correspondence), associating the number $2^{n_1}.3^{n_2}.\dots.p_k^{n_k}$, where p_k denotes the kth prime number (in order of increasing magnitude), with the sequence $n_1, n_2, \dots, n_k$. A natural number ⟦out of a certain subset⟧ is thus assigned one-to-one not only to every primitive sign but also to every finite sequence of such signs. We denote by $\Phi(a)$ the number assigned to the primitive sign (or to the sequence of primitive signs) a. Now let some relation (or class) $R(a_1, a_2, \dots, a_n)$ between [or of] primitive signs or sequences of primitive signs be given. With it we associate the relation (or class) $R'(x_1, x_2, \dots, x_n)$ between [or of] natural numbers that obtains between $x_1, x_2, \dots, x_n$ if and only if there are some $a_1, a_2, \dots, a_n$ such that $x_i = \Phi(a_i)$ $(i = 1, 2, \dots, n)$ and $R(a_1, a_2, \dots, a_n)$ hold. The relations between (or classes of) natural numbers that in this manner are associated with the metamathematical notions defined so far, for example, "variable", "formula", "sentential formula", "axiom", "provable formula", and so on, will be denoted by the same words in SMALL CAPITALS. The proposition that there are undecidable problems in the system P, for example, reads thus: There are SENTENTIAL FORMULAS a such that neither a nor the NEGATION of a is a PROVABLE FORMULA.

We now insert a parenthetic consideration that for the present has nothing to do

[24] The rule of substitution is rendered superfluous by the fact that all possible substitutions have already been carried out in the axioms themselves. (This procedure was used also by *von Neumann 1927*.)

with the formal system P. First we give the following definition: A number-theoretic function[25] $\varphi(x_1, x_2, \ldots, x_n)$ is said to be *recursively defined in terms of* the number-theoretic functions $\psi(x_1, x_2, \ldots, x_{n-1})$ and $\mu(x_1, x_2, \ldots, x_{n+1})$ if

$$\begin{aligned} \varphi(0, x_2, \ldots, x_n) &= \psi(x_2, \ldots, x_n), \\ \varphi(k+1, x_2, \ldots, x_n) &= \mu(k, \varphi(k, x_2, \ldots, x_n), x_2, \ldots, x_n) \end{aligned} \tag{2}$$

hold for all $x_2, \ldots, x_n, k$.[26]

A number-theoretic function φ is said to be *recursive* if there is a finite sequence of number-theoretic functions $\varphi_1, \varphi_2, \ldots, \varphi_n$ that ends with φ and has the property that every function φ_k of the sequence is recursively defined in terms of two of the preceding functions, or results from any of the preceding functions by substitution,[27] or, finally, is a constant or the successor function $x + 1$. The length of the shortest sequence of φ_i corresponding to a recursive function φ is called its *degree*. A relation $R(x_1, \ldots, x_n)$ between natural numbers is said to be recursive[28] if there is a recursive function $\varphi(x_1, \ldots, x_n)$ such that, for all $x_1, x_2, \ldots, x_n$,

$$R(x_1, \ldots, x_n) \sim [\varphi(x_1, \ldots, x_n) = 0].^{29}$$

The following theorems hold:

I. *Every function (relation) obtained from recursive functions (relations) by substitution of recursive functions for the variables is recursive; so is every function obtained from recursive functions by recursive definition according to schema (2);*

II. *If R and S are recursive relations, so are $\bar{R}$ and $R \vee S$ (hence also R & S);*

III. *If the functions $\varphi(\mathfrak{x})$ and $\psi(\mathfrak{y})$ are recursive, so is the relation $\varphi(\mathfrak{x}) = \psi(\mathfrak{y})$;*[30]

IV. *If the function $\varphi(\mathfrak{x})$ and the relation $R(x, \mathfrak{y})$ are recursive, so are the relations S and T defined by*

$$S(\mathfrak{x}, \mathfrak{y}) \sim (Ex)[x \leqq \varphi(\mathfrak{x}) \;\&\; R(x, \mathfrak{y})]$$

and

$$T(\mathfrak{x}, \mathfrak{y}) \sim (x)[x \leqq \varphi(\mathfrak{x}) \rightarrow R(x, \mathfrak{y})],$$

as well as the function ψ defined by

$$\psi(\mathfrak{x}, \mathfrak{y}) = \varepsilon x[x \leqq \varphi(\mathfrak{x}) \;\&\; R(x, \mathfrak{y})],$$

where $\varepsilon x F(x)$ means the least number x for which $F(x)$ holds and 0 in case there is no such number.

[25] That is, its domain of definition is the class of nonnegative integers (or of n-tuples of nonnegative integers) and its values are nonnegative integers.

[26] In what follows, lower-case italic letters (with or without subscripts) are always variables for nonnegative integers (unless the contrary is expressly noted).

[27] More precisely, by substitution of some of the preceding functions at the argument places of one of the preceding functions, for example, $\varphi_k(x_1, x_2) = \varphi_p[\varphi_q(x_1, x_2), \varphi_r(x_2)]$ $(p, q, r < k)$. Not all variables on the left side need occur on the right side (the same applies to the recursion schema (2)).

[28] We include classes among relations (as one-place relations). Recursive relations R, of course, have the property that for every given n-tuple of numbers it can be decided whether $R(x_1, \ldots, x_n)$ holds or not.

[29] Whenever formulas are used to express a meaning (in particular, in all formulas expressing metamathematical propositions or notions), Hilbert's symbolism is employed. See *Hilbert and Ackermann 1928*.

[30] We use German letters, $\mathfrak{x}$, $\mathfrak{y}$, as abbreviations for arbitrary n-tuples of variables, for example, $x_1, x_2, \ldots, x_n$.

Theorem I follows at once from the definition of "recursive". Theorems II and III are consequences of the fact that the number-theoretic functions

$$\alpha(x), \quad \beta(x, y), \quad \gamma(x, y),$$

corresponding to the logical notions $\overline{}$, $\vee$, and $=$, namely,

$$\alpha(0) = 1, \ \alpha(x) = 0 \quad \text{for } x \neq 0,$$

$$\beta(0, x) = \beta(x, 0) = 0, \quad \beta(x, y) = 1 \quad \text{when } x \text{ and } y \text{ are both } \neq 0,$$

$$\gamma(x, y) = 0 \quad \text{when } x = y, \qquad \gamma(x, y) = 1 \quad \text{when } x \neq y,$$

are recursive, as we can readily see. The proof of Theorem IV is briefly as follows. By assumption there is a recursive $\rho(x, \mathfrak{y})$ such that

$$R(x, \mathfrak{y}) \sim [\rho(x, \mathfrak{y}) = 0].$$

We now define a function $\chi(x, \mathfrak{y})$ by the recursion schema (2) in the following way:

$$\chi(0, \mathfrak{y}) = 0,$$

$$\chi(n + 1, \mathfrak{y}) = (n + 1).a + \chi(n, \mathfrak{y}).\alpha(a),$$ [31]

where $a = \alpha[\alpha(\rho(0, \mathfrak{y}))].\alpha[\rho(n + 1, \mathfrak{y})].\alpha[\chi(n, \mathfrak{y})]$. Therefore $\chi(n + 1, \mathfrak{y})$ is equal either to $n + 1$ (if $a = 1$) or to $\chi(n, \mathfrak{y})$ (if $a = 0$).[32] The first case clearly occurs if and only if all factors of a are 1, that is, if

$$\bar{R}(0, \mathfrak{y}) \ \& \ R(n + 1, \mathfrak{y}) \ \& \ [\chi(n, \mathfrak{y}) = 0]$$

holds. From this it follows that the function $\chi(n, \mathfrak{y})$ (considered as a function of n) remains 0 up to ⟦but not including⟧ the least value of n for which $R(n, \mathfrak{y})$ holds and, from there on, is equal to that value. (Hence, in case $R(0, \mathfrak{y})$ holds, $\chi(n, \mathfrak{y})$ is constant and equal to 0.) We have, therefore,

$$\psi(\mathfrak{x}, \mathfrak{y}) = \chi(\varphi(\mathfrak{x}), \mathfrak{y}),$$

$$S(\mathfrak{x}, \mathfrak{y}) \sim R[\psi(\mathfrak{x}, \mathfrak{y}), \mathfrak{y}].$$

The relation T can, by negation, be reduced to a case analogous to that of S. Theorem IV is thus proved.

The functions $x + y$, $x.y$, and x^y, as well as the relations $x < y$ and $x = y$, are recursive, as we can readily see. Starting from these notions, we now define a number of functions (relations) 1–45, each of which is defined in terms of preceding ones by the procedures given in Theorems I–IV. In most of these definitions several of the steps allowed by Theorems I–IV are condensed into one. Each of the functions (relations) 1–45, among which occur, for example, the notions "FORMULA", "AXIOM", and "IMMEDIATE CONSEQUENCE", is therefore recursive.

1. $x/y \equiv (Ez)[z \leqq x \ \& \ x = y.z]$,[33]
x is divisible by y.[34]

[31] We assume familiarity with the fact that the functions $x + y$ (addition) and $x \,.\, y$ (multiplication) are recursive.

[32] a cannot take values other than 0 and 1, as can be seen from the definition of α.

[33] The sign $\equiv$ is used in the sense of "equality by definition"; hence in definitions it stands for either $=$ or $\sim$ (otherwise, the symbolism is Hilbert's).

[34] Wherever one of the signs (x), (Ex), or εx occurs in the definitions below, it is followed by a bound on x. This bound merely serves to ensure that the notion defined is recursive (see Theorem IV). But in most cases the *extension* of the notion defined would not change if this bound were omitted.

2. $\text{Prim}(x) \equiv \overline{(Ex)}[z \leqq x \;\&\; z \neq 1 \;\&\; z \neq x \;\&\; x/z] \;\&\; x > 1$,
x is a prime number.

3. $0 \; Pr \; x \equiv 0$,
$(n + 1) \; Pr \; x \equiv \varepsilon y[y \leqq x \;\&\; \text{Prim}(y) \;\&\; x/y \;\&\; y > n \; Pr \; x]$,
$n \; Pr \; x$ is the nth prime number (in order of increasing magnitude) contained in x.[34a]

4. $0! \equiv 1$,
$(n + 1)! \equiv (n + 1).n!$.

5. $Pr(0) \equiv 0$,
$Pr(n + 1) \equiv \varepsilon y[y \leqq \{Pr(n)\}! + 1 \;\&\; \text{Prim}(y) \;\&\; y > Pr(n)]$,
$Pr(n)$ is the nth prime number (in order of increasing magnitude).

6. $n \; Gl \; x \equiv \varepsilon y[y \leqq x \;\&\; x/(n \; Pr \; x)^y \;\&\; \overline{x/(n \; Pr \; x)^{y+1}}]$,
$n \; Gl \; x$ is the nth term of the number sequence assigned to the number x (for $n > 0$ and n not greater than the length of this sequence).

7. $l(x) \equiv \varepsilon y[y \leqq x \;\&\; y \; Pr \; x > 0 \;\&\; (y + 1) \; Pr \; x = 0]$,
$l(x)$ is the length of the number sequence assigned to x.

8. $x{*}y \equiv \varepsilon z\{z \leqq [Pr(l(x) + l(y))]^{x+y} \;\&\; (n)[n \leqq l(x) \rightarrow n \; Gl \; z = n \; Gl \; x] \;\&$
$(n)[0 < n \leqq l(y) \rightarrow (n + l(x)) \; Gl \; z = n \; Gl \; y]\}$,
$x{*}y$ corresponds to the operation of "concatenating" two finite number sequences.

9. $R(x) \equiv 2^x$,
$R(x)$ corresponds to the number sequence consisting of x alone (for $x > 0$).

10. $E(x) \equiv R(11){*}x{*}R(13)$,
$E(x)$ corresponds to the operation of "enclosing within parentheses" (11 and 13 are assigned to the primitive signs "(" and ")", respectively).

11. $n \; \text{Var} \; x \equiv (Ez)[13 < z \leqq x \;\&\; \text{Prim}(z) \;\&\; x = z^n] \;\&\; n \neq 0$,
x is a VARIABLE OF TYPE n.

12. $\text{Var}(x) \equiv (En)[n \leqq x \;\&\; n \; \text{Var} \; x]$,
x is a VARIABLE.

13. $\text{Neg}(x) \equiv R(5){*}E(x)$,
$\text{Neg}(x)$ is the NEGATION of x.

14. $x \; \text{Dis} \; y \equiv E(x){*}R(7){*}E(y)$,
$x \; \text{Dis} \; y$ is the DISJUNCTION of x and y.

15. $x \; \text{Gen} \; y \equiv R(x){*}R(9){*}E(y)$,
$x \; \text{Gen} \; y$ is the GENERALIZATION of y with respect to the VARIABLE x (provided x is a VARIABLE).

16. $0 \; N \; x \equiv x$,
$(n + 1) \; N \; x \equiv R(3){*}n \; N \; x$,
$n \; N \; x$ corresponds to the operation of "putting the sign 'f' n times in front of x".

17. $Z(n) \equiv n \; N \; [R(1)]$,
$Z(n)$ is the NUMERAL denoting the number n.

18. $\text{Typ}_1'(x) \equiv (Em, n)\{m, n \leqq x \;\&\; [m = 1 \vee 1 \; \text{Var} \; m] \;\&\; x = n \; N \; [R(m)]\}$,[34b]
x is a SIGN OF TYPE 1.

[34a] For $0 < n \leqq z$, where z is the number of distinct prime factors of x. Note that $n \; Pr \; x = 0$ for $n = z + 1$.

[34b] $m, n \leqq x$ stands for $m \leqq x \;\&\; n \leqq x$ (similarly for more than two variables).

19. $\mathrm{Typ}_n(x) \equiv [n = 1 \,\&\, \mathrm{Typ}_1'(x)] \vee [n > 1 \,\&$
$(Ev)\{v \leqq x \,\&\, n \,\mathrm{Var}\, v \,\&\, x = R(v)\}]$,

x is a SIGN OF TYPE n.

20. $Elf(x) \equiv (Ey, z, n)[y, z, n \leqq x \,\&\, \mathrm{Typ}_n(y) \,\&$
$\mathrm{Typ}_{n+1}(z) \,\&\, x = z{*}E(y)]$,

x is an ELEMENTARY FORMULA.

21. $Op(x, y, z) \equiv x = \mathrm{Neg}(y) \vee x = y \,\mathrm{Dis}\, z \vee (Ev)[v \leqq x \,\&\, \mathrm{Var}(v) \,\&$
$x = v \,\mathrm{Gen}\, y]$.

22. $FR(x) \equiv (n)\{0 < n \leqq l(x) \rightarrow Elf(n \,Gl\, x) \vee (Ep, q)[0 < p, q < n \,\&$
$Op(n \,Gl\, x, p \,Gl\, x, q \,Gl\, x)]\} \,\&\, l(x) > 0$,

x is a SEQUENCE OF FORMULAS, each of which either is an ELEMENTARY FORMULA or results from the preceding FORMULAS through the operations of NEGATION, DISJUNCTION, or GENERALIZATION.

23. $\mathrm{Form}(x) \equiv (En)\{n \leqq (Pr[l(x)^2])^{x\cdot[l(x)]^2} \,\&\, FR(n) \,\&\, x = [l(n)] \,Gl\, n\}$,[35]

x is a FORMULA (that is, the last term of a FORMULA SEQUENCE n).

24. $v \,\mathrm{Geb}\, n, x \equiv \mathrm{Var}(v) \,\&\, \mathrm{Form}(x) \,\&\, (Ea, b, c)[a, b, c \leqq x \,\&$
$x = a{*}(v \,\mathrm{Gen}\, b){*}c \,\&\, \mathrm{Form}(b) \,\&\, l(a) + 1 \leqq n \leqq l(a) + l(v \,\mathrm{Gen}\, b)]$,

the VARIABLE v is BOUND in x at the nth place.

25. $v \,Fr\, n, x \equiv \mathrm{Var}(v) \,\&\, \mathrm{Form}(x) \,\&\, v = n \,Gl\, x \,\&\, n \leqq l(x) \,\&\, \overline{v \,\mathrm{Geb}\, n, x}$,

the VARIABLE v is FREE in x at the nth place.

26. $v \,Fr\, x \equiv (En)[n \leqq l(x) \,\&\, v \,Fr\, n, x]$,

v occurs as a FREE VARIABLE in x.

27. $Su\, x\binom{n}{y} \equiv \varepsilon z\{z \leqq [Pr(l(x) + l(y))]^{x+y} \,\&\, [(Eu, v)\, u, v \leqq x \,\&$
$x = u{*}R(n \,Gl\, x){*}v \,\&\, z = u{*}y{*}v \,\&\, n = l(u) + 1]\}$,

$Su\, x\binom{n}{y}$ results from x when we substitute y for the nth term of x (provided that $0 < n \leqq l(x)$).

28. $0 \,St\, v, x \equiv \varepsilon n\{n \leqq l(x) \,\&\, v \,Fr\, n, x \,\&\, \overline{(Ep)}[n < p \leqq l(x) \,\&\, v \,Fr\, p, x]\}$,
$(k + 1) \,St\, v, x \equiv \varepsilon n\{n < k \,St\, v, x \,\&\, v \,Fr\, n, x \,\&\, \overline{(Ep)}[n < p < k \,St\, v, x$
$\&\, v \,Fr\, p, x]\}$,

$k \,St\, v, x$ is the $(k + 1)$th place in x (counted from the right end of the FORMULA x) at which v is FREE in x (and 0 in case there is no such place).

29. $A(v, x) \equiv \varepsilon n\{n \leqq l(x) \,\&\, n \,St\, v, x = 0\}$,

$A(v, x)$ is the number of places at which v is FREE in x.

30. $Sb_0(x^v_y) \equiv x$,
$Sb_{k+1}(x^v_y) \equiv Su\,[Sb_k(x^v_y)]\binom{k \,St\, v, x}{y}$.

31. $Sb(x^v_y) \equiv Sb_{A(v,x)}(x^v_y)$,[36]

$Sb(x^v_y)$ is the notion SUBST $a\binom{v}{b}$ defined above.[37]

32. $x \,\mathrm{Imp}\, y \equiv [\mathrm{Neg}(x)] \,\mathrm{Dis}\, y$,
$x \,\mathrm{Con}\, y \equiv \mathrm{Neg}\{[\mathrm{Neg}(x)] \,\mathrm{Dis}\, [\mathrm{Neg}(y)]\}$,

[35] That $n \leqq (Pr([l(x)]^2))^{x\cdot[l(x)]^2}$ provides a bound can be seen thus: The length of the shortest sequence of formulas that corresponds to x can at most be equal to the number of subformulas of x. But there are at most $l(x)$ subformulas of length 1, at most $l(x) - 1$ of length 2, and so on, hence altogether at most $l(x)(l(x) + 1)/2 \leqq [l(x)]^2$. Therefore all prime factors of n can be assumed to be less than $Pr([l(x)]^2)$, their number $\leqq [(lx)]^2$, and their exponents (which are subformulas of x) $\leqq x$.

[36] In case v is not a VARIABLE or x is not a FORMULA, $Sb(x^v_y) = x$.

[37] Instead of $Sb[Sb(x^v_y)^w_z]$ we write $Sb(x^{v\,w}_{y\,z})$ (and similarly for more than two VARIABLES).

$x \text{ Aeq } y \equiv (x \text{ Imp } y) \text{ Con } (y \text{ Imp } x)$,
$v \text{ Ex } y \equiv \text{Neg}\{v \text{ Gen } [\text{Neg}(y)]\}$.

33. $n \; Th \; x \equiv \varepsilon y\{y \leqq x^{(x^n)} \; \& \; (k)[k \leqq l(x) \rightarrow (k \; Gl \; x \leqq 13 \; \& \; k \; Gl \; y = k \; Gl \; x) \vee (k \; Gl \; x > 13 \; \& \; k \; Gl \; y = k \; Gl \; x . [1 \; Pr \; (k \; Gl \; x)]^n)]\}$,

$n \; Th \; x$ is the nth TYPE ELEVATION of x (in case x and $n \; Th \; x$ are FORMULAS).

Three specific numbers, which we denote by z_1, z_2, and z_3, correspond to the Axioms I, 1–3, and we define

34. $Z\text{-}Ax(x) \equiv (x = z_1 \vee x = z_2 \vee x = z_3)$.

35. $A_1\text{-}Ax(x) \equiv (Ey)[y \leqq x \; \& \; \text{Form}(y) \; \& \; x = (y \text{ Dis } y) \text{ Imp } y]$,

x is a FORMULA resulting from Axiom schema II, 1 by substitution. Analogously, $A_2\text{-}Ax$, $A_3\text{-}Ax$, and $A_4\text{-}Ax$ are defined for Axioms [rather, Axiom Schemata] II, 2–4.

36. $A\text{-}Ax(x) \equiv A_1\text{-}Ax(x) \vee A_2\text{-}Ax(x) \vee A_3\text{-}Ax(x) \vee A_4\text{-}Ax(x)$,

x is a FORMULA resulting from a propositional axiom by substitution.

37. $Q(z, y, v) \equiv \overline{(En, m, w)[n \leqq l(y) \; \& \; m \leqq l(z) \; \& \; w \leqq z \; \& \; w \equiv m \; Gl \; z \; \& \; w \text{ Geb } n, y \; \& \; v \; Fr \; n, y]}$

z does not contain any VARIABLE BOUND in y at a place at which v is FREE.

38. $L_1\text{-}Ax(x) \equiv (Ev, y, z, n)\{v, y, z, n \leqq x \; \& \; n \text{ Var } v \; \& \; \text{Typ}_n(z) \; \& \; \text{Form}(y) \; \& \; Q(z, y, v) \; \& \; x = (v \text{ Gen } y) \text{ Imp } [Sb(y^v_z)]\}$,

x is a FORMULA resulting from Axiom schema III, 1 by substitution.

39. $L_2\text{-}Ax(x) \equiv (Ev, q, p)\{v, q, p \leqq x \; \& \; \text{Var}(v) \; \& \; \text{Form}(p) \; \& \; \overline{v \; Fr \; p} \; \& \; \text{Form}(q) \; \& \; x = [v \text{ Gen } (p \text{ Dis } q)] \text{ Imp } [p \text{ Dis } (v \text{ Gen } q)]\}$,

x is a FORMULA resulting from Axiom schema III, 2 by substitution.

40. $R\text{-}Ax(x) \equiv (Eu, v, y, n)[u, v, y, n \leqq x \; \& \; n \text{ Var } v \; \& \; (n + 1) \text{ Var } u \; \& \; \overline{u \; Fr \; y} \; \& \; \text{Form}(y) \; \& \; x = u \text{ Ex } \{v \text{ Gen } [[R(u)*E(R(v))] \text{ Aeq } y]\}]$,

x is a FORMULA resulting from Axiom schema IV, 1 by substitution.

A specific number z_4 corresponds to Axiom V, 1, and we define:

41. $M\text{-}Ax(x) \equiv (En)[n \leqq x \; \& \; x = n \; Th \; z_4]$.

42. $Ax(x) \equiv Z\text{-}Ax(x) \vee A\text{-}Ax(x) \vee L_1\text{-}Ax(x) \vee L_2\text{-}Ax(x) \vee R\text{-}Ax(x) \vee M\text{-}Ax(x)$,

x is an AXIOM.

43. $Fl(x, y, z) \equiv y = z \text{ Imp } x \vee (Ev)[v \leqq x \; \& \; \text{Var}(v) \; \& \; x = v \text{ Gen } y]$,

x is an IMMEDIATE CONSEQUENCE of y and z.

44. $Bw(x) \equiv (n)\{0 < n \leqq l(x) \rightarrow Ax(n \; Gl \; x) \vee (Ep, q)[0 < p, q < n \; \& \; Fl(n \; Gl \; x, p \; Gl \; x, q \; Gl \; x)]\} \; \& \; l(x) > 0$,

x is a PROOF ARRAY (a finite sequence of FORMULAS, each of which is either an AXIOM or an IMMEDIATE CONSEQUENCE of two of the preceding FORMULAS.

45. $x \; B \; y \equiv Bw(x) \; \& \; [l(x)] \; Gl \; x = y$,

x is a PROOF of the FORMULA y.

46. $\text{Bew}(x) \equiv (Ey)y \; B \; x$,

x is a PROVABLE FORMULA. (Bew(x) is the only one of the notions 1–46 of which we cannot assert that it is recursive.)

The fact that can be formulated vaguely by saying: every recursive relation is definable in the system P (if the usual meaning is given to the formulas of this system), is expressed in precise language, *without* reference to any interpretation of the formulas of P, by the following theorem:

Theorem V. *For every recursive relation* $R(x_1, \ldots, x_n)$ *there exists an n-place*

RELATION SIGN r *(with the* FREE VARIABLES[38] $u_1, u_2, \ldots, u_n$) *such that for all n-tuples of numbers* $(x_1, \ldots, x_n)$ *we have*

$$R(x_1, \ldots, x_n) \rightarrow \text{Bew}[Sb(r_{Z(x_1)\ldots Z(x_n)}^{u_1 \ldots u_n})], \tag{3}$$

$$\bar{R}(x_1, \ldots, x_n) \rightarrow \text{Bew}[\text{Neg}(Sb(r_{Z(x_1)\ldots Z(x_n)}^{u_1 \ldots u_n}))]. \tag{4}$$

We shall give only an outline of the proof of this theorem because the proof does not present any difficulty in principle and is rather long.[39] We prove the theorem for all relations $R(x_1, \ldots, x_n)$ of the form $x_1 = \varphi(x_2, \ldots, x_n)$[40] (where φ is a recursive function) and we use induction on the degree of φ. For functions of degree 1 (that is, constants and the function $x + 1$) the theorem is trivial. Assume now that φ is of degree m. It results from functions of lower degrees, $\varphi_1, \ldots, \varphi_k$, through the operations of substitution or recursive definition. Since by the induction hypothesis everything has already been proved for $\varphi_1, \ldots, \varphi_k$, there are corresponding RELATION SIGNS, $r_1, \ldots, r_k$, such that (3) and (4) hold. The processes of definition by which φ results from $\varphi_1, \ldots, \varphi_k$ (substitution and recursive definition) can both be formally reproduced in the system P. If this is done, a new RELATION SIGN r is obtained from $r_1, \ldots, r_k$,[41] and, using the induction hypothesis, we can prove without difficulty that (3) and (4) hold for it. A RELATION SIGN r assigned to a recursive relation[42] by this procedure will be said to be recursive.

We now come to the goal of our discussions. Let κ be any class of FORMULAS. We denote by Flg(κ) (the set of consequences of κ) the smallest set of FORMULAS that contains all FORMULAS of κ and all AXIOMS and is closed under the relation "IMMEDIATE CONSEQUENCE". κ is said to be ω-consistent if there is no CLASS SIGN a such that

$$(n)[Sb(a_{Z(n)}^{v}) \ \varepsilon \ \text{Flg}(\kappa)] \ \& \ [\text{Neg}(v \ \text{Gen} \ a)] \ \varepsilon \ \text{Flg}(\kappa),$$

where v is the FREE VARIABLE of the CLASS SIGN a.

Every ω-consistent system, of course, is consistent. As will be shown later, however, the converse does not hold.

The general result about the existence of undecidable propositions reads as follows:

Theorem VI. *For every ω-consistent recursive class κ of* FORMULAS *there are recursive* CLASS SIGNS r *such that neither* v Gen r *nor* Neg(v Gen r) *belongs to* Flg(κ) (where v is the FREE VARIABLE of r).

Proof. Let κ be any recursive ω-consistent class of FORMULAS. We define

$$Bw_\kappa(x) \equiv (n)[n \leq l(x) \rightarrow Ax(n \ Gl \ x) \lor (n \ Gl \ x) \ \varepsilon \ \kappa \ \lor$$
$$(Ep, q)\{0 < p, q < n \ \& \ Fl(n \ Gl \ x, p \ Gl \ x, q \ Gl \ x)\}] \ \& \ l(x) > 0 \tag{5}$$

[38] The VARIABLES $u_1, \ldots, u_n$ can be chosen arbitrarily. For example, there always is an r with the FREE VARIABLES 17, 19, 23, ..., and so on, for which (3) and (4) hold.

[39] Theorem V, of course, is a consequence of the fact that in the case of a recursive relation R it can, for every n-tuple of numbers, be decided *on the basis of the axioms of the system P* whether the relation R obtains or not.

[40] From this it follows at once that the theorem holds for every recursive relation, since any such relation is equivalent to $0 = \varphi(x_1, \ldots, x_n)$, where φ is recursive.

[41] When this proof is carried out in detail, r, of course, is not defined indirectly with the help of its meaning but in terms of its purely formal structure.

[42] Which, therefore, in the usual interpretation expresses the fact that this relation holds.

(see the analogous notion 44),

$$x\ B_\kappa\ y \equiv Bw_\kappa(x)\ \&\ [l(x)]\ Gl\ x = \tag{6}$$

$$\mathrm{Bew}_\kappa(x) \equiv (Ey)y\ B_\kappa\ x \tag{6.1}$$

(see the analogous notions 45 and 46).

We obviously have

$$(x)[\mathrm{Bew}_\kappa(x) \sim x\ \varepsilon\ \mathrm{Flg}(\kappa)] \tag{7}$$

and

$$(x)[\mathrm{Bew}(x) \rightarrow \mathrm{Bew}_\kappa(x)]. \tag{8}$$

We now define the relation

$$Q(x, y) \equiv \overline{x\ B_\kappa\ [Sb(y^{19}_{Z(y)})]}. \tag{8.1}$$

Since $x\ B_\kappa\ y$ (by (6) and (5)) and $Sb(y^{19}_{Z(y)})$ (by Definitions 17 and 31) are recursive, so is $Q(x, y)$. Therefore, by Theorem V and (8) there is a RELATION SIGN q (with the FREE VARIABLES 17 and 19) such that

$$\overline{x\ B_\kappa\ [Sb(y^{19}_{Z(y)})]} \rightarrow \mathrm{Bew}_\kappa[Sb(q^{17}_{Z(x)}\ {}^{19}_{Z(y)})], \tag{9}$$

and

$$x\ B_\kappa\ [Sb(y^{19}_{Z(y)})] \rightarrow \mathrm{Bew}_\kappa[\mathrm{Neg}(Sb(q^{17}_{Z(x)}\ {}^{19}_{Z(y)}))]. \tag{10}$$

We put

$$p = 17\ \mathrm{Gen}\ q \tag{11}$$

(p is a CLASS SIGN with the FREE VARIABLE 19) and

$$r = Sb(q^{19}_{Z(p)}) \tag{12}$$

(r is a recursive CLASS SIGN[43] with the FREE VARIABLE 17).
Then we have

$$Sb(p^{19}_{Z(p)}) = Sb([17\ \mathrm{Gen}\ q]^{19}_{Z(p)}) = 17\ \mathrm{Gen}\ Sb(q^{19}_{Z(p)}) = 17\ \mathrm{Gen}\ r \tag{13}$$

(by (11) and (12));[44] furthermore

$$Sb(q^{17}_{Z(x)}\ {}^{19}_{Z(p)}) = Sb(r^{17}_{Z(x)}) \tag{14}$$

(by (12)). If we now substitute p for y in (9) and (10) and take (13) and (14) into account, we obtain

$$\overline{x\ B_\kappa\ (17\ \mathrm{Gen}\ r)} \rightarrow \mathrm{Bew}_\kappa[Sb(r^{17}_{Z(x)})], \tag{15}$$

$$x\ B_\kappa\ (17\ \mathrm{Gen}\ r) \rightarrow \mathrm{Bew}_\kappa[Neg(Sb(r^{17}_{Z(x)}))]. \tag{16}$$

This yields:

1. 17 Gen r is not κ-PROVABLE.[45] For, if it were, there would (by (6.1)) be an n such

[43] Since r is obtained from the recursive RELATION SIGN q through the replacement of a VARIABLE by a definite number, p. [Precisely stated the final part of this footnote (which refers to a side remark unnecessary for the proof) would read thus: "REPLACEMENT of a VARIABLE by the NUMERAL for p."]

[44] The operations Gen and Sb, of course, can always be interchanged in case they refer to different VARIABLES.

[45] By "x is *κ-provable*" we mean $x\ \varepsilon\ \mathrm{Flg}(\kappa)$, which, by (7), means the same thing as $\mathrm{Bew}_\kappa(x)$.

that $n\ B_\kappa$ (17 Gen r). Hence by (16) we would have $\mathrm{Bew}_\kappa[\mathrm{Neg}(Sb(r^{17}_{Z(n)}))]$, while, on the other hand, from the κ-PROVABILITY of 17 Gen r that of $Sb(r^{17}_{Z(n)})$ follows. Hence, κ would be inconsistent (and a fortiori ω-inconsistent).

2. Neg(17 Gen r) is not κ-PROVABLE. Proof: As has just been proved, 17 Gen r is not κ-PROVABLE; that is (by (6.1)), $(n)\overline{n\ B_\kappa\ (17\ \mathrm{Gen}\ r)}$ holds. From this, $(n)\mathrm{Bew}_\kappa[Sb(r^{17}_{Z(n)})]$ follows by (15), and that, in conjunction with Bew_κ[Neg(17 Gen r)], is incompatible with the ω-consistency of κ.

17 Gen r is therefore undecidable on the basis of κ, which proves Theorem VI.

We can readily see that the proof just given is constructive;[45a] that is, the following has been proved in an intuitionistically unobjectionable manner: Let an arbitrary recursively defined class κ of FORMULAS be given. Then, if a formal decision (on the basis of κ) of the SENTENTIAL FORMULA 17 Gen r (which [for each κ] can actually be exhibited) is presented to us, we can actually give

1. A PROOF of Neg(17 Gen r);
2. For any given n, a PROOF of $Sb(r^{17}_{Z(n)})$.

That is, a formal decision of 17 Gen r would have the consequence that we could actually exhibit an ω-inconsistency.

We shall say that a relation between (or a class of) natural numbers $R(x_1, \ldots, x_n)$ is *decidable* ⟦*entscheidungsdefinit*⟧ if there exists an n-place RELATION SIGN r such that (3) and (4) (see Theorem V) hold. In particular, therefore, by Theorem V every recursive relation is decidable. Similarly, a RELATION SIGN will be said to be *decidable* if it corresponds in this way to a decidable relation. Now it suffices for the existence of undecidable propositions that the class κ be ω-consistent and decidable. For the decidability carries over from κ to $x\ B_\kappa\ y$ (see (5) and (6)) and to $Q(x, y)$ (see (8.1)), and only this was used in the proof given above. In this case the undecidable proposition has the form v Gen r, where r is a decidable CLASS SIGN. (Note that it even suffices that κ be decidable in the system enlarged by κ.)

If, instead of assuming that κ is ω-consistent, we assume only that it is consistent, then, although the existence of an undecidable proposition does not follow [by the argument given above], it does follow that there exists a property (r) for which it is possible neither to give a counterexample nor to prove that it holds of all numbers. For in the proof that 17 Gen r is not κ-PROVABLE only the consistency of κ was used (see p. 608). Moreover from $\overline{\mathrm{Bew}_\kappa}$(17 Gen r) it follows by (15) that, for every number x, $Sb(r^{17}_{Z(x)})$ is κ-PROVABLE and consequently that $\mathrm{Neg}(Sb(r^{17}_{Z(x)}))$ is not κ-PROVABLE for any number.

If we adjoin Neg(17 Gen r) to κ, we obtain a class of FORMULAS κ' that is consistent but not ω-consistent. κ' is consistent, since otherwise 17 Gen r would be κ-PROVABLE. However, κ' is not ω-consistent, because, by $\overline{\mathrm{Bew}_\kappa}$(17 Gen r) and (15), $(x)\mathrm{Bew}_\kappa Sb(r^{17}_{Z(x)})$ and, a fortiori, $(x)\mathrm{Bew}_{\kappa'} Sb(r^{17}_{Z(x)})$ hold, while on the other hand, of course, $\mathrm{Bew}_{\kappa'}$[Neg(17 Gen r)] holds.[46]

We have a special case of Theorem VI when the class κ consists of a finite number of FORMULAS (and, if we so desire, of those resulting from them by TYPE ELEVATION).

[45a] Since all existential statements occurring in the proof are based upon Theorem V, which, as is easily seen, is unobjectionable from the intuitionistic point of view.

[46] Of course, the existence of classes κ that are consistent but not ω-consistent is thus proved only on the assumption that there exists some consistent κ (that is, that P is consistent).

Every finite class κ is, of course, recursive.[46a] Let a be the greatest number contained in κ. Then we have for κ

$$x \,\varepsilon\, \kappa \sim (Em, n)[m \leqq x \,\&\, n \leqq a \,\&\, n \,\varepsilon\, \kappa \,\&\, x = m\, Th\, n].$$

Hence κ is recursive. This allows us to conclude, for example, that, even with the help of the axiom of choice (for all types) or the generalized continuum hypothesis, not all propositions are decidable, provided these hypotheses are ω-consistent.

In the proof of Theorem VI no properties of the system P were used besides the following:

1. The class of axioms and the rules of inference (that is, the relation "immediate consequence") are recursively definable (as soon as we replace the primitive signs in some way by natural numbers);
2. Every recursive relation is definable (in the sense of Theorem V) in the system P.

Therefore, in every formal system that satisfies the assumptions 1 and 2 and is ω-consistent there are undecidable propositions of the form $(x)F(x)$, where F is a recursively defined property of natural numbers, and likewise in every extension of such a system by a recursively definable ω-consistent class of axioms. As can easily be verified, included among the systems satisfying the assumptions 1 and 2 are the Zermelo-Fraenkel and the von Neumann axiom systems of set theory,[47] as well as the axiom system of number theory consisting of the Peano axioms, recursive definition (by schema (2)), and the rules of logic.[48] Assumption 1 is satisfied by any system that has the usual rules of inference and whose axioms (like those of P) result from a finite number of schemata by substitution.[48a]

3

We shall now deduce some consequences from Theorem VI, and to this end we give the following definition:

A relation (class) is said to be *arithmetic* if it can be defined in terms of the notions $+$ and $.$ (addition and multiplication for natural numbers)[49] and the logical constants $\vee$, $\overline{}$, (x), and $=$, where (x) and $=$ apply to natural numbers only.[50] The notion "arithmetic proposition" is defined accordingly. The relations "greater than" and "congruent modulo n", for example, are arithmetic because we have

$$x > y \sim \overline{(Ez)}[y = x + z],$$
$$x \equiv y \pmod{n} \sim (Ez)[x = y + z.n \vee y = x + z.n].$$

[46a] [[On page 190, lines 21, 22, and 23, of the German text the three occurrences of α are misprints and should be replaced by occurrences of κ.]]

[47] The proof of assumption 1 turns out to be even simpler here than for the system P, since there is just one kind of primitive variables (or two in von Neumann's system).

[48] See Problem III in *Hilbert 1928a*.

[48a] As will be shown in Part II of this paper, the true reason for the incompleteness inherent in all formal systems of mathematics is that the formation of ever higher types can be continued into the transfinite (see *Hilbert 1925*, p. 184 [[above, p. 387]]), while in any formal system at most denumerably many of them are available. For it can be shown that the undecidable propositions constructed here become decidable whenever appropriate higher types are added (for example, the type ω to the system P). An analogous situation prevails for the axiom system of set theory.

[49] Here and in what follows, zero is always included among the natural numbers.

[50] The definiens of such a notion, therefore, must consist exclusively of the signs listed, variables for natural numbers, $x, y, \ldots$, and the signs 0 and 1 (variables for functions and sets are not permitted to occur). Instead of x any other number variable, of course, may occur in the prefixes.

We now have

Theorem VII. *Every recursive relation is arithmetic.*

We shall prove the following version of this theorem: every relation of the form $x_0 = \varphi(x_1, \ldots, x_n)$, where φ is recursive, is arithmetic, and we shall use induction on the degree of φ. Let φ be of degree s ($s > 1$). Then we have either

1. $\varphi(x_1, \ldots, x_n) = \rho[\chi_1(x_1, \ldots, x_n), \chi_2(x_1, \ldots, x_n), \ldots, \chi_m(x_1, \ldots, x_n)]$[51]

(where ρ and all χ_1 are of degrees less than s) or

2. $\varphi(0, x_2, \ldots, x_n) = \psi(x_2, \ldots, x_n)$,

$$\varphi(k+1, x_2, \ldots, x_n) = \mu[k, \varphi(k, x_2, \ldots, x_n), x_2, \ldots, x_n]$$

(where ψ and μ are of degrees less than s).

In the first case we have

$$x_0 = \varphi(x_1, \ldots, x_n) \sim (Ey_1, \ldots, y_m)[R(x_0, y_1, \ldots, y_m) \; \& \; S_1(y_1, x_1, \ldots, x_n) \; \& \ldots \& \; S_m(y_m, x_1, \ldots, x_n)],$$

where R and S_i are the arithmetic relations, existing by the induction hypothesis, that are equivalent to $x_0 = \rho(y_1, \ldots, y_m)$ and $y = \chi_i(x_1, \ldots, x_n)$, respectively. Hence in this case $x_0 = \varphi(x_1, \ldots, x_n)$ is arithmetic.

In the second case we use the following method. We can express the relation $x_0 = \varphi(x_1, \ldots, x_n)$ with the help of the notion "sequence of numbers" (f)[52] in the following way:

$$x_0 = \varphi(x_1, \ldots, x_n) \sim (Ef)\{f_0 = \psi(x_2, \ldots, x_n) \; \& \; (k)[k < x_1 \rightarrow f_{k+1} = \mu(k, f_k, x_2, \ldots, x_n)] \; \& \; x_0 = f_{x_1}\}.$$

If $S(y, x_2, \ldots, x_n)$ and $T(z, x_1, \ldots, x_{n+1})$ are the arithmetic relations, existing by the induction hypothesis, that are equivalent to $y = \psi(x_2, \ldots, x_n)$ and $z = \mu(x_1, \ldots, x_{n+1})$, respectively, then

$$x_0 = \varphi(x_1, \ldots, x_n) \sim (Ef)\{S(f_0, x_2, \ldots, x_n) \; \& \; (k)[k < x_1 \rightarrow T(f_{k+1}, k, f_k, x_2, \ldots, x_n)] \; \& \; x_0 = f_{x_1}\}. \qquad (17)$$

We now replace the notion "sequence of numbers" by "pair of numbers", assigning to the number pair n, d the number sequence $f^{(n,d)}$ ($f_k^{(n,d)} = [n]_{1+(k+1)d}$), where $[n]_p$ denotes the least nonnegative remainder of n modulo p.

We then have

Lemma 1. If f is any sequence of natural numbers and k any natural number, there exists a pair of natural numbers, n, d such that $f^{(n,d)}$ and f agree in the first k terms.

Proof. Let l be the maximum of the numbers $k, f_0, f_1, \ldots, f_{k-1}$. Let us determine an n such that

$$n \equiv f_i [\mathrm{mod}(1 + (i+1)l!)] \quad \text{for } i = 0, 1, \ldots, k-1,$$

which is possible, since any two of the numbers $1 + (i+1)l!$ ($i = 0, 1, \ldots, k-1$)

[51] Of course, not all $x_1, \ldots, x_n$ need occur in the χ_1 (see the example in footnote 27).

[52] f here is a variable with the [infinite] sequences of natural numbers as its domain of values. f_k denotes the $(k+1)$th term of a sequence f (f_0 denoting the first).

are relatively prime. For a prime number contained in two of these numbers would also be contained in the difference $(i_1 - i_2)l!$ and therefore, since $|i_1 - i_2| < l$, in $l!$; but this is impossible. The number pair n, $l!$ then has the desired property.

Since the relation $x = [n]_p$ is defined by

$$x \equiv n \pmod{p} \;\&\; x < p$$

and is therefore arithmetic, the relation $P(x_0, x_1, \ldots, x_n)$, defined as follows:

$$P(x_0, \ldots, x_n) \equiv (En, d)\{S([n]_{d+1}, x_2, \ldots, x_n) \;\&\; (k)\,[k < x_1 \rightarrow T([n]_{1+d(k+2)}, k, [n]_{1+d(k+1)}, x_2, \ldots, x_n)] \;\&\; x_0 = [n]_{1+d(x_1+1)}\},$$

is also arithmetic. But by (17) and Lemma 1 it is equivalent to $x_0 = \varphi(x_1, \ldots, x_n)$ (the sequence f enters in (17) only through its first $x_1 + 1$ terms). Theorem VII is thus proved.

By Theorem VII, for every problem of the form $(x)F(x)$ (with recursive F) there is an equivalent arithmetic problem. Moreover, since the entire proof of Theorem VII (for every particular F) can be formalized in the system P, this equivalence is provable in P. Hence we have

Theorem VIII. *In any of the formal systems mentioned in Theorem VI*[53] *there are undecidable arithmetic propositions.*

By the remark on page 610, the same holds for the axiom system of set theory and its extensions by ω-consistent recursive classes of axioms.

Finally, we derive the following result:

Theorem IX. *In any of the formal systems mentioned in Theorem VI*[53] *there are undecidable problems of the restricted functional calculus*[54] (that is, formulas of the restricted functional calculus for which neither validity nor the existence of a counterexample is provable).[55]

This is a consequence of

Theorem X. *Every problem of the form $(x)F(x)$ (with recursive F) can be reduced to the question whether a certain formula of the restricted functional calculus is satisfiable* (that is, for every recursive F we can find a formula of the restricted functional calculus that is satisfiable if and only if $(x)F(x)$ is true.

By formulas of the restricted functional calculus (r. f. c.) we understand expressions formed from the primitive signs $\overline{}$, $\vee$, (x), $=$, $x, y, \ldots$ (individual variables), $F(x), G(x, y), H(x, y, z), \ldots$ (predicate and relation variables), where (x) and $=$ apply to individuals only.[56] To these signs we add a third kind of variables, $\varphi(x)$, $\psi(x, y)$,

[53] These are the ω-consistent systems that result from P when recursively definable classes of axioms are added.

[54] See *Hilbert and Ackermann 1928.*

In the system P we must understand by formulas of the restricted functional calculus those that result from the formulas of the restricted functional calculus of *PM* when relations are replaced by classes of higher types as indicated on page 599.

[55] In *1930a* I showed that every formula of the restricted functional calculus either can be proved to be valid or has a counterexample. However, by Theorem IX the existence of this counterexample is *not* always provable (in the formal systems we have been considering).

[56] Hilbert and Ackermann (*1928*) do not include the sign $=$ in the restricted functional calculus. But for every formula in which the sign $=$ occurs there exists a formula that does not contain this sign and is satisfiable if and only if the original formula is (see *Gödel 1930a*).

$\kappa(x, y, z)$, and so on, which stand for object-functions ⟦Gegenstandsfunktionen⟧ (that is, $\varphi(x)$, $\psi(x, y)$, and so on denote single-valued functions whose arguments and values are individuals).[57] A formula that contains variables of the third kind in addition to the signs of the r. f. c. first mentioned will be called a formula in the extended sense (i. e. s.).[58] The notions "satisfiable" and "valid" carry over immediately to formulas i. e. s., and we have the theorem that, for any formula A i. e. s., we can find a formula B of the r. f. c. proper such that A is satisfiable if and only if B is. We obtain B from A by replacing the variables of the third kind, $\varphi(x)$, $\psi(x, y)$, ..., that occur in A with expressions of the form $(\imath z)F(z, x)$, $(\imath z)G(z, x, y)$, ..., by eliminating the "descriptive" functions by the method used in *PM* (I, *14), and by logically multiplying[59] the formula thus obtained by an expression stating about each $F, G, \ldots$ put in place of some $\varphi, \psi, \ldots$ that it holds for a unique value of the first argument [for any choice of values for the other arguments].

We now show that, for every problem of the form $(x)F(x)$ (with recursive F), there is an equivalent problem concerning the satisfiability of a formula i. e. s., so that, on account of the remark just made, Theorem X follows.

Since F is recursive, there is a recursive function $\Phi(x)$ such that $F(x) \sim [\Phi(x) = 0]$, and for Φ there is sequence of functions, $\Phi_1, \Phi_2, \ldots, \Phi_n$, such that $\Phi_n = \Phi$, $\Phi_1(x) = x + 1$, and for every Φ_k $(1 < k \leqq n)$ we have either

1. $$(x_2, \ldots, x_m)[\Phi_k(0, x_2, \ldots, x_m) = \Phi_p(x_2, \ldots, x_m)],$$

$$(x, x_2, \ldots, x_m)\{\Phi_k[\Phi_1(x), x_2, \ldots, x_m] = \Phi_q[x, \Phi_k(x, x_2, \ldots, x_m), x_2, \ldots, x_m]\}, \quad (18)$$

with $p, q < k$,[59a]

or

2. $$(x_1, \ldots, x_m)[\Phi_k(x_1, \ldots, x_m) = \Phi_r(\Phi_{i_1}(\mathfrak{x}_1), \ldots, \Phi_{i_s}(\mathfrak{x}_s))],^{60} \quad (19)$$

with $r < k$, $i_v < k$ (for $v = 1, 2, \ldots, s$),

or

3. $$(x_1, \ldots, x_m)[\Phi_k(x_1, \ldots, x_m) = \Phi_1(\Phi_1, \ldots, \Phi_1\,(0))]. \quad (20)$$

We then form the propositions

$$(x)\overline{\Phi_1(x) = 0}\ \&\ (x, y)[\Phi_1(x) = \Phi_1(y) \rightarrow x = y], \quad (21)$$

$$(x)[\Phi_n(x) = 0]. \quad (22)$$

In all of the formulas (18), (19), (20) (for $k = 2, 3, \ldots, n$) and in (21) and (22) we now replace the functions Φ_i by function variables φ_i and the number 0 by an

[57] Moreover, the domain of definition is always supposed to be the *entire* domain of individuals.

[58] Variables of the third kind may occur at all argument places occupied by individual variables, for example, $y = \varphi(x)$, $F(x, \varphi(y))$, $G(\psi(x, \varphi(y)), x)$, and the like.

[59] That is, by forming the conjunction.

[59a] [The last clause of footnote 27 was not taken into account in the formulas (18). But an explicit formulation of the cases with fewer variables on the right side is actually necessary here for the formal correctness of the proof, unless the identity function, $I(x) = x$, is added to the initial functions.]

[60] The $\mathfrak{x}_i$ $(i = 1, \ldots, s)$ stand for finite sequences of the variables $x_1, x_2, \ldots, x_m$; for example, x_1, x_3, x_2.

individual variable x_0 not used so far, and we form the conjunction C of all the formulas thus obtained.

The formula $(Ex_0)C$ then has the required property, that is,

1. If $(x)[\Phi(x) = 0]$ holds, $(Ex_0)C$ is satisfiable. For the functions $\Phi_1, \Phi_2, \ldots, \Phi_n$ obviously yield a true proposition when substituted for $\varphi_1, \varphi_2, \ldots, \varphi_n$ in $(Ex_0)C$);

2. If $(Ex_0)C$ is satisfiable, $(x)[\Phi(x) = 0]$ holds.

Proof. Let $\psi_1, \psi_2, \ldots, \psi_n$ be the functions (which exist by assumption) that yield a true proposition when substituted for $\varphi_1, \varphi_2, \ldots, \varphi_n$ in $(Ex_0)C$. Let $\mathfrak{J}$ be their domain of individuals. Since $(Ex_0)C$ holds for the functions ψ_i, there is an individual a (in $\mathfrak{J}$) such that all of the formulas (18)–(22) go over into true propositions, (18′)–(22′), when the Φ_i are replaced by the ψ_i and 0 by a. We now form the smallest subclass of $\mathfrak{J}$ that contains a and is closed under the operation $\psi_1(x)$. This subclass ($\mathfrak{J}'$) has the property that every function ψ_i, when applied to elements of $\mathfrak{J}'$, again yields elements of $\mathfrak{J}'$. For this holds of ψ_1 by the definition of $\mathfrak{J}'$, and by (18′), (19′), and (20′) it carries over from ψ_i with smaller subscripts to ψ_i with larger ones. The functions that result from the ψ_i when these are restricted to the domain $\mathfrak{J}'$ of individuals will be denoted by ψ_i'. All of the formulas (18)–(22) hold for these functions also (when we replace 0 by a and Φ_i by ψ_i').

Because (21) holds for ψ_1' and a, we can map the individuals of $\mathfrak{J}'$ one-to-one onto the natural numbers in such a manner that a goes over into 0 and the function ψ_1' into the successor function Φ_1. But by this mapping the functions ψ_i' go over into the functions Φ_i, and, since (22) holds for ψ_n' and a, $(x)[\Phi_n(x) = 0]$, that is, $(x)[\Phi(x) = 0]$, holds, which was to be proved.[61]

Since (for each particular F) the argument leading to Theorem X can be carried out in the system P, it follows that any proposition of the form $(x)F(x)$ (with recursive F) can in P be proved equivalent to the proposition that states about the corresponding formula of the r. f. c. that it is satisfiable. Hence the undecidability of one implies that of the other, which proves Theorem IX.[62]

4

The results of Section 2 have a surprising consequence concerning a consistency proof for the system P (and its extensions), which can be stated as follows:

Theorem XI. *Let κ be any recursive consistent*[63] *class of* FORMULAS; *then the* SENTENTIAL FORMULA *stating that κ is consistent is not* κ-PROVABLE; in particular, the consistency of P is not provable in P,[64] provided P is consistent (in the opposite case, of course, every proposition is provable [in P]).

The proof (briefly outlined) is as follows. Let κ be some recursive class of FORMULAS chosen once and for all for the following discussion (in the simplest case it is the

[61] Theorem X implies, for example, that Fermat's problem and Goldbach's problem could be solved if the decision problem for the r. f. c. were solved.

[62] Theorem IX, of course, also holds for the axiom system of set theory and for its extensions by recursively definable ω-consistent classes of axioms, since there are undecidable propositions of the form $(x)F(x)$ (with recursive F) in these systems too.

[63] "κ is consistent" (abbreviated by "Wid(κ)") is defined thus: $\mathrm{Wid}(\kappa) \equiv (Ex)(\mathrm{Form}(x) \;\&\; \overline{\mathrm{Bew}_\kappa(x)})$.

[64] This follows if we substitute the empty class of FORMULAS for κ.

empty class). As appears from 1, page 608, only the consistency of κ was used in proving that 17 Gen r is not κ-PROVABLE;[65] that is, we have

$$\mathrm{Wid}(\kappa) \rightarrow \overline{\mathrm{Bew}_{\kappa}}(17 \text{ Gen } r), \tag{23}$$

that is, by (6.1),

$$\mathrm{Wid}(\kappa) \rightarrow (x)\, \overline{x\, B_{\kappa}\, (17 \text{ Gen } r)}.$$

By (13), we have

$$17 \text{ Gen } r = Sb(p^{19}_{Z(p)}),$$

hence

$$\mathrm{Wid}(\kappa) \rightarrow (x)\, \overline{x\, B_{\kappa}\, Sb(p^{19}_{Z(p)})},$$

that is, by (8.1),

$$\mathrm{Wid}(\kappa) \rightarrow (x)Q(x), p). \tag{24}$$

We now observe the following: all notions defined (or statements proved) in Section 2,[66] and in Section 4 up to this point, are also expressible (or provable) in P. For throughout we have used only the methods of definition and proof that are customary in classical mathematics, as they are formalized in the system P. In particular, κ (like every recursive class) is definable in P. Let w be the SENTENTIAL FORMULA by which $\mathrm{Wid}(\kappa)$ is expressed in P. According to (8.1), (9), and (10), the relation $Q(x, y)$ is expressed by the RELATION SIGN q, hence $Q(x, p)$ by r (since, by (12), $r = Sb(q^{19}_{Z(p)})$), and the proposition $(x)Q(x\, p)$ by 17 Gen r.

Therefore, by (24), w Imp (17 Gen r) is provable in P[67] (and a fortiori κ-PROVABLE). If now w were κ-PROVABLE, then 17 Gen r would also be κ-PROVABLE, and from this it would follow, by (23), that κ is not consistent.

Let us observe that this proof, too, is constructive; that is, it allows us to actually derive a contradiction from κ, once a PROOF of w from κ is given. The entire proof of Theorem XI carries over word for word to the axiom system of set theory, M, and to that of classical mathematics,[68] A, and here, too, it yields the result: There is no consistency proof for M, or for A, that could be formalized in M, or A, respectively, provided M, or A, is consistent. I wish to note expressly that Theorem XI (and the corresponding results for M and A) do not contradict Hilbert's formalistic viewpoint. For this viewpoint presupposes only the existence of a consistency proof in which nothing but finitary means of proof is used, and it is conceivable that there exist finitary proofs that *cannot* be expressed in the formalism of P (or of M or A).

Since, for any consistent class κ, w is not κ-PROVABLE, there always are propositions (namely w) that are undecidable (on the basis of κ) as soon as Neg(w) is not κ-PROVABLE; in other words, we can, in Theorem VI, replace the assumption of ω-consistency by the following: The proposition "κ is inconsistent" is not κ-PROVABLE. (Note that there are consistent κ for which this proposition is κ-PROVABLE.)

[65] Of course, r (like p) depends on κ.

[66] From the definition of "recursive" on page 602 to the proof of Theorem VI inclusive.

[67] That the truth of w Imp (17 Gen r) can be inferred from (23) is simply due to the fact that the undecidable proposition 17 Gen r asserts its own unprovability, as was noted at the very beginning.

[68] See *von Neumann 1927*.

In the present paper we have on the whole restricted ourselves to the system P, and we have only indicated the applications to other systems. The results will be stated and proved in full generality in a sequel to be published soon.[68a] In that paper, also, the proof of Theorem XI, only sketched here, will be given in detail.

Note added 28 August 1963. In consequence of later advances, in particular of the fact that due to A. M. Turing's work[69] a precise and unquestionably adequate definition of the general notion of formal system[70] can now be given, a completely general version of Theorems VI and XI is now possible. That is, it can be proved rigorously that in *every* consistent formal system that contains a certain amount of finitary number theory there exist undecidable arithmetic propositions and that, moreover, the consistency of any such system cannot be proved in the system.

[68a] [[This explains the "I" in the title of the paper. The author's intention was to publish this sequel in the next volume of the *Monatshefte*. The prompt acceptance of his results was one of the reasons that made him change his plan.]]

[69] See *Turing 1937*, p. 249.

[70] In my opinion the term "formal system" or "formalism" should never be used for anything but this notion. In a lecture at Princeton (mentioned in *Princeton University 1946*, p. 11 [[see *Davis 1965*, pp. 84–88]]) I suggested certain transfinite generalizations of formalisms, but these are something radically different from formal systems in the proper sense of the term, whose characteristic property is that reasoning in them, in principle, can be completely replaced by mechanical devices.

ON COMPLETENESS AND CONSISTENCY

(1931a)

Let Z be the formal system that we obtain by supplementing the Peano axioms with the schema of definition by recursion (on one variable) and the logical rules of the *restricted* functional calculus. Hence Z is to contain no variables other than variables for individuals (that is, natural numbers), and the principle of mathematical induction must therefore be formulated as a rule of inference. Then the following hold:

1. Given any formal system S in which there are finitely many axioms and in which the sole principles of inference are the rule of substitution and the rule of implication, if S contains[1] Z, S is incomplete, that is, there are in S propositions (in

[1] That a formal system S contains another formal system T means that every proposition expressible (provable) in T is expressible (provable) also in S.

[[Remark by the author, 18 May 1966:]]

[This definition is not precise, and, if made precise in the straightforward manner, it does not yield a sufficient condition for the nondemonstrability in S of the consistency of S. A sufficient condition is obtained if one uses the following definition: "S contains T if and only if every meaningful formula (or axiom or rule (of inference, of definition, or of construction of axioms)) of T *is a* meaningful formula (or axiom, and so forth) of S, that is, if S is an extension of T".

Under the weaker hypothesis that Z is recursively one-to-one translatable into S, with demonstrability preserved in this direction, the consistency, even of very strong systems S, *may* be provable in S and even in primitive recursive number theory. However, what can be shown to be unprovable in S is the fact that the rules of the equational calculus applied to equations, between primitive recursive terms, demonstrable in S yield only correct numerical equations (provided that S possesses the property that is asserted to be unprovable). Note that it is necessary to prove this "outer" consistency of S (which for the usual systems is trivially equivalent with consistency) in order to "justify", in the sense of Hilbert's program, the transfinite axioms of a

particular, propositions of Z) that are undecidable on the basis of the axioms of S, provided that S is ω-consistent. Here a system is said to be ω-consistent if, for no property F of natural numbers, $(Ex)\overline{Fx}$ as well as all the formulas $F(i)$, $i = 1, 2, \ldots,$ are provable.

2. In particular, in every system S of the kind just mentioned the proposition that S is consistent (more precisely, the equivalent arithmetic proposition that we obtain by mapping the formulas one-to-one on natural numbers) is unprovable.

Theorems 1 and 2 hold also for systems in which there are infinitely many axioms and in which there are other principles of inference than those mentioned above, provided that when we enumerate the formulas (in order of increasing length and, for equal length, in lexicographical order) the class of numbers assigned to the axioms is definable and decidable ⟦entscheidungsdefinit⟧ in the system Z, and that the same holds of the following relation $R(x_1, x_2, \ldots, x_n)$ between natural numbers: "the formula with number x_1 follows from the formulas with numbers $x_2, \ldots, x_n$ by a single application of one of the rules of inference". Here a relation (class) $R(x_1, x_2, \ldots, x_n)$ is said to be decidable in Z if for every n-tuple $(k_1, k_2, \ldots, k_n)$ of natural numbers either $R(k_1, k_2, \ldots, k_n)$ or $\bar{R}(k_1, k_2, \ldots, k_n)$ is provable in Z. (At present no decidable number-theoretic relation is known that is not definable and decidable already in Z.)

If we imagine that the system Z is successively enlarged by the introduction of variables for classes of numbers, classes of classes of numbers, and so forth, together with the corresponding comprehension axioms, we obtain a sequence (continuable into the transfinite) of formal systems that satisfy the assumptions mentioned above, and it turns out that the consistency (ω-consistency) of any of those systems is provable in all subsequent systems. Also, the undecidable propositions constructed for the proof of Theorem 1 become decidable by the adjunction of higher types and the corresponding axioms; however, in the higher systems we can construct other undecidable propositions by the same procedure, and so forth. To be sure, all the propositions thus constructed are expressible in Z (hence are number-theoretic propositions); they are, however, not decidable in Z, but only in higher systems, for example, in that of analysis. In case we adopt a type-free construction of mathematics, as is done in the axiom system of set theory, axioms of cardinality (that is, axioms postulating the existence of sets of ever higher cardinality) take the place of the type extensions, and it follows that certain arithmetic propositions that are undecidable in Z become decidable by axioms of cardinality, for example, by the axiom that there exist sets whose cardinality is greater than every α_n, where $\alpha_0 = \aleph_0$, $\alpha_{n+1} = 2^{\alpha_n}$.

system S. ("Rules of the equational calculus" in the foregoing means the two rules of substituting primitive recursive terms for variables and substituting one such term for another to which it has been proved equal.)

The last-mentioned theorem and Theorem 1 of the paper remain valid for much weaker systems than Z, in particular for primitive recursive number theory, that is, what remains of Z if quantifiers are omitted. With insignificant changes in the wording of the conclusions of the two theorems they even hold for any recursive translation into S of the equations between primitive recursive terms, under the sole hypothesis of ω-consistency (or outer consistency) of S in this translation.]

On the consistency of arithmetic

JACQUES HERBRAND

(*1931b*)

Dated "Göttingen, 14 July 1931", this paper was sent by Herbrand for publication (to the *Journal für die reine und angewandte Mathematik*) just before he left for a vacation trip in the Alps. The paper was received on 27 July and on that day Herbrand was killed in a fall.

The paper presents a consistency proof for a segment of arithmetic and was intended, no doubt, to be a contribution to the realization of the Hilbert school's program. Consistency is proved for an arithmetic in which the well-formed formula that can be substituted in the induction schema does not contain any bound variable (or, if it does, does not contain any function but the successor function). Consistency proofs likewise requiring some restriction on the induction axiom schema had been presented by Ackermann (*1924*) and von Neumann (*1927*). Herbrand's proof remains relatively simple and straightforward because he has at his disposal his powerful fundamental theorem (*1930*). § 1 consists of a very clear presentation of this theorem.

Throughout his work Herbrand applies the word "intuitionistic" to the methods that he considers admissible in metamathematics, and, although there may be some variations in the meaning that he gives to the term, this meaning is on the whole much closer to that of Hilbert's word "finitary" ("finit") than to "intuitionistic" as applied to Brouwer's doctrine. Herbrand writes, for instance, that Hilbert had undertaken to solve metamathematical problems "exclusively through intuitionistic arguments" ("uniquement par des raisonnements intuitionnistes", *1931a*, p. 187), and in the paper below (p. 622, line 1u) he writes that the statement and the proof of his fundamental theorem are "intuitionistic". The identification of "intuitionistic" with "finitary" was then current among members of the Hilbert school (see, for example, *von Neumann 1927*, p. 2; the distinction between the two notions was to be made explicit a few years later (*Hilbert and Bernays 1934*, pp. 34 and 43, *Bernays 1934a*, *1935*, and *1938*).

A key part of Herbrand's proof is the elimination of the induction axiom schema (with no apparent variables) through the introduction of functions; the definition conditions for each of these functions are such that, for every set of arguments, a well-determined number can be proved in a finitary way to be the value of the function (Group C of hypotheses). The functions are, in fact, (general) recursive functions, and here is the first appearance of the notion of recursive (as opposed to primitive recursive) function. It is interesting to see how, a few months earlier, Herbrand had been led to this notion by his conception of "intuitionism". (Let us note in passing that, it seems, Herbrand's knowledge of Brouwer's ideas is not first-hand, but derived from Hilbert.) He writes (*1931a*, p. 187): "In its extreme form this theory ⟦intuitionism⟧ allows only arguments dealing with the integers (or with objects that can actually be numbered by means of integers) and satisfying the following conditions: all the functions that are introduced must be actually computable for all values of their arguments, by means of operations that are completely described in advance; every time that we come to say 'A proposition is true for every integer', this means 'We can actually verify it for every integer'; every time that we come to say 'There exists an integer x that has such and such a property', this means implicitly 'In what precedes we gave a means of constructing such an x'".

Gödel (*1934*, p. 26) gives, as "suggested by Herbrand", the following definition of a recursive function: "If φ denotes an unknown function and $\psi_1, \ldots, \psi_k$ are known functions, and if the ψ's and the φ are substituted in one another in the most general fashions and certain pairs of the resulting expressions are equated, then, if the resulting set of functional equations has one and only one solution for φ, φ is a recursive function".

Answering a query of the editor's about the history of Herbrand's suggestion, Professor Gödel wrote in a letter dated 23 April 1963:

"I have never met Herbrand. His suggestion was made in a letter in 1931, and it was formulated *exactly* as on page 26 of my lecture notes, that is, without any reference to computability. However, since Herbrand was an intuitionist, this definition for him evidently meant that there exists a *constructive* proof for the existence and unicity of φ. He probably believed that such a proof can be given only by exhibiting a computational procedure. (Note that, if Church's thesis is correct, it is *true* that, if $(\exists!\varphi)A(\varphi)$ holds intuitionistically, then $(\iota\varphi)A(\varphi)$ is general recursive, although, in order to obtain the computational procedure for φ, it may be necessary to add some equations to those contained in $A(\varphi)$.) So I don't think there is any discrepancy between his two definitions as he meant them. What he failed to see (or to make clear) is that the computation, for *all* computable functions, proceeds by *exactly the same rules*. It is this fact that makes a precise definition of general recursiveness possible. Unfortunately I do not find Herbrand's letter among my papers. It probably was lost in Vienna during World War II, as many other things were. But my recollection is very distinct and was still very fresh in 1934."

Herbrand's consistency proof for a fragment of arithmetic still belongs to the period that preceded Gödel's famous result (*1931*). He probably started to write his paper before Gödel's paper reached him. But he had ample opportunity to examine Gödel's result and he wrote a last section dealing with it. Herbrand explains very clearly why this result does not hold for the fragment of arithmetic that he considers: whatever functions may be introduced by definition conditions of Group E, these functions never include a function that enumerates them (this is shown by a diagonal argument), hence the metamathematical description of the system cannot be projected into the system.

The translation is by the editor, and it is printed here by permission of Walter de Gruyter and Co.

When published, Herbrand's text was preceded by a bibliographical note written by Hans Hasse in German. Here is its translation:

"Jacques Herbrand, born on 12 February 1908 in Paris, had a fatal accident on 27 July 1931 while mountain climbing in the Alps. The same day the editors received the manuscript published below.

"He spent the last six months of his life at German universities, in close contact with a number of German mathematicians and engaged in a lively exchange of ideas with them. They were all profoundly impressed by his noble personality, which was endowed with rich scientific gifts. With him an extraordinarily gifted mind has perished in the bloom of youth. The beautiful and important results that he found in the field of number theory and of mathematical logic and the fruitful ideas that he expressed in mathematical conversations justified the greatest hopes. The science of mathematics has suffered a severe and irreparable loss by his untimely death."

In two previous papers[1] we stated and proved a general theorem that makes it possible to solve many problems in metamathematics, and we gave some applications of it. In the present paper we wish to apply it to the problem of the consistency of arithmetic more completely than we did in D I. The fundamental result is stated at the beginning of § 3.

§ 1. Statement of a general theorem

Let us first recall the statement of the theorem in question.

(**a**) We use Russell's signs, with slight changes: $\vee$, $\sim$, $\times$ (and), $\rightarrow$ (for implication), $\equiv$, (Ex) (instead of Russell's inverted E), and (x),[2] as well as Russell's rules of inference, which we have slightly modified. In particular, we shall not use Hilbert's logical ε function in propositions, and instead of his transfinite axiom we shall use the axioms and rules of Russell that are equivalent to it. We complete Russell's formal system by allowing the use of functions (in the ordinary sense of the word); the functions of 0 variable are the constants. For more details see D I, chaps. 1 and 3.

In a given theory we have certain elementary propositions and functions, and all propositions considered within the theory are combinations of logical signs with elementary propositions and functions. We have, moreover, certain propositions called *hypotheses*, and the propositions true in the theory are those that we can obtain from the hypotheses by applying the rules of inference.

If a proposition can be proved by application of these rules without the use of any hypothesis, we call it an *identity*. If a proposition P is true in a given theory, there exist certain hypotheses $H_1, H_2, \ldots, H_n$ such that $H_1 \times H_2 \times \cdots \times H_n . \rightarrow P$ is an identity; and conversely (see D I, chap. 3, 2.4), if this is the case, P is true in the theory in question.

The part of a proposition on which a logical sign operates is called the *scope* of the

[1] *1930*, *1931*. In what follows these papers are denoted by D I and D II. We take advantage of the present opportunity to make the following two remarks about them.

(a) The theorem in chap. 3, 3.3, of D I is correct only if $A(x)$ contains no restricted variables (only under this condition is the expansion of $(x)A(x)$ true).

(b) In D II, at the bottom of p. 32, we point out that the elements of a domain can be "equated". This calls for a clarification, since the values of the index functions are not well-determined in the new domain. Among all elements α_1 equal to a given one, we choose one, β. These β form the new domain. In this domain the value of a function $f(\beta_1, \beta_2, \ldots, \beta_n)$ (which may be either an elementary descriptive function or an index function) will be the β equal to the value of $f(\beta_1, \beta_2, \ldots, \beta_n)$ in the former domain. We thus see that every proposition true in the former domain remains true in the new one.

[2] We recall that the signs $\times$, $\rightarrow$, and $\equiv$ can be defined by means of the signs $\vee$ and $\sim$, these alone being taken as primitive. Let us recall their meaning. $\vee$, $\sim$, $\times$, $\rightarrow$, $\equiv$, (Ex), (x) mean: or, not, and, implies, is equivalent to, there exists an x such that, for all x, respectively.

sign. Thus in $(x) \sim A(x, y) . \vee . (Ey) \sim B(y)$ the scope of (x) is $\sim A(x, y)$, that of (Ey) is $\sim B(y)$, that of the first $\sim$ is $A(x, y)$, that of the second $\sim$ is $B(y)$. Instead of the scope of (x) or (Ey) we shall also speak of the scope of x or y.

The variables that occur in signs of the form (x) or (Ex) are said to be *apparent*; the others are said to be *real*.

An apparent variable x occurring either in a sign (Ex) that is within the scope of an even number of signs $\sim$ or in a sign (x) within the scope of an odd number of signs $\sim$ is said to be *restricted*. Otherwise, it is said to be *general*. Thus in $\sim (x)A(x, x) . \vee . (Ey) \sim \sim (z)A(y, z)$, x and y are restricted, z is general.

(b) When a proposition contains neither apparent variables nor functions, there is a simple criterion that enables us to decide whether or not it is an identity.

Let us consider the elementary propositions occurring in this proposition (they may or may not contain variables; of course, we consider those that contain the same variables in the same order to be identical); let us assign to each of them a *logical value*, that is, let us associate with each of them one of the signs V or F, and let us agree that to all propositions without apparent variables formed from these propositions we assign a logical value according to the following rules (see footnote 2):

(α) If F (or V) is associated with P, then V (or F) is associated with $\sim P$;

(β) F is associated with $P \vee Q$ only if F is associated with P and with Q.

It can then be proved (D I, chap. 1, 5.21) that P is an identity if and only if P has the logical value V whatever may be the logical values assigned to the elementary propositions. Hence, if P contains n distinct elementary propositions, we need 2^n trials to decide the truth of P.

(c) Let us consider an arbitrary proposition P (we assume that all the variables are represented by different letters). We shall apply the remarks that follow to the proposition $(x)(Ey) . A(x, y)$, which we shall take as an example (A is a proposition with no variable other than x and y, formed from elementary propositions and functions).

We shall consider functions that include, first, the elementary functions occurring in P and, then, new functions obtained as follows (it should be clear that by a function we understand a mere logical sign and that we are not concerned, at least for the time being, with its "values"): with every restricted variable ⟦x⟧ of P we associate a function whose arguments are the real variables of P and the general variables having a scope containing a quantifier in x, (x) or (Ex); if there are no such real or general[2a] variables, the function has 0 argument (and is a constant). For example, for $(x)(Ey) . A(x, y)$ we have the function $\varphi(x)$ associated with y.

Let us now consider all the distinct elementary propositions occurring in P (here two elementary propositions differing only in their variables are considered identical), denoting them by $A_i(x_1, x_2, \ldots, x_{n_i})$; let us also consider all distinct functions (elementary functions as well as those just introduced, with the same convention), denoting them by $f_i(x_1, x_2, \ldots, x_{m_i})$. For example, if $A(x, y)$ is $B(x, x) \times B(y, y) . \vee . \sim B(x, y)$, we have $B(x, y)$ as the only elementary proposition and $\varphi(x)$ as the only function.

Let us now consider finite sets of the letters $a_0, a_1, \ldots, a_n$; these sets we shall call

[2a] ⟦Instead of "générales" the original text has "restreintes", which is an oversight.⟧

domains; let us assume that we assign a "value" to each $f_i(a_{\alpha_1}, a_{\alpha_2}, \ldots, a_{\alpha_{n_i}})$, that is, we associate one of the a_i with this collection of signs. Let us assume that this has been done for some of the $f_i(a_{\alpha_1}, a_{\alpha_2}, \ldots, a_{\alpha_{n_i}})$ (not necessarily for all).

From our elementary functions we can construct other functions by "substitution"; for example, from the functions $f_1(x)$, $f_2(x)$, and $f_3(x, y)$ we can construct $f_3(f_2(x), f_1(y))$, and so on. What we should understand by the value of such a function is clear. Let us denote by $F_i(x_1, x_2, \ldots, x_{p_i})$ any one of these functions (old or new).

To every function we assign a *height* in the following way: elementary functions have height 1; if the maximum of the heights of $F_{u_1}, F_{u_2}, \ldots, F_{u_{m_i}}$ (the variables are omitted) is h, then the height of $f_i(F_{u_1}, F_{u_2}, \ldots, F_{u_{m_i}})$ is $h + 1$.

If in the domain there exist values of a sufficient number of the $f_i(a_{\alpha_1}, a_{\alpha_2}, \ldots, a_{\alpha_{m_i}})$ for us to compute the values of all $F_i(a_0, a_0, \ldots, a_0)$ of at most height h and if h is the greatest number having this property, the domain is said to be of order h, and a_0 is the *initial element* (this element is not always uniquely determined).

In our example let us set $\varphi_1(x) = \varphi(x)$, $\varphi_2(x) = \varphi(\varphi(x)), \ldots, \varphi_{n+1}(x) = \varphi(\varphi_n(x)), \ldots$; then the functions of at most height h are $\varphi_1(x), \varphi_2(x), \ldots, \varphi_h(x)$. A domain of order h will be formed by the letters $a_0, a_1, a_2, \ldots, a_h$ if to $\varphi(a_i)$ we assign the value a_{i+1} (to express this we shall often write $\varphi(a_i) = a_{i+1}$. We see that $\varphi(a_h)$ has no value in the domain.

Let us now associate a logical value with every $A_i(a_{\alpha_1}, a_{\alpha_2}, \ldots, a_{\alpha_{n_i}})$.

Let us consider P, delete in it all signs of the form (x) and (Ex), and then replace each restricted variable by the associated function. We thus obtain a certain proposition without apparent variables (for example, $A(x, \varphi(x))$.

Let us replace the variables by some of the a_i in any way whatsoever and, in the expression thus obtained, let us replace the functions of the form $F_i(a_{\alpha_1}, a_{\alpha_2}, \ldots, a_{\alpha_{n_i}})$ by their values. (This is not possible for every choice of the a_i; in our example we obtain $A(a_i, a_{i+1})$, and the operation is impossible if we replace x by a_h.)

Since the $A_i(a_{\alpha_1}, a_{\alpha_2}, \ldots, a_{\alpha_{n_i}})$ have a logical value, the same holds of the proposition obtained. If this logical value is always V, in whatever way the variables are replaced by the a_i, then P is said to be true in the domain of order h considered. (We see that this domain is characterized by the values of the functions and the logical values of the $A_i(a_{\alpha_1}, a_{\alpha_2}, \ldots, a_{\alpha_{n_i}}$.)

We then have the following fundamental theorem:

(α) *If P is an identity, we can find, given the proof of P, a number h such that $\sim P$ is not true in any domain of order h;*

(β) *If P is not an identity, for every h we can construct a domain of order h in which $\sim P$ is true.*

The statement and the proof of this theorem are intuitionistic.[3]

[3] By an intuitionistic argument we understand an argument satisfying the following conditions: in it we never consider anything but a given finite number of objects and of functions; these functions are well-defined, their definition allowing the computation of their value in a univocal way; we never state that an object exists without giving the means of constructing it; we never consider the totality of all the objects x of an infinite collection; and when we say that an argument (or a theorem) is true for all these x, we mean that, for each x taken by itself, it is possible to repeat the general argument in question, which should be considered to be merely the prototype of these particular arguments.

(**d**) If for every h we can construct a domain of order h in which P is true, we say that there is an *infinite domain* in which P is *true*. We say that P is *false* in an infinite domain if $\sim P$ is true in it. (This expression "infinite domain" is merely an abbreviation; in order to have the intuitionistic meaning of the statements we have to return to the precise statements (α) and (β).) We can then state:

P is not an identity if and only if there exists an infinite domain in which P is false.

In this form the theorem is a more precise statement of the well-known Löwenheim-Skolem theorem;[4] in fact, if there exists, in our sense, an infinite domain in which P is true, we can easily obtain from it by nonintuitionistic procedures a denumerable set in which P is true for certain values of the functions, that is, a model ⟦réalisation⟧ of P, and conversely. But

(α) Our statement is completely intuitionistic; we even have the means, once P has been proved, of finding a number h such that in the "partial" domain of order h, that is, at the hth step in the construction of the infinite domain, it is impossible to make P false;

(β) We do not restrict ourselves to propositions in canonical form, as Löwenheim and, even more, Skolem do;

(γ) Löwenheim and Skolem implicitly assume that, once we have a model of $\sim P$, P cannot be proved; but this means that the consistency of mathematics (or at least of arithmetic) is implicitly assumed; our theorem, on the contrary, will allow us to investigate the consistency of arithmetic.

(**e**) Let us now consider a theory with given hypotheses.

If we can determine an infinite domain such that all the hypotheses of the theory are true in it, the theory is not contradictory.

For, if the theory were contradictory, there would be n hypotheses $H_1, H_2, \ldots, H_n$ such that $\sim . H_1 \times H_2 \times \cdots \times H_n$ would be an identity (see D I, chap. 5, 6.5 ⟦above, p. 561⟧). But, since $H_1, H_2, \ldots, H_n$ are true in the infinite domain, $H_1 \times H_2 \times \cdots \times H_n$ would also be true in it (this is easy to see); hence $\sim . H_1 \times H_2 \times \cdots \times H_n$ would be false in it, which is impossible if it is an identity.

The foregoing theorems have a large number of applications (see D I and especially D II). In particular they allow us

(1) To solve many special cases of the decision problem;

(2) To prove rigorously (intuitionistically, which Löwenheim did not do) that the decision problem can always be reduced to the case in which there are only *three elementary propositions of two variables* or *one of three variables.*

By means of them we shall deal with the consistency of arithmetic (see already D I, chap. 5, 6.8 ⟦above, pp. 563–567⟧).

§ 2. The axioms of arithmetic

The theory that we shall study, which is the formal translation of classical arithmetic, has only one elementary proposition (of two variables), $x = y$; it has one constant, 0, one function of one variable, $x + 1$, and other functions that we shall indicate later.

[4] See *Löwenheim 1915* and *Skolem 1920.*

The hypotheses are the following:

$$\left.\begin{array}{l} x = x, \\ x = y \,.\!\rightarrow\!.\, y = x, \\ x = y \times y = z \,.\!\rightarrow\!.\, x = z, \\ x = y \,.\!\equiv\!.\, x + 1 = y + 1, \\ \sim . x + 1 = 0. \end{array}\right\} \quad \text{(Group A)}$$

Then come all the hypotheses obtained when in

$$\Phi(0) \times :(x) . \Phi(x) \rightarrow \Phi(x + 1) :\rightarrow. (x)\Phi(x)$$

$\Phi(x)$ is replaced by any proposition containing the variable x (Group B, mathematical induction).

We can also introduce any number of functions $f_i(x_1, x_2, \ldots, x_{n_i})$ together with hypotheses such that

(a) *The hypotheses contain no apparent variables*;

(b) *Considered intuitionistically,*[5] *they make the actual computation of the* $f_i(x_1, x_2, \ldots, x_{n_i})$ *possible for every given set of numbers, and it is possible to prove intuitionistically that we obtain a well-determined result* (Group C).

As a specific example, let us assume that we have already introduced a certain number of such functions; then a new one, $f(x)$, can be introduced, together with the new hypotheses

$$f(0) = \alpha,$$
$$f(x + 1) = \beta(f(x)),$$

α and $\beta(y)$ being functions formed from previously introduced functions. This is the ordinary schema of definition by recursion.

But we can introduce much more complicated definition schemata, for example, multiple recursions on several variables. Thus we can introduce Hilbert's function[6] together with the hypotheses

$$\varphi(n + 1, a, b) = \varphi(n, a, \varphi(n + 1, a, b - 1)),$$
$$\varphi(n, a, 1) = a,$$
$$\varphi(0, a, b) = a + b.$$

Let us finally introduce, following Hilbert,[7] a fourth group of hypotheses, which does not belong to classical arithmetic:

Let $A(x)$ be a proposition without apparent variables; if it can be proved by intuitionistic procedures that this proposition, intuitionistically considered,[5] *is true for every x, then we add $(x)A(x)$ to the hypotheses.* (Group D)

§ 3. The problem of consistency

The fundamental result is the following:

The hypotheses A, B, C, *and* D *yield a consistent theory if we assume that, in the hypotheses* B, $\Phi(x)$ *contains no apparent variables.*

[5] This expression means: when they are translated into ordinary language, considered as a property of integers and not as a mere symbol.

[6] See *Hilbert 1925* [[above, p. 388]].

[7] See *Hilbert 1930*, p. 491.

The proof is very simple.

For every $\Phi(x)$ without apparent variables we introduce a new function $\varepsilon(x, y_1, y_2, \ldots, y_n)$, the y_i being the variables of $\Phi(x)$ that are distinct from x. We write $\varepsilon(x)$ for short, omitting the y_i. We introduce the following hypotheses:

$$\varepsilon(0) = 0, \quad (1)$$

$$\Phi(0) . \sim \Phi(x + 1) . \varepsilon(x) = 0 .\rightarrow. \varepsilon(x + 1) = x + 1, \quad (2)$$

$$\sim[\Phi(0) . \sim \Phi(x + 1) . \varepsilon(x) = 0] :\rightarrow. \varepsilon(x + 1) = \varepsilon(x), \quad (3)$$

$$\varepsilon(x) = y + 1 .\rightarrow: \varepsilon(y + 1) = y + 1 \times \varepsilon(y) = 0. \quad (4)$$

(Group E)

Translated into ordinary language, they mean that, if a is the least number for which Φa is false, $\varepsilon(x) = 0$ for $x < a$ and $\varepsilon(x) = a$ for $x \geqq a$. If the number a does not exist, $\varepsilon(x) = 0$ for every x.

We can then prove that

(a) *The hypotheses* B *follow from these new hypotheses.*

To see that, we have merely to formalize the following argument.

Let us assume that $\Phi(0)$ and $(x) . \Phi(x) \rightarrow \Phi(x + 1)$ are true and also that there exists an x such that $\sim \Phi(x + 1)$ is true, then it follows from E(2) and E(3) that

$$\varepsilon(x + 1) = x + 1 \quad \text{if } \varepsilon(x) = 0,$$

$$\varepsilon(x + 1) = \varepsilon(x) \quad \text{if } \varepsilon(x) \neq 0.$$

Let us then take $\varepsilon(x + 1) = y + 1$. By E(4) we have

$$\varepsilon(y + 1) = y + 1,$$

$$\varepsilon(y) = 0.$$

Because of E(2) and E(3), this entails that $\sim \Phi(y + 1)$ is true (since $\Phi(0)$ is assumed to be true). We have $y \neq 0$, since otherwise $\Phi(0)$ and $\sim \Phi(0 + 1)$ would be true, which is contrary to the assumption made. Hence we can take $y = z + 1$. By this assumption, $\sim \Phi(z + 1)$ and $\Phi(0)$ are true. E(2), together with the fact that $\varepsilon(z + 1) \neq z + 1$, entails that $\varepsilon(z) \neq 0$; then E(3) entails that $\varepsilon(z) = \varepsilon(z + 1)$, which leads to a contradiction.

(b) *The new hypotheses are of the form* C *if* $\Phi(x)$ *contains no apparent variables.*

For in this case we can actually determine whether $\Phi(x)$ is true or not, and the hypotheses E allow us to actually compute the $\varepsilon(x)$ in a univocal and consistent way.

It is therefore sufficient to prove the consistency of the hypotheses A, C, and D. But this follows immediately from the general theorem. We can indeed construct, by intuitionistic procedures, a domain of order h that consists of ordinary integers and in which the hypotheses are true; this domain consists of the values (in the intuitive sense of the word) of the functions of at most height h that are constructed from the elementary functions, here the latter being merely the function $x + 1$ and the functions introduced in the hypotheses C; 0 is the initial element. According to what has been said above, it can then be proved by intuitionistic procedures that all the hypotheses are true in that domain.

The foregoing provides a procedure that enables us to construct, for every h, a domain of order h in which all the hypotheses are true. These hypotheses are therefore true in an infinite domain, and the theory cannot be inconsistent (compare D I, chap. 5, 6.8 ⟦above, pp. 563–567⟧; the present argument is simpler because we reduce the hypotheses B to the hypotheses C).

We see how simple the proof is, once we have the general theorem. However, the question is not entirely solved. Still to be considered is the case in which, in Group B, $\Phi(x)$ contains apparent variables. We can only say:

In case $\Phi(x)$ *contains apparent variables we can add the hypotheses* B *without introducing a contradiction if the only function occurring in* $\Phi(x)$ (*besides the constant* 0) *is the function* $u + 1$.

For we can introduce the function $\delta(x)$ together with the hypotheses

$$\delta(0) = 0,$$

$$\delta(x + 1) = x.$$

From this it follows that the proposition

$$x = 0 \,.\vee.\, (Ey)(x = y + 1)$$

is true. However, we have shown in D I (chap. 4, 8.1) that the propositions of the form $x \neq x + 1 + 1 + \cdots + 1$ follow from some hypotheses of Group B in which the $\Phi(x)$ have no apparent variables. Finally, we have proved in D I (chap. 4, 8.1) that in the case under consideration the propositions above, together with the hypotheses A, entail the hypotheses B. Q. e. d.

We would have a still more general problem by admitting other schemata of definition by recursion. For example, we could introduce a function $f(x)$ together with the hypotheses

$$A(f(0)),$$

$$B(f(x), f(x + 1)),$$

$(Ex)A(x)$ and $(x)(Ey) . B(x, y)$ being true propositions, proved without the use of the new function.[8]

§ 4. Comparison with a theorem of Gödel's

Following Gödel's argument, we can prove that[9]

The consistency of a theory cannot be proved by arguments formalizable in the theory, whenever the theory contains arithmetic.

The exact meaning of this proposition is easy to understand. For the details one

[8] Bernays drew our attention to the fact that, if an arithmetic containing hypotheses A, B, and, as functions, those occurring in $A(x)$ and $B(x, y)$, as well as addition and multiplication, were proved consistent, it would follow that we could add the new hypotheses without contradiction, for any argument carried out with the new function $f(x)$ could be carried out in the arithmetic just described.

[9] See *Gödel 1931*. To understand this proposition, one has to imagine that all signs occurring in metamathematics are represented by objects of the theory being considered, for example, by integers in arithmetic; the properties of these signs and the relations among them will then be represented by certain propositions of the theory; every argument in the theory in question carried out with these objects and these propositions will correspond to a metamathematical argument, of which it will be, in a way, the translation.

can consult Gödel's paper, where the theorem is proved for the system of Russell and Whitehead. Let us recall briefly the essence of his argument.

We can intuitionistically number all the propositions containing just one real variable and all the proofs of the theory considered. Let $P(x, y, z)$ be a proposition, intuitionistically defined, whose meaning is: proof number x is a proof of proposition number y for the value z of its variable. (In his paper Gödel effectively constructs this function for the theory he considers.) Let us observe that the computation that we have to do in order to check whether $P(x, y, z)$ is true or false for given x, y, and z—a computation that is intuitionistically carried out—can also be formally carried out in the theory; hence, if $P(x, y, z)$ is true, it is provable in the theory. Let us assume that the theory is consistent. Then, if β is the number of the proposition $(x) . \sim P (x, y, y)$, the proposition $P(x, \beta, \beta)$ cannot be true, for it would mean that proof number x is a proof of $(x) . \sim P(x, \beta, \beta)$; hence $\sim P(x, \beta, \beta)$ could be proved in the theory, and we would have a contradiction. Moreover, $(x) . \sim P(x, \beta, \beta)$ is not provable in the theory, for, if proof number y were a proof of it, $P(y, \beta, \beta)$ would be true (by definition); hence, because of what we have said, it would be provable in theory, and we would again have a contradiction. If we formalize the foregoing considerations in the theory (and they are intuitionistic), we obtain the following result: w being the translation[10] of the proposition "The theory is consistent", $w \rightarrow \sim P(x, \beta, \beta)$ is a true proposition of the theory (x being a variable). If w were provable in the theory so would be $\sim P(x, \beta, \beta)$, hence also $(x) . \sim P(x, \beta, \beta)$, and we have just seen that this would make the theory inconsistent. Hence *w is unprovable*.

Let us apply this method to the arithmetic that contains the hypotheses A, B, and C (leaving out the hypotheses D for the sake of simplicity). Our arguments, as well as those necessary to establish the fundamental theorem, are formalizable in this theory. But, if we want to carry out Gödel's argument, we have to consider a definite theory. In Group C we merely have a description of the hypotheses that may be introduced. We must consider a definite group of schemata of hypotheses of type C, say, Group C′. We shall see that, if we wish to carry out Gödel's argument, we must introduce other schemata for the definition of functions; it is therefore impossible to apply his considerations to our arithmetic.

One could object that it is perhaps possible to describe at once all the schemata that are included in C for the construction of functions. This would mean that it is possible to describe outright all intuitionistic procedures for the construction of number-theoretic functions. But this is impossible. For, if this were the case, we could, by an intuitionistic procedure, number all the functions of just one variable, $f_1(x), f_2(x), \ldots, f_n(x), \ldots$; then $f_x(x) + 1$ would be an intuitionistically defined function that would not be among the functions previously listed, hence a contradiction.

Let us apply this result to all functions definable by C′ and those obtained by compounding these functions with one another and with $x + 1$. The function $f_x(x) + 1$ cannot be among these functions. But, to carry out Gödel's argument, we have to number all objects occurring in proofs; we are thus led to construct the function of two variables $f_y(x)$; this justifies what we were saying above, namely, that it is impossible, in an arithmetic containing the hypotheses C′, to formalize Gödel's argument about this arithmetic.

[10] See footnote 9.

Let us make two more remarks.

(a) Every intuitionistic argument can always be carried out with ordinary integers, because we can always replace the objects considered in the argument by ordinary integers, for example, if we number them. It seems to us almost certain that every intuitionistic argument can then be carried out in an arithmetic containing the hypotheses A, B, and C [[only]]. But the justification of this point (a *proof*, in the mathematical sense of the word, is obviously impossible, just as it is for any statement involving the totality of *all* intuitionistic arguments) would lead us to a detailed discussion, which we shall perhaps undertake some other time. It would show that the hypotheses of Group D follow from those that precede them.

(b) It seems to us impossible, contrary to Gödel's opinion,[11] that there could be intuitionistic arguments not formalizable in ordinary analysis. Thus we believe that the arithmetic containing the hypotheses A, B, C, and D is a part of ordinary analysis. It seems to us probable that the question stands as follows: it will never be possible to give an example of an intuitionistic argument not formalizable in ordinary analysis, and also it will always be impossible to prove that there exists no such argument; for we have shown that it is impossible to describe all intuitionistic procedures of reasoning (since it is impossible to describe all procedures for the construction of functions). *There might be some sort of logical postulate here.* It would follow that it is impossible to prove the consistency of ordinary analysis.

It is not even impossible that every intuitionistic argument could be carried out in an arithmetic containing the hypotheses A and B and allowing in C only ordinary addition and multiplication.[12] If this were so, the consistency of ordinary arithmetic would already be unprovable.

[11] See *Gödel 1931*, p. 197 [[above, p. 615]].

[12] See *Gödel 1931*, Theorem VII [[p. 611, above]], and also footnote 8 above.

References

Throughout this volume, an author's name followed by a year number, both in italics, denotes an entry in the present list of references. Thus "*Herbrand 1928*" denotes the first publication listed under Herbrand in this list. When the context leaves no doubt as to the author, the year number alone is used. For papers that originally were addresses or lectures, the year indicated is that in which they were delivered; for communications to learned societies, it is that in which they were made; for other papers, it is that of the complete volume of the periodical in which they appear (there are a few exceptions, with irregular publications—any compiler of a bibliography knows that no rule is without exception). For one author, additional titles in the same year are distinguished by *a*, *b*, and so on; an attempt has been made to follow the actual chronological order. A star in front of an entry indicates that the publication is included in the present volume. When a book had several editions or when a paper was subsequently reprinted, the fact is indicated only when this indication is considered a real service to the reader.

Aanderaa, Stål

See Dreben, Burton, and Stål Aanderaa; also Dreben, Burton, Peter Andrews, and Stål Aanderaa.

Ackermann, Wilhelm

1924 Begründung des "tertium non datur" mittels der Hilbertschen Theorie der Widerspruchsfreiheit, *Mathematische Annalen 93*, 1–36.

1928 *Zum Hilbertschen Aufbau der reellen Zahlen, *ibid. 99*, 118–133 (493–507 of the present volume).

1928a Über die Erfüllbarkeit gewisser Zählausdrücke, *ibid. 100*, 638–649.

1940 Zur Widerspruchsfreiheit der Zahlentheorie, *ibid. 117*, 162–194.

1954 *Solvable cases of the decision problem* (North-Holland, Amsterdam); 2nd printing 1962.

See Hilbert, David, and Wilhelm Ackermann.

Addison, John West

1958 Separation principles in the hierarchies of classical and effective descriptive set theory, *Fundamenta mathematicae 46*, 123–135.

——— Leon Henkin, and Alfred Tarski

1963 (eds.) *The theory of models, Proceedings of the 1963 International Symposium at Berkeley* (North-Holland, 1965).

Aleksandrov, Pavel Sergeievich (Павел Сергеевич Александров)

1916 Sur la puissance des ensembles mesurables B, *Comptes rendus hebdomadaires des séances de l'Académie des sciences* (Paris) *162*, 323–325.

Andrews, Peter

See Dreben, Burton, Peter Andrews, and Stål Aanderaa.

Artin, Emil, and Otto Schreier

1926 Algebraische Konstruktion reeller Körper, *Abhandlungen aus dem mathematischen Seminars der Hamburgischen Universität 5* (1927), 85-99.

Asser, Günter

1957 Theorie der logischen Auswahlfunktionen, *Zeitschrift für mathematische Logik und Grundlagen der Mathematik 3*, 30–68.

Baire, René
See Borel, Emile, *1905a*.
Bar-Hillel, Yehoshua
See Fraenkel, Abraham A., and Yehoshua Bar-Hillel.
Bauer-Mengelberg, Stefan
1965 Review of *Gödel 1962*, *The journal of symbolic logic 30*, 359–362.
1966 Review of Elliott Mendelson's translation of *Gödel 1931* in *Davis 1965*, *ibid. 31*, 486–489.
Behmann, Heinrich
1922 Beiträge zur Algebra der Logik, insbesondere zum Entscheidungsproblem, *Mathematische Annalen 86*, 163–229.
Belinfante, Maurits Joost
1929 Über einen Grenzwertsatz aus der Theorie der unendlichen Folgen, *Mathematische Annalen 101*, 312–315.
1929a Zur intuitionistischen Theorie der unendlichen Reihen, *Sitzungsberichte der Preussischen Akademie der Wissenschaften*, 639–660.
1930 Über eine besondere Klasse von non-oszillierenden Reihen, *Koninklijke Akademie van wetenschappen te Amsterdam, Proceedings of the section of sciences 33*, 1170–1179.
1930a Absolute Konvergenz in der intuitionistischen Mathematik, *ibid.*, 1180–1184.
1931 Die Hardy-Littlewoodsche Umkehrung des Abelschen Stetigkeitssatzes in der intuitionistischen Mathematik, *ibid. 34*, 401–412.
1931a Über die Elemente der Funktionentheorie und die Picardschen Sätze in der intuitionistischen Mathematik, *ibid.*, 1395–1397.
1932 Über den intuitionistischen Beweis der Picardschen Sätze, *Verhandlungen des Internationalen Mathematiker-Kongresses Zürich 1932* (Orell Füssli, Zurich and Leipzig), vol. 2, 345–346.
1939 Das Riemannsche Umordnungsprinzip in der intuitionistischen Theorie der unendlichen Reihen, *Compositio mathematica 6*, 118–123.
1939a Der Lévysche Umordnungssatz und seine intuitionistische Übertragung, *ibid.*, 124–135.
1941 Elemente der intuitionistischen Funktionentheorie, *Nederlandse Akademie van Wetenschappen, Proceedings of the section of sciences 44*, 173–185, 276–285, 420–425, 563–567, 711–717; also *Indagationes mathematicae 3*, 64–76, 90–99, 170–175, 226–230, 322–328.
Benacerraf, Paul, and Hilary Putnam (eds.)
1964 *Philosophy of mathematics: selected readings* (Prentice-Hall, Englewood Cliffs, New Jersey).
Bentham, Jeremy
1843 *The works of Jeremy Bentham*, edited by John Bowring (London); reprinted 1962 (Russell and Russell, New York).
Berkeley, Edmund Collis
1954 The algebra of states and events, *The scientific monthly 78*, 232–242.
Bernays, Paul
1922 Über Hilberts Gedanken zur Grundlegung der Arithmetik, *Jahresbericht der Deutschen Mathematiker-Vereinigung 31*, 1st section, 10–19.
1923 Erwiderung auf die Note von Herrn Aloys Müller: "Über Zahlen als Zeichen", *Mathematische Annalen 90*, 159–163; reprinted in *Annalen der Philosophie und philosophischen Kritik 4* (1924), 492–497.
1926 Axiomatische Untersuchung des Aussagen-Kalkuls der "Principia mathematica", *Mathematische Zeitschrift 25*, 305–320.
1927 *Zusatz zu Hilberts Vortrag über "Die Grundlagen der Mathematik", *Abhandlungen aus dem mathematischen Seminar der Hamburgischen Universität 6*, 89–92 (485–489 of the present volume); reprinted in *Hilbert 1928*, 25–28.

Bernays, Paul—*contd.*

1932 Methoden des Nachweises von Widerspruchsfreiheit und ihre Grenzen, *Verhandlungen des Internationalen Mathematiker-Kongresses Zürich 1932* (Orell Füssli, Zurich and Leipzig), vol. 2, 342–343.

1934 Sur le platonisme dans les mathématiques, *L'enseignement mathématique 34*, (1935), 52–69.

1934a Quelques points essentiels de la métamathématique, *ibid.*, 70–95.

1935 Hilberts Untersuchungen über die Grundlagen der Arithmetik, in *Hilbert 1935*, 196–216.

1936 *Logical calculus*, lecture notes 1935–1936 (The Institute for Advanced Study, Princeton, New Jersey).

1937–1954 A system of axiomatic set theory, *The journal of symbolic logic 2*, 65–77; *6* (1941), 1–17; *7* (1942), 65–89, 133–145; *8* (1943), 89–106; *13* (1948), 65–79; *19*, 81–96.

1938 Sur les questions méthodologiques actuelles de la théorie hilbertienne de la la démonstration, in *Gonseth 1938*, 144–152; Discussion, 153–161.

1940 Review of *Gödel 1939*, *The journal of symbolic logic 5*, 117–118.

1940a Review of *Löwenheim 1940*, *ibid.*, 127–128.

1941 See *1937–1954*.

1954 Zur Beurteilung der Situation in der beweistheoretischen Forschung, *Revue internationale de philosophie 8*, 9–13; Discussion, 15–21.

1957 Über eine natürliche Erweiterung des Relationenkalkuls, in *Heyting 1957*, 1–14.

1958 *Axiomatic set theory*, with a historical introduction by Abraham A. Fraenkel (North-Holland, Amsterdam).

1961 Zur Frage der Unendlichkeitsschemata in der axiomatischen Mengenlehre, *Essays on the foundations of mathematics dedicated to A. A. Fraenkel on his seventieth anniversary* (Magnes Press, Jerusalem; distributed by North-Holland, Amsterdam).

See Hilbert, David, and Paul Bernays.

——— and Moses Schönfinkel

1928 Zum Entscheidungsproblem der mathematischen Logik, *Mathematische Annalen 99*, 342–372.

Bernoulli, Jakob

1686 Demonstratio rationum etc., *Acta eruditorum*, 360–361; reprinted in *Bernoulli 1744*, 282–283.

1744 *Opera*, vol. 1 (Geneva).

Bernstein, Felix

1904 Bemerkung zur Mengenlehre, *Nachrichten von der Königlichen Gesellschaft der Wissenschaften zu Göttingen*, Mathematisch-physikalische Klasse, 557–560.

1905 Über die Reihe der transfiniten Ordnungszahlen, *Mathematische Annalen 60*, 187–193.

1905a Die Theorie der reellen Zahlen, *Jahresbericht der Deutschen Mathematiker-Vereinigung 14*, 447–449.

Beth, Evert Willem

1953 On Padoa's method in the theory of definition, *Koninklijke Nederlandse Akademie van wetenschappen, Proceedings of the section of sciences 56*, series A, Mathematical sciences, 330–339; also *Indagationes mathematicae 15*, 330–339.

1955 *Semantic entailment and formal derivability* (Mededelingen der Koninklijke Nederlandse Akademie van wetenschappen, afd. letterkunde, new series, vol. 18, no. 13, N. V. Noord-Hollandsche Uitgevers Maatschappij, Amsterdam).

1956 *L'existence en mathématiques* (Gauthier-Villars, Paris; Nauwelaerts, Louvain).

1956a *La crise de la raison et la logique* (Gauthier-Villars, Paris; Nauwelaerts, Louvain; 1957).

1959 *The foundations of mathematics* (North-Holland, Amsterdam).

Blumenthal, Otto

1935 Lebensgeschichte, in *Hilbert 1935*, 388–429.

Boole, George

1847 *The mathematical analysis of logic, being an essay toward a calculus of deductive reasoning* (London and Cambridge, England).

1848 The calculus of logic, *The Cambridge and Dublin mathematical journal 3*, 183–198.

1854 *An investigation of the laws of thought, on which are founded the mathematical theories of logic and probabilities* (London).

Borel, Emile

1905 Quelques remarques sur les principes de la théorie des ensembles, *Mathematische Annalen 60*, 194–195.

1905a Cinq lettres sur la théorie des ensembles, *Bulletin de la Société mathématique de France 33*, 261–273 (Hadamard to Borel, Baire to Hadamard, Lebesgue to Borel, Hadamard to Borel, Borel to Hadamard); reprinted in Borel, *Leçons sur la théorie des functions*, 2nd ed. (Gauthier-Villars, Paris, 1914), 150–160, and in subsequent editions.

Bradley, Francis Herbert

1883 *The principles of logic* (London).

Brocard, Henri

1906 Réponse à la note 2549, *L'intermédiaire des mathématiciens 13*, 18.

Brouwer, Luitzen Egbertus Jan

1907 *Over de grondslagen der wiskunde* (Maas and van Suchtelen, Amsterdam and Leipzig; Noordhoff, Groningen).

1908 De onbetrouwbaarheid der logische principes, *Tijdschrift voor wijsbegeerte 2*, 152–158; reprinted in *Brouwer 1919a*, 5–12.

1912 *Intuitionisme en formalisme* (Noordhoff, Groningen); reprinted in *Brouwer 1919a*, 5–29.

1912a Intuitionism and formalism, *Bulletin of the American Mathematical Society 20* (1913), 81–96; reprinted in *Benacerraf and Putnam 1964*, 66–77.

1914 Review of *Schoenflies 1913*, *Jahresbericht der Deutschen Mathematiker-Vereinigung 23*, 2nd section, 78–83.

1917 Addenda en corrigenda over de grondslagen der wiskunde, *Koninklijke Akademie van wetenschappen te Amsterdam, Verslagen van de gewone vergaderingen der wis- en natuurkundige afdeeling 25*, 1418–1423; reprinted in *Nieuw archief voor wiskunde*, 2nd series, *12* (1918), 439–445.

1918 Begründung der Mengenlehre unabhängig vom logischen Satz vom ausgeschlossenen Dritten. Erster Teil: Allgemeine Mengenlehre, *Verhandelingen der Koninklijke Akademie van wetenschappen te Amsterdam*, 1st section, *12*, no. 5; errata on last page of *Brouwer 1919*.

1919 Begründung der Mengenlehre unabhängig vom logischen Satz vom ausgeschlossenen Dritten. Zweiter Teil: Theorie der Punktmengen, *ibid.*, no. 7.

1919a *Wiskunde, waarheid, werkelijkheid* (Noordhoff, Groningen).

1919b Intuitionistische Mengenlehre, *Jahresbericht der Deutschen Mathematiker-Vereinigung 28*, 1st section, 203–208; reprinted in *Koninklijke Akademie van wetenschappen te Amsterdam, Proceedings of the section of sciences 23* (1922), 949–954.

1920 Besitzt jede reelle Zahl eine Dezimalbruchentwickelung?, *Koninklijke Akademie van wetenschappen te Amsterdam, Verslag van de gewone vergaderingen der wis- en natuurkundige afdeeling 29* (1921), 803–812; also *Proceedings of the section of sciences 23* (1922), 955–964; reprinted in *Mathematische Annalen* 83 (1921), 201–210.

1923 Begründung der Funktionenlehre unabhängig vom logischen Satz vom ausgeschlossenen Dritten. Erster Teil: Stetigkeit, Messbarkeit, Derivierbarkeit, *Verhandelingen der Koninklijke Akademie van wetenschappen te Amsterdam*, 1st section, *13*, no. 2.

Brouwer, Luitzen Egbertus Jan—*contd.*

1923a Over de rol van het principium tertii exclusi in de wiskunde, in het bijzonder in de functietheorie, *Wis- en natuurkundig tijdschrift 2*, 1–7.

1923b *Über die Bedeutung des Satzes vom ausgeschlossenen Dritten in der Mathematik, insbesondere in der Funktionentheorie, *Journal für die reine und angewandte Mathematik 154*, 1–7 (334–341 of the present volume).

1923c Die Rolle des Satzes vom ausgeschlossenen Dritten in der Mathematik, *Jahresbericht der Deutschen Mathematiker-Vereinigung 33* (1925), 2nd section, 67–68 (abstract of *Brouwer 1923b*).

1923d Intuitionistische Zerlegung mathematischer Grundbegriffe, *ibid.*, 1st section, 251–256.

1924 Bewijs, dat iedere volle functie gelijkmatig continu is, *Koninklijke Akademie van wetenschappen te Amsterdam, Verslag van de gewone vergaderingen der wis- en natuurkundige afdeeling 33*, 189–193.

1924a Beweis, dass jede volle Funktion gleichmässig stetig ist, *Koninklijke Akademie van wetenschappen te Amsterdam, Proceedings of the section of sciences 27*, 189–193.

1924b Opmerkingen aangaande het bewijs der gelijkmatige continuiteit van volle functies, *Koninlijke Akademie van wetenschappen te Amsterdam, Verslag van de gewone vergaderingen der wis- en natuurkundige afdeeling 33*, 646–648.

1924c Bemerkungen zum Beweise der gleichmässigen Stetigkeit voller Funktionen, *Koninklijke Akademie van wetenschappen te Amsterdam, Proceedings of the section of sciences 27*, 644–646.

1925 Zur Begründung der intuitionistischen Mathematik I, *Mathematische Annalen 93*, 244–257; Berichtigung, *ibid. 95*, 472.

1925a Zur Begründung der intuitionistischen Mathematik II, *ibid. 95*, 453–472; Berichtigung, *ibid. 96*, 488.

1926 Zur Begründung der intuitionistischen Mathematik III, *ibid. 96*, 451–488.

1926a De intuitionistische vorm van het theorema van Heine-Borel, *Koninklijke Akademie van wetenschappen te Amsterdam, Verslag van de gewone vergaderingen der afdeeling natuurkunde 35*, 677–678.

1926b Die intuitionistische Form des Heine-Borelschen Theorems, *Koninklijke Akademie van wetenschappen te Amsterdam, Proceedings of the section of sciences 29*, 866–867.

1927 *Über Definitionsbereiche von Funktionen, *Mathematische Annalen 97*, 60–75 (446–463 of the present volume).

1927a *Intuitionistische Betrachtungen über den Formalismus, *Koninklijke Akademie van wetenschappen te Amsterdam, Proceedings of the section of sciences 31* (1928), 374–379 (490–492 of the present volume); also, with slight variations, *Sitzungsberichte der Preussischen Akademie der Wissenschaften, Physikalisch-mathematische Klasse* (1928), 48–52.

1927b Intuitionistische Betrachtungen über den Formalismus, *Koninklijke Akademie van wetenschappen te Amsterdam, Verslag van de gewone vergaderingen der afdeeling natuurkunde 36*, 1189 (abstract of *Brouwer 1927a*).

1928 Mathematik, Wissenschaft und Sprache, *Monatshefte für Mathematik und Physik 36* (1929), 153–164.

1928a *Die Struktur des Kontinuums* (Gistel, Vienna).

1942 Zum freien Werden von Mengen und Funktionen, *Nederlandsche Akademie van wetenschappen, Proceedings of the section of sciences 45*, 322–323; also *Indagationes mathematicae 4*, 107–108.

1947 Richtlijnen der intuïtionistische wiskunde, *Koninklijke Nederlandsche Akademie van wetenschappen, Proceedings of the section of sciences 50*, 339; also *Indagationes mathematicae, 9*, 197.

1948 Consciousness, philosophy, and mathematics, *Proceedings of the Tenth International Congress of Philosophy (Amsterdam, August 11–18, 1948)* (North-Holland, Amsterdam, 1949), 1235–1249.

Brouwer, Luitzen Egbertus Jan—*contd.*

1948a Essentieel-negatieve eigenschappen, *Koninklijke Nederlandsche Akademie van wetenschappen, Proceedings of the section of sciences 51,* 963–964; also *Indagationes mathematicae 10,* 322–323.

1948b Opmerkingen over het beginsel van het uitgesloten derde en over negatieve asserties, *Koninklijke Nederlandsche Akademie van wetenschappen, Proceedings of the section of sciences 51,* 1239–1243; also *Indagationes mathematicae 10,* 383–387.

1949 De non-aequivalentie van de constructieve en de negatieve orderelatie in het continuum, *Koninklijke Nederlandsche Akademie van wetenschappen, Proceedings of the section of sciences 52,* 122–124; also *Indagationes mathematicae 11,* 37–39.

1949a Contradictoriteit der elementaire meetkunde, *Koninklijke Nederlandsche Akademie van wetenschappen, Proceedings of the section of sciences 52,* 315–316; also *Indagationes mathematicae 11,* 89–90.

1950 Remarques sur la notion d'ordre, *Comptes rendus hebdomadaires des séances de l'Académie des sciences* (Paris) *230,* 263–265.

1950a Sur la possibilité d'ordonner le continu, *ibid.,* 349–350.

1951 On order in the continuum, and the relation of truth to non-contradictority, *Koninklijke Nederlandse Akademie van wetenschappen, Proceedings of the section of sciences 54,* series A, Mathematical sciences, 357–358; also *Indagationes mathematicae 13,* 357–358.

1952 Historical background, principles and methods of intuitionism, *South African journal of science 49,* 139–146.

1952a Over accumulatiekernen van oneindige kernsoorten, *Koninklijke Nederlandse Akademie van wetenschappen, Proceedings of the section of sciences 55,* series A, Mathematical sciences, 439–441; also *Indagationes mathematicae 14,* 439–441.

1953 Points and spaces, *Canadian journal of mathematics 6* (1954), 1–17.

1954 *Addenda en corrigenda over de rol van het principium tertii exclusi in de wiskunde, *Koninklijke Nederlandse Akademie van wetenschappen, Proceedings of the section of sciences 57,* series A, Mathematical sciences, 104–105 (341–342 of the present volume); also *Indagationes mathematicae 16,* 104–105.

1954a *Nadere addenda en corrigenda over de rol van het principium tertii exclusi in de wiskunde, *Koninklijke Nederlandse Akademie van wetenschappen, Proceedings of the section of sciences,* series A, Mathematical sciences, 109–111 (342–345 of the present volume); also *Indagationes mathematicae 16,* 109–111.

1954b Intuitionistische differentieerbaarheid, *Koninklijke Nederlandse Akademie van wetenschappen, Proceedings of the section of sciences,* series A, Mathematical sciences, 201–204; also *Indagationes mathematicae 16,* 201–204.

1954c An example of contradictority in classical theory of functions, *Koninklijke Nederlandse Akademie van wetenschappen, Proceedings of the section of sciences,* series A, Mathematical sciences, 204–206; also *Indagationes mathematicae 16,* 204–206.

Burali-Forti, Cesare

1894 Sulle classi ordinate e i numeri transfiniti, *Rendiconti del Circolo matematico di Palermo 8,* 169–179.

1894a *Logica matematica* (Milan).

1896 Le classi finite, *Atti dell'Accademia di Torino 32,* 34–52.

1896a Sopra un teorema del sig. G. Cantor, *ibid.,* 229–237.

1897 *Una questione sui numeri transfiniti, *Rendiconti del Circolo matematico di Palermo 11,* 154–164 (104–111 of the present volume).

1897a *Sulle classi ben ordinate, *ibid.,* 260 (111–112 of the present volume).

Cantor, Georg

1878 Ein Beitrag zur Mannigfaltigkeitslehre, *Journal für die reine und angewandte Mathematik 84,* 242–258; reprinted in *Cantor 1932,* 119–133.

Cantor, Georg—*contd.*

1883 Ueber unendliche, lineare Punktmannichfaltigkeiten, *Mathematische Annalen 21*, 545–591; also printed as *Grundlagen einer allgemeinen Mannichfaltigkeitslehre, ein mathematisch-philosophischer Versuch in der Lehre des Unendlichen* (Leipzig); reprinted in *Cantor 1932*, 165–209.

1887 Mitteilungen zur Lehre vom Transfiniten, *Zeitschrift für Philosophie und philosophische Kritik*, new series, *91*, 81–125, 252–270; *92* (1888), 240–265; reprinted in *Cantor 1932*, 378–439.

1895 Beiträge zur Begründung der transfiniten Mengenlehre (Erster Artikel), *Mathematische Annalen 46*, 481–512; reprinted in *Cantor 1932*, 282–311.

1895a Contribuzione al fondamento della teoria degli insiemi transfiniti, *Rivista di matematica 5*, 129–162.

1897 Beiträge zur Begründung der transfiniten Mengenlehre (Zweiter Artikel), *Mathematische Annalen 49*, 207–246; reprinted in *Cantor 1932*, 312–356.

1899 *Cantor an Dedekind, in *Cantor 1932*, 443–447, 451 (113–117 of the present volume).

1932 *Gesammelte Abhandlungen mathematischen und philosophischen Inhalts*, edited by Ernst Zermelo (Springer, Berlin); reprinted 1962 (Olms, Hildesheim).

See Noether, Emmy, and Jean Cavaillès.

Cavaillès, Jean

See Noether, Emmy, and Jean Cavaillès.

Chang, Chen Chung

1964 Some new results in definability, *Bulletin of the American Mathematical Society 70*, 808–813.

Church, Alonzo

1936 A note on the Entscheidungsproblem, *The journal of symbolic logic 1*, 40–41; correction, *ibid.*, 101–102; reprinted in *Davis 1965*, 110–115.

1956 *Introduction to mathematical logic* (Princeton University Press, Princeton, New Jersey), vol. 1.

1955 Review of *Berkeley 1954*, *The journal of symbolic logic 20*, 286–287.

1957 Binary recursive arithmetic, *Journal de mathématiques pures et appliquées*, series 9, *36*, 39–55.

1957a Application of recursive arithmetic to the problem of circuit synthesis, *Summaries of talks presented at the Summer Institute for symbolic logic, Cornell University 1957*, 2nd ed. (Institute for Defense Analyses, Princeton, New Jersey, 1960), 3–50.

Cohen, Paul

1963 *The independence of the axiom of choice* (mimeographed, Stanford University).

1963a The independence of the continuum hypothesis, *Proceedings of the National Academy of Sciences of the U.S.A. 50*, 1143–1148.

1964 The independence of the continuum hypothesis, II, *ibid. 51*, 105–110.

Couturat, Louis

1899 La logique mathématique de M. Peano, *Revue de métaphysique et de morale 7*, 616–646.

1905 *L'algèbre de la logique* (Gauthier-Villars, Paris).

1906 Pour la logistique (Réponse à M. Poincaré), *Revue de métaphysique et de morale 14*, 208–250.

Craig, William

1956 Review of *Beth 1953*, *The journal of symbolic logic 21*, 194–195.

1957 Linear reasoning. A new form of the Herbrand-Gentzen theorem, *ibid. 22*, 250–268.

1957a Three uses of the Herbrand-Gentzen theorem in relating model theory and proof theory, *ibid.*, 269–285.

1960 Bases for first-order theories and subtheories, *The journal of symbolic logic 25*, 97–142.

Craig, William—*contd.*

1963 An implicit definition of satisfaction in n-th order theory, *ibid. 28*, 301; abstract of a paper presented at the 28 December 1963 session of a meeting of the Association for Symbolic Logic.

Curry, Haskell Brooks

1929 An analysis of logical substitution, *American journal of mathematics 51*, 363–384.

1930 Grundlagen der kombinatorischen Logik, *ibid. 52*, 509–536, 789–834.

1941 A formalization of recursive arithmetic, *American journal of mathematics 63*, 263–282.

——— and Robert Feys

1958 *Combinatory logic* (North-Holland, Amsterdam), vol. 1.

Davis, Martin (ed.)

1965 *The undecidable. Basic papers on undecidable propositions, unsolvable problems, and computable functions* (Raven Press, Hewlett, New York).

de Bouvère, Karel Louis

1959 *A method in proofs of undefinability* (North-Holland, Amsterdam).

Dedekind, Richard

1888 *Was sind und was sollen die Zahlen?* (Brunswick).

1890 Über den Begriff des Unendlichen, unpublished manuscript in the Niedersächsische Staats- und Universitätsbibliothek, Göttingen (Cod. Ms. Dedekind 3, folder 1), dated 8 February 1890.

1890a *Letters to Keferstein, 9 February, 27 February (the latter in the present volume, 98–103), 1 April, 23 December 1890, copies in Dedekind's hand in the Niedersächsische Staats- und Universitätsbibliothek, Göttingen (Cod. Ms. Dedekind 13).

1893 2nd ed. of *Dedekind 1888*, unchanged but for an additional preface.

1911 3rd ed. of *Dedekind 1888*, unchanged but for a third preface.

1932 *Gesammelte mathematische Werke*, edited by Robert Fricke, Emmy Noether, Öystein Ore (Vieweg, Brunswick), vol. 3.

See Noether, Emmy, and Jean Cavaillès.

Denton, John

1963 Applications of the Herbrand theorem (thesis, Harvard University).

See Dreben, Burton, and John Denton.

Dijkman, Jacobus Gerhardus

1948 Recherche de la convergence négative dans les mathématiques intuitionnistes, *Koninklijke Nederlandsche Akademie van Wetenschappen, Proceedings of the section of sciences 51*, 681–692; also *Indagationes mathematicae 10*, 232–243.

Dixon, Alfred Cardew

1906 On "well-ordered" aggregates, *Proceedings of the London Mathematical Society*, 2nd series, *4*, 18–20.

Dreben, Burton

1952 On the completeness of quantification theory, *Proceedings of the National Academy of Sciences of the U.S.A. 38*, 1047–1052.

1961 Solvable Surányi subclasses: An introduction to the Herbrand theory, *Proceedings of a Harvard symposium on digital computers and their applications, 3–6 April 1961* (The annals of the Computation Laboratory of Harvard University *31*; Harvard University Press, Cambridge, Massachusetts), 32–47 (1962).

1963 Corrections to Herbrand, *American Mathematical Society, Notices 10*, 285; abstract presented by title at the 29 April–3 May 1963 meeting of the American Mathematical Society.

——— and Stål Aanderaa

1964 Herbrand analyzing functions, *Bulletin of the American Mathematical Society 70*, 697–698.

Dreben, Burton—*contd.*
——— Peter Andrews, and Stål Aanderaa
1963 Errors in Herbrand, *American Mathematical Society, Notices 10,* 285; abstract presented by title at the 29 April–3 May 1963 meeting of the American Mathematical Society.
1963a False lemmas in Herbrand, *Bulletin of the American Mathematical Society 69,* 699–706.
——— and John Denton
1966 A supplement to Herbrand, *The journal of symbolic logic 31,* 393–398.
1967 *The Herbrand theorem and the consistency of number theory,* forthcoming.

Feys, Robert
See Curry, Haskell Brooks, and Robert Feys.

Finsler, Paul
1923 Gibt es Widersprüche in der Mathematik?, *Jahresbericht der Deutschen Mathematiker-Vereinigung 34* (1925), 1st section, 143–155.
1926 *Formale Beweise und die Entscheidbarkeit, *Mathematische Zeitschrift 25,* 676–682 (438–445 of the present volume).
1927 Über die Lösung von Paradoxien, *Philosophischer Anzeiger 2,* 183–192.
1944 Gibt es unentscheidbare Sätze?, *Commentarii mathematici helvetici 16,* 310–320.

Fraenkel, Abraham A.
1921 Über die Zermelosche Begründung der Mengenlehre, *Jahresbericht der Deutschen Mathematiker-Vereinigung 30,* 2nd section, 97–98.
1922 Axiomatische Begründung der transfiniten Kardinalzahlen I, *Mathematische Zeitschrift 13,* 153–188.
1922a Zu den Grundlagen der Cantor-Zermeloschen Mengenlehre, *Mathematische Annalen 86,* 230–237.
1922b *Der Begriff "definit" und die Unabhängigkeit des Auswahlsaxioms, *Sitzungsberichte der Preussischen Akademie der Wissenschaften, Physikalisch-mathematische Klasse,* 253–257 (284–289 of the present volume).
1922c Zu den Grundlagen der Mengenlehre, *Jahresbericht der Deutschen Mathematiker-Vereinigung 31,* 2nd section, 101–102.
1923 *Einleitung in die Mengenlehre,* 2nd ed. (Springer, Berlin).
1923a Die Axiome der Mengenlehre, *Scripta Universitatis atque Bibliothecae Hierosolymitanarum, Mathematica et physica 1,* VI.
1923b Die neueren Ideen zur Grundlegung der Analysis und Mengenlehre, *Jahresbericht der Deutschen Mathematiker-Vereinigung 33* (1924), 1st section, 97–103.
1925 Untersuchungen über die Grundlagen der Mengenlehre, *Mathematische Zeitschrift 22,* 250–273.
1927 *Zehn Vorlesungen über die Grundlegung der Mengenlehre* (Teubner, Leipzig and Berlin).
1935 Sur la notion d'existence dans les mathématiques, *L'enseignement mathématique 34,* 18–32.
1937 Ueber eine abgeschwaechte Fassung des Auswahlsaxioms, *The journal of symbolic logic 2,* 1–25.
See Bernays, Paul, *1958.*
——— and Yehoshua Bar-Hillel
1958 *Foundations of set theory* (North-Holland, Amsterdam).

Frege, Gottlob
1873 *Ueber eine geometrische Darstellung der imaginären Gebilde in der Ebene,* Inaugural-Dissertation der philosophischen Facultät zu Göttingen zur Erlangung der Doctorwürde (Jena).
1874 *Rechnungsmethoden, die sich auf eine Erweiterung des Grössenbegriffes gründen,* Dissertation zur Erlangung der venia docendi bei der philosophischen Fakultät in Jena (Jena).

Frege, Gottlob—*contd.*

1878 Ueber eine Weise, die Gestalt eines Dreiecks als complexe Grösse aufzufassen, *Sitzungsberichte der Jenaischen Gesellschaft für Medicin und Naturwissenschaft für das Jahr 1878* (1879), 18.

1879 **Begriffsschrift, eine der arithmetischen nachgebildete Formelsprache des reinen Denkens* (Halle) (1–82 of the present volume); reprinted in *Frege 1964.*

1879a Anwendungen der Begriffsschrift, *Sitzungsberichte der Jenaischen Gesellschaft für Medicin und Naturwissenschaft für das Jahr 1879*, 29–33; reprinted in *Frege 1964*, 89–93.

1881 Ueber den Briefwechsel Leibnizens und Huygens mit Papin, *Sitzungsberichte der Jenaischen Gesellschaft für Medicin und Naturwissenschaft für das Jahr 1881*, 29–32; reprinted in *Frege 1964*, 93–96.

1882 Ueber den Zweck der Begriffsschrift, *Sitzungsberichte der Jenaischen Gesellschaft für Medicin und Naturwissenschaft für das Jahr 1882* (1883), 1–10; reprinted in *Frege 1964*, 97–106.

1882a Ueber die wissenschaftliche Berechtigung einer Begriffschrift, *Zeitschrift für Philosophie und philosophische Kritik*, new series, *81*, 48–56; reprinted in *Frege 1962*, 89–95, and in *Frege 1964*, 106–114.

1883 Geometrie der Punktpaare in der Ebene, *Sitzungsberichte der Jenaischen Gesellschaft für Medicin und Naturwissenschaft für das Jahr 1883* (1884), 98–102.

1884 *Die Grundlagen der Arithmetik, eine logisch-mathematische Untersuchung über den Begriff der Zahl* (Breslau); reprinted 1934 (Marcus, Breslau) and in *Frege 1950.*

1885 Ueber formale Theorien der Arithmetik, *Sitzungsberichte der Jenaischen Gesellschaft für Medicin und Naturwissenschaft für das Jahr 1885*, 94—104.

1891 *Function und Begriff. Vortrag gehalten in der Sitzung vom 9. Januar 1891 der Jenaischen Gesellschaft für Medicin und Naturwissenschaft* (Jena); reprinted in *Frege 1962*, 16–37; English translation in *Frege 1952*, 21–41.

1892 Ueber Begriff und Gegenstand, *Vierteljahrschrift für wissenschaftliche Philosophie 16*, 192–205; reprinted in *Frege 1962*, 64–78; English translation in *Frege 1952*, 42–55.

1892a Über Sinn und Bedeutung, *Zeitschrift für Philosophie und philosophische Kritik*, new series, *100*, 25–50; reprinted in *Frege 1962*, 38–63; English translation in *Frege 1952*, 56–78.

1893 *Grundgesetze der Arithmetik, begriffsschriftlich abgeleitet* (Jena), vol. 1; reprinted 1962 (Olms, Hildesheim); partial translation in *Frege 1964a.*

1895 Kritische Beleuchtung einiger Punkte in E. Schröder's Vorlesungen über die Algebra der Logik, *Archiv für systematische Philosophie 1*, 433–456; reprinted in *Frege 1966*, 92–112.

1896 Ueber die Begriffsschrift des Herrn Peano und meine eigene, *Berichte über die Verhandlungen der Königlich Sächsischen Gesellschaft der Wissenschaften zu Leipzig, Mathematisch-physikalische Klasse 48*, 361–378.

1896a Lettera del sig. G. Frege all'editore, *Revue de mathématiques 6* (1898), 53–59; reprinted in *Peano 1958*, 288–294.

1902 *Letter to Russell (first published in the present volume, 126–128).

1903 *Grundgesetze der Arithmetik, begriffsschriftlich abgeleitet* (Pohle, Jena), vol. 2; reprinted 1962 (Olms, Hildesheim); partial translation in *Frege 1964a.*

1950 *The foundations of arithmetic, A logico-mathematical enquiry into the concept of number*, English translation of *Frege 1884*, with the German text, by John Langshaw Austin (Blackwell, Oxford; Philosophical Library, New York); 2nd revised ed., 1953; reprinted, but without the German text, 1960 (Harper, New York).

1952 *Translations from the philosophical writings of Gottlob Frege*, edited by Peter Geach and Max Black (Blackwell, Oxford); 2nd ed., 1960.

Frege, Gottlob—*contd.*

1962 *Funktion, Begriff, Bedeutung. Fünf logische Studien*, edited by Günther Patzig (Vandenhoeck and Ruprecht, Göttingen); 2nd ed., revised, 1965.

1964 *Begriffsschrift und andere Aufsätze*, edited by Ignacio Angelelli (Olms, Hildesheim); contains reprints of *Frege 1879, 1879a, 1881, 1882*, and *1882a*.

1964a *The basic laws of arithmetic. Exposition of the system*, translated and edited, with an introduction, by Montgomery Furth (University of California Press, Berkeley and Los Angeles, California).

1966 *Logische Untersuchungen*, edited, with an introduction, by Günther Patzig (Vandenhoeck and Ruprecht, Göttingen).

Gentzen, Gerhard

1934 Untersuchungen über das logische Schliessen, *Mathematische Zeitschrift 39*, 176–210, 405–431.

1936 Die Widerspruchsfreiheit der reinen Zahlentheorie, *Mathematische Annalen 112*, 493–565.

1938 Neue Fassung des Widerspruchsfreiheitsbeweises für die reine Zahlentheorie, *Forschungen zur Logik und zur Grundlegung der exakten Wissenschaften*, new series, no. 4, 19–44.

1943 Beweisbarkeit und Unbeweisbarkeit von Anfangsfällen der transfiniten Induktion in der reinen Zahlentheorie, *Mathematische Annalen 119*, 140–161.

Gergonne, Joseph Diez

1817 Essai de dialectique rationnelle, *Annales de mathématiques pures et appliquées 7* (1816–1817), 189–228.

Gödel, Kurt

1930 Über die Vollständigkeit des Logikkalküls (thesis, University of Vienna).

1930a *Die Vollständigkeit der Axiome des logischen Funktionenkalküls, *Monatshefte für Mathematik und Physik 37*, 349–360 (582–591 of the present volume).

1930b *Einige metamathematische Resultate über Entscheidungsdefinitheit und Widerspruchsfreiheit, *Anzeiger der Akademie der Wissenschaften in Wien, Mathematisch-naturwissenschaftliche Klasse 67*, 214–215 (communicated on 23 October 1930 by Hans Hahn; 595–596 of the present volume).

1931 *Über formal unentscheidbare Sätze der Principia mathematica und verwandter Systeme I, *Monatshefte für Mathematik und Physik 38*, 173–198 (596–616 of the present volume); English translations in *Gödel 1962*, 35–72, and in *Davis 1965*, 5–38 (but see *Bauer-Mengelberg 1965* and *1966*).

1931a *Über Vollständigkeit und Widerspruchsfreiheit, *Ergebnisse eines mathematischen Kolloquiums 3* (1932), 12–13 (616–617 of the present volume).

1934 *On undecidable propositions of formal mathematical systems*, lecture notes by Stephen Cole Kleene and John Barkley Rosser (The Institute for Advanced Study, Princeton, New Jersey); reprinted with corrections, emendations, and a postscript in *Davis 1965*.

1934a Über die Länge von Beweisen, *Ergebnisse eines mathematischen Kolloquiums 7* (1936), 23–24; English translation in *Davis 1965*, 82–83.

1938 The consistency of the axiom of choice and of the generalized continuum-hypothesis, *Proceedings of the National Academy of Sciences 24*, 556–557.

1939 Consistency-proof for the generalized continuum-hypothesis, *ibid. 25*, 220–224.

1940 *The consistency of the continuum hypothesis* (Princeton University Press, Princeton, New Jersey); 2nd printing 1951; 3rd printing 1953. ("In the second and third printings several misprints and slight errors in the first printing have been corrected and some new Notes have been added at the end of the book.")

1947 What is Cantor's continuum problem?, *The American mathematical monthly, 54*, 515–525; revised and expanded version in *Benacerraf and Putnam 1964*, 258–273.

1958 Über eine bisher noch nicht benützte Erweiterung des finiten Standpunktes, *Dialectica 12*, 280–287.

Gödel, Kurt—*contd.*

1962 *On formally undecidable propositions of Principia mathematica and related systems*; English translation of *Gödel 1931* by B. Meltzer, with an introduction by R. B. Braithwaite (Oliver and Boyd, Edinburgh and London).

Gonseth, Ferdinand (ed.)

1938 *Les entretiens de Zurich sur les fondements et la méthode des sciences mathématiques, 6–9 décembre 1938* (Leemann, Zurich, 1941).

Goodman, Nelson

1942 Sequences, *The journal of symbolic logic 6*, 150–153.

Goodstein, Reuben Louis

1941 Function theory in an axiom-free equation calculus, *Proceedings of the London Mathematical Society*, 2nd series, *48* (1945), 401–434.

1957 *Recursive number theory. A development of recursive arithmetic in a logic-free equation calculus* (North-Holland, Amsterdam).

Grassmann, Hermann

1861 *Lehrbuch der Arithmetik für höhere Lehranstalten* (Berlin).

Hadamard, Jacques

1897 Sur certaines applications possibles de la théorie des ensembles, *Verhandlungen des ersten Internationalen Mathematiker-Kongresses in Zürich vom 9. bis 11. August 1897* (Leipzig, 1898), 201–202.

See Borel, Emile, *1905a*.

Hamel, Georg

1905 Eine Basis aller Zahlen und die unstetigen Lösungen der Funktionalgleichung: $f(x + y) = f(x) + f(y)$, *Mathematische Annalen 60*, 459–462.

Hardy, Godfrey Harold

1904 A theorem concerning the infinite cardinal numbers, *The quarterly journal of pure and applied mathematics 35*, 87–94.

Hausdorff, Felix

1908 Grundzüge einer Theorie der geordneten Mengen, *Mathematische Annalen 65*, 435–505.

1914 *Grundzüge der Mengenlehre* (Veit, Leipzig); reprinted 1949 (Chelsea, New York).

Henkin, Leon

1949 The completeness of the first-order functional calculus, *The journal of symbolic logic 14*, 159–166.

See Addison, John West, Leon Henkin, and Alfred Tarski.

Herbrand, Jacques

1928 Sur la théorie de la démonstration, *Comptes rendus hebdomadaires des séances de l'Académie des sciences* (Paris) *186*, 1274–1276.

1929 Non-contradiction des axiomes arithmétiques, *ibid. 188*, 303–304.

1929a Sur quelques propriétés des propositions vraies et leurs applications, *ibid.*, 1076–1078.

1929b Sur le problème fondamental des mathématiques, *ibid. 189*, 554–556, 720.

1930 **Recherches sur la théorie de la démonstration*, Thesis at the University of Paris; also Prace Towarzystwa Naukowego Warszawskiego, Wydział III, no. 33 (chap. 5, 525–581 of the present volume).

1930a Les bases de la logique hilbertienne, *Revue de métaphysique et de morale 37*, 243–255.

1931 Sur le problème fondamental de la logique mathématique, *Sprawozdania z posiedzeń Towarzystwa Naukowego Warszawskiego, Wydział III, 24*, 12–56.

1931a Unsigned note on *Herbrand 1930*, *Annales de l'Université de Paris 6*, 186–189.

1931b *Sur la non-contradiction de l'arithmétique, *Journal für die reine angewandte Mathematik 166*, 1–8 (618–628 of the present volume).

Hessenberg, Gerhard

1906 Grundbegriffe der Mengenlehre, *Abhandlungen der Fries'schen Schule*, new series, *1*, 479–706; separate reprint 1906 (Vandenhoeck and Ruprecht, Göttingen).

Hessenberg, Gerhard—*contd.*

1910 Kettentheorie und Wohlordnung, *Journal für Mathematik 135*, 81–133.

Heyting, Arend

1929 De telbaarheidspraedicaten van Prof. Brouwer, *Nieuw archief voor wiskunde*, 2nd series, *16*, 47–58.

1934 *Mathematische Grundlagenforschung. Intuitionismus. Beweistheorie* (Springer, Berlin).

1955 *Les fondements des mathématiques. Intuitionnisme. Théorie de la démonstration* (Gauthier-Villars, Paris; Nauwelaerts, Louvain); French translation, with additions, of *Heyting 1934*.

1956 *Intuitionism. An introduction* (North-Holland, Amsterdam).

1957 (ed.) *Constructivity in mathematics, Proceedings of the colloquium held at Amsterdam, 1957* (North-Holland, Amsterdam, 1959).

Hilbert, David

1896 Zur Theorie der aus *n* Haupteinheiten gebildeten complexen Größen, *Nachrichten von der Königlichen Gesellschaft der Wissenschaften zu Göttingen*, Mathematisch-physikalische Klasse, 179–183.

1899 *Grundlagen der Geometrie* (Leipzig).

1900 Über den Zahlbegriff, *Jahresbericht der Deutschen Mathematiker-Vereinigung 8*, 180–194; reprinted in *Hilbert 1909*, 256–262; *1913*, 237–242; *1922b*, 237–242; *1923*, 237–242; *1930a*, 241–246.

1900a Mathematische Probleme. Vortrag, gehalten auf dem internationalen Mathematiker-Kongress zu Paris 1900, *Nachrichten von der Königlichen Gesellschaft der Wissenschaften zu Göttingen*, 253–297; reprinted with some additions as *Hilbert 1900b* and in *Hilbert 1935*, 290–329; French translation, with emendations and additions, in *Compte rendu du Deuxième congrès international des mathématiciens tenu á Paris du 6 au 12 août 1900* (Gauthier-Villars, Paris, 1902), 58–114.

1900b Mathematische Probleme. Vortrag, gehalten auf dem internationalen Mathematiker-Kongress zu Paris 1900, *Archiv der Mathematik und Physik*, 3rd series, *1* (1901), 44–63, 213–237.

1903 *Grundlagen der Geometrie*, 2nd ed. (Teubner, Leipzig).

1904 *Über die Grundlagen der Logik und der Arithmetik, *Verhandlungen des Dritten Internationalen Mathematiker-Kongresses in Heidelberg vom 8. bis 13. August 1904* (Teubner, Leipzig, 1905), 174–185 (129–138 of the present volume); reprinted in *Hilbert 1909*, 263–279; *1913*, 243–258; *1922b*, 243–258; *1923*, 243–258; *1930a*, 247–261.

1905 On the foundations of logic and arithmetic, *The monist 15*, 338–352; English translation of *Hilbert 1904* by George Bruce Halsted.

1909 *Grundlagen der Geometrie* (Teubner, Leipzig and Berlin), 3rd ed.

1913 ——— 4th ed.

1917 Axiomatisches Denken, *Mathematische Annalen 78* (1918), 405–415; reprinted in *Hilbert 1935*, 146–156.

1922 Neubegründung der Mathematik (Erste Mitteilung), *Abhandlungen aus dem mathematischen Seminar der Hamburgischen Universität 1*, 157–177; reprinted in *Hilbert 1935*, 157–177.

1922a Die logischen Grundlagen der Mathematik, *Mathematische Annalen 88* (1923), 151–165; reprinted in *Hilbert 1935*, 178–191.

1922b *Grundlagen der Geometrie* (Teubner, Leipzig and Berlin), 5th ed.

1923 ——— 6th ed.

1925 *Über das Unendliche, *Mathematische Annalen 95* (1926), 161–190 (367–392 of the present volume); reprinted in abbreviated form as *Hilbert 1925a* and also in abbreviated form, but with minor emendations and additions, in *Hilbert 1930a*, 262–288; partial translation in *Benacerraf and Putnam 1964*, 134–151.

Hilbert, David—*contd.*

1925a Über das Unendliche, *Jahresbericht der Deutschen Mathematiker-Vereinigung 36* (1927), 1st section, 201–215.

1927 *Die Grundlagen der Mathematik, *Abhandlungen aus dem mathematischen Seminar der Hamburgischen Universität 6* (1928), 65–85 (464–479 of the present volume); reprinted in *Hilbert 1928*, 1–21; *1930a*, 289–312.

1928 *Die Grundlagen der Mathematik*, mit Zusätzen von Hermann Weyl and Paul Bernays, Hamburger Mathematische Einzelschriften *5* (Teubner, Leipzig); contains reprints of *Hilbert 1927*, *Weyl 1927*, and *Bernays 1927*.

1928a Probleme der Grundlegung der Mathematik, *Atti del Congresso internazionale dei matematici, Bologna 3–10 settembre 1928* (Bologna, 1929), vol. 1, 135–141; reprinted, with emendations and additions, in *Mathematische Annalen 102* (1929), 1–9, and in *Hilbert 1930a*, 313–323.

1930 Die Grundlegung der elementaren Zahlenlehre, *Mathematische Annalen 104* (1931), 485–494; reprinted in part in *Hilbert 1935*, 192–195.

1930a *Grundlagen der Geometrie* (Teubner, Leipzig and Berlin), 7th ed.

1930b Naturerkennen und Logik, *Die Naturwissenschaften 18*, 959–963.

1931 Beweis des Tertium non datur, *Nachrichten von der Gesellschaft der Wissenschaften zu Göttingen*, Mathematisch-physikalische Klasse, 120–125.

1935 *Gesammelte Abhandlungen* (Springer, Berlin), vol. 3.

——— and Wilhelm Ackermann

1928 *Grundzüge der theoretischen Logik* (Springer, Berlin).

1938 ——— 2nd ed.

——— and Paul Bernays

1934 *Grundlagen der Mathematik* (Springer, Berlin), vol. 1.

1939 ——— vol. 2.

Hintikka, K. Jaakko J.

1955 Form and content in quantification theory, *Acta philosophica fennica 8*, 7–55.

Hobson, Ernest William

1906 On the arithmetic continuum, *Proceedings of the London Mathematical Society*, 2nd series, *4*, 21–28.

1921 *The theory of functions of a real variable and the theory of Fourier's series*, 2nd ed. (Cambridge University Press, Cambridge, England), vol. 1.

1926 ——— vol. 2.

Howard, William A., and Georg Kreisel

1966 Transfinite induction and bar induction of types zero and one, and the role of continuity in intuitionistic analysis, *The Journal of symbolic logic 31*, 325–358.

Huntington, Edward Vermilye

1904 Sets of independent postulates for the algebra of logic, *Transactions of the American Mathematical Society 5*, 288–309.

Jevons, William Stanley

1879 *The principles of science, a treatise on logic and scientific method*, 3rd ed. (London).

1883 ——— stereotyped reprint.

Johansson, Ingebrigt

1936 Der Minimalkalkül, ein reduzierter intuitionistischer Formalismus, *Compositio mathematica 4*, 119–136.

Jourdain, Philip Edward Bertrand

1904 On the transfinite cardinal numbers of well-ordered aggregates, *Philosophical magazine*, series 6, *7*, 61–75.

1904a On the transfinite cardinal numbers of number classes in general, *ibid.*, 294–303.

1905 On transfinite cardinal numbers of the exponential form, *ibid. 9*, 42–56.

1905a On a proof that every aggregate can be well-ordered, *Mathematische Annalen 60*, 465–470.

1912 The development of the theories of mathematical logic and the principles of mathematics, *The quarterly journal of pure and applied mathematics 43*, 219–314.

Kahr, Andrew Seth, Edward Forrest Moore, and Hao Wang
1961 Entscheidungsproblem reduced to the ∀∃∀ case, *Proceedings of the National Academy of Sciences of the United States of America 48* (1962), 365–377.

Kalmár, László
1932 Ein Beitrag zum Entscheidungsproblem, *Acta litterarum ac scientiarum Regiae Universitatis Hungaricae Francisco-Josephinae, Sectio scientiarum mathematicarum 5*, 222–236.
1932a Zum Entscheidungsproblem der mathematischen Logik, *Verhandlungen des Internationalen Mathematiker-Kongresses Zürich 1932* (Orell Füssli, Zurich and Leipzig), vol. 2, 337–338.
1934 Über einen Löwenheimschen Satz, *Acta litterarum ac scientiarum Regiae Universitatis Hungaricae Francisco-Josephinae, Sectio scientiarum mathematicarum 7*, 112–121.
1936 Zurückführung des Entscheidungsproblem auf den Fall von Formeln mit einer einzigen, binären, Funktionsvariablen, *Compositio mathematica 4* (1937), 137–144.
1951 Contributions to the reduction theory of the decision problem. Fourth paper. Reduction to the case of a finite set of individuals, *Acta mathematica Academiae scientiarum hungaricae 2*, 125–142.

Keferstein, Hans
1890 Über den Begriff der Zahl, *Festschrift herausgegeben von der Mathematischen Gesellschaft in Hamburg anlässlich ihres 200jährigen Jubelfestes 1890, Zweiter Teil: Wissenschaftliche Abhandlungen* (Teubner, Leipzig), 119–125 (this book is considered volume 2 of *Mitteilungen der Mathematischen Gesellschaft in Hamburg*).
1890a Letters to Dedekind dated 14 February, 19 March, 19 December 1890; originals in the Niedersächsische Staats- und Universitätsbibliothek, Göttingen (Cod. Ms. Dedekind 13).
1890b Erwiderung auf vorstehende Abhandlung, unpublished reply dated 17 November 1890 to *Dedekind 1890*, copy in Dedekind's hand in the Niedersächsische Staats- und Universitätsbibliothek, Göttingen (Cod. Ms. Dedekind 13).

Kleene, Stephen Cole
1936 General recursive functions of natural numbers, *Mathematische Annalen 112*, 727–742; for an erratum and a simplification see *Kleene 1938*, footnote 4, *Péter 1937a*, and *The journal of symbolic logic 4*, iv; for an addendum see *Davis 1965*, 253. Reprinted in *Davis 1965*, 237–253.
1938 On notation for ordinal numbers, *The journal of symbolic logic 3*, 150–155.
1944 On the forms of the predicates in the theory of constructive ordinals, *American journal of mathematics 66*, 41–58.
1952 *Introduction to metamathematics* (Van Nostrand, New York and Toronto; North-Holland, Amsterdam; Noordhoff, Groningen).
1955 On the forms of the predicates in the theory of constructive ordinals (Second paper), *American journal of mathematics 77*, 405–428.
1955a Arithmetical predicates and function quantifiers, *Transactions of the American Mathematical Society 79*, 312–340.
1958 Extension of an effectively generated class of functions by enumeration, *Colloquium mathematicum 6*, 67–78.

——— and Richard Eugene Vesley
1965 *The foundations of intuitionistic mathematics, especially in relation to recursive functions* (North-Holland, Amsterdam).

Kneale, William, and Martha Kneale
1962 *The development of logic* (Oxford University Press, Oxford).

König, Dénes
1926 Sur les correspondances multivoques des ensembles, *Fundamenta mathematicae 8*, 114–134.

König, Dénes—*contd.*

1927 Über eine Schlussweise aus dem Endlichen ins Unendliche, *Acta litterarum ac scientiarum Regiae Universitatis Hungaricae Francisco-Josephinae, Sectio scientiarum mathematicarum 3*, 121–130.

1936 *Theorie der endlichen und unendlichen Graphen* (Akademische Verlagsgesellschaft, Leipzig); reprinted 1950 (Chelsea, New York).

König, Julius (Gyula)

1904 Zum Kontinuum-Problem, *Verhandlungen des Dritten Internationalen Mathematiker-Kongresses in Heidelberg vom 8. bis 13. August 1904* (Teubner, Leipzig, 1905), 144–147.

1905 Zum Kontinuum-Problem, *Mathematische Annalen 60*, 177–180; Berichtigung, *ibid.*, 462.

1905a *Über die Grundlagen der Mengenlehre und das Kontinuumproblem, *ibid. 61*, 156–160 (145–149 of the present volume).

1906 Über die Grundlagen der Mengenlehre und das Kontinuumproblem (Zweite Mitteilung), *ibid. 63*, 217–221.

1914 *Neue Grundlagen der Logik, Arithmetik und Mengenlehre* (Veit, Leipzig).

Kolmogorov, Andrei Nikolaevich (Андрей Николаевич Колмогоров)

1925 *О принципе tertium non datur, *Математический сборник 32*, 646–667 (414–437 of the present volume).

Kreisel, Georg

1953 Note on arithmetic models for consistent formulae of the predicate calculus II, *Actes du XIème Congrès international de philosophie, Bruxelles, 20–26 août 1953* (North-Holland, Amsterdam; Nauwelaerts, Louvain), vol. 14, 39–49.

1958 Review of *Beth 1956a*, *The journal of symbolic logic 23*, 35–37.

1963 *Reports of the seminar on foundations of analysis, Stanford University, Summer 1963* (mimeographed, Stanford University, Stanford, California); contains contributions by others.

1964 ——— vol. 2, Introduction.

See Howard, William A., and Georg Kreisel.

Kummer, Ernst Eduard

1835 Über die Convergenz und Divergenz der unendlichen Reihen, *Journal für die reine und angewandte Mathematik 13*, 171–184.

Kuratowski, Kazimierz

1921 Sur la notion d'ordre dans la théorie des ensembles, *Fundamenta mathematicae 2*, 161–171.

Langford, Cooper Harold

1926 Some theorems on deducibility, *Annals of mathematics*, 2nd series, *28* (1927), 16–40.

1926a Theorems on deducibility (Second paper), *ibid.*, 459–471.

Lebesgue, Henri

1905 Sur les fonctions représentables analytiquement, *Journal de mathématiques pures et appliquées*, 6th series, *1*, 139–216.

See Borel, Emile, *1905a*.

Lennes, Nels Johann

1922 On the foundations of the theory of sets, *Bulletin of the American Mathematical Society 28*, 300; abstract of a paper read at a session of the 14–15 April 1922 meeting of the American Mathematical Society.

Levi, Beppo

1902 Intorno alla teoria degli aggregati, *Reale Istituto lombardo di scienze e lettere, Rendiconti*, 2nd series, *35*, 863–868.

Lévy, Paul

1964 Remarques sur un théorème de Paul Cohen, *Revue de métaphysique et de morale 69*, 88–94.

Lewis, Clarence Irving

1918 *A survey of symbolic logic* (University of California Press, Berkeley); reprinted 1960 with the omission of chaps. V and VI (Dover, New York).

Lindenbaum, Adolf

See Tarski, Alfred, and Adolf Lindenbaum.

——— and Andrzej Mostowski

1938 Über die Unabhängigkeit des Auswahlaxioms und einiger seiner Folgerungen, *Sprawozdania z posiedzeń Towarzystwa Naukowego Warszawskiego, Wydział III, 31*, 27–32.

Löwenheim, Leopold

1910 Über die Auflösung von Gleichungen im logischen Gebietekalkul, *Mathematische Annalen 68*, 169–207.

1915 *Über Möglichkeiten im Relativkalkül, *ibid. 76*, 447–470 (228–251 of the present volume).

1940 Einkleidung der Mathematik in Schröderschen Relativkalkul, *The journal of symbolic logic 5*, 1–15.

Lukasiewicz, Jan

1920 O logice trójwartościowej, *Ruch filozoficzny 5*, 170–171.

1929 O znaczeniu i potrzebach logiki matematycznej, *Nauka polska 10*, 604–620.

1929a *Elementy logiki matematycznej*, authorized lecture notes prepared by Mojżesz Presburger (Warsaw University, Warsaw).

1958 ——— 2nd ed. (PWN, Warsaw).

1963 *Elements of mathematical logic*, English translation of *Łukasiewicz 1958* by Olgierd Wojtasiewicz (PWN and Pergamon, Oxford; Macmillan, New York).

——— and Alfred Tarski

1930 Untersuchungen über den Aussagenkalkül, *Sprawozdania z posiedzeń Towarzystwa Naukowego Warszawskiego, Wydział III, 23*, 30–50; English translation in *Tarski 1956*, 38–59.

Luzin, Nikolai Nikolaevich (Lusin, Николай Николаевич Лузин)

1927 Sur les ensembles analytiques, *Fundamenta mathematicae 10*, 1–95.

1928 Sur les voies de la théories des ensembles, *Atti del Congresso internazionale dei matematici, Bologna 3–10 settembre 1928* (Zanichelli, Bologna, 1929) vol. 1, 295–299.

1930 *Leçons sur les ensembles analytiques et leurs applications* (Gauthier-Villars, Paris).

1933 Sur les classes des constituantes des complémentaires analytiques, *Annali della Reale Scuola normale superiore di Pisa, Scienze fisiche e matematiche*, 2nd series, *2*, 269–282.

1935 Sur les ensembles analytiques nuls, *Fundamenta mathematicae 25*, 109–131.

1953 *Лекции об аналитических множествах и их приложениях*, edited with a preface and notes by L. V. Keldysh and P. S. Novikov (Государственное издательство, Moscow).

1958 *Собрание сочинений* (Издательство Академии Наук СССР, Moscow), vol. 1.

1958a Лекции об аналитических множествах и их приложениях, in *Luzin 1958*, 9–269.

1958b О путях развития теории множеств, in *Luzin 1958*, 464–469.

1958c О классах конституант аналитических дополнений, in *Luzin 1958*, 627–641.

1958d О пустых аналитических множествах, in *Luzin 1958*, 662–680.

——— and Wacław Sierpiński

1918 Sur quelques propriétés des ensembles (*A*), *Bulletin international de l'Académie des sciences de Cracovie, Classe des sciences mathématiques et naturelles, série A: sciences mathématiques*, April–May, 35–48.

1958 О некоторых свойствах *A*-множеств, in *Luzin 1958*, 273–284.

Lyndon, Roger Conant

1950 The representation of relational algebras, *Annals of mathematics*, 2nd series, *51*, 707–729.

MacColl, Hugh

1877 The calculus of equivalent statements and integration limits, *Proceedings of the London Mathematical Society 9*, 9–20.

1878 The calculus of equivalent statements (Second paper), *ibid.*, 177–186.

1878a The calculus of equivalent statements (Third paper), *ibid. 10*, 16–28.

1880 The calculus of equivalent statements (Fourth paper), *ibid. 11*, 113–121.

McKinsey, John Charles Chenoweth

1935 On the independence of undefined ideas, *Bulletin of the American Mathematical Society 41*, 291–297.

1940 Postulates for the calculus of binary relations, *The journal of symbolic logic 5*, 85–97.

Makkai, M.

1964 On a generalization of a theorem of E. W. Beth, *Acta mathematicae Academiae scientiarum hungaricae 15*, 227–235.

Maltsev, Anatolii Ivanovich (Malcev, Анатолий Иванович Мальцев)

1936 Untersuchungen aus dem Gebiete der mathematischen Logik, *Математический сборник 1*, 323–336.

Mendelson, Elliott

1956 The independence of a weak axiom of choice, *The journal of symbolic logic 21*, 350–366.

Menger, Karl

1928 Bemerkungen zu Grundlagenfragen, *Jahresbericht der Deutschen Mathematiker-Vereinigung 37*, 1st section, 213–226.

Mirimanoff, Dimitry

1917 Les antinomies de Russell et de Burali-Forti et le problème fondamental de la théorie des ensembles, *L'enseignement mathématique 19*, 37–52.

Montague, Richard

1956 Zermelo-Fraenkel set theory is not a finite extension of Zermelo set theory, *Bulletin of the American Mathematical Society 62*, 260; abstract of a paper presented at the 25 February 1956 meeting of the American Mathematical Society.

1961 Fraenkel's addition to the axioms of Zermelo, *Essays on the foundations of mathematics dedicated to A. A. Fraenkel on his seventieth anniversary* (Magnes Press, Jerusalem; distributed by North-Holland, Amsterdam), 91–113.

Moore, Edward Forrest

See Kahr, Andrew Seth, Edward Forrest Moore, and Hao Wang.

Mostowski, Andrzej

1939 Über die Unabhängigkeit des Wohlordnungssatzes vom Ordnungsprinzip, *Fundamenta mathematicae 32*, 201–252.

1948 On the principle of dependent choices, *ibid. 35*, 127–130.

1955 A formula with no recursively enumerable model, *ibid. 42*, 125–140.

See Lindenbaum, Adolf, and Andrzej Mostowski; also Tarski, Alfred, *1953*.

Müller, Eugen

See Schröder, Ernst, *1909, 1910*.

Nicod, Jean

1916 A reduction in the number of primitive propositions of logic, *Proceedings of the Cambridge Philosophical Society 19* (1917–1920), 32–41 (received 30 October 1916).

Noether, Emmy, and Jean Cavaillès (eds.)

1937 *Briefwechsel Cantor-Dedekind* (Hermann, Paris).

Padoa, Alessandro

1898 *Conférences sur la logique mathématique* (Brussels).

1900 *Essai d'une théorie algébrique des nombres entiers, précédé d'une introduction logique à une théorie déductive quelconque, *Bibliothèque du Congrès international de philosophie, Paris, 1900* (Armand Colin, Paris, 1901), vol. 3, 309–365 (118–123 of the present volume).

Padoa, Alessandro—*contd.*

1900a Un nouveau système irréductible de postulats pour l'algèbre, *Compte rendu du Deuxième congrès international des mathématiciens tenu à Paris du 6 au 12 août 1900* (Gauthier-Villars, Paris, 1902), 249–256.

1903 Le problème no. 2 de M. David Hilbert, *L'enseignement mathématique 5*, 85–91.

Peano, Giuseppe

1888 *Calcolo geometrico secondo l'Ausdehnungslehre di H. Grassmann, preceduto dalle Operazioni della logica deduttiva* (Turin); partly reprinted in *Peano 1958*, 3–19.

1889 **Arithmetices principia, nova methodo exposita* (Turin) (83–97 of the present volume); reprinted in *Peano 1958*, 20–55.

1890 Démonstration de l'intégrabilité des équations différentielles ordinaires, *Mathematische Annalen 37*, 182–288; reprinted in *Peano 1957*, 119–170.

1891 Principii di logica matematica, *Rivista di matematica 1*, 1–10; reprinted in *Peano 1958*, 92–101.

1891a Formole di logica matematica, *Rivista di matematica 1*, 24–31, 182–184; reprinted in *Peano 1958*, 102–113.

1891b Sul concetto di numero, *Rivista di matematica 1*, 87–102, 256–267; reprinted in Peano *1959*, 80–109.

1894 *Notations de logique mathématique. Introduction au formulaire de mathématique* (Turin).

1895 *Formulaire de mathématiques* (Turin), vol. 1.

1897 ——— vol. 2, § 1.

1897a Studii di logica matematica, *Atti della Reale Accademia delle scienze di Torino 32*, 565–583; reprinted in *Peano 1958*, 201–217.

1898 Riposta, *Revue de mathématiques 6*, 60–61; reprinted in *Peano 1958*, 295–296.

1898a *Formulaire de mathématiques* (Turin), vol. 2, § 2.

1899 ——— § 3.

1901 ——— vol. 3.

1903 Note 2549, *L'intermédiaire des mathématiciens 10*, 70; see Brocard, Henri, *1906*.

1903a *Formulaire de mathématiques* (Bocca, Turin; Clausen, Turin), vol. 4.

1906 Super theorema de Cantor-Bernstein, *Rendiconti del Circolo matematico di Palermo 21*, 360–366; also *Revista de mathematica 8*, 136–143; reprinted in *Peano 1957*, 337–344.

1906a Additione, *Revista de mathematica 8*, 143–157; reprinted in *Peano 1957*, 344–358.

1957 *Opere scelte* (Edizioni cremonese, Rome), vol. 1.

1958 ——— vol. 2

1959 ——— vol. 3.

Peirce, Charles Sanders

1870 Description of a notation for the logic of relatives, resulting from an amplification of the conceptions of Boole's calculus of logic, *Memoirs of the American Academy of arts and sciences 9*, 317–378; reprinted in *Peirce 1933*, 27–98.

1880 On the algebra of logic, *American journal of mathematics 3*, 15–57; reprinted in *Peirce 1933*, 104–157.

1882 Brief description of the algebra of relatives, in *Peirce 1933*, 180–186.

1883 A theory of probable inference. Note B. The logic of relatives, *Studies in logic by members of the Johns Hopkins University* (Boston), 187–203; reprinted in *Peirce 1933*, 195–209.

1885 On the algebra of logic: a contribution to the philosophy of notation, *American journal of mathematics 7*, 180–202; reprinted in *Peirce 1933*, 210–238.

1933 *Collected papers of Charles Sanders Peirce*, edited by Charles Hartshorne and Paul Weiss (Harvard University Press, Cambridge, Massachusetts), vol. 3.

Péter, Rózsa (Politzer)

1932 Rekursive Funktionen, *Verhandlungen des Internationalen Mathematiker-Kongresses Zürich 1932* (Orell Füssli, Zurich and Leipzig), vol. 2, 336–337.

Péter, Rózsa (Politzer)—*contd.*

1934 Über den Zusammenhang der verschiedenen Begriffe der rekursiven Funktion, *Mathematische Annalen 110*, 612–632.

1935 Konstruktion nichtrekursiver Funktionen, *ibid. 111*, 42–60.

1937 Über die mehrfache Rekursion, *ibid. 113*, 489–527.

1937a Review of *Kleene 1936*, *The journal of symbolic logic 2*, 38.

1951 *Rekursive Funktionen* (Verlag der Ungarischen Akademie der Wissenschaften, Budapest).

1951a Probleme der Hilbertschen Theorie der höheren Stufen von rekursiven Funktionen, *Acta mathematica Academiae scientiarum hungaricae 2*, 247–274.

1953 Rekursive Definitionen, wobei frühere Funktionswerte von variabler Anzahl verwendet werden, *Publicationes mathematicae 3*, 33–70.

1956 Die beschränkt-rekursiven Funktionen und die Ackermannsche Majorisierungsmethode, *ibid. 4*, 362–375.

1957 2nd, enlarged ed. of *Péter 1951*.

Pieri, Mario

1906 Sur la compatibilité des axiomes de l'arithmétique, *Revue de métaphysique et de morale 14*, 196–207.

Poincaré, Henri

1905 Les mathématiques et la logique, *Revue de métaphysique et de morale 13*, 815–835; *14* (1906), 17–34.

1906 Les mathématiques et la logique, *ibid.*, 294–317.

1906a A propos de la logistique, *ibid.* 866–868.

1908 *Science et méthode* (Flammarion, Paris).

Post, Emil Leon

1920 Introduction to a general theory of elementary propositions, *Bulletin of the American Mathematical Society 26*, 437; abstract of a paper presented at the 24 April 1920 meeting of the American Mathematical Society.

1920a Determination of all closed systems of truth tables (abstract of a paper presented at the 24 April 1920 meeting of the American Mathematical Society), *ibid.*

1921 *Introduction to a general theory of elementary propositions, *American journal of mathematics 43*, 163–185 (264–283 of the present volume).

1921a On a simple case of deductive systems (abstract of a paper read by title at the 23 April 1921 meeting of the American Mathematical Society), *Bulletin of the American Mathematical Society 27*, 396–397.

1936 Finite combinatory processes, formulation I, *The journal of symbolic logic 1*, 103–105; reprinted in *Davis 1965*, 288–291.

1941 *The two-valued iterative systems of mathematical logic* (Annals of mathematics series, no. 5, Princeton University Press, Princeton, New Jersey).

1941a Absolutely unsolvable problems and relatively undecidable propositions. Account of an anticipation, in *Davis 1965*, 340–433.

1943 Formal reductions of the general combinatorial decision problem, *American journal of mathematics 65*, 197–215.

Presburger, Mojżesz

1929 Über die Vollständigkeit eines gewissen Systems der Arithmetik ganzer Zahlen, in welchem die Addition als einzige Operation hervortritt, *Sprawozdanie z I Kongresu matematyków krajów słowiańskich, Warszawa 1929* (Warsaw, 1930), 92–101, 395.

Princeton University

1946 *Problems of mathematics*, Princeton University bicentennial conferences, series 2, conference 2.

Pringsheim, Alfred

1916 *Vorlesungen über Zahlen- und Funktionenlehre* (Teubner, Leipzig and Berlin), vol. 1, part 2.

Putnam, Hilary
1956 Arithmetic models for consistent formulas of quantification theory, *The journal of symbolic logic 22* (1957), 110–111; abstract of a paper presented at the 27 December 1956 meeting of the Association for Symbolic Logic.
1961 *Trial and error predicates and the solution to a problem of Mostowski's* (New York University, New York).
1965 Trial and error predicates and the solution to a problem of Mostowski, *The journal of symbolic logic 30*, 49–57.
See Benacerraf, Paul, and Hilary Putnam.

Quine, Willard Van Orman
1940 *Mathematical logic* (Harvard University Press, Cambridge, Massachusetts); revised edition 1951; reprinted 1962 (Harper, New York).
1945 On ordered pairs, *The journal of symbolic logic 10*, 95–96.
1950 *Methods of logic* (Holt, New York); see *Quine 1955b*.
1955 On Frege's way out, *Mind 64*, 145–159; reprinted in *Quine 1966*, 146–158.
1955a A proof procedure for quantification theory, *The journal of symbolic logic 20*, 141–149; reprinted in *Quine 1966*, 196–204.
1955b *Appendix. Completeness of quantification theory. Löwenheim's theorem*, enclosed as a pamphlet with part of the third printing (1955) of *Quine 1950* and incorporated in the revised edition (1959), 253–260.
1956 Unification of universes in set theory, *The journal of symbolic logic 21*, 267–279.
1960 Variables explained away, *Proceedings of the American Philosophical Society 104*, 343–347; reprinted in *Quine 1966*, 227–235.
1963 *Set theory and its logic* (Harvard University Press, Cambridge, Massachusetts).
1966 *Selected logic papers* (Random House, New York).

Ramsey, Frank Plumpton
1925 The foundations of mathematics, *Proceedings of the London Mathematical Society*, series 2, *25*, 338–384, reprinted in *Ramsey 1931*, 1–61.
1931 *The foundations of mathematics and other logical essays*, edited by Richard Bevan Braithwaite (Paul, Trench, Trubner, London; Harcourt, Brace, New York); reprinted 1950 (Humanities Press, New York).

Rasiowa, Helena, and Roman Sikorski
1950 A proof of the completeness theorem of Gödel, *Fundamenta mathematicae 37*, 193–200.

Richard, Jules
1905 *Les principes des mathématiques et le problème des ensembles, *Revue générale des sciences pures et appliquées 16*, 541 (142–144 of the present volume).
1906 Lettre à Monsieur le rédacteur de la Revue générale des sciences, *Acta mathematica 30*, 295–296.
1907 Sur un paradoxe de la théorie des ensembles et sur l'axiome Zermelo, *L'enseignement mathématique 9*, 94–98.

Robinson, Abraham
1955 A result on consistency and its application to the theory of definition, *Koninklijke Nederlandse Akademie van wetenschappen, Proceedings*, series A, Mathematical sciences *59* (1956), 47–58; also *Indagationes mathematicae 18*, 47–58.

Robinson, Raphael Mitchel
1937 The theory of classes, a modification of von Neumann's system, *The journal of symbolic logic 2*, 29–36.
1947 Primitive recursive functions, *Bulletin of the American Mathematical Society 53*, 925–942.
1948 Recursion and double recursion, *ibid. 54*, 987–993.
See Tarski, Alfred, *1953*.

Rose, Harvey Ernest
1961 On the consistency and undecidability of recursive arithmetic, *Zeitschrift für mathematische Logik und Grundlagen der Mathematik 7*, 124–135.

Rose, Harvey Ernest—*contd.*

1962 Ternary recursive arithmetic, *Mathematica scandinavica 10*, 210–216.

Rosser, John Barkley

1936 Extensions of some theorems of Gödel and Church, *The journal of symbolic logic 1*, 87–91; reprinted in *Davis 1965*, 231–235.

Russell, Bertrand

1902 *Letter to Frege (first published in the present volume, 124–125).

1903 *The principles of mathematics*, vol. 1 (Cambridge University Press, Cambridge, England; 2nd ed., Norton, New York, no date).

1905 On denoting, *Mind*, new series, *14*, 479–493; reprinted in *Russell 1956*, 41–56.

1905a On some difficulties in the theory of transfinite numbers and order types, *Proceedings of the London Mathematical Society*, 2nd series, *4* (1907), 29–53.

1906 The theory of implication, *American journal of mathematics 28*, 159–202.

1906a Les paradoxes de la logique, *Revue de métaphysique et de morale 14*, 627–650.

1908 Mr. Haldane on infinity, *Mind*, new series, *17*, 238–242.

1908a *Mathematical logic as based on the theory of types, *American journal of mathematics 30*, 222–262 (150–182 of the present volume); reprinted in *Russell 1956*, 59–102.

1919 *Introduction to mathematical philosophy* (Allen and Unwin, London; Macmillan, New York).

1944 My mental development, *The philosophy of Bertrand Russell*, edited by Paul Arthur Schilpp (Tudor, New York), 3–20.

1956 *Logic and knowledge, Essays 1901–1950*, edited by Robert Charles Marsh (Allen and Unwin, London).

See Whitehead, Alfred North, and Bertrand Russell.

Schönfinkel, Moses

1924 *Über die Bausteine der mathematischen Logik, *Mathematische Annalen 92*, 305–316 (355–366 of the present volume).

See Bernays, Paul, and Moses Schönfinkel.

Schoenflies, Arthur

1905 Über wohlgeordnete Mengen, *Mathematische Annalen 60*, 181–186.

1913 *Allgemeine Theorie der unendlichen Mengen und Theorie der Punktmengen* (Teubner, Leipzig and Berlin), published as first half of *Entwickelung der Mengenlehre und ihrer Anwendungen* gemeinsam mit Hans Hahn herausgegeben von Artur Schoenflies (second half never published).

1921 Zur Axiomatik der Mengenlehre, *Mathematische Annalen 83*, 173–200.

1922 Bemerkung zur Axiomatik der Größen und Mengen, *ibid. 85*, 60–64.

1922a Zur Erinnerung an Georg Cantor, *Jahresbericht der Deutschen Mathematiker-Vereinigung 31*, 1st section, 97–106.

1928 Georg Cantor, *Mitteldeutsche Lebensbilder 3*, 548–563.

Schreier, Otto

See Artin, Emil, and Otto Schreier.

Schröder, Ernst

1873 *Lehrbuch der Arithmetik und Algebra für Lehrer und Studirende. Erster Band: Die sieben algebraischen Operationen* (Leipzig).

1877 *Der Operationskreis des Logikkalkuls* (Leipzig).

1880 Review of *Frege 1879*, *Zeitschrift für Mathematik und Physik 25*, Historisch-literarische Abtheilung, 81–94.

1890 *Vorlesungen über die Algebra der Logik* (*exakte Logik*) (Leipzig), vol. 1; reprinted in *Schröder 1966*.

1891 ——— vol. 2, part 1; reprinted in *Schröder 1966*.

1895 ——— vol. 3, *Algebra und Logik der Relative*, part 1; reprinted in *Schröder 1966*.

1909 *Abriss der Algebra der Logik*, edited by Eugen Müller (Teubner, Leipzig and Berlin), part 1, *Elementarlehre*; reprinted in *Schröder 1966*.

1910 ——— part 2, *Aussagentheorie, Funktionen, Gleichungen und Ungleichungen*; reprinted in *Schröder 1966*.

1966 *Vorlesungen ueber die Algebra der Logik* (Chelsea, New York).

Schütte, Kurt

1951 Beweistheoretische Erfassung der unendlichen Induktion in der Zahlentheorie, *Mathematische Annalen 122*, 369–389.

1960 *Beweistheorie* (Springer Verlag, Berlin, Göttingen, and Heidelberg).

1962 Der Interpolationssatz der intuitionistischen Prädikatenlogik, *Mathematische Annalen 148*, 192–200.

Schwabhäuser, Wolfram

1954 Zur Definition des geordneten Paares von Mengen beliebiger Stufe, *Mathematische Nachrichten 11*, 81–84.

Sheffer, Henry Maurice

1913 A set of five independent postulates for Boolean algebras, with applications to logical constants, *Transactions of the American Mathematical Society 14*, 481–488.

Shepherdson, John Cedric

1951–1953 Inner models for set theory, *The journal of symbolic logic 16*, 161–190; *17* (1952), 225–237; *18*, 145–167.

1963 Non-standard models for fragments of number theory, in *Addison, Henkin, and Tarski 1963*, 342–358.

Sierpiński, Wacław

1958 *Cardinal and ordinal numbers* (Państwowe wydawnictwo naukowe, Warsaw); 2nd ed. 1965.

See Luzin, Nikolai Nikolaevich, and Wacław Sierpiński.

Sigwart, Christoph

1904 *Logik*, 3rd ed. (Mohr, Tübingen), vol. 1.

1908 *Логика*, vol. 1, Russian translation of *Sigwart 1904* by I. A. Davidov (Общественная Польза and Провинція, Saint Petersburg).

Sikorski, Roman

See Rasiowa, Helena, and Roman Sikorski.

Skolem, Thoralf

1919 Untersuchungen über die Axiome des Klassenkalkuls und über Produktations- und Summationsprobleme, welche gewisse Klassen von Aussagen betreffen, *Videnskapsselskapets skrifter, I. Matematisk-naturvidenskabelig klasse*, no. 3.

1920 *Logisch-kombinatorische Untersuchungen über die Erfüllbarkeit oder Beweisbarkeit mathematischer Sätze nebst einem Theoreme über dichte Mengen, *ibid.*, no. 4 (§ 1, 252–263 of the present volume).

1922 *Einige Bemerkungen zur axiomatischen Begründung der Mengenlehre, *Matematikerkongressen i Helsingfors den 4–7 Juli 1922, Den femte skandinaviska matematikerkongressen, Redogörelse* (Akademiska Bokhandeln, Helsinki, 1923), 217–232 (290–301 of the present volume).

1923 *Begründung der elementaren Arithmetik durch die rekurrierende Denkweise ohne Anwendung scheinbarer Veränderlichen mit unendlichem Ausdehnungsbereich, *Videnskapsselskapets skrifter, I. Matematisk-naturvidenskabelig klasse*, no. 6 (302–333 of the present volume).

1928 *Über die mathematische Logik, *Norsk matematisk tidsskrift 10*, 125–142 (508–524 of the present volume).

1928a Review of *Ackermann 1928*, *Jahrbuch über die Fortschritte der Mathematik 54* (for 1928, published 1932), 56–57.

1928b Review of *Ackermann 1928a*, *ibid.*, 57.

1929 Über einige Grundlagenfragen der Mathematik, *Skrifter utgitt av Det Norske Videnskaps-Akademi i Oslo, I. Matematisk-naturvidenskapelig klasse*, no. 4.

1929a Über die Grundlagendiskussionen in der Mathematik, *Den syvende skandinaviske matematikerkongress i Oslo 19–22 August 1929* (Brøggers, Oslo, 1930), 3–21.

1930 Einige Bemerkungen zu der Abhandlung von E. Zermelo: "Über die Definitheit in der Axiomatik", *Fundamenta mathematicae 15*, 337–341.

Skolem, Thoralf—*contd.*

1933 Über die Unmöglichkeit einer vollständigen Charakterisierung der Zahlenreihe mittels eines endlichen Axiomensystems, *Norsk matematisk forenings skrifter*, series 2, no. 10, 73–82.

1934 Über die Nicht-charakterisierbarkeit der Zahlenreihe mittels endlich oder abzählbar unendlich vieler Aussagen mit ausschliesslich Zahlenvariablen, *Fundamenta mathematicae 23*, 150–161.

1935 Ein Satz über Zählausdrücke, *Acta litterarum ac scientiarum Regiae Universitatis Hungaricae Francisco-Josephinae, Sectio scientiarum mathematicarum 7*, 193–199.

1935a Über die Erfüllbarkeit gewisser Zählausdrücke, *Skrifter utgitt av Det Norske Videnskaps-Akademi i Oslo, I. Matematisk-naturvidenskapelig klasse*, no. 6.

1936 Utvalgte kapitler av den matematiske logikk, *Chr. Michelsens Institutt for Videnskap og Åndsfrihet, Beretninger 6*, no. 6.

1938 Sur la portée du théorème de Löwenheim-Skolem, in *Gonseth 1938*, 25–47; Discussion, 47–52.

1946 Den rekursive aritmetikk, *Norsk matematisk tidsskrift 28*, 1–12.

1946a The development of recursive arithmetic, *Den 10. skandinaviske matematikerkongress i København 26.–30. August 1946*, 1–46.

1950 Some remarks on the foundation of set theory, *Proceedings of the International Congress of Mathematicians, Cambridge, Massachussets, U.S.A., August 30–September 6, 1950* (American Mathematical Society, Providence, Rhode Island, 1952), vol. 1, 695–704.

1962 *Abstract set theory* (University of Notre Dame, Notre Dame, Indiana).

Specker, Ernst

1957 Zur Axiomatik der Mengenlehre (Fundierungs- und Auswahlaxiom), *Zeitschrift für mathematische Logik und Grundlagen der Mathematik 3*, 173–210.

Spector, Clifford

1955 Recursive well-orderings, *The journal of symbolic logic 20*, 151–163.

1961 Provably recursive functionals of analysis: A consistency proof of analysis by an extension of principles formulated in current intuitionistic mathematics, *Recursive function theory, Proceedings of symposia in pure mathematics* (American Mathematical Society, Providence, Rhode Island, 1962), vol. 5, 1–27.

Sudan, Gabriel

1927 Sur le nombre transfini ω^ω, *Bulletin mathématique de la Société roumaine des sciences 30*, 11–30.

Suslin, Mikhail Iakovlevich (Souslin, Михаил Яковлевич Суслин)

1917 Sur une définition des ensembles mesurables B sans nombres transfinis, *Comptes rendus hebdomadaires des séances de l'Académie des sciences* (Paris) *164*, 88–91 (session of 8 January 1917).

Svenonius, Lars

1959 A theorem on permutations in models, *Theoria* (Lund) *25*, 173–178.

Tarski, Alfred

1934 Z badán metodologicznych nad definiowalnością terminów, *Przegląd filozoficzny 37*, 438–460) (see Bibliographical Note in *Tarski 1956*, 296).

1935 Einige methodologische Untersuchungen über die Definierbarkeit der Begriffe, *Erkenntnis 5*, 80–100; English translation in *Tarski 1956*, 296–319.

1941 On the calculus of relations, *The journal of symbolic logic 6*, 73–89.

1948 *A decision method for elementary algebra and geometry* (Rand Corporation, Santa Monica, California).

1948a Axiomatic and algebraic aspects of two theorems on sums of cardinals, *Fundamenta mathematicae 35*, 79–104.

1951 2nd ed., revised, of *Tarski 1948* (University of California Press, Berkeley and Los Angeles).

Tarski, Alfred—*contd.*

1953 *Undecidable theories*, in collaboration with Andrzej Mostowski and Raphael M. Robinson (North-Holland, Amsterdam).

1956 *Logic, semantics, metamathematics* (Oxford University Press, Oxford).

See Addison, John West, Leon Henkin, and Alfred Tarski; also Łukasiewicz, Jan, and Alfred Tarski.

——— and Adolf Lindenbaum

1926 Sur l'indépendance des notions primitives dans les systèmes mathématiques, *Annales de la Société polonaise de mathématique 5*, 111–113.

——— and Robert Lawson Vaught

1956 Arithmetical extensions of relational systems, *Compositio mathematica 13*, 81–102.

Trendelenburg, Adolf

1867 *Historische Beiträge zur Philosophie* (Berlin), vol. 3, Vermischte Abhandlungen.

Turing, Alan Mathison

1937 On computable numbers, with an application to the Entscheidungsproblem, *Proceedings of the London Mathematical Society*, 2nd series, *42*, 230–265; correction, *ibid. 43*, 544–546; reprinted in *Davis 1965*, 116–154.

Ulam, Stanislaw

1958 John von Neumann, *Bulletin of the American Mathematical Society 64*, no. 3, part 2, 1–49.

Vaught, Robert Lawson

See Tarski, Alfred, and Robert Lawson Vaught.

Veblen, Oswald

1904 A system of axioms for geometry, *Transactions of the American Mathematical Society 5*, 343–384.

Vesley, Richard Eugene

See Kleene, Stephen Cole, and Richard Eugene Vesley.

Vieler, Heinrich

1926 Untersuchungen über Unabhängigkeit und Tragweite der Axiome der Mengenlehre in der Axiomatik Zermelos und Fraenkels (thesis, University of Marburg).

Vivanti, Giulio

1893 Lista bibliografica della teoria degli aggregati, *Rivista di matematica 3*, 189–192.

von Neumann, John

1923 *Zur Einführung der transfiniten Zahlen, *Acta litterarum ac scientiarum Regiae Universitatis Hungaricae Francisco-Josephinae, Sectio scientiarum mathematicarum 1*, 199–208 (346–354 of the present volume); reprinted in *von Neumann 1961*, 24–33.

1925 *Eine Axiomatisierung der Mengenlehre, *Journal für die reine und angewandte Mathematik 154*, 219–240; Berichtigung, *ibid. 155*, 128 (393–413 of the present volume); reprinted in *von Neumann 1961*, 34–56.

1926 Az általános nalmazelmélet axiomatikus folépitése (thesis, University of Budapest).

1927 Zur Hilbertschen Beweistheorie, *Mathematische Zeitschrift 26*, 1–46; reprinted in *von Neumann 1961*, 256–300.

1928 Über die Definition durch transfinite Induktion und verwandte Fragen der allgemeinen Mengenlehre, *Mathematische Annalen 99*, 373–391; reprinted in *von Neumann 1961*, 320–338.

1928a Die Axiomatisierung der Mengenlehre, *Mathematische Zeitschrift 27*, 669–752; reprinted in *von Neumann 1961*, 339–422.

1929 Über eine Widerspruchsfreiheitsfrage in der axiomatischen Mengenlehre, *Journal für die reine und angewandte Mathematik 160*, 227–241; reprinted in *von Neumann 1961*, 494–508.

1961 *Collected works* (Pergamon Press, New York), vol. 1.

Wang, Hao

1949 On Zermelo's and von Neumann's axioms for set theory, *Proceedings of the National Academy of Sciences of the U. S. A. 35*, 150–155.

1950 Remarks on the comparison of axiom systems, *ibid. 36*, 448–453.

1951 Arithmetic models for formal systems, *Methodos 3*, 217–232.

1952 Truth definitions and consistency proofs, *Transactions of the American Mathematical Society 73*, 243–275; reprinted with revisions as Chapter 18 of *Wang 1962*.

1955 Undecidable sentences generated by semantic paradoxes, *The journal of symbolic logic 20*, 31–43; reprinted with revisions as Chapter 22 of *Wang 1962*.

1957 The axiomization of arithmetic, *The journal of symbolic logic 22*, 145–157; reprinted as Chapter 4 of *Wang 1962*.

1962 *A survey of mathematical logic* (Science Press, Peking; distributed by North-Holland, Amsterdam).

See Kahr, Andrew Seth, Edward Forrest Moore, and Hao Wang.

Weierstrass, Karl

1884 Zur Theorie der aus *n* Haupteinheiten gebildeten complexen Grössen, *Nachrichten von der Königlichen Gesellschaft der Wissenschaften zu Göttingen*, 395–419.

Wernick, William

1942 Complete sets of logical functions, *Transactions of the American Mathematical Society 51*, 117–132.

Weyl, Hermann

1910 Über die Definitionen der mathematischen Grundbegriffe, *Mathematisch-naturwissenschaftliche Blätter 7*, 93–95, 109–113.

1918 *Das Kontinuum. Kritische Untersuchungen über die Grundlagen der Analysis* (Veit, Leipzig); reprinted 1932 (de Gruyter, Berlin and Leipzig) and without date (Chelsea, New York).

1920 Über die neue Grundlagenkrise der Mathematik, *Mathematische Zeitschrift 10* (1921), 39–79.

1927 *Diskussionsbemerkungen zu dem zweiten Hilbertschen Vortrag über die Grundlagen der Mathematik, *Abhandlungen aus dem Mathematischen Seminar der Hamburgischen Universität 6* (1928), 86–88 (480–484 of the present volume); reprinted in *Hilbert 1928*, 22–24.

1949 *Philosophy of mathematics and natural science* (Princeton University Press, Princeton, New Jersey).

Whitehead, Alfred North

1898 *A treatise of universal algebra with applications* (Cambridge University Press, Cambridge, England), vol. 1.

1902 On cardinal numbers, *American journal of mathematics 24*, 367–394.

1906 On mathematical concepts of the material world, *Philosophical transactions of the Royal Society of London*, series A, *205*, 465–525.

——— and Bertrand Russell

1910 **Principia mathematica* (Cambridge University Press, Cambridge, England), vol. 1 (sec. (1) of chap. III, 216–223 of the present volume).

1912 ——— vol. 2.

1913 ——— vol. 3.

1925 ——— 2nd ed., vol. 1.

1927 ——— vol. 2.

1927a ——— vol. 3.

Wiener, Norbert

1913 A comparison between the treatment of the algebra of relatives by Schroeder and that by Whitehead and Russell (thesis, Harvard University).

1914 *A simplification of the logic of relations, *Proceedings of the Cambridge Philosophical Society 17*, 387–390 (224–227 of the present volume).

1953 *Ex-prodigy, my childhood and youth* (Simon and Schuster, New York).

Young, William Henry, and Grace Chisholm Young

1929 Review of *Hobson 1921* and *1926*, Second notice, *The mathematical gazette 14*, 98–104.

Zermelo, Ernst

1904 *Beweis, daß jede Menge wohlgeordnet werden kann, *Mathematische Annalen 59*, 514–516 (139–141 of the present volume).

1908 *Neuer Beweis für die Möglichkeit einer Wohlordnung, *ibid. 65*, 107–128 (183–198 of the present volume).

1908a *Untersuchungen über die Grundlagen der Mengenlehre I, *ibid.*, 261–281 (199–215 of the present volume).

1908b Ueber die Grundlagen der Arithmetik, *Atti del IV Congresso internazionale dei matematici, Roma, 6–11 Aprile 1908* (Accademia dei Lincei, Rome, 1909), vol. 2, 8–11.

1909 Sur les ensembles finis et le principe de l'induction complète, *Acta mathematica 32*, 183–193.

1929 Über den Begriff der Definitheit in der Axiomatik, *Fundamenta mathematicae 14*, 339–344.

1930 Über Grenzzahlen und Mengenbereiche, *ibid. 16*, 29–47.

See Cantor, Georg, *1932*.

Żyliński, Eustachy

1925 Some remarks concerning the theory of deduction, *Fundamenta mathematicae 7*, 203–209.

Index